Ion Exchange Technology I

Inamuddin • Mohammad Luqman
Editors

Ion Exchange Technology I

Theory and Materials

Springer

Editors
Inamuddin
Assistant Professor
Department of Applied Chemistry
Faculty of Engineering & Technology
Aligarh Muslim University
Aligarh, India

Mohammad Luqman
Assistant Professor
Chemical Engineering Department
College of Engineering
King Saud University
Riyadh, Kingdom of Saudi Arabia

ISBN 978-94-017-8172-5 ISBN 978-94-007-1700-8 (eBook)
DOI 10.1007/978-94-007-1700-8
Springer Dordrecht Heidelberg New York London

Printed on acid-free paper

Springer is part of Springer Science+Business Media (www.springer.com)

Preface

Ion exchange is a process of exchanging ions between stationary and mobile phases. It is a natural process and has been in practice for ages. Since commercial development took place in the last century, both academic and industrial research have been improving technology to find new applications.

This edition covers the introduction, principle, instrumental and theoretical fundamentals, structure, synthesis and characterization, kinetic and equilibrium, simulation and computer modeling studies of ion exchange materials in addition to the preparation and properties of ion exchange membranes for electrodialysis and fuel cells.

Chapter 1 covers the basic fundamentals of ion exchange kinetics and equilibrium and discusses the various applications that utilize ion exchange processes. Chapter 2 reviews the selectivity coefficient as well as the exchange isotherm diffusion and transport in terms of thermodynamics, equilibria and ion exchange kinetics. Chapter 3 examines the various conditions of ion exchange equilibrium with important theories developed in literature and reviews ion exchange kinetics and mass transport processes based on semi empirical models, Fick's law and derived expressions. Chap. 4, presents fundamentals of ion exchange fixed bed operations. Chapter 5 deals with the performance of ion exchange membrane electrodialysis for saline water desalination. The desalination performance of a practical-scale electrodialyzer is discussed using computer simulation. Chapter 6 is devoted to the structure, synthesis and properties of organic ion exchange materials. Preparation, properties and application of ion exchange membranes are discussed in Chap. 7. Chapter 8 focuses on the synthesis, structure, properties and applications of synthetic ion exchange materials. Chapter 9 reviews the most important aspects such as: synthesis, physical and chemical properties, equilibria and kinetics, as well as of sorption processes, possible and real field applications of fibrous ion exchangers. Fibrous catalysts, color-changing sorbents and hybrid fibrous sorbents impregnated with nanoparticles of inorganic substances are also described. The structure, coordination chemistry and applications of most commonly employed chelating ion exchangers are discussed in Chap. 10. Chapter 11 focuses on the recent advances in the field of ion exchanger–based voltammetric sensors, whose widespread use has instigated a new electroanalytical technique named ion exchange voltammetry. Chapter 12 discusses the properties of sulfonated poly(ether ether ketone) (SPEEK) as a promising membrane material for polymer electrolyte fuel cell. Chapter 13 reviews the preparation and use of organic–inorganic hybrid ion exchangers in organic reaction catalysis. An introduction to the ion exchange technique in solid matter, mainly optical glasses, to fabricate wave guides telecommunications is reviewed in Chap. 14. Network simulation of electrical response using Nernst–Planck and Poisson equations is used to describe the ionic transport processes through a cation-exchange membrane and two diffusion boundary layers on both sides of the membrane in Chap. 15. Chapter 16 reviews the authors' work on the mathematical and computer modeling of ion exchangers on styrene–divinylbenzene matrix, a mathematical model based on the concept of the influence of neighbouring exchange sites on the properties of each other. Such a model allowed to explain the dependence of selectivity and additive properties of the ion exchange system on the degree of ion exchange.

Acknowledgments

We are most indebted to the grace of the Almighty "One Universal Being," who inspires entire Humanity to knowledge, and who blessed us with the needed favor to complete this work.

This book gathers the remarkable contributions from international leading experts in the field of ion exchange technology and provides a comprehensive review and research work. We are thankful to all the authors for their esteemed contribution to this book. We would also like to thank all the publishers and authors who granted us permission to use their copyright material. Although sincere efforts were made to obtain copyright permissions from the respective owners and to include citations with the reproduced materials, we would like to offer our sincere apologies to any copyright holder whose rights may have been unknowingly infringed.

Dr. Inamuddin would like to express his deep sense of gratitude to Profs. Syed Ashfaq Nabi, Ishtiyaq Ahmad, Rakesh Kumar Mahaja, Seon Jeon Kim, Kenneth I. Ozoemena, Saleem-ur-Rahman, S.M.J. Zaidi, Gaber E. Eldesoky, Zeid A. AL-Othman, Sheikh Raisuddin, Byong-Hun Jeon, and A.I. Yahya, and to Drs. B.D. Malhotra and Raju Khan for their valuable suggestions, guidance, and constant inspiration.

He would also like to thank his departmental colleagues, Profs. M. Mobin, Asif Ali Khan, and R.A.K. Rao, and Drs. M.Z.A. Rafiqui, Abu Nasar, Rais Ahmad, and Yasser Azim. Without their continuous encouragement, this book would have not been brought to its final form. Dr. Inamuddin is at a loss of words to express his gratitude to his friends and colleagues, Drs. M.M. Alam, Amir-Al-Ahmad, Zafar Alam, Mu. Naushad, Salabh Jain, Hemendra Kumar Tiwari, Adesh Bhadana, Shakeel Ahmad Khan, Satish Singh, and others, for their timely help, good wishes, encouragement, and affection.

Dr. Luqman would like to acknowledge the lively environment provided by Samsung Cheil Industries during the time he spent with them. He is also very grateful to his colleagues, more specifically to the Chairman of the Chemical Engineering Department, King Saud University, and other departmental staff for their help, support, and encouragement throughout the preparation of this book.

Last but not least, we would like to express our heartfelt gratitude to our family members for their constant inspiration and gracious support.

Inamuddin
Mohammad Luqman
Editors

Editors' Bios

Dr. Inamuddin is currently working as Assistant Professor in the Department of Applied Chemistry, Aligarh Muslim University (AMU), India. He received his Master of Science degree in Organic Chemistry from Chaudhry Charan Singh (CCS) University, Meerut, India, in 2002 and his Master of Philosophy and Doctor of Philosophy degrees in Applied Chemistry from AMU in 2004 and 2007, respectively. He has extensive research experience in the multidisciplinary fields of Analytical Chemistry, Material Chemistry, and Electrochemistry and, more specifically, Renewable Energy and Environment. He has worked in different projects funded by the University Grants Commission (UGC), Government of India, and Council of Scientific and Industrial Research (CSIR), Government of India. He has received Fast Track Young Scientist Award of Science and Technology, Government of India, to work in the area of bending actuators and artificial muscles. He has published 30 research articles and 4 book chapters of international repute. Recently, he has edited the book entitled *Advanced Organic–inorganic Composites: Materials, Devices and Allied Applications* and co-edited the books entitled *Green Solvents I: Properties and Applications in Chemistry* and *Green Solvents II: Properties and Applications of Ionic Liquids* published by Nova Science Publishers, Inc, and Springer, United Kingdom, respectively. He is presently working as editor in chief of the Journal of Chemical and Environmental Research published from The Muslim Association for the Advancement of Science (MAAS), India. He has worked as a Postdoctoral Fellow, leading a research team at Creative Research Initiative Center for Bio-Artificial Muscle, Hanyang University, South Korea, in the field of renewable energy, especially biofuel cells. He has also worked as Postdoctoral Fellow at Center of Research Excellence in Renewable Energy, King Fahd University of Petroleum and Minerals, Saudi Arabia, in the field of polymer electrolyte membrane fuel cells and computer fluid dynamics of polymer electrolyte membrane fuel cells. He is a life member of the *Journal of the Indian Chemical Society*.

Mohammad Luqman, Ph.D., is serving as an Assistant Professor in Chemical Engineering Department, King Saud University (KSU), Saudi Arabia. He teaches Polymer/Material Science and Engineering courses. Before joining the KSU, he served as an Assistant Manager in Samsung Cheil Industries, Korea. He worked on the development of heat-resistant polymers, organic glass, and block copolymers as impact modifiers and compatibilizers for engineering alloys. He served as a post-doctoral fellow at Artificial Muscle Research Center, Konkuk University, Korea, in the field of Ionic Polymer Metal Composites. He was awarded the Ph.D. degree in 2007, in the field of Ionomers, by Chosun University, Korea. Dr. Luqman has published numerous papers and book chapters in the field of Ionomers, Ion-exchange Polymers, and Polymer Nanocomposites. One of his papers, published in "*Polymer*", Elsevier, was Ranked the "first" among the "Top 25 Hottest Articles" in "all" chemistry journals during April–June 2008. Dr. Luqman is acting as a Regional (Middle East) Editor of the International Journal *Chemical and Environmental Research*, Published from MAAS, India. He also serves as a Section Editor of *Journal of Industrial Research & Technology* by HATAM Publishers, Malaysia. He has delivered few invited lectures at International podiums and very recently, has served as an Invited Editor of a reference book entitled *Recent Advances in Plasticizers*.

Contents

Contributors

Amir-Al-Ahmed Center of Research Excellence in Renewable Energy, King Fahd University of Petroleum & Minerals, Dhahran, Saudi Arabia

Simão P. Cardoso CICECO/Department of Chemistry, University of Aveiro, Aveiro, Portugal

Dilson Cardoso Chemical Engineering Department, Federal University of São Carlos – UFSCar, São Carlos, SP, Brazil

Vassilis J. Inglezakis SC European Focus Consulting srl, Bacau, Romania

Mahadevappa Y. Kariduraganavar P. G. Department of Studies in Chemistry, Karnatak University, Dharwad, KA, India

Arjumand A. Kittur Department of Chemistry, SDM College of Engineering & Technology, Dharwad, KA, India

E.G. Kosandrovich Institute of Physical Organic Chemistry of the National Academy of Sciences of Belarus, Minsk, Republic of Belarus

Srikant S. Kulkarni Department of Chemistry, JSS Degree College, Dharwad, KA, India

Patrícia F. Lito CICECO/Department of Chemistry, University of Aveiro, Aveiro, Portugal

José M. Loureiro LSRE/Department of Chemical Engineering, School of Engineering, University of Oporto, Porto, Portugal

Leandro Martins Institute of Chemistry, Universidade Estadual Paulista – UNESP, Araraquara, SP, Brazil

Ligia M. Moretto Department of Molecular Sciences and Nanosystems, University Ca' Foscari of Venice, Venice, Italy

A.A. Moya Departamento de Física, Universidad de Jaén, Jaén, Spain

Tulsi Mukherjee Analytical Chemistry Division, Chemistry Group, Bhabha Atomic Research Centre, Trombay, Mumbai, India

Mohamed Mahmoud Nasef Institute of Hydrogen Economy, Universiti Teknologi Malaysia, Kuala Lumpur, Malaysia

Jin-Soo Park Department of Environmental Engineering, College of Engineering, Sangmyung University, Dongnam-gu, Cheonan, Republic of Korea

Jayshree Ramkumar Analytical Chemistry Division, Chemistry Group, Bhabha Atomic Research Centre, Trombay, Mumbai, India

Victor Anthony Garcia Rivera Institute of Physics and National Institute of Optical and Photonic (INOF) of São Carlos, University of São Paulo, São Carlos, SP, Brazil

Wolfgang Schmidt Max-Planck-Institut für Kohlenfoschung, Mülheim an der Ruhr, Germany

Carlos Manuel Silva CICECO/Department of Chemisty, University of Aveiro, Aveiro, Portugal

V.S. Soldatov Institute of Physical Organic Chemistry of the National Academy of Sciences of Belarus, Minsk, Republic of Belarus

Lublin University of Technology, Lublin, Poland

Dhiraj Sud Department of Chemistry, Sant Longowal Institute of Engineering and Technology, (Deemed to be University), Longowal, India

Abdullah S. Sultan Center for Petroleum & Minerals, King Fahd University of Petroleum & Minerals, Dhahran, Saudi Arabia

Yoshinobu Tanaka IEM Research, Ushiku-shi, Ibaraki, Japan

Paolo Ugo Department of Molecular Sciences and Nanosystems, University Ca' Foscari of Venice, Venice, Italy

Zaini Ujang Institute of Environmental and Water Resources Management, Universiti Teknologi Malaysia, Skudai, Johor, Malaysia

S.M. Javaid Zaidi Center of Research Excellence in Renewable Energy, King Fahd University of Petroleum & Minerals, Dhahran, Saudi Arabia

Department of Chemical Engineering, King Fahd University of Petroleum & Minerals, Dhahran, Saudi Arabia

V.M. Zelenkovskii Institute of Physical Organic Chemistry, Belarus National Academy of Sciences, Minsk, Republic of Belarus

Antonis Zorpas Institute of Environmental Technology and Sustainable Development, Department of Research and Development, Laboratory of Environmental Friendly Technology, Paralimni, Cyprus

Abbreviations

α_B^A	Separation factor
12MR	Twelve-membered ring
4MR	Four-membered ring
4-VP	4-Vinylpyridine
6MR	Six-membered ring
8MR	Eight-membered ring
$a_{Az}{}^{b}$, $a_{Bs}{}^{a}$	Activities
APAS	Aminophosphonic acid silica
ASV	Anodic stripping voltammetry
BEA	Zeolite *Beta*
BT	Breakthrough
BV	Bed volumes
CD	Cyclodextrin
CEC	Cation-exchange capacity
CMX	Cation exchange membranes
CPE	Carbon paste electrode
CsEBS	Cesium salts of ethylbenzenesulfoacid
CSV	Cathodic stripping voltammetry
CV	Cyclic voltammetry
Cyt *c*	Cytochrome *c*
D	Diffusion coefficient
D4R	Double four-membered ring
D6R	Double six-membered ring
D_{app}	Apparent diffusion coefficient
DEA	Diethylamine
DEAPA	Diethylaminopropylamine
DETA	Diethylenetriamine

DFs	Decontamination factors
DL	Detection limit
DMAPA	Dimethylaminopropylamine
DMFC	Direct methanol fuel cell
DMG	Dimethylglyoxime
DP-ASV	Differential pulse anodic stripping voltammetry
DPV	Differential pulse voltammetry
DS	Degree of sulfonation
D_s	Diffusion coefficient in the solution phase
DVB	Di-vinyl benzene
E(OC)	Open circuit electrochemical potential
$E_{1/2}$	Half-wave potential
E_{appl}	Applied potential
ECL	Electrochemiluminescence
ED	Electrodialysis
EDA	Ethylenediamine
EDR	Electrodialysis reversal
EDTA	Ethylenediaminetetraacetic acid
E_i	Initial potential
EIS	Electrochemical impedance spectroscopy
ENM	Electrospin nano-fiber membrane
Ep_b	Backward peak potential
EPBI(DMG)	Epoxidized polybenzimidazole(Dimethylglyoxime)
Ep_f	Forward peak potential
ETSS	Ethyl styrene sulfonate
EW	Equivalent weight
F	Faraday constant
FAU	*Fau*jasite
Fc	Ferrocene
Fc^+	Ferricinium cation
FCC	Fluid catalytic cracking
FS	Full scale
GCE	Glassy carbon electrode
GIS	*Gis*mondine
GME	*Gme*linite
HASB	Hard soft acid base
HPA	Hydrated tungstophosphoric acid
HPCIC	High performance chelation ion chromatography
IDA	Iminodiacetic acid
IEC	Ion exchange capacity
IEV	Ion-exchange voltammetry
IO	Integrated-optic

Ip_P	Peak current for analytes n the polymer phase
Ip_s	Peak current for analytes in the solution phase
IS	Iontosorb salicyl
ITO	Indium thin oxide
K_a	Equilibrium constant
k_D	Distribution coefficient
k_{ex}	Site to site electron exchange rate constant
$K_X{}^M$	Selectivity coefficient
LB	Langmuir-Blodgett
LBL	Layer-by-layer
LDPE	Low density polyethylene
LS	Lab scale
LTA	*L*inde *T*ype *A*
m- DVB	metha-Divinylbenzene
M_{As}, M_{Bs}	Molarities
m_{Az}, m_{Bz}	Molalities
MEL	Maximum exchange level
MFI	*ZSM*-5 (*fi*ve)
MHL	Metal proton ligand
MINI, MIDI(d), 3-21G*	Basis sets for non-empirical calculations.
MMT	Montmorillonite
MOR	*Mor*denite
MP2/3-21G* and MP2/MIDI(d)	Level of theory of non-empirical calculations with using basis sets 3-21G* and MIDI(d) and with accounting for electronic correlation in the frame of the second order Moeller-Plesset perturbation theory.
MSWV	Multiple square wave voltammetry
MTA	Methylthriamyl ether
MTB	Methylthributhyl ether
MV	Methyl viologen
MWW	*M*CM-22 (t*w*enty-t*w*o)
N	Noise
N-DC	N, N′ di(caroxymethyl)dithiocarbamate
NMP	N-methyl-2-pyrrolidinon
NMR	Nuclear magnetic resonance
PA	Polyamide
PAMAM	Polyamidoamine
PAN	Polyacrylonitrile
PBI	Poly(benzimidazole)
PBI	Polybenzimidazole
PC	Polycarbonate
PDDMACl	Poly(diallyldimethylammonium chloride)
PDDPCl	Poly(1,1-dimethyl-3,5-dimethylenepiperidinium chloride)

PEEK	Poly(ether ether ketone)
PEI	Poly(ether imine)
PEK	Poly ether ketone
PEKEKK	Poly(ether ketone ether ketone ketone)
PEM	Proton exchange membrane
PEMFC	Polymer electrolyte membrane fuel cell
PES	Polyether sulphone
PET	Poly(ethyleneterphthalate)
PFSA	Perfluorosulfonic acid
PI	Polyimides
PILC	Pillared clay
PLE's	Polymeric ligand exchangers
PMA	Poly mtharcylate
PMeT	Poly(3-methylthiophene)
PP	Polypropylene
PPO	poly(phenylene oxide)
PP-ST-DVB	Polypropylene with grafted polystyrene with divinylbenzene
PPy	Polypyrrole
PS-DVB copolymer	PSDC
PS-DVB	Polystyrene divinylbenzene
PSS	Poly(4-styrene sulfonate)
PSSH	Poly(styrenesulfonic acid)
PSSNa	Poly(sodium styrenesulfonate)
PSU	Polysulphone udel
PSU-NH_2	Aminated polysulfone udel
PTFE	Poly(tetrafluoroethylene)
PV	Pervaporation
PVA	Poly(vinyl alcohol)
PVC	Poly(vinyl chloride)
PVP	Polyvinyl pyrollidone
Q	Ion exchange capacity
R	Gas constant
REC	Real exchange level
RHF	Restricted Hartree-Fock method for closed shalls.
RO	Reverse osmosis
ROHF	Restricted open shall Hartree-Fock method.
S	Signal
SCF MO LCAO	Model, in which a molecular orbital (MO) is represented as a linear combination of atomic orbitals (LCAO), are examined in light of ab initio self-consistent field (SCF) computations with bases of various sizes.

s-IPNs	Semi-interpenetrating polymer networks
SMM	Surface modifying macromolecules
SPE	Screen printed electrode
SPEEK	Sulfonated poly(ether ether ketone)
SPI	Sulfonated polyimide
SPME	Solid phase microextraction
SPSU	Ortho-sulfonesulfonated poly (ethersulfone)
ST	Polystyrene
ST-DVB Matrix	Styrene – divinylbenzene matrix
ST-DVB	Styrene–divinylbenzene
SWV	Square wave voltammetry
T	Temperature
t	Time
TCB	Phenol-trichlorobenzene
TEC	Theoretical exchange level
TETA	Triethylenetetraamine
T_g	Glass transition temperature
THF	Tetrahydrofuran
TMFE	Thin mercury film electrode
TPA	Tripropylamine
TPABr	Tetrapropylammonium bromide
UF	Ultra filtration
v	Scan rate
WKB method	Wentzel-Kramers-Brillouin method
XAD	Commercial polystrene divinylbenzene resin
Z_A, Z_B, S_A, S_B	Equivalent fractions
ZrP	Zirconium phosphate
ΔG^0	Free energy change
ΔH	Enthalpy change
ΔS	Entropy change

Nomenclature

a	Minimum approximation distance between ions
A	External particle surface area
A_γ	Debye-Huckel constant
A'_{ij}, A'_{ji}	Margules parameters
$A^{z_A}, B^{z_B}, C^{z_C}$	Counter ions with valences z_A, z_B, z_C
$\bar{A}^{z_A}, \bar{B}^{z_B}$	Counter ions with valences z_A, z_B inside the exchanger
a_i	Activity of species i in solution
$\bar{a}_i$	Activity of species i in exchanger
$A_i^{z_i}$	Generic counter ion i with valence z_i
a_p	External surface area per unit particle volume
B	Second Virial coefficient
$B_{j,i}$	Langmuir constant

C_b	Solute concentration at breakthrough time
$C_{ef,i}$	Concentration of sorbate in the effluent
$C_{F,i}$	Concentration of species *i* in the feed
C_i	Molar concentration of species *i* in solution
C_i^*	Molar concentration of species *i* at the exchanger/film interface
$C_{N,i}$	Normality of species *i*
$C_{N,t}$	Total normality of solution
$C_{p,i}$	Molar concentration of species *i* inside the pores
$\bar{C}_{p,i}$	Average concentration of species *i* inside the pores
C_{sat}	Saturation concentration
C_t	Total molar concentration of ionic species in solution
d	Particle diameter
D_A, D_B	Self-diffusion coefficients of species *A* and *B*
D_{AB}	Interdiffusion coefficient
$D_{eff,p,i}$	Effective diffusion coefficient of species *i* in macropores
$D_{eff,s,i}$	Effective diffusion coefficient of species *i* in micropores
D_f	Diffusion coefficient in the film
D_i	Diffusion coefficient of species *i*
	MS surface diffusivity of the pair $i–j$
	MS surface diffusivity corresponding to the interaction between *i* and the fixed ionic charges
D_L	Axial dispersion coefficient
e	Electron charge
$E_{i,j}$	Energy of adsorption of ion *i* on site *j*
$\bar{E}_i$	Average adsorption of ion *i*
F	Faraday constant
F_i	Fractional attainment of equilibrium of species *i*
g_{ij}	Energy parameter characteristic of the $i - j$ interaction
I	Ionic strength
J_i	Diffusion flux of species *i*
k	Boltzmann's constant
k_1	Rate constant of the first order sorption
k_2	Rate constant of the second order sorption
K_{aB}^{A}	Corrected selectivity coefficient
k_{AB}	Bohart and Adams rate constant
K_B^A	Thermodynamic (equilibrium) constant
K_C	Selectivity coefficient
K_D	Distribution coefficient
k_f	Convective mass transfer coefficient
K_{LDF}	Linear driving force coefficient
$K_S^{M_xM_m}$	Stability constant
k_{Th}	Thomas rate constant
k_{YN}	Yoon-Nelson rate constant
L	Column length

M^{m+}	Cation
m_i	Molality of species *i*
m_t	Total molality of ionic species
n	Freundlich constant, number of ionic species in solution
N_0	Avogadro's constant
n_c	Number of counter ions
n_f	Number of functional groups
N_i	Molar flux of species *i*
$N_{p,i}$	Diffusion fluxes of species *i* through the macropores
$N_{s,i}$	Diffusion fluxes of species *i* through the micropores
n_w	Number of water molecules in the zeolite
$n_x + n_y$	Total number of tetrahedral in the unit cell of zeolite
q_i	Molar concentration of ionic species *i* in exchanger
p	Parameter in binomial distribution
p_j	Equivalent fraction of exchanger site of type *j*
$\bar{q}_i$	Average loading of ionic species *i* in exchanger
q_i^*	Resinate concentration in equilibrium with the fluid concentration
Q_i	Equivalent ionic concentration of species *i* in exchanger
Q_i^e	Surface excess of ion *i*
$Q_{j,i}$	Equivalent ionic concentration of species *i* on exchanger site *j*
q_M	Kusik-Meissner parameter
q_{max}	Maximum sorbate concentration in the solid phase
q_s	Molar concentration of ionic fixed groups in exchanger
q_t	Total molar concentration of ionic species in exchanger
Q_t	Ion exchange capacity (in equivalents)
r	Radial position
R	Particle radius
$\Re$	Universal gas constant
t	Time
T	Absolute temperature
$t_{1/2}$	Time required for 50% sorbate breakthrough; stoichiometric time
t_b	Breakthrough time
U_0	Superficial velocity
u_i	Electrochemical mobility of species *i*, velocity of diffusing species *i*
V_{ef}	Volume of effluent
V_L	Volume of fluid phase
V_s	Volume of solid phase
V_{ZLC}	Volume of the ZLC column
W_{exch}	Mass of exchanger
W_{ij}	Weighting factor
x_i	Ionic fraction of species *i* in solution
X_i	Equivalent ionic fraction of species *i* in solution
X^{x-}	Anion
y_i	Ionic fraction of species *i* in exchanger

Y_i	Equivalent ionic fraction of species *i* in exchanger
y_s	Mole fraction of ionic fixed groups in exchanger
z_i	Valence of ionic species *i*

Subscripts

0	Initial condition
e	Equilibrium
f	Free
s	Solid, fixed ionic groups of the exchanger
t	Total
intra	Intraparticle

Greek Letters

α_B^A	Separation factor
$\bar{\alpha}_j^i$	Average separation factor
δ	Film thickness
ε	Dielectric constant
ε_b	Bed porosity
ε_p	Particle porosity
φ	Electric potential
γ_i	Activity coefficient of species *i* in solution
$\bar{\gamma}_i$	Activity coefficient of species *i* in exchanger
Γ_{ij}	Thermodynamic factor
Γ	Reduced activity coefficient of Meissner and Kusik
λ_i	Distribution coefficient of species *i*
$\Lambda_{ij}, \Lambda_{ji}$	Wilson parameters
μ_i	Chemical potential of species *i* in solution
$\bar{\mu}_i$	Surface chemical potential of species *i*
v	Volumetric flow rate, number of site types
v_c	Number of cations per electrolyte
v_i	Pure-component molar volume
v_a	Number of anions per electrolyte
ρ_w	Density of pure solvent
σ_i	Standard deviation of energy distribution
τ_d	Time constant for intraparticle diffusion
$\tau_{d,m}$	Maximum value of τ_d
$\tau_{d,i}$	Minimum value of τ_d

Chapter 1
Introduction to Ion Exchange Processes

Mohamed Mahmoud Nasef and Zaini Ujang

Abstract Ion exchange technology remains the workhorse of various chemical, petrochemical, food, power, and pharmaceutical industries. The success of ion exchange process depends literally on understanding of its basic principles and applying them in a way suiting the nature of the treated feed. This chapter reviews the basic fundamentals and key components of ion exchange process taking into consideration the latest progress taking place in the field. The variation in the ion exchange materials, their nature, forms, and functions are reviewed. The kinetics, sorption equilibrium, operating modes, and engineering configurations for ion exchange processes are also discussed. A brief encounter for the various applications utilizing ion exchange processes is also presented.

1.1 Introduction

Ion exchange is a technology that has been ever receiving growing attention in various industries for several decades. This technology is commonly used to purify solutions by removing the dissolved ions by electrostatic sorption into ion exchange materials of various physical forms. The removed ions are replaced with equivalent amounts of other ions of the same charge in the solutions. The use of ion exchange reaction allows either all ions to be removed from a solution or particular ions to be selectively separated. Therefore, both selective removal of ionic contamination and

M.M. Nasef (✉)
Institute of Hydrogen Economy, Universiti Teknologi Malaysia, International City Campus, Jalan Semarak, 54100 Kuala Lumpur, Malaysia
e-mail: mahmoureithar@cheme.utm.my

Z. Ujang
Institute of Environmental and Water Resources Management, Universiti Teknologi Malaysia, 81310 Skudai, Johor, Malaysia
e-mail: zaini@utm.my

Inamuddin and M. Luqman (eds.), *Ion Exchange Technology I: Theory and Materials*, DOI 10.1007/978-94-007-1700-8_1,

complete deionization can be distinguished. The selection between both depends mainly on the composition of the solution and the extent of decontamination required [1].

The applications of ion exchange are numerous and cover wide ranges of industries and households. These applications mainly cover purification purposes; however, ion exchange is also widely implemented in the separation and extraction of valuable substances such as uranium and plutonium from the nuclear industry waste [2]. Deionization (demineralization) of water and water softening are known to be the most common applications. However, the spectrum of other applications varies from large-scale extraction of metals in hydrometallurgical and metal finishing processes to recovery of precious metals [3, 4]. The applications of ion exchange are also extended to food and beverages, petrochemical and chemical, pharmaceutical, sugar and sweeteners, industrial wastewater, ground and potable water, semiconductor, production power soil remedy, and pulp and paper industry.

In principle, ion exchange is a process involving an exchange of ions between an electrolyte solution (aqueous phase) and similarly charged ions immobilized in an ion exchange material (solid phase), which takes place through a stoichiometric reversible ion exchange reaction. Ion exchange materials represent the heart of ion exchange processes that fall into various categories: polymeric and mineral, cationic and anionic, and resins and membranes depending on their classification. Engineering systems of various configurations meeting the requirements for industrial application are available and vary depending on the morphology of the ion exchange materials. Batch and column systems are the most common configurations to accomplish ion exchange processes using resins, whereas plate and frame modules/cells are favored upon using membrane/sheet forms. Currently, a large number of commercial resins and membranes are available giving high possibility for more than one technically effective solution that allows the utilization of custom-designed ion exchange process. However, having a robust system design requires a thorough knowledge of all available resin types along with a clear understanding of basic fundamentals and economics of ion exchange to ensure highly efficient and cost-effective operations.

This chapter provides an intensive review for the basic fundamentals of ion exchange process covering its essential ion exchange materials, reaction kinetics and sorption equilibrium, operating modes, system configurations, process economy, and industrial applications.

1.2 Historical Perspective

Ion exchange phenomena have been known for many years. The first examples of this phenomena were discovered by Thompson and Way (1850) [5, 6] during their investigations concerning the way in which soluble manures were retained for long periods in the soil, instead of being washed out by rainwater. The importance of this discovery (in ion exchange terms) was not fully understood until later in that decade

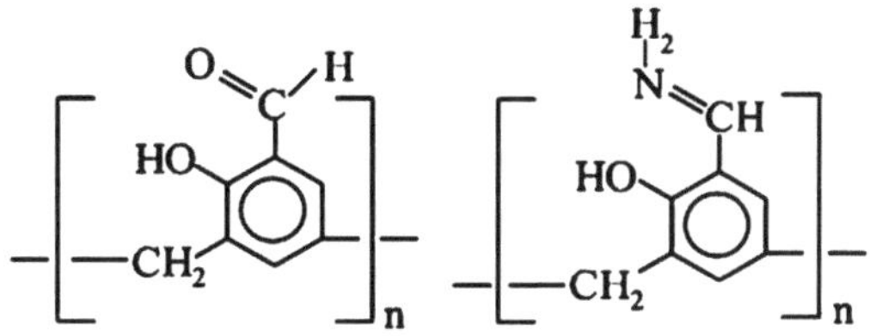

Fig. 1.1 Phenol formaldehyde ion exchange resins

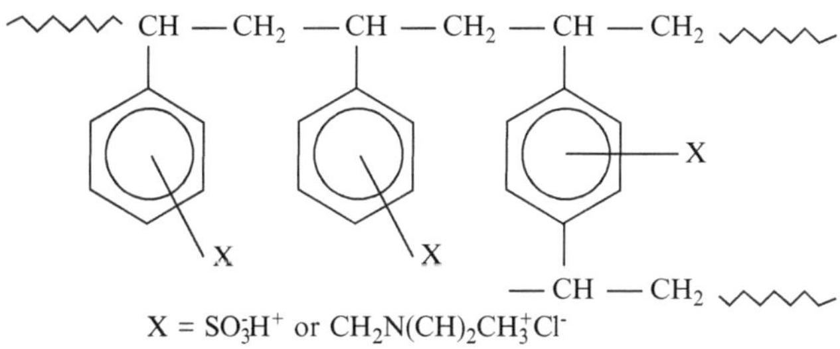

Fig. 1.2 Strong cation- and anion-exchange resins based on polystyrene divinylbenzene copolymer resins

when this reaction was found to be reversible. This phenomenon was caused by certain minerals in the soil as released in the latter half of the nineteenth century. These minerals, called resins, are based on tetrahedron structure of silicon and aluminum compounds called zeolites. In 1905, synthetic zeolites were manufactured and utilized for water treatment in a form of water-softening agent ever since [7]. Synthetic cation-exchange resins were developed during the 1930s using certain types of coal treated with sulfuric acid [8, 9]. This was an important evolution due to the fact that the sulfonated coal would operate in a greater pH range, 1–10. This made the sulfonated coal more versatile for the use in many more industrial applications. However, these resins were found to have serious deficiency caused by their lower exchange capacity compared to the zeolites. A few years later, the phenol formaldehyde polymer resin from the type shown in Fig. 1.1 was synthesized [10]. This polymer was sulfonated forming strong acid ion exchange resin. Using the same base polymer only functionalized with an amine (NH_2) produced the first weak base ion exchange resins. The major development for the power industry came in USA in 1944 when strong acid and strong base resins from the types shown in Fig. 1.2 were produced based on divinylbenzene cross-linked polystyrene, which was treated with sulfuric acid to make a strongly acidic resins or chloromethylated and subsequently aminated to produce strongly basic resins [11–13]. These resins possess much better characteristics than earlier phenol/formaldehyde resins. These new resins are now used almost exclusively in water demineralization plants for high pressure boilers. By the year 1950, weakly acidic ion exchange resins shown in Fig. 1.3 based on polymerization of methacrylic acid and divinylbenzene were developed. Eventually, the macroporous methyl methacrylate and divinylbenzene resins were synthesized with various functionalities (weakly basic, strongly basic,

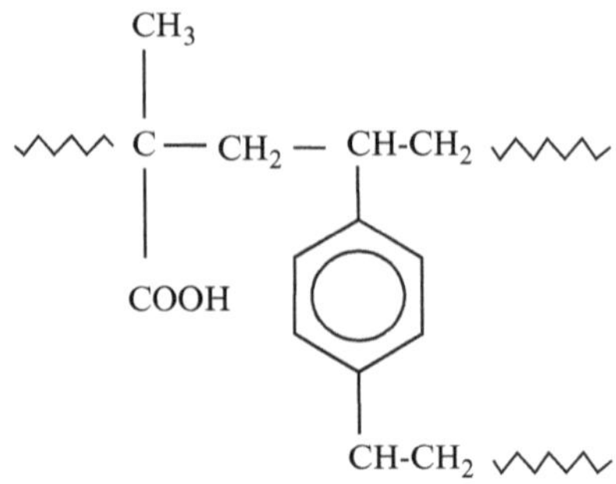

Fig. 1.3 Weak cation-exchange resins based on polymethylmethacrylate divinylbenzene copolymer resins

Table 1.1 The most important milestones in the development of ion exchange resins

Year	Milestones
1850	Agricultural chemists Harry Thompson and John Way discovered ion exchange phenomena
1858	German Chemist Eichom reported that ion exchange is a reversible reaction
1905	Robert Gans introduced first process to soften water using zeolite (sodium aluminosilicate)
1913	American company (Permutit) introduced first commercial zeolites
1935	English chemists Adams and Holmes prepared first synthetic polymer cation and anion exchangers (phenol formaldehyde and polyamine formaldehyde)
1944	D'Alelio developed cation-exchange resins based on polymerization of styrene and divinylbenzene
1946	Anion-exchange resins based on polymerization of styrene and divinylbenzene were developed
1950	Weakly acidic cation-exchange resins based on polymerization of methacrylic acid and divinylbenzene were developed
1965	Weakly basic resins based on polymerization of methyl methacrylate and divinylbenzene were developed
>1965	Bifunctional resins based on polymerization of methyl methacrylate and divinylbenzene were developed

and bipolar), by the year 1965 and above, with each resin having its own niche application in the water treatment industry [3]. Table 1.1 shows the most important milestones in the development of ion exchange resins. Today, hundreds of resins of various types, chemical groups, structures, and morphologies are available in the market with many more materials being researched to introduce more tolerance, cost-effectiveness, and new applications to ion exchange processes.

Ion exchange membranes are another significant class of materials that have been explored since the discovery of ion exchange phenomena. A significant development in ion exchange membranes was started by studies on ion-permeable membranes, collodion-type membranes which were carried out by Michaelis [14] who recognized the effect of membrane charge on ion permeation through the membrane. The theory of membrane potential was proposed by Meyer et al. [15]

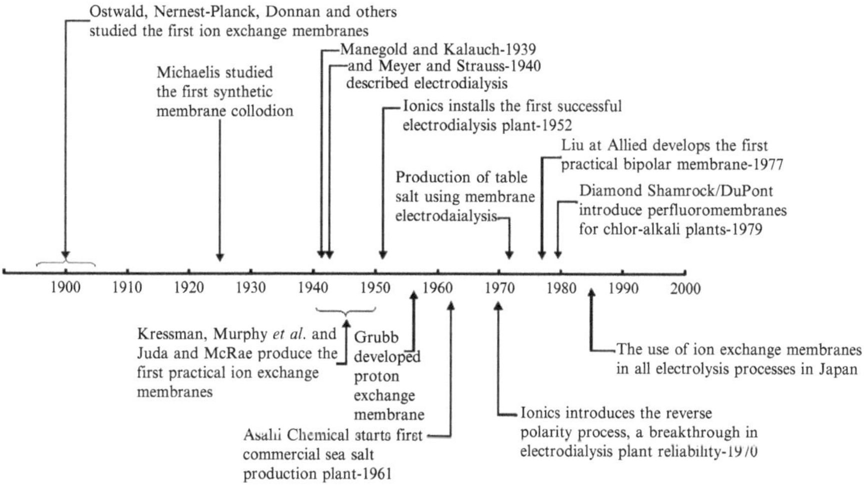

Fig. 1.4 Milestones of the development of ion exchange membranes and related processes

paving the way for the modified collodion membrane to be the first charged artificial membrane. The synthesis of ion exchange membranes was reported in 1950 by Juda and MacRae [16]. The preparation of proton exchange membrane and its use in fuel cells was first reported by Grubb in 1955. Later, in 1961 [17], Asahi Chemicals installed a membrane electrodialysis plant for the production of edible salt in Japan, and as a result, electrolysis in Japan was totally converted from the mercury method to a process using the ion exchange membranes by 1986. In 1977, the membrane chlor-alkali industry was introduced by Asahi Chemicals. This was followed by the introduction of Nafion (perfluorinated sulfonic acid membrane) for chlor-alkali industry by Diamond Shamrock and DuPont in 1979 [18]. The milestones in the development of ion exchange membranes are schematized in Fig. 1.4.

1.3 Ion Exchange Materials

Ion exchangers are a class of functional materials that display ion exchange properties owing to existence of fixed ionic sites bonded to their framework, which is held together by chemical bonds or lattice energy and can be called polyions. Oppositely charged ions move throughout the framework and can be replaced by ions of similar charge. Ion exchange materials are available in different forms and structures varying in their classifications depending on origin, physical form (morphology), immobilized functional group, and their functions, as shown in Fig. 1.5. The mechanism of ion exchange is dictated by various parameters related to the ion exchange materials such as the nature and type of fixed functional groups, the physical forms, and the origin of the ion exchange material [19].

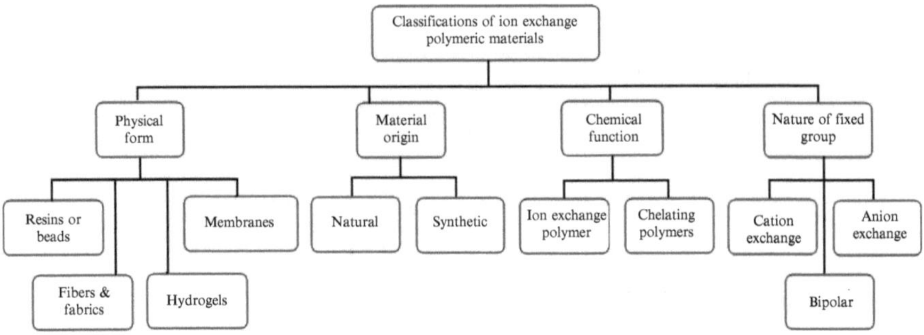

Fig. 1.5 Various classifications of ion exchange materials

1.3.1 Classifications of Ion Exchange Materials

On the basis of origin, there are two general types of ion exchange materials, that is, organic and minerallic; the former majority are synthetic polymers available in cationic and anionic forms whereas the latter exists in cation-exchange form only (e.g., zeolites and betonites). Thus, organic ion exchange materials can be cationic, anionic, and combined cationic/anionic (amphoteric) exchangers considering the nature of fixed ion exchange sites (functional groups).

Since ion exchangers act in a similar way to conventional acids and bases, the main classes of these materials, that is, cation and anion exchangers, can be further classified depending on the type of the functional group into several types: strongly acidic, strongly basic, weakly acidic, and weakly basic materials. Ion exchange materials containing sulfonate ($–SO_3^-$) and phosphorate acid ($–PO_3^-$) groups and those containing tetraammonium ($–NR_3^+$) basic groups are strongly acidic and strongly basic exchangers, respectively. On the other hand, materials containing phenolic (–OH) groups and primary amine ($–NH_2$) and secondary amine (–NRH) groups are weakly acidic and weakly basic exchangers, respectively. Carboxyl groups ($–COO^-$) and tertiary amine ($–NR_2$) groups take a medium position between strong and weak acidic and basic exchangers, respectively.

Practically, most strong acid exchangers contain sulfonate groups, which are active over the entire pH range. Unlikely, most weak acid exchangers have carboxylic groups, which are not active at pH values below 4–6. However, such exchangers often have higher ion exchange capacities than sulfonate exchangers together with other specific advantages [4]. Similarly, strong basic exchangers are active over the entire pH range unlike weak base exchangers which are not active at alkaline pH. A summary of the common functional groups and their negative logarithm of the dissociation constant (pK) are presented in Table 1.2. It can be clearly seen that each of these major resin classes has several physical or chemical variations within the class. Such variations impart different operating properties to the resin. Thus, the terms strong and weak in the ion exchange world do not refer to the strength of binding; it rather reflects the extent of variation of ionization with pH

Table 1.2 Common functional groups of polymeric ion exchange materials and their respective pK values

Anion-exchange materials		Cation-exchange materials	
Fixed ionic groups	pK	Fixed ionic group	pK
$\equiv N^+$	1–2	$-SO_3H$	1–2
$=N$	4–6	$-PO_3H_2$	2–5
$=NH$	6–8	–COOH	4–6
–NH2	8–10	–OH	9–10

of the medium solution. Each of these major resin classes has several physical or chemical variations within the class. Strongly acidic resins are commonly available in Na^+ form or H^+ form with different degrees of cross-linking to meet the requirements in various applications, whereas strongly basic resins are available in Cl^- or OH^- forms.

Physically, organic (polymeric) ion exchange materials are available in various morphologies related to the polymer framework carrying the functional groups. This includes beads, fibers, and membranes. Such variation in the physical forms brings about wide differences in chemical and physical properties of these ion exchangers. The majority of these ionic forms have synthetic polymer structures and mainly exist in a resin form represented by a wide number of commercial resins with polystyrene divinylbenzene backbone. A smaller class of biosorbents obtained from modified natural polymer sources including alginate, chitosan, and cellulose is also under development. Ion exchange resins fall into two main categories: cation- and anion-exchange forms. A combination of cation- and anion-exchange groups can be used to obtain a bipolar form of the resins that can replace mixed bed in ion exchange column. More details on the various types of ion exchange materials can be found in Refs. [4] and [20].

1.3.2 *Ion Exchange Resins*

Considering the separation mechanism, ion exchangers can be further classified into various categories including ion exchange resins, chelating adsorbents, hydrogels, affinity polymers, and ion exchange membranes. Among all, ion exchange resins, which are covalently cross-linked insoluble polyions supplied as spherical beads (particles), represent the major class of exchanger being commercially produced as stated earlier.

1.3.2.1 Preparation of Ion Exchange Resins

Commercial ion exchange resins that are available in market today are commonly produced by suspension polymerization, polycondensation, or polymer-analogous transformations [19]. Resins based on styrene-divinylbenzene copolymers as a building block involve the preparation of a cross-linked bead copolymer followed

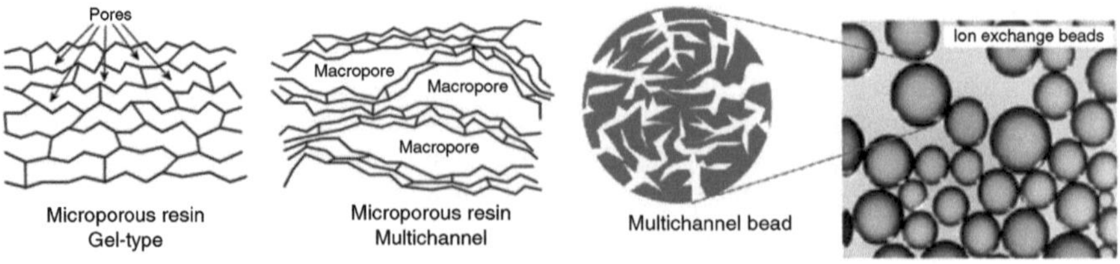

Fig. 1.6 Macroporous ion exchange resin bead with multichannel structure

by chemical treatment with a sulfonating agent in the case of strong acid cation-exchange resins, or chloromethylation and subsequent amination of the copolymer for obtaining anion-exchange resins. The degree of cross-linking achieved in the resin beads depends on the proportions of the styrene monomer and the cross-linking agent (divinylbenzene) used in the polymerization step. Practical ranges of cross-linking are in the range of 4–16%; however, 8% cross-linking level is preferred [4]. It is not straightforwardly known which one of the two resin structures, that is, strongly or weakly, is advantageous without knowing specific operating conditions of the treatment site. This is because the operating properties for acrylic resins differ from those of the corresponding sulfonic acid resin. The preferred resin is the one that has operating properties matching up best with the site's operating parameters, thus maximizing operating efficiency and cost-effectiveness.

1.3.2.2 Physical Characteristics of Ion Exchange Resins

Physically, ion exchange beads have either a dense internal structure with no or minimal discrete pores, that is, gel (microporous) (0.5–20 nm) resins or macroporous (macroreticular) (20–200 nm) resins with multichanneled structure, as schematized in Fig. 1.6. The type of internal structure of the resins beads dictates the selection of an ion exchanger for a particular application. Macroporous resins, with their tridimensional matrix with high effective surface area, obtained by adding porogen (phase extender) during their production followed by its washing out leaving voids in the structure, facilitate the ion exchange process. Also, multichannel structure gives an access to the exchange sites for larger ions allowing the use of almost any solvent, irrespective of whether it is a good for resin solvation together with little or no change in volume upon solvent loss. This confers more rigidity to the resin beads and facilitates their removal from the reaction system. On the other hand, the presence of no discrete pores in the microporous beads leads to a limitation in the solute ions diffusion hindering the interaction with fixed ionic sites and leading to a slowdown in the reaction rates. However, these resins offer certain advantages such as less fragility, less required care in handling, faster in functionalization and applications reactions, and possessing higher loading capacities [20].

Cross-linking of the resins during the preparation step is an essential step toward inheriting toughness and insolubility to the resin. The level of cross-linking plays an important role in affecting the physical and chemical properties of the resins. For example, resins with very low degree of cross-linking tend to adsorb more moisture and change dimensions markedly depending on the type of ionic sites bound to polymer backbone. The main physical properties that get influenced by cross-linking mainly include (1) moisture content and (2) particle size, together with other associated properties such as pore size and morphology, surface area, and partial volume of the resins in a swollen state [4]. The moisture content, which is primarily a hydration of the fixed ionic groups, is a function of the degree of cross-linking and the type of functional group. Low degree of cross-linking in gel resins having sulfonic acid or quaternary ammonium groups allows absorption of large amounts of water resulting in remarkable resin swelling. Water swelling of resins leads to variation in their volume, which varies in turn with degree of hydration that depends on the attached counter (incoming) ion [21]. However, frequent swelling and subsequent contraction during drying may shorten the resins lifetime.

On the other hand, resin particle size causes an impact on their performance. For instance, smaller particles improve the kinetics of the ion exchange reaction but cause an increase in the water pressure drop leading to a decrease of the flow rate [19]. The chemical properties of resins involve ion exchange capacity, type of resin's matrix, and type and concentration of functional groups attached to backbone of the resin.

The stability (chemical, physical, and mechanical) and the behavior of the ion exchange resins depend primarily not only on the structure and the degree of cross-linking of the resin matrix but also on the nature and number of fixed ionic groups. The former determines the porosity of the matrix and accordingly the degree of swelling of the resin and the mobility of the counter ions through it, which in turn controls the rates of ion exchange in the resin. Furthermore, highly cross-linked resins are harder and more resistant to mechanical breakdown but with lesser swelling and counter ion accessibility. However, higher and lower cross-linked resins are also available with less or more water sorption capabilities. Therefore, the most desired combinations of properties in ideal ion exchange resins required of industrial applications are:

1. Fast kinetics, that is, speedy rate of exchange
2. High chemical stability
3. Physical stability in terms of mechanical strength and resistance to wear
4. Reasonable degree of cross-linking
5. Hydrophilic structure of regular particle size and reproducible form
6. Effective and lasting ion exchange capacity
7. Consistent particle size
8. Effective surface area compatible with the hydraulic design requirements for large-scale plant

Finally, it can be stated that the performance of ion exchange resins in terms of kinetics and sorption equilibrium depends on the physical and chemical properties of the resins.

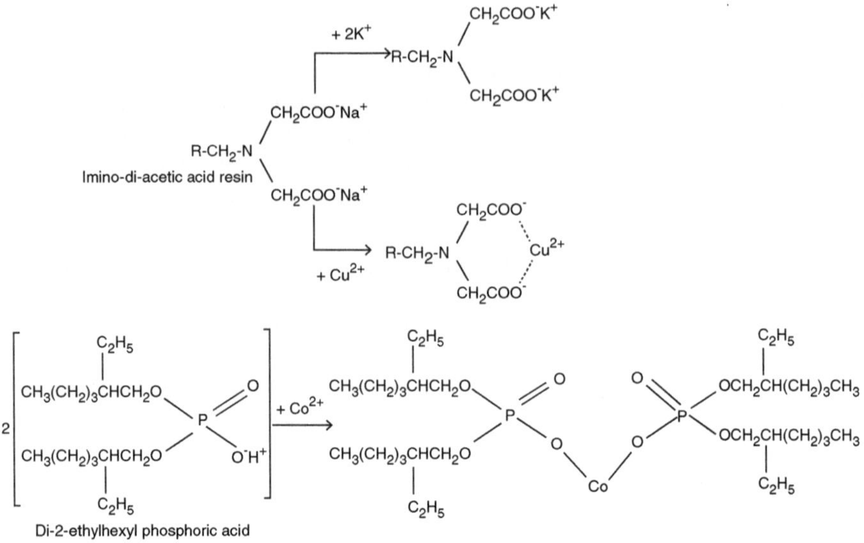

Fig. 1.7 Mechanisms of ion exchange in chelating resins

1.3.3 Special Ion Exchange Materials

1.3.3.1 Chelating Resins

Chelating or ligand exchange resins are considered as subgroup of normal ion exchange resins that bind counter ions by coordinate bonds interaction and sometimes by electrostatic interactions, as shown in the mechanisms in Fig. 1.7. Similar to common ion exchange resins, the chelating resins are insoluble in water as they have cross-linked matrix and are characterized by two components: polymer matrix, which provides the stability, and the functional groups, representing the ligands required for metal complexation. The most common coordinating atoms present in the main or side chains are N, O, P, and S in the forms of functional groups such as iminodiacetic acid and aminophosphoric acid, amidoxime, and thiol making the metal complexes, which their stability is a function of pH of the sorption medium. The principal active groups of chelating ion exchangers and their chemical formula are shown in Table 1.3.

Chelating resins have the same bead form and cross-linked polymer matrix as in normal ion exchange resins. They are mainly based on divinylbenzene-cross-linked polystyrene [22]. However, they differ from ordinary resins in their high selectivity toward certain metal ions in the sorption processes. Hence, it is possible to make chelating resins that have selective absorption capacity for certain metal ions by fixing desired ligand groups on the polymer backbone. For example, chelating polymers containing amidoxime groups were found to have high selectivity for uranyl ions and proposed for the recovery of uranium from seawater [23–26].

Table 1.3 Principal active groups of chelating ion exchangers and their chemical formula [76]

Chemical groups	Formula	Selectivity
Iminodiacetate	$-CH_2N(CH_2COOH)_2$	$Hg^+ > Cu^{2+} > UO_2^{2+} > Pb^{2+} > Fe^{3+} > Al^{3+} > Cr^{3+} > Ni^{2+} > Zn^{2+} > Ag^+ > Co^{2+} > Cd^{2+} > Fe^{2+} > Mn^{2+} > Ba^{2+} > Ca^{2+} > Sr^{3+} > Mg^{2+} > Na^+$
Aminophosphonic acid	$-CH_2NHCH_2CH_2PO_3H$	$Th^{4+} > U^{4+} > UO_2^{2+} > Fe^{3+} > Be^{2+} >$ Rare earth $> H^+ > Ag^+ > Cd^{2+} > Zn^{2+} > Cu^{2+} > Ni^{2+} > Co^{2+} > Mn^{2+} > Ca^{2+} > Na^+$
Thiol	$-SH$	$Ag^+ > Cu^+ > > Pb^{2+} > Cd^{2+} >> Zn^{2+} > Ni^{2+} > Fe^{3+} > Ca^{2+}$
Amidoxime	$-\underset{NH_2}{\underset{\vert}{C}}=NOH$	Cu^{2+}, Ru^{6+}, Au^{3+}, Rh^{3+}, V^{4+}, Pd^{2+}, U^{6+}, Pt^{2+}, Fe^{3+}, $Mo^{6+} > Cu^{2+} > Ni^{2+} > Co^{2+} > Zn^{2+} > Mn^{2+}$
N-methylglucamine	$-CH_2\underset{CH_3}{\underset{\vert}{N}}-(CHOH)_4CH_2OH$	BO_3^{2-}
N-bis-picolylamine	$-CH_2N(CH_2-C_5H_4N)_2$	$Ni^{2+} > Co^{2+}$ at low pH
Triethylammonium	$-N(C_2H_5)_3^+OH^-$	NO_3^-
Thiourea	$-NH-\overset{S}{\overset{\Vert}{C}}-NH_2$	$Hg^+ < Cd^{2+}$

Upon use, the selectivity of chelating resins is based on the different stabilities of metal complexes at the appropriate pH values. Chelating polymers are very useful for the purpose of selective adsorption of certain metal ions from their mixtures, removal of metal ions, and recovery of soluble metal ions [27–29]. Adsorption of toxic metal ions using chelating polymers has the advantage of high efficiency, ease of handling, availability of different adsorbents, and cost-effectiveness. Thus, chelating polymers has a great importance in environmental applications including wastewater treatment [30, 31].

Chelating polymers can be divided into two classes: graft adsorbents and graft biosorbents. The former have side chains carrying functional groups attached to a synthetic matrix, whereas the functional groups are attached to biopolymer matrix (e.g., chitosan, alginate, and starch) in the latter [19]. Recently, a great attention has been paid for developing cheaper and effective sorbents based on various natural polymers to overcome the high cost of the synthetic resins. The recent progress in the development of biosorbents from various natural polymers and their applications in heavy metal and color removal from industrial wastewater can be found on other reviews [32, 33].

1.3.3.2 Affinity Ion Exchange Materials

Affinity beads are porous particles having selective affinity ligands including immobilized biological or synthetic molecules attached to their pores. These beads have found a widespread application in separation and purification of biological macromolecules in both analytical and preparative scales in a variety of bioprocesses [34]. Among all, affinity (adsorption) chromatography, which provides an alternative technique to conventional chromatography, has proven its efficiency and time stability [35]. Such adsorption chromatographic techniques are often used for the isolation and purification of fermentation products [36]. They are also used to eliminate toxic effects on growth, reduce feedback repression of production, and ultimately extend fermentation time and product yield [37]. Column operations are commonly adopted for those applications, where affinity resins (beads) are used as chromatographic separation media. The selectivity of beads is largely a function of the three-dimensional geometry of the pores of the beads and occluded functional/ligand groups; therefore, both pore and chemical characteristics of the beads dictate the separation and adsorption capacity of the target compounds [38]. However, conventional packed beds have several major limitations: (1) the pressure drop across a packed bed is generally high and (2) the dependence on intraparticle diffusion transport of biomolecules such as proteins to their binding sites within the pores of such beads. Various chromatographic modes and techniques using beads selective for specific proteins have been introduced to enhance the protein separation and purification processes [39]. More details on the use of affinity resins and their recent progress in protein separation and purification can be found elsewhere [40, 41].

1.3.3.3 Amphoteric Ion Exchange Materials

Amphoteric or bipolar ion exchange materials are ion exchangers that have the ability to exchange cations and anions simultaneously with stoichiometric sorption of the whole dissolved salts. These polyions contain both cationic and anionic groups, or a combinations of groups such as $-SO_3^-H^+$ and $-N(CH_3)_2$ and –COOH and $-N^+(CH_3)_3OH^-$ laying back to back on a polymer backbone and also known as *ampholytic* polymers [42]. When ionic groups of opposite signs are incorporated into the same pendant groups, the material is called, depending on the structure of the pendant groups, a zwitterionic polymers or polybetaines. An example for preparation of cross-linked polyzwitterionic carboxybetaines based on the 4-vinylpyridine and divinylbenzene macromolecular supports can be found in Neagu et al. 2010 [43]. Zwitterionic polymers have high selectivity toward specific ionic species depending on the fixed ionic sites and are commonly used in selective separation as their corresponding amphoteric resins. The various trends in synthesis of functional polymer and the corresponding selective ion exchange polymers were early reviewed in Refs. [44] and [45].

1.3.3.4 Snake in Cage Ion Exchange Resins

Snake in cage represents a class bipolar polymeric ion exchangers having network structure in three dimensions with fixed charges and trapped linear polyelectrolytes of

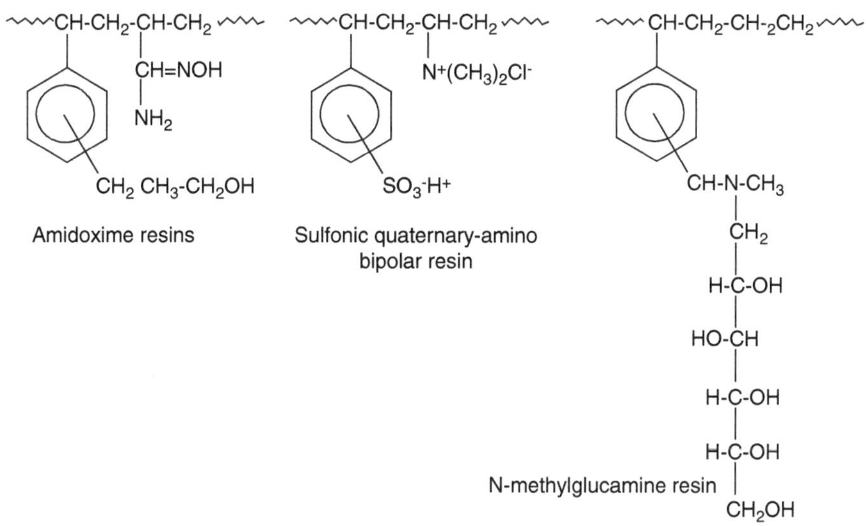

Fig. 1.8 Some special ion-selective resins

the opposite charge, which have equal values and tend to neutralize each other. These ion exchangers can absorb both anions and cations of equivalent amounts simultaneously. A typical commercial example for this resin is *Retardion* 11-A-8 (Dow Chemicals Co.) [46]. The resin of such type is characterized by having the ionic groups of counter ions not attached to the matrix; thus, the changes in the counter ions and that of the matrix have more freedom to move relative to one another and hence may neutralize one another. As a consequence, it is not necessary for the resin to contain mobile ions to remain electrically neutral provided that the changes of the fixed ionic groups and counter ions are properly balanced [47]. These resins are excellent reversible sorbents of electrolytes. They are used for separation by ion retardation [4, 20]. Examples of special ion-selective resins are shown in Fig. 1.8. N-methylglucamine-containing resins (e.g., Amberlite IRA 743, Dowex XUS 43594, Purolite S 108, and Diaion WA 30) are boron selective [48–50], whereas amidoxime resin is selective for uranyl ions [51–53]. The latest developments and achievements of various ion-selective resins together with ordinary resins in terms of materials and applications have been reviewed by Alexandratos (2009) [54].

1.3.4 Ion Exchange Membranes

Ion exchange membranes is a class of ion exchange materials in a form of highly swollen polymer film carrying fixed charged chemical groups covalently bonded to the polymer backbone that can exchange their mobile ions for ions of similar charge from the surrounding medium [55]. Ion exchange membranes can be divided into (1) cation-exchange membranes, which contain negatively charged groups ($-SO_3^-$, $-COO^-$, $-PO_3^{2-}$, $-PO_3H^-$ and $-C_6H_4O^-$),

(2) anion-exchange membranes, which contain positively charged groups ($-NH_3^+$, $-NRH_2^+$, $-NR_2H^+$, $-NR_3^+$, $-PR_3^+$ *and* $-SR_2^+$), and (3) bipolar (amphoteric) membranes containing both negatively and positively charged groups fixed to the polymer matrix in film/sheet forms.

The presence of fixed ionic groups (anionic or/and cationic) in the dense sheet-form structures confers these membranes their essential properties and infringes various interesting phenomena. The former is represented by ion exchange capacity and water uptake (swelling), both of which dictate the electrical resistance. The latter is marked by permselectivity of cations or anions through the membrane, generation of membrane potential and bi-ionic potential, ion conductivity, diffusion of electrolyte and nonelectrolyte through the membrane, water transport involving osmotic water and electroosmotic water, and salt rejection in the presence of hydraulic pressure [56]. Thus far, the separation behavior (cations from anions or vice versa from aqueous solution) of each type of these membranes is determined by the chemical nature of attached charged group, and thus, these membranes possess high permselectivity, that is, high preferential transport number for counter ions [57, 58].

Ion exchange membranes are also broadly classified into strong or weak acid cation-exchange membranes and strong or weak base anion-exchange membranes depending on the strength of the functional groups as explained in the corresponding ion exchange resins. In the amphoteric membranes, one side contains weak or strong anion-exchange layer and the other side possesses weak or strong cation-exchange layer. Ion exchange membranes can be also divided depending on the degree of heterogeneity into (1) homogeneous membranes, (2) interpolymer membranes, (3) microheterogeneous graft- and block-copolymer membranes, (4) chelating membrane, and (5) heterogeneous ion exchange membranes [59, 60]. Ion exchange membrane materials can be further divided into three large classes: (1) high molecular weight membranes based on polymer containing ionic moieties, (2) inorganic membranes, and (3) hybrid membranes composed of organic/inorganic materials [61].

Like their corresponding resins, affinity polymers in the form of membranes are also available. These membranes are capable of overcoming the shortcomings (e.g., fouling) associated with membrane operation based on sieving mechanism such as microfiltration and ultrafiltration. Affinity membranes have selective functional groups immobilized in the porous structure of a film that allow high flow rates to be achieved at low pressures, enabling the processing of large volumes (such as cell culture fluids) within short processing times compared to affinity beads [62]. Ideal film substrates used for affinity membranes should fulfill the following conditions: (1) proper pore structures and mechanical strength for use at high flow rates and low back pressure in rapid processing, (2) availability of reactive groups such as *–OH*, *–NH*, *–SH*, and *–COOH* for the further coupling of spacer arms or ligands, (3) chemical and physical stability under harsh conditions of high temperature or chemical sterilization, and (4) hydrophilic surfaces to avoid any unspecific interaction between biomolecules and the membrane surface [63]. This has made affinity membrane, which combines a high processing rate with a selective affinity adsorption, proven to be more efficient than the classical column affinity method in chromatography [64].

Ion exchange membranes have been widely used in various industrial applications related to separation processes and ion-selective transport in many electrochemical processes and solid-state devices. The most famous membrane separation systems are utilizing diffusion dialysis, electrodialysis, pervaporation, and reverse osmosis, which operate based on concentration, potential, or pressure gradient as a driving force. Electrochemical devices using ion exchange membranes including fuel cells, batteries, and sensors are also available. To meet the requirements for such applications, ion exchange membranes have to essentially possess (1) low electrical resistance, (2) high ionic permselectivity, (3) moderate degree of swelling, (4) reasonable degree of cross-linking, (5) thermal resistance, (6) chemical stability, and (7) high mechanical strength. The various characteristics of ion exchange membranes and the diversity of their applications have been reviewed in literature on several occasions [60, 65–69].

1.4 Fundamentals of Ion Exchange Reactions

1.4.1 Ion Exchange Equilibrium and Selectivity

Ion exchange process depends on the mechanism by which mobile ions from an external solution are exchanged in the opposite direction for an equivalent number of ions that are electrostatically bound to the functional groups contained within a solid matrix of the ion exchange material.

The preference of one ionic species over another by the ion exchangers can be attributed to several causes [20]:

1. The electrostatic interaction between the charged framework and the counter ions depends on the size and valence of the counter ion.
2. Other effective interactions between ions and their environment.
3. Large counter ions may be sterically excluded from the narrow pores of the ion exchanger.

All these effects depend on the nature of the counter ion and thus may lead to preferential uptake of species by the ion exchanger. The ability of the ion exchanger to distinguish between the various counter ion species is called selectivity.

Metal ions initially contained in an aqueous solution are exchanged with ions initially contained in ion exchange materials. This process represents cation-exchange reaction that can be illustrated by the reaction:

$$\overline{R^{-}H^{+}} + A^{+} \rightarrow \overline{R^{-}H^{+}} + H^{+} \tag{1.1}$$

where R is ion exchanger and A^+ is a positively charged metal ion. The bars indicate the phase of the ion exchanger. The equilibrium constant (K) is given by the following formula:

$$K = \frac{[RA]}{[RH]} x \frac{[H^+]}{[A^+]} \tag{1.2}$$

The distribution of the two ions between the two solid and liquid phases (ion exchange material and external solution) depends on the reaction constant value. A similar reaction involving anions is called anion exchange:

$$\overline{R^+Y^-} + B^- \rightarrow \overline{R^+B^-} + Y^- \tag{1.3}$$

where B^- and Y^- are anions or negatively charged metal ions. The equilibrium constant (K) is given by the following formula:

$$K = \frac{[RB]}{[RY]} x \frac{[Y^-]}{[A^-]} \tag{1.4}$$

Ion exchange reactions are reversible and follow stoichiometry, as seen in the following reaction:

$$2(R - SO_3H) + NiSO_4 \leftrightarrow (R - SO_3)_2\text{Ni} + \text{H}_2\text{SO}_4 \tag{1.5}$$

To achieve the removal of both positively and negatively charged ions from solution, a mixture of cation and anion exchanger in a mixed bed system is often used. Alternative, bipolar bed system can replace the cation and anion exchanger in a mixed bed system.

Unlike many other chemical separations, reactions 1 and 3 can be successfully used even if they are shifted to the opposite direction. By taking advantage of the fact that, under certain conditions, ion exchange media has a greater affinity for certain ionic species than for others, a separation of these species can be made. Chemical selectivity of reactions 1 and 3 is desirable but is not a requirement [4]. The selectivity (ion exchange) coefficient for reaction 1 expresses the relative distribution of the ions when a resin in the H^+ form is placed in a solution containing A^+ ions, whereas the selectivity coefficient for reaction 3 expresses the relative distribution of the ions when a resin in Y^- form is placed in a solution containing B^- ions. The selectivity coefficients for reaction (1) ($K_{H^+}^{A^+}$) and (3) ($K_{Y^-}^{B^-}$) are given by:

$$K_{H^+}^{A^+} = \frac{[A^+]_{\text{Re sin}}}{[H^+]_{\text{Re sin}}} \times \frac{[H^+]_{Solution}}{[A^+]_{Solution}} \quad \text{and} \quad K_{Y^-}^{B^-} = \frac{[B^-]_{\text{Re sin}}}{[Y^-]_{\text{Re sin}}} \times \frac{[Y^-]_{Solution}}{[B^-]_{Solution}} \tag{1.6}$$

The selectivity coefficients in ion exchange systems can describe the ion exchange equilibrium and provide a means of determining what to expect when various ions are involved. However, they are not constant and vary with the operation conditions such as types and properties of the resins, concentration of the feed solution, pH of the solution, temperature, and the presence of other competing ions in the solution [70].

When two ions have similar charge and ionic radii, the selectivity stemmed from the properties of the ion exchange material (acidity, basicity, and the degree of cross-linking) is not sufficient for performing effective separation. Therefore, an appropriate complexing agent has to be added to the aqueous phase to allow the selectivity to be attained through either the difference in the stability constants, the difference in charges, or the variation in the structures of the complexes formed. As a result, an increased selectivity can be achieved. However, the determination of selectivity coefficients is practically complicated task and is ordinarily not undertaken in the design of ion exchange processes such as wastetreatment systems. In fact most of these parameters can be extracted from manufacturers' data or research literature [1].

The rules of thumb that have to be considered for general design purposes when cationic polymeric ion exchange resins at low concentrations and normal waste processing, the selectivity typically increases with increasing charge on the exchanging cation in the order: $Li^+ < H^+ < Na^+ < K^+ < Cs^+ < Rb^+ < Ag^+ < Be^{+2} < Mg^{2+} < Cu^{2+} < Ni^{2+} < Co^{2+} < Ca^{2+} < Sr^{2+} < Ba^{2+} < Ce^{3+} < La^{3+} < Th^{4+}$. Selectivity also increases with increasing the atomic number (decreasing hydrated ionic radii) of the exchanging cation in the order of $Li^+ < H^+ < Na^+ < K^+ < Cs^+$ (Li^+ is an exception, owing to its high hydration energy).

For anions, a typical series of selectivity is as follows: $F^- \approx OH^- < HCO_3^- < CH_3COO^- < Cl^- < HSO_3^- < Br^- < CrO_4^{2-} < NO_3^- < SCN^- < I^- < ClO_4^- < C_2\,O_4^{2-} < SO_4^{2-}$.

The ion exchange equilibrium can be also described in terms of the ion exchange isotherm, the thermodynamic equilibrium constant, and the distribution coefficient.

1.4.2 Ion Exchange Thermodynamics

The distribution coefficient (K_d) is a measure of the effectiveness of the ion exchange resins to remove particular ionic species (metal ion) from a liquid solution. It is defined as the ratio of metal ion concentration in the solid to metal concentration in the liquid. The K_d is generally determined with batch contacts between the exchanger and solution and is calculated according to the following formula:

$$K_d = \frac{(C_i - C_f)}{C_i} \cdot \frac{V_s}{m_e} \tag{1.7}$$

where C_i is the initial concentration (before contact), C_f is the final concentration of metal in solution, m_s is the mass of exchanger used, and V_s is the volume of solution. The K_d is specific to the temperature and concentration of other ions in the solution. Systematic information of the K_d forms fundamental base from ion exchange process to be planned. High values of the distribution coefficient are always desirable. Changes in the physical parameters and the concentration of functional groups affect the K_d and can affect the driving force for the ion exchange process.

The K_d value is used in the following equation to determine the Gibbs free energy of adsorption (ΔG^o):

$$\Delta G^o = -RT \ . \ \ln \ . \ K_d \tag{1.8}$$

where ΔG^{o} is standard free energy change (J), R is the universal gas constant (8.314 J mol^{-1} K^{-1}), and T is the absolute temperature (K).

1.4.3 Ion Exchange Kinetics

Kinetics of the ion exchange reaction represents the speed with which reaction takes place. Upon designing an ion exchange process system, the rate of exchange, which is subjected to mass-transfer resistances in both liquid and solid phases, is highly significant in the design and operation of the ion exchange process. The exchange rate has impact on the size of the ion exchange columns required and/or its flow rate. The kinetics of the ion exchange reaction is affected by the type and nature of the exchanger, solution concentration, and temperature at which the exchanger is operated.

When an ion exchange bead is brought into contact with a solution, a static liquid film of a thickness varying in the range of 10–100 nm depending on the rate of flow of liquid passes the particle, is formed around it. The ion exchange process taking place between the resin particle and the solution involves six distinct steps; any combination thereof can be the rate controlling step. They are [4, 20]:

1. Diffusion of counter ions through the bulk solution surrounding the bead
2. Diffusion of counter ions through the hydrated film of solution that is in close contact with the bead matrix
3. Diffusion of counter ions within the bead matrix
4. The actual ion exchange reaction
5. Diffusion of the exchanged species out of the ion exchange bead
6. Diffusion of the exchanged species from the bead surface particle into the bulk solution

It is noteworthy mentioning that the concentration differences that might occur in the bulk solution are constantly leveled out by agitation, which disturbs the necessary transfer by convection. Nonetheless, agitation neither affects the interior of the beads nor the thin liquid film, which adheres to the bead surfaces. The ion transfer takes place within the beads and through the thin film only by diffusion, and as a result, the actual ion exchange reaction occurs very quickly and is not generally considered to be rate limiting. Thus, the rate-determining step is either the interdiffusion of counter ions within the ion exchanger itself (particle diffusion) or interdiffusion of counter ions in the adherent film (film diffusion). In practice, either step can be the rate-determining step. However, the slower of these two steps controls the overall reaction rate. In intermediate cases, the rate may be affected by both steps [4]. Moreover, the particle size of the resin being used is also a determining factor. Uniform particle-sized resins exhibit enhanced kinetic performance compared to conventional polydispersed resins due to the absence of kinetically slow larger beads. Recent studies showed that ion exchange resins in fibrous forms have improved the kinetics allowing greater adsorption of toxic and precious metals [71, 72].

It is important to mention that in systems with ion exchange through organic solvents, the transfer of the counter ions in the resin is usually lower than in aqueous systems because swelling is less pronounced and electrostatic interactions with the fixed charges are stronger. Particle diffusion, thus, is relatively slow and, hence, usually controls the rate-controlling mechanism [2]. Finally, the actual rates of ion exchange can vary over a wide range, requiring a few seconds to several months to reach equilibrium.

1.4.4 Exchange Capacity

Capacity of the ion exchange materials is a significant chemical characteristic that is defined as the number of counter ion equivalents that can be adsorbed in a specified amount of ion exchange material. The ion exchange capacity and related data are primarily used for (1) describing the total available exchange capacity of ion exchange materials and (2) the use in the numerical predesign calculations of ion exchange processes. This value is expressed in milliequivalents per gram (meq/g) or milliequivalents per milliliter (meq/mL) based on the dry and wet weights of the material in a given form (H^+ or Cl^-), respectively. The capacity of the resins can be described in various ways depending on the conditions. For example, maximum ion exchange capacity (meq/g dry H^+ or Cl^- form) is represented by the total static exchange capacity when it is determined under static conditions, whereas the dynamic or volume-exchange capacity (eq/L packed bed in H^+ or Cl^- form and fully water-swollen) is determined by passing a solution through a bed of the exchanger. The extent of the use of the total exchange capacity depends on the level of ionization of the functional groups of the exchanger together with the chemical and physical conditions of the process. The total dynamic capacity can be also expressed as breakthrough capacity, which characterizes the column operation and depends on (1) column design and operating parameters, (2) the concentration of the ions being removed, and (3) the effects of interference from other competing ions.

Under dynamic conditions, in which the ion exchange resin is used in a packed bed or a column, the distribution coefficient K_d can be used to calculate the theoretical maximum capacity of the ion exchange bed in terms of weight. If the bed includes m (kg) of ion exchange resin, the total capacity of the bed V_{tot} can be calculated as:

$$V_{tot} = K_d \times m \tag{1.9}$$

In water purification, the volume of each type of the resins can be determined as a function of the operating time chosen between two regenerations using the calculated operating capacities:

$$V = (Q \,.\, t \,.\, S)/C \tag{1.10}$$

where

V = resin volume, m^3
Q = flow rate, m^3/h
t = operating time, h
S = salinity to be adsorbed by the resin, eq/m^3 of water or meq/L
C = operating capacity of the resin, eq/m^3 of resin

In column operation, the total capacity or breakthrough capacity generally refers to the volume of the solution that can be treated before a sharp increase in the effluent concentration of the ionic species being removed is observed. At this point, the ion exchange medium is considered to be exhausted and must be replaced or regenerated. This trend can be clearly seen in the column schematic representation shown in Fig. 1.9. Part I refers to the resin, which has finished exchange reaction; part II refers to the exchange zone in the resin, which is working until it moves downward with the continuous flow of Y^- ions (ion to be removed); and part III represents the zone in which the resin is exhausted and Y^- ions completely break through the resins [73].

Thus, the performance of ion exchanger is measured by comparing the solution concentration or conductivity of both influent and effluent, which can be monitored in terms of ion-exchanger effectiveness known as *decontamination factor (DF)*, which is defined as a ratio of the concentration (or activity) of the fluid at the inlet compared to the concentration (or activity) at the effluent (Eq. 1.11), which expresses the effectiveness of an ion exchange process.

$$DF = \frac{\text{Influent concentration or conductivity}}{\text{Effluent concentration or conductivity}} \tag{1.11}$$

In practice, the operator of the ion exchange system has to have a requirement for DF. For example, if the decontamination factor requirement is 100, the ion exchange bed has to be replaced when 1% breakthrough occurs. Therefore, the breakthrough capacity is practically the most interesting parameter in the design of a column type ion exchange system and is generally given as the number of bed volumes (the ratio of the volume of the liquid processed before the breakthrough point to the volume of the settled bed of the exchanger).

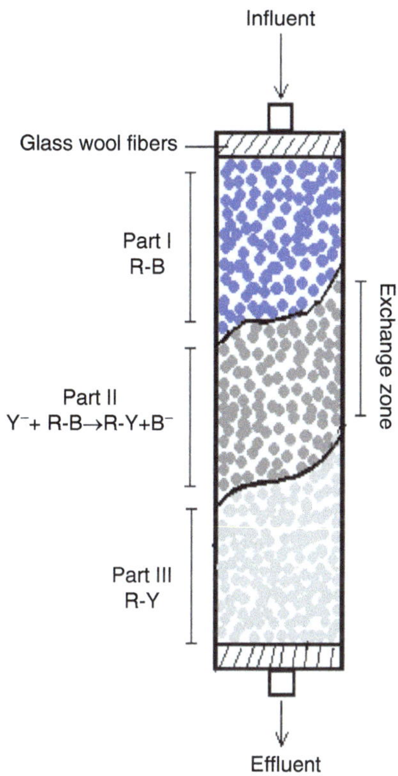

Fig. 1.9 Schematic diagram of ion exchange column

A typical breakthrough curve is shown in Fig. 1.10. The points C_o and C_x are the concentrations of Y^- in the influent and effluent, respectively. The breakthrough begins at point *c*, and the extent of breakthrough increases until point *e* (end point) beyond which no further ion exchange takes place (i.e., a complete exhaustion point of the column). The breakthrough capacity is proportional to the area *abcd* and so is the overall capacity to the area *aecd*. The shape of the breakthrough curve and the breakthrough capacity (i.e., the total liquid volume treated up to a particular percentage breakthrough) depend on the operating parameters.

Some important parameters that affect the breakthrough capacity include [1, 4, 20, 73]:

1. The nature of the functional group on the exchanger
2. The ion exchange medium grain size
3. The degree of cross-linking of the resins
4. The concentration and type of ions in the solution to be treated
5. The flow rate of the feed solution
6. The rate of percolation
7. Depth of the resin bed
8. Type, concentration, and quantity of regenerant

In practical operations, it is desirable to have the breakthrough curve as steep as possible to increase the extent of column utilization (i.e., bringing the ratio of the

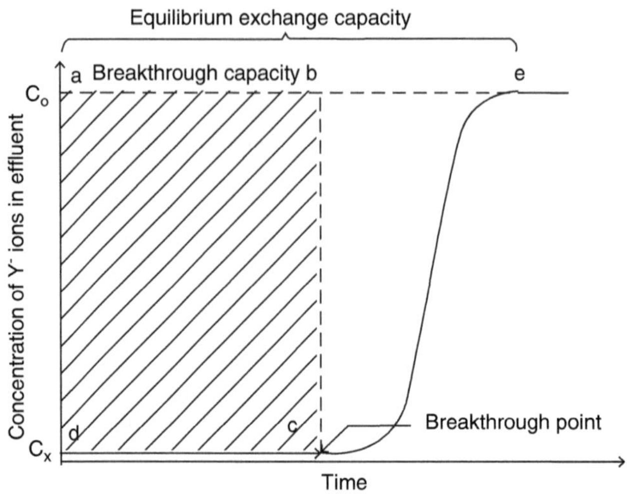

Fig. 1.10 Schematic representation of breakthrough curve

breakthrough capacity to the total capacity closer to unity). It is worth noting that upon requirement of high DF, the degree of column utilization can vary considerably for different ion exchange resins.

1.5 Ion Exchange Process Equipment and Operation

The feasibility and success of ion exchange processes depend on the nature of the ion exchange material and its process operating conditions. For instance, the morphology and type of the ion exchanger dictate the selection of the process configuration and its operating mode. As mentioned earlier, the basic morphologies of ion exchange materials include resins or beads, membranes, fibers, and hydrogels. To bring such polymer forms into applications, they have to be hosted in a proper engineering system that provides a set of adjustable operating parameters to control their performance. Accordingly, the ion exchange systems take various configurations depending primarily on the physical form of ion exchange material, as indicated in Fig. 1.11 which depicts a schematic representation of various morphologies of ion exchangers and their respective operating systems. In addition, the volume of the solution to be treated and the concentration of the ionic species to be separated also play a role in selection of system configuration but to lesser extent.

1.5.1 Cycle of Ion Exchange Process

When ion exchangers are in a resin form, the separation is conventionally accomplished by either batch or column method. Most commonly, the ion

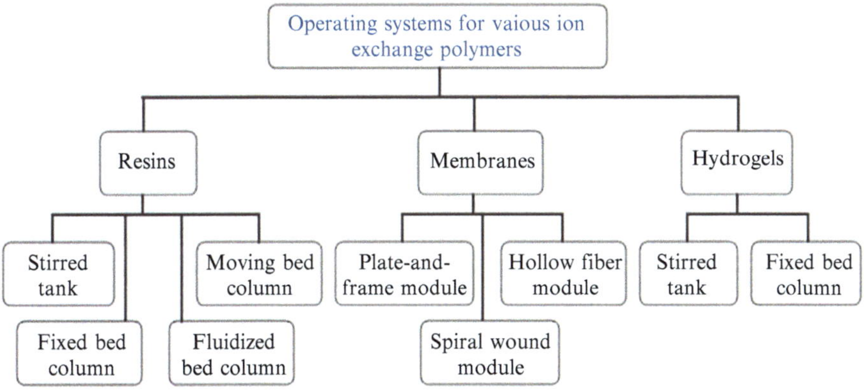

Fig. 1.11 Various operating configurations for different morphologies of ion exchange materials

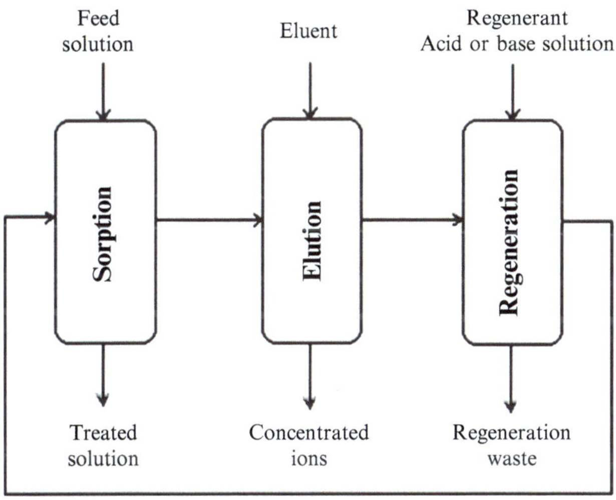

Fig. 1.12 Schematic representation of ion exchange operation cycle

exchange is performed in cyclic operations. Each cycle is divided into three main stages: (1) sorption, (2) elution, and (3) regeneration, as shown in the ion operation cycle schematized in Fig. 1.12. Details of ion operation cycle are as follows:

1. Sorption: The solution containing the targeted ions is passed slowly through to reaction vessel. The ions bind into the resins. Ions initially contained in the exchanger are released.
2. Elution (stripping): The target ions are subsequently stripped from the loaded resin with a small volume of an eluent. The eluent replaces and hence also releases the target ions from the resin into the solution phase.
3. Regeneration: The principle of displacement of selectively binding ions by less selective binding ion is the basis for each regeneration procedure. Depending on

the type of the ion exchanger and the stripping agent, most of the ion exchanger can be regenerated by acids (excess of H^+ ions), salt-brines (excess of sodium or chloride ions), or alkali (excess of OH^- ions). During regeneration, the adsorbed ions are removed and replaced by suitable ions. For example, if the sorption step uses a cation exchanger loaded with H^+ ions but the elution leaves Na^+ ions in the exchanger phase, the material has to be protonated. A strong acid could be applied in order to convert (regenerate) the exchanger in the initial state. It is important to mention that designing an optimized cyclic ion exchange process requires efficient utilization of bed through selection of resins having high selectivity toward the target ions.

1.5.2 Operating Configurations of Ion Exchange Processes with Resins

Upon utilization of ion exchange process, solution (influent) is pumped through a vessel loaded with the ion exchange resin and represents one of four basic operating configurations (1) continuous stirred (batch) tank, (2) fixed bed column, (3) fluidized bed column, and (4) a moving bed (closed loop) system.

In a batch method, the resin and solution are mixed in a batch tank, the exchange reaction is allowed to come to equilibrium, and then the resin is separated from the solution. The degree to which the exchange takes place is limited by the preference the resin exhibits for the target ions in solution. Consequently, the use of the resin's exchange capacity is limited unless the selectivity for the ions in solution is far greater than for the exchangeable ion attached to the resin. Because batch regeneration of the resin is chemically inefficient, batch processing by ion exchange has limited potential for industrial applications and often used in laboratories for fundamental sorption equilibrium and kinetic studies [74].

The fixed bed column operation is the most usable ion exchange configuration in industrial applications. It has a basic component of which the bead column is considered to be analogous of several batch reactors in series [1, 4]. The purpose of the column operation is to work around the limitation of selectivity of the resin. To favor high selectivity, more stages can be used (multicolumn system). Three types of column operation modes are available: (1) down flow, (2) up flow, and (3) counterflow. Most column beds operate with down flow operation, as shown in Fig. 1.13 that depicts a schematic representation of fixed bed ion exchange column system with operation and regeneration modes. This is where feed and beads pass down through the resin bed. On the contrary, up flow operation is used when the feed and resin are raised through a bed. The final flow is counterflow, and it consists of the feed flowing down from the top and the regenerate flows up from the bottom. Most industrial applications of ion exchange use fixed bed column systems, the basic component of which is the resin column.

Ion exchange column is typically applied in a pressure vessel, as shown in Fig. 1.14. It is equipped with appropriate internal plumbing that has two purposes:

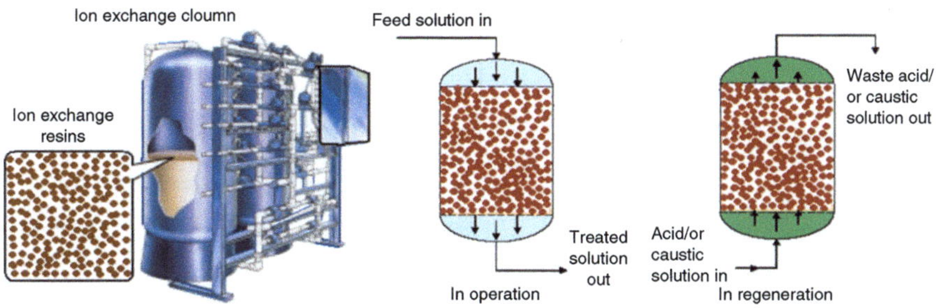

Fig. 1.13 Schematic representation of fixed bed ion exchange column system showing operation and regeneration modes

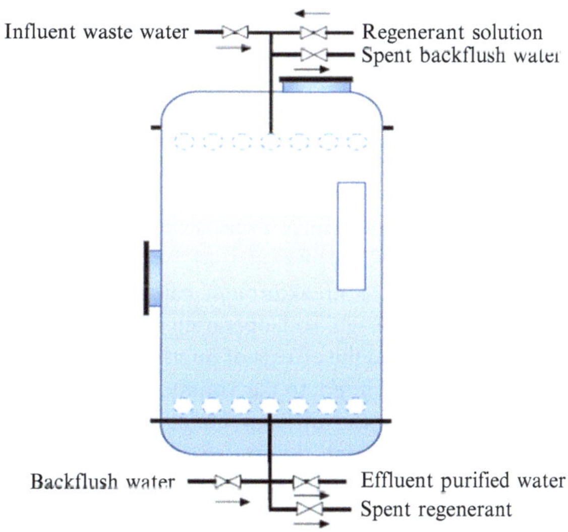

Fig. 1.14 Typical structure of ion exchange column [Reproduced from Ref. [1] with kind permission of © Wiley-VCH Verlag GmbH & Co (2005)]

(1) it prevents the ion exchange resin from being washed out of the vessel and (2) it provides an appropriate distribution of liquid flow through the ion exchange resin bed. The internal system is a corrosion-resistant screen mounted above a porous backing plate. The pressure vessel is also equipped with slit nozzles (strainers), piping, and fittings to maintain low pressure drop. During the operation, the pressure vessel is filled to only half of its volume with the empty space above the ion exchange resin bed (called freeboard) to allow free volume for swelling and shrinking during operations of exhaustion and regeneration.

In summary, the column design must consider the following [1]:

1. Contain and support the ion exchange resin.
2. Uniformly distribute the service and regeneration flow through the resin bed.
3. Provide space to allow the resins to swell, shrink, and be fluidized during backwash.
4. Include the piping, valves, and instruments needed to regulate flow of feed, regenerant, and backwash solutions.

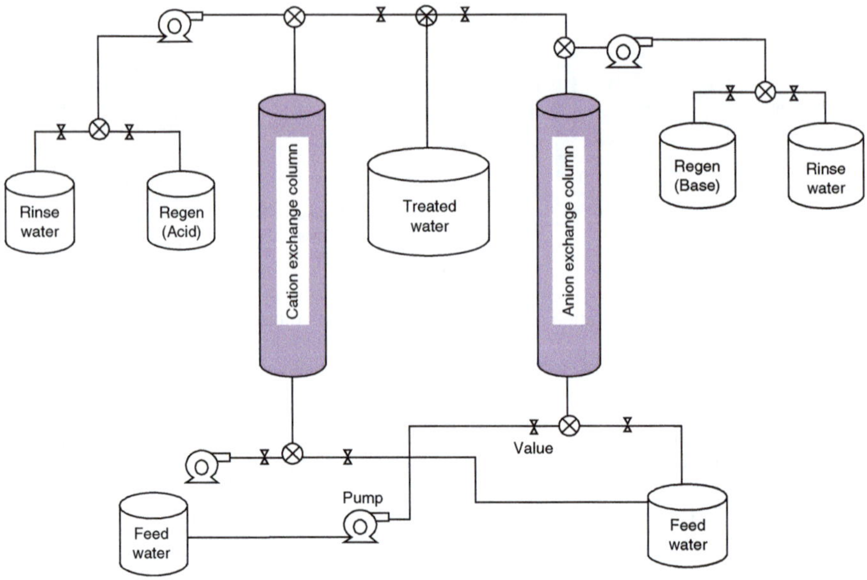

Fig. 1.15 Schematic diagram of a laboratory scale ion exchange column system

The operating or breakthrough capacity of an ion exchange column system depends on its design and operating parameters, the concentration of the ions being removed, and the effects of interference from other ions. In a column system, capacity generally refers to the volume of the solution that can be treated before a sharp increase in the effluent concentration of the species being removed is observed. At this point, the ion exchange medium is considered to be spent and must be replaced or regenerated. The operating or breakthrough capacity is the most significant parameter in the design of a column-type ion exchange system and is generally given as the number of bed volumes (the ratio of the volume of liquid processed before the breakthrough point to the volume of the settled bed of the exchanger). Some important parameters that affect the breakthrough capacity include (1) the nature of the functional group on the exchanger, (2) the degree of cross-linking of the resins, (3) the concentration of the solution to be treated, (4) the valence and the size of ions to be exchanged, and (5) the operating temperature. Figures 1.15 and 1.16 show a schematic diagram and picture, respectively, of a laboratory-scale fixed bed column ion exchange system. In practical operations, it is desirable to have the breakthrough curve as steep as possible. A steeper breakthrough curve will increase the extent of column utilization, that is, to bring the ratio of the breakthrough capacity to the total capacity closer to unity as early mentioned. If the required decontamination factor is high, the degree of column utilization can vary considerably for different ion exchange media.

The fluidized bed column operation is used when the raw feed contains suspended solids. The column is partially filled with the ion exchange beds. The feed solution flows down up through the bed with a speed sufficient to push up the

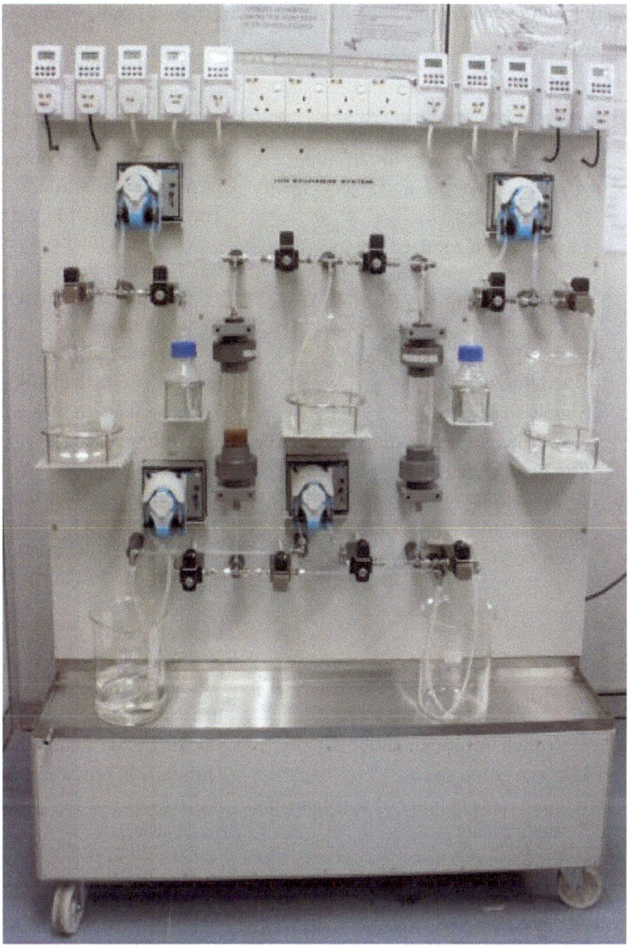

Fig. 1.16 Picture of a laboratory scale ion exchange column system schematized in Fig. 1.15

beads up into the column causing them to float and swirl around. The floating of the ion exchange beads reduces the hydrodynamic resistance of the column. The presence of free-pass ways between beads allows the solid suspended particles to be transferred through the column giving fluidized bed column the advantage of processing solid-containing solution without any pretreatment. Therefore, fluidized beds are commonly used for biochemical filtration such as fermentation broth where treated solutions are highly contained with suspended solids.

The moving beds are the most economical beds of ion exchange systems. The principle of this operation relies on bringing the beads and the solution to flow through the system. The beads are contacted countercurrently with the exhausting and regenerated streams. The advantage of this operation is featured by continuous product of uniform quality at less space, capital, and labor. However, the

Table 1.4 Comparison between the advantages and disadvantages of various ion exchange systems

System configuration	Advantages	Disadvantages
Batch system	Simple to construct and operate Good for small-scale applications A wide variety if ion exchange media can be used Easy to be customized for specific treatment problems Operate with resins of various shapes	Manual operation may be cumbersome to operate with large volumes of waste The separation of liquid and ion exchange media is required Can only be operated at atmospheric pressures and ambient temperatures Once through use only
Conventional column system	Good throughput Simple to operate A wide variety of media are available Can be operated at elevated temperatures and pressures High decontamination factors are possible	Large equipment can be costly The regeneration of media may require extra equipment There are difficulties in transporting inorganic ion exchangers through pipelines Cannot treat solid-containing solutions and thus prefiltration is necessary Limited part of the bed particle is involved in ion exchange in the same time
Moving bed system	The process is continuous Very high decontamination factors Less operation space Requires less number of supplementary equipments Low pressure drop Can treat solid-containing solution without bed plugging	Instability of the process Longitudinal mixing problem Difficulty in providing circular motion Extensive mechanical deterioration of resin beads
Membrane system	Can be used either as a waste treatment or fluid concentration technique Prefiltration is not necessary Does not need granular materials High decontamination factors are possible	Equipment can be costly to construct and operate on a large scale

complexity of the design poses a problem during operation. Comparison between the advantages and disadvantages of various ion exchange systems is presented in Table 1.4.

With the wide number of resin types available in the marketplace, it is probable that there is more than one technically effective solution that will meet all the system's design specifications. This is where experience and knowledge are required to help selecting a system design suitable for the separation expected by the customer under the conditions existing at the desired site. This experience includes a thorough knowledge of all available resin types along with their various advantages and disadvantages so they can be applied in ion exchange systems that are both technically sound and cost-effective. Sound equipment design, effective use of available resins, and readily available services that meet all the expectations of the industrial customer will lead to a win/win long-term customer relationship that benefits both the supplier and the customer.

1.5.3 Operating Configurations of Ion Exchange Membrane Processes

Putting ion exchange membrane into practical application requires the membrane to be housed in a vessel called module. Membrane module provides support and protection against operating pressures (if any) and daily wear and tear of a production environment and allows applying control strategy over the system performance. More detail on the requirements for efficient module designs to house various membrane types can be found elsewhere [75]. The most common configurations of membrane modules are (1) plate and frame, (2) spiral wound, (3) tubular, and (4) hollow fiber. Among all, plate-and-frame module is the most common module used in ion exchange membrane processes.

Plate-and-frame module consists of layers of membranes separated by corrugated structural sheets. Corrugations run at right angles in alternating layers with feed material flowing in and retentate flowing out in one direction, while a carrier fluid flows in and permeate out in the other direction. Advantages of this system include ability to accommodate low levels of suspended solids and viscous fluids and the ability to change out membranes if needed (though it is not difficult). Disadvantages of this module include relatively low packing density, high initial cost, and difficulty of cleaning. The common uses for this module configuration are in electrodialysis and electrochemical membranes; other membrane types may be differently configured [65, 68]. Based on their applications, ion exchange membrane processes can be classified into five categories:

1. Mass separation processes such as electrodialysis, diffusional dialysis, and Donnan dialysis
2. Chemical synthesis processes such as chlorine-alkaline electrodialysis and production of H_2 and O_2 by water electrolysis
3. Catalysis processes such as inversion of sugar and production of biodiesel
4. Energy conversion and storage processes such as fuel cell and batteries
5. Sensing and stimuli-response such as sensors and actuators

Potential applications of novel ion exchange membrane-based processes, their state of development, and possible advantages and experienced problems were summarized in Table 1.5. More details on various ion exchange membrane processes and their fundamental operation and essential process parameters can be found in Refs. [65] and [67].

Table 1.5 Potential applications of novel ion exchange membrane-based process, their state of development, and possible advantages and experienced problems [65, 67]

Applications	State of process development	Potential advantages	Problems related to application
Electrodialysis with bipolar membranes: Production of mineral acids and bases from corresponding salts	Pilot plant operation	Lower energy consumption	Contamination of products and low current utilization due to poor membrane permselectivity
Recovering/producing of organic acids from fermentation processes	Commercial and pilot plant operation	Simple integrated process, lower costs	Unsatisfactory membrane stability and fouling application experience
Removal of SO_2 from flue gas	Extensive pilot plant test	Decreased salt production, reduced salt disposal costs	High investment costs, long-term membrane stability
Recovering and recycling of H_2SO_4 and NaOH from wastewaters, such as the rayon production effluent	Laboratory and pilot plant tests, some commercial plants	Purity of the recovered products is not critical, savings in chemicals and sludge disposal costs	No long-term experience, membrane stability under operating conditions, membrane fouling, high investment costs
Recycling of HF and HNO_3 from steel pickling solutions	Commercial plants	Cost savings due to recovered acids and decreased salt disposal	Relatively complex process, high investment costs
In food industry	Laboratory and pilot plant tests	Fewer by-products, less chemicals, and salt production and disposal	Application experience, process costs, investment costs
Energy storage: Battery separator	Commercial plant	Economical advantages, enhanced efficiency, and button battery production	Uses only radiation-induced grafting method and process cost
Energy conversion: Polymer electrolyte membrane fuel cell Direct method fuel cell	Extensive laboratory and pilot tests	High efficiency and power density	High cost of membranes and performance at 80°C

Production of sodium methoxide from methanol	Laboratory tests	More economic than conventional production process	No long-term experience
Elector-deionization and ultra-pure water production	Commercial plants	Continuous process without by-products, high efficiency	Higher investment costs, waste disposal, care pretreatment
Electrodialysis and membrane electrolysis	Commercial plants	Continuous process, high efficiency	High investment costs, membrane stability and selectivity
Electrodialysis and reactor: Denitrification of drinking water, fermentation process	Pilot plant tests, some commercial plants	Continuous process, high efficiency	Membrane stability and selectivity, relatively complex process
Electrodialysis and filtration: Waste recovery	Commercial plants	Continuous process, more compacted process	Relatively complex process, connection with care for each process

1.6 Selection of Ion Exchange Materials and Systems

Selection of ion exchange materials and system is a crucial step toward achieving the economic and technical goals of treatment. To summarize the process of selection of ion exchange materials, system configuration, and operating mode, three major factors have to be carefully taken into account: (1) Waste characteristics where the concentration of total suspended solids in the waste should be less than 4 mg/L and the total dissolved salts content is less than 1–2 g/L. (2) Ion exchange material and system where the selection of appropriate materials (resin, hydrogel, or fibers) depends on the needs of the system. The ion exchange media must be compatible with the chemical nature of the waste (such as the pH and type of ionic species present) and the operating parameters, notably temperature and pressure. However, if there are large concentrations of chemically similar ions in the waste, the process of selection becomes more difficult. (3) Cost consideration where the total cost of operating a process is the sum of the capital costs, the initial cost of the ion exchange media, the operating costs, and the costs associated with the treatment and disposal of the exhausted ion exchange resins. The total costs can be reduced by a frequent regeneration of the exhausted ion exchange resin instead of using it once through process.

A good ion exchange system designer not only will design the system to meet all design specifications but also will utilize resins that will allow the system to operate at peak efficiency and maximum cost-effectiveness. The presence of a wide number of commercial resins available in the marketplace gives high possibility for finding more than one technically effective solution to meet all the system's design specifications.

1.7 Advantages of Ion Exchange Processes

The widespread applications of ion exchange processes in various industrial aspects are supported by a combination of various advantages including the following:

- Proven ability to remove variety of impurities from various volumes with the availability of a wide number of resins.
- Tolerance for fluctuating feed flow rates.
- Low energy consumption.
- Accumulated experience that provides technically effective solutions that meet all system's design specifications.
- Large varieties of specific resins are available from suppliers. Each resin is effective in removing specific contaminants.
- Fast reaction and simple process operation.
- Can be operated at a high flow rate.
- The discharged effluents can achieve regulator acceptance.

- Cost-effectiveness which can be further improved by technical innovation including introducing cheap and highly tolerant ion exchange materials.
- The regenerant chemicals are cheap, and if well maintained, resin beds can last for many years before replacement is needed.

1.8 Limitations of Ion Exchange Process

Despite the extendable and diverse uses of the ion exchange process applications, there are a number of limitations which must be taken into consideration very carefully during the design stages. They include:

- High levels of suspended solids (greater than 10 ppm) and oil together with grease in wastewater may cause clogging of nonselective resins.
- Waste brine from regeneration step requires treatment and disposal, though waste volume can be reduced.
- Spent nonselective resins require frequent replacement and careful disposal.
- Competitive uptake by other ions may limit the effectiveness of nonselective exchange resins.
- Effectiveness of treatment is strongly influenced by water chemistry of the site (e.g., the presence of competing ions and pH of the water source).
- Oxidants present in the ground water may damage the ion exchange resin.
- Usually not feasible with high levels of total dissolved solids (TDS).
- Pretreatment required for most surface water treatments.
- Wastewater is generated during the regeneration step, and it requires additional treatment and disposal.

1.9 Economy of Ion Exchange Processes

Ion exchange has long been proven to be capable technology for removing many dissolved contaminants from various streams despite concerns over capital cost. Capital and operational costs for ion exchange systems vary depending on a number of factors, such as the effluent discharge requirements, the volume of water to be treated, contaminant concentration, the presence of other contaminants, type of utilized resin and regenerant utilization, brine disposal, and site-specific hydrological and geochemical conditions. Among all, the key cost factors include (1) pretreatment requirements, (2) discharge requirements and resin utilization, (3) regenerant used, and (4) efficiency.

The economy of the ion exchange process in an application can be optimized by [4]:

1. Selection of the right process based on the feed composition and the requirement of effluent to be discharged

2. Selection of the type of ion exchange materials based on raw solution characteristics and optimum combinations of the operating parameters
3. Selection of sorption and elution conditions taking into consideration that

   (a) Sharper sorption front allows more efficiency in bed utilization
   (b) Higher pumping rates allow higher productivity per unit time
   (c) Higher concentration in the effluent causes more regeneration time

4. Optimization of regeneration including the type of regeneration agent and the degree of regeneration and the reuse of incompletely exhausted regenerating solution
5. Utilization of waste solution

The economy of the ion exchange processes can be further improved by process innovation. Particularly, introducing new ion exchange materials with low cost and high tolerance to feed concentrations are key issues in boosting cost reduction and expanding the usage of the technology to new frontiers. Also, introducing computer software and automation help very much to enhance the ion exchange process efficiency.

1.10 Applications of Ion Exchange Processes

The applications for ion exchange processes are numerous and cover wide range of industries and house appliances. The purpose of separation dictates the selection of the type of the ion exchange materials, their physical form, and system configuration for practical application and thus forms the basis of a large number of ion exchange processes, which can be functionally divided into three main categories [1]:

1. Substitution: A valuable ion (e.g., copper and silver) can be recovered from solution and replaced by worthless one. Toxic ion such as cyanide can be similarly removed from solution and replaced by nontoxic one.
2. Separation: A solution containing a number of different ions passes through a column containing beads of an ion exchange resin. The ions are separated and emerge following the order of their increasing affinity for the resin.
3. Removal: By using combination of cation-exchange resin (in the H^+ form) and an anion-exchange resin (in the OH^- form) or bipolar resins, all ions are removed and replaced by water (H^+OH^-). The solution is thus demineralized.

Despite the diversity of ion exchange processes, their chief application of today is still the treatment of water with the principle offering of unlimited possibilities in other fields. Commercial ion exchange installations are serving in water and water treatment, food, and chemical industries include processes such as purification of sugar solutions, separation and purification of drugs and fine chemicals, purification of waste effluents, and the recovery of valuable wastes, for example, in the metallurgical industries, the extraction and quantitative separation of elements and metallic complexes. The diverse applications of ion exchange processes can be classified into the following categories [4, 20, 60, 65–69]:

Water production, for example:

- Desalination of seawater
- Desalination of brackish water
- Production of pure and ultrapure water
- Water softening

Wastewater treatment, for example:

- Removal of heavy metals from industrial streams
- Removal of dyes and colors
- Removal of nitrate and ammonia
- Removal of fluoride
- Dealkalization

Radioactive waste treatment, for example:

- Waste decontamination
- Storage of radioactive waste

Isotopes separation and radioactive industry, for example:

- Separation of uranium isotopes
- Separation of lithium isotopes
- Separation of boron isotopes

Recovery of metals, for example:

- Recovery of precious metals such as silver, gold, and platinum
- Recovery of rare earth metals
- Recovery of transition metals

Purification and separation in chemical industries, for example:

- Production of caustic soda in chlor-alkali industry
- Production of sulfuric acid
- Recovery of chromic acid
- Synthesis of ethylbenzene
- Production of table salt
- Dehydration of solvents
- Removal of inorganic salts from waste streams in pulp and paper industry
- Electrolysis of water
- Catalysis in petroleum refineries (zeolites)

Separation and purification in food industry, for example:

- Purification of sugars
- Inversion of sugar
- Removal of tastes and odors
- Recovery of glutamic acid

- Deacidification of fruit juice
- Extraction of lactoperoxidase and purification of casein in dairy industries

Biotechnological applications, for example:

- Production of organic acids such as citric, L-glutamic, and amino acids
- Separation of lactic acid from fermentative broth
- Recovery of proteins
- Enzyme immobilization
- Recovery of enzymes
- Production of biodiesel

Biomedical and pharmaceutical applications, for example:

- Production of antibiotics
- Extraction of vitamins
- Drug delivery systems
- Gene and hormone delivery systems
- Ultrapure water for dilution of medicine

Soil remedy and improvement, for example:

- Artificial soil
- Soil decontamination
- Improve water retention in sandy soil
- Determination of nitrogen content in soil

Chemical analysis, for example:

- Ion chromatography

Energy conversion and solid state applications:

- Battery separators
- Solid polymer electrolyte in fuel cells
- Sensors and actuators

Details of potential and fast-growing applications and their basic principles of operation together with the latest progress in the materials and accommodating process are discussed in the next chapters of this book.

Acknowledgment The authors wish to acknowledge the financial support by the Malaysian Ministry of Science, Technology and Innovation (MOSTI) under the Science Fund program (vote # 79283).

References

1. Dardel F, Arden TV (2005) Ion exchangers. In: Ullmann F, Gerhartz W, Yamamoto YS, Campbell FT, Pfefferkorn R, Rounsaville JF (eds) Ullmann's encyclopedia of industrial chemistry. Wiley-VCH/GmbH & Co. KGaA, Weinheim

2. Badawy SM (2003) Uranium isotope enrichment by complexation with chelating polymer adsorbent. Radiat Phys Chem 66:67–71
3. Harland CE (1994) Ion exchange: theory and practice. Royal Society of Chemistry, Cambridge
4. Zagorondni AA (2006) Ion exchange materials: properties and application. Elsevier, Amsterdam
5. Thompson HS (1850) On the absorbent power of soils. Roy Agric Soc Engl 11:68–74
6. Way JT (1850) On the power of soils to absorb manure. J Roy Agric Soc Engl 11:68–74
7. Gans R (1905) Zeolithe und ahnliche Verbindungen, ihre Konstitution und Bedeutung fiir Technik und Landwiertschafi. Jahrb Preuss Geol Landesandstadt 26:179–21
8. Liebknecht O (1940) Carbonaceous zeolite and the preparation thereof. US Patent 2,191,060
9. Smith P (1940) Manufacture of absorbent and ion exchanging materials. US Patent 2,191,063
10. Adams BA, Holmes EL (1935) Absorptive properties of synthetic resins. Part I. J Soc Chem Ind 54:1T–9T
11. D' Alelio GF (1944) Production of synthetic polymeric compositions comprising sulphonated polymerizates of poly-vinyl aryl compounds and treatment of liquid media therewith. US Patent 2,366,007
12. D' Alelio GF (1944) Production of synthetic polymeric compositions comprising aminated polymerizates of poly-vinyl aryl compounds and treatment of liquid media therewith. US Patent 2,366,008
13. McBurney CH (1952) Resinous insoluble reaction products of tertiary amines with haloalkylated vinyl aromatic hydrocarbon copolymers. US Patent 2,591,573
14. Michaelis L (1926) Die Permeabilitate von Membranen. Nutunvissen schuften 14:33–42
15. Meyer KH, Sievers JF (1936) La permeabilite des membranes I. Theorie de la permeabilite ionique. Helv Chim Acta 19:649–665
16. Juda W, McRae WA (1950) Coherent ion-exchange gels and membranes. J Am Chem Soc 72:1044–1053
17. Grubb WT (1959) Fuel cell. US Patent 2,913,511
18. Mauritz KA, Moore RB (2004) State of understanding of nafion. Chem Rev 104:4535–4585
19. Nasef MM (2008) In separation and purification. In: Battacharya A, Rawlins JW, Ray P (eds) Grafting and crosslinking of polymers. Wiley, New Jersey
20. Helfferich F (1995) Ion exchange. General Publishing Company, Toronto
21. De Silva F (1999) Essentials of ion exchange. In: An article presented at the 25th annual WQA conference. Lisle, USA
22. Garg BS, Sharma RK, Bhojak N, Mittal S (1999) Chelating resins and their applications in the analysis of trace metal ions. Microchem J 61:94–114
23. Goto A, Kusakabe K, Morooka S, Kago T (1993) A test of uranium recovery from seawater with a packed bed of amidoxime fiber adsorbent. Sep Sci Technol 28:1273–1285
24. Egawa H, Kabay N, Jyo A, Hirono M, Shuto T (1994) Recovery of uranium from seawater I. Development of amidoxime resins with high sedimentation velocity for passively driver fluidized bed adsorbers. Ind Eng Chem Res 33:657–661
25. Das S, Pandey AK, Athawale A, Kumar V, Bhardwaj YK, Sabharwal S, Manchanda VK (2008) Chemical aspects of uranium recovery from seawater by amidoximated electron-beam-grafted polypropylene membranes. Desalination 232:243–253
26. Nilchi A, Rafiee R, Babalou AA (2008) Adsorption behavior of metal ions by amidoxime chelating resins. Macromol Symp 274:101–108
27. Agrawal A, Sahu KK (2006) Separation and recovery of lead from a mixture of some heavy metals using Amberlite IRC 718 chelating resin. J Hazard Mater 133:299–303
28. Janin A, Blais J-F, Mercier G, Drogu P (2009) Selective recovery of Cr and Cu in leachate from chromated copper arsenate treated wood using chelating and acidic ion exchange resins. J Hazard Mater 169:1099–1105
29. Hubicki Z, Wołowicz A (2009) A comparative study of chelating and cationic ion exchange resins for the removal of palladium(II) complexes from acidic chloride media. J Hazard Mater 164:1414–1419

30. Kantipuly C, Katragadda S, Chow A, Gesser HD (1990) Chelating polymers and related supports for separation and preconcentration of trace metals. Talanta 37:491–517
31. Chen CY, Chiang CL, Huang PC (2006) Absorptions of heavy metal ions by a magnetic chelating resin containing hydroxyl and iminodiacetate groups. Sep Purif Technol 50:15–21
32. Crini G (2005) Recent developments in polysaccharide-based materials used as adsorbents in waste water treatment. Prog Polym Sci 30:38–70
33. Bolto B, Gregory J (2007) Organic polyelectrolytes in water treatment. Water Res 41:2301–2324
34. Casey JT, O'Cleirigh C, Walsh PK, O' Shea DG (2004) Development of a robust microtiter plate-based assay method for assessment of bioactivity. J Microbiol Methods 58:327–334
35. Guo W, Ruckenstein E (2003) Cross-linked glass fiber affinity membrane chromatography and its application to fibronectin separation. J Chromatogr B 795:61–72
36. Gordon NF, Moore CMV, Cooney CL (1990) An overview of continuous protein purification processes. Biotechnol Adv 8:471–762
37. Voser W (1982) Isolation of hydrophilic fermentation products by adsorption chromatography. J Chem Technol Biotechnol 32:109–118
38. Tadashi A, Isobe E (2004) Fundamental characteristics of synthetic adsorbents intended for industrial chromatographic separations. J Chromatogr 1063:33–44
39. Leonard M (1997) New packing materials for protein chromatography. J Chromatogr B 699:3–27
40. Kawai T, Saito K, Lee W (2003) Protein binding to polymer brush, based on ion-exchange, hydrophobic, and affinity interactions. J Chromatogr B 790:131–142
41. Banki MR, Wood DW (2005) Inteins and affinity resin substitutes for protein purification and scale up. Microb Cell Factor 4:32–38
42. Hess M, Jones RG, Kahovec J, Kitayama T, Kratochvíl P, Kubisa P, Mormann W, Stepto RFT, Tabak D, Vohlídal J, Wilks ES (2006) Terminology of polymers containing ionizable or ionic groups and of polymers containing. Pure Appl Chem 78:2067–2074
43. Neagu V, Vasiliu S, Racovita S (2010) Adsorption studies of some inorganic and organic salts on new zwitterionic ion exchangers with carboxybetaine moieties. Chem Eng J 162:965–973
44. Knaebel KS, Cobb DD, Shih TT, Pigford RL (1979) Ion-exchange rates in bifunctional resins. Ind Eng Chem Fundam 18:175–180
45. Warshawsky P (1982) Selective ion exchange polymers. Die Angew Makromolekul Chemie 10:171–196
46. Bernahl WE, Rossow CE (2002) Current trends in ion exchange operations-what the resins are telling us. In: Proceedings of the 63rd annual international water conference, Pittsburgh, 20–24 Oct 2002
47. Biswas M, Packirisamy S (1985) Synthetic ion-exchange resins. Adv Polym Sci 70:71–118
48. Kabay N, Yilmaz I, Bryjak M, Yüksel M (2004) Removal and recovery of boron from geothermal wastewater by selective ion-exchange resins- field tests. Desalination 167:427–438
49. Wang L, Qi T, Zhang Y (2006) Synthesis of novel chelating adsorbents for boron uptake from aqueous solutions. Chinese J Process Eng 6:375–379
50. Liu H, Ye X, Li Q, Kim T, Qing B, Guo M, Ge F, Wu Z, Lee K (2009) Boron adsorption using a new boron-selective hybrid gel and the commercial resin D564. Colloids Surf A Physicochem Eng Asp 341:118–126
51. Vernon F, Shah T (1983) The extraction of uranium from seawater by poly(amidoxime)/poly (hydroxamic acid) resins and fiber. React Polym Ion Exch Sorb 1:301–308
52. Seko N, Katakai A, Tamada M, Sugo T, Yoshii F (2004) Fine fibrous amidoxime adsorbent synthesized by grafting and uranium adsorption-elution cyclic test with seawater. Sep Sci Technol 39:3753–3767
53. Pal S, Satpati SK, Hareendran KN, Kumar SA, Thalor KL, Roy SB, Tewari PK (2010) Recovery and pre-concentration of uranium from secondary effluent using novel resin. Int J Nucl Desalination 4:28–36
54. Alexandratos SD (2009) Ion-exchange resins: a retrospective from industrial and engineering chemistry research. Ind Eng Chem Res 48:388–398

55. Nasef MM, Saidi H, Nor HM (1999) Radiation-induced graft copolymerization for preparation of cation exchange membranes: a review. J Nucl Sci Malaysia 1:39–54
56. Sata T (2004) Ion exchange membranes: preparation, characterization, modification and application. The Royal Society of Chemistry, Cambridge
57. Hideo K, Tsuzura K, Shimizu H (1991) Ion exchange membranes. In: Dorfner K (ed) Ion exchangers. Walter de Gruyter Berlin, New York
58. Strathmann H (1995) Electrodialysis and related processes. In: Nobe RD, Stern SA (eds) Membrane separation technology-principles and applications. Elesevier Science B.V, Amsterdam/The Netherlands
59. Nasef MM, Hegazy EA (2004) Ion exchange membranes by radiation induced graft copolymerization of polar monomers onto non-polar films: preparations and applications. Prog Polym Sci 29:499–561
60. Nagarale RK, Gohil GS, Shahi VK (2006) Recent developments on ion-exchange membranes and electro-membrane processes. Adv Colloid Interface Sci 119:97–130
61. Yaroslavtsev AB, Nikonenko VV (2009) Ion-exchange membrane materials: properties, modification, and practical application. Nanotechnol in Russia 4:137–159
62. Klein E (2000) Affinity membranes: a 10 year review. J Membr Sci 179:1–27
63. Zou H, Lou Q, Zhou D (2001) Affinity membrane chromatography for the analysis and purification of proteins. J Biochem Biophys Methods 49:199–240
64. Ruckenstei E, Guo W (2004) Cellulose and glass fiber affinity membranes for the chromatographic separation of biomolecules. Biotechnol Prog 20:13–25
65. Strathmann H (2004) Ion-exchange membrane separation processes. Elsevier, Amsterdam
66. Nasef MM, Saidi H, Ujang Z (2007) Ion exchange technology for water and wastewater treatment: principles and progress in materials development. In: Manan ZA, Nasef MM, Setapar SHM (eds) Advances in separation processes. UTM press, Johor
67. Xu T (2005) Ion exchange membranes: state of the development and their perspective. J Membr Sci 263:1–29
68. Kariduraganavar MY, Nagarale RK, Kittur AA, Kulkarni SS (2006) Ion-exchange membranes preparative methods for electrodialysis and fuel cell applications. Desalination 197:225–246
69. Tanaka Y (2007) Ion exchange membranes: fundamentals and applications. Elsevier, Amsterdam/The Netherlands
70. Wheaton RM, Lefevre LJ (1981) Ion exchange. In: Kir-Othmer (ed) Encyclopedia of chemical technology, vol 13, 3rd edn. Wiley, New York
71. Seko N, Tamada M (2005) Current status of adsorbent for metal ions with radiation grafting and crosslinking techniques. Nucl Instr Meth B 236:21–29
72. Seko N, Bang LT, Tamada M (2007) Syntheses of amine-type adsorbents with emulsion graft polymerization of glycidyl methacrylate. Nucl Instr Meth B 265:146–149
73. Li Y-S, Dong Y-L (2004) Determination of anion-exchange resin performance based on facile chloride-ion monitoring by FIA-spectrophotometry with applications to water treatment operation. Anal Sci 20:831–835
74. Nasef MM, Saidi H, Ujang Z, Dahlan KZM (2011) Adsorption of Co(II), Cu(II), Ni(II), Pb(II) and Ag(I) ions from aqueous solutions using crosslinked polyethylene-*graft*-polystyrene sulfonic acid membrane prepared by radiation grafting. J Chil Chem Soc 55:421–427
75. Baker R (2004) Membrane technology and applications. Wiley, Chichester
76. Tamada M (2009) In: 2nd RCM meeting on development of novel adsorbents and membranes by radiation grafting for environmental and industrial applications, International Atomic Energy Agency, Aargau

Chapter 2
Principles of Ion Exchange Equilibria

Jayshree Ramkumar and Tulsi Mukherjee

Abstract The old axiom of chemists, namely, "Corpura non agunt nisi fluida sive soluta", used to give a general feeling that reactions do not occur unless the components are in fluid state or dissolved. However, this is no longer true as many fundamental reactions in nature, like the transport of ions by soil and also the geochemical distributions of various ions, show that reactions do occur in the heterogeneous phase. This is the basis for the understanding of ion exchange process which is not a new phenomenon at all. The basic principles of ion exchange like ion exchange equilibria and capacity are useful for understanding the process of ion exchange. The selectivity coefficient and exchange isotherm have been discussed in terms of thermodynamics and equilibria. The diffusion and transport have been illustrated within the context of exchange kinetics.

2.1 Introduction

There is always a general feeling that reactions do not occur unless the components are in fluid state or dissolved as per the old general principle of "Copura non agunt nisi fluida sive soluta". However, this old axiom of chemists is now being considered as not true. This is because the fundamental reactions in nature, like the transport of ions by soil and also the geochemical distributions of various ions, make it easy to realize that reactions do occur in the heterogeneous phase. This is the basis for the understanding of ion exchange process. The discovery of ion exchange is not recent but is dated to as early as times of the ancient Bible. Ion exchange properties were found in nature itself to. Soil is one of the best known ion exchange materials.

J. Ramkumar (✉)
Analytical Chemistry Division, Trombay, Mumbai 400085, India
e-mail: jrk@barc.gov.in

T. Mukherjee
Chemistry Group Bhabha Atomic Research Centre, Trombay, Mumbai 400085, India

Inamuddin and M. Luqman (eds.), *Ion Exchange Technology I: Theory and Materials*, DOI 10.1007/978-94-007-1700-8_2,

In this chapter, the basic principles of ion exchange like ion exchange equilibria and capacity will be reviewed with a view to understand the concept of ion exchange. The selectivity coefficient and exchange isotherm have been discussed in terms of thermodynamics and equilibria. The diffusion and transport have been illustrated within the context of exchange kinetics.

2.2 Ion Exchange and Ion Exchangers

Ion exchange is defined as the process in which ions in the liquid phase (electrolyte solution) exchange with the ions present in the solid phase, known as ion exchangers. Ion exchangers are insoluble solid materials with fixed ion exchange sites and contain exchangeable cations or anions which can be exchanged with stoichiometric equivalent amount of other ions of the same charge when placed in contact with aqueous solutions. Ion exchangers which can exchange cations are known as cation exchangers, whereas those with replaceable anions are known as anion exchangers. Amphoteric ion exchangers can exchange both cations and anions. The four main types of ion exchange resins are described in Table 2.1.

In general, the ion exchange process is an equilibrium-controlled process. For instance, when a cation exchanger with counter-ions A (RA) is placed in an electrolyte containing the counter-cations B and coions C, the distribution of cations occurs till equilibrium is reached, as given by the following equation. The ion exchange process is shown in Fig. 2.1.

$$\bar{R}A + BC \Leftrightarrow \bar{R}B + AC$$

Table 2.1 Classification of ion exchangers and various associated properties

	Cation exchanger		Anion exchanger	
Parameter	Strong acid	Weak acid	Strong base	Weak base
Functional group	SO_3H	COOH	NR_4^+ Type 1: R = methyl Type 2: R = methyl and ethanol	Polyamine
Operating *pH*	0–14	7–14	10–14	1–7
Regenerant	Acid		Base	
	(% of stoichiometric amount)			
	200–300	100	200–300	100
Application	Softening	Used along with strong cation resin	Type: total anion removal; for high alkalinity and silica waters Type: in dealkalizers; without high silica content	Used for water with high concentration of sulphate, chloride

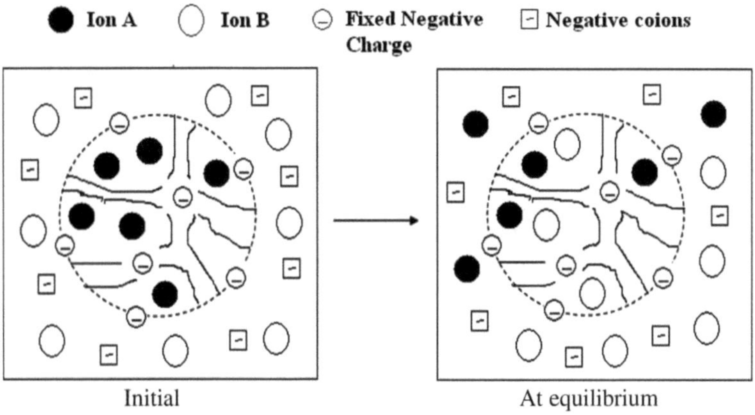

Fig. 2.1 Schematic representation of the ion exchange equilibrium process

Ion exchangers owe their properties to their structure. The framework is held together by chemical bonds and carries a surplus positive or negative charge which is then charge neutralized by ions of opposite charge known as counter-ions. The counter-ions are free to move within the matrix and can also be replaced by ions of same charge present in solution. To put it simply, the ion exchanger may be considered as a sponge with counter-ions floating in the pores. When the sponge is immersed in a solution, the counter-ions leave the pores and float out. To maintain electro-neutrality, stoichiometric amount of ions from the solution fill the pores of the sponge. However, this simple model does not help in the actual understanding of the selectivity of ion exchange. The electrostatic interactions are well understood by the sponge model. Capacity and related data are used for two fundamental applications, first to characterize the ion exchangers and secondly to use in numerical calculations for use in actual applications. Capacity is a term that should be defined in such a manner that it is characteristic constant of the material and independent of experimental conditions. Capacity is defined as the number of ionogenic groups present per specified amount of ion exchanger. The scientific weight capacity has units of meq/g of dry resin (H^+ form for cationic exchanger and Cl^- for anion exchanger). For technical volume capacity, the units are eq/L of packed column bed of resin. Thus, capacity is a constant used to characterize the ion exchangers [1]. Apparent capacity or effective capacity is the number of exchangeable counter-ions per amount of ion exchanger. The unit is meq/g of dry resin (H^+ form for cationic exchanger and Cl^- for anion exchanger). It is lower than the maximum capacity and depends on experimental condition like *pH*, solution concentration, etc. Sorption capacity is the amount of solute taken up by sorption per specified amount of sorbent rather than by ion exchange alone, and this depends on the experimental conditions. Useful capacity is defined as the capacity utilized when equilibrium is not attained and depends on experimental conditions. The values can be determined by titration. The capacity unit expressed in terms of weight q_w can be converted into volume basis q_v, by using the following expression, wherein β is the fraction void volume of the packing, ρ the density of the swollen resin in g/cm^3 and ω is the water content given in percentage:

$$q_v = (1 - \beta)\rho(1 - \omega)q_w.$$

For a weak acid resin, the apparent pK_a value is used to obtain the apparent weight capacity factor: $q_{app} = aq_w$.

Under experimental conditions, these values are found to be very useful. For a weak acid resin (RH), the dissociation constant K and degree of dissociation α are given by the equations

$$K = [R][H]/[RH] \text{ and}$$

$$\alpha = \overline{[R^-]}\Big/\left(\overline{[R^-]} + \overline{[RH]}\right).$$

Thus, $pK = \overline{pH}$ ($\alpha = 0.5$) and the pH of the resin phase should be related to the pH of the solution as a first approximation. At 50% conversion of the HR resin into the Na^+ form, the concentration of Na^+ in the resin phase is given by $\overline{Na} = (\overline{X})/2$, where $(\overline{X}) = \overline{[R^-]} + \overline{[RH]}$ is the total concentration of the dissociated and undissociated functional groups. Thus, substituting the various parameters, the final relation between pH, pK and α can be written as $\log[(1-\alpha)/\alpha^2] = pK - pH + \log(\overline{X}) - \log(A^-)$.

2.3 Selectivity Coefficient and Ion Exchange Isotherm

For any ion exchange reaction, the equilibrium is represented as follows:

$$m\overline{BR}_n + nA^{m+} \Leftrightarrow n\overline{AR}_m + mB^{n+}$$

$$Q = \frac{(\overline{AR}_m)^n \,.\, (B^{n+})^m}{(\overline{BR}_n)^m \,.\, (A^{m+})^n}.$$

When n and $m = 1$ and $Q = \alpha$, the separation factor can be rewritten as

$$\alpha = \frac{(\overline{AR}_m)/(A^{m+})}{(\overline{BR}_n)/(B^{n+})}.$$

If $n = m$ and $Q = \alpha^n$, then after incorporating the corrections for non-ideality of solution,

$$K = Q(\gamma_B/\gamma_A)^n = [\alpha(\gamma_B/\gamma_A)]^n.$$

K is known as the selectivity coefficient, and α, the separation factor, refers to the ratio of the two metal ions A and B. Selectivity coefficient depends on various factors [2]:

1. Higher valence ions are more preferred:
   Tetravalent > Trivalent > Divalent > Monovalent

$$Th^{4+} > Fe^{3+} > Ca^{2+} > K^{+}$$
$$PO_4{}^{3-} > SO_4{}^{2-} > Cl^{-}$$

2. Least hydrated ion is the most preferred:
   The higher the ionic radius, the lower is the radius of the hydrated ion.

$$Cs^{+} > Rb^{+} > K^{+} > Na^{+} > Li^{+}$$
$$SCN^{-} > ClO_4{}^{-} > I^{-} > NO_3{}^{-} > Br^{-} > CN^{-}$$

3. Ions with greater ability to polarize are preferred:

$$NO_3{}^{-} > Br^{-} > NO_2{}^{-} > Cl^{-}$$

4. Ions with least tendency to form complexes in solution:

$$Cl^{-} > HCO_3{}^{-} > CH_3COO^{-} > OH^{-} > F^{-}$$

For a two component system using a cation exchange number,

$$A^{n+} + nRB \Leftrightarrow nB^{+} + R_n^{-}A^{n+}$$

Using the ionic concentrations, the selectivity coefficient *K* can be written as

$$K_B^A = (\overline{B^+})^n (A^{n+}) / (\overline{A^{n+}}) (B^+)^n$$

The equilibrium constant *K* can also be expressed in terms of activity as given below:

$$K = \frac{\bar{a}_M \times a_H}{\bar{a}_H \times a_M}.$$

In the “rational approach”, the exchanger is considered as a solid solution of two forms of the swollen resin and the standard states are chosen as monoionic H^+ and M^+ forms of the exchanger [3]. The rational thermodynamics equilibrium constant is given by

$$K = \frac{\overline{[M]}[H]}{\overline{[H]}[M]} \times \frac{\gamma_H}{\gamma_M} \times \frac{f_M}{f_H} = K_c \times \frac{\gamma^2_{\pm HCl}}{\gamma^2_{\pm MCl}} \times \frac{f_M}{f_H},$$

where $K_c = \frac{\overline{N}_M \times N_H}{\overline{N}_H \times N_M}$.

The quantities in brackets are the appropriate stoichiometric concentrations of the ions in the two phases; γ_M and γ_H are the solution phase activity coefficients of the two ions and f_M and f_H are the exchanger phase activity coefficient; K_c is the stoichiometric concentration product (selectivity coefficient) which can be estimated using equilibrium composition in solution; N and N are the equivalent fraction of ions in the solution and exchanger phase, respectively; $\gamma_\pm$ is the mean molal activity coefficient for both MCl and HCl in aqueous electrolyte mixture [4]. Further, an approximation is made that the ionic coefficients are a function of the total ionic strength only and are independent of ionic composition of the solution [which is valid for uni-univalent electrolyte where the ionic strength (μ) ≤ 0.1].

The mean activity coefficient of the electrolyte in solution phase can be obtained from the literature. However, one cannot directly determine the activity coefficient (f) of the different ionic forms in the exchanger. In this rational approach, the f's are included into the equilibrium constant, that is, $K'_c = K_c \times \frac{f_M}{f_H}$.

As a result, the equilibrium constant K_c will vary with the resin phase composition, $\overline{N}_M$(equivalent fraction in the resin phase). It has been shown that

$$\log K'_c = \int_0^1 \log K_c d\overline{N}_M.$$

K'_c can be obtained by finding the area under the curve of the plot of log K_c versus $\overline{N}_M$. The equilibrium constant, K, is given by

$$\log K = \int_0^1 \log K_c d\overline{N}_M + 2 \log \frac{\gamma_{\pm HCl}}{\gamma_{\pm MCl}}.$$

The term $2 \log \gamma_{\pm \mathrm{HCl}}/\gamma_{\pm \mathrm{MCl}}$ refers to the activity coefficients of a mixture of MCl and HCl, keeping the ionic strength constant. Application of Harned's rule to these systems of mixed electrolytes reveals that this term reduces to $\log \gamma_{\pm HCl}/\gamma_{\pm MCl}$, the $\gamma_\pm$ values now refer to the pure electrolyte at the same concentration [5]. Then

$$\log K = \int_0^1 \log K_c d\overline{N}_M + 2 \log \frac{\gamma_{\pm \mathrm{HCl}}}{\gamma_{\pm \mathrm{MCl}}}.$$

Ion exchange is an adsorption process which occurs when there is equilibrium between a fixed number of ions on the resin phase and ions in solution. The concentrations in the solution and resin phases can be expressed in the form of mole fractions. The solution phase concentration can be expressed in terms of C which is the total ionic concentration of the solution in equivalents per litre:

$$X_{A^{n+}} = (A^{n+})/C \quad \text{and} \quad X_{B^+} = (B^+)/C.$$

Similarly, for the resin phase, $X_{A^{n+}} = (A^{n+})/\overline{C}$ and $X_{B^+} = (B^+)/\overline{C}$, wherein $\overline{C}$ is the total ion exchange capacity in equivalents per litre.

Also, $X_{A^{n+}} + X_{B^+} = 1$ and $\overline{X}_{A^{n+}} + \overline{X}_{B^+} = 1$.

Therefore, $\frac{\overline{X}_{A^{n+}}}{(1-\overline{X}_{A^{n+}})} = K_B^A \left(\frac{\overline{C}}{C}\right)^{n-1} \frac{X_{A^{n+}}}{(1-X_{A^{n+}})}$.

When considering monovalent exchange, $n = 1$ and therefore $C, \overline{C}$ do not appear in the equation. However, for ion exchange of different valence of ions, these terms become very important.

Being an adsorption process, the data can be fitted to both Langmuir and Freundlich models of adsorption [6]. The linearized form of Freundlich equation is given below, where q_e is the amount adsorbed at equilibrium (mg/g) and C_e is the equilibrium concentration (mg/L). K_f and n are constants indicative of adsorption capacity and adsorption intensity, respectively.

$$\ln\ q_e = \ln\ K_f + \frac{1}{n} \ln\ C_e$$

The Langmuir equation assumes that (1) the solid surface presents a finite number of identical sites which are energetically uniform; (2) there are no interactions between adsorbed species, meaning that the amount adsorbed has no influence on the rate of adsorption and (3) a monolayer is formed when the solid surface reaches saturation. The linear form of Langmuir equation is given below, where q_e is the amount adsorbed at equilibrium (mg/g), C_e is the equilibrium concentration (mg/L), b is a constant and Q^0 is the mass of adsorbed solute required to saturate a unit mass of adsorbent (mg/g). Q^0 represents a practical limiting adsorption capacity when the surface is fully covered with ion:

$$\frac{C_e}{q_e} = \frac{1}{Q^0 * b} + \frac{C_e}{Q^0}$$

The isotherms can be derived from the equilibrium process. For an exchange reaction, given below,

$$\frac{C_e}{q_e} - \frac{1}{Q^0 * b} + \frac{C_e}{Q^0}$$

The isotherms can be obtained from the equilibrium process. Consider the following equilibrium process, $A^+ + \overline{BR} \Leftrightarrow \overline{AR} + B^+$. If b is the maximum capacity of the resin and C_O is the initial concentration of A^+ ion in solution, then $\overline{BR} \Leftrightarrow b - \overline{AR}$ and $B^+ = C_0 - A^+$:

$$Q = \frac{(\overline{AR})[C_0 - (A^+)]}{[b - (\overline{AR})](A^+)}.$$

After rearrangement, $\frac{(\overline{AR})}{b} = \frac{Q}{\left[\frac{(A^+)}{C_O}\right]^{-1} + (Q-1)}$.

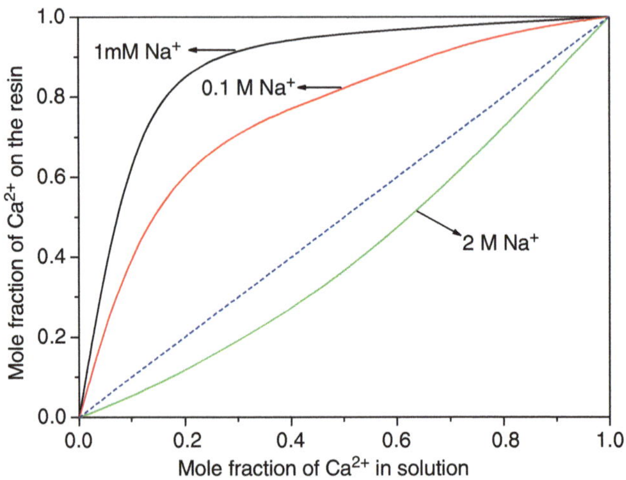

Fig. 2.2 Ion exchange isotherm of Ca^{2+} in presence of varying concentrations of Na^{+}

Hence, we have a relation between the concentration of the cation A^+ in the resin phase $\frac{(\overline{AR})}{b}$ and the concentration in solution phase $\left[\frac{(A^+)}{C_o}\right]$ for different values of Q. The plot of concentration in the resin phase versus the concentration in the solution phase is known as ion exchange isotherm. Above is an example of an equilibrium process between a divalent cation, say Ca^{2+} and a monovalent cation Na^+. In the Fig. 2.2, the blue line at 45° represents no selectivity to either of the phase. For dilute solutions of Na^+ ions, it is seen that the resin phase prefers the divalent cation with increase in selectivity observed with decrease in the concentration of Na^+ ions. However, at higher concentration, there exists a selectivity reversal [7]. This phenomenon is very helpful for the application of regeneration of used ion exchange resins using high concentrations of salt solutions.

2.4 Ion Exchange Equilibria and Kinetics

Different theories have been proposed for understanding the ion exchange kinetics. The *crystal lattice theory* attempted to account for the fixed sites of ions, and for many minerals, the crystal lattice ions can be interchangeable. The *double layer theory* was postulated for understanding the ion exchange phenomenon in colloidal systems to taken into consideration the *pH* of the external solution and concentration effects. For ion exchange membranes, *Donnan membrane theory* was postulated to understand the transport of certain ions in preference to the others which are not allowed to diffuse through. However, all the theories postulated come to a common understanding that ion exchange process has certain characteristic features irrespective of the nature of the resin or condition of use. It is understood that ion exchange reaction is a stoichiometric reversible process following mass law. It occurs by adsorption process, and transport and diffusion processes are allowed. Ion exchange is a diffusion-controlled process wherein the ions get redistributed by diffusion. The rate-determining step in the ion exchange process is the interdiffusion of the exchanged counter-ions within the ion exchanger (*particle diffusion*) or with an adjacent liquid film that is not affected by agitation of solution (*film diffusion*) [8].

Film diffusion is favoured by a resin with cross-linking, low concentration of exchanging ions in solution and high exchange capacity, and the rate of change of concentration is given by

$$-dC\Big/dt = k_f(C - C_e)\frac{a^o}{v},$$

where k_f is the film diffusion coefficient, a^o is the effective area for mass transfer, v is the volume of the solution and C_e is the concentration of the metal ion at equilibrium.

Acknowledgements The authors thank Dr. A.V.R. Reddy, Head, Analytical Chemistry Division (BARC), for constant support and encouragement.

References

1. Juda W, McRae WA (1950) Coherent ion-exchange gels and membranes. J Am Chem Soc 72:1044
2. Kunin R (1951) Ion exchange. Ind Eng Chem 43:102–108
3. Slabaugh WH (1954) Cation exchange properties of bentonite. J Phys Chem 58:162–165
4. Soldano B, Larson QV (1955) Osmotic behavior of anion and cation exchangers. J Am Chem Soc 77:1331–1334
5. Boyd GE, Bunzl K (1967) The donnan equilibrium in cross-linked polystyrene cation and anion exchangers. J Am Chem Soc 89:1776–1780
6. Boyd GE, Soldano BA (1953) Self-diffusion of cations in and through sulfonated polystyrene cation-exchange polymers. J Am Chem Soc 75:6091–6099
7. Muraveiv D, Gorshkov VI, Warshawsky A (1999) Ion exchange. CRC press, Taylor and Francias
8. Helferich FG (1995) Ion exchange. Corier Dover, New York

Chapter 3
Ion Exchange Equilibria and Kinetics

Patrícia F. Lito, Simão P. Cardoso, José M. Loureiro, and Carlos M. Silva

Abstract The accurate modelling of equilibrium and kinetics of ion exchange is fundamental for economic and safe design of industrial units, particularly to carry out the delicate scale-up studies and simulations.

With regard to equilibrium, this chapter covers the following topics: (1) ion association phenomena; (2) activity coefficients in solution and exchanger phases; (3) the milestones works of Gaines and Thomas, Argersinger et al., and Ioannidis et al. to the thermodynamic treatment of ion exchange; and (4) a deep discussion of the three most important theories in the literature (homogeneous mass action models, ion adsorption and related models, and heterogeneous mass action models).

Concerning ion exchange kinetics and mass transport processes, the chapter reviews semiempirical models, Fick's law and derived expressions, Nernst–Planck equations, and the Maxwell–Stefan formulation.

The chapter ends with the general modelling approaches to the omnipresent batch and fixed bed applications.

Modelling and simulation are essential tools to study and design chemical plants. It is fundamental to be able to predict the dynamic behaviour, optimise the operating conditions, and scale up any unit from laboratory to industrial scale. Such goals may be accomplished by computer simulation in order to reduce the number of indispensable experiments. A model can be validated through the fitting of experimental results, but its usefulness comes mainly from its ability to predict the behaviour of a process under operating conditions different from those used to obtain its parameters.

P.F. Lito • S.P. Cardoso • C.M. Silva (✉)
CICECO/Department of Chemistry, University of Aveiro,
Campus Universitário de Santiago, Aveiro 3810-193, Portugal
e-mail: carlos.manuel@ua.pt

J.M. Loureiro
LSRE/Department of Chemical Engineering, School of Engineering,
University of Oporto, Rua do Dr. Roberto Frias, 4200-465 Porto, Portugal

Inamuddin and M. Luqman (eds.), *Ion Exchange Technology I: Theory and Materials*, DOI 10.1007/978-94-007-1700-8_3,

This chapter focuses the equilibrium and kinetic modelling of ion exchange, with the aim to provide accurate and theoretically sound approaches and theories to describe these phenomena. Topics of equilibria, kinetics, and flow patterns are fundamental for computer-aided design of several ion exchange applications as, for instance, softening and deionisation of water, waste treatment, catalysis, chemicals purification, plating, food, and pharmaceutical uses.

The chapter starts with two preliminary sections (Sects. 3.1 and 3.2) where some introductory concepts and notes about ion exchangers are given. Then, Sect. 3.3 is devoted to the equilibrium modelling, being divided into four subsections: Sect. 3.3.1 deals with mass action law principles; Sect. 3.3.2 describes activity coefficients models for ions in solution or in exchanger phases; Sect. 3.3.3 analyses ion association in solution; Sect. 3.3.4 presents the most important approaches for equilibrium modelling, namely, the homogeneous mass action models for ideal and non-ideal systems (group I), the ion adsorption and other related models like those based on statistical thermodynamics (group II), and the heterogeneous mass action models (group III). The subsequent Sect. 3.4 covers ion exchange kinetics along four subsections, where semiempirical models, Fick's law and derived expressions, Nernst–Planck equations, and Maxwell–Stefan formulation are discussed. Finally, Sect. 3.5 describes some applications of the previous models to batch, fixed bed, and zero-length-column operations.

3.1 Introduction

Ion exchangers may be considered insoluble electrolytes which possess the property to exchange ionic species with those of the same charge sign existent in a surrounding electrolyte solution. Solids with mobile cations, like H^+ in resins containing sulphonic acid functional groups $\left(-SO_3^- H^+\right)$, are named cation exchangers, and solids with mobile anions, like Cl^- in resins with quaternary ammonium group $\left(-CH_2N^+(CH_3)_3Cl^-\right)$, are called anion exchangers. The exchangeable ions are named counter ions, whether in the solid or in the external solution. The fixed ions on the solid framework are the co-ions, and the same designation is attributed to the species of the same charge sign in solution.

Ion exchange is commonly treated as a chemical reaction because it is a stoichiometric process in order to ensure electroneutrality in both phases. Accordingly, the exchange of counter ions A^{z_A} and B^{z_B}, with general valences z_A and z_B, respectively, may be represented by the general equation:

$$z_B A^{z_A} + z_A \overline{B^{z_B}} \leftrightarrow z_B \overline{A^{z_A}} + z_A B^{z_B} \tag{3.1}$$

where the top bar was drawn to distinguish the solid phase domain and the modulus was omitted in the stoichiometric coefficients for simplicity. The total concentration of ionic charges in the solid is established by the abundance of co-ions in the framework and defines its exchange capacity, Q_t, which is frequently expressed in equivalents/kg or equivalents/m3.

In the following, the main concentration scales and conversions found in the field of ion exchange are given for the case of two counter ions, A^{z_A} and B^{z_B}. Their extension to n_c counter ions is straightforward.

In the external fluid phase, the molar concentration, molality, and normality of species i are denoted by C_i, m_i, and $C_{N,i}$. The concentrations of all counter ions in solution are:

$$\begin{aligned} m_t &= m_A + m_B \\ C_t &= C_A + C_B \\ C_{N,t} &= C_{N,A} + C_{N,B} \end{aligned} \tag{3.2}$$

Then, the ionic fraction and the equivalent ionic fraction of A^{z_A} are calculated by, respectively:

$$\begin{aligned} x_A &= \frac{C_A}{C_t} = \frac{m_A}{m_t} = \frac{C_{N,A}/z_A}{C_{N,A}/z_A + C_{N,B}/z_B} \\ X_A &= \frac{C_{N,A}}{C_{N,t}} = \frac{z_A C_A}{z_A C_A + z_B C_B} = \frac{z_A m_A}{z_A m_A + z_B m_B} \end{aligned} \tag{3.3}$$

The next relationships must be adhered to:

$$\sum_{i=1}^{n_c} x_i = 1 \quad \text{and} \quad \sum_{i=1}^{n_c} X_i = 1 \tag{3.4}$$

$$x_A = \frac{X_A/z_A}{X_A/z_A + X_B/z_B} \quad \text{and} \quad X_A = \frac{z_A x_A}{z_A x_A + z_B x_B} \tag{3.5}$$

For the ion exchanger phase, the molar and equivalent concentrations of counter ion i are denoted by q_i and Q_i. Hence, the ionic fraction and the equivalent ionic fraction of A^{z_A} are computed by, respectively:

$$\begin{aligned} y_A &= \frac{q_A}{q_A + q_B} = \frac{Q_A/z_A}{Q_A/z_A + Q_B/z_B} \\ Y_A &= \frac{Q_A}{Q_A + Q_B} = \frac{z_A q_A}{z_A q_A + z_B q_B} \end{aligned} \tag{3.6}$$

It is to be noted that:

$$q_t = q_A + q_B \tag{3.7}$$

$$Q_t = Q_A + Q_B \tag{3.8}$$

$$\sum y_i = 1 \text{ and } \sum Y_i = 1 \tag{3.9}$$

The above definitions are essential for the next sections, particularly those relative to equilibrium and kinetic modelling. In the next section, a brief discussion about ion exchange materials is presented.

3.2 Ion Exchange Materials

The ion exchangers may be either organic or inorganic. The most common materials are organic polymers (resins) with ionisable groups placed along their chains. The well-established backbone for ion exchange resins is polystyrene. It is relatively easy to incorporate ionic groups into the polymer, usually on the benzene of polystyrene. To make the resin essentially insoluble in water, even with the large concentration of polar sites, and provide the necessary structural strength, the polystyrene is cross-linked with divinylbenzene.

The exchangers may be classified as strong or weak. In the first ones, the active sites are fully ionised or dissociated, being available for ion exchange. This is the case of sulphonic acid groups $\left(-SO_3^-H^+\right)$ in cation exchangers or quaternary ammonium groups $\left(-CH_2N^+(CH_3)_3Cl^-\right)$ in anion exchangers. In weak exchangers, the functional groups are not fully ionised. Examples of weak cation resins are copolymers of divinylbenzene and acrylic or methacrylic acid. When ternary or even secondary amines are built into the polymer, weak anion resins are obtained. It is important to refer that although weak exchangers often exhibit lower ion exchange capacity owing to being partially ionised, they are easier to regenerate than the stronger ones.

Some exchangers may be prepared with more than one type of functional group, being designated by polyfunctional exchangers. For instance, resins obtained by sulfonation of acrylic and methacrylic acid polymers contain both sulphonic and carboxylic acid groups.

Molecular sieves, such as zeolites and titanosilicates, are well-known types of inorganic ion exchangers. A minute discussion about the structure of zeolites will be presented in order to explain the origin of the ion exchange ability of these materials. Zeolites have a regular framework formed by a three-dimensional combination of SiO_4 and AlO_4 tetrahedrons building blocks connected with each other by shared oxygen atoms. An isolated SiO_4 group contains a charge of −4, while it is neutral in a solid containing an O/Si ratio of 2 since each oxygen is connected with two Si atoms. Taking into account the AlO_4 unit charge is −1, the global structure of zeolites carries a negative charge, which is neutralised by exchangeable extra-framework cations. The structural formula of zeolites may be expressed for the crystallographic unit cell as:

$$\mathrm{M}_{nx/z_\mathrm{M}}\left[(\mathrm{AlO_2})_{nx}(\mathrm{SiO_2})_{ny}\right]n_\mathrm{w}\mathrm{H_2O}$$

where M^{z_M} is a cation of valence z_M which balances the negative charge of AlO_4, n_w is the number of water molecules in the zeolite, and $nx + ny$ is the total number of tetrahedra in the unit cell. The channels and cavities of zeolites framework are occupied by the cations and water molecules. These cations are able to exchange with surrounding counter ions.

3.3 Ion Exchange Equilibrium

In this section, the ion exchange equilibrium is focused in detail. Firstly, the most important models for estimating activity coefficients of ions in solution and in exchanger phases are described. These properties are essential for the accurate modelling of the equilibrium, particularly under the scope of the mass action law approach. Furthermore, activity coefficients play an important role in mass transport at high concentrations or in non-ideal liquid solutions. This aspect will be emphasised in Sect. 3.4.5, when discussing Maxwell–Stefan diffusion equations. The ion association phenomenon is important at high concentrations because it reduces the number of free ions available for ion exchange. It will be interpreted here by the Kester and Pytkowicz approach. The milestones works of Gaines and Thomas, Argersinger and collaborators, and Ioannidis and coworkers on the thermodynamic treatment of the equilibrium are essential topics in this section. Finally, these tools will be called to model the equilibrium, whose main objective is the determination of isotherms. The prediction of multicomponent systems on the basis of parameters optimised only from binary data is most desirable. The most significant and theoretically sound models existent in the literature for non-ideal mixtures are explained and reviewed.

3.3.1 Mass Action Law Fundamentals

The thermodynamic constant corresponding to the equilibrium of Eq. 3.1 depends only on the temperature and is rigorously expressed as function of the activities a_i and $\bar{a}_i$ of the two ions by:

$$K_{\mathrm{B}}^{\mathrm{A}}(T) = \frac{\bar{a}_{\mathrm{A}}^{z_{\mathrm{B}}}\, a_{\mathrm{B}}^{z_{\mathrm{A}}}}{a_{\mathrm{A}}^{z_{\mathrm{B}}}\, \bar{a}_{\mathrm{B}}^{z_{\mathrm{A}}}} \tag{3.10}$$

For simplicity, the valences of both particles are omitted in the subscripts. Once more, the top bar identifies the exchanger phase. The activities of both species, whether in solution or exchanger, are calculated by the product of their concentrations by the corresponding activity coefficients on the same concentration scale. Generally, mole fractions, y_i, are adopted for the ion exchanger and molalities, m_i, for the solution:

$$\begin{aligned} \bar{a}_i &= \bar{\gamma}_i \times y_i \\ a_i &= \gamma_i \times m_i \end{aligned} \tag{3.11}$$

The activity coefficient γ_i is commonly calculated in asymmetrical convention, which implies $\gamma_i \rightarrow 1$ when $m_i \rightarrow 0$. On the contrary, in the solid phase, the symmetric convention is mostly adopted, by which the limit $\bar{\gamma}_i \rightarrow 1$ when $y_i \rightarrow 1$ is observed. In Sect. 3.3.2, the most frequently used models to represent non-ideal behaviour in both phases are listed.

The equilibrium constant is related with the standard Gibbs energy change of *reaction* by $Ln\ K_{\mathrm{B}}^{\mathrm{A}} = -\Delta G^{\mathrm{o}}/\Re T$, where $\Re$ is the universal gas constant and T is the absolute temperature. The influence of temperature upon $K_{\mathrm{B}}^{\mathrm{A}}$ is expressed by the classical thermodynamic relation involving the standard enthalpy change of *reaction* by $d\,Ln\ K_{\mathrm{B}}^{\mathrm{A}}/dT = \Delta H^{\mathrm{o}}/\Re T^2$ [1, 2].

Some authors, like originally Dranoff and Lapidus [3] and Pieroni and Dranoff [4], assumed the system is ideal, i.e. the activity coefficients are unitary in both phases, what is expected to be correct only for diluted solutions. It should be emphasised that in this case, the quantity used to represent equilibrium is not a true thermodynamic constant. It is called selectivity coefficient and is calculated by:

$$K_{\mathrm{c}}(T, m_i, y_i) = \frac{y_{\mathrm{A}}^{z_{\mathrm{B}}}\ m_{\mathrm{B}}^{z_{\mathrm{A}}}}{m_{\mathrm{A}}^{z_{\mathrm{B}}}\ y_{\mathrm{B}}^{z_{\mathrm{A}}}} \tag{3.12}$$

By comparing Eqs. 3.10 and 3.12, one concludes that K_{c} depends simultaneously on the temperature and on the composition of both phases due to the effect of the activity coefficients involved:

$$K_{\mathrm{c}} = K_{\mathrm{B}}^{\mathrm{A}} \times \frac{\bar{\gamma}_{\mathrm{B}}^{z_{\mathrm{A}}} \gamma_{\mathrm{A}}^{z_{\mathrm{B}}}}{\bar{\gamma}_{\mathrm{A}}^{z_{\mathrm{B}}} \gamma_{\mathrm{B}}^{z_{\mathrm{A}}}} \tag{3.13}$$

When the deviations to the ideal behaviour are taken into account in the liquid, a corrected selectivity coefficient, $K_{\mathrm{aB}}^{\mathrm{A}}$, dependent on temperature and on the composition on the ion exchanger results:

$$K_{\mathrm{aB}}^{\mathrm{A}}(T, y_i) = K_{\mathrm{B}}^{\mathrm{A}} \times \left(\frac{\bar{\gamma}_{\mathrm{B}}^{z_{\mathrm{A}}}}{\bar{\gamma}_{\mathrm{A}}^{z_{\mathrm{B}}}} \right) = \frac{y_{\mathrm{A}}^{z_{\mathrm{B}}} a_{\mathrm{B}}^{z_{\mathrm{A}}}}{a_{\mathrm{A}}^{z_{\mathrm{B}}} y_{\mathrm{B}}^{z_{\mathrm{A}}}} \tag{3.14}$$

Another quantity of interest in practical applications, particularly for trace levels of counter ions, is the distribution coefficient, which establishes the ratio between the concentrations in both regions:

$$\lambda_i = y_i/m_i \quad \text{or} \quad \lambda_i = y_i/x_i \tag{3.15}$$

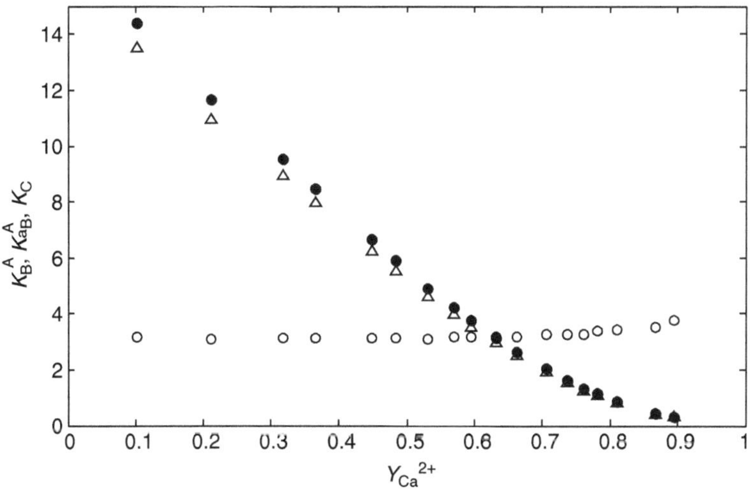

Fig. 3.1 Calculated selectivity coefficient (Δ), corrected selectivity coefficient (●), and equilibrium constant (○) for the system Ca^{2+}/Mg^{2+}/Cl^-/Amberjet 1200 H at 298 K and 0.10 N [15]

The separation factor, α_B^A, analogous to the well-known relative volatility in distillation, corresponds to the quotient between the distribution coefficients of two counter ions:

$$\alpha_B^A = \frac{\lambda_A}{\lambda_B} = \frac{y_A\, m_B}{m_A\, y_B} = \frac{y_A\, x_B}{x_A\, y_B} \tag{3.16}$$

and may be taken as a shorthand method of comparing the easy separation of different ions in solution, i.e. the selectivity of the ion exchanger. When $\alpha_B^A = 1$, the exchange is then non-selective. Attending to the previous discussion about K_c, it is obvious that λ_i and α_B^A are also functions of temperature and concentration.

In order to illustrate the impact of exchanger composition upon K_c and K'^{A}_{aB} due to non-idealities, their plots are shown in Fig. 3.1 for the binary system Ca^{2+}/Mg^{2+} 0.10 N in Amberjet 1500 H along with the true equilibrium constant for comparison. As the figure shows, the former two constants are very sensitive to the concentration in the solid, whereas K_B^A remains practically invariable. In Sect. 3.3.4.1, the most important methods that may be applied to extract the equilibrium constant from experimental data are presented and discussed in detail.

The previous developments refer to binary systems A^{z_A}/B^{z_B}. Nonetheless, in practical applications, ion exchangers are usually exposed to more than two exchangeable species, and so more equations are needed to represent equilibrium. For instance, for the ternary system $A^{z_A}/B^{z_B}/C^{z_C}$, the following reactions are now implied:

$$z_B A^{z_A} + z_A \overline{B^{z_B}} \leftrightarrow z_B \overline{A^{z_A}} + z_A B^{z_B} \tag{3.17}$$

$$z_B C^{z_C} + z_C \overline{B^{z_B}} \leftrightarrow z_B \overline{C^{z_C}} + z_C B^{z_B} \tag{3.18}$$

$$z_C A^{z_A} + z_A \overline{C^{z_C}} \leftrightarrow z_C \overline{A^{z_A}} + z_A C^{z_C} \tag{3.19}$$

whose equilibrium constants are:

$$K_B^A = \frac{\bar{a}_A^{z_B} a_B^{z_A}}{a_A^{z_B} \bar{a}_B^{z_A}} \tag{3.20}$$

$$K_B^C = \frac{\bar{a}_C^{z_B} a_B^{z_C}}{a_C^{z_B} \bar{a}_B^{z_C}} \tag{3.21}$$

$$K_C^A = \frac{\bar{a}_A^{z_C} a_C^{z_A}}{a_A^{z_C} \bar{a}_C^{z_A}} \tag{3.22}$$

It may be easily shown that only two chemical equations are independent, once the third one can be obtained by linear combination of the other two. Equation 3.19 arises when Eq. 3.18 is multiplied by $-z_A/z_C$, added to Eq. 3.17, and finally multiplied by z_C/z_B. Therefore, the three equilibrium constants are also interrelated, and it is now straightforward to show that they satisfy the so-called triangle rule, $\left(K_B^C\right)^{z_A}\left(K_C^A\right)^{z_B}\left(K_B^A\right)^{-z_C} = 1$, which we prefer to recast as:

$$\left(K_B^A\right)^{z_C} = \left(K_C^A\right)^{z_B} / \left(K_C^B\right)^{z_A} \tag{3.23}$$

Akin to binary systems, if concentrations are used in Eqs. 3.20–3.22 in spite of activities, the corresponding K_{aj}^i and K_c are obtained, and the corresponding triangle rules express themselves as:

$$\left(K_{aB}^C\right)^{z_A}\left(K_{aC}^A\right)^{z_B}\left(K_{aB}^A\right)^{-z_C} = 1 \quad \text{and} \quad \left(K_{c,C/B}\right)^{z_A}\left(K_{c,A/C}\right)^{z_B}\left(K_{c,A/B}\right)^{-z_C} = 1 \tag{3.24}$$

These results may be extended to any ion exchange system involving $n_c>3$ counter ions, $A_i^{z_i}$. The complete description of equilibrium only needs $n_c - 1$ binary chemical equations:

$$z_i A_j^{z_j} + z_j \overline{A_i^{z_i}} \leftrightarrow z_i \overline{A_j^{z_j}} + z_j A_i^{z_i} \tag{3.25}$$

where $i<j$, $i = 1, 2, ..., n_c - 1$, and $j = 2, 3, ..., n_c$.

If selectivity coefficients or equilibrium constants are available for a series of ions of interest on the ion exchanger under study, then a reference counter ion C^{z_C} may be chosen to provide $K_{c,j/C}$ or K_C^j for the various pairs j^{z_j}/C^{z_C}. After that, the desired K_c or K_B^A for the system A^{z_A}/B^{z_B} can be estimated by Eqs. 3.23 and 3.24.

3.3.2 *Activity Coefficients Models*

The deviations from ideality in electrolyte solutions are important even at low concentrations since ions interact strongly with each other and with the solvent or exchanger through their electric charges. In this section, the most utilised models for estimating the activity coefficients of ions in the solution and exchanger phases are summarily presented. In the following, they are divided into these two main groups.

3.3.2.1 Solution Phase Modelling

In solution, molal unsymmetrical activity coefficients are mostly adopted. Several models may be found in literature, namely, Debye–Hückel, Hückel, Bromley, Guggenheim, Pitzer, Meissner and Kusik, etc. (e.g. [2, 5]).

Debye–Hückel Model

The Debye–Hückel limiting law was formulated in 1923 for extremely diluted electrolyte solutions. The activity coefficient depends on the ionic strength, I, defined as:

$$I = \frac{1}{2} \sum_{i=1}^{n} z_i^2 m_i \tag{3.26}$$

where n is the total number of ionic species in solution. It is expressed by:

$$Ln\ \gamma_i = -A_\gamma\, z_i^2 \sqrt{I} \tag{3.27}$$

and provides accurate results for ionic strengths less than 0.001 molal. Constant A_ϕ is the well-known Debye–Hückel constant given by:

$$A_\gamma = \left(\frac{e^2}{\varepsilon kT}\right)^{3/2} \sqrt{\frac{2\pi\, \rho_w N_0}{1000}} \tag{3.28}$$

where N_0 is the Avogadro constant, ρ_w and ε are density (g cm^{-3}) and dielectric constant ($C^2N^{-1}\ m^{-2}$) of pure solvent, $k = 1.38066 \times 10^{-23}$ J K^{-1} is Boltzmann constant, T is absolute temperature (K), and $e = 1.60206 \times 10^{-19}$ C is the electron charge. Pitzer et al. [6] published A_γ values for aqueous solution in the range 0–300°C; at 298.15 K, $A_\gamma = 1.174$ $kg^{1/2}mol^{-1/2}$ for natural logarithm.

In order to extend the previous limiting law to higher concentrations, a corrective term has been introduced to take into account ions size and short-range interactions. The model includes a minimum approximation distance equal for all ions, *a*, and can be applicable for ionic strengths up to 0.1 molal:

$$Ln\ \gamma_i = -\frac{A_\gamma z_i^2 \sqrt{I}}{1 + \beta a \sqrt{I}} \tag{3.29}$$

$$\beta = \sqrt{\frac{8\,\pi\, e^2 N_0 \rho}{1000\,\varepsilon\, k\, T}} \tag{3.30}$$

By setting the standard value $a = 3.04\text{Å}$, the quantity βa in the denominator of Eq. 3.29 becomes unity.

Hückel Model

Hückel tried to expand Eq. 3.29 to higher concentrations by summing one term related with the reduction of the dielectric constant raised by the increasing ionic concentration:

$$Ln\ \gamma_i = -\frac{A_\gamma z_i^2 \sqrt{I}}{1 + \beta \mathrm{a} \sqrt{I}} + \mathrm{c}I \tag{3.31}$$

Here, a and c are ion specific. Robinson and Stokes [7] obtained reliable results for binary systems up to 1 molal, but failed to describe concentrated multicomponent solutions.

Bromley Model

Bromley [8] based his model on the Debye–Hückel theory by adding a correction term F_i which includes electrostatic contributions characteristic of Coulomb forces besides those of long range. For a generic ion *i* in a multicomponent mixture, it is given by:

$$Ln\ \gamma_i = -\frac{A_\gamma z_i^2 \sqrt{I}}{1 + \sqrt{I}} + F_i \tag{3.32}$$

$$F_i = \sum_j \dot{B}_{ij} z_{ij}^2 m_j \tag{3.33}$$

$$z_{ij} = \frac{z_i + z_j}{2} \tag{3.34}$$

$$\dot{B}_{ij} = \frac{0.06 + 0.6 B_{ij}}{\left(1 + 1.5 I / \left|z_i z_j\right|\right)^2} \left|z_i z_j\right| + B_{ij} \tag{3.35}$$

In these equations, when i is a cation, j represent anions; on the other hand, if i is an anion, j refers to cations. B_{ij} is Bromley parameter for ij pair. This equation may give good results up to 6 molal.

Guggenheim Model

Similarly to the previous models, Guggenheim extended the Debye–Hückel expression to improve its behaviour at higher concentrations by summing additional terms. The activity coefficients of a particular cation, γ_c, and a particular anion, γ_a, are given by:

$$Ln\,\gamma_c = -\frac{A_\gamma z_c^2\sqrt{I}}{1+\sqrt{I}} + \sum_j B_{cj}\,m_j \tag{3.36}$$

$$Ln\,\gamma_a = -\frac{A_\gamma z_a^2\sqrt{I}}{1+\sqrt{I}} + \sum_i B_{ia}\,m_i \tag{3.37}$$

where subscripts i and j refer to all cations and anions in solution, respectively; c and a denote the specific cation and anion under study.

Pitzer Model

The Pitzer equations may give good results up to 6 molal. The activity coefficients of a particular cation, γ_c, and a particular anion, γ_a, are given by [2, 9]:

$$\begin{aligned} Ln\,\gamma_c = {} & z_c^2F + \sum_j m_j\left\{2B_{cj} + \left(2\sum_i m_i z_i\right)C_{cj}\right\} + \sum_i m_i\left(2\Theta_{ci} + \sum_j m_j\Psi_{cij}\right) \\ & + \sum_i\sum_j m_i m_j\left(z_c^2B'_{ij} + |z_c|C_{ij}\right) + \frac{1}{2}\sum_j\sum_{j'} m_j m_{j'}\Psi_{cjj'} \end{aligned} \tag{3.38}$$

$$\begin{aligned} Ln\,\gamma_a = {} & z_a^2F + \sum_i m_i\left\{2B_{ia} + \left(2\sum_j m_j z_j\right)C_{ia}\right\} + \sum_j m_j\left(2\Theta_{aj} + \sum_i m_i\Psi_{aji}\right) \\ & + \sum_i\sum_j m_i m_j\left(z_a^2B'_{ij} + |z_a|C_{ij}\right) + \frac{1}{2}\sum_i\sum_{i'} m_i m_{i'}\Psi_{ii'a} \end{aligned} \tag{3.39}$$

$$F = -\frac{A_\gamma}{3}\left[\frac{\sqrt{I}}{1+1.2\sqrt{I}} + \frac{Ln\left(1+1.2\sqrt{I}\right)}{0.6}\right] \tag{3.40}$$

In these equations, subscripts i and j are intended to span all cations and anions in solution, respectively; c and a identify the specific cation and anion under calculation.

The term F includes the Debye–Hückel contribution to take the electrostatic far field interactions between ions into account. The second virial coefficients, B, represent the specific binary, near-field interactions between pairs ij. The corresponding ternary interactions ijk are described by the third virial coefficient, Ψ. For the limiting case of diluted solutions, the second and higher terms vanish, and Pitzer model reduces to Debye–Hückel:

$$B_{ij} = \beta_{ij}^{(0)} + \beta_{ij}^{(1)} g\left(\alpha_1 \sqrt{I}\right) + \beta_{ij}^{(2)} g\left(\alpha_2 \sqrt{I}\right) \tag{3.41}$$

$$B'_{ij} = \frac{\beta_{ij}^{(1)} g'\left(\alpha_1 \sqrt{I}\right) + \beta_{ij}^{(2)} g'\left(\alpha_2 \sqrt{I}\right)}{I} \tag{3.42}$$

For any salt containing a monovalent ion, $\alpha_1 = 2$ and $\alpha_2 = 0$, while for 2–2 electrolytes and higher valence types, the corresponding values are $\alpha_1 = 1.4$ and $\alpha_2 = 12$. The functions $g(x)$ and $g'(x)$ are given by:

$$g(x) = \frac{2[1 - (1 + x)\exp(-x)]}{x^2} \tag{3.43}$$

$$g'(x) = \frac{-2[1 - (1 + x + 0.5x^2)\exp(-x)]}{x^2} \tag{3.44}$$

where $x = \alpha\sqrt{I}$. Coefficients C_{ij} are defined by Pitzer [162] as:

$$C_{ij} = \frac{C_{ij}^{(0)}}{2\left|z_i z_j\right|^{0.5}} \tag{3.45}$$

For binary systems, the activity coefficients depend only on the interaction parameters $\beta_{ij}^{(0)}$, $\beta_{ij}^{(1)}$, $\beta_{ij}^{(2)}$, and $C_{ij}^{(0)}$ to compute the implied variables B and C. For multicomponent mixtures, it is also necessary to include parameters Θ and Ψ: the parameters Θ account for cation–cation and anion–anion interactions, while parameters Ψ call cation–cation–anion and anion–anion–cation interactions.

Meissner and Kusik Model

The mean activity coefficient of an electrolyte ca, γ_{ca}, was defined by Meissner and Kusik [10, 11] in terms of a reduced activity coefficient Γ_{ca}:

$$Ln\, \gamma_{ca} = \left|z_c z_a\right| Ln \Gamma_{ca} \tag{3.46}$$

The mean ionic activity coefficient may be calculated from the individual ionic coefficients, γ_c and γ_a, taking into account their stoichiometry (v_c and v_a):

$$\gamma_{ca} = \left(\gamma_c^{v_c} \times \gamma_a^{v_a}\right)^{\frac{1}{v_c + v_a}} \tag{3.47}$$

For a multicomponent solution, Γ_{ca} is calculated at 25°C by:

$$\Gamma_{ca} = [1 + B(1 + 0.1I)^{q_{ca,mix}} - B]\Gamma^* \tag{3.48}$$

where

$$B = 0.75 - 0.065 q_{ca,mix} \tag{3.49}$$

$$Ln\ \Gamma^* = -\frac{A_\gamma \sqrt{I}}{1 + C\sqrt{I}} \tag{3.50}$$

$$C = 1 + 0.055\, q_{ca,mix} \exp\left(-0.023\, I^3\right) \tag{3.51}$$

The Meissner interaction parameter $q_{ca,mix}$ comprehends the contributions of all couples ia and cj, where i and j span all cations and all anions present in the multicomponent solution, respectively. Hence:

$$q_{ca,mix} = \frac{\sum_j I_j q^{o}_{cj}}{I} + \frac{\sum_i I_i q^{o}_{ia}}{I} \tag{3.52}$$

where q^{o}_{ij} is the Meissner parameter for electrolyte ij in pure solution, and $I_k = 0.5 m_k z_k^2$ is the partial ionic strength of ion k.

The temperature effect upon q^{o}_{ij} may be estimated by:

$$\frac{q^{o}_{ij}(T) - q^{o}_{ij}(25°C)}{T - 25} = \frac{0.0027 q^{o}_{ij}(25°C)}{|z_i\, z_j|} \tag{3.53}$$

Figure 3.2 presents a comparison between experimental and estimated mean activity coefficients of the well-known binary system NaCl in aqueous solution at 25°C. The models of Debye–Hückel, Bromley, Guggenheim, Pitzer, and Meissner and Kusik were selected to perform the predictions. It is clear from the figure that Debye–Hückel and Guggenheim provide unacceptable results along almost all concentration range. The remaining models behave similarly, but the accurate description accomplished by Meissner and Kusik and particularly Pitzer models should be emphasised.

3.3.2.2 Exchanger Phase Modelling

The equations applied to compute solid phase activity coefficients are those originally developed for vapour–liquid equilibrium. One may cite, for instance, the models of Wilson, NRTL, Margules, van Laar, UNIQUAC, etc. For strongly non-ideal binary mixtures, the Wilson equation is the most adopted since it contains only

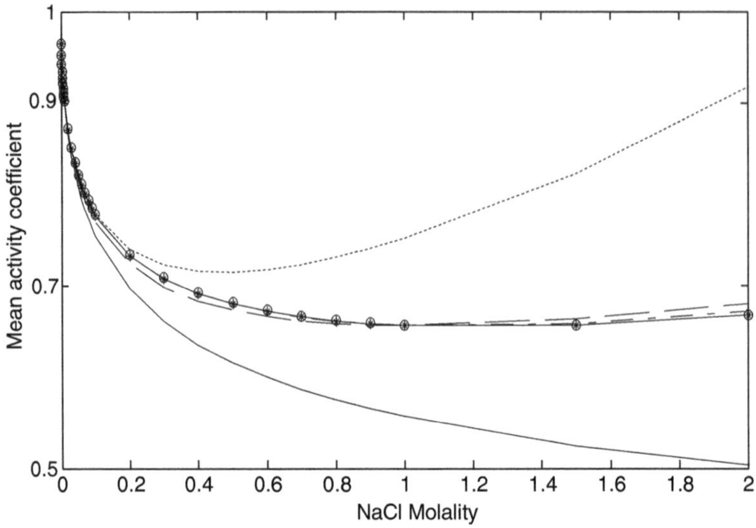

Fig. 3.2 Comparison between experimental [2] and estimated mean activity coefficients of NaCl in aqueous solution at 25°C. Models: (—) Debye–Hückel, (···) Guggenheim, (- – – –) Bromley, (- -) Meissner and Kusik, and (–*–) Pitzer

two adjustable parameters while NRTL possesses three, and the UNIQUAC expression is more complex. Such facts sponsor its frequent utilisation for multicomponent mixtures too because prediction is based on binary parameters only. In the following, the activity coefficients of Wilson, NRTL, and Margules models are summarily presented due to being the most adopted in ion exchange research. The remaining models, as well as others not mentioned, may be consulted elsewhere (e.g. [5, 12]). Recently, Aniceto et al. [163] compared the performance of Wilson, NRTL, Margules and UNIQUAC models to represent the non-idealities in the exchanger phase for binary systems, and concluded that the Wilson, NRTL and Margules activity coefficient models provide similar representations and predictions, whereas UNIQUAC performed worse. In the case of multicomponent systems, NRTL and Wilson provided the best results [164].

Wilson Model

The Wilson equation takes into account the combined effects of differences in molecular size and intermolecular forces. For a solution of n_c species, it requires only parameters obtained from binary data:

$$Ln\,\bar{\gamma}_i = 1 - Ln\left(\sum_{j=1}^{n_c} y_j\Lambda_{ij}\right) - \sum_{k=1}^{n_c} \frac{y_k\Lambda_{ki}}{\sum_{j=1}^{n_c} y_j\Lambda_{kj}} \tag{3.54}$$

$\Re$ is the ideal gas constant, and Λ_{ij} and Λ_{ji} are binary interaction temperature-dependent parameters defined such that $i \neq j$, being calculated as function of the pure-component molar volumes (v_i,v_j) and characteristic energy difference ($\lambda_{ij} - \lambda_{ii}$) by:

$$\Lambda_{ij} = \frac{v_j}{v_i} \exp\left(-\frac{\lambda_{ij} - \lambda_{ii}}{\Re T} \right) \tag{3.55}$$

For a binary ij system, the model simplifies to:

$$Ln\, \bar{\gamma}_i = -Ln\left(y_i + y_j \Lambda_{ij} \right) + y_j \left[\frac{\Lambda_{ij}}{y_i + y_j \Lambda_{ij}} + \frac{\Lambda_{ji}}{y_i \Lambda_{ji} + y_j} \right] \tag{3.56}$$

NRTL Model

The multicomponent NRTL expression contains only parameters obtained from binary data and is given by:

$$Ln\, \bar{\gamma}_i = \frac{\sum_{j=1}^{n_c} y_j \tau_{ji} G_{ji}}{\sum_{l=1}^{n_c} y_l G_{li}} + \sum_{j=1}^{n_c} \frac{y_j G_{ij}}{\sum_{l=1}^{n_c} y_l G_{lj}} \left(\tau_{ij} - \frac{\sum_{r=1}^{n_c} y_r \tau_{rj} G_{rj}}{\sum_{l=1}^{n_c} y_l G_{lj}} \right) \tag{3.57}$$

where

$$\tau_{ij} = \frac{g_{ij} - g_{jj}}{\Re T}, \quad G_{ij} = \exp\left(-\alpha_{ij} \tau_{ij} \right), \quad \left(\alpha_{ji} = \alpha_{ij} \right) \tag{3.58}$$

g_{ij} is an energy parameter characteristic of the $i - j$ interaction, whose significance is similar to that of λ_{ij} in Wilson equation; α_{ij} is related to the nonrandomness in the mixture (when $\alpha_{ij} = 0$, the binary mixture is completely random and the model reduces to the two-suffix Margules equation). α_{ij} varies frequently in the range 0.20–0.47 and can be set arbitrarily when data are scarce (e.g. $\alpha_{ij} = 0.3$).

For binary ij mixtures, Eq. 3.57 shortens to:

$$Ln\, \bar{\gamma}_i = y_j^2 \left[\tau_{ji} \left(\frac{G_{ji}}{y_i + y_j G_{ji}} \right)^2 + \frac{\tau_{ij} G_{ij}}{\left(y_j + y_i G_{ij} \right)^2} \right] \tag{3.59}$$

Margules Model

The simplest Margules expressions for binary systems are the two-suffix equations that contain only one parameter ($Ln\ \bar{\gamma}_i = A^{'}\ y_i^2$) and are applicable for mixtures of very similar (size, shape, and chemical structure) components. The one-parameter Gibbs excess function is symmetrical with composition and exhibits an extremum at $y_i = y_j = 0.5$. To represent the skewness of real binary data, it is necessary to include two parameters, $A^{'}_{ij}$ and $A^{'}_{ji}$, giving rise to the well-known three-suffix Margules equations:

$$Ln\ \bar{\gamma}_i = y_j^2 \left[A^{'}_{ij} + 2\left(A^{'}_{ji} - A^{'}_{ij} \right) y_i \right] \tag{3.60}$$

The three-suffix Margules equations may be derived for multicomponent mixtures by Wohl method [5, 13]. For instance, for a ternary mixture *ijk*, the activity coefficient of component *i* is:

$$\begin{aligned} Ln\,\bar{\gamma}_i = {} & A^{'}_{ij}y_j^2(1-2y_i) + 2A^{'}_{ji}y_iy_j(1-y_i) + A^{'}_{ik}y_k^2(1-2y_i) + 2A^{'}_{ki}y_iy_k(1-y_i) \\ & - 2A^{'}_{ik}y_jy_k^2 - 2A^{'}_{kj}y_j^2y_k + \left[\frac{1}{2}\left(A^{'}_{ij} + A^{'}_{ji} + A^{'}_{ik} + A^{'}_{jk} + A^{'}_{kj} \right) - W^{'} \right] \left(y_jy_k - 2y_iy_jy_k \right) \end{aligned} \tag{3.61}$$

The $A^{'}_{ij}$ constants refer to pure binary *ij* systems, while $W^{'}$ is determined from ternary data (if these data are not available, one may set $W^{'} = 0$ as a first approximation). To get $\bar{\gamma}_j$ from Eq. 3.61, it is sufficient to substitute *i* by *j*, *j* by *k*, and *k* by *i*. In the case of $\bar{\gamma}_k$, replace *i* by *k*, *j* by *i*, and *k* by *j*.

3.3.3 *Ion Association*

The correct description of ion exchange should take into account the real concentrations of the various ionic species present in solution. When ions of opposite sign are close together, the energy of their mutual electrical attraction may be considerably greater than their thermal energy, so they may constitute a virtually new stable entity in solution. With unsymmetrical electrolytes, new ionic species of a charge type not previously present may result, while no net charge components appear in the case of symmetrical electrolytes, notwithstanding they should exhibit a dipole moment.

When a generic salt M_xX_m is introduced into an aqueous solution, the following equilibrium is established:

$$M_xX_m \leftrightarrow xM^{m+} + mX^{x-} \tag{3.62}$$

The corresponding stability constant, $K_S^{M_xX_m}$, is:

$$K_S^{M_xX_m} = \frac{[M_xX_m]}{[M^{m+}]_f^x [X^{x-}]_f^m} \tag{3.63}$$

where $[M^{m+}]_f$ and $[X^{x-}]_f$ are the free concentrations of cation M^{m+} and anion X^{x-} in solution. As mentioned above, some of the ions are not available for ion exchange since they are not present as free species but rather as ion pairs. Even simple salts are weak electrolytes in solvents of low dielectric constants, for example, KBr in acetic acid or ammonia; CsCl in ethanol; KI in acetone, n-propanol, or pyridine; and NaI in ethylenediamine.

The method of Kester and Pytkowicz [14] can be successfully applied to determine the real ionic concentrations in solution. For instance, consider the five component system studied by Vo and Shallcross [15], where K^+, Na^+, H^+, Ca^{2+}, and Mg^{2+} are present with co-ion Cl^- in solution. Some of the particles will exist as KCl, NaCl, $MgCl^+$, and $CaCl^+$; it is assumed that all $CaCl_2$ and $MgCl_2$ dissociate to $CaCl^+$ and Cl^-, and $MgCl^+$ and Cl^-, respectively. The material balance to potassium is:

$$[K]_t = [K^+]_f + [KCl] \tag{3.64}$$

where $[K]_t$ is the total concentration of potassium present in solution in whatever form. Taking into account the stability constant, $[KCl] = K_S^{KCl}[K^+]_f[Cl^-]_f$, and Eq. 3.64, the concentration of free K^+ can be obtained after rearrangement:

$$[K^+]_f = \frac{[K]_t}{1 + K_S^{KCl}[Cl^-]_f} \tag{3.65}$$

The same approach may be applied to the remaining ionic species, giving rise to:

$$[Na^+]_f = \frac{[Na]_t}{1 + K_S^{NaCl}[Cl^-]_f} \tag{3.66}$$

$$[H^+]_f = \frac{[H]_t}{1 + K_S^{HCl}[Cl^-]_f} \tag{3.67}$$

$$[Ca^{2+}]_f = \frac{[Ca]_t}{1 + K_S^{CaCl^+}[Cl^-]_f} \tag{3.68}$$

$$[Mg^{2+}]_f = \frac{[Mg]_t}{1 + K_S^{MgCl^+}[Cl^-]_f} \tag{3.69}$$

$$[Cl^-]_f = \frac{[Cl]_t}{1 + K_S^{KCl}[K^+]_f + K_S^{NaCl}[Na^+]_f + K_S^{HCl}[H^+]_f + K_S^{CaCl^+}[Ca^{2+}]_f + K_S^{MgCl^+}[Mg^{2+}]} \tag{3.70}$$

All free concentrations can be now calculated from Eqs. 3.65–3.70, once the total concentrations and stability constants are known.

Kester and Pytkowicz [14] proposed a simple relation between the stability constant and the solution ionic strength, I:

$$Ln\left(K_{S}^{M_xX_m}\right) = A' + B'I \tag{3.71}$$

$K_{S}^{M_xX_m}$ values or correlations are available in the literature for several salts and complexes [14, 16–20]. The method of Kester and Pytkowicz [14] for the partial dissociation of salts in solution has been applied to ion exchange processes by several authors, as, for instance, Mehablia et al. [21], Vo and Shallcross [15], Mumford et al. [22], and Aniceto et al. [163, 164].

3.3.4 Modelling Ion Exchange Equilibrium

An ion exchange isotherm characterises equilibrium by representing the concentrations of counter ions in exchanger as function of their concentrations in solution at constant temperature. It is desirable that they cover all possible experimental conditions with accuracy in order to be used in the design and optimisation of ion exchange processes.

The number of research works in literature concerning binary equilibria is rather large in comparison to those on multicomponent systems, which results from the large complexity arisen from the various background processes causing non-idealities and coupling of their models. The nature of the systems, the interactions of counter ions in the solution and in the solid phases, the interactions between them and the exchanger, the heterogeneity of the exchanger surface, the clustering and dissociation of exchangeable ions, the limited solubility of counter ions, and the synergistic effect of competitive ions may be cited, for instance [23]. This scenario explains why no fully theoretical model has been developed to date to predict ion exchange equilibrium without at least some experimental data.

In order to simplify multicomponent analysis, the general approach adopted by researchers is the development of good semi-theoretical models based on binary ion exchange data. The models are firstly applied to binary equilibrium data to optimise the corresponding binary parameters. Afterwards, these are used in the extended model formulation to predict the multicomponent equilibrium of interest. This has been the primary goal of wide variety of models published during decades. For example, the ternary system $A^{z_A}/B^{z_B}/C^{z_C}$ may be represented by the three constitutive binaries A^{z_A}/B^{z_B}, A^{z_A}/C^{z_C}, and B^{z_B}/C^{z_C}. It is assumed that these parameters are independent of the presence of any other ionic species.

Shallcross [24] proposed the following series of requirements that any proposed ion exchange equilibrium model should meet: (1) Model parameters must be independent of solution phase concentration. (2) Model parameters must remain constant in the presence of different nonexchanging counter ions. (3) Internal consistency of the model is essential. Most models are overspecified, with more equations to be solved than unknowns present. The parameters' values must not be significantly dependent on which of these equations are used in the solution procedure. (4) The model must be applicable up to quaternary systems.

The models adopted to represent and/or predict ion exchange equilibrium may be divided into four groups.

In the first group, ion exchange equilibrium is described in terms of the law of mass action, i.e. it is treated as a chemical reaction. The principles of this approach were already exposed in Sect. 3.3.1, while the most important and significant mathematical treatments used to determine the embodied parameters are presented in this section. The exchanger is homogeneous, and the non-idealities are introduced via the activity coefficients of ions in solution and in the solid phases. Hence, the models of group I are named homogeneous mass action models.

The second group comprehends models where ion exchange is treated as an adsorption process. The first attempts to explain and predict equilibrium utilised the well-known Freundlich and Langmuir isotherms, whereas, presently, more thorough models have appeared. In this method, the deviations from ideal behaviour are explained in terms of energetic heterogeneity of the functional groups of the ion exchanger. Accordingly, these models may be called heterogeneous adsorption models.

The third group collects models derived from the original work of Melis et al. [25]. It is based also on the mass action law, but differs from the first group because it takes into account the heterogeneity of ion exchange sites. This is the reason why they are frequently called heterogeneous mass action models.

In the last group, we included purely empirical models, which generally possess limited theoretical interest and small extrapolation ability, being used for correlative purposes only. For this reason, they will not be analysed in this chapter.

In view of their importance, the former three groups will be discussed in separate subsections.

3.3.4.1 Group I: Homogeneous Mass Action Models

Group I describes ion exchange in terms of the law of mass action, as it may be seen in Eqs. 3.10 and 3.20–3.22. Several authors adopted this approach, for example, Dranoff and Lapidus [3], Pieroni and Dranoff [4], Klein et al. [26], Smith and Woodburn [27], Sengupta and Paul [28], Shallcross et al. [29], Martínez et al. [30], Mehablia et al. [21], Ioannidis et al. [31], Vo and Shallcross [15], Mumford et al. [22], Borba et al. [32], and Aniceto et al. [163, 164]. In the following, the modelling principles of ion exchange equilibrium for ideal and real binary and multicomponent systems according to mass action law are presented and discussed in detail.

Ideal Systems Modelling

As has been mentioned above, some early publications assumed the ion exchange systems are ideal (e.g. [3, 4, 26], Pieroni and Dranoff [4], Klein et al. [26]), i.e. activity coefficients are unitary in both phases, which means that selectivity coefficients may be used in calculations. It is worth noting that only dilute solutions are expected to fulfil this behaviour. In this case, binary isotherms are very simple and explicit, whether for equal or distinct valence counter ions. With the aid of Eqs. 3.3, 3.4, and 3.9, Eq. 3.12 may be rewritten as:

$$y_\mathrm{A} = \frac{K_\mathrm{c}^{1/z} x_\mathrm{A}}{1+\left(K_\mathrm{c}^{1/z}-1\right)x_\mathrm{A}}, \quad \text{for} \quad z_\mathrm{A} = z_\mathrm{B} = z \tag{3.72}$$

$$\frac{y_\mathrm{A}}{(1-y_\mathrm{A})^2} = \frac{K_\mathrm{c}}{m_t}\frac{x_\mathrm{A}}{(1-x_\mathrm{A})^2}, \quad \text{for} \quad z_\mathrm{A} = 2 \text{ and } z_\mathrm{B} = 1 \tag{3.73}$$

From Eq. 3.73, the important role played by the solution concentration, m_t, when $z_\mathrm{A} \neq z_\mathrm{B}$, is clear.

For multicomponent ideal systems, when selectivity coefficients $K_{\mathrm{c},i/j}$ are known, the mole fraction of any counter ion $\mathrm{A}_j^{z_j}$ in exchanger may be expressed in terms of y_i by:

$$y_j = \frac{x_j}{K_{\mathrm{c},i/j}^{1/z_i} m_t^{z_j/z_i-1}}\left(\frac{y_i}{x_i}\right)^{z_j/z_i} \tag{3.74}$$

For any component i, Eq. 3.9 gives rise to a polynomial in y_i when the remaining $j \neq i$ fractions are substituted:

$$y_i + \sum_{j\neq i}\frac{x_j}{K_{\mathrm{c},i/j}^{1/z_i} m_t^{z_j/z_i-1}}\left(\frac{y_i}{x_i}\right)^{z_j/z_i} = 1 \tag{3.75}$$

These n_c equations allow us to calculate the exchanger phase whenever the solution is known. For the external solution, the fundamental relations may be derived analogously to give:

$$x_i + \sum_{j\neq i}\frac{y_j}{K_{\mathrm{c},i/j}^{1/z_i} m_t^{z_j/z_i-1}}\left(\frac{x_i}{y_i}\right)^{z_j/z_i} = 1 \tag{3.76}$$

Non-ideal Systems Modelling

The behaviour of real systems, where the selectivity coefficients vary with both exchanger and solution compositions, cannot be accurately described by the previous equations (e.g. [15, 22, 32, 33, 163, 164]). This statement is clearly illustrated in Fig. 3.1, where K_c exhibits a strong dependence on the resin composition. The combination of activity coefficients and ion association approaches are undoubtedly necessary to achieve good equilibrium representations and predictions. The solution non-idealities may be usually introduced by activity coefficients estimated by the Debye–Hückel, Hückel, Bromley, Guggenheim, Pitzer, or Meissner and Kusik models (Sect. 3.3.2.1), while those for the solid phase may be typically Wilson, NRTL, or Margules (Sect. 3.3.2.2). Very recently, Aniceto et al. [163, 164] compared the performance of Wilson, NRTL, Margules and UNIQUAC models for binary [163] and multicomponent systems [164]. Results showed that the Wilson, NRTL and Margules activity coefficient models provide similar representations and predictions, whereas UNIQUAC performed worse. The validation of these conclusions was accomplished with a large database containing 26 binary systems, 20 multicomponent systems, and 1546 data points totally.

Kataoka and Yoshida [34] assumed that only the exchanger phase was ideal, and used corrected selectivity coefficients, K^i_{aj}. Elprince and Babcock [35] proposed the Wilson model to estimate the $\bar{\gamma}_i$. Later, Smith and Woodburn [27] used the same model for the solid and the extended Debye–Hückel equation for the solution with the aim to study the equilibrium behaviour of $SO_4^{2-}/NO_3^-/Cl^-$ in Amberlite IR-400. Exactly the same recipe has been considered later by Shehata et al. [36] to study the ternary system $Sr^{2+}/Cs^+/Na^+$ on chabazite. Shallcross et al. [29] proposed an identical model for the equilibrium of $Ca^{2+}/Mg^{2+}/Na^+$ on Amberlite 252, but estimated the aqueous phase activity coefficients by Pitzer theory inasmuch as the extended Debye–Hückel equation did not take into account the nature of any other ions present in the mixture, except through their effect upon the ionic strength (see Eqs. 3.26 and 3.29). de Lucas et al. [37] studied the Na^+/K^+/Amberlite IR - 120 ion exchange equilibrium on different organic and mixed solvents, making use of the Debye–Hückel and Wilson models to compute γ_i and $\bar{\gamma}_i$, respectively. Vamos and Haas [38] used the Margules equations to represent the resin phase activity coefficients. The statistical analysis of data fits indicated that although the Wilson and Margules equations were able to model adequately the resin phase non-idealities, the Wilson model provided superior data fits. Pabalan and Bertetti [39] also chose the Margules expressions to represent the deviations from ideality of Na^+/Sr^{2+}, K^+/Sr^{2+}, and K^+/Ca^{2+} in clinoptilolite. Recently, Borba et al. [32] selected the Bromley and the Wilson models for solution and solid non-idealities, respectively, being able to predict the ternary system $Cu^{2+}/Zn^{2+}/Na^+$ in Amberlite IR-120.

Several works have investigated the influence of temperature upon ion exchange equilibria, for instance, those of Mumford et al. [22, 40], Carmona et al. [41], and de Lucas et al. [42]. Generally, the influence of the temperature on the equilibrium constant is interpreted by thermodynamic or semiempirical relations. Regarding its effect on the activity coefficients, it is taken into account by the proper relations of each model as, for instance, Eqs. 3.28, 3.30, 3.53, 3.55, and 3.58. For the particular case of the Wilson model, which is the most adopted for ion exchange calculation, other approaches appear, as the Robinson and Gilliland relationship [41–43].

Most publications estimate solution non-idealities with Pitzer or Debye–Hückel type models. de Lucas et al. [44] and Martínez et al. [30] are two essays where the Meissner and Kusik electrolyte solution theory has been selected instead. Vo and Shallcross [45] analysed binary and multicomponent ion exchange equilibria involving H^+, Na^+, K^+, Mg^{2+}, and Ca^{2+} ions in Amberjet 1200 H. They adopted the Wilson model for the resin and compared the performance of three models for the non-ideal behaviour of the solution, which may be arranged in order of increasing accuracy as ideal < extended Debye–Hückel < Meissner and Kusik < Pitzer. Nonetheless, it has been demonstrated that the ion association effects contribute more significantly to more accurate results than the employment of sophisticated models to account for the solution non-idealities [45].

The models of Smith and Woodburn [27], Shallcross et al. [29], de Lucas et al. [37], and Borba et al. [32] require three parameters per binary system, namely, the equilibrium constant K_B^A and the Wilson interaction parameters Λ_{AB} and Λ_{BA}, which have been simultaneously regressed from the binary equilibrium data. In the case of n_c counter ions in solution, this type of models involves $n_c(n_c-1)$ Wilson parameters to describe the interactions between $n_c(n_c-1)/2$ different pairs of ions. These parameters, along with the equilibrium constants, should be evaluated from experimental data for the corresponding $n_c(n_c-1)/2$ binary systems. In order to eliminate one adjustable Wilson parameter per binary, the following reciprocity relationship investigated by Allen et al. [46] and Allen and Addison [47] may be applied:

$$\Lambda_{AB}\,\Lambda_{BA} = 1 \tag{3.77}$$

In the multicomponent case, the connection between Λ_{ij} values is referred to as the Hála constraint [48, 49]. For ternary systems, it writes:

$$\Lambda_{AB}\,\Lambda_{BC}\Lambda_{CA} = 1 \tag{3.78}$$

Notwithstanding these equations, as well as other triangle relations to reduce model complexity, they do not adequately represent experimental data. Allen et al. [46] applied their modified version of the Smith and Woodburn [27] model, which contains only parameters K_B^A and Λ_{AB} per binary system, but both optimised values varied significantly with solution concentration. Other authors have drawn the same conclusions [21, 47, 50].

The works of Smith and Woodburn [27], Shallcross et al. [29], and Shehata et al. [36] demonstrate the predictive capability of mass action models. Nonetheless, the simultaneously correlated constants that they obtained for each constitutive binary system (K_B^A, Λ_{AB}, and Λ_{BA}) exhibit a disturbing interdependence, specifically K_B^A values very sensitive to minor changes in the values of the Wilson parameters, Λ_{AB} and Λ_{BA}. Hence, it is vital to apply and develop alternative approaches to decouple the model parameters in order to get equilibrium constants independent of the solution concentration. Fortunately, there are several methods available which may be successfully applied with that purpose. The next topics under discussion are the approaches of Argersinger et al. [51], Gaines and Thomas [52], and Ioannidis et al. [31] that will be presented below in detail due to their importance in the field of ion exchange equilibrium.

Another point of concern in the modelling of multicomponent ion exchange equilibrium by mass action law principles is the overspecified nature of the problem. As has been mentioned before, the number of parameters available from binary subsystems exceeds the number required to describe the corresponding multicomponent system. In general, for n_c counter ions, we may form $n_c(n_c-1)/2$ binary systems, though just $n_c - 1$ are truly independent, which means it is necessary to choose the most appropriate subsystems for modelling. Various methods have been used to select this optimal choice. The most accurate one in the majority of cases is the use of those binary subsystems containing the most strongly exchanging component which may be seen as the reference component in this context [21, 47, 50, 53].

Thermodynamic Treatment of Equilibrium

The problem of the simultaneous fitting of the equilibrium constant and parameters of the activity coefficient models in the solid phase from experimental data is the high degree of correlation between the two sets of parameters. It is fundamental to apply thermodynamically consistent models that provide estimates of equilibrium constants independently of the activity coefficients in the sorbed phase. Here we present the most utilised and powerful approaches in the literature: Argersinger et al. [51], Gaines and Thomas [52], and Ioannidis et al. [31]. They use only thermodynamic relations and consist essentially in the calculation of the equilibrium constant and activity coefficients in exchanger from corrected selectivity coefficients data under various conditions. With this purpose, the Gibbs–Duhem equation is used in the derivation, as it is shown below.

Argersinger–Davidson–Bonner and Gaines and Thomas Approaches

Both methods provide very similar equations for the equilibrium constant, K_B^A, and allow the calculation of the activity coefficients of counter ions in the exchanger, $\bar{\gamma}_i$. Nonetheless, Argersinger et al. [51] adopted ionic fraction scale, y_i, while Gaines and Thomas [52] used equivalent ionic fractions, Y_i.

For the sake of simplicity, the derivation of the Gaines and Thomas [52] equations will be accomplished here in terms of the less elaborate approach, by neglecting solvent and electrolyte sorption effects.

The Gibbs–Duhem equation applied to the ion exchanger relates the chemical potentials, $\bar{\mu}_i$, of both counter ions:

$$n_A d\bar{\mu}_A + n_B d\bar{\mu}_B = 0 \qquad (3.79)$$

After expressing the chemical potentials in terms of activity coefficients and equivalent ionic fractions, Eq. 3.79 gives rise to:

$$dLn\,\bar{\gamma}_A = -\frac{y_B}{y_A} dLn\,\bar{\gamma}_B - dLn\,Y_A - \frac{y_B}{y_A} dLn\,Y_B \qquad (3.80)$$

Furthermore, by applying logarithms to Eq. 3.14 followed by differentiation, one gets:

$$dLn\,K_{aB}^A = z_A dLn\,\bar{\gamma}_B - z_B dLn\,\bar{\gamma}_A \qquad (3.81)$$

The previous equations may be used to eliminate $dLn\,\bar{\gamma}_B$ so that:

$$dLn\,\bar{\gamma}_A = -\frac{Y_B}{z_B} dLn\,K_{aB}^A - dY_A - \frac{z_A}{z_B} dY_B \tag{3.82}$$

Since $Y_B\,dLn\,K_{aB}^A = d\left[Y_B Ln\,K_{aB}^A\right] - Ln\,K_{aB}^A\,dY_B$, Eq. 3.82 may be recast as:

$$dLn\,\bar{\gamma}_A = -\frac{1}{z_B} d\left[Y_B Ln\,K_{aB}^A\right] + \frac{1}{z_B} Ln\,K_{aB}^A dY_B - dY_A - \frac{z_A}{z_B} dY_B \tag{3.83}$$

which after partial integration under initial mono-ionic A form condition yields:

$$Ln\,\bar{\gamma}_A = -\frac{1}{z_B} Y_B Ln\,K_{aB}^A + \left(1 - \frac{z_A}{z_B}\right) Y_B + \frac{1}{z_B} \int_0^{Y_B} Ln\,K_{aB}^A\,dY_B \tag{3.84}$$

By an analogous procedure, one obtains an expression for $Ln\,\bar{\gamma}_B$:

$$Ln\,\bar{\gamma}_B = \frac{1}{z_A} Y_A\,Ln\,K_{aB}^A + \left(1 - \frac{z_B}{z_A}\right) Y_A - \frac{1}{z_A} \int_0^{Y_A} Ln\,K_{aB}^A dY_A \tag{3.85}$$

By substituting these two relations into Eq. 3.14, one finally gets the Gaines and Thomas model:

$$Ln\,K_B^A = (z_B - z_A) + \int_0^1 Ln\,K_{aB}^A\,dY_A \tag{3.86}$$

According to the previous equation, the equilibrium constant can be calculated by integrating the natural logarithm of the corrected selectivity coefficient, $Ln\,K_{aB}^A$, from the mono-ionic B form ($Y_A = 0$) to the mono-ionic A form ($Y_A = 1$) of the ion exchanger. The activity coefficients can also be calculated using Eqs. 3.84 and 3.85. Data is previously regressed to obtain an expression for $Ln\,K_{aB}^A$ as function of Y_A, and then the integration is accomplished. Polynomials and rational functions are frequently used.

For illustration, the graphical representation of $Ln\,K_{aB}^A$ against $Y_{Ca^{2+}}$ is shown in Fig. 3.3 for the binary system Ca^{2+}/Mg^{2+}/Amberjet 1200 H at 25°C and 0.10 N, along with the polynomial fitted to the experimental data. The Pitzer model (main Eqs. 3.38–3.40) was selected to compute the activity coefficients for the solution. The polynomial obtained is:

$$Ln\,K_{aB}^A = -5.8486\,Y_{Ca^{2+}}^4 + 7.4164\,Y_{Ca^{2+}}^3 - 4.5334\,Y_{Ca^{2+}}^2 + 0.1515\,Y_{Ca^{2+}} + 1.3371 \tag{3.87}$$

which after integration between 0 and 1 according to Eq. 3.86 yields the true equilibrium constant, $K_B^A = 1.7969$.

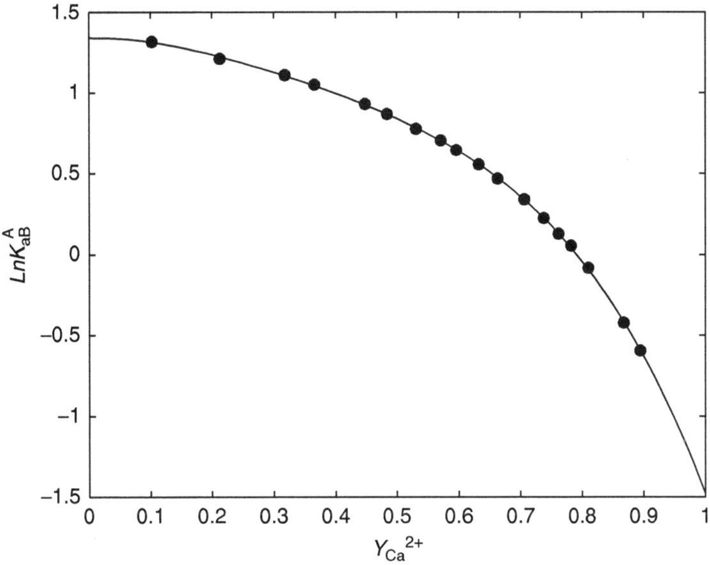

Fig. 3.3 Gains and Thomas approach for the calculation of the equilibrium constant of the system $Ca^{2+}/Mg^{2+}/Cl^{-}$/Amberjet 1200 H at 25°C and 0.10 N [15]

As has been mentioned above, Argersinger et al. [51] adopted the ionic fraction scale instead of equivalents. Their final expression is:

$$Ln\,K_{\mathrm{B}}^{\mathrm{A}} = \int_0^1 Ln\,K_{\mathrm{aB}}^{\mathrm{A}} dy_{\mathrm{A}} \tag{3.88}$$

The procedure to implement the model of Argersinger et al. [51] is totally similar to that of Gaines and Thomas, though Eq. 3.88 is now implicated.

Particular attention must be paid when equilibrium data are not available in the limits of the integral ($y_{\mathrm{A}}, Y_{\mathrm{A}} \to 0$ or $y_{\mathrm{A}}, Y_{\mathrm{A}} \to 1$), and extrapolations are necessary to span all range. In such cases, large errors may result, which requires an alternative approach. Ioannidis et al. [31, 54] solved this problem, providing new expressions that do not require extrapolations outside the experimental data range (derivation given below).

Fletcher and Townsend [55] and Soldatov and Bychkova [56] extended the methodologies of Argersinger et al. [51] and Gaines and Thomas [52] to multicomponent systems.

Ioannidis–Anderko–Sanders Approach

Ioannidis and Anderko [57] and Ioannidis et al. [31] developed a mathematical treatment analogous to that of Gaines and Thomas [52] and Argersinger et al. [51], which avoids the need of experimental data over the full range of the ionic fraction to determine the equilibrium constant, in contrast with previous Eqs. 3.86 and 3.88. This is fundamental for works where it is impossible to cover all concentration range or data is not available.

The Gibbs–Duhem equation is once more applied to the ion exchanger but includes now the solvent contribution:

$$n_A d\bar{\mu}_A + n_B d\bar{\mu}_B + n_s d\bar{\mu}_s = 0 \tag{3.89}$$

Expressing the chemical potentials in terms of activity coefficients and ionic fractions, one obtains after manipulation:

$$Y_A dLn(y_A \bar{\gamma}_A)^{z_B} + Y_B dLn(y_B \bar{\gamma}_B)^{z_A} + n_s z_A z_B dLn\, a_s = 0 \tag{3.90}$$

From Eq. 3.81, originally developed from Eq. 3.14, one may easily get:

$$Y_B z_A dLn\, \bar{\gamma}_B = Y_B dLn\, K^A_{aB} - Y_B z_B dLn\, \bar{\gamma}_A \tag{3.91}$$

$$Y_A z_B dLn \bar{\gamma}_A = Y_A z_A dLn \bar{\gamma}_B - Y_A dLn K^A_{aB} \tag{3.92}$$

Substituting Eqs. 3.91 and 3.92 individually into 3.90 yields, respectively:

$$z_B d\, Ln\, \bar{\gamma}_A = -Y_B d \ln K^A_{aB} - Y_A z_B d\, Ln\, y_A - Y_B z_A d\, Ln\, y_B - n_s z_A z_B d \ln a_s \tag{3.93}$$

$$z_A d\, Ln\, \bar{\gamma}_B = Y_A d\, Ln\, K^A_{aB} - Y_A z_B d\, Ln\, y_A - Y_B z_A d\, Ln\, y_B - n_s z_A z_B d\, Ln\, a_s \tag{3.94}$$

The integration of Eq. 3.93 between two points of the isotherm, labelled θ_1 and θ_2, may be written as:

$$z_B\, Ln \frac{\bar{\gamma}_A(\theta_1)}{\bar{\gamma}_A(\theta_2)} = \int_{\theta_1}^{\theta_2} Y_B d Ln K^A_{aB} + z_B \int_{\theta_1}^{\theta_2} Y_A d Ln y_A + z_A \int_{\theta_1}^{\theta_2} Y_B d Ln y_B + z_A z_B \int_{\theta_1}^{\theta_2} n_s d Ln\, a_s \tag{3.95}$$

The three former integrals on the right-hand side can be treated and calculated to give:

$$\int_{\theta_1}^{\theta_2} Y_B d\, Ln\, K^A_{aB} = \int_{\theta_1}^{\theta_2} \left[d\left(Y_B\, Ln\, K^A_{aB}\right) - Ln\, K^A_{aB}\, dY_B \right]$$

$$= \left[Y_B\, Ln\, K^A_{aB} \right]_{\theta_1}^{\theta_2} - \int_{\theta_1}^{\theta_2} Ln\, K^A_{aB}\, dY_B \tag{3.96}$$

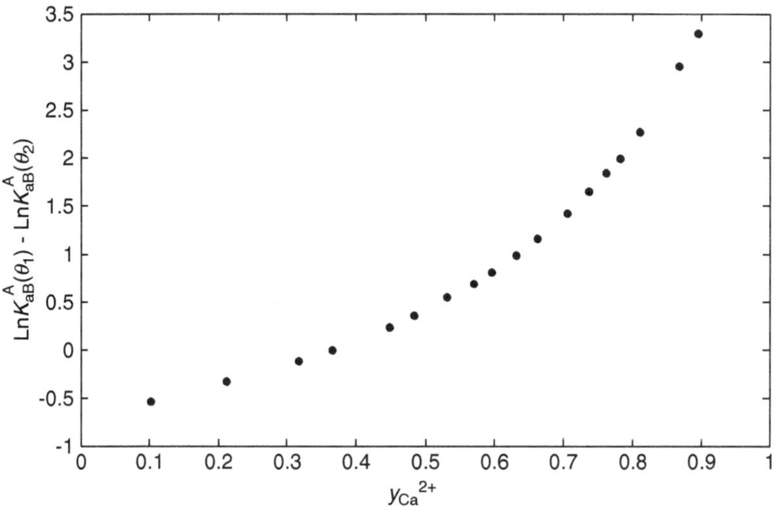

Fig. 3.4 Ioannidis et al. [31] plot for the $Ca^{2+}/Mg^{2+}/Cl^-$/Amberjet 1200 H at 25°C and 0.10 N [15]

$$z_B \int_{\theta_1}^{\theta_2} Y_A d\,Ln\,y_A = \frac{1}{z_B^{-1} - z_A^{-1}} Ln\left(\frac{y_A}{z_B} + \frac{y_B}{z_A}\right)\Bigg]_{\theta_1}^{\theta_2} \tag{3.97}$$

$$z_A \int_{\theta_1}^{\theta_2} Y_B d\,Ln\,y_B = \frac{1}{z_A^{-1} - z_B^{-1}} Ln\left(\frac{y_A}{z_B} + \frac{y_B}{z_A}\right)\Bigg]_{\theta_1}^{\theta_2} \tag{3.98}$$

Similarly, integrating Eq. 3.94 between the same isotherm points, θ_1 and θ_2, an expression for $z_A\,Ln\,[\bar{\gamma}_B(\theta_1)/\gamma_B(\theta_2)]$ can be derived. Subtracting both equations to eliminate the terms containing the solvent activity, the final relation of Ioannidis et al. arises:

$$Ln\frac{\bar{\gamma}_A^{z_B}(\theta_2)}{\bar{\gamma}_A^{z_B}(\theta_1)} + Ln\frac{\bar{\gamma}_B^{z_A}(\theta_1)}{\bar{\gamma}_B^{z_A}(\theta_2)} = Ln\,K_{aB}^A(\theta_1) - Ln\,K_{aB}^A(\theta_2) \tag{3.99}$$

The calculation procedure is quite different from those of Argersinger et al. [51] and Gaines and Thomas [52]. Now, from a set of experimental data, one chooses a reference point (θ_1) and calculates $Ln\,K_{aB}^A$ for all points (θ_2). Then, the differences $Ln\,K_{aB}^A(\theta_1) - Ln\,K_{aB}^A(\theta_2)$ are computed, and the parameters of the activity coefficient model adopted for $\bar{\gamma}_i$ are optimised by minimising the sum of residuals of Eq. 3.99.

Figure 3.4 displays such a plot for the same system studied above, $Ca^{2+}/Mg^{2+}/Cl^-$/Amberjet 1200 H at 25°C and 0.10 N. The activity coefficients for the solution were estimated by Pitzer model (main Eqs. 3.38–3.40), and Wilson was selected for

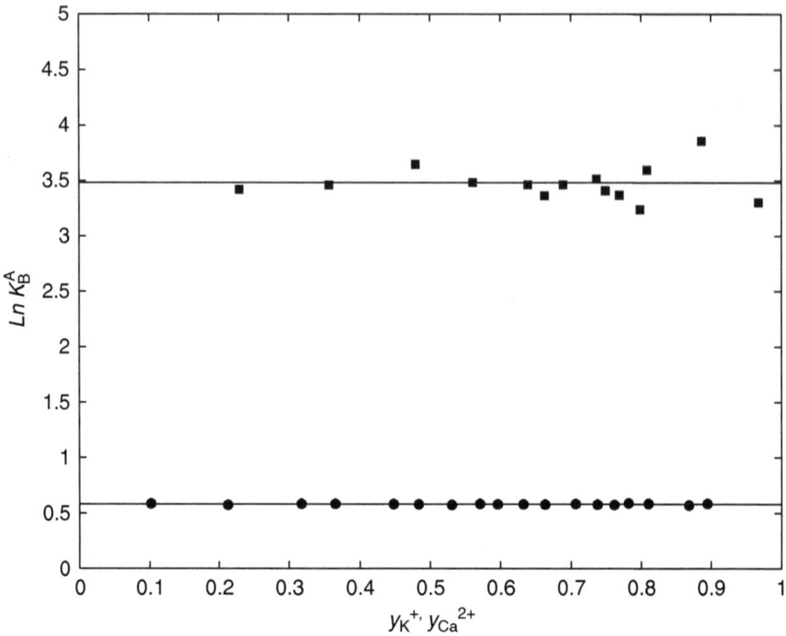

Fig. 3.5 Equilibrium constants calculated by Ioannidis et al. approach for $Ca^{2+}/Mg^{2+}/Cl^-$/Amberjet 1200 H at 25°C and 0.10 N (●) and $K^+/Na^+/Cl^-$/clinoptilolite at 25°C and 0.05 N (■) [15, 59]

the resin (Eq. 3.56). The reference point is the number four, for which the difference $Ln\,K_{aB}^{A}(\theta_1) - Ln\,K_{aB}^{A}(\theta_2)$ vanishes (see Fig. 3.4). The equilibrium constant determined by Ioannidis et al. approach is $K_B^A = 1.7873$, which differs by only 0.54% from the value offered by the method of Gaines and Thomas. It is worth noting that all calculations were performed between $y_{Ca^{2+}} = 0.1$ and 0.9, i.e. without extrapolations to $y_{Ca^{2+}} = 0$ and $y_{Ca^{2+}} = 1$. In Fig. 3.5, the equilibrium constants calculated by Eq. 3.14 using the optimised Wilson parameters, Λ_{12} and Λ_{21}, and Pitzer model are plotted. The invariance of the values found is remarkable. For illustration, an equivalent plot for another system ($K^+/Na^+/Cl^-$/clinoptilolite at 25°C and 0.05 N) was superimposed.

Application of Modelling Techniques

As has been mentioned above, it is fundamental to study ion exchange equilibrium on the basis of thermodynamically consistent models that provide estimates of equilibrium constants independently of the activity coefficients of the resinates. Furthermore, not only the non-ideal behaviour of both phases (Sect. 3.3.2) should be taken into account but also the incomplete ionic dissociation/ionisation in solution (Sect. 3.3.3).

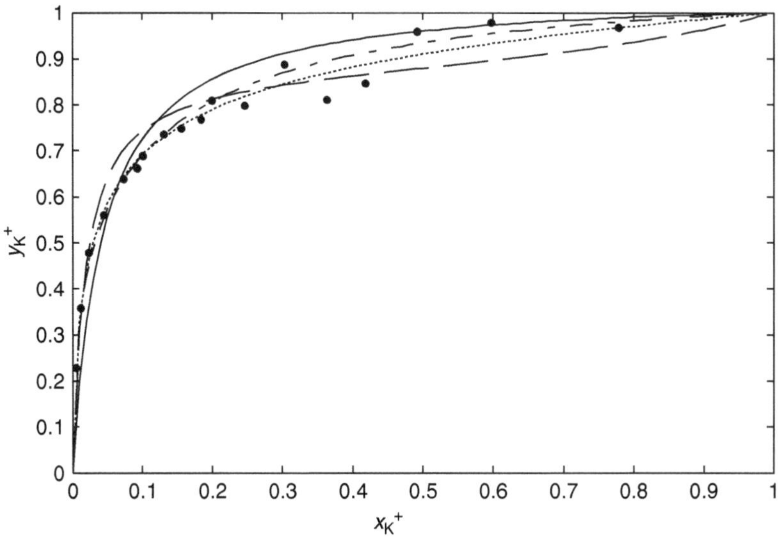

Fig. 3.6 Modelling the ion exchange equilibrium of $K^+/Na^+/Cl^-$/clinoptilolite at 25°C and 0.05 N. Models: (–) ideal, (···) Smith and Woodburn approach, (– – –) Gaines and Thomas approach, and (- -) Melis et al. model [59]

The research carried out by Mehablia and coworkers [21, 58] has been essential in the field of ion exchange modelling of multicomponent systems. Their model accounted for the non-idealities of the solution and exchanger phases by applying Pitzer's electrolyte solution theory and the Wilson model, respectively; the ionic pairing effect has been also considered by the equilibrium-based approach of Kester and Pytkowicz [14]. By employing the Gaines and Thomas [52] approach, they decoupled the calculation of the binary equilibrium constants from that of the Wilson parameters. The same approach has been applied recently to 26 binary [163] and 20 multicomponent [164] systems, where the Wilson, NRTL and UNIQUAC models were adopted for the exchanger phase.

Let us compare modelling results accomplished by techniques discussed up till now under the scope of mass action law. With this purpose, the ion exchange equilibrium of the binary system $K^+/Na^+/Cl^-$/clinoptilolite at 25°C and 0.05 N studied by Pabalan [59] was taken. As Fig. 3.6 points out, the ideal model (Eq. 3.72) provides poor results, in contrast to the isotherm achieved using Gaines and Thomas [52] approach combined with Pitzer and Wilson models. It is also plotted the correlation achieved by fitting the equilibrium constant and the two binary interaction parameters simultaneously; the Pitzer model was used for the solution. It is important to observe the split of the y_{K^+} provided by the last two methods – due to Gaines and Thomas [52] and Smith and Woodburn [27] – since the embodied equations are exactly the same: mass action law, Pitzer and Wilson. This fact denunciates the interdependence of the three parameters under analysis. The calculations achieved for the same system using the Ioannidis et al. approach [31, 54] are displayed in Fig. 3.7, where ideal, Pitzer and Debye–Hückel equations were

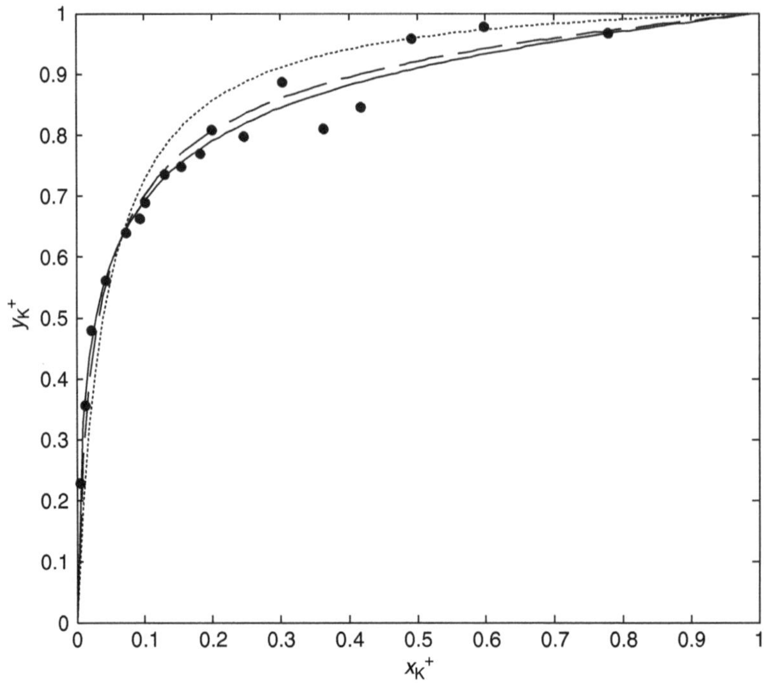

Fig. 3.7 Modelling ion exchange equilibrium of $K^+/Na^+/Cl^-$/clinoptilolite at 25°C and 0.05 N by Ioannidis et al. approach. Activity coefficient models for the liquid phase: (···) ideal, (–) Debye–Hückel, and (- -) Pitzer [59]

adopted for the liquid phase. Once more, the ideal model exhibits the worse behaviour, whereas Pitzer and Debye–Hückel give rise to identical results since they are inside their range of validity ($C_{N,t} = 0.05$ N, $I = 0.05$ molal).

In Fig. 3.8, the correlation of three binary isotherms achieved with Gaines and Thomas method combined with the activity coefficients models of Pitzer and Wilson is presented. The systems under study are Ca^{2+}/Na^+, Mg^{2+}/Na^+, and Ca^{2+}/Mg^{2+} at 25°C and 0.10 N, with co-ion Cl^- and Amberjet 1500 H. The results depicted emphasise the good performance of this approach. Afterwards, the optimised binary parameters were used to estimate the corresponding ternary system $Ca^{2+}/Mg^{2+}/Na^+/Cl^-$ in Amberjet 1500 H. The prediction accomplished was very successful, as Fig. 3.9 illustrates.

The effect of total concentration in solution is fundamental in ion exchange. Figure 3.10 exhibits the isotherms of the aqueous system $Ca^{2+}/Na^+/Cl^-$/ clinoptilolite at 25°C for 0.05 and 0.5 N. Once again, we emphasise the good performance assured by the thermodynamic treatment of the ion exchange equilibrium, in this case via Gains and Thomas approach, together with the Pitzer equation and Wilson model for activity coefficients of ions in solution and exchanger phases, respectively.

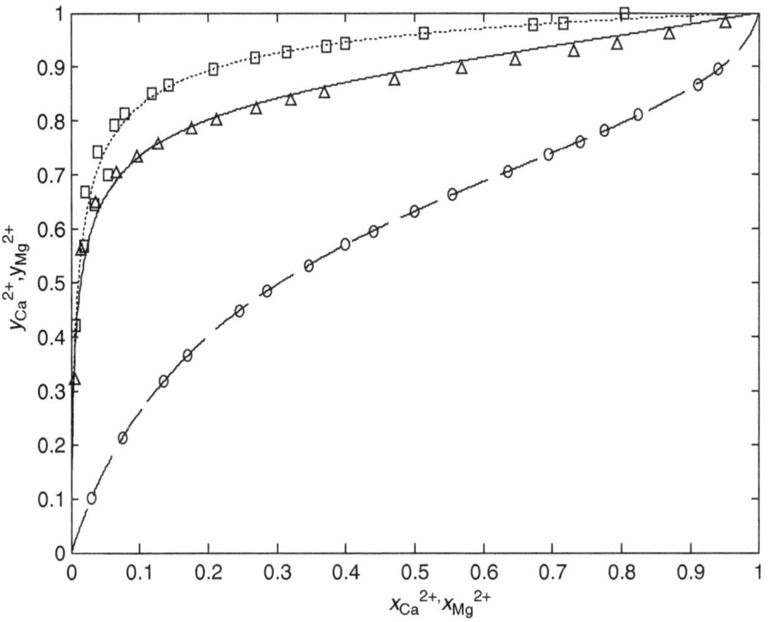

Fig. 3.8 Comparison of modelling results for the ion exchange equilibrium of (Δ) Ca^{2+}/Na^{+}, (□) Mg^{2+}/Na^{+}, and (○) Ca^{2+}/Mg^{2+}, with co-ion Cl^{-}, on Amberjet 1500 H at 25°C and 0.10 N [15]

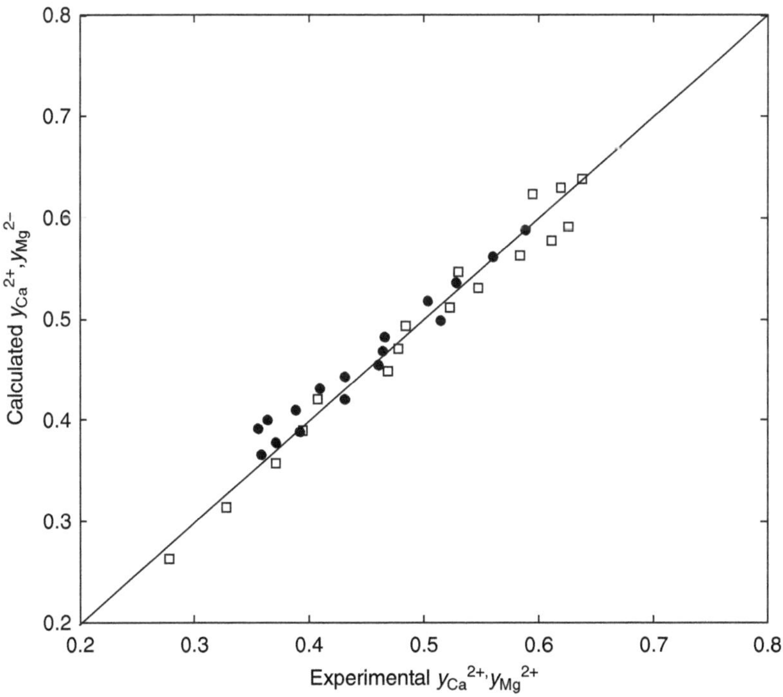

Fig. 3.9 Comparison of model predictions to experimental data for the ion exchange equilibrium $Ca^{2+}/Mg^{2+}/Na^{+}/Cl^{-}$ in Amberjet 1500 H at 25°C and 0.05 N [15]

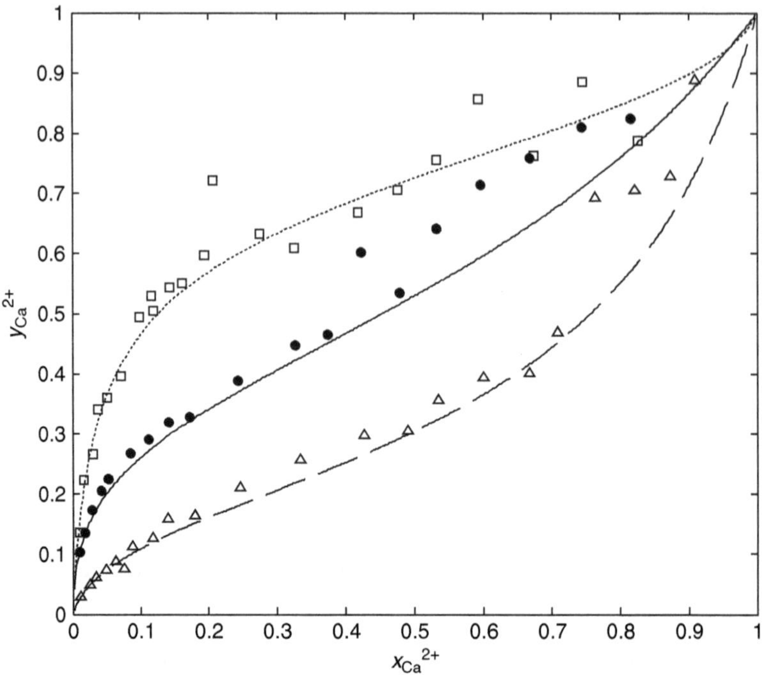

Fig. 3.10 Isotherms of $Ca^{2+}/Na^{+}/Cl^{-}$/clinoptilolite at 25°C for (□) $C_{N,t} = 0.005N$, (●) $C_{N,t} = 0.05N$, and (Δ) $C_{N,t} = 0.5N$ [39]

3.3.4.2 Group II: Ion Adsorption and Related Models

Models of group II describe ion exchange as phase equilibrium akin to an adsorption process. Under this group, one may consider the empirical or semiempirical application of classical isotherms, such as Langmuir or Freundlich [60–72], as well as more theoretically sound models, where deviations from ideal adsorption are generally explained in terms of the functional groups heterogeneity of the ion exchanger [73]. Another reliable and successful approach is based on the principles of statistical thermodynamics, as well as on the Metropolis Monte Carlo numerical method [53, 74–77].

Application of Adsorption Isotherms

Altin et al. [68] presented a review of adsorption isotherms that successfully represent binary equilibrium data between heavy metal and clay minerals. Some applications may be cited. Petrus and Warchol [69] utilised the Langmuir and competitive Langmuir models to study ternary and quaternary ion exchange equilibria of $Pb^{2+}/Cd^{2+}/Cu^{2+}$solutions with Na-form of clinoptilolite. Ku et al. [70] used Langmuir and Freundlich models to describe the equilibria of aqueous solutions of phenols with Purolite A-510 in chloride form, a strong basic anionic resin. They considered that the uptake occurs only on the active sites of the resin by either adsorption or ion exchange. pH affects significantly equilibria since phenols act as weak acids; consequently, the species distribution has been taken into account by including the corresponding ionisation reactions and acidity constants. Later, Carmona et al. [71] studied phenol removal from aqueous solution with Amberlite IRA-420 and developed an improved theoretical model to define the contribution of the adsorption and ion exchange mechanisms in the overall uptake. In their analysis, Langmuir equations were adopted for both phenomena. Recently, Langmuir and Freundlich models have been also adopted to represent equilibria of phenol from aqueous solution on Dowex XZ (strong anion exchange resin) and AuRIX 100 (weak anion exchange) [72]. Langmuir and Freundlich isotherms have been used to represent ion exchange equilibria of Hg^{2+} and Cd^{2+} aqueous solutions on titanosilicates ETS-4 and ETS-10 [60–67].

Fundamental Works of Myers and Collaborators

In the original work of Novosad and Myers [74], the thermodynamic description of adsorption from liquid solutions was used as a basis for a rigorous thermodynamic treatment of ion exchange. The authors defined the surface excess of ion i, Q_i^{e}, as the difference between the concentration of adsorbed ion and that for a non-selective exchanger, i.e. equal to the bulk liquid:

$$Q_i^{\mathrm{e}} = Q_{\mathrm{t}}\,(Y_i - X_i) \tag{3.100}$$

This quantity may be expressed solely in terms of the equivalent ionic fraction in the liquid at equilibrium and of the initial concentrations and amounts of solution and exchanger:

$$Q_i^{\mathrm{e}} = \frac{W_{\mathrm{exch}} Q_{\mathrm{t}} + C_{\mathrm{N,t}} V_{\mathrm{L}}}{W_{\mathrm{exch}}} \left(X_{i,0} - X_i\right) \tag{3.101}$$

where W_{exch} is the mass of ion exchanger and V_{L} is the solution volume. From Eq. 3.100, one concludes that:

$$Q_{\mathrm{A}}^{\mathrm{e}} + Q_{\mathrm{B}}^{\mathrm{e}} = 0 \tag{3.102}$$

$$Q_{\mathrm{A}} = Q_{\mathrm{A}}^{\mathrm{e}} + Q_{\mathrm{t}} X_{\mathrm{A}} \tag{3.103}$$

The relationship between the surface excess and the separation factor follows from Eqs. 3.16 (recast in equivalent ionic fractions), 3.8, and 3.100:

$$Q_{\mathrm{A}}^{\mathrm{e}} = \frac{Q_{\mathrm{t}} X_{\mathrm{A}} X_{\mathrm{B}} \left(\alpha_{\mathrm{B}}^{\mathrm{A}} - 1\right)}{\alpha_{\mathrm{B}}^{\mathrm{A}} X_{\mathrm{A}} + X_{\mathrm{B}}} \tag{3.104}$$

Novosad and Myers [74] also devised an integral thermodynamic consistency test for an ion exchange cycle involving three pairs of 1–1 counter ions, A/B, B/C, and C/A, formed from the set A/B/C, in analogy with the thermodynamic consistency test for adsorption at liquid/solid interface [75]:

$$\begin{aligned} \mathrm{B} + \bar{\mathrm{A}} &\leftrightarrow \bar{\mathrm{B}} + \mathrm{A} \\ \mathrm{C} + \bar{\mathrm{B}} &\leftrightarrow \bar{\mathrm{C}} + \mathrm{B} \\ \mathrm{A} + \bar{\mathrm{C}} &\leftrightarrow \bar{\mathrm{A}} + \mathrm{C} \end{aligned} \tag{3.105}$$

Considering the isofugacity principle of equilibrium and the Gibbs–Duhem relationship [5], such test can be derived at constant normality and temperature to give the following triangle rule [74]:

$$\int_{X_A=0}^{1} Q_A^e d\left(Ln\frac{a_{AX}}{a_{BX}}\right) + \int_{X_B=0}^{1} Q_B^e d\left(Ln\frac{a_{BX}}{a_{CX}}\right) + \int_{X_C=0}^{1} Q_C^e d\left(Ln\frac{a_{CX}}{a_{AX}}\right) = 0 \tag{3.106}$$

where X is the co-ion in solution and AX, BX, and CX denote the dissolved electrolytes. Equation 3.106 is rigorous and is evaluated with experimentally accessible variables: surface excess and solution activities. It can be simplified if Harned's rule [76] is obeyed for the activity coefficients of the electrolytes in solution, so activities are replaced by ionic fractions:

$$\int_{X_A=0}^{1} \frac{Q_A^e}{X_A X_B} dX_A + \int_{X_B=0}^{1} \frac{Q_B^e}{X_B X_C} dX_B + \int_{X_C=0}^{1} \frac{Q_C^e}{X_C X_A} dX_C = 0 \tag{3.107}$$

If it is further assumed that the total capacity of exchanger is the same for all counter ions, then substituting Eqs. 3.102 and 3.104 into 3.107 yields:

$$\bar{\alpha}_B^A \times \bar{\alpha}_C^B \times \bar{\alpha}_A^C = 1 \tag{3.108}$$

where $\bar{\alpha}_j^i$ is an average relative separation factor given by:

$$Ln\,\bar{\alpha}_j^i = \int_{X_i=0}^{1} \frac{\alpha_j^i - 1}{\alpha_j^i X_i + X_j} dX_i \tag{3.109}$$

If separation factors are constant, i.e. $\alpha_j^i = k$, Eq. 3.109 reduces to:

$$\bar{\alpha}_j^i = \alpha_j^i = k \tag{3.110}$$

Four years later, Myers and Byington [73] published the milestone model of group II. In that essay, they assume the ion exchange non-idealities are due to energetic heterogeneity of the adsorption sites. The non-uniformities were simulated through the binomial distribution of $v = n + 1$ site types, each having a characteristic energy level $E_{i,j}$ for the adsorption of ion i on site j. The fraction of sites of type j is:

$$P_j = \binom{n}{j} p^j (1-p)^{n-j} \tag{3.111}$$

with the following energy level:

$$E_{j,i} = \bar{E}_i + \frac{j - np}{\sqrt{np(1-p)}}\sigma_i \tag{3.112}$$

where $0<p<1$ is the skewness parameter (for a symmetric distribution, $p = 1/2$), $\bar{E}_i$ is the average adsorption energy of counter ion i, and σ_i is the standard deviation of the distribution. It is worth to note that $P_j = Q_j/Q_t$, where Q_j represents the ion exchanger capacity in terms of j functional sites.

The adsorption process of ion i on site j has been described by the ideal Langmuir isotherm [73]:

$$Q_{j,i} = Q_t \frac{B_{j,i} X_i C_{N,t}}{1 + \sum_{l=1}^{n_c} B_{j,l} X_l C_{N,t}} \quad (j = 0, \ldots, n) \tag{3.113}$$

where $B_{j,i}$ is the Langmuir constant of adsorption calculated by:

$$B_{j,i} = \bar{B}_i \exp\left(\frac{E_{j,i} - \bar{E}_i}{\Re T}\right) \tag{3.114}$$

It is assumed that the total capacity of ion exchanger, Q_t, is the same for all ionic species; otherwise, Langmuir equation fails.

The concentration of ion i in the sorbent is computed by carrying out the summation over all sites, $Q_i = \sum_{j=0}^{n} Q_{j,i} P_j$. From Eq. 3.113, one gets:

$$\frac{Q_i}{X_i} = Q_t \sum_{j=0}^{n} \frac{P_j B_{j,i} C_{N,t}}{1 + \sum_{l=1}^{n_c} B_{j,l} X_l C_{N,t}} \tag{3.115}$$

The general separation factor for pair $i - j$ is obtained by combining Eqs. 3.16 (recast in equivalent ionic fractions) and 3.115:

$$\alpha_k^i = \frac{Q_i/X_i}{Q_k/X_k} = \frac{\sum_{j=0}^{n} P_j B_{j,i} / (1+\tau_j)}{\sum_{j=0}^{n} P_j B_{j,k} / (1+\tau_j)}, \quad \tau_j \equiv \sum_{l=1}^{n_c} B_{j,l} X_l C_{N,t} \tag{3.116}$$

The parameters of the model are the number of sites, $v = n + 1$; the skewness of the distribution, p; the average adsorption energy, $\bar{E}_i$; and standard deviation, σ_i, for each counter ion. Input variables are the solution normality, $C_{\mathrm{N,t}}$; equivalent ionic fractions in solution, X_i; and temperature, T.

At high normality, we may consider $1 + \tau_j \approx \tau_j$, which simplifies Eqs. 3.115 and 3.116. For instance, Eq. 3.115 reduces to:

$$\frac{Q_i}{X_i} = Q_{\mathrm{t}} \sum_{j=0}^{n} \frac{P_j B_{j,i}}{\sum_{l=1}^{n_c} B_{j,l} X_l} \tag{3.117}$$

Saunders et al. [77] used this result without introducing the $C_{\mathrm{N,t}} \to \infty$ approximation since they noted that for completely ionised functional groups, it is more appropriate to describe the single site exchange equilibrium by Eq. 3.118 below, instead of Eq. 3.113:

$$Q_{j,i} = Q_{\mathrm{t}} \frac{B_{j,i} X_i}{\sum_{l=1}^{n_c} B_{j,l} X_l} \quad (j = 0, ..., n) \tag{3.118}$$

For the simplest case of two counter ions, $n_{\mathrm{c}} = 2$, on a two-site exchanger, $v = n + 1 = 2$, the separation factor at high normality reduces to:

$$\alpha_2^1 = \bar{\alpha} \frac{\bar{\alpha} W^{U+V} X_1 + \left[W^U (1-p) + W^V p\right] X_2}{\bar{\alpha}\left[W^V (1-p) + W^U p\right] X_1 + X_2} \tag{3.119}$$

where

$$\bar{\alpha} = \exp\left(\frac{\bar{E}_1 - \bar{E}_2}{\Re T}\right) \quad \text{and} \quad W = \exp\left(\frac{\sigma_1 - \sigma_2}{\Re T}\right) \tag{3.120}$$

$$U = -\frac{p}{\sqrt{p(1-p)}} \quad \text{and} \quad V = \frac{1-p}{\sqrt{p(1-p)}} \tag{3.121}$$

In this case, the model comprehends three parameters. If additionally the distribution is symmetric, $p = 1/2$, the resultant two-parameter model may be rewritten as:

$$\alpha_2^1 = \bar{\alpha} \frac{\eta + X_1(\bar{\alpha} - \eta)}{1 + X_1(\bar{\alpha}\eta - 1)} \tag{3.122}$$

where

$$\eta = \cosh\left(\frac{\sigma_1 - \sigma_2}{\Re T}\right) \quad \text{and} \quad \bar{\alpha} = \exp\left(\frac{\bar{E}_1 - \bar{E}_2}{\Re T}\right) \tag{3.123}$$

In both models, Eqs. 3.119 and 3.122, when $\sigma_1 = \sigma_2$, the separation factor is constant, i.e. $\alpha_2^1 = \bar{\alpha}$. Furthermore, it is to note that substituting Eq. 3.122 into 3.109 and integrating, one obtains $\bar{\alpha}_2^1 = \bar{\alpha}$, which implies that this model satisfies the triangular rule given by Eq. 3.108, being thermodynamically consistent [73].

The form of the multicomponent model of Myers and Byington [73] usually adopted in applications may be expressed by:

$$\alpha_r^i = \bar{\alpha}_r^i \frac{\sum_{k=1}^{n_c} \left\{ \bar{\alpha}_r^k X_k W_{k,r}^{U+V} \left[(1-p) W_{i,k}^U + p W_{i,k}^V \right] \right\}}{\sum_{k=1}^{n_c} \left\{ \bar{\alpha}_r^k X_k W_{k,r}^{U+V} \left[(1-p) W_{r,k}^U + p W_{r,k}^V \right] \right\}} \tag{3.124}$$

with U and V given by Eq. 3.121, and

$$\bar{\alpha}_r^i = \exp\left(\frac{\bar{E}_i - \bar{E}_r}{\Re T} \right) \quad \text{and} \quad W_{i,r} = \exp\left(\frac{\sigma_i - \sigma_r}{\Re T} \right) \tag{3.125}$$

Here, subscript r denotes a reference counter ion. Note that $\alpha_r^i = 1/\alpha_i^r$, $\alpha_k^i = \alpha_r^i \alpha_k^r = \alpha_r^i/\alpha_r^k$, and $\alpha_i^i = 1$. The concentration of species i in the exchanger phase is calculated by:

$$Y_i = \frac{X_i \alpha_r^i}{\sum_{k=1}^{n_c} X_k \alpha_r^k} \tag{3.126}$$

The models of Myers and Byington have been applied to correlate binary data of amino acids and subsequent prediction of the corresponding multicomponent equilibria, taking into account the ionisation equilibria involved [77, 78]. Later, Zammouri et al. [79] studied the equilibrium uptake of amino acids by strongly basic anion exchange resins in the hydroxyl form (Amberlite IRA458 and Amberlite IRA900B) using the same modelling approach. In that work, the authors proved the importance of energetic heterogeneities in ion exchangers. The same model has been applied to describe the equilibria of dipeptides on cation resins with varying degrees of cross-linking, and rather satisfactory results were obtained [80]. The ion exchange equilibria of phenylalanine and tyrosine on a strong-acid cation exchanger resin (PK220) and on a strong-base anion exchanger resin (PA316) were investigated by Moreira and Ferreira [81], and their experimental binary and multicomponent data were successfully represented by Myers and Byington [73] model. In Fig. 3.11, the isotherm for the binary equilibrium $Tyr^{2-}/OH^-/PA316$ obtained at NaOH concentrations of 10 and 100 mM is shown, where Tyr^{2-} is the divalent form of tyrosine. In this range, the ionic fraction of its monovalent form (Tyr^-) is negligible. As the figure points out, the Myers and Byington model provides very good representation of the experimental data.

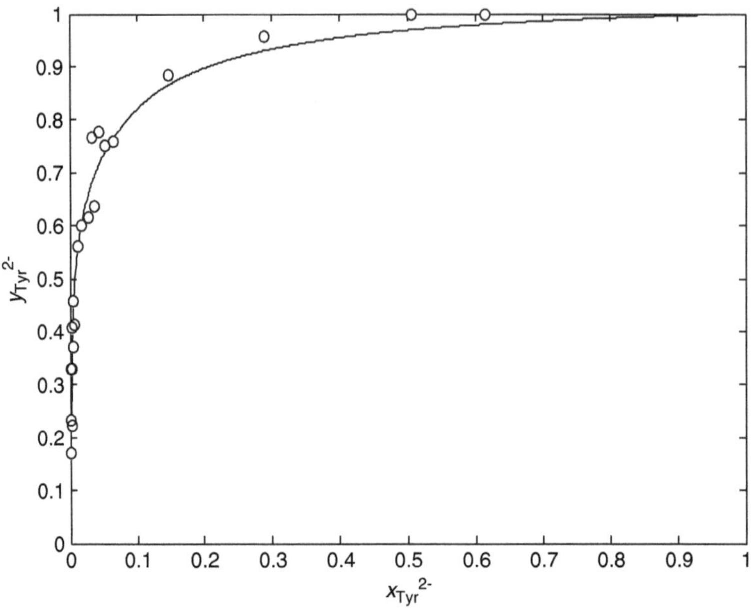

Fig. 3.11 Experimental [73] and calculated ion exchange equilibrium for the binary system Tyrosine^{2-}/OH^-/PA316 (model by Myers and Byington [73])

Models Based on the Principles of Statistical Thermodynamics

In last years, theoretical models based on the principles of statistical thermodynamics as well as on the Metropolis Monte Carlo numerical method have been published. The model was originally developed by de Kock and van Deventer [82], refined and extended by Lukey et al. [83, 84], and recently adapted by Provis et al. to generate a single-parameter model for binary [85] and multicomponent [53] ion exchange equilibria.

The main assumptions made in the formulation of the original model by de Kock and van Deventer [82] are as follows: (1) The solvent is an additional species (denoted by 0). Hence, a binary system (two counter ions) is considered a three component competitive sorption. (2) The ion exchanger is initially loaded by the solvent only. This means that for the first iteration, the concentration of any counter ion in the sorbent is zero. (3) The mean sorption energy of any species i, E_i, is measured relative to the solvent. Consequently, the mean sorption energy of the solvent is zero, $E_0 = 0$. Moreover, E_i is taken as negative because total system energy diminishes after ion exchange. (4) The probability $\Pr(I_{\text{load}})$ that an active site is loaded with species i is given by:

$$\Pr(I_{\text{load}}) = Y_i \tag{3.127}$$

where Y_i is the equivalent fraction of all active sites occupied by i. It is implied in this equation that the whole surface is available for sorption. In cases of highly non-ideal selective sorption, it is necessary to include selectivity effects, modifying the form of Eq. 3.127 [82–84]. (5) Ion exchange is assumed to occur by collision between dissolved exchanging species with active sites occupied by counter ions or solvent molecules. The probability of resinate species i being replaced by dissolved species j is computed by:

$$\Pr(I \rightarrow J) = \Pr(J_{\mathrm{col}}) \Pr(I_{\mathrm{load}}) \Pr(I \rightarrow J)_{\mathrm{col}} \tag{3.128}$$

where $\Pr(J_{\mathrm{col}})$ is the probability that species j in the bulk phase collides with an active site, and $\Pr(I \rightarrow J)_{\mathrm{col}}$ is the probability that species i will be replaced by j if a collision between them occurs. (6) The Boltzmann criterion is utilised to calculate $\Pr(I \rightarrow J)_{\mathrm{col}}$ in terms of the difference between the dimensionless sorption energies of i and j by:

$$\Pr(I \rightarrow J)_{\mathrm{col}} = \min\left[\exp\left(-\Delta E^{*}_{\mathrm{I}\rightarrow\mathrm{J}}\right), 1\right] \tag{3.129}$$

where $\Delta E^{*} = \Delta E/(kT)$, k is Boltzmann constant, and $T = 294$ K is the reference temperature adopted. (7) The ratio between collisions of different species from solution is equal to the ratio of their concentrations in the bulk phase. This assumption implies that the solution exhibits ideal properties – Provis et al. [85] utilised activities instead. (8) At equilibrium, the probability of an increase or a decrease in the loading of any species i must be equal:

$$\sum_{j=0}^{n_c} \Pr(I \rightarrow J) = \sum_{j=0}^{n_c} \Pr(J \rightarrow I) \tag{3.130}$$

(Note that summation includes solvent, 0, and all n_c counter ions.)

The multicomponent model can be now developed. Substituting Eq. 3.128 into both sides of Eq. 3.130 yields:

$$\Pr(I_{\mathrm{load}}) \sum_{j=0}^{n_c} \left[\Pr(J_{\mathrm{col}}) \Pr(I \rightarrow J)_{\mathrm{col}}\right] = \Pr(I_{\mathrm{col}}) \sum_{j=0}^{n_c} \left[\Pr(J_{\mathrm{load}}) \Pr(J \rightarrow I)_{\mathrm{col}}\right] \tag{3.131}$$

which after assumptions 4 and 7 gives rise to:

$$Y_i \sum_{j=0}^{n_c} \left[X_j \Pr(I \rightarrow J)_{\mathrm{col}}\right] = X_i \sum_{j=0}^{n_c} \left[Y_j \Pr(J \rightarrow I)_{\mathrm{col}}\right] \tag{3.132}$$

Attending to the fact that the solvent fraction is $Y_0 = 1 - \sum_{j=1}^{n_c} Y_j$, Eq. 3.132 can be rewritten as:

$$Y_i \sum_{j=0}^{n_c} \left[\frac{X_j}{X_i} \Pr(I \rightarrow J)_{\mathrm{col}}\right] + \sum_{j=1}^{n_c} \left\{ Y_j \left[\Pr(0 \rightarrow I)_{\mathrm{col}} - \Pr(J \rightarrow I)_{\mathrm{col}}\right]\right\} = \Pr(0 \rightarrow I)_{\mathrm{col}} \tag{3.133}$$

This equation reduces to a set of linear equations with the following matrix formulation:

$$\left(\overline{\overline{M}}+\overline{\overline{N}}\right)\,\overline{S}=\overline{D} \tag{3.134}$$

where $\overline{\overline{M}}$ is a square matrix, $\overline{\overline{N}}$ is a diagonal matrix, $\overline{S}$ is a column vector containing the exchanger phase loadings, and column vector $\overline{D}$ contains the probabilities that the sorbed solvent will be replaced by counter ion I:

$$M(i,j)=\Pr\left(0\rightarrow I\right)_{\mathrm{col}}-\Pr\left(J\rightarrow I\right)_{\mathrm{col}} \tag{3.135}$$

$$N(i,i)=\sum_{j=0}^{n_c}\frac{X_j}{X_i}\Pr(I\rightarrow J)_{\mathrm{col}} \tag{3.136}$$

$$S(i)=Y_i \tag{3.137}$$

$$D(i)=\Pr\left(0\rightarrow I\right)_{\mathrm{col}} \tag{3.138}$$

Attending to assumption 3, $\overline{D}=[1]$ in most cases. It must also be noted that $\Pr\left(I\rightarrow I\right)_{\mathrm{col}}=1$ since this imposes no change in system configuration (see Eq. 3.129).

For the simplest case of a binary system containing the solvent and counter ions A and B, the model reduces to:

$$\overline{\overline{M}}=\begin{bmatrix}\Pr\left(0\rightarrow A\right)_{\mathrm{col}}-1 & \Pr\left(0\rightarrow A\right)_{\mathrm{col}}-\Pr\left(B\rightarrow A\right)_{\mathrm{col}}\\ \Pr\left(0\rightarrow B\right)_{\mathrm{col}}-\Pr\left(A\rightarrow B\right)_{\mathrm{col}} & \Pr\left(0\rightarrow B\right)_{\mathrm{col}}-1\end{bmatrix} \tag{3.139}$$

$$\overline{\overline{N}}=\begin{bmatrix}1+\sum\limits_{j=0,B}\frac{X_j}{X_{\mathrm{A}}}\Pr(A\rightarrow J)_{\mathrm{col}} & 0\\ 0 & 1+\sum\limits_{j=0,A}\frac{X_j}{X_{\mathrm{B}}}\Pr(B\rightarrow J)_{\mathrm{col}}\end{bmatrix} \tag{3.140}$$

$$\overline{\mathrm{S}}=\begin{bmatrix}Y_{\mathrm{A}}\\ Y_{\mathrm{B}}\end{bmatrix} \tag{3.141}$$

$$\overline{\mathrm{D}}=\begin{bmatrix}\Pr\left(0\rightarrow A\right)_{\mathrm{col}}\\ \Pr\left(0\rightarrow B\right)_{\mathrm{col}}\end{bmatrix}=\begin{bmatrix}1\\ 1\end{bmatrix} \tag{3.142}$$

This model can be applied to dilute systems containing equally sized particles that non-selectively and reversibly sorb onto a homogeneous surface of an ion exchanger. Furthermore, no interactions between adjacent particles occur. Lukey et al. [83, 84] incorporated non-ideal properties to solve these limitations, such as surface heterogeneity, selective and irreversible sorption, and lateral interactions in the solid.

The heterogeneous nature of the ion exchanger was modelled by dividing the surface into n_p equally sized and homogeneous patches, each of a different energy. The energy distribution over the patches was characterised by a unimodal or bimodal Gaussian distribution. The more the surface is divided $(n_p \rightarrow \infty)$, the better the representation of the real exchanger can be made. In practice, 30 patches are sufficient in most cases since superior numbers require additional computation time for minimal improvement in model results [53, 83–85].

The selective sorption was modelled by dividing the homogeneous patch into $2^{n_c} - 1$ different areas, whose fractions are initially pre-assigned and then determined by nested calculations. Here, the solvent is excluded as a species.

The lateral interactions between sorbed species and the irreversible ion exchange are modelled by the following equation:

$$E^*_{i,\text{load}} = \frac{E^*_{i,\text{load},0}}{\zeta_i} + z_{\text{adj}} \sum_{j=0}^{n_c} Y_j L_{i,j} \tag{3.143}$$

$E^*_{i,\text{load}}$ is the dimensionless observed sorption energy of species i, $E^*_{i,\text{load},0}$ is the dimensionless sorption energy of i on a specific site, ζ_i is its irreversible parameter $(0<\zeta_i \leq 1)$, Y_j is the equivalent fraction of species j on patch, z_{adj} is the number of interacting adjacent sites, and $L_{i,j}$ is the lateral interaction factor between species i and j, which is calculated based on their values for similar species:

$$L_{i,j} = \sqrt{L_{i,i} \times L_{j,j}}\left(1 - \beta_{i,j}\right) \tag{3.144}$$

$\beta_{i,j}$ may be taken as a deviation to the classical combining rule.

Prior to evaluating the model, the following parameters and initial values must be specified: initial surface composition, number of patches into which the surface will be divided, and the statistical distribution used to model the energy distribution of the surface experienced by each species.

The parameters used to describe the exchanger surface are chosen using prior knowledge of the sorption system; they are the energy distribution of each species, the distribution fraction if a bimodal distribution is adopted, the distribution mean and standard deviation, and the selective sorption parameter.

Realistic estimates of parameters are necessary to initially solve the model: mean sorption energy of each species between -5 and -100 and interaction energy of each species between -5 and 5.

Lukey et al. [84] noticed that the model was able to successfully describe the multicomponent equilibrium sorption of gold cyanide and copper cyanide onto an ion exchange resin containing trimethylammonium functional groups in non-saline as well as saline solutions attaining 0.5 or 1.0 M KCl.

In another work, Lukey et al. [83] used this model to describe the same equilibria onto distinct resins – Amberjet 4400, dimethylamine, and tripropylamine resins – in order to determine the influence of functional group structure, hydrophilicity, and ionic density of the resin on specific model parameters.

The former applications of the model did not include any solution phase non-idealities. Hence, subsequent works by Provis et al. [53, 85] used its simplest possible form in conjunction with accurate description of solution behaviour to provide the most accurate results based on a minimal set of fitted parameters. Accordingly, Provis et al. incorporated the ion association following the approach of Kester and Pytkowicz (Sect. 3.3.3) and the Pitzer model for the activity coefficients in solution (Sect. 3.3.2.1), but disregarded the complex exchanger phase effects cited above (lateral interactions, selective sorption, irreversible sorption). Consequently, a single-parameter model was derived, where only $\Delta E^{*}_{I \rightarrow J}$ is required to represent a binary ion exchange system.

Provis et al. [85] studied the behaviour of all binary combinations of Ca^{2+}, Mg^{2+}, Na^{+}, K^{+}, and H^{+} ions in the presence of Cl^{-} with the commercial resin Amberjet 1200 H. They obtained acceptable fits to all data sets, and the optimised parameters for each binary system were independent of the solution concentration.

Later, Provis et al. [53] continued the analysis of their one-parameter approach by detailing its extension to ternary and quaternary ion exchange systems. They used the parameters fitted to binary systems and accomplished predictions for the following systems: $Ca^{2+}/Mg^{2+}/H^{+}/Cl^{-}$/Amberjet 1200 H, $Ca^{2+}/H^{+}/Na^{+}/NO_3^{-}$/Dowex HSR-S(E), $Ca^{2+}/H^{+}/Na^{+}/SO_4^{2-}$/Dowex HSR-S(E), $Ca^{2+}/K^{+}/Na^{+}/H^{+}/Cl^{-}$/Dowex HSR-S(E), and $Ca^{2+}/Mg^{2+}/Na^{+}/H^{+}/Cl^{-}$/Amberjet 1200 H. The modelling results were highly accurate in many cases; those systems that were not predicted so accurately from binary data were able to be described by recalculating one parameter using the multicomponent data.

3.3.4.3 Group III: Heterogeneous Mass Action Models

Group III accounts for the heterogeneous nature of the ion exchanger in the framework of the equilibrium models based on the mass action law discussed in Sects. 3.3.1 and 3.3.4.1. The milestone models are due to Barrer and Meier [86] and Melis et al. [25]. The first authors recognised the existence of two different sites in a zeolite and described the equilibrium of binary systems of monovalent cations through the superimposition of two ideal mass action laws. Later, Melis et al. [25] adopted a similar approach to interpret the behaviour of systems involving ions of distinct valences, giving rise to the most significant model of group III.

The model of Melis et al. [25] assumes that both the solution and the exchanger phases behave ideally, in contrast to the majority of the expressions belonging to group I. The effect of mixture non-idealities is supposed to be smaller than that caused by the heterogeneity of the exchanger phase, in a similar way to the idea of Myers and Byington [73] discussed above in Sect. 3.3.4.2. Nonetheless, the activity coefficients of the ionic species in solution can be perfectly incorporated in the calculations by using one of the models of Sect. 3.3.2.1.

In the simplest situation, the resin is constituted by only two functional groups ($n_f = 2$). Accordingly, its total capacity corresponds to the sum of the individual resin capacities with respect to functional groups of types 1 and 2:

$$Q_t = \sum_{j=1}^{n_f=2} Q_j = Q_1 + Q_2 \tag{3.145}$$

and the fraction of the active site $j = 1, 2$ is given by $P_j = Q_j/Q_t$. Following this notation, the equilibrium constant for the ion exchange on the functional group j is:

$$K_{B,j}^{A} = \frac{\bar{a}_{j,A}^{z_B}\, a_B^{z_A}}{\bar{a}_{j,B}^{z_A}\, a_A^{z_B}} \quad (j = 1, 2) \tag{3.146}$$

The corresponding standard free energy change is ΔG_j^0, which means the average value for the global ion exchange process is:

$$\overline{\Delta G^0} = \sum_{j=1}^{n_f=2} P_j\, \Delta G_j^0 = P_1\, \Delta G_1^0 + P_2\, \Delta G_2^0 \tag{3.147}$$

From thermodynamics, the equilibrium constant can be estimated as function of temperature if the standard free energy change is known: $Ln\, K = -\Delta G^o/\Re T$ [5]. Accordingly, the average equilibrium constant must be related to the individual constants by:

$$\bar{K}_B^A = \exp\left(-\frac{\overline{\Delta G^0}}{\Re T}\right) = \left(K_{B,1}^A\right)^{P_1} \left(K_{B,2}^A\right)^{P_2} \tag{3.148}$$

If all activity coefficients are unitary, the activities in Eq. 3.146 are replaced by concentrations, for example, equivalent fractions and molalities. In this case, the selectivity coefficient for the functional group j is:

$$K_{c,j} = \frac{Y_{j,A}^{z_B}\, X_B^{z_A}}{Y_{j,B}^{z_A}\, X_A^{z_B}} \left(\frac{Q_j}{C_{N,t}}\right)^{z_B - z_A} \quad (j = 1, 2) \tag{3.149}$$

where the following stoichiometric restrictions must be adhere to:

$$Y_{j,A} + Y_{j,B} = 1 \quad (j = 1, 2) \tag{3.150}$$

The overall resin composition is obtained by considering, for each counter ion i, the quantities present on each type of functional group:

$$Y_i = \sum_{j=1}^{n_f=2} P_j\, Y_{j,i} \quad (i = \text{A, B}) \tag{3.151}$$

With this purpose, for given values of $K_{c,1}$ and $K_{c,2}$, Eq. 3.149 is solved in conjunction with Eq. 3.150 for the unknowns $Y_{j,A}$ and $Y_{j,B}$, for each active site j independently. The overall compositions are finally computed through Eq. 3.151.

For a multicomponent system containing n_c exchangeable ions, there exist $n_c - 1$ independent equilibrium relations between a reference counter ion and the remaining ones. The n_c^{th} ionic species is frequently taken as reference. Consequently, similarly to Eq. 3.149, the selectivity coefficients for the equilibrium between each counter ion $i = 1, 2, ..., (n_c - 1)$ and reference ion $r = A_{n_c}^{z_{nc}}$ on sites of type j are:

$$K_{c,j|i/r} = \frac{Y_{j,i}^{z_r} X_r^{z_i}}{Y_{j,r}^{z_i} X_i^{z_r}} \left(\frac{Q_j}{C_{N,t}}\right)^{z_r - z_i} \quad (j = 1, 2; i = 1, 2, ..., n_c - 1; r = A_{n_c}^{z_{nc}}) \tag{3.152}$$

which, along with the stoichiometric restrictions:

$$\sum_{i=1}^{n_c} Y_{j,i} = 1 \quad (j = 1, 2) \tag{3.153}$$

provide the n_c necessary equations to compute the concentrations $Y_{j,i}$ on active site j. Then, the overall exchanger loads can be obtained by summing up the corresponding concentrations of the individual active sites, analogously to Eq. 3.151:

$$Y_i = \sum_{j=1}^{n_f=2} P_j Y_{j,i} \quad (i = 1, 2, ..., n_c) \tag{3.154}$$

It is worth noting that for multicomponent systems, the model requires $n_f(n_c - 1) = 2(n_c - 1)$ equilibrium constants and the value of the parameter P_1, which is the same for all binary systems (remember that $P_2 = 1 - P_1$). Once more, the average equilibrium constants may be calculated by:

$$\bar{K}_r^i = \exp\left(-\frac{\overline{\Delta G_{i,r}^0}}{\Re T}\right) = \left(K_{r,1}^i\right)^{P_1} \left(K_{r,2}^i\right)^{P_2} \tag{3.155}$$

Melis et al. [87] tested their model with a large database collected from the literature. Its reliability in describing binary uptake data and in predicting the behaviour of multicomponent systems (using the parameters of the corresponding $n_c - 1$ couples firstly optimised) has been proved. Valverde et al. [88] compared this model with the homogeneous mass action law (group I) where the non-idealities of the solution and solid phases have been included. They studied ten binary systems and five ternary systems, combining Ca^{2+}, Mg^{2+}, K^+, Na^+, and H^+ on the same resin (Amberlite IR-120), and observed that, for binary systems with counter ions of different valences and for ternary systems, the homogeneous approach provided better results than the heterogeneous model developed by Melis et al. [87].

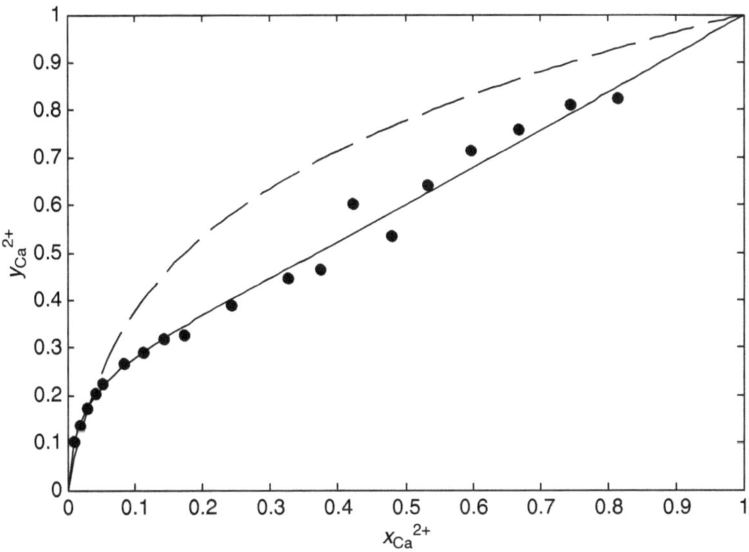

Fig. 3.12 Representation of the ion exchange equilibrium of the aqueous system $Ca^{2+}/Na^{+}/Cl^{-}$/ clinoptilolite at 25°C and 0.05 N, by the models of (–) Melis et al. [87] and (- -) ideal mass action law [59]

In Fig. 3.6 above, a comparison between ideal and non-ideal mass action law models for the binary system $K^{+}/Na^{+}/Cl^{-}$/clinoptilolite is presented. Superimposed, it is also the correlation achieved by Melis et al. model, which is reliable when compared to the remaining ones.

In Fig. 3.12, the excellent correlation achieved by the model of Melis et al. [87] for the aqueous system $Ca^{2+}/Na^{+}/Cl^{-}$clinoptilolite at 25°C is shown. The ideal mass action law representation is also graphed for comparison and provides rather poor results. However, it is worth noting that the first model contains three parameters ($K^{A}_{B,1}$, $K^{A}_{B,2}$, p_1) in contrast to the single-parameter ideal model.

3.4 Ion Exchange Kinetics

This section focuses the modelling of ion exchange kinetics. First, a brief overview of the classical semiempirical pseudo-first and pseudo-second order equations is presented. Then, a more detailed description around fundamental equations based on the Fick's law, where electrochemical gradients are not taken into account, will be given. Finally, particular attention will be devoted to models that accurately describe ion exchange in the presence of an electrical field, namely, the Nernst–Planck and the Maxwell–Stefan equations. In particular, a detailed model based on the Maxwell–Stefan formulation to describe ion exchange in microporous materials is presented.

3.4.1 General Overview

Apparently based on the original work of Lagergren in 1898, the kinetic performance of an ion exchange process is frequently interpreted by semiempirical pseudo-first and pseudo-second order equations [64, 66, 89–103]. Besides, the Ritchie and Elovich kinetic models have been also commonly applied to model ion exchange kinetics [104–110]. However, such models have no theoretical background, which limits their application and extrapolation.

Some authors represent experimental data with analytical models devised specifically for particle diffusion control or film diffusion control under well-defined conditions, as, for instance, isotopic exchange at infinite solution volume or constant diffusion rate and radial diffusion coefficient. However, such conditions generally never hold in practice [111–114].

Mass transport in dilute ionic systems can be effectively described by the Nernst–Planck (NP) equations [111, 115–126]. In an electrolyte solution, the electric field induced by the different mobility of counter ions produces an additional force also responsible for the transference of ions. The Nernst–Planck equations account for both concentration and electric potential gradients. However, the interdiffusion coefficients of counter ions are composition dependent and the ionic interactions (ion–ion, ion–solvent, and ion–ion exchanger) are lumped into effective diffusivities [111]. Furthermore, the Nernst–Planck model does not take into account non-ideality effects and the pressure gradient [111, 127]. Such weak points are solved by the Maxwell–Stefan approach of mass transport, which is particularly advantageous at high concentrations.

3.4.2 Semiempirical Models

The pseudo-first order rate equation of Lagergren [128] is one of the most widely used kinetic models to express the ion exchange mechanism. The Lagergren equation was the first rate equation for sorption of liquid/solid systems based on the solid capacity. It may be represented by:

$$\frac{d\bar{q}_i}{dt} = k_1 \left(\bar{q}_{i,\mathrm{e}} - \bar{q}_i \right) \tag{3.156}$$

where k_1 is the rate constant of the first order sorption and $\bar{q}_{i,\mathrm{e}}$ is the sorbed solute concentration at equilibrium. The top bars denote average concentrations. By partial integration of Eq. 3.156, from $t = 0$ and $\bar{q}_i = 0$, one obtains after manipulation:

$$Ln\left(\bar{q}_{i,\mathrm{e}} - \bar{q}_i \right) = Ln\left(\bar{q}_{i,\mathrm{e}} \right) - k_1\, t \tag{3.157}$$

In order to fit Eq. 3.157 to experimental batch data, $\bar{q}_{i,\mathrm{e}}$ must be known. The $\bar{q}_i$ values can be obtained for each time t from material balance to the experimental batch system. Then, constant k_1 can be determined by plotting $Ln(\bar{q}_{i,e} - \bar{q}_i)$ versus t. The value of $\mathrm{Ln}(\bar{q}_{i,e})$ is expected to be equal to the intercept of the $\mathrm{Ln}(\bar{q}_{i,e} - \bar{q}_i)$ versus t representation; however, this is often not the case.

The pseudo-second order rate equation is also based on the sorption capacity of the solid and can be written as:

$$\frac{d\bar{q}_i}{dt} = k_2\left(\bar{q}_{i,e} - \bar{q}_i\right)^2 \tag{3.158}$$

where k_2 is the rate constant of the second order sorption. After integration, it can be expressed as:

$$\frac{t}{q_i} = \frac{1}{k_2\bar{q}_{i,e}^2} + \frac{1}{\bar{q}_{i,e}}t \tag{3.159}$$

and k_2 can be determined by liner fitting.

Since the ion concentration in the solid is calculated from a material balance to the batch vessel, both the k_1 and k_2 constants of the pseudo-first and pseudo-second order models depend on the relative amounts of solid and fluid in the experiments used to extract them. This prevents their further use to simulate any other system, namely, the operation of open systems; in this context, these models are useless.

3.4.3 Fick's Law–Based Models

The counter ions diffusion is frequently the rate-determining step in ion exchange processes [111], occurring simultaneously in both the ion exchanger and in the solution. In particular, kinetics of ion exchange can be governed by either interparticle diffusion (film diffusion) or intraparticle diffusion (particle diffusion) or, in intermediate cases, by a combination of both steps.

Particle diffusion may be described by homogeneous models assuming that the particle has homogeneous structure, being the ion transport controlled by solid phase diffusion. However, ion exchangers generally consist of small microporous crystals embedded in a macroporous composite pellet, in which two distinct resistances to ion transfer may be distinguished: the micropore resistance of the solid phase and macropore resistance of the pellet (see Fig. 3.13). In this case, the ion exchange kinetics is more accurately described by heterogeneous diffusion models.

When a solid ion exchanger is in contact with an ionic solution, a diffusion process occurs by transference of ions from and into the solid particle. Neglecting the electrochemical gradients, the flux of exchangeable ions through the pellet, J_i, may be described by the Fick's first law, which for homogeneous pellets writes:

$$J_i = -D_i\,\nabla q_i \tag{3.160}$$

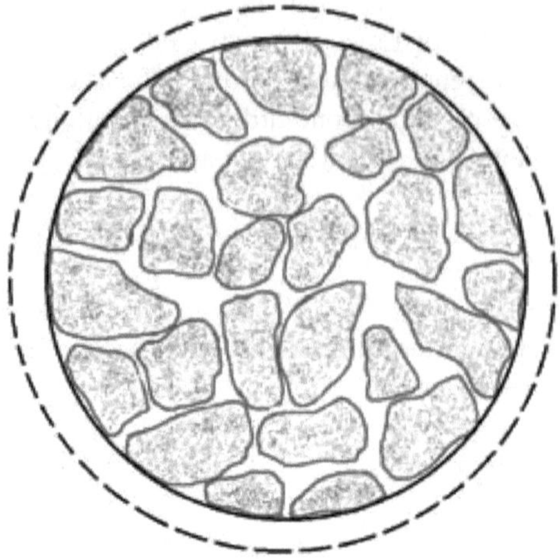

Fig. 3.13 External film, macropores, and microporous crystals where mass transfer limitations may occur in a composite pellet

where D_i is the diffusion coefficient. The evolution of concentration along time may be determined accordingly to the Fick's second law:

$$\frac{\partial q_i}{\partial t} = -\nabla \bullet J_i \tag{3.161}$$

which for spherical coordinates and constant diffusion coefficients yields:

$$\frac{\partial q_i}{\partial t} = D_i \left(\frac{\partial^2 q_i}{\partial r^2} + \frac{2}{r} \frac{\partial q_i}{\partial r} \right) \tag{3.162}$$

where r is the radial position.

Numerous solutions for Eq. 3.162 could be obtained for a variety of initial and boundary conditions, giving rise to several single-particle methods frequently used to evaluate the implied diffusion coefficients. In the following, some analytical solutions are briefly presented.

3.4.3.1 Expression for the *Infinite Solution Volume Condition*

When the volume of the external solution is much higher than the exchanger quantity (batch method), the fluid concentration remains approximately invariable throughout the process – *infinite solution volume* condition.

For particle diffusion rate control, the concentrations at exchanger surface are the same as in the bulk solution since there is no film. The corresponding initial and boundary conditions for particles initially clean are:

$$t = 0, \quad q_i = q_{i,0} \tag{3.163}$$

$$r = R, \quad q_i = 0 \tag{3.164}$$

$$r = 0, \quad \left(\frac{\partial q_i}{\partial r}\right) = 0 \tag{3.165}$$

being R the particle radius. By integration and averaging, one obtains [111, 113, 129, 130]:

$$F_i(t) = 1 - \frac{6}{\pi^2}\sum_{i=1}^{\infty}\frac{1}{i^2}\exp\left(-\frac{D_i t\pi^2 i^2}{R^2}\right) \tag{3.166}$$

where $F_i(t)$ is the fractional attainment of equilibrium, defined as $F_i(t) = \left(q_{i,o} - q_i(t)\right)/\left(q_{i,o} - q_{i,e}\right)$. In this equation, $\tau_d = R^2/D_i$ can be interpret as a time constant for intraparticle diffusion.

3.4.3.2 Diffusion Model According to Vermeulen's Approximation

For small values of t/τ_d, Eq. 3.166 does not converge rapidly. As an alternative, Vermeulen's approximation [131] can be adopted for isotopic exchange under infinite solution volume conditions. Less accurate than the exact solution, Vermeulen's approximation may be more suitable for practical use, fitting the entire range $0 \leq F_i(t) \leq 1$:

$$F_i(t) = \left[1 - \exp\left(-\frac{D_i t\pi^2}{R^2}\right)\right]^{1/2} = \left[1 - \exp\left(-\pi^2 t/\tau_d\right)\right]^{1/2} \tag{3.167}$$

The diffusion coefficient can be determined by linear regression of previous equation recast as [111, 113, 129, 132]:

$$Ln\left(1 - F_i(t)^2\right) = -\frac{D_i t\pi^2}{R^2} \tag{3.168}$$

3.4.3.3 Parabolic Diffusion Model

This model was derived for diffusion in a cylinder, assuming that diffusion is the controlling mechanism (rate limiting) and that the surface concentration is constant. Assuming only radial diffusion, i.e. neglecting axial diffusion (semi-infinite cylinder), the model equation is:

$$\frac{\partial q_i}{\partial t} = r\frac{\partial}{\partial r}\left(rD_i\frac{\partial q_i}{\partial r}\right) \tag{3.169}$$

If the cylinder is initially at constant concentration, the analytical solution for short times is [113, 130, 133, 134]:

$$F_i(t) = \frac{4}{\pi^{1/2}}\left(\frac{D_i t}{R^2}\right)^{1/2} - \frac{D_i t}{R^2} - \frac{1}{3\pi^{1/2}}\left(\frac{D_i t}{R^2}\right)^{3/2} \tag{3.170}$$

Neglecting the third term of this equation, for small values of t/τ_d, the diffusion coefficient may be determined from the slope of the plot F_i/t versus $1/t^{1/2}$.

3.4.3.4 Another Homogeneous Diffusion Model

Another homogeneous model has been developed from the second Fick's law for the case of constant diffusion rate, radial diffusion coefficient, and surface concentration [113, 133, 134]:

$$F_i(t) = 1 - \sum_{i=0}^{\infty} b_i \exp\left(-\frac{c_i t}{\tau_d}\right) \tag{3.171}$$

where b_i and c_i are parameters dependent on the exchanger geometry. For spherical particles and short times, Eq. 3.171 simplifies to:

$$F_i(t) = \frac{6}{\pi^{1/2}}\frac{D_i^{1/2}}{R}t^{1/2} \tag{3.172}$$

The diffusion coefficient may be easily calculated from the slope of $F_i(t)$ versus $t^{1/2}$. For long times, Eq. 3.171 gives rise to, after manipulation:

$$Ln(1 - F_i(t)) = Ln\left(\frac{6}{\pi^{1/2}}\right) - \frac{\pi^2 D_i}{R^2}t \tag{3.173}$$

3.4.3.5 Film Diffusion Control

When intraparticle diffusion is much faster than external diffusion, the film becomes the rate-controlling step [111, 114, 135].

In the following treatment, we assume that the film is planar and quasi-stationary, i.e. diffusion across the film is faster than the concentration changes at the boundary. According to Fick's first law for isotopic exchange under constant diffusivity, the flux through the film is thus:

$$J_i = -D_i\,\frac{\Delta C_i}{\delta} \tag{3.174}$$

where δ is film thickness.

The time dependence of the solution concentration is obtained from lumped material balance:

$$\frac{dC_i}{dt} = -J_i \frac{A_\mathrm{p}}{\mathrm{V_L}} \tag{3.175}$$

A_p is the external surface area of the solid particles. Assuming clean solution and uniform distribution in the solid at beginning, the proper conditions are [111]:

$$\begin{aligned} &r = R,\ C_i^* = \frac{q_{i,\mathrm{o}} C_\mathrm{t}}{q_\mathrm{t}} \\ &r \geq R + \delta, C_i = 0 \end{aligned} \tag{3.176}$$

where C_t and q_t are total concentrations in solution and in ion exchanger and C_i^* is the concentration of species i at the interface. Under the infinite solution volume approximation, the boundary condition should be:

$$r \geq R + \delta,\ t \geq 0 \qquad C_i = 0 \tag{3.177}$$

Accordingly, after integration and manipulation of previous equations, one gets under infinite solution volume restriction:

$$F(t) = 1 - \exp\left(-\frac{3 D_i C_i t}{R \delta q_\mathrm{t}}\right) \tag{3.178}$$

For the case of finite solution volume, the interface concentration varies along time, being necessary to include a global material balance (both phases) to follow their bulk concentration and compute ΔC_i. The final analytical solution is:

$$F(t) = 1 - \exp\left(-3D_i \frac{V_\mathrm{s} q_\mathrm{t} + V_\mathrm{L} C_\mathrm{t}}{R \delta V_\mathrm{L} q_\mathrm{t}} t\right) \tag{3.179}$$

3.4.3.6 Heterogeneous Diffusion Model

The heterogeneous diffusion model assumes that processes with distinct diffusion coefficients can occur simultaneously inside particle [134]. In this case, different τ_d's should thus be assigned to each individual process, being the overall sorption determined by:

$$q = \int_{\tau_{\mathrm{d},i}}^{\tau_{\mathrm{d,m}}} q_{\tau,\mathrm{e}} \left[q_\tau(t)/q_{\tau,\mathrm{e}}\right] d\tau_\mathrm{d} \tag{3.180}$$

where $\tau_{d,m}$ and $\tau_{d,i}$ are the maximum and minimum values of τ_d, respectively. After appropriate approximations for small and large values of t, the kinetic equation takes the form [134]:

$$\frac{d\left[q_i(t)/q_{i,e}\right]}{dLn(t)} = \frac{1}{Ln\left(\tau_{d,m}/\tau_{d,i}\right)}\left[1 - \left(\frac{4t}{\pi\tau_{d,m}}\right)^{1/2} - \frac{8}{\pi^2}\exp\left(-\frac{\pi^2 t}{4\tau_{d,i}}\right)\right] \tag{3.181}$$

For small values of t, the second term on the right-hand side of Eq. 3.181 is negligible, while the third term can be neglected at large t.

3.4.4 Nernst–Planck Model

An electric field in an electrolyte solution, caused by the different mobilities of counter ions, produces an additional force responsible for the transference of ions. In contrast to the classical Fick's law analysed above, the Nernst–Planck model does account for the influence of this electric field.

In the absence of concentration gradients, the transference of a species i in the direction of the current is proportional to the gradient of the electric potential ϕ and to its concentration and electrochemical valence:

$$N_i = -u_i z_i C_i \nabla\phi \tag{3.182}$$

where u_i is the electrochemical mobility, defined by the Nernst–Einstein equation:

$$u_i = \frac{D_i F}{\Re T} \tag{3.183}$$

where F is the Faraday constant. In solutions with concentration gradients, both electrical and diffusion transferences must be taken into account. Hence, the flux of each counter ion in dilute ionic solutions may be described by the Nernst–Planck equations [111]:

$$N_A = -D_A\left(\frac{\partial q_A}{\partial r}\right) - D_A z_A q_A \frac{F}{\Re T}\left(\frac{\partial\phi}{\partial r}\right) \tag{3.184}$$

$$N_B = -D_B\left(\frac{\partial q_B}{\partial r}\right) - D_B z_B q_B \frac{F}{\Re T}\left(\frac{\partial\phi}{\partial r}\right) \tag{3.185}$$

being D_A and D_B are the self-diffusion coefficients of species A and B. In the following development, the particle is assumed to be subjected to the restrictions of electroneutrality and absence of electric current, mathematically represented by:

$$q_A z_A + q_B z_B = Q_t \tag{3.186}$$

$$z_A N_A + z_B N_B = 0 \tag{3.187}$$

The electric potential term in the transport equations may be eliminated by substituting Eqs. 3.184 and 3.185 in Eq. 3.187 and then combining with Eq. 3.186:

$$\frac{F}{\Re T}\frac{\partial \phi}{\partial r} = \frac{z_A(D_B - D_A)}{z_A q_A(z_A D_A - z_B D_B) + D_B z_B Q_t}\frac{\partial q_A}{\partial r} \tag{3.188}$$

After substitution of Eq. 3.188 into 3.184, the general expression for the molar flux of A is obtained:

$$N_A = -\frac{D_A D_B(z_B^2 q_B + z_A^2 q_A)}{D_A z_A^2 q_A + D_B z_B^2 q_B}\left(\frac{\partial q_A}{\partial r}\right) \tag{3.189}$$

This equation may be recast as a special form of the Fick's first law, where a coupled interdiffusion coefficient, D_{AB}, appears:

$$N_A = -D_{AB}\left(\frac{\partial q_A}{\partial r}\right), \quad \text{and} \quad D_{AB} \equiv \frac{D_A D_B\left(z_A^2 q_A + z_B^2 q_B\right)}{D_A z_A^2 q_A + D_B z_B^2 q_B} \tag{3.190}$$

D_{AB} depends on D_A, D_B, and on the ionic concentration in the ion exchanger, which varies throughout the ion exchange process. Accordingly, D_{AB} takes the value of D_A when $y_A \to 0$, while it assumes D_B for vanishing values of y_B. Such tendency, depicted in Fig. 3.14 for univalent exchange, elucidates that the ion present in smaller concentration mainly controls the interdiffusion rate [111].

A simplified solution for multicomponent ion exchange kinetics based on the Nernst–Planck model has been developed by Pátzay [136]. The good results obtained prove its applicability for multispecies systems.

3.4.5 *Maxwell–Stefan Approach*

An alternative approach to describe an ion exchange process deals with the application of the Maxwell–Stefan (MS) equations, due to their well-documented advantages in mass transport [137–139]. However, in the case of porous structures, such formulation has been essentially applied to non-ionic systems, particularly in the gas phase [140, 141]. In the last years, some papers focus on the application of MS theory to mass transfer in membrane electrolysis process (e.g. works of van der Stegen [142] and Hogendoorn et al. [143]) and electrodialysis [144]. The MS equations do take into account and distinguish both ion–ion and ion–solid interactions, and they have been proved to be more effective in predicting ion exchange than the above mentioned Nernst–Planck relationships [127, 145–147]. Furthermore, under the scope of MS formalism, one diffusivity coefficient

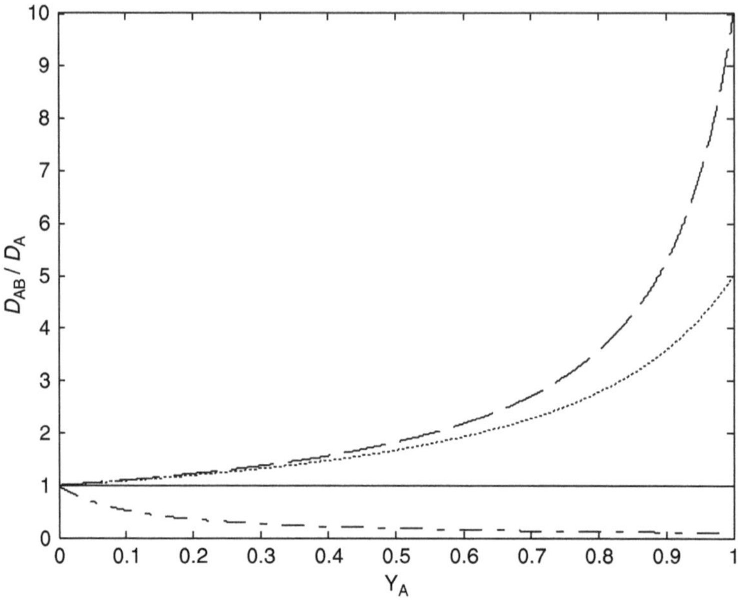

Fig. 3.14 Plot of the ratio between the interdiffusion and self-diffusion coefficients against the ionic composition inside the solid for different values of D_B/D_A, for univalent exchange. *Dashed lines* $D_B/D_A = 10$, *dotted lines* $D_B/D_A = 5$, *solid lines* $D_B/D_A = 1$, and *dash-dotted lines* $D_B/D_A = 0.1$

is defined for each pair of components, being dependent on their properties only. In addition, these coefficients are only weakly dependent on composition.

In the following, the intraparticle resistance to ion exchange in microporous materials will be described using the generalised MS equations (e.g. Krishna and Wesselingh [137]) subjected to the following assumptions: (1) Surface diffusion is the only transport mechanism since due to the very small pore diameters, ions never escape from the force field of the matrix co-ions, mainly due to the strong and long-range nature of the electrostatic interactions. (2) The solid matrix is considered to be a uniform distribution of fixed ionic groups corresponding to the $(n_c + 1)$th component, through which counter ions diffuse. Therefore, $u_{n_c+1} = 0$, as in the well-known Dusty Gas Model [137, 148, 149].

Considering the abovementioned assumptions, the MS transport equation for species i in multicomponent ionic systems is:

$$- \nabla \bar{\mu}_i - F z_i \nabla \phi = \sum_{\substack{j=1 \\ j \neq i}}^{n_c} \frac{y_j \Re T (u_i - u_j)}{Đ_{ij}} + \frac{y_s \Re T u_i}{Đ_{is}} \tag{3.191}$$

where $\nabla \bar{\mu}_i$ is the surface chemical potential gradient of i, $Đ_{ij}$ is the common MS surface diffusivity of the pair i - j, $Đ_{is}$ is the MS surface diffusivity corresponding to the interaction between i and the fixed ionic groups (subscript s stands for solid); u_i and u_j are the velocities of species i and j, $y_j = q_j/(q_t + q_s)$ is the mole fraction of counter ion j, q_s being the molar concentration of ionic fixed groups (i.e. co-ions) and q_t the molar concentration of all counter ions (extension of Eq. 3.7); and $y_s = q_s/(q_t + q_s)$ is mole fraction of ionic fixed groups. It should be noted that $\sum_{i=1}^{n_c+1} y_i = 1$ and q_t is not constant whenever counter ions have different electrochemical valences. Taking into account the definition of molar flux of the ionic species j:

$$N_j = q_t y_j u_j \tag{3.192}$$

Equation 3.191 may be rewritten as:

$$-\frac{y_i}{\Re T}\nabla\bar{\mu}_i - y_i z_i \frac{F}{\Re T}\nabla\phi = \sum_{\substack{j=1\\ j\neq i}}^{n_c} \frac{y_j N_i - y_i N_j}{(q_t + q_s)Đ_{ij}} + \frac{y_s N_i}{(q_t + q_s)Đ_{is}} \tag{3.193}$$

Assuming equilibrium between exchanger and bulk solution, (i.e. $\bar{\mu}_i = \mu_i$), $\nabla\bar{\mu}_i$ can be expressed in terms of the mole fraction gradients of counter ions in the particle by:

$$\frac{y_i}{\Re T}\nabla\bar{\mu}_i = \sum_{j=1}^{n_c} \Gamma_{ij}\nabla y_j, \quad \text{with} \quad \Gamma_{ij} \equiv y_i \frac{\partial \ln(\gamma_i x_i)}{\partial y_j} \tag{3.194}$$

where γ_i is activity coefficient of counter ion i in solution and Γ_{ij} is the thermodynamic factor. It is worth noting that in Eq. 3.194, x_i and y_i are related by the equilibrium isotherm. This fact emphasises the importance of good isotherms (discussed in Sect. 3.3.4) even in kinetic equations.

Equation 3.193 can be expressed in n-dimensional matrix notation. Introducing the matrix of thermodynamic factors, $[\Gamma]$, given by Eq. 3.194, one obtains:

$$-(q_t + q_s)[\Gamma](\nabla y) - (q_t + q_s)(\nabla\xi) = [B](N) \tag{3.195}$$

or

$$(N) = -(q_t + q_s)[B]^{-1}[\Gamma](\nabla y) - (q_t + q_s)[B]^{-1}(\nabla\xi) \tag{3.196}$$

where

$$B_{ii} = \frac{y_s}{Đ_{is}} + \sum_{\substack{j=1\\ j\neq i}}^{n_c} \frac{y_j}{Đ_{ij}}, \quad B_{ij} = -\frac{y_i}{Đ_{ij}} \tag{3.197}$$

$$\nabla\xi_i = y_i z_i \frac{F}{\Re T}\nabla\phi \tag{3.198}$$

Under conditions of electroneutrality and absence of electric current, the following relations are adhered to:

$$\sum_{i=1}^{n_c+1} q_i z_i = 0 \tag{3.199}$$

$$\sum_{i=1}^{n_c+1} z_i N_i = 0 \tag{3.200}$$

Equations 3.196 and 3.200 may be combined to eliminate $\nabla\phi$ from the generalised MS equations. As a result, the mass transport process in a mixture of $n_c + 1$ components can be described by the following set of equations:

1. $n_c - 1$ Maxwell–Stefan equations for components 1 to $n_c - 1$ (Eq. 3.196)
2. One reference condition (bootstrap relation):

$$N_{n_c+1} = 0 \quad \text{or} \quad u_{n_c+1} = 0 \tag{3.201}$$

3. The no current relationship, Eq. 3.200

The electric potential gradient obtained from Eqs. 3.196 and 3.200 is:

$$\frac{F}{\Re T}\nabla\phi = \frac{-\sum_{i=1}^{n_c} z_i \left(\sum_{j=1}^{n} L_{ij}\nabla y_j\right)}{\sum_{i=1}^{n_c} y_i z_i \left(\sum_{i=1}^{n} z_j L_{ji}\right)} \tag{3.202}$$

which gives for a binary system ($n_c = 2$):

$$\frac{F}{\Re T}\nabla\phi = \frac{\begin{array}{c}-z_A[L_{11}(\Gamma_{11}\nabla y_A + \Gamma_{12}\nabla y_B) + L_{12}(\Gamma_{21}\nabla y_A + \Gamma_{22}\nabla y_B)] \\ -z_B[L_{21}(\Gamma_{11}\nabla y_A + \Gamma_{12}\nabla y_B) + L_{22}(\Gamma_{21}\nabla y_A + \Gamma_{22}\nabla y_B)]\end{array}}{y_A z_A^2 L_{11} + y_B z_A z_B L_{12} + y_A z_A z_B L_{21} + y_B z_B^2 L_{22}} \tag{3.203}$$

where $[L] = [B]^{-1}$.

3.5 Batch and Fixed Bed Applications

The time dependence of ion concentrations in a phase can be calculated by solving the appropriate material balances combined with equilibrium relations, kinetic equations, and specific initial and boundary conditions. In this section, the general modelling guidelines of the omnipresent batch and fixed bed applications are presented. A minute description of the zero-length column method is given at the end.

3.5.1 Batch Operation

Most commercially available resins used in batch experiments consist of small microporous phases into a macroporous pellet (see Fig. 3.13). Accordingly, both macropore and micropore diffusion resistances may control the ion exchange kinetics in a composite pellet. For a perfectly stirred tank, the material balances to the spherical particle and reservoir are:

$$(1-\varepsilon_{\mathrm{p}})\frac{\partial q_i}{\partial t}+\varepsilon_{\mathrm{p}}\frac{\partial C_i}{\partial t}=-\frac{1}{r^2}\frac{\partial}{\partial r}\left(r^2 N_{i,\mathrm{intra}}\right) \tag{3.204}$$

$$\frac{dC_i}{dt}=-\frac{V_{\mathrm{s}}}{V_{\mathrm{L}}}\frac{d}{dt}\left(\bar{q}_i+\varepsilon_{\mathrm{p}}\bar{C}_{\mathrm{p},i}\right) \tag{3.205}$$

where V_{s}, V_{L}, and ε_{p} are the volume of the solid and fluid phases and particle porosity, respectively. The intraparticle diffusion flux is calculated by:

$$N_{\mathrm{intra},i}=N_{\mathrm{p},i}+N_{\mathrm{s},i} \tag{3.206}$$

where $N_{\mathrm{p},i}$ and $N_{\mathrm{s},i}$ are the diffusion fluxes through the macropore and micropore regions, respectively, which according to the most common Nernst–Planck approach may be calculated as:

$$N_{\mathrm{p},i}=-D_{\mathrm{eff,p},i}\frac{\partial C_{\mathrm{p},i}}{\partial r} \tag{3.207}$$

$$N_{\mathrm{s},i}=-D_{\mathrm{eff,s},i}\frac{\partial q_i}{\partial r} \tag{3.208}$$

being $D_{\mathrm{eff,p},i}$ and $D_{\mathrm{eff,s},i}$ the corresponding effective diffusion coefficients. For a binary system A/B, the $D_{\mathrm{eff,s},i}$ is calculated using the interdiffusion coefficient given by Eq. 3.190; a similar approach is used for $D_{\mathrm{eff,p},i}$ also. The average loading and the average pore concentration are calculated by:

$$\bar{q}_i=\frac{3}{R^3}\int_0^R r^2 q_i\,dr \tag{3.209}$$

$$\bar{C}_{\mathrm{p},i}=\frac{3}{R^3}\int_0^R r^2 C_{\mathrm{p},i}\,dr \tag{3.210}$$

and the boundary and initial conditions for exchanger are:

$$t=0,\quad C_{\mathrm{p},i}=q_i=0,\quad \text{and}\quad C_i=C_{i,0} \tag{3.211}$$

$$r=R,\quad C_{\mathrm{p},i}=C_{i,\mathrm{s}},\quad \text{and}\quad q_i=q_{i,\mathrm{s}} \tag{3.212}$$

$$r = 0, \quad \left(\frac{\partial C_{p,i}}{\partial r}\right) = \left(\frac{\partial q_i}{\partial r}\right) = 0 \tag{3.213}$$

In Eq. 3.211, it is assumed that the solid is initially clean. The concentration at the interface is uniquely determined by equalising internal diffusion and film diffusion fluxes, i.e.:

$$N_{\text{intra},i}\big|_{r=R} = \left(N_{p,i} + N_{s,i}\right)\big|_{r=R} = k_f\left(C_i - C_{i,s}\right) \tag{3.214}$$

where k_f is the external mass transfer coefficient by film diffusion.

For homogeneous particles, i.e. $\varepsilon_p = 0$, the model equations reduce to:

$$\frac{\partial q_i}{\partial t} = -\frac{1}{r^2}\frac{\partial}{\partial r}\left(r^2 N_{s,i}\right) \tag{3.215}$$

$$\frac{dC_i}{dt} = -\frac{V_s}{V_L}\frac{d\bar{q}_i}{dt} \tag{3.216}$$

together with Eqs. 3.208, 3.209, and 3.214. Now, the effective diffusivity is the interdiffusion coefficient in the solid phase. The initial and boundary conditions become:

$$t = 0, \quad q_i = 0, \quad \text{and} \quad C_i = C_{i,0} \tag{3.217}$$

$$r = R, \quad q_i = q_{i,s} \tag{3.218}$$

$$r = 0, \quad \left(\frac{\partial q_i}{\partial r}\right) = 0 \tag{3.219}$$

For well-established agitated systems, it is possible to estimate k_f using correlations which depend generally on the Reynolds, Schmidt, and Power numbers and on geometrical parameters such as the ratio of impeller to tank diameter, the specific geometry of the impeller, and the geometry of baffling, if any, used to inhibit vortex formation in the vessel. Several correlations may be found elsewhere [135, 150–153].

In the following, a comparison between Nernst–Planck (NP) and Maxwell–Stefan (MS) models' ability to correlate and predict a batch ion exchange system where ETS-4 is used to uptake Cd^{2+} from aqueous solution at room temperature will be presented. The hypotheses of the model are (1) there exist film and intraparticle mass transfer resistances; (2) spherical solid particles; (3) isothermal operation; (4) liquid and solid volume changes are neglected; (5) co-ions are excluded from the ion exchanger particles (Donnan exclusion); and (6) ideal solution.

In Fig. 3.15, the experimental data and calculated results are shown, being possible to conclude that both MS (—) and NP (—) flux equations provide similar representations and predictions, because the corresponding curves are almost indistinguishable. It is worth noting that from the four sets of experimental data, only one

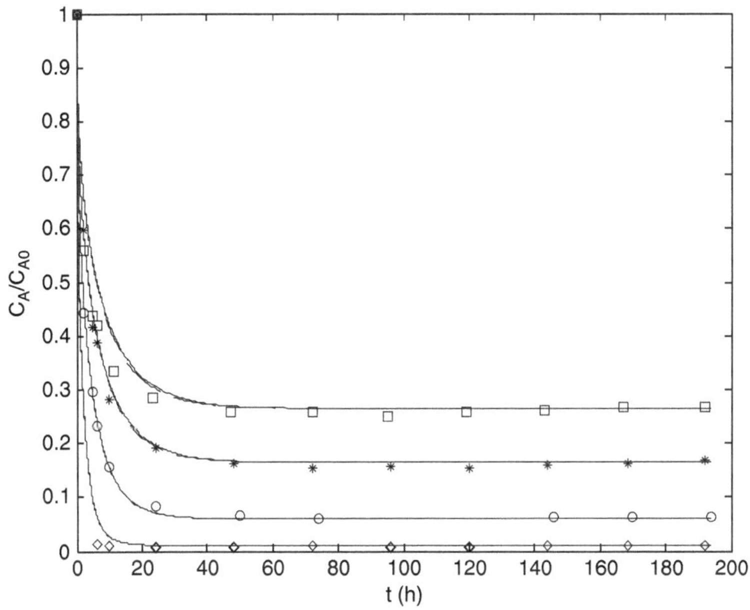

Fig. 3.15 Prediction and correlation of ion exchange uptake curves by Maxwell–Stefan (—) and Nernst–Planck (—) based models. System: Cd^{2+}/Na^{+}/ETS-4 [62]: □, Exp. 1 ($C_o = 0.51 \times 10^{-3}$ kg m^{-3} and $W_{exch} =$ 44 mg); *, Exp. 2 ($C_o = 0.64 \times 10^{-3}$ kg m^{-3} and $W_{exch} =$ 50 mg); ○, Exp. 3 ($C_o = 0.84 \times 10^{-3}$ kg m^{-3} and $W_{exch} =$ 51 mg); ◇, Exp. 4 ($C_o = 0.62 \times 10^{-3}$ kg m^{-3} and $W_{exch} = 100$ mg)

(Exp. 3) has been used to fit the model parameters (diffusivities and k_f). The remaining three curves (Exp. 1, Exp. 2, and Exp. 4) have been purely predicted. Such findings prove that, under the particular conditions of these experiments, both approaches can be equally applied notwithstanding the above mentioned concerns about NP. In fact, since in this case one deals with dilute solutions, MS do not import any additional accuracy to the final solution.

Figure 3.16 shows, as an example, the normalized concentration of cadmium(II) in the ETS-4 particles, calculated using the MS based model, as function of time and radial position, for Exp.3 of Figure 3.15. This plot illustrates an interesting time evolution of the concentration at surface normalized with the final equilibrium concentration, $q_A(t; r = R)/q_{A\infty}$. Accordingly, an initial sudden rise of surface concentration is so pronounced that it goes through a maximum and then decreases gradually until equilibrium is reached, while far from surface a monotonic behaviour is found instead. With no film resistance, the initial particle concentration at surface would be expected to raise suddenly from $q_A(t = 0^-,\ r = R) = 0$ to $q_A(t = 0^+, r = R) = q_A(C_{A,0})$, which is the concentration in equilibrium with bulk solution. Then, for $t > 0, q_A(t, r = R)$ would decrease monotonically until final system equilibration. This figure illustrates such behaviour, but the existence of external diffusion smoothes the ideal trend identified with such initial step increase.

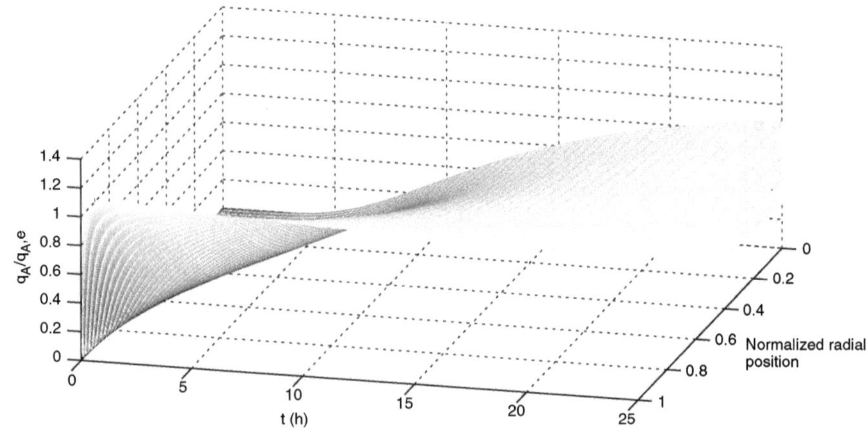

Fig. 3.16 Simulation results obtained for Exp.3 (Fig. 3.15) graphed as normalized Cd^{2+} concentration inside the ETS-4 particle as function of time and normalized radial position.

3.5.2 Fixed Bed Column

In practice, most ion exchange processes are carried out in percolation columns, where the solution flows through a packed bed of exchanger beads. Depending on the system and operating conditions, micropores or macropores intraparticle resistances and/or external film resistance may be controlling.

For isothermal operation and negligible concentration radial gradients, the material balance to the column is:

$$\frac{\partial C_i}{\partial t} + u_i \frac{\partial C_i}{\partial z} - D_{\mathrm{L}} \frac{\partial^2 C_i}{\partial z^2} = -\frac{(1-\varepsilon_{\mathrm{b}})}{\varepsilon_{\mathrm{b}}} \frac{d}{dt} \left(\varepsilon_{\mathrm{p}} \bar{C}_{\mathrm{p},i} + \bar{q}_i \right) \tag{3.220}$$

where u_i is the interstitial fluid velocity, D_{L} is the axial dispersion coefficient, and ε_{b} is the bed porosity. For a bed initially free of solute, Eq. 3.220 is subjected to the following initial and boundary conditions:

$$t = 0, \quad C_i = \bar{C}_{\mathrm{p},i} = \bar{q}_i = 0 \tag{3.221}$$

$$z = 0, \quad u_i C_{\mathrm{F},i} = u_i C_i\big|_{z=0^+} - D_{\mathrm{L}} \frac{\partial C_i}{\partial z}\Big|_{z=0^+} \tag{3.222}$$

$$z = L, \quad \frac{\partial C_i}{\partial z} = 0 \tag{3.223}$$

where $C_{\mathrm{F},i}$ is the feed concentration and L is the column length.

The representation of breakthrough curves is essential to attain successful design of a column sorption process. However, the complexity and diversity of the previous models require the development of simplifications and approximations for practical column modelling. As a result, simplified expressions are frequently adopted to describe mass transfer as, for instance, the linear driving force (LDF) approximation.

The exact solution of the fixed bed column is obtained combining Eq. 3.220 with the material balance and averages to the particle given by Eqs. 3.204 and 3.206–3.210 and with an isotherm. Moreover, the implied fluxes are calculated according to Fick, Nernst–Planck, or Maxwell–Stefan formalism. This gives rise to a very complex system of algebraic and partial differential equations, whose solution necessitates the application of reliable numerical methods.

Alternatively, the LDF approach provides an excellent compromise between accuracy and simplification. According to LDF, the material balance to the particle expresses now as:

$$\frac{d}{dt}\left(\varepsilon_{\mathrm{p}}\bar{C}_{\mathrm{p},i}+\bar{q}_i\right)=K_{\mathrm{LDF},i}\left(q_i^*-\bar{q}_i\right) \tag{3.224}$$

where the global linear driving force coefficient is computed by:

$$\frac{1}{K_{\mathrm{LDF},i}}=\frac{R^2}{15D_{\mathrm{eff},i}}+\frac{1}{k_{\mathrm{f}}a_{\mathrm{p}}} \tag{3.225}$$

q_i^* is the resinate concentration in equilibrium with the external solution, and a_{p} is external surface area per unit particle volume. For homogeneous particles, $\varepsilon_{\mathrm{p}}=0$, or negligible pore accumulation, this equation simplifies to:

$$\frac{d\bar{q}_i}{dt}=K_{\mathrm{LDF},i}\left(q_i^*-\bar{q}_i\right) \tag{3.226}$$

Townsend and Harjula [154] have proposed a simplified approach based on the plate concept by considering the number of "transfer units" or "effective plates" as a measure of the column efficiency. Accordingly, the following relation has been obtained for film diffusion–controlled exchange:

$$N_i=A(D_i)^{1/2}d^{-3/2}Lu_{i,0}{}^{-1/2} \tag{3.227}$$

where d is the particle diameter and D_i is the diffusion coefficient in the film.

Their expression for particle diffusion–controlled exchange is:

$$N_i=BK_{\mathrm{D}}D_id^{-2}Lu_{i,0}{}^{-1} \tag{3.228}$$

In previous equations, A and B are empirical factors and K_{D} the distribution coefficient.

Additionally, a number of analytical models have been developed to describe fixed bed sorption, as, for instance, the Thomas [155], Bohart–Adams [156], Clark [157], and Yoon and Nelson [158] models. The first two expressions are representative of chemical kinetic type models, where rate laws are represented by rate equations of chemical reactions. Nonetheless, it is well known that several equivalences may be found between apparently distinct models. For instance, the expressions of Bohart–Adams [156] and Wolborska [159] are equivalent, notwithstanding the last one comprehends the concepts of axial dispersion and external mass transfer resistance. A short description of each model is given in the following.

The Thomas model [155] is one of the most general and widely used kinetic models to study the column performance. It is based on the Langmuir kinetics of adsorption–desorption with no axial dispersion and assumes the rate driving force to follow a second order reversible reaction kinetics. Besides, the separation factor is assumed constant, although being appropriate to either favourable or unfavourable isotherms. It takes the form:

$$\frac{C_{\mathrm{ef},i}}{C_{\mathrm{F},i}} = \frac{1}{1 + \exp\left[\frac{k_{\mathrm{Th}}}{v}\left(q_{\max} W_{\mathrm{exch}} - C_{\mathrm{F},i} V_{\mathrm{ef}}\right)\right]} \tag{3.229}$$

where k_{Th} is the Thomas rate constant, $C_{\mathrm{ef},i}$ is the concentration of sorbate in the effluent, $q_{\max}$ is the maximum sorbate concentration in the solid phase, V_{ef} is the volume of effluent, and v is the volumetric flow rate.

Bohart and Adams model [156] was originally developed to describe the adsorption of chloride on charcoal in a continuous flowing system, being nowadays used to quantitatively describe other systems [160]. It was derived based on the surface reaction theory considering that equilibrium is not instantaneous. Accordingly, the sorption rate was assumed proportional to both the residual capacity of the exchanger and the concentration of the solute species [161]. The Bohart–Adams equation can be written as:

$$\frac{C_{\mathrm{ef},i}}{C_{\mathrm{F},i}} = \exp\left[k_{\mathrm{AB}} C_{\mathrm{F},i} t - k_{\mathrm{AB}} C_{\mathrm{sat}} \left(\frac{L}{U_0}\right)\right] \tag{3.230}$$

where the k_{AB} is the rate constant, C_{sat} is the saturation concentration, and U_0 is the superficial velocity.

In 1987, Clark [157] has developed a new relation for the breakthrough curve by using the mass transfer coefficient in combination with the Freundlich isotherm. The following relation was obtained:

$$\frac{C_{\mathrm{ef},i}}{C_{\mathrm{F},i}} = \left[\frac{1}{1 + \beta \exp(-\theta t)}\right]^{1/n-1} \tag{3.231}$$

which can be rearranged to the following linear form:

$$Ln\left(\frac{C_{\mathrm{F},i}}{C_{\mathrm{ef},i}}\right)^{n-1} - 1 = -\theta t + Ln(\beta) \tag{3.232}$$

where n is the Freundlich constant and C_{b} is the solute concentration at breakthrough time, t_{b}. The values of β and θ can be determined using the above equation to predict the breakthrough curve.

The model proposed by Yoon and Nelson [158] is simpler than the previous ones and requires no detailed data concerning the characteristics of sorbate, the type of sorbent, and the physical properties of the bed [161]. It was developed to describe the sorption and breakthrough curves of gases in activated charcoal, assuming the rate of decrease in the probability of sorption for each sorbate molecule to be proportional to the probability of sorbate sorption and sorbate breakthrough on the sorbent. The Yoon–Nelson model for a single component system is expressed as:

$$\frac{C_{\mathrm{ef},i}}{C_{\mathrm{F},i} - C_{\mathrm{ef},i}} = \exp\left(k_{\mathrm{YN}} t_{1/2} - \tau k_{\mathrm{YN}}\right) \tag{3.233}$$

where k_{YN} is the Yoon–Nelson rate constant and $t_{1/2}$ is the time required for 50% sorbate breakthrough, i.e. the stoichiometric time.

3.5.3 ZLC Bed

Beyond batch, shallow bed, or single-particle methods, the zero-length column (ZLC) provides a simple and effective technique for measuring intraparticle diffusion coefficients in adsorbents. The column consists in a differential bed of particles initially saturated with a solution of ions, assumed as a well-mixed cell. The system is exposed to a solution containing a different counter ion, while desorption of the ion initially present in the particles is followed along time. Assuming spherical particles, the mass balance for the ZLC is:

$$V_{\mathrm{ZLC}} \varepsilon_{\mathrm{b}} \frac{\partial C_i}{\partial t} + V_{\mathrm{ZLC}} (1 - \varepsilon_{\mathrm{b}}) \frac{\partial q_i}{\partial t} + v C_i = 0 \tag{3.234}$$

being V_{ZLC} the volume of the ZLC column. The initial conditions are:

$$t = 0, \quad q_i(r) = q_{i,\mathrm{o}}, \quad C_i = C_{i,\mathrm{o}} \tag{3.235}$$

References

1. Smith JM, Ness HCV, Abbott MM (2000) Introduction to chemical engineering thermodynamics, 6th edn. McGraw-Hill, Singapore
2. Zemaitis JF, Clark DM, Rafal M, Scrivner NC (1986) Handbook of aqueous electrolyte thermodynamics. ed. A.I.o.C. Engineers, New York, pp 64–66
3. Dranoff J, Lapidus L (1957) Equilibrium in ternary ion exchange systems. Ind Eng Chem 49 (8):1297–1302
4. Pieroni LJ, Dranoff JS (1963) Ion exchange equilibria in a ternary system. Aiche J 9(1):42–45
5. Prausnitz JM, Lichtenthaler RN, Azevedo EG (1999) Molecular thermodynamics of fluid-phase equilibria. Prentice Hall, New Jersey
6. Pitzer KS, Peiper JC, Busey RH (1984) Thermodynamic properties of aqueous sodium chloride solutions. J Phys Chem Ref Data 13(1):1–102

7. Robinson RA, Stokes RH (1923) Electrolyte solutions. Butterworths, London
8. Bromley LA (1973) Thermodynamic properties of strong electrolytes in aqueous solutions. Aiche J 19(2):313–320
9. Pitzer KS (1973) Thermodynamics of electrolytes. I. Theoretical basis and general equations. J Phys Chem 77(2):268–277
10. Meissner HP, Kusik CL (1972) Activity coefficients of strong electrolytes in multicomponent aqueous solutions. Aiche J 18(2):294–298
11. Kusik CL, Meissner HP (1978) Electrolyte activity coefficients in inorganic processing. AIChE Symp Ser 74(173):14–20
12. Poling BE, Prausnitz JM, O'Connel JP (2000) The properties of gases and liquids, 5th edn. McGraw-Hill, Singapore
13. Wohl K (1946) Thermodynamic evaluation of binary and ternary liquid systems. Trans Am Inst Chem Eng 42(2):215–249
14. Kester DR, Pytkowicz RM (1975) Theoretical model for the formation of ion-pairs in seawater. Mar Chem 3(4):365–374
15. Vo BS, Shallcross DC (2005) Ion exchange equilibria data for systems involving H^+, Na^+, K^+, Mg^{2+}, and Ca^{2+} ions. J Chem Eng Data 50(3):1018–1029
16. Majer V, Stulík K (1982) A study of the stability of alkaline-earth metal complexes with fluoride and chloride ions at various temperatures by potentiometry with ion-selective electrodes. Talanta 29(2):145–148
17. Perrin DD, Högfeldt E, Sillen LG, Martell AE (1971) Stability constants of metal-ion complexes. Pergamon, Oxford, Supplement
18. Johnson KS, Pytkowicz RM (1979) Ion association of chloride and sulphate with sodium, potassium, magnesium and calcium in seawater at 25°C. Mar Chem 8(1):87–93
19. Atlas E, Culberson C, Pytkowicz RM (1976) Phosphate association with Na^+, Ca^{2+} and Mg^{2+} in seawater. Mar Chem 4(3):243–254
20. De Robertis A, Rigano C, Sammartano S, Zerbinati O (1987) Ion association of Cl^- with Na^+, K^+, Mg^{2+} and Ca^{2+} in aqueous solution at $10 \leq T \leq 45$°C and $0 \leq I \leq 1$ mol L^{-1}: a literature data analysis. Thermochimica Acta 115:241–248
21. Mehablia MA, Shallcross DC, Stevens GW (1994) Prediction of multicomponent ion exchange equilibria. Chem Eng Sci 49(14):2277–2286
22. Mumford KA, Shallcross DC, Snape I, Stevens GW (2008) Application of a temperature-dependent semiempirical thermodynamic ion-exchange model to a multicomponent natural zeolite system. Ind Eng Chem Res 47(21):8347–8354
23. Velayudhan A, Horváth C (1994) Adsorption and ion-exchange isotherms in preparative chromatography. J Chromatogr A 663(1):1–10
24. Shallcross DC (2003) Modelling multicomponent ion exchange equilibrium behaviour. J Ion Exch 14(Supp):5–8
25. Melis S, Cao G, Morbidelli M (1995) A new model for the simulation of ion exchange equilibria. Ind Eng Chem Res 34(11):3916–3924
26. Klein G, Tondeur D, Vermeulen T (1967) Multicomponent ion exchange in fixed beds. General properties of equilibrium systems. Ind Eng Chem Fundam 6(3):339–351
27. Smith RP, Woodburn ET (1978) Prediction of multicomponent ion exchange equilibria for the ternary system SO_4^{2-}- NO_3^-- Cl^- from data of binary systems. Aiche J 24(4):577–587
28. Sengupta M, Paul TB (1985) Multicomponent ion exchange equilibria. I. Zn^{2+}–Cd^{2+}–H^+ and Cu^{2+}–Ag^+–H^+ on Amberlite IR 120. React Polym Ion Exch Sorb 3(3):217–229
29. Shallcross DC, Herrmann CC, McCoy BJ (1988) An improved model for the prediction of multicomponent ion exchange equilibria. Chem Eng Sci 43(2):279–288
30. De Martínez AL, Cañizares P, Díaz JZ (1993) Binary ion exchange equilibrium for Ca^{2+}, Mg^{2+}, K^+, Na^+ and H^+ ions on amberlite IR-120. Chem Eng Technol 16(1):35–39
31. Ioannidis S, Anderko A, Sanders SJ (2000) Internally consistent representation of binary ion exchange equilibria. Chem Eng Sci 55(14):2687–2698

32. Borba CE, Silva EA, Spohr S, Santos GHF, Guirardello R (2010) Ion exchange equilibrium prediction for the system Cu^{2+}–Zn^{2+}–Na^{+}. J Chem Eng Data 55(3):1333–1341
33. Helfferich F (1995) Ion exchange. Dover, New York
34. Kataoka T, Yoshida H (1980) Ion exchange equilibria in ternary systems. J Chem Eng Japan 13(4):328–330
35. Elprince AM, Babcock KL (1975) Prediction of ion-exchange equilibria in aqueous systems with more than two counter-ions. Soil Sci 120(5):332–338
36. Shehata FA, El-Kamash AM, El-Sorougy MR, Aly HF (2000) Prediction of multicomponent ion-exchange equilibria for a ternary system from data of binary systems. Sep Sci Technol 35 (12):1887–1900
37. de Lucas A, Valverde JL, Romero MC, Gómez J, Rodríguez JF (2002) The ion exchange equilibria of Na^{+}/K^{+} in nonaqueous and mixed solvents on a strong acid cation exchanger. Chem Eng Sci 57(11):1943–1954
38. Vamos RJ, Haas CN (1994) Reduction of ion-exchange equilibria data using an error in variables approach. Aiche J 40(3):556–569
39. Pabalan RT, Bertetti FP (1999) Experimental and modeling study of ion exchange between aqueous solutions and the zeolite mineral clinoptilolite. J Sol Chem 28(4):367–393
40. Mumford KA, Northcott KA, Shallcross DC, Snape I, Stevens GW (2008) Comparison of amberlite IRC-748 resin and zeolite for copper and ammonium ion exchange. J Chem Eng Data 53(9):2012–2017
41. Carmona M, Warchoł J, Ad L, Rodriguez JF (2008) Ion-exchange equilibria of Pb^{2+}, Ni^{2+}, and Cr^{3+} ions for H^{+} on amberlite IR-120 resin. J Chem Eng Data 53(6):1325–1331
42. de Lucas A, Rodriguez L, Sanchez P, Lobato J (2003) Retention capacity of the builder δ-$Na_2Si_2O_5$. Modeling the $Ca^{2+}/Mg^{2+}/Na^{+}$ equilibrium. Ind Eng Chem Res 42(14):3257–3262
43. Robinson C, Gilliland E (1950) Elements of fractional distillation. McGraw-Hill, New York
44. De Lucas A, Zarca J, Cañizares P (1992) Ion-exchange equilibrium of Ca^{2+}, Mg^{2+}, K^{+}, Na^{+}, and H^{+} ions on amberlite IR-120: experimental determination and theoretical prediction of the ternary and quaternary equilibrium data. Sep Sci Technol 27(6):823–841
45. Vo BS, Shallcross DC (2005) Modeling solution phase behavior in multicomponent ion exchange equilibria involving H^{+}, Na^{+}, K^{+}, Mg^{2+}, and Ca^{2+} ions. J Chem Eng Data 50 (6):1995–2002
46. Allen RM, Addison PA, Dechapunya AH (1989) The characterization of binary and ternary ion exchange equilibria. Chem Eng J 40(3):151–158
47. Allen RM, Addison PA (1990) Ion exchange equilibria for ternary systems from binary exchange data. Chem Eng J 44(3):113–118
48. Stewart WE, Liou CT, Lim HC, Weigand WA, Berger AJ, Hála E (1972) Letters to the editor. AIChE J 18(4):875–876
49. Hála E (1972) Liquid–vapor equilibrium. LII. On boundary conditions between constants of Wilson and NRTL equations in three- and more component systems. Collect Czechoslovak Chem Commun 37:2817–2819
50. Bajpai RK, Gupta AK, Rao MG (1973) Binary and ternary ion-exchange equilibriums. Sodium-cesium-manganese-Dowex 50W-X8 and cesium-manganese-strontium-Dowex 50W-X8 systems. J Phys Chem 77(10):1288–1293
51. Argersinger WJ, Davidson AW, Bonner OD (1950) Thermodynamics and ion exchange phenomena. Kansas Acad Trans 53:404–410
52. Gaines JGL, Thomas HC (1953) Adsorption studies on clay minerals. II. A formulation of the thermodynamics of exchange adsorption. J Chem Phys 21(4):714–718
53. Provis JL, Lukey GC, Shallcross DC (2005) Modeling multicomponent ion exchange: application of the single-parameter binary system model. Ind Eng Chem Res 44 (7):2250–2257
54. Ioannidis S, Anderko A (2001) Equilibrium modeling of combined ion-exchange and molecular adsorption phenomena. Ind Eng Chem Res 40(2):714–720
55. Fletcher P, Townsend RP (1981) Ternary ion exchange in Zeolites. Part 1. – problem of predicting equilibrium compositions. J Chem Soc Faraday Trans 2:77

56. Soldatov VS, Bychkova VA (1971) Calculation of activity coefficients of components of the ion exchanger phase in multicomponent systems. Russ J Phys Chem 45(5):707–709
57. Ioannidis S, Anderko A (2000) Equilibrium modeling of combined ion-exchange and molecular adsorption phenomena. Ind Eng Chem Res 40(2):714–720
58. Mehablia MA, Shallcross DC, Stevens GW (1996) Ternary and quaternary ion exchange equilibria. Sol Extract Ion Exch 14(2):309–322
59. Pabalan RT (1994) Thermodynamics of ion exchange between clinoptilolite and aqueous solutions of Na^+/k^+ and Na^+/Ca^{2+}. Geochim Cosmochim Acta 58(21):4573–4590
60. Barreira LD, Lito PF, Antunes BM, Otero M, Lin Z, Rocha J, Pereira E, Duarte AC, Silva CM (2009) Effect of pH on cadmium (II) removal from aqueous solution using titanosilicate ETS-4. Chem Eng J 155(3):728–735
61. Camarinha ED, Lito PF, Antunes BM, Otero M, Lin Z, Rocha J, Pereira E, Duarte AC, Silva CM (2009) Cadmium(II) removal from aqueous solution using microporous titanosilicate ETS-10. Chem Eng J 155(1–2):108–114
62. Ferreira TR, Lopes CB, Lito PF, Otero M, Lin Z, Rocha J, Pereira E, Silva CM, Duarte A (2009) Cadmium(II) removal from aqueous solution using microporous titanosilicate ETS-4. Chem Eng J 147(2–3):173–179
63. Lopes CB, Lito PF, Otero M, Lin Z, Rocha J, Silva CM, Pereira E, Duarte AC (2008) Mercury removal with titanosilicate ETS-4: batch experiments and modelling. Micropor Mesopor Mater 115(1–2):98–105
64. Lopes CB, Otero M, Coimbra J, Pereira E, Rocha J, Lin Z, Duarte A (2007) Removal of low concentration Hg^{2+} from natural waters by microporous and layered titanosilicates. Micropor Mesopor Mater 103(1–3):325–332
65. Lopes CB, Otero M, Lin Z, Silva CM, Pereira E, Rocha J, Duarte AC (2010) Effect of pH and temperature on Hg^{2+} water decontamination using ETS-4 titanosilicate. J Hazard Mater 175 (1–3):439–444
66. Lopes CB, Otero M, Lin Z, Silva CM, Rocha J, Pereira E, Duarte AC (2009) Removal of Hg^{2+} ions from aqueous solution by ETS-4 microporous titanosilicate – kinetic and equilibrium studies. Chem Eng J 151(1–3):247–254
67. Otero M, Lopes CB, Coimbra J, Ferreira TR, Silva CM, Lin Z, Rocha J, Pereira E, Duarte AC (2009) Priority pollutants (Hg^{2+} and Cd^{2+}) removal from water by ETS-4 titanosilicate. Desalination 249(2):742–747
68. Altin O, Özbelge HÖ, Dogu T (1998) Use of general purpose adsorption isotherms for heavy metal-clay mineral interactions. J Colloid Interface Sci 198(1):130–140
69. Petrus R, Warchol JK (2005) Heavy metal removal by clinoptilolite. An equilibrium study in multi-component systems. Water Res 39(5):819–830
70. Ku Y, Lee K-C, Wang W (2005) Removal of phenols from aqueous solutions by purolite A-510 resin. Sep Sci Technol 39(4):911–923
71. Carmona M, Lucas AD, Valverde JL, Velasco B, Rodríguez JF (2006) Combined adsorption and ion exchange equilibrium of phenol on Amberlite IRA-420. Chem Eng J 117(2):155–160
72. Caetano M, Valderrama C, Farran A, Cortina JL (2009) Phenol removal from aqueous solution by adsorption and ion exchange mechanisms onto polymeric resins. J Colloid Interface Sci 338(2):402–409
73. Myers AL, Byington S (1986) Thermodynamics of ion exchange: prediction of multicomponent equilibria from binary data. In: Rodrigues AE (ed) Ion exchange: science and technology. Martinus Nijhoff, Dordrecht, pp 119–145
74. Novosad J, Myers AL (1982) Thermodynamics of ion exchange as an adsorption process. Can J Chem Eng 60(4):500–503
75. Sircar S, Myers AL (1971) A thermodynamic consistency test for adsorption from binary liquid mixtures on solids. AIChE J 17(1):186–190
76. Harned HS, Owen BB (1958) The physical chemistry of electrolytic solutions, 3rd edn. Reinhold, New York
77. Saunders MS, Vierow JB, Carta G (1989) Uptake of phenylalanine and tyrosine by a strong-acid cation exchanger. AIChE J 35(1):53–68

78. Dye SR, Decarli JP, Carta G (1990) Equilibrium sorption of amino-acids by a cation-exchange resin. Ind Eng Chem Res 29(5):849–857
79. Zammouri A, Chanel S, Muhr L, Grevillot G (2000) Ion-exchange equilibria of amino acids on strong anionic resins. Ind Eng Chem Res 39(5):1397–1408
80. Jones IL, Carta G (1993) Ion exchange of amino acids and dipeptides on cation resins with varying degree of crosslinking. 1. Equilibrium. Ind Eng Chem Res 32(1):107–117
81. Moreira MJA, Ferreira LMGA (2005) Equilibrium studies of phenylalanine and tyrosine on ion-exchange resins. Chem Eng Sci 60(18):5022–5034
82. de Kock FP, van Deventer JSJ (1995) Statistical thermodynamic model for competitive ion exchange. Chem Eng Commun 135(1):21–45
83. Lukey GC, Van Deventer JSJ, Shallcross DC (2000) Equilibrium model for the selective sorption of gold cyanide on different ion-exchange functional groups. Minerals Eng 13 (12):1243–1261
84. Lukey GC, Van Deventer JSJ, Shallcross DC (2001) Equilibrium model for the sorption of gold cyanide and copper cyanide on trimethylamine ion exchange resin in saline solutions. Hydrometallurgy 59(1):101–113
85. Provis JL, Lukey GC, Shallcross DC (2004) Single-parameter model for binary ion-exchange equilibria. Ind Eng Chem Res 43(24):7870–7879
86. Barrer RM, Meier WM (1959) Exchange equilibria in a synthetic crystalline exchanger. Trans Faraday Soc 55:130–141
87. Melis S, Markos J, Cao G, Morbidelli M (1996) Multicomponent equilibria on ion-exchange resins. Fluid Phase Equilibria 117(1–2):281–288
88. Valverde JL, de Lucas A, Rodriguez JF (1998) Comparison between heterogeneous and homogeneous MASS action models in the prediction of ternary ion exchange equilibria. Ind Eng Chem Res 38(1):251–259
89. Namasivayam C, Senthilkumar S (1998) Removal of arsenic(V) from aqueous solution using industrial solid waste: adsorption rates and equilibrium studies. Ind Eng Chem Res 37 (12):4816–4822
90. Ho YS, McKay G (1999) Pseudo-second order model for sorption processes. Process Biochem 34(5):451–465
91. Ho YS, McKay G (1999) The sorption of lead(II) ions on peat. Water Res 33(2):578–584
92. Reddad Z, Gerente C, Andres Y, Le Cloirec P (2002) Adsorption of several metal ions onto a low cost biosorbent: kinetic and equilibrium studies. Environ Sci Technol 36(9):2067–2073
93. Yardim MF, Budinova T, Ekinci E, Petrov N, Razvigorova M, Minkova V (2003) Removal of mercury (II) from aqueous solution by activated carbon obtained from furfural. Chemosphere 52(5):835–841
94. Chiron N, Guilet R, Deydier E (2003) Adsorption of Cu(II) and Pb(II) onto a grafted silica: isotherms and kinetic models. Water Res 37(13):3079–3086
95. Aksu Z (2005) Application of biosorption for the removal of organic pollutants: a review. Process Biochem 40(3–4):997–1026
96. Zhang F-S, Nriagu JO, Itoh H (2005) Mercury removal from water using activated carbons derived from organic sewage sludge. Water Res 39:389–395
97. Chen S, Yue Q, Gao B, Xu X (2010) Equilibrium and kinetic adsorption study of the adsorptive removal of Cr(VI) using modified wheat residue. J Colloid Interface Sci 349 (1):256–264
98. Edebali S, Pehlivan E (2010) Evaluation of amberlite IRA96 and Dowex 1x8 ion-exchange resins for the removal of Cr(VI) from aqueous solution. Chem Eng J 161(1–2):161–166
99. Faghihian H, Kabiri-Tadi M (2010) Removal of zirconium from aqueous solution by modified clinoptilolite. J Hazard Mater 178(1–3):66–73
100. Ofomaja AE, Naidoo EB, Modise SJ (2010) Biosorption of copper(II) and lead(II) onto potassium hydroxide treated pine cone powder. J Environ Manage 91(8):1674–1685
101. Wahab MA, Jellali S, Jedidi N (2010) Effect of temperature and pH on the biosorption of ammonium onto Posidonia oceanica fibers: equilibrium, and kinetic modeling studies. Bioresour Technol 101(22):8606–8615

102. Wang L, Zhang J, Zhao R, Li Y, Li C, Zhang C (2010) Adsorption of Pb(II) on activated carbon prepared from *Polygonum orientale* Linn.: kinetics, isotherms, pH, and ionic strength studies. Bioresour Technol 101(15):5808–5814
103. Wang XM, Huang JH, Huang KL (2010) Surface chemical modification on hyper-cross-linked resin by hydrophilic carbonyl and hydroxyl groups to be employed as a polymeric adsorbent for adsorption of p-aminobenzoic acid from aqueous solution. Chem Eng J 162 (1):158–163
104. Chien SH, Clayton WR (1980) Application of Elovich equation to the kinetics of phosphate release and sorption on soils. Soil Sci Amer J 44:265
105. Sparks DL (1986) Kinetics of reaction in pure and mixed systems. In: Sparks DL (ed) Soil physical chemistry. CRC press, Boca Raton
106. Anirudhan TS, Radhakrishnan PG (2010) Uptake and desorption of nickel(II) using polymerised tamarind fruit shell with acidic functional groups in aqueous environments. Chem Ecol 26(2):93–109
107. Davila-Rangel JI, Solache-Rios M, Badillo-Almaraz VE (2005) Comparison of three Mexican aluminosilicates for the sorption of cadmium. J Radioanal Nuc Chem 267(1):139–145
108. Ngah WSW, Hanafiah M (2008) Adsorption of copper on rubber (*Hevea brasiliensis*) leaf powder: kinetic, equilibrium and thermodynamic studies. Biochem Eng J 39(3):521–530
109. Sivasankara V, Ramachandramoorthy T, Chandramohan A (2010) Fluoride removal from water using activated and MnO_2-coated Tamarind Fruit (*Tamarindus indica*) shell: batch and column studies. J Hazard Mater 177(1–3):719–729
110. Ritchie AG (1977) Alternative to the Elovich equation for the kinetics of adsorption of gases on solids. J Chem Soc Faraday Trans 73:1650–1653
111. Helfferich F (1962) Ion exchange. McGraw-Hill, New York
112. Liberti L, Boari G, Passino R (1978) Chloride-sulfate exchange on anion resins – kinetic investigations .2. Particle diffusion rates. Desalination 25(2):123–134
113. Trgo M, Peric J, Medvidovic NV (2006) A comparative study of ion exchange kinetics in zinc/lead – modified zeolite-clinoptilolite systems. J Hazard Mater 136(3):938–945
114. Trgo M, Peric J, Medvidovic NV (2006) Investigations of different kinetic models for zinc ions uptake by a natural zeolitic tuff. J Environ Manage 79(3):298–304
115. Cincotti A, Mameli A, Locci AM, Orru R, Cao G (2006) Heavy metals uptake by Sardinian natural zeolites: experiment and modeling. Ind Eng Chem Res 45(3):1074–1084
116. Varshney KG, Gupta PA, Tayal N (2003) Kinetics of ion exchange of alkaline earth metal ions on, acrylamide cerium(IV) phosphate: a fibrous ion exchanger. Colloids Surf B Biointerfaces 28(1):11–16
117. Patzay G (1995) A simplified numerical solution method for the Nernst-Planck multicomponent ion exchange kinetics model. React Funct Polym 27(1):83–89
118. Smith TG, Dranoff JS (1964) Film diffusion-controlled kinetics in binary ion exchange. Ind Eng Chem Fundam 3(3):195–200
119. Chanda M, Rempel GL (1995) Sorption of sulfide on a macroporous, quaternized poly (4-vinyl pyridine) in alkaline medium. React Polym 24(3):203–212
120. Dolgonosov AM, Khamizov RK, Krachak AN, Prudkovsky AG (1995) Macroscopic model for multispecies ion-exchange kinetics. React Funct Polym 28(1):13–20
121. Rodriguez JF, Valverde JL, Rodrigues AE (1998) Measurements of effective self-diffusion coefficients in a gel-type cation exchanger by the zero-length-column method. Ind Eng Chem Res 37(5):2020–2028
122. Samson E, Marchand J (1999) Numerical solution of the extended Nernst-Planck model. J Colloid Interface Sci 215(1):1–8
123. Varshney KG, Pandith AH (1999) Forward and reverse ion-exchange kinetics for some alkali and alkaline earth metal ions on amorphous zirconium(IV) aluminophosphate. Langmuir 15(22):7422–7425
124. Rodriguez JF, de Lucas A, Leal JR, Valverde JL (2002) Determination of intraparticle diffusivities of Na^+/K^+ in water and water/alcohol mixed solvents on a strong acid cation exchanger. Ind Eng Chem Res 41(12):3019–3027

125. Valverde JL, De Lucas A, Carmona M, Gonzalez M, Rodriguez JF (2004) A generalized model for the measurement of effective diffusion coefficients of heterovalent ions in ion exchangers by the zero-length column method. Chem Eng Sci 59(1):71–79
126. Valverde JL, De Lucas A, Carmona M, Gonzalez M, Rodriguez JF (2005) Model for the determination of diffusion coefficients of heterovalent ions in macroporous ion exchange resins by the zero-length column method. Chem Eng Sci 60(21):5836–5844
127. Wesselingh JA, Vonk P, Kraaijeveld G (1995) Exploring the Maxwell-Stefan description of ion-exchange. Chem Eng J Biochem Eng J 57(2):75–89
128. Lagergren S (1898) About the theory of so-called adsorption of soluble substances Kungliga Svenska Vetenskapsakademiens. Handlingar 24:1–39
129. Ruthven DM (1984) Principles of adsorption and adsorption processes. Wiley, New York
130. Crank J (1999) The mathematics of diffusion, 2nd edn. Oxford University Press, Great Britain
131. Vermeulen T (1953) Theory for irreversible and constant-pattern solid diffusion. Ind Eng Chem 45(8):1664–1670
132. Inglezakis VJ, Grigoropoulou HP (2001) Applicability of simplified models for the estimation of ion exchange diffusion coefficients in zeolites. J Colloid Interface Sci 234(2):434–441
133. Aharoni C, Sparks DL (1991) Kinetics of soil chemical-reactions – a theoretical treatment. Rates Soil Chem Process 27:1–18
134. Aharoni C, Sparks DL, Levinson S, Ravina I (1991) Kinetics of soil chemical-reactions – relationships between empirical equations and diffusion-models. Soil Sci Soc Am J 55(5):1307–1312
135. Slater MJ (1991) Principles of ion exchange technology. Butterworth-Heinemann, Great Britain
136. Patzay G (1995) A simplified numerical solution method for the Nernst-Planck multicomponent ion exchange kinetics model. React Funct Polym 27:83–89
137. Krishna R, Wesselingh JA (1997) Review article number 50 – the Maxwell-Stefan approach to mass transfer. Chem Eng Sci 52(6):861–911
138. Krishna R (1990) Multicomponent surface-diffusion of adsorbed species – a description based on the generalized Maxwell-Stefan equations. Chem Eng Sci 45(7):1779–1791
139. Krishna R (1993) Problems and pitfalls in the use of the Fick formulation for intraparticle diffusion. Chem Eng Sci 48(5):845–861
140. Kapteijn F, Moulijn JA, Krishna R (2000) The generalized Maxwell-Stefan model for diffusion in zeolites: sorbate molecules with different saturation loadings. Chem Eng Sci 55(15):2923–2930
141. van de Graaf JM, Kapteijn MF, Moulijn JA (1999) Modeling permeation of binary mixtures through zeolite membranes. Aiche J 45(3):497–511
142. van der Stegen JHG et al (1999) Application of the Maxwell-Stefan theory to the transport in ion-selective membranes used in the chloralkali electrolysis process. Chem Eng Sci 54 (13–14):2501–2511
143. Hogendoorn JA, Veen AJVd, Stegen JHGVd, Kuipers JAM, Versteeg GF (2001) Application of the Maxwell-Stefan theory to the membrane electrolysis process: model development and simulation. Comput Chem Eng 25:1251–1265
144. Wesselingh JA, Vonk P, Kraaijeveld G (1995) Exploring the Maxwell-Stefan description of ion-exchange. Chem Eng J Biochem Eng J 57(2):75–89
145. Graham EE, Dranoff JS (1982) Application of the Stefan-Maxwell equations to diffusion in ion-exchangers .1. Theory. Ind Eng Chem Fundam 21(4):360–365
146. Graham EE, Dranoff JS (1982) Application of the Stefan-Maxwell equations to diffusion in ion-exchangers .2. Experimental results. Ind Eng Chem Fundam 21(4):365–369
147. Pinto NG, Graham EE (1987) Characterization of ionic diffusivities in ion-exchange resins. Ind Eng Chem Res 26(11):2331–2336
148. Jackson R (1977) Transport in porous catalysts. Elsevier, Amsterdam
149. Mason EA, Malinauskas AP (1983) Gas transport in porous media: the dusty gas model. Elsevier, Amsterdam/The Netherlands

150. Treybal RE (1981) Mass-transfer operations, 3rd edn. McGraw-Hill, Singapore
151. Misic DM, Sudo Y, Suzuki M, Kawazoe K (1982) Liquid-to-particle mass-transfer in a stirred batch adsorption tank with non-linear isotherm. J Chem Eng Jpn 15(1):67–70
152. Kulov NN, Nikolaishvili EK, Barabash VM, Braginski LN, Malyusov VA, Zhavoronkov NM (1983) Dissolution of solid particles suspended in agitated vessels. Chem Eng Commun 21 (4–6):259–271
153. Miller SA, Amber CM, Bennet RC, Dahlstrom DA, Darji JD, Emmet RC, Gray JB, Gurnham CF, Jacobs LJ, Klepper RP, Michalson AW, Oldshue JY, Silverblatt CE, Smith JC, Todd DB Liquid-solid systems, In: Perry RH, Green D (eds) Perry's chemical engineers' handbook. McGrow-Hill, Singapore
154. Townsend R, Harjula R (2002) Ion exchange in molecular sieves by conventional techniques, in post-synthesis modification I. Springer, Berlin/Heidelberg, pp 1–42
155. Thomas HC (1944) Heterogeneous ion exchange in a flowing system. J Am Chem Soc 66:1664–1666
156. Bohart GS, Adams EQ (1920) Some aspects of the behavior of charcoal with respect to chlorine. J Am Chem Soc 42:523–544
157. Clark RM (1987) Evaluating the cost and performance of field-scale antigranulocytes activated carbon systems. Environ Sci Technol 21(6):573–580
158. Yoon YH, Nelson JH (1984) Application of gas-adsorption kinetics .1. A theoretical-model for respirator cartridge service life. Am Ind Hyg Assoc J 45(8):509–516
159. Wolborska A (1989) Adsorption on activated carbon of p-nitrophenol from aqueous solution. Water Res 23(1):85–91
160. Baral SS, Das N, Ramulu TS, Sahoo SK, Das SN, Chaudhury GR (2009) Removal of Cr(VI) by thermally activated weed Salvinia cucullata in a fixed-bed column. J Hazard Mater 161 (2–3):1427–1435
161. Han R, Wang Y, Zhao X, Wang Y, Xie F, Cheng J, Tang M (2009) Adsorption of methylene blue by phoenix tree leaf powder in a fixed-bed column: experiments and prediction of breakthrough curves. Desalination 245(1–3):284–297
162. Pitzer KS (1991) Ion interaction approach: theory and data correlation. In: Pitzer KS (ed) Activity coefficients in electrolyte solutions. CRC Press, Boca Raton, pp 75–153
163. Aniceto JPS, Cardoso SP, Faria TL, Lito PF, Silva CM (2012) Modeling ion exchange equilibrium: analysis of exchanger phase non-ideality. Desalination 290:43–53
164. Aniceto JPS, Lito PF, Silva CM (2012b) Modeling sorbent phase non-ideality for accurate prediction of multicomponent ion exchange equilibrium with homogeneous mass action law. J Chem Eng Data doi:10.1021/je300156H

Chapter 4
Fundamentals of Ion Exchange Fixed-Bed Operations

Vassilis J. Inglezakis and Antonis Zorpas

Abstract As a physical phenomenon, ion exchange is basically a subject of physical chemistry, while fixed bed is a common reactor studied in chemical engineering. Furthermore, it is a subject bringing together the experimentalists and the modeling scientists, who in the typical case have different views on the same system. This interrelation of several sciences and approaches is by no means a particularity of ion exchange fixed-beds field. The experience of the authors and the study of the related published research in journals for many years indicate a clear gap in many scientific publications, where authors are frequently missing aspects which are well known to other scientists working on the same subject. This chapter comes to fill the gap and in the same time to balance between all these different views of scientists with different approaches and background. It should be mentioned that the principal target of this work is to analyze the fundamentals of the operation (fixed bed) and not the physical chemistry of the phenomenon (ion exchange), which is a subject of other chapters in the present book. From this point of view, the authors have followed the most common approach, in which ion exchange is treated as a sorption process when it comes to the design of fixed-bed processes. Consequently, the authors consider this chapter as an added value for the related literature.

V.J. Inglezakis (✉)
SC European Focus Consulting srl, Banatului 16, Bacau 600 276, Romania
e-mail: inglezakis@efcon-group.ro; v_inglezakis@yahoo.com

A. Zorpas
Cyprus Open University, Faculty of Pure and Applied Science,
Environmental Conservation and Management, Latsia, Cyprus

Institute of Environmental Technology and Sustainable Development,
Laboratory of Environmental Friendly Technology, Paralimni Cyprus,
e-mail: antonis.zorpas@ouc.ac.cy; antoniszorpas@envitech.org; antoniszorpas@yahoo.com;
envitech@cytanet.com.cy

Inamuddin and M. Luqman (eds.), *Ion Exchange Technology I: Theory and Materials*, DOI 10.1007/978-94-007-1700-8_4,

4.1 Introduction

4.1.1 General

Ion exchange is a process in which mobile ions from an external solution are exchanged for ions that are electrostatically bound to the functional groups contained within a solid matrix. These functional groups and the ions carry opposite charge and electroneutrality of the matrix and solution is always preserved because ion exchange is a stoichiometric process. Ion exchange is of the most common processes used for wastewater treatment and other separation operations and in the most of the cases takes place in fixed beds [1, 2]. A large number of predictive models exist for the description of the dynamics of ion exchange fixed-bed operations. However, the mathematical complexity and the need of knowing many parameters from separate experiments make them rather inconvenient for practical use. This is the reason why adsorption models are frequently used for ion exchange fixed-bed operations as is true that the two phenomena share many common features, of the most basic being the transfer and resulting equilibrium distribution of one or more solutes between a fluid phase and particles [3, 4]. In the related literature, ion exchange and adsorption are grouped together as sorption for a unified treatment.

4.1.2 Ion Exchange, Adsorption and Sorption

Sorption is a separation process involving two phases between which certain components are distributed. There are three types of sorption, classified according to the type of bonding involved [1]:

- Physical sorption or adsorption, where no exchange of electrons occurs, rather intermolecular attractions between sorption sites and is therefore independent of the electronic properties of the molecules involved.
- Chemical sorption or chemisorptions, where an exchange of electrons between specific surface sites and solute molecules occurs, which results in the formation of a chemical bond. In general, chemisorption exhibits stronger adsorption energy than physical adsorption.
- Electrostatic sorption or ion exchange, where coulombic attractive forces between ions and charged functional groups occur.

Ion exchange shares many common characteristics with adsorption, such as mass transfer from the fluid to the solid phase, which is a diffusion-driven process as is presented in Sect. 4.2.1.1. However, there are some significant differences, the most important being the nature of the sorbed species, which are ions in ion exchange and electrically neutral substances in adsorption. In ion exchange, the ions removed from the liquid phase are replaced by ions from the solid phase, and thus a two-way traffic (exchange) occurs in contrast to adsorption, where one-way removal of species takes place. This exchange of ions between the two phases (liquid and solid) is such that the total charge sorbed and desorbed is exactly the same as imposed by the electroneutrality principle [5]. All these factors constitute the quantitative treatment of ion exchange much more complicated than adsorption. Although the differences, adsorption and ion exchange are grouped together as sorption for the purposes of a unified treatment in practical applications and in the related literature most of the mathematical theories and approaches developed originally for adsorption are used with minor modifications for ion exchange. As a general rule of thumb, the applicability of a simplified theory hinges on the mode of operation rather than on the particular mechanism of solute uptake [5]. As the principal target of this chapter is to analyze the fundamentals of the operation (fixed bed) and not the physicochemistry of the phenomenon (ion exchange), it is based on the simplified approach as presented above.

4.1.3 Fixed-Bed Operations

Fixed bed is the most frequently used reactor for ion exchange and adsorption operations. In Figs. 4.1 and 4.2, a typical fixed-bed operation and its breakthrough curve are presented.

As is visible in Fig. 4.1, the operation is upflow, i.e., the liquid phase is introduced in the bottom of the bed. This mode of operation is preferred as it provides better flow quality and complete wetting of the material. In Fig. 4.2, the dimensionless exit concentration C/C_o versus effluent volume V_{eff} is plotted, i.e., the breakthrough curve of the operation. The breakpoint is the moment when the operation is stopped, typically when the exit concentration reaches a certain predefined level. This is the case in all wastewater treatment operations, where a maximum concentration is selected above which the effluent becomes unacceptable for disposal. In the optimum fixed-bed operation, the breakthrough curve is a single step, a vertical line, but in the most actual cases is a sigmoidal curve, and the steeper the breakthrough curve, the higher is the fixed-bed utilization. High degree of utilization means that the ratio of the breakthrough capacity to the total capacity is closer to unity [3].

During sizing fixed beds, the main aim is to predict the time (or total effluent volume) until the effluent exceeds the desired breakpoint concentration. The corresponding breakpoint effluent volume is the most important information which could be derived from experiments conducted in laboratory fixed beds and is used during scale-up in order to design the large-scale unit. Fixed-bed operation is influenced by equilibrium (isotherm and capacity), kinetic (diffusion and convection coefficients) and hydrodynamic (liquid holdup, geometric analogies, and maldistribution) factors, constituting the design a complicated task and in some cases, for complicated systems the full process modeling is

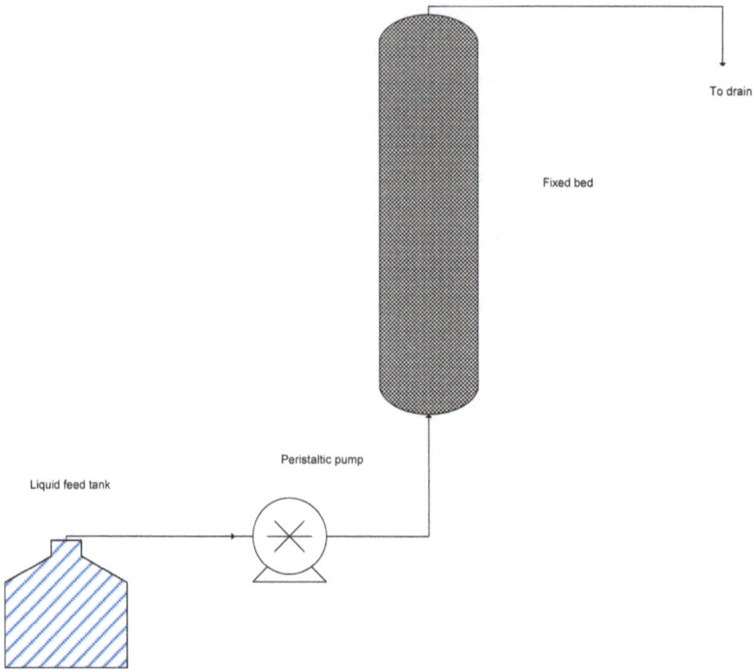

Fig. 4.1 Fixed-bed upflow operation

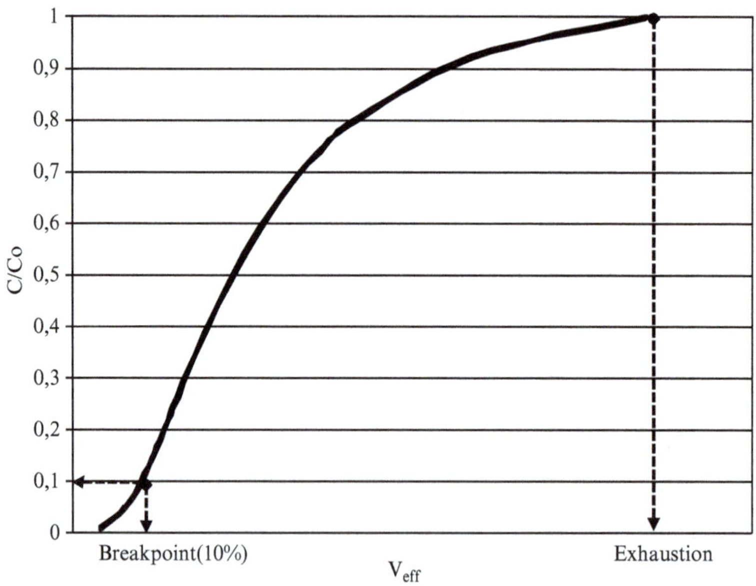

Fig. 4.2 Breakthrough curve

practically impossible [3, 5]. A characteristic example is the simple case of the treatment of mono-component solutions in ion exchange fixed beds where it is experimentally found that equilibrium parameters and/or solid diffusion coefficients seem to be unexpectedly influenced by contact time, making the modeling of the operation a complex task [6]. Furthermore, it is true that the effectiveness of a fixed-bed operation is heavily dependent on its hydrodynamic performance, a fact that in many cases is overlooked and inevitably leads many operations to failure. Failures are frequently a result of the competitive effects of kinetics and fluid dynamics: internal mixing, dead spaces, channeling, and bypasses, phenomena that radically alter the performance of a reactor, especially when compared to the "ideal" or theoretical presentation [7]. In liquid–solid systems, these phenomena are termed as "liquid maldistribution," and they are related to inadequate liquid distribution and low liquid holdup, especially in downflow operation. Liquid maldistribution is a critical factor in laboratory-, pilot-, and commercial-scale reactors, and scale-up has always to deal with such problems. As scale-up is always a complicated procedure, in most cases, engineers utilize empirical rules by selecting the essential design variables and taking into consideration several safety or overdesign factors [3, 8]. It is not surprising that even today many decisions in commercial-scale operations are not always sufficiently supported by adequate experimental evidences or modeling but are taken based on experience and approximate methods.

4.2 Theoretical Fundamentals and Modeling

4.2.1 Ion Exchange Kinetics and Equilibrium

4.2.1.1 Kinetics

When an ion-exchanger particle is brought in contact with a solution, a static adherent liquid film is formed around it, the thickness of which may vary between 10 and 100 μm, depending on the flow conditions around the particle. Ion exchange involves five distinct steps [1]:

(a) Transport of ions through the bulk solution in order to reach the ion-exchanger particle
(b) Diffusion of ions through the liquid film surrounding the particle
(c) Diffusion of the ion across the film–particle interface
(d) Diffusion of the ion through the particle
(e) The exchange of ions on the active sites

Unless the concentration of ions in the solution is extremely low, steps (a), (c), and (e) are very fast, and the overall ion exchange rate is determined by step (b) or step (d), or a combination of these two steps. In other words, the overall rate is controlled by mass-transfer resistance, rather than by the intrinsic sorption kinetics [5, 9]. Thus, the phenomenon is a diffusion process and Fick's first law is applicable [5, 10]. Fick's first law relates the diffusive flux to the concentration field, by postulating that the flux goes from regions of high concentration to regions of low concentration, with a magnitude that is proportional to the concentration gradient (in one dimension, this gradient is equal to $\frac{\partial C_i}{\partial x}$):

$$J_i = -D \cdot grad(C_i) = -D \cdot \nabla C_i, \tag{4.1}$$

where J_i is the flux of the diffusing species i, C_i is its concentration, and D is the diffusion coefficient. The negative sign in Eq. 4.1 arises because diffusion occurs in direction opposite to that of increasing concentration. In some cases, for example, diffusion in dilute solutions, D can be reasonably taken as constant, while in others, it depends on the concentration [10]. Several authors showed that internal (particle) diffusivity increases with particle loading, and various relationships have been proposed as for example exponential expressions [11]. However, it is true that in the most cases for practical applications the diffusion coefficient is assumed to be a constant.

As mentioned above, an electric field is developed in ion exchange systems and thus, the flux because of the electric potential should be added in Eq. 4.1. Exchange between counter ions A in solid and counter ions B may be represented by the Nernst–Planck equation as [12]:

$$J_i = -D_i \cdot \nabla C_i - u_i \cdot z_i \cdot C_i \cdot \nabla \varphi = -D_i \cdot \left(\nabla C_i - \frac{z_i \cdot C_i \cdot F}{R \cdot T} \cdot \nabla \varphi \right), \tag{4.2}$$

where i = A or B, u is the electrochemical mobility, φ is the electric potential, z is the ion charge, T is the temperature, R is the ideal gas constant, and F is the Faraday constant. The quantity φ represents the electrical potential generated by the difference in mobility. The requirements of electroneutrality and absence of net electric current reduce Eq. 4.2 to

$$J_i = -D_{AB} \cdot \nabla C_i, \tag{4.3}$$

where [13]

$$D_{AB} = \frac{D_A \cdot D_B \cdot \left(z_A^2 \cdot C_A + z_B^2 \cdot C_B\right)}{z_A^2 \cdot C_A \cdot D_A + z_B^2 \cdot C_B \cdot D_B}. \tag{4.4}$$

This is the interdiffusion coefficient and depends on the individual diffusion coefficients, on the local concentration of both species and therefore on the radial position and time. Thus, the effect of the electric field is expressed by the variant interdiffusion coefficient. For ions of equal mobility, D_{AB} becomes equal to the self-diffusion coefficient of each ion and Eq. 4.3 reduces to Eq. 4.1. Furthermore, as it can be proved, an average constant value could be used under certain conditions valid for many practical applications [13, 14]. As mentioned above, the concentration dependence is a general fact for diffusion-driven phenomena and is not a particularity of ion exchange. Again, the simplified approach of a constant diffusion coefficient is very common in the related literature, and although in batch systems, models incorporating the electric field are frequently used, for fixed-bed systems, the ordinary diffusion-driven potential is the most common approach. It is not by chance that in *Perry's Chemical Engineers Handbook* ion exchange and adsorption are grouped together as sorption for a unified treatment when dealing with application in batch and fixed-bed processes [4]. The same approach is followed in *Coulson and Richardson's Chemical Engineering* book, where the design of fixed-bed ion exchangers shares a common theory with fixed-bed adsorbers [12]. Clearly, this approach reduces the physical meaning of the models providing however simplicity on the interpretation of fixed-bed operation, a process considerably more complicated than a batch system. A more detailed analysis on the applicability of simplified models for ion exchange in batch systems is given elsewhere [13, 15].

Focusing on the solid-phase diffusion, a substantial fraction of the particle volume is occupied by the solid matrix, which may consist of small microporous crystals formed into a macroporous structure. In this case, two distinct diffusion resistances exist: the macropore (pore diffusion) and the micropore (solid diffusion) resistances, frequently grouped under the single term "intraparticle diffusion" [4, 5, 9]. Pore diffusion is the diffusion of solutes in fluid-filled pores, which are so large enough for the diffusing species to escape the adsorbent surface force field [4]. On the other hand, solid diffusion is the diffusion in the adsorbent surface itself, where the pores are so small that the diffusing species never escapes the force field of the adsorbent surface. Pore and solid diffusion act in parallel, and thus the dominant transport process is the faster one. For the purpose of this chapter, Helfferich's approach on the kinetics is followed: the particle, consisting of pore fluid and a solid matrix, is treated as a single quasi-homogeneous phase, regardless of its inhomogeneities in molecular or colloidal dimensions and its particular geometrical structure [15]. In solid diffusion-controlled process, the corresponding diffusion coefficient is an "effective" kinetic parameter characterizing the ease of movement of the solutes in the particle pores. Furthermore, the diffusion coefficient considered to be constant, or to be more realistic, is an average value as it can be dependent on the solute solid-phase concentration. Following this simplified approach, the time dependence of the concentration is related with the flux by the Fick's second law [10]:

$$\frac{\partial C_i}{\partial t} = -div(J_i) = D \cdot \nabla^2 C_i. \tag{4.5}$$

Fick's second law is derived from first law and the mass balance and predicts how diffusion causes the concentration field to change with time. Combining Eqs. 4.1 and 4.5, the mass-transfer process for spherical particles is described by the following diffusion equation, written in spherical coordinates [5]:

$$\frac{\partial C_i}{\partial t} = D_s \cdot \left(\frac{\partial^2 C_i}{\partial r^2} + \frac{2}{r} \cdot \frac{\partial C_i}{\partial r} \right), \tag{4.6}$$

Table 4.1 Factors affecting ion exchange kinetics (subscript s denotes solid phase and f fluid phase)

Parameter	Particle diffusion control	Film diffusion control
Counter ion mobility		
Solid phase	Proportional to D_s	No effect
Liquid phase	No effect	Proportional to D_f
Particle size	Proportional to $1/r^2$	Proportional to $1/r$
Capacity of the ion exchanger	No effect	Inverse proportional to the concentration of functional ionic groups
Degree of cross-linking	Decreases with increasing cross-linking	No effect
Selectivity of the ion exchanger	Preferred counter ion is generally taken up at higher rate and released in lower rate	
Concentration of the solution	No effect	Proportional to the concentration
Temperature	Increases with temperature	
Rate of agitation or flow	No effect	Proportional to the agitation or flow rate

where D_s is the solid diffusion coefficient and C_i is the solid-phase concentration of the solute (in the following paragraphs, this concentration is denoted by q). Equation 4.6 must be solved under the appropriate initial and boundary conditions.

For the mass transfer associated with the transport of solutes through the fluid layer surrounding the particle, the driving force is the concentration difference across the boundary layer, and the flux at the particle surface is [5]

$$J_i = k_f \cdot \left(C_{i,f} - C_{i,s}\right). \tag{4.7}$$

where k_f is the mass-transfer coefficient and $C_{i,f}$, $C_{i,s}$ are the concentrations of solute in the bulk fluid and at the particle surface phase, respectively.

The ion exchange rate can vary over a wide range, requiring a few seconds to several months to reach equilibrium, and it is affected by a number of parameters, such as the properties and nature of the exchanger and counter ions, the extent of agitation and flow conditions, the concentration of the cations and anions, etc. The main factors affecting the exchange rate are given in Table 4.1 [1].

4.2.1.2 Equilibrium

Ion exchange reactions are stoichiometric and reversible, and in that way, they are similar to other solution phase reactions. A common way to represent the equilibrium in ion exchange systems is the equilibrium isotherm, which represents the distribution of the adsorbed material between the adsorbed phase and the solution phase at equilibrium. The basic difference between adsorption and ion exchange is that while there is only one isotherm at a specified temperature for adsorption, more than one isotherm can exist at a specified temperature for different normality of the solution in ion exchange of ions of different valences due to the concentration–valence effect [5]. Thus, a specific ion exchange system presents one equilibrium curve (isotherm) only under constant temperature and normality.

Ion exchange equilibrium has been studied for many decades and numerous theories and models have been proposed. Although the sound theoretical basis and proven quality of such models, it is true that in many cases they fail to incorporate several interfering phenomena encountered in an ion exchange system, like the hydrolysis of ion exchangers and cations and the sorption of electrolytes [16]. Another weakness of rigorous complex models is the errors associated to several experimental parameters and the unavailability of such parameters. An alternative has always been the development and use of simplified equations, both theoretical and empirical in nature, which in the most of the cases are valid as approximations under certain assumptions and experimental conditions. Nevertheless, such simplified equations would probably serve best the purpose of equilibrium modeling and in a further step of a commercial-scale operation design [16]. In the following paragraph, the two most common isotherms used for liquid-phase adsorption and ion exchange systems are presented, namely, the Langmuir and Freundlich isotherms.

Langmuir equation is applicable to a manifold of simple binary ion exchange systems. The underlying theory assumes the adsorbents to contain fixed individual sites, each of which adsorbs equally only one molecule, forming thus a monolayer [4]:

$$\frac{q_e}{Q_M} = \frac{K \cdot C_e}{1 + K \cdot C_e}. \tag{4.8}$$

where q_e is the solid-phase concentration in equilibrium with liquid-phase concentration C_e, Q_M is the ultimate sorptive capacity, and K is an equilibrium constant. For $C_e = C_o$, Eq. 4.8 becomes

$$\frac{q_{\max}}{Q_M} = \frac{K \cdot C_o}{1 + K \cdot C_o}. \tag{4.9}$$

where q_{max} is the solid-phase concentration in equilibrium with C_o. Dividing Eq. 4.8 with Eq. 4.9, we have

$$\frac{q_e}{q_{\max}} = \frac{C_e}{C_o} \cdot \frac{1 + K \cdot C_o}{1 + K \cdot C_e}. \tag{4.10}$$

The equilibrium parameter La is defined as

$$La = \frac{1}{1 + K \cdot C_o}. \tag{4.11}$$

Then, the dimensionless form of Langmuir Eq. 4.10 is

$$Y = \frac{X}{La + (1 - La) \cdot X}. \tag{4.12}$$

where

$$Y = \frac{q_e}{q_{\max}} \tag{4.13}$$

and

$$X = \frac{C_e}{C_o}. \tag{4.14}$$

By using the dimensionless equilibrium relationship (Eq. 4.12), *La* can also be expressed as

$$La = \frac{X \cdot (1 - Y)}{Y \cdot (1 - X)}. \tag{4.15}$$

Freundlich isotherm is an empirical equation and assumes the adsorbents to have energetically heterogeneous surface consisting of sites with different adsorption potentials. The equation is [4]

$$\frac{q_e}{Q_M} = k \cdot C_e^{Fr}. \tag{4.16}$$

where q_e is the solid-phase concentration in equilibrium with liquid-phase concentration C_e, Q_M is the ultimate sorptive capacity, and k, Fr are equilibrium constants. Another form of the Freundlich isotherm is the following:

$$q_e = K_F \cdot C_o^{Fr}. \tag{4.17}$$

where

$$K_F = Q_M \cdot k. \tag{4.18}$$

Applying Eq. 4.16 for $C_e = C_o$:

$$\frac{q_{\max}}{Q_M} = k \cdot C_o^{Fr}. \tag{4.19}$$

here q_{max} is the solid-phase concentration in equilibrium with C_o. Dividing Eqs. 4.16 and 4.19,

$$Y = X^{Fr}. \tag{4.20}$$

At this point, it is important to underline the difference between q_{max} and Q_M, the first being the solid-phase concentration in equilibrium with the initial fluid-phase concentration, while Q_M is the maximum adsorption capacity. For an ion exchange system, q_{max} corresponds to the maximum exchange level (*MEL*), while Q_M corresponds to the real exchange capacity (*REC*) [17]:

- Real exchange capacity refers to the amount of the actual exchangeable cations of the solid, measured by ion exchange methods, and is a characteristic constant of the ion exchanger independent of the experimental conditions. *REC* is the upper limit of the actual solid capacity and, in general, is lower than the theoretical exchange capacity (*TEC*) calculated by chemical analysis of the solid.
- Maximum exchange level is a characteristic property of the specific ion exchange system. An ion exchange system is defined a system consisting of a material and a liquid solution which contains the ion to be exchanged, under a specified temperature and normality. *MEL* is straightforwardly related to the equilibrium behavior of the specific ion exchange system. It is measured by repeated equilibrations or by determinations using equilibrium data. *MEL* and *REC* could be equal for "ideal" ion exchange systems, i.e., systems where complete exchange is achieved. However, generally, MEL is only a fraction of REC and this is the case of partial exchange.

The empirical constants *La* and *Fr* are related to the particular system under investigation and are obtained from laboratory experiments [5]. Generally, in *X* versus *Y* plot, an isotherm is favorable for the incoming ion if its shape is convex upward $\frac{\partial^2 Y}{\partial X^2}>0$ and unfavorable for the incoming ion if its shape is concave upward $\frac{\partial^2 Y}{\partial X^2}<0$. In Fig. 4.3, the characteristic isotherm shapes are presented.

The parameter *La* is also called separation factor and is $La = 0$ for irreversible, $La < 1$ for favorable, $La = 1$ for linear, and $La > 1$ for unfavorable isotherm type. The same holds for *Fr* in Freundlich isotherm. More isotherm types can be found elsewhere [5].

The discussion of equilibrium is completed by presenting some basic relationships and parameters involved in ion exchange equilibrium. Let us suppose that the solid phase is initially in the B form and that the liquid-phase ion is A [5]:

$$A + \overline{B} \leftrightarrow \overline{A} + B.$$

The separation factor α_{A-B} is defined as

$$a_{A-B} = \frac{q_A \cdot C_B}{q_B \cdot C_A} = \frac{Y_A \cdot X_B}{Y_B \cdot X_A}. \tag{4.21}$$

where

$$Y_i = \frac{q_i}{q_{\max}} \tag{4.22}$$

$$X_i = \frac{C_i}{C_o}. \tag{4.23}$$

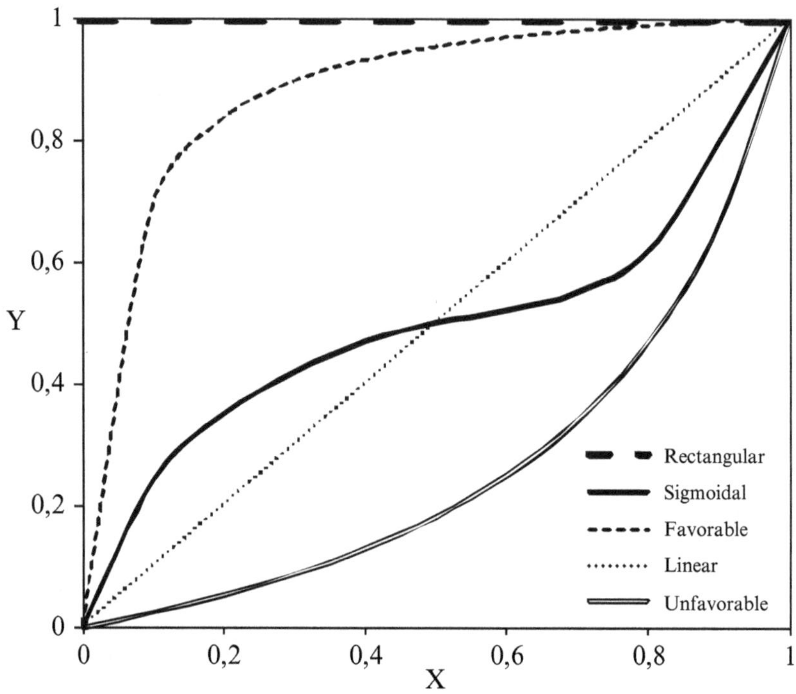

Fig. 4.3 Basic isotherm types

where i is A or B. In Eq. 4.22, q_{max} is the maximum exchange level and only in the case of complete exchange is equal to the real exchange capacity (*REC*). A complete analysis of the ion exchange capacity types and its measurement is given elsewhere [17]. Furthermore, C_o is the initial concentration (normality) of the solution phase, which for the case of ion exchange is constant throughout the exchanging process. If the ion A is preferred, the separation factor is larger than unity, whereas if ion B is preferred, it is smaller than unity. Expressing all concentrations in equivalents:

$$X_A + X_B = 1 \Rightarrow X_B = 1 - X_A \tag{4.24}$$

$$Y_A + Y_B = 1 \Rightarrow Y_B = 1 - Y_A. \tag{4.25}$$

Then,

$$a_{A-B} = \frac{Y_A \cdot (1 - X_A)}{X_A \cdot (1 - Y_A)}. \tag{4.26}$$

Recalling Eq. 4.15 is easy to show that α_{A-B} and La are related as follows:

$$\alpha_{A-B} = \frac{1}{La}. \tag{4.27}$$

This relation is the reason why La is also called as "separation factor" in the related literature. A constant separation factor α_{A-B} in an ion exchange system means that this system obeys a Langmuir equilibrium isotherm, in which La is constant. Finally, the distribution coefficient λ_i and the selectivity coefficient K_{A-B} are defined as follows:

$$\lambda_i = \frac{q_i}{C_i} \tag{4.28}$$

$$K_{A-B} = \frac{{q_A}^{Z_B} \cdot {C_B}^{Z_A}}{{q_B}^{Z_A} \cdot {C_A}^{Z_B}}. \tag{4.29}$$

where z_i is the absolute value of the ions charge. If the ion A is preferred, then the selectivity coefficient is larger than unity, whereas if ion B is preferred, it is smaller than unity. It is obvious that due to the exponents, different units result in different values of the selectivity coefficient.

4.2.2 Fixed-Bed Mass Balances

In a fixed bed, the continuity equation around a segment of the solid phase between the height x and $x + dx$ and during time dt is [2–5, 9]

$$-\rho_b \cdot \frac{\partial \overline{q}}{\partial t} = \varepsilon \cdot \frac{\partial C}{\partial t} + u_s \cdot \frac{\partial C}{\partial x} - D_{ax} \cdot \varepsilon \cdot \frac{\partial^2 C}{\partial x^2}. \tag{4.30}$$

where ε and ρ_b are bed porosity and bulk density, respectively, u_s is the superficial liquid velocity, D_{ax} is the axial dispersion coefficient, C is the solution concentration, and $\overline{q}$ is the mean concentration of solute in the solid phase. For the common case of dilute solutions, u_s is considered constant throughout the fixed bed, but under high concentration levels, u_s is variant, and this term should be incorporated into the derivative. Such effects are usually significant only for adsorption in gaseous systems [9].

Dividing and multiplying the second and third term of Eq. 4.30 by the bed height Z, we have

$$-\rho_b \cdot \frac{\partial \overline{q}}{\partial t} = \varepsilon \cdot \frac{\partial C}{\partial t} + \frac{u_s}{Z} \cdot \frac{\partial C}{\partial \left(\frac{x}{Z}\right)} - \frac{D_{ax} \cdot \varepsilon}{Z^2} \cdot \frac{\partial^2 C}{\partial \left(\frac{x}{Z}\right)^2}. \tag{4.31}$$

and rearranging

$$-\rho_b \cdot \frac{Z}{u_s} \cdot \frac{\partial \overline{q}}{\partial t} = \varepsilon \cdot \frac{Z}{u_s} \cdot \frac{\partial C}{\partial t} + \frac{\partial C}{\partial \left(\frac{x}{Z}\right)} - \frac{D_{ax} \cdot \varepsilon}{u_s \cdot Z} \cdot \frac{\partial^2 C}{\partial \left(\frac{x}{Z}\right)^2}. \tag{4.32}$$

the term $\varepsilon \cdot Z/u_s$ is the liquid residence time (or contact time), while the term $\frac{D_{ax} \cdot \varepsilon}{u_s \cdot Z}$ is the inversed bed Peclet number. Both C and q are dependent on time t and height Z, and Eq. 4.32 is a partial differential equation. As the solute uptake is controlled by the diffusion rates, the rate equations to be used are [3–5]

$$\frac{\partial q}{\partial t} = \frac{k_f \cdot a_u}{\rho_b} \cdot (C - C*) \tag{4.33}$$

$$\frac{\partial q}{\partial t} = \frac{D_s}{\rho_b} \cdot \left(\frac{\partial^2 q}{\partial r^2} + \frac{2}{r} \cdot \frac{\partial q}{\partial r}\right) = \frac{D_s}{\rho_b \cdot r_p^2} \cdot \left(\frac{\partial^2 q}{\partial \left(\frac{r}{r_p}\right)^2} + \frac{2}{\left(\frac{r}{r_p}\right)} \cdot \frac{\partial q}{\partial \left(\frac{r}{r_p}\right)}\right). \tag{4.34}$$

where C^* is the concentration of the solute in the fluid phase, which is assumed to be in equilibrium with q and α_u is the total external solid surface area per unit bed volume [3]:

$$a_u = \frac{6 \cdot (1 - \varepsilon)}{d_p}. \tag{4.35}$$

The mean solid concentration $\overline{q}$ is [18, 19]

$$\overline{q} = \frac{3}{r_p^3} \cdot \int_0^R q \cdot r^2 \cdot dr. \tag{4.36}$$

where r is the distance from particle center and d_p, r_p are the particle diameter and radius, respectively. Equation 4.33 holds for liquid film control, and Eqs. 4.34 and 4.36 for solid diffusion control. For liquid film control, $\overline{q}=q$. For combined resistances, Eqs. 4.33, 4.34, and 4.36 should be used simultaneously. The solid diffusion coefficient D_s is theoretically independent of the flow type and rate, and it can be calculated using appropriate diffusion models in batch reactors. Liquid mass-transfer coefficient is a flow-dependent parameter, which can be evaluated from several correlations found in the related literature (see Sect. 4.3.5.1). Experimental methods for the determination of several model parameters are provided in Sect. 4.3.5.3.

As in the case of adsorption, the physical process of ion exchange is considered to be so fast relative to diffusion steps that in and near the solid particles a local equilibrium exists, and thus, the equilibrium isotherm $q = f(C^*)$ relates the stationary- and mobile-phase concentrations at equilibrium [17]. For the purposes of simplified fixed-bed modeling, Langmuir and Freundlich equilibrium equations are used [9, 18, 19]. The empirical constants *La* and *Fr* are related to the particular system under investigation and are obtained from laboratory experiments [5]. The continuity, rate and equilibrium equations are solved simultaneously using the appropriate initial and boundary conditions [9]. This system consists of four equations and four unknowns, i.e., C, $\overline{q}$, q, and C^*.

4.2.3 *Fixed-Bed Hydrodynamics*

Following the study of the mass balance, kinetics and thermodynamics of the system, the hydrodynamic behavior of the fixed-bed reactor should be investigated. This investigation basically involves the evaluation of liquid holdup, quality of the flow and bed geometry. During this step, physical mechanisms that are sensitive to size are investigated separately from chemical (kinetic or equilibrium) studies [20]. Here, the fixed bed is examined in respect to its flow patterns, as a vessel, irrespective of the specific chemical reaction or physical phenomenon that takes place in it.

4.2.3.1 Peclet Number and Liquid Holdup

A fixed bed is an array of voids into which fluid flows at relatively high velocity, accelerating in the ports created by particle–particle intersections and decelerating upon entering the voids, leading to a certain degree of mixing. In the ideal case of plug flow in fixed beds, axial mixing between the several cross sections of the bed is minimal, whereas radial mixing in each section is maximal [3, 5]. Ideal flow is studied and represented using the classic dispersion or dispersed plug-flow model of Levenspiel [21]. Recall the material balance of a fixed-bed reactor (Eq. 4.31):

$$-\rho_b \cdot \frac{Z}{u_s} \cdot \frac{\partial \overline{q}}{\partial t} = \varepsilon \cdot \frac{Z}{u_s} \cdot \frac{\partial C}{\partial t} + \frac{\partial C}{\partial \left(\frac{x}{Z}\right)} - \frac{D_{ax} \cdot \varepsilon}{u_s \cdot Z} \cdot \frac{\partial^2 C}{\partial \left(\frac{x}{Z}\right)^2}.$$

The last term of this equation is zero in both cases of plug and mixed flow: in plug flow, $D_{ax} = 0$, and in mixed flow, $\frac{\partial^2 C}{\partial \left(\frac{x}{Z}\right)^2} = 0$. The physical parameter relevant to the flow quality is the axial dispersion coefficient, incorporated into the particle Peclet number:

$$Pe_p = \frac{u \cdot d_p}{D_{ax}}. \tag{4.37}$$

where d_p is the particle diameter, D_{ax} is the axial dispersion coefficient, and u is the interstitial fluid velocity (equal to u_s/ε). Multiplying this number with the term Z/d_p, where Z is the fixed-bed length, we obtain the bed Peclet number. In the case of fixed beds, the higher the vessel Peclet number, the better flow quality, and if close or higher than about 100, the flow is considered to be ideal (plug flow) [3].

Another critical parameter in fixed beds is the liquid holdup, representing the part of the bed occupied (wetted) by the liquid phase. This is important especially in downflow operation, as in upflow operation the liquid occupies practically the whole external free void volume of the bed. Total liquid holdup h_t consists of two parts: static h_s and dynamic holdup h_d. Static holdup is related to the volume of liquid which is adherent to the particles surface, whereas dynamic holdup is related to the flowing part of the liquid [5]. Liquid holdup (h_t, h_d, or h_s) based on the total volume of the bed occupied by liquid h_e and liquid holdup based the void volume of the bed occupied by liquid h_v are related as follows:

$$h_v = \frac{V_{liquid}}{\varepsilon \cdot V_{totalbed}} = \frac{1}{\varepsilon} \cdot \left(\frac{V_{liquid}}{V_{totalbed}} \right) = \frac{h_e}{\varepsilon}. \tag{4.38}$$

Then, $0 \leq h_e \leq \varepsilon$ while $0 \leq h_v \leq 1$. Several correlations for the determination of Peclet number and liquid holdup are given in Sect. 4.3.5.2. It is worthwhile to mention that in the case of partial wetting of the bed ($h_e \leq \varepsilon$), h_e should be used instead of ε in all model equations.

4.2.3.2 Fixed-Bed Sizing and Geometry

Fixed-bed geometric analogies are essential for a good hydrodynamic behavior in both laboratory- and full-scale units, especially during scale-up where in order to "transfer" experimental data from the small to the large unit, maldistribution and liquid holdup should be kept in a satisfactory level and similar in both scales. In particular, the following analogies are of the most important for avoiding large-scale flow maldistribution [3, 5]:

$$\frac{Z}{D} \geq 5 \tag{4.39}$$

$$\frac{D}{d_p} \geq 12 - 30 \tag{4.40}$$

$$\frac{Z}{d_p} \geq 50 - 150. \tag{4.41}$$

where D is the bed diameter, Z is the bed height, and d_p is the particle diameter. Relationship (4.40) is relevant to the bed porosity, which is practically constant for low d_p/D (< 0.1).

In order to avoid extended friction between the packing material in both down and upflow operations, a maximum linear velocity is defined [9]. This velocity is 0.8 times the minimum fluidization velocity for upflow operation and 1.8 times the same velocity for downflow operation. The minimum fluidization velocity can be evaluated using the following equation [4]:

$$\frac{150 \cdot \mu \cdot u_{fm}}{\Phi_s{}^2 \cdot d_p{}^2} \cdot \frac{1-\varepsilon}{\varepsilon^3} + \frac{1.75 \cdot \rho \cdot u_{fm}{}^2}{\Phi_s \cdot d_p} \cdot \frac{1}{\varepsilon^3} = g \cdot (\rho_p - \rho). \tag{4.42}$$

where d_p is the particles diameter, ρ_p and ρ are the density of the particles and of the liquid, respectively, g is the gravity acceleration constant (9.81 m^2/s), μ is the

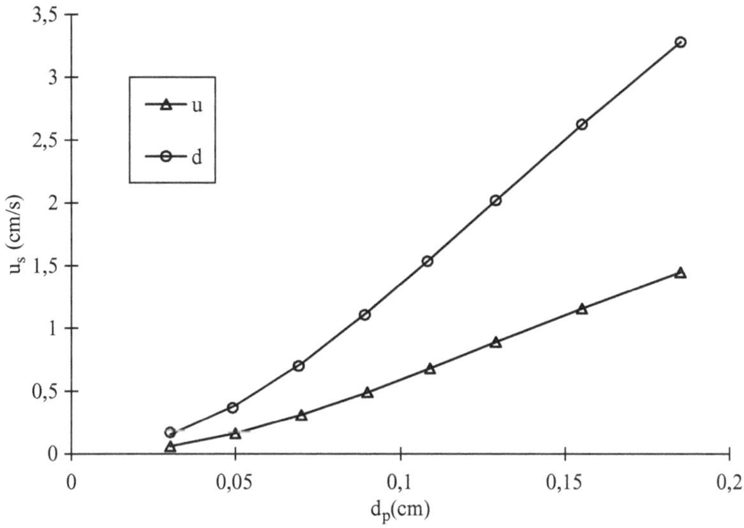

Fig. 4.4 Maximum linear velocity versus particle diameter (u upflow, d downflow) for $\rho_p = 2.08$ g/cm^3, $\mu = 0.9\ cP$, $\rho = 1\ g/cm^3$, $\varepsilon = 0.5$, and $\Phi = 0.65$ (for sand-type granular materials)

dynamic viscosity of the liquid, ε is the bed voidage, u_{fm} is the minimum fluidization velocity, and Φ_S is the sphericity of the particle. For granular particles, sphericity is between 0.6 and 0.95, while for spherical particles, it is equal to 1 [22]. In Fig. 4.4, the maximum linear velocity versus particle diameter is presented.

In the typical system shown in Fig. 4.4, it is clear that the smaller the particles, the lower the permissible operating liquid velocity, especially in the upflow mode. The operating velocity could be higher under the condition that larger particles are used. However, as is well known, by increasing particle size, the adsorption and ion exchange rates are decreased [3, 9, 15]. In laboratory fixed beds, the use of small particles and low velocity is not a problem, but during scale-up, in order to keep the same contact time and particle size in the large bed, inevitably the liquid velocity is increased, and there is the possibility to exceed the maximum permissible limit. In this case, it could be the case of operating in downflow mode as the maximum permissible velocity for the same particle size is higher, leading to a safer operation. Concluding, it is clear that particle size should be chosen considering kinetics and hydrodynamics in the same time for both scales (laboratory- and large-scale beds).

Another problem associated with small particle sizes is the increased pressure drop, which in downflow mode is possible to cause flooding. In ion exchange operations where pretreatment and/or regeneration steps are in use, by using the maximum permissible velocity and the contact time during pretreatment/regeneration (which is generally higher than the treatment one), the maximum bed height is determined as follows [3]:

$$Q_{FS} = a \cdot V_{FS} \Rightarrow u_{FS} = \frac{Q_{FS}}{A_{FS}} = \frac{a \cdot V_{FS}}{A_{FS}} \leq u_{\max,FS} \Rightarrow Z_{FS} \leq \frac{u_{\max,FS}}{a}. \tag{4.43}$$

where a is the number of bed volumes per hour (BV/h), subscript FS is for full-scale bed, u_{max} is the maximum allowable linear velocity, and A is the bed cross-sectional area. For safety, 85% of this height should be used. The fixed-bed height is also influencing the force applied on the particles in the bed bottom, which exhibit a certain level of particle strength. These particles are under the pressure of the upper layers as well as the hydraulic pressure of the liquid, and thus, a maximum permissible bed height is allowed [23]:

$$Z \leq \frac{F_c \cdot d_p \cdot \rho_b}{g \cdot M_k \cdot (\rho_b + \varepsilon \cdot \rho)}. \tag{4.44}$$

where Z is the bed height, ρ_b is the bed bulk density, M_k is the particle weight, and F_c is the maximum force that can be applied to a single particle without breaking it (particle strength).

4.3 The Use of Predictive Models

4.3.1 Models for Single Controlling Step

Predictive models could be used in order to model the process and determine the controlling step, which for the most cases in ion exchange and liquid-phase adsorption is the particle diffusion within the solid matrix. Nevertheless, the controlling step depends on liquid velocity because in low velocities, the liquid film diffusion step could influence the process. In order to use simplified models, the following assumptions should be met [3]:

(a) Plug flow. In this case, the last term in Eq. 4.32 is neglected. This assumption holds only if the bed Peclet number is greater than 100 [21]. Generally, in upflow operation, the quality of the flow is better, especially at low velocities. For complete analysis of fixed-bed hydrodynamics, see Sect. 4.2.3.

(b) Constant-pattern condition. With a favorable isotherm and a mass-transfer resistance or axial dispersion, the concentration front approaches a constant pattern, which is an asymptotic shape beyond which the wave will not spread. The wave is said to be "self-sharpening" [4]. In contrast, if the concentration front is seen to spread out more and more as it propagates toward the column outlet, it is a dispersive or spreading behavior (Figs. 4.5 and 4.6). Constant-pattern condition reduces the continuity Eq. 4.32 to the simple relation: $C/C_o = \overline{q}/q_{max}$, or, to put it in words, the dimensionless profiles of concentration and exchanger loadings are identical. This means that all points on the breakthrough curve are moving in the bed under the same velocity, and thus a constant shape of this curve is established [5]. Practically, the constant-pattern assumption holds if the

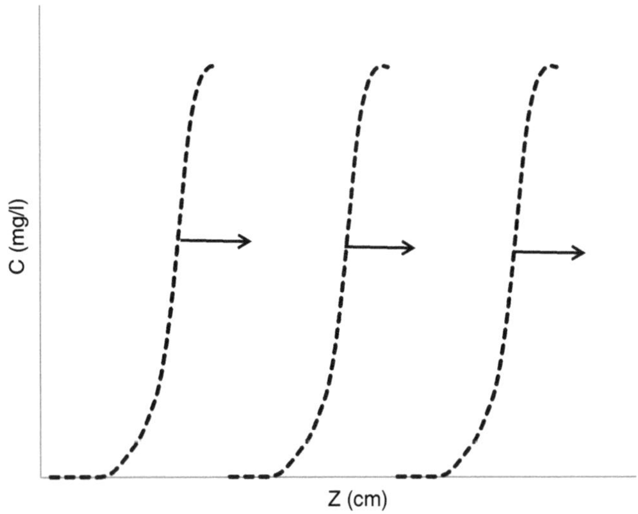

Fig. 4.5 Self-sharpening behavior

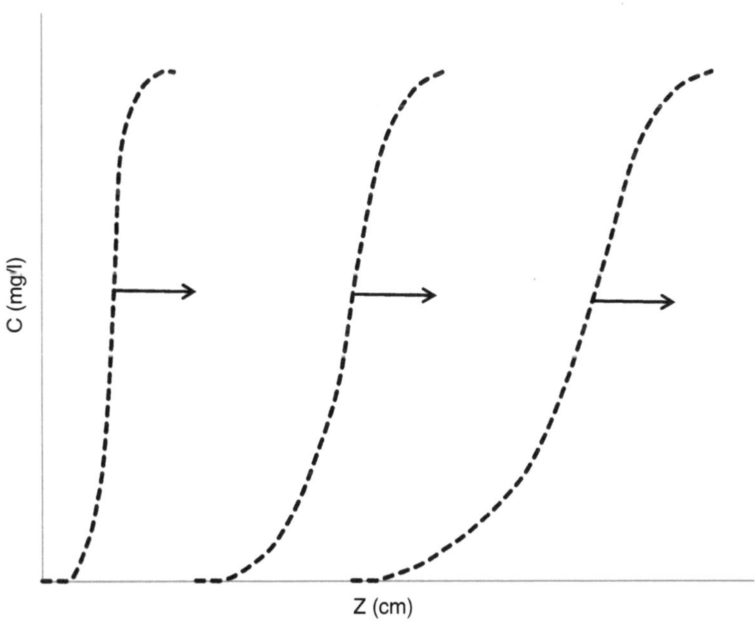

Fig. 4.6 Spreading behavior

equilibrium is favorable and under high residence times, i.e., deep-bed and low superficial velocity [4]. However, constant-pattern assumption is weak if the system exhibits very slow kinetics [24].

Simplified models under the above assumptions have been proposed and analyzed in the related literature and are in the form of either arithmetic or analytical solutions [5, 9]. The following dimensionless parameters are defined [4]:

$$\Lambda = \frac{\rho_b \cdot q_{\max}}{C_o} \tag{4.45}$$

$$T = \frac{t - \left(\frac{V_o \varepsilon}{Q}\right)}{\left(\frac{\Lambda \cdot V_o}{Q}\right)} \tag{4.46}$$

$$N_f = \frac{k_f \cdot a_u \cdot V_o}{Q} \tag{4.47}$$

$$N_s = \frac{15 \cdot D_s \cdot \Lambda \cdot V_o}{{r_p}^2 \cdot Q} \tag{4.48}$$

$$N_p = \frac{15 \cdot D_p \cdot (1 - \varepsilon) \cdot V_o}{r_p^2 \cdot Q}. \tag{4.49}$$

where t is time, V_o is the total bed volume, ε is the bed voidage, Q is the volumetric flow rate, ρ_b is the bed density, and q_{max} is the operating capacity, which should be evaluated from bed data in dynamic experiments [6]. Subscript f refers to the fluid, s to the solid, and p to the pore phase resistance. The following equations are approximate solutions of fixed-bed model under the constant-pattern and plug-flow assumption for favorable Langmuir isotherm and linear driving forces [3, 4]:

$$N_f \cdot (T - 1) = \frac{\ln(X) - La \cdot \ln(1 - X)}{1 - La} + 1 \tag{4.50}$$

$$N_s \cdot (T - 1) = \frac{1}{\Psi_S} \cdot \left[\frac{La \cdot \ln(X) - \ln(1 - X)}{1 - La} - 1\right] \tag{4.51}$$

$$\Psi_S = \frac{0.894}{1 - 0.106 \cdot La^{0.5}}. \tag{4.52}$$

Equation 4.50 is for liquid film diffusion control and Eq. 4.51 for solid diffusion control. In Fig. 4.7, Eq. 4.51 is plotted for two different values of the Langmuir constant ($La = 0.1$ and 0.7), and in Fig. 4.8, both Eqs. 4.50 and 4.51 are plotted for $La = 0.3$.

From Figs. 4.7 and 4.8, it is evident that the more favorable the isotherm (lower value of La), the steeper the breakthrough curve (the performance of the fixed bed is better) while a tail is formed, either at the start (liquid film control) or at the end (solid diffusion control) of the curve.

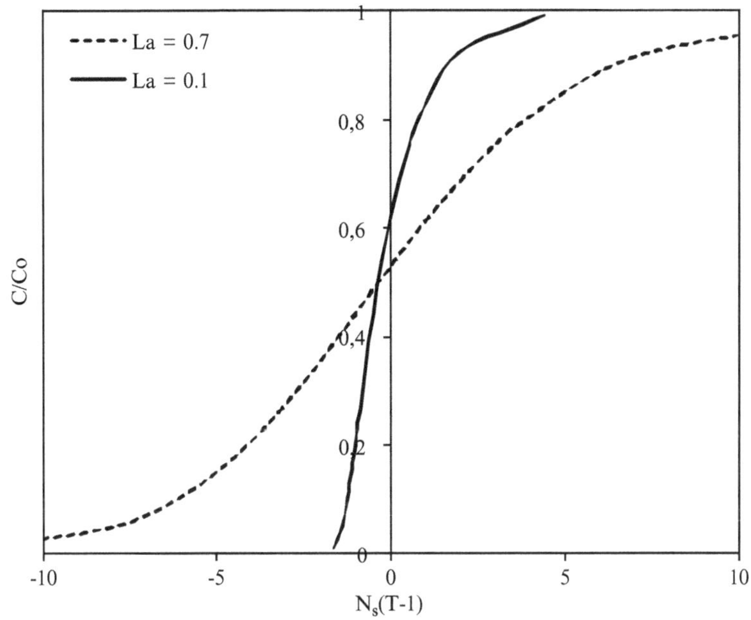

Fig. 4.7 Solid diffusion control model for *La* 0.1 and 0.7

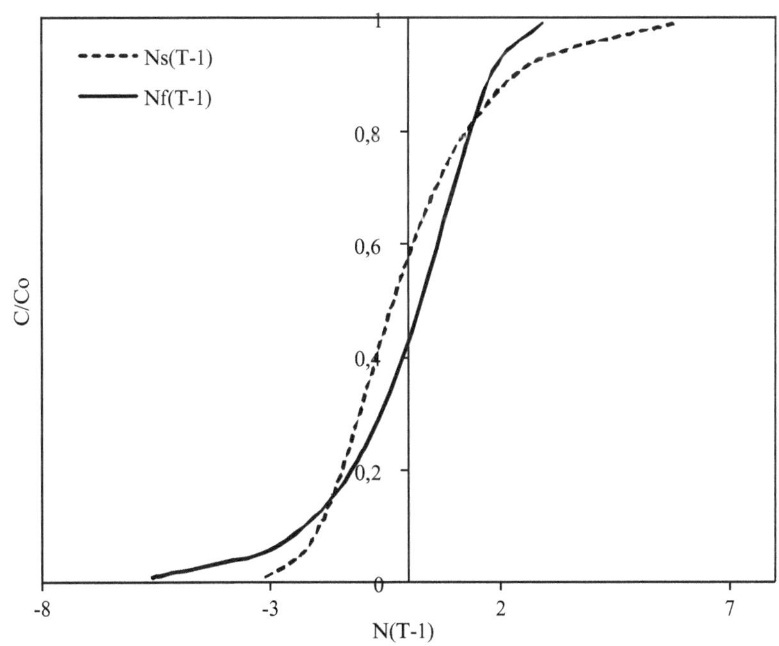

Fig. 4.8 Solid and film diffusion control models for *La* = 0.3

The following equations are based on nonlinear driving forces and Langmuir isotherm, the first for pore diffusion control and the second for solid diffusion control. For the case of pore diffusion control, the equations are [3, 5]

$$N_p \cdot (T-1) = \frac{1}{\psi_{pore}} \cdot \left(-2 \cdot \sqrt{1-(1-La)\cdot X} - \frac{La}{1-La} \cdot \ln\rho_1 + \frac{\sqrt{La}}{1-La} \cdot \ln\rho_2\right) + I_c \tag{4.53}$$

$$\rho_1 = \frac{1+\sqrt{1-(1-La)\cdot X}}{1-\sqrt{1-(1-La)\cdot X}} \tag{4.54}$$

$$\rho_2 = \frac{\sqrt{1-(1-La)\cdot X}+\sqrt{La}}{\sqrt{1-(1-La)\cdot X}-\sqrt{La}} \tag{4.55}$$

$$I_c = 2.44 - 2.15 \cdot La \tag{4.56}$$

$$\psi_{pore} = \frac{1}{La^2 + 1.83 \cdot (1-La)^{0.92}}. \tag{4.57}$$

For solid diffusion control, the equations are [5]

$$N_s \cdot (T-1) = \frac{1}{\psi_{solid}} \cdot \left[\frac{1}{1-La^2} \cdot (\lambda_1 + \lambda_2 + \lambda_3)\right] + I_c \tag{4.58}$$

$$\lambda_1 = \ln\left[\frac{1}{(1+La) - 2 \cdot La \cdot X - (1-La) \cdot X^2}\right] \tag{4.59}$$

$$\lambda_2 = La \cdot \ln\left(\frac{X + \frac{1+La}{1-La}}{1-X}\right) \tag{4.60}$$

$$\lambda_3 = 2 \cdot La^2 \cdot \ln X \tag{4.61}$$

$$\psi_{solid} = \frac{1}{La^{1.5} + 1.688 \cdot (1-La)} \tag{4.62}$$

$$I_c = -106.67 \cdot La^5 + 177.82 \cdot La^4 - 114.38 \cdot La^3 + 31.604 \cdot La^2 - 4.5323 \cdot La - 0.8408. \tag{4.63}$$

Both models are satisfactory approximations of the respective numerical solutions, especially when the interest is in the first part of the breakthrough curve, which is of great importance in industrial applications [23]. Finally, the following equation is a solution of the fixed-bed model under the constant-pattern and plug-flow assumption, for fluid-film diffusion control and favorable Freundlich isotherm [5]:

$$N_f(T-1) = 1 + \ln(X) - \frac{Fr}{1-Fr} \cdot \ln\left(1 - X^{\frac{1-Fr}{Fr}}\right) + \omega \tag{4.64}$$

$$\omega = \frac{Fr}{Fr-1} \cdot \sum_{k=1}^{\infty} \frac{Fr}{k \cdot [k \cdot (1-Fr) + Fr]}. \tag{4.65}$$

4.3.2 Models for Combined Controlling Steps

For combined diffusion resistances (fluid film and solid diffusion), the concentration in the bulk liquid phase is different than this on interface due to the effect of the fluid-film resistance. The following equations can be used for Langmuir and Freundlich equilibrium equations as published by Miura and Hashimoto [3, 5, 25].

Langmuir isotherm

$$\theta_\tau - X_\tau = \frac{1}{1+\zeta} \cdot \varphi_1 + \frac{\zeta}{1+\zeta} \cdot \frac{1}{\eta} \cdot \varphi_2 \tag{4.66}$$

$$\varphi_1 = \frac{1}{1-La} \cdot \ln X_i - \frac{La}{1-La} \cdot \ln(1-X_i) - \ln[La + (1-La) \cdot X_i] - \frac{La}{1-La} \cdot \ln La + 1 \tag{4.67}$$

$$\varphi_2 = \frac{La}{1-La} \cdot \ln X_i - \frac{1}{1-La} \cdot \ln(1-X_i) - 1 \tag{4.68}$$

$$\eta = 1 - 0.192 \cdot (1-La)^3. \tag{4.69}$$

Freundlich isotherm

$$\theta_\tau - X_\tau = \frac{1}{1+\zeta} \cdot \omega_1 + \frac{\zeta}{1+\zeta} \cdot \frac{1}{\eta} \cdot \omega_2 + \frac{1}{1+\zeta} \cdot \frac{\zeta}{\zeta+\eta} \cdot \omega_3 \tag{4.70}$$

$$\omega_1 = \frac{Fr}{Fr-1} \cdot \ln\left(X_i^{Fr-1} - 1\right) + 1 + \frac{\eta}{\zeta+\eta} \cdot \frac{Fr^2}{Fr-1} \cdot I_A \tag{4.71}$$

$$\omega_2 = \frac{1}{Fr-1} \cdot \ln\left(1 - X_i^{1-Fr}\right) + \frac{\zeta}{\zeta+\eta} \cdot \frac{1}{Fr-1} \cdot I_B \tag{4.72}$$

$$\omega_3 = Fr - 1 + \frac{Fr}{Fr - 1} \cdot (I_A + I_B) \tag{4.73}$$

$$\eta = 0.808 + 0.192 \cdot Fr \tag{4.74}$$

$$I_A = \sum_{n=1}^{\infty} \frac{1}{n \cdot [n \cdot (1 - Fr)] + Fr} \tag{4.75}$$

$$I_B = \sum_{n=1}^{\infty} \frac{1}{n \cdot [n \cdot (1 - Fr)] + 1} \tag{4.76}$$

where

$$\theta_\tau = \frac{k_s \cdot a_u}{\rho_b \cdot (1 + \frac{1}{\zeta})} \cdot \left(t - \frac{\varepsilon \cdot Z}{u_s} \right) \tag{4.77}$$

$$X_\tau = \frac{k_s \cdot a_u \cdot \gamma}{1 + \frac{1}{\zeta}} \cdot \frac{Z}{u_s} \tag{4.78}$$

$$k_s \cdot a_u = \frac{15 \cdot D_s \cdot \rho_b}{r_p^2} \tag{4.79}$$

$$\gamma = \frac{q_{\max}}{C_o} \tag{4.80}$$

$$\zeta = \frac{k_f \cdot a_u}{k_s \cdot a_u \cdot \gamma} \tag{4.81}$$

$$a_u = \frac{3}{r_p} \cdot (1 - \varepsilon). \tag{4.82}$$

where q_{max} is in mass of solute per unit mass of solid and C_o is in mass of solute per unit volume of fluid. For practical use, the infinite series I_A and I_B are shown in Fig. 4.9.

The interface relative concentration X_i is related to the respective bulk relative concentration X as follows [5]:

Langmuir

$$X = \frac{(\zeta \cdot La + \eta) \cdot X_i + \zeta \cdot (1 - La) \cdot X_i^2}{(\zeta + \eta) \cdot [La + (1 - La) \cdot X_i]}. \tag{4.83}$$

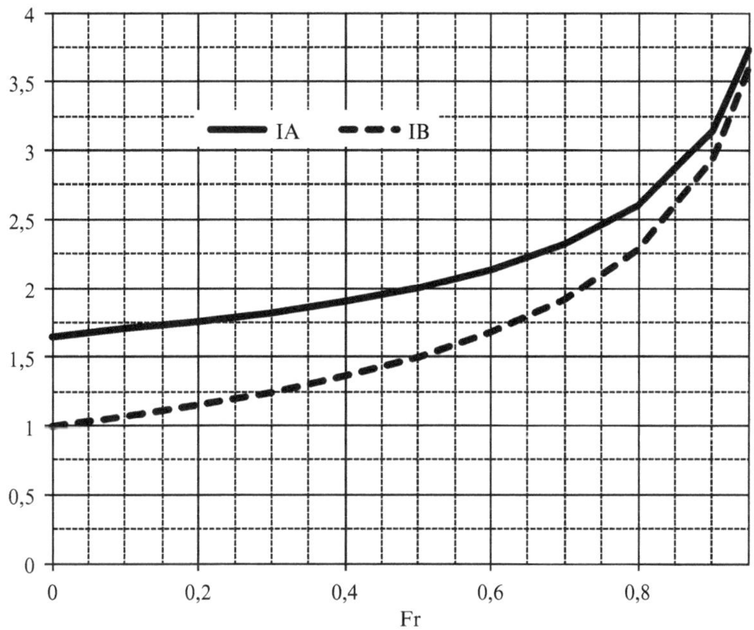

Fig. 4.9 Values of (I_A) and (I_B)

Freundlich

$$X = \frac{\zeta \cdot X_i + \eta \cdot X_i^{Fr}}{\zeta + \eta}. \tag{4.84}$$

The equations provide only the X_i versus θ_τ-X_τ relation, while for a direct relationship between X and θ_τ-X_τ, X_i should be explicitly expressed by using only X. However, there is not an explicit relationship for Freundlich isotherm and, in this case, an iteration procedure is needed. For Langmuir isotherm,

$$X_i = \frac{-b + \sqrt{b^2 - 4 \cdot a \cdot c}}{2 \cdot a} \tag{4.85}$$

where

$$a = \zeta \cdot (1 - La) \tag{4.86}$$

$$b = (\zeta \cdot La + \eta) - (\zeta + \eta) \cdot (1 - La \cdot X) \tag{4.87}$$

$$c = -La \cdot (\zeta + \eta) \cdot X. \tag{4.88}$$

The same equations can be used as approximations for pore diffusion control, by setting $k_s \cdot a_u$ equal to $\frac{15 \cdot D_p \cdot (1-\varepsilon)}{\gamma \cdot r_p^2}$.

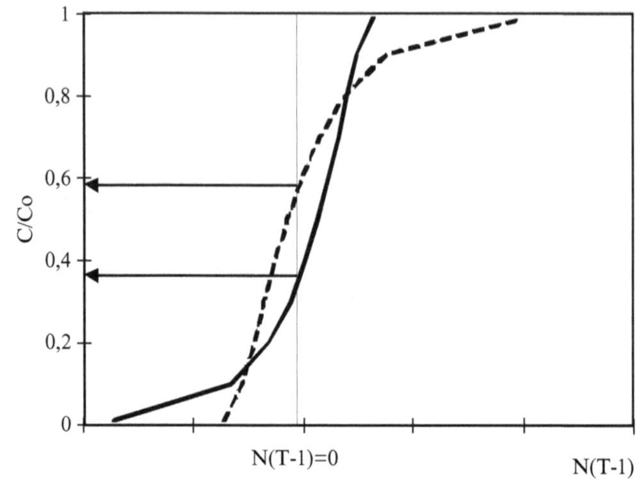

Fig. 4.10 Characteristic C/C_o versus *N·(T-1)* curves for solid diffusion control (*dotted line*) and liquid film diffusion control ($La = 0.05$)

4.3.3 Approximate Methods for the Determination of the Controlling Mechanism

If the target is the determination of the controlling mechanism, another useful method is the following [3, 23]. This method is approximate and is based on the experimental breakthrough curve. Based on these data, it is easy to construct the *T-1* versus C/C_o curve, where C is the exit concentration, C_o the inlet concentration, and T is the dimensionless time modulus. If the experimental data are plotted in *N·(T-1)* versus C/C_o graph, the C/C_o at which $N \cdot (T\text{-}1)$ equals to zero is called stoichiometric point and is independent of volumetric flow rate [4]. The characteristic C/C_o versus *N·(T-1)* curves are shown in Fig. 4.10.

The stoichiometric point is always lower than about $C/C_o = 0.7$, regardless of the controlling mechanism for both Langmuir and Freundlich isotherms [3, 4, 18, 19]. Furthermore, the $(C/C_o)_{stoich}$ values are between 0.51 and 0.70 for solid diffusion control, and the more favorable equilibrium, the higher the $(C/C_o)_{stoich}$. The opposite holds for liquid film control where the corresponding $(C/C_o)_{stoich}$ values are between 0.31 and 0.5. Using the approximate models (Eqs. 4.50, 4.51), the X_{stoich} versus La is shown in Fig. 4.11.

As an alternative method, Miura and Hashimoto defined the mechanical parameter ζ, as follows [25]:

$$\zeta = \frac{k_f \cdot a_u}{K \cdot \left(\frac{q_{\max}}{C_o}\right)} \tag{4.89}$$

$$K = \frac{15 \cdot D_s \cdot \rho_b}{r_p^2}. \tag{4.90}$$

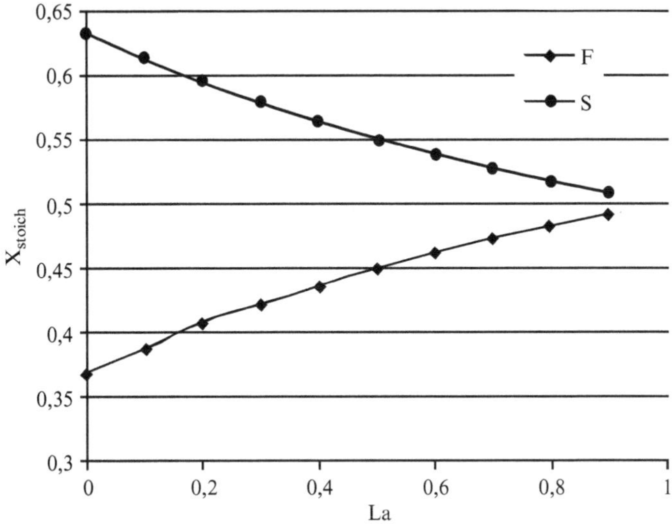

Fig. 4.11 Stoichiometric point curves (S solid diffusion control, F liquid film diffusion control)

If ζ is zero (<< 1), then liquid film diffusion is controlling the process rate, while if ζ is infinite (>> 1), then the solid diffusion is controlling the process rate. Essentially, the mechanical parameter represents the ratio of the diffusion resistances (solid and liquid film). This equation can be used irrespective of the constant-pattern assumption.

If modeling or other approximate methods are not applicable, then a series of experimental runs are conducted in order to examine the effect of liquid velocity on the bed performance and on the breakpoint in particular. Then, by keeping the same contact time and particle size, the effect of liquid velocity can be examined by changing the bed length: liquid velocity has no effect in case of solid diffusion control while the opposite holds for film diffusion control. That way, a minimum liquid velocity is calculated, above which the controlling mechanism is solid diffusion control. Thus, for higher liquid velocity, which is expected in large-scale units, it is guaranteed that the controlling step is solid diffusion, unaltered in all bed sizes. This procedure is useful only if the flow conditions are close to ideal plug flow as in a different case, changes in the liquid velocity are affecting the flow quality and consequently the bed performance.

4.3.4 Equilibrium Limited Systems

In the case of unfavorable equilibrium, the local equilibrium analysis could be applied. Local equilibrium is a condition under which the mass transport step is neglected as negligible and equilibrium is established between the liquid and solid phase. In this case, equation of continuity reduces to [3, 4]

$$\frac{dY^*}{dX} = T \Leftrightarrow \frac{dF(X)}{dX} = T. \tag{4.91}$$

where $Y^* = F(X)$ is the equilibrium relationship between Y and X. The limits of validity of Eq. 4.91 can be found setting $X = 0$ (for T minimum) and $X = 1$ (for T maximum). Finally, using the local equilibrium analysis, the isotherm of a system can be found from breakthrough experiments using the following equation:

$$Y^* = \int_0^1 T \cdot dX. \tag{4.92}$$

In the case of favorable equilibrium, local equilibrium analysis predicts that at $T = 1$, the concentration X will rise instantly from 0 to 1 (ideal step change). This situation is ideal and does not correspond to real situations, as when a system exhibits favorable equilibrium, mass transfer always is controlling the rate.

4.3.5 Derivation of Basic Model Parameters

4.3.5.1 External Mass-Transfer Coefficient

Liquid mass-transfer coefficient or convection coefficient is a flow-dependent parameter, which is evaluated by use of empirical and semiempirical correlations. One of the most well-known correlations for use in water treatment situations is the following [3, 26]:

$$Sh = \left(2 + 0.644 \cdot \mathrm{Re}^{\frac{1}{2}} \cdot Sc^{\frac{1}{3}}\right) \cdot [1 + 1.5 \cdot (1 - \varepsilon)]. \tag{4.93}$$

In adsorption processes for wastewater treatment by use of granular activated carbon, the Williamson correlation is proposed [27]:

$$k_f = \frac{2.4 \cdot u_s}{Sc^{0.58} \cdot \mathrm{Re}^{0.66}}. \tag{4.94}$$

Kataoka correlation has been used in adsorption systems from liquid phase [28]:

$$k_f = 1.85 \cdot (\mathrm{Re} \cdot Sc)^{-\frac{2}{3}} \cdot \frac{u_s}{\varepsilon} \cdot \left(\frac{1 - \varepsilon}{\varepsilon}\right)^{\frac{-1}{3}}. \tag{4.95}$$

Finally, for very low Reynolds number (<0.01), the Wilson–Geankoplis correlation has been used in ion exchange as well as in adsorption from liquid phase [11, 29, 30]:

$$Sh = \frac{1.09}{\varepsilon} \cdot Sc^{\frac{1}{3}} \cdot \mathrm{Re}^{\frac{1}{3}}. \tag{4.96}$$

where

$$Sh = \frac{k_f \cdot d_p}{D_f} \tag{4.97}$$

$$\mathrm{Re} = \frac{d_p \cdot u_s}{v} \tag{4.98}$$

$$Sc = \frac{v}{D_f}. \tag{4.99}$$

where *Re* is the particle Reynolds number, *Sh* is the Sherwood number, *Sc* is the Schmidt number, D_f is the diffusion coefficient of the solute in the liquid phase, d_p is the particle diameter, and v is the liquid kinematic viscosity.

4.3.5.2 Peclet Number and Liquid Holdup

For materials that are frequently used in adsorption and ion exchange systems in wastewater treatment applications, as for zeolites and similar particles of irregular shape, the following equations can be used [3, 5, 31]:

$$Pe_p = L \cdot \mathrm{Re}^k \tag{4.100}$$

$$\%h_{v,t} = 21 + 99.72 \cdot u_s^{0.28}. \tag{4.101}$$

where v is the kinematic viscosity of the liquid phase, L is 0.52 for upflow and 0.05 for downflow, and k is -0.65 for upflow and 0.48 for downflow. In Eq. 4.101, total liquid holdup is expressed as percentage of the total void bed volume, while superficial velocity is in *cm/s*. Peclet correlation (Eq. 4.100) holds for $0.6 < Re < 8.5$ and liquid holdup correlation (Eq. 4.101) for particle size of 1.2–1.3 *mm*. For different particle size, the dynamic liquid holdup is evaluated using the following analogy:

$$\frac{\%h(d_1)}{\%h(d_2)} \cong \left(\frac{d_2}{d_1}\right)^m. \tag{4.102}$$

where m is 0.72. Several correlations for the Peclet number and liquid holdup can be found elsewhere [3, 32, 33]. In the equation above, the bed voidage is considered to be the same for different particle sizes, which is true for low d_p/D

values (see scale-up analysis), while the value proposed here is for activated carbon particles, which are of similar shape as zeolites [34, 35]. The constant part in liquid holdup correlation (Eq. 4.101) is the static liquid holdup. This part is between $0.11/\varepsilon$ and $0.072/\varepsilon$ for particle size between 0.2 and 2 *mm*, where ε is the bed voidage [23]. In the typical case of $\varepsilon = 0.4$–0.5, static holdup is between 14% and 27%. More correlations for liquid holdup are given in [5].

The bed Peclet number is evaluated as follows [21]:

$$Pe_L = Pe_p \cdot \frac{Z}{d_p}. \tag{4.103}$$

Generally, for spherical and other irregular-shaped particles (intalox saddles, raschig rings, berl saddles), particle Peclet number is found to be between 0.3 and 0.8 for Reynolds number between 0.01 and 150. For high Reynolds number, the following expression holds [5]:

$$Pe_p = \frac{1}{\varepsilon} \cdot \left(0.2 + 0.011 \cdot \mathrm{Re}^{0.48}\right). \tag{4.104}$$

This expression holds for $25 < Re < 320$. This correlation is derived using glass beads, aluminum beads, and steel beads, and it has been used for ion exchange systems by use of resins beads [11]. Furthermore, according to a literature review presented in [32], the particle Peclet number is between 0.06 and 0.3, showing no particular trend, for $0.01 < Re < 10$ and is steadily increased for $Re > 10$. More correlations for Peclet are given in [5]. For the specific case of irregular-shaped particles, a review is published by Inglezakis [33].

4.3.5.3 Experimental Data

In any design problem in engineering, preliminary experiments in laboratory-scale beds are always essential. In the laboratory, a series of measurements are collected for all mechanisms and phenomena that are independent of size. This applies in particular to everything related to thermodynamics and chemical kinetics [20]. The laboratory reactor should not necessarily be similar to the industrial one but has to be designed in order to give the best information [7].

Figures 4.12–4.15 show the relationships between areas determined by a breakthrough curve and the axes, plotted in terms of concentration versus effluent volume. These areas represent different elements of an overall material balance on a complete exchange operation and imply no assumption of equilibrium or kinetics [4, 15].

(a) The amount in *mg* of solute that has been leaked out of the column between effluent volumes 0 and V in excess of the initial interstitial solute is represented by the hatched area on Fig. 4.12 and given by the integral

$$L = \int_0^V C \cdot dV. \tag{4.105}$$

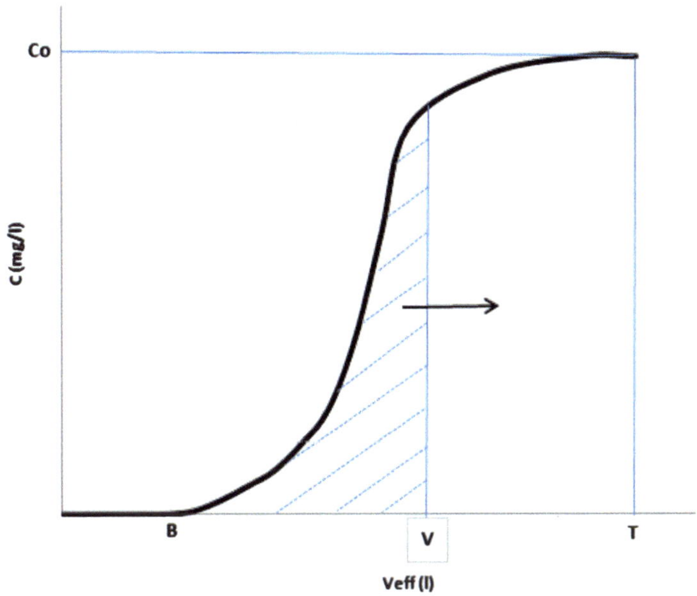

Fig. 4.12 Breakthrough curve

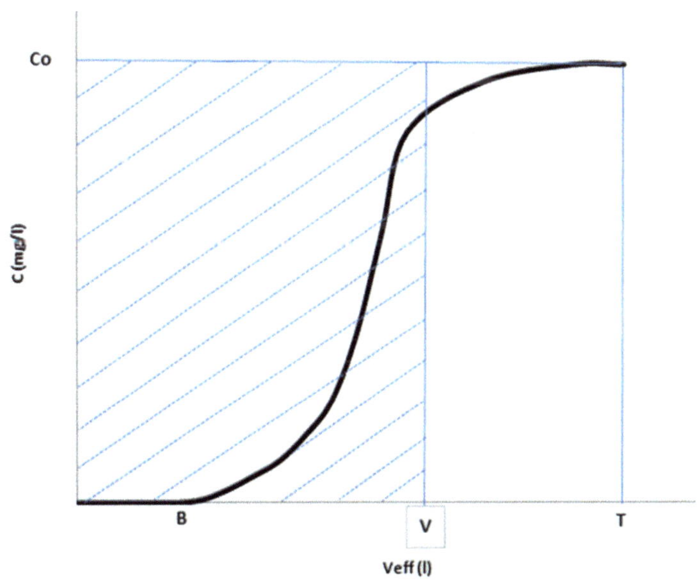

Fig. 4.13 Breakthrough curve

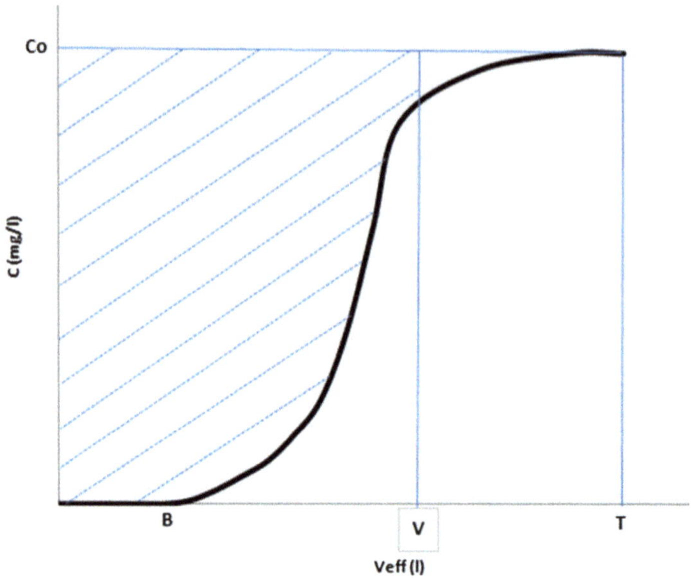

Fig. 4.14 Breakthrough curve

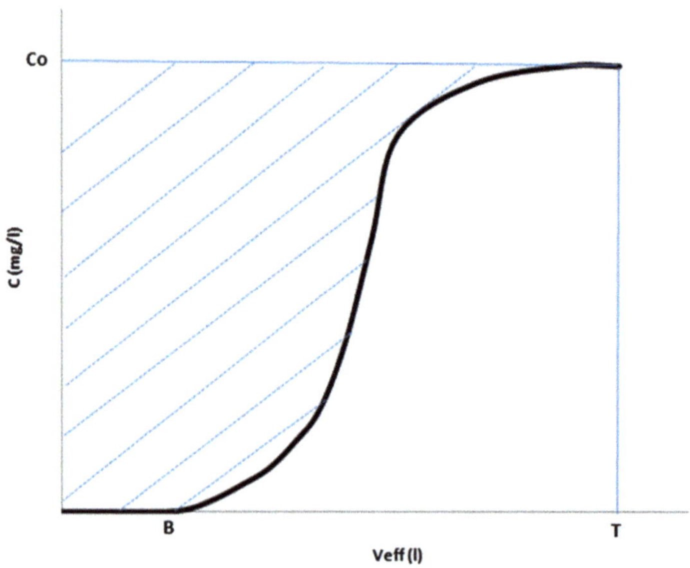

Fig. 4.15 Breakthrough curve

(b) The amount of solute in *mg* introduced into the column in excess of the initial content is simply (Fig. 4.13)

$$I = C_o \cdot V. \tag{4.106}$$

(c) The amount of solute in *mg* stored in the bed in this interval is the difference of the two preceding quantities (Fig. 4.14):

$$S = I - L = Co \cdot V - \int_0^V C \cdot dV. \tag{4.107}$$

(d) The exchange capacity available is the maximum amount (mg) of solute the column can store, accounting for its initial state and feed concentration C_o. This is obtained by letting V becoming very large (Fig. 4.15):

$$A = \int_0^\infty (Co - C) \cdot dV \cong Co \cdot V_T - \int_0^T C \cdot dV. \tag{4.108}$$

Then, the operational ion exchange capacity in *mg*/*g* is

$$q_{\max} = \frac{Co \cdot V_T - \int_0^T C \cdot dV}{\rho_b \cdot V_o}. \tag{4.109}$$

During experiments, it is essential to minimize the flow non-idealities; otherwise, the underlying phenomena could be masked. The most well-known method for hydrodynamics studies, which is basically the determination of the Peclet number and liquid holdup, is the tracing technique, and details are provided elsewhere [31, 33, 36]. It is always safer to run experimental laboratory-scale fixed beds in upflow mode as in small-scale beds downflow operation frequently suffers from severe liquid maldistribution and low liquid holdup [31]. Another advantage of upflow operation is that liquid holdup is near 100%, and thus the theoretical contact time equals the real contact time of the reactor. The minimal set of data needed for a shortcut design is the bed diameter D_{LS}, particle diameter d_p, relative volumetric flow rates *BV*/*h* for treatment β and pretreatment/regeneration α, and the achieved breakpoint volume (bed volumes, *BV*) for the selected volumetric flow rate (treatment) and breakpoint concentration. Obviously, fixed-bed operation is stopped at this point, and the performance of the unit is based on the achieved treated volume (breakpoint volume). Simple correlations between contact time and breakpoint volume could be valuable information in scale-up procedure. Breakthrough volume is considered to be a primary parameter for design purposes. In experimental section, it could be valuable to determine the following: the maximum adsorption capacity of the solid q_{max}, the diffusion coefficients of the solutes in the solid phase D_s, and the equilibrium isotherm. Generally, all the above measurements are made in batch systems using simplified kinetics models. The maximum adsorption (or ion exchange) capacity of the solid q_{max} is frequently measured separately by the "repeated equilibrations method" in batch systems [6, 37].

Theoretically, for a specific ion exchange or adsorption system, the equilibrium isotherm, ion exchange capacity, and solid diffusion coefficient are independent of the experimental method used for their determination and independent of flow conditions [5, 15]. However, a number of experiments have shown that the maximum adsorption capacity, the equilibrium isotherm, and the diffusion coefficient measured in fixed beds and batch systems could be different, and they might be flow dependent [6, 11, 38–41]. All these observations seem to be a result of the limiting (low) contact time in fixed beds and of the different concentration gradients in different reactor arrangements, and it is true that batch reactors do not approximate the hydrodynamic and contaminant removal patterns of fixed-bed reactors [40]. In case resins are used as packing material, another explanation could be the lack of adequate mechanical strength of particles and permeability of particles to fluid flow [42]. The dependence of q_{max} on residence time could be expressed as follows [23, 38]:

$$q^t_{\max} = q^0_{\max} \cdot [1 - \exp(-\omega \cdot \tau)] \tag{4.110}$$

$$q^t_{\max} = q^0_{\max} \cdot \left[1 - \exp\left(-\omega \cdot \tau^{0.5}\right)\right]. \tag{4.111}$$

where $q^t_{\max}$ and $q^o_{\max}$ are the operational bed capacity and the maximum exchange level (measured in batch reactor systems), τ is the residence time, and ω is a system-specific constant. It is obvious that if residence time is infinite, the operational bed maximum capacity is equal to *MEL*, which is theoretically expected, as noticed elsewhere [6].

Concluding, solid-phase capacity should be experimentally measured in fixed-bed experiments, and if models are applicable, diffusion coefficients should also be determined in fixed-bed apparatus. However, equilibrium isotherms require extended experimental data and are quite difficult to be conducted in fixed-bed reactors, and by this point of view, it is more practical to be studied in batch reactor systems. In the case that such isotherms are used in fixed-bed models, the equilibrium discrepancy (if exists) can be compensated by a different estimate for the solid diffusion coefficient [6, 38].

4.4 Scale-Up

4.4.1 General Analysis

As is well known, while the rate (chemical kinetics) of a given system is independent of the size and the structure of the reactor, the physical processes involved in the overall rate, for example, mass and heat transfer, are usually controlled by these factors [4]. Nevertheless, it is true that scale-up has better chances of success if laboratory- and large-scale units are carried out in the same type of reactor, for example, fixed bed, fluidized bed and batch reactors [43, 44]. Then, by keeping the same reactor structure, the critical parameter is the reactor size, which incorporates its geometrical analogies and dimensions. In the particular case of wastewater treatment in ion exchange and adsorption fixed beds, no real chemical reaction or heat transfer occurs; it is a crucial subject as these operations are essentially isothermal. Thus, scale-up mainly considers the mass-transfer phenomena and the hydrodynamic performance of the reactor, i.e., liquid distribution, small-scale dispersion of the flow, and liquid holdup.

From a practical point of view, a distinction should be made between a system that can be fully modeled mathematically, for example, treatment of one-component solutions in fixed beds, and a system that such a modeling is not practically feasible, for example, treatment of a complicated multicomponent solution in the same operation. In the case where full modeling is applicable, scale-up can be based on the model, keeping adequate flow conditions and geometric analogies in order to avoid extensive particle attrition, partial fluidization, and large-scale maldistribution of the flow as channeling, dead spaces, and fluid recirculation [43].

Even in one-component systems because of the nonlinearity associated with the equilibrium expressions, a full solution of the resulting partial differential equations of the models requires complicated numerical solutions. In our days with the aim of computers and sophisticated software, these solutions are easily found. However, to obtain a priori prediction of the breakthrough curve, independent experiments and/or reliable engineering correlations are required to estimate the numerous equilibrium, transport, and sorption kinetic parameters. As a result, some of the model parameters are evaluated by fitting the model to experimental breakthrough curves. Such multiparameter fitting may reduce the physical significance of the parameters [43, 45]. By this point of view, scale-up is often following other routes.

In the case where modeling is not applicable, apart from flow conditions and geometric considerations, scale-up should also be based on some additional rules, called similitude rules. Then, under specified conditions, the data of a laboratory-scale unit can be used for the evaluation of the large-scale unit performance. The similitude rules can be withdrawn using the continuity equation, rates equations, and hydrodynamic equations (for $\%h$ and Pe_p). From continuity equation, it is clear that for the same ion exchange or adsorption system and the same C_o, q_{max}, the exit concentration C at each time interval t will be the same for any bed size if the equation terms are the same. These terms are shown in Table 4.2.

In many cases, not all of the parameters in Table 4.2 are critical. In Table 4.3, the most critical comments are summarized.

Table 4.2 Parameters derived from the analysis of model equations [5, 43]

Parameters	Equation	Parameter is a function of
ε	Continuity equation (4.32)	D/d_p
Z/u_s		–
Pe_L		Pe_p, Z/d_p or Z/d_p, Re_p
D_s	Rate equation for solid diffusion (4.34)	–
R (or d_p)		–
k_f	Rate equation for liquid film diffusion (4.33)	ε and Re_p
a_u		d_p, ε, and particle shape
$h\%$	Liquid holdup equations (e.g., 4.101)	u_s, d_p, ε
Equilibrium behavior	Isotherm (–)	Possibly influenced by contact time Z/u_s

Table 4.3 Comments on the critical design parameters [5, 43]

Parameter	Comments
Z/u_s	Critical
Z/d_p	Minimal effect if is higher than 150
D/d_p	Minimal effect if is higher than 30
Re_p	Minimal effect if solid diffusion is the controlling mechanism and the bed is operated in near plug-flow mode (or for Z/d_p higher than 150)
d_p	Critical
E	Minimal effect if d_p/D is lower than 0.1
u_s	Minimal effect if solid diffusion is the controlling mechanism and the unit is operated in upflow mode. The same for downflow mode if the liquid holdup is 100%

4.4.2 Fixed-Bed Scale-Up

There are some basic scale-up rules for each operation. In fixed bed, adsorption and ion exchange operations are scaled-up from laboratory tests in small-diameter bed, and the large unit is designed for the same superficial velocity and particle size [22, 43]. Particle size should be kept the same, as it is well known that it affects the rate of the adsorption and ion exchange [5, 9, 15]. Liquid velocity is considered as an important parameter also in other studies. As is pointed out, due to geometric considerations it is not possible to keep linear velocity constant during scale-up, and only if the pilot plant has a very high ratio of height to diameter, compared to the large unit, the linear velocity could be similar in the two units. However, contact time is more critical than superficial velocity and is incorporated in the continuity equation, representing the physical contact time allowed for the two phases (liquid/solid) to interact. The physical meaning of this is that contact time is the time allowed for the system to react. However, it is known that in some systems, contact time is possibly influencing the equilibrium, leading to a "partial" equilibrium behavior in the bed, at least in reversible and of low kinetics liquid-phase adsorption and ion exchange processes [6]. This effect represents another "non ideality" not incorporated in the continuity Eq. 4.32. Concluding, fixed-bed scale-up is based on the same contact time and particle size [43].

Constant contact time has been used as an essential factor for scale-up of adsorption and ion exchange fixed beds, provided the degree of maldistribution of the flow is very small [46, 47]. Contact time is also used as the basic scale-up factor in two-phase flow (gas–liquid) fixed beds in the common case of liquid reactant limited systems [48, 49]. Again, the full-scale fixed bed, being taller than the experimental pilot or laboratory one, at the same contact time (or space velocity) has a much larger liquid mass velocity which causes improved contacting efficiency.

Contact time (or space time) is determined by the linear velocity as follows [43]:

$$\tau = \frac{\varepsilon}{I} = \frac{V_o \cdot \varepsilon}{Q} = \frac{Z \cdot \varepsilon}{u_s}. \tag{4.112}$$

where I is the relative volumetric flow rate commonly expressed in empty bed volumes per hour (BV/h). Essentially, I is equal to the reactor space velocity [21].

From a hydrodynamics perspective by keeping the same contact time liquid velocity is much higher in the large fixed bed. As a consequence, the higher velocity minimizes the diffusion resistance in the liquid film, leading in parallel the system closer to solid diffusion control and results in higher liquid holdup (for downflow operation) and higher Pe_L, provided that liquid dispersion in the top of the fixed bed is well designed. Under these conditions, the performance of the bed is expected better, i.e., higher breakthrough volume will be achieved. In systems that solid diffusion controls the uptake rate, as in the case of adsorption and ion exchange, the influence of superficial velocity on the bed performance is expected minimal, provided that the flow is near plug flow and the liquid holdup is near to 100% [43]. In these cases, if scale-up is based on the same contact time, the experimental results from laboratory-scale bed can be directly used in large bed. For example, if contact time-breakpoint volume relationships derived from the laboratory bed experiments are available, the relationship can be safely used to predict the performance of the large bed. Nevertheless, when modeling is applicable in laboratory-scale bed, the possible change in the controlling mechanism should be taken into account, and different model type is likely to be needed in large-scale bed [43].

Considering the liquid distribution in downflow operation, several liquid distributors could be used as weir-through, simple slotted-pipe, and perforated-pipe distributor [4, 43, 50]. For the later, the design procedure is rather simple and is presented in the literature [51]. Finally, reproducible and appropriate bed voidage should be ensured in large beds in order to avoid dead volumes. A suitable method called "snow storm filling" is presented in the literature [52]. Furthermore, in order to have near-ideal plug flow in small-scale bed, particle Peclet number should be high enough to have a high bed Peclet number. In contrast, in large-scale unit particle, Peclet number is of minimal importance since it is multiplied by Z/d_p, and thus the bed Peclet number is expected high enough (see Eq. 4.103). In order to ignore the geometrical similarity between the different unit sizes, Eqs. 4.39–4.41 and the following one should be satisfied [5, 8, 43]:

$$\frac{D_{FS}}{D_{LS}} \leq 10. \tag{4.113}$$

Using the above relationships, the limits of volumetric flow rates could be evaluated. Here, I is the relative volumetric flow rate in BV/h, β is for treatment, and α is for pretreatment/regeneration (generally $\alpha > \beta$):

$$Q = I \cdot V = I \cdot Z \cdot A \Rightarrow A = \frac{Q}{I \cdot Z} \Rightarrow D = \left(\frac{4 \cdot Q}{I \cdot \pi \cdot Z}\right)^{0.5}. \tag{4.114}$$

Then the following equations hold [43]:

$$D \le \frac{Z}{5} \Rightarrow \left(\frac{4 \cdot Q}{I \cdot \pi \cdot Z}\right)^{0.5} \le \frac{Z}{5} \Rightarrow Q \le \left(\frac{Z}{5}\right)^2 \cdot \frac{I \cdot \pi \cdot Z}{4} \tag{4.115}$$

$$D \ge 30 \cdot d_p \Rightarrow \left(\frac{4 \cdot Q}{I \cdot \pi \cdot Z}\right)^{0.5} \ge 30 \cdot d_p \Rightarrow Q \ge \left(30 \cdot d_p\right)^2 \cdot \frac{I \cdot \pi \cdot Z}{4} \tag{4.116}$$

$$D_{FS} \le 10 \cdot D_{LS} \Rightarrow \left(\frac{4 \cdot Q_{FS}}{I \cdot \pi \cdot Z_{FS}}\right)^{0.5} \le 10 \cdot D_{LS} \Rightarrow Q_{FS} \le (10 \cdot D_{LS})^2 \cdot \frac{I \cdot \pi \cdot Z_{FS}}{4}, \tag{4.117}$$

where subscript LS is for laboratory-scale bed and FS for the large unit. Using Eqs. 4.115–4.117, the appropriate limits of working volumetric flow rates can be evaluated. Then, the full-scale bed diameter can be evaluated using Eq. 4.114 according to the chosen working volumetric flow rate [43].

4.4.3 Bed Voidage Considerations

In order to use safely the above analysis, the bed voidage should be approximately the same for both units. Bed voidage is a function of d_p/D [53]. However, for small d_p/D values (typically lower than 0.1), the bed voidage is practically constant. As in practical applications d_p/D is kept lower than 0.1, the bed voidage could be considered the same in both scales. For example, using particles of 2 *mm* diameter, the bed diameter should be greater than 2 *cm*. This is critical then only in laboratory-scale units. In Fig. 4.16, Dixon's correlations are presented for spheres and cylinders, in the case of $d_p/D < 0.5$ [53]. For cylinders, d_p is equal to diameter of equal volume sphere.

From Fig. 4.16, it is clear that for $d_p/D < 0.1$, the differences of bed voidage are small, while for $d_p/D > 0.1$, the bed voidage is greatly affected by particle/bed diameter ratio. Although such data are available for regular-shaped particles, for irregular ones, measurements should be conducted in order to evaluate bed voidage and its dependence on d_p/D. Finally, as pressure drop is very sensitive to the bed

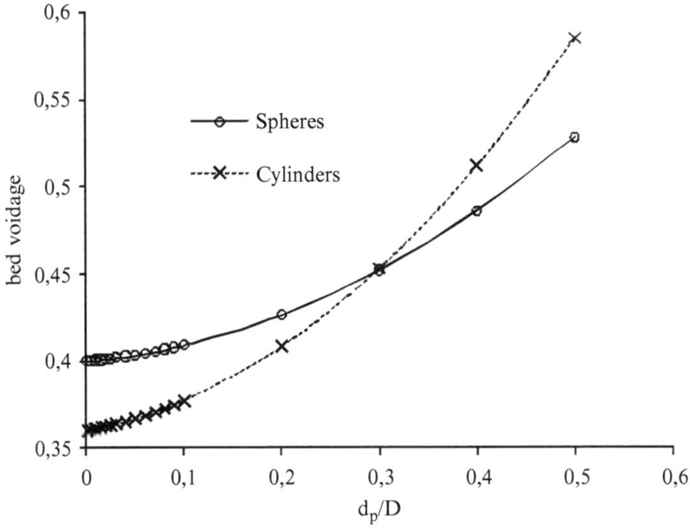

Fig. 4.16 Bed voidage versus d_p/D

voidage, the ratio d_p/D greatly influences pressure drop across the bed in the case of $d_p/D > 0.1$ [52, 54]. Again this happens because for $d_p/D > 0.1$, bed voidage changes considerably (Fig. 4.16).

References

1. International Atomic Energy Agency (2002) Application of ion exchange processes for the treatment of radioactive waste and management of spent ion exchangers. Technical reports series no 408, Vienna
2. Babel S, Kurniawan T (2003) Low-cost adsorbents for heavy metals uptake from contaminated water: a review. J Hazard Mater B97:219–243
3. Inglezakis VJ (2010) Ion exchange and adsorption fixed bed operations for wastewater treatment – Part I: modeling fundamentals and hydraulics analysis. J Eng Stud Res 16:29–41
4. Perry RH, Green D (1997) Perry's chemical engineers' handbook. McGraw-Hill, International Editions, New York
5. Inglezakis V, Poulopoulos S (2006) Adsorption ion exchange and catalysis: design of operations and environmental applications. Elsevier, International Edition, Amsterdam
6. Inglezakis V, Grigoropoulou H (2003) Modeling of ion exchange of Pb^{2+} in fixed-beds of clinoptilolite. Micropor Mesopor Mater 61:273–282
7. Donati G, Paludetto R (1997) Scale up of chemical reactors. Catal Today 34:483–533
8. Peters M, Timmerhaus K (1968) Plant design and economics for chemical engineers. McGraw-Hill International Edition, New York
9. Ruthven D (1984) Principles of adsorption and adsorption processes. Wiley, New York
10. Crank J (1975) The mathematics of diffusion. Clarendon, Oxford
11. Ben-Shebil S, Alkan-Sungur A, Özdural A (2007) Fixed-bed ion exchange columns operating under non-equilibrium conditions: estimation of mass transfer properties via non-equilibrium modeling. React Funct Polym 67:1540–1547

12. Coulson J, Richardson J (2002) Coulson and Richardson's chemical engineering, vol 2, Particle technology and separation processes. Butterworth-Heinemann, International Edition, Oxford
13. Inglezakis V, Grigoropoulou H (2001) Applicability of simplified models for the estimation of ion exchange diffusion coefficients in zeolites. J Colloid Interf Sci 234:434–441
14. Ruthven D (2008) Fundamentals of adsorption equilibrium and kinetics in microporous solids. In: Karge GH, Weitkamp J (eds) Molecular sieves science and technology, adsorption and diffusion, vol 7. Springer, Berlin/Heidelberg
15. Helfferich F (1962) Ion exchange. Dover, New York
16. Inglezakis V (2007) Solubility-normalized Dubinin-Astanhov adsorption isotherm for ion exchange systems. Micropor Mesopor Mater 103:72–81
17. Inglezakis V (2005) The concept of "capacity" in zeolite ion exchange systems. J Colloid Interf Sci 281:68–79
18. Hall K, Eagleton L, Acrivos A, Vermeulen T (1966) Pore- and solid-diffusion kinetics in fixed bed adsorption under constant-pattern conditions. Ind Eng Chem Fund 5:212–223
19. Fleck R, Kirwan D, Hall K (1973) Mixed-resistance diffusion kinetics in fixed bed adsorption under constant pattern conditions. Ind Eng Chem Fund 12:95–99
20. Trambouze P (1990) Reactor scaleup methodology. Chem Eng Prog 2:23–31
21. Levenspiel O (1972) Chemical reaction engineering. Wiley Eastern Limited, International Edition, New York
22. McCabe W, Smith J, Harriott P (1983) Unit operations of chemical engineering. McGraw-Hill, International Edition
23. Inglezakis V (2002) Design of ion exchange column using natural minerals as packing materials. Ph.D. Thesis, National Technical University of Athens, Athens
24. Wevers C (1959) Research on ion exchange by means of the percolation technique. Chem Eng Sci 10:171–189
25. Miura K, Hashimoto K (1977) Analytical solutions for the breakthrough curves of fixed-bed adsorbers under constant pattern and linear driving force approximations. J Chem Eng Jpn 10:490–493
26. Chern J, Chien Y (2002) Adsorption of nitrophenol onto activated carbon: isotherms and breakthrough curves. Water Res 36:647–655
27. Crittenden J, Hand D, Arora H, Lykins B Jr (1987) Design considerations for GAC treatment of organic chemicals. J AWWA J 1:74–82
28. Zulfadhly Z, Mashitah M, Bhatia S (2001) Heavy metals removal in fixed-bed column by the macro fungus *Pycnoporus sanguineus*. Environ Pollut 112:463–470
29. Xiu G, Li P (2000) Prediction of breakthrough curves for adsorption of lead(II) on activated carbon fibers in a fixed bed. Carbon 38:975–981
30. Chen J, Wang L (2004) Characterization of metal adsorption kinetic properties in batch and fixed-bed reactors. Chemosphere 54:397–404
31. Inglezakis V, Lemonidou M, Grigoropoulou H (2001) Liquid holdup and dispersion in zeolite packed beds. Chem Eng Sci 56:5049–5057
32. Chung S, Wen C (1968) Longitudinal dispersion of liquid flowing through fixed and fluidized beds. AIChE J 14:857–866
33. Inglezakis V (2010) Non ideal flow in liquid–solid fixed beds of irregular-shaped particles: a critical review. Int J Chem Reactor Eng 8:R3
34. Specchia V, Baldi G (1977) Pressure drop and liquid holdup for two-phase concurrent flow in packed beds. Chem Eng Sci 32:515–523
35. Colombo A, Baldi G (1976) Solid–liquid contacting effectiveness in trickle bed reactors. Chem Eng Sci 31:1101–1108
36. Inglezakis V, Zorpas A, Grigoropoulou H, Arapoglou D (2002) Liquid holdup measurements using tracing techniques in zeolite packed beds. In: Proceedings of the 5th international scientific and technical conference "Water Supply and Water Quality", Poznan, pp 430–440
37. Inglezakis V, Papadeas C, Loizidou M, Grigoropoulou H (2001) Effects of pretreatment on physical and ion exchange properties of natural clinoptilolite. Environ Technol 22:75–83

38. Weber W, Wang C (1987) A microscale system for estimation of model parameters for fixed bed adsorbers. Environ Sci Technol 21:1096–1102
39. Barros M, Silva E, Arroyo P, Tavares C, Schneider R, Suszek M, Sousa-Aguiar E (2004) Removal of Cr (III) in the fixed bed column and batch reactors using as adsorbent zeolite NaX. Chem Eng Sci 59:5959–5966
40. Weber W Jr, Smith E (1987) Simulation and design models for adsorption processes. Environ Sci Technol 21:1040–1050
41. Ko D, Porter J, McKay G (2003) Fixed bed studies for the sorption of metal ions onto peat. Trans I Chem Eng Part B 81:73–86
42. Fernandez M, Laughinghouse W, Carta G (1996) Characterization of protein adsorption by composite silica – polyacrylamide gel anion exchangers. II. Mass transfer in packed columns and predictability of breakthrough behavior. J Chromatogr A 746:185–198
43. Inglezakis VJ (2010) Ion exchange and adsorption fixed bed operations for wastewater treatment – Part II: scale-up and approximate design methods. J Eng Stud Res 16:42–50
44. Smith J (1981) Chemical engineering kinetics. McGraw-Hill, International Edition, New York
45. Inglezakis V (2007) The physical significance of simplified models in engineering. Model Optim Mach Build Field 13:338–341
46. Chen J, Buege J, Cunningham F, Northam J (1968) Scale-up of a column adsorption process by computer simulation. Ind Eng Chem Proc DD 7:26–31
47. McCabe D, Hassan N, King W, Steimke J, Norano M, Hamm L, Oji L, Johnson M (2001) Comprehensive scale testing of the ion exchange removal of cesium and technetium from Hanford tank wastes. Waste management conference (WM'01), University of Arizona, Tucson, 25 Feb–1 Mar 2001
48. Dudukovic M (1999) Trends in catalytic reaction engineering. Catal Today 48:5–15
49. Dudukovic M, Larachi F, Mills P (1999) Multiphase reactors – revisited. Chem Eng Sci 54:1975–1995
50. Treybal R (1980) Mass transfer operations. McGraw-Hill, International Edition, New York
51. Feintuch H (1997) Distillation distributor design. Hydrocarb Process 10:150–151
52. Afandizadeh S, Foumeny E (2001) Design of packed bed reactors: guides to catalyst shape, size, and loading selection. Appl Therm Eng 21:669–682
53. Dixon A (1988) Correlations for wall and particle shape effects on fixed bed bulk voidage. Can J Chem Eng 66:705–708
54. Fumeny E, Kulkarni A, Roshani S, Vatani A (1996) Elucidation of pressure drop in packed-bed systems. Appl Therm Eng 16:192–202

Chapter 5
Principles of Ion Exchange Membrane Electrodialysis for Saline Water Desalination

Yoshinobu Tanaka

Abstract Irreversible thermodynamics is the fundamental principle in ion exchange membrane electrodialysis. The mechanism of saline water desalination is explained based on irreversible thermodynamics. The overall mass transport equation is developed on the basis of the electrodialysis experiments. The phenomenological equation appearing in irreversible thermodynamics is substantially identical to the overall mass transport equation. The overall membrane pair characteristics appearing in the overall mass transport equation are expressed by functions of the overall hydraulic permeability of the membrane pair. In an electrodialyzer, solution velocities in desalting cells vary between the cells. Salt concentrations are decreased along the flow-passes in desalting cells. These events give rise to electric resistance distribution and current density distribution in the electrodialyzer and exert an influence on the limiting current density of the electrodialyzer. The electrodialysis process is classified into a continuous (one-pass flow), a batch, and a feed-and-bleed process. The performance of each process is discussed using computer simulation (electrodialysis program) and by applying the principles of mass transport, current density distribution, cell voltage, energy consumption, and limiting current density. An electrodialysis program is operated for desalinating saline water, and the performance of a practical-scale electrodialyzer is discussed with figures created using computer simulation. The program aims to work as a pilot plant operation.

5.1 Introduction

Ion exchange membrane electrodialysis is a process to separate or extract ionic species in an electrolyte solution. The largest scale application of the electrodialysis is the demineralization of saline water. Other applications are desalination and

Y. Tanaka (✉)
IEM Research, 1-46-3 Kamiya, Ushiku-shi, Ibaraki 300-1216, Japan
e-mail: fwis1202@mb.infoweb.ne.jp

Inamuddin and M. Luqman (eds.), *Ion Exchange Technology I: Theory and Materials*, DOI 10.1007/978-94-007-1700-8_5,

reuse of sewage or industrial waste, recovery of useful components from a metal surface treatment process, refining of amino acid solution, desalination of milk, whey, sugar liquor, concentration of seawater, organic acid, etc. Electrodialysis is the fundamental technology based on the ion exchange membrane, and it is applied to the succeeding technology such as bipolar membrane electrodialysis, electrodeionization, electrolysis, diffusion dialysis, fuel cell, etc.

Looking back at the development of the history of ion exchange membrane electrodialysis, the notable achievement is the theory of mass transport, the Nernst–Planck equation developed by Planck [1]. Donnan presented the membrane equilibrium theory [2]. Michaelis and Fujita studied permselectivity of biological membranes using colloidal membranes [3]. Teorell [4] and Meyer and Sievers [5] discussed membrane phenomena and revealed the mechanism of membrane potential, electric conductivity, transport number, etc. Meyer and Straus electrodialyzed a KCl solution and realized doubled KCl concentration [6]. Wyllie and Patnode [7] and Juda and McRae [8] invented artificial ion exchange membranes, and the membranes were applied to electrodialysis for desalting saline water. Ion exchange membrane electrodialysis is now one of the basic technologies in saline water desalination industry. This chapter discusses "saline water desalination" from fundamental and practical viewpoints.

5.2 Principles

5.2.1 Irreversible Thermodynamics

Mass transport across the membrane must be discussed fundamentally on the basis of thermodynamics because the thermodynamics describes the rule of energy changes inevitably generating in the mass transport. Many phenomena (including membrane phenomena) generating in the natural world are far away from reversible states, so the reversible thermodynamics was ineffective to discuss their mechanism. However, the irreversible thermodynamics (nonequilibrium thermodynamics) succeeded in discussing the mass transport by introducing the concept of "time" in its system.

When two kinds of ions are transported with solvent in a solution, their fluxes influence each other, because the fluxes and driving forces are not independent but are coupled together. The irreversible thermodynamics describes such an interaction by the following phenomenological equation [9]:

$$J_i = L_{i1}X_1 + L_{i2}X_2 + - - - + L_{in}X_n = \sum_{k=1}^{n} L_{ik}X_k \quad (i = 1, 2, - - - n) \qquad (5.1)$$

where X_i is the force and J_i is the flux. L_{ik} is the phenomenological coefficients, in which i and k are components. The following Onsager's reciprocal theorem [10] suggests that the matrix composed of the phenomenological coefficients is symmetrical:

$$L_{ik} = L_{ki}(i, k = 1, 2, - - - - - - \mathrm{n}) \qquad (5.2)$$

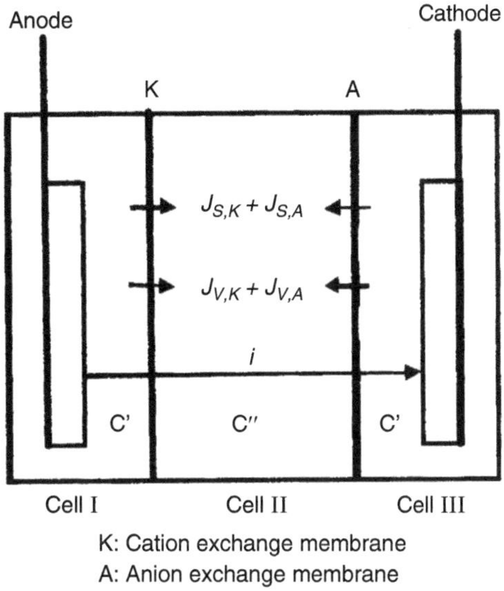

Fig. 5.1 Three-cell electrodialysis system

Based on the irreversible thermodynamics, an electric current, I, volume flow of a solution, J, and a mass flux of component, i, J_i, are introduced as the functions of a potential difference $\Delta\psi$, a pressure difference ΔP, and a chemical potential difference $\Delta\mu_i$ as follows [11]:

$$I = L_E\Delta\psi + L_{EP}\Delta P + \sum_i L_{Ei}\Delta\mu_i \tag{5.3}$$

$$J = L_{EP}\Delta\psi + L_P\Delta P + \sum_i L_{Pi}\Delta\mu_i \tag{5.4}$$

$$J_i = L_{iE}\Delta\psi + L_{iP}\Delta P + \sum_i L_{ik}\Delta\mu_i \tag{5.5}$$

Further, the phenomenological equation is also introduced as follows:

$$J_i = \sum_i \left(z_k L_{ik}\Delta\psi + v_k L_{ik}\Delta P + L_{ik}\Delta\mu_{ik}\right) \tag{5.6}$$

where z_k and v_k are the electric charge number and partial volume of component k, respectively.

Hause [12] discussed an electrokinetic phenomenon based on the approaches of Kedem and Katchalsky [11]. Referring to this suggestion, Schultz [13] discussed ionic transport in an electrodialysis process and introduced the phenomenological equation. The mechanism of saline water desalination is presented as follows [14] based on the equation developed by Schultz.

A salt solution dissolving monovalent cations and monovalent anions is assumed to be electrodialyzed using a three-cell electrodialysis system (Fig. 5.1) consisting of central cell II and electrode cells I and III placed on both outsides of the cell II.

A cation exchange membrane (K) is placed between cell I and cell II, and an anion exchange membrane (A) is placed between cell II and cell III. Supplying a salt solution (electrolyte concentration C') into cells I and III, constant current density i is applied, and the salt solution being collected in cell II is taken out from the system until the electrolyte concentration in cell II reaches steady constant (C''). Salt accumulation $J_{S,K} + J_{S,A}$ and solution accumulation $J_{V,K} + J_{V,A}$ in cell II in the steady state are given by the following equations introduced from the phenomenological equation:

$$J_{S,K} + J_{S,A} = (t_K + t_A - 1)\frac{i}{F} - RT\left[(\omega_K + \omega_A) - \{L_{P,K}\sigma_K(1-\sigma_K) + L_{P,A}\sigma_A(1-\sigma_A)\}C_S^*\right]\Delta C \quad (5.7)$$

$$J_{V,K} + J_{V,A} = (\beta_K + \beta_A)i + RT(\sigma_K L_{P,K} + \sigma_A L_{P,A})\Delta C \quad (5.8)$$

Here, t is the transport number, ω is the solute permeability, L_P is the hydraulic conductivity, σ is the reflection coefficient, β is the electro-osmotic permeability, F is the Faraday constant, R is the gas constant, T is the absolute temperature, and ΔC ($= C'' - C'$) is the concentration difference between cell I and cell II. In an ion exchange membrane, the reflection coefficient is generally assumed to be 1; $\sigma = 1$. C_S* is the logarithmic mean concentration defined by:

$$C_S^* = \frac{C'' - C'}{\ln(C''/C')} \quad (5.9)$$

5.2.2 Overall Mass Transport

Irreversible thermodynamics described in Sect. 5.2.1 is the classical and basic theory of the transport phenomena across an ion exchange membrane. However, it cannot elucidate the mechanism of phenomena occurring in a practical-scale electrodialyzer because the phenomena generating in the electrodialyzer include too many parameters not being considered in the irreversible thermodynamics. The overall mass transport equation is developed to understand the mechanism of saline water desalination process on the basis of the following electrodialysis experiments. This equation is in agreement with the phenomenological equations based on irreversible thermodynamics [14].

An electrodialyzer was set up incorporating with commercially available ion exchange membranes. Saline water was supplied to desalting cells at linear velocity u, and electric potential was applied between electrodes. After electrolyte concentration in concentrating cells reached stable, the fluxes of ions J_S (eq/cm^{2}s) and a solution J_V (cm^3/cm^{2}s) across an ion exchange membrane pair, electrolyte concentration in desalting cells C' (eq/cm^3) and in concentrating cells C'' were measured. Changing

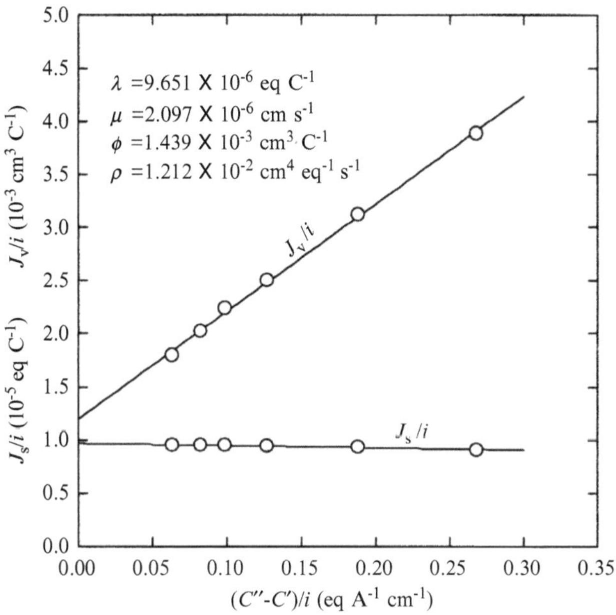

Fig. 5.2 J_S/i versus $\Delta C/i$ plot and J_V/i versus $\Delta C/i$ plot

current density i (A/cm^2) incrementally, electrodialysis was repeated. Plotting J_S/i and J_V/i against $(C'' - C')/i = \Delta C/i$ produced linear lines as exemplified in Fig. 5.2.

These lines are expressed by the following overall mass transport equation:

$$J_S = C''J_V = \lambda i - -\mu(C'' - C') = \lambda i - \mu\Delta C \tag{5.10}$$

$$J_V = \phi i + \rho(C'' - C') = \phi i + \rho\,\Delta C \tag{5.11}$$

λ (eq C^{-1}) is the overall transport number, μ (cm s^{-1}) is the overall solute permeability, ϕ (cm^{3}C^{-1}) is the overall electro-osmotic permeability, and ρ (cm^{4}eq^{-1} s^{-1}) is the overall hydraulic permeability, and these are overall membrane pair characteristics. The term "overall" means that the coefficients express the contributions of a cation and an anion exchange membrane. It also means that the coefficients express the contributions of many kinds of ions dissolving in an electrolyte solution. Terms λi and $\mu \Delta C$ in Eq. 5.10 stand for migration and diffusion of ions, respectively. Terms ϕi and $\rho \Delta C$ in Eq. 5.11 correspond to electro-osmosis and hydraulic osmosis.

Based on the saline water electrodialysis experiment described above, ρ versus λ, μ, and ϕ plots are presented by the following empirical equations:

$$\lambda = 9.208 \times 10^{-6} + 1.914 \times 10^{-5}\rho \tag{5.12}$$

$$\mu = 2.005 \times 10^{-4}\rho \tag{5.13}$$

$$\phi = 3.768 \times 10^{-3}\rho^{0.2} - 1.019 \times 10^{-2}\rho \tag{5.14}$$

Further ρ versus alternating current electric resistance of ion exchange membrane pair $R_{alter} = r_{alter,K} + r_{alter,A}$ is presented by the following equation:

$$R_{alter} = r_{alter,K} + r_{alter,A} = 1.2323\rho^{-(1/3)} \tag{5.15}$$

The salt concentration in a concentrating cells C'' after arriving at steady state and current efficiency η in the above electrodialysis are given by the following equations:

$$C'' = \frac{1}{2\rho}\left(\sqrt{A^2 + 4\rho B} - A\right) \tag{5.16}$$

$$A = \phi i + \mu - \rho C' \tag{5.17}$$

$$B = \lambda i + \mu C' \tag{5.18}$$

$$\eta = \frac{J_S}{i/F} \tag{5.19}$$

The phenomenological equation introduced in irreversible thermodynamics Eqs. 5.7 and 5.8 describes the mass transport in a pair membrane system. The overall mass transport equation, Eqs. 5.10 and 5.11, also presents the mass transport across a pair of membranes. Both equations are substantially identical, so the overall mass transport coefficients are presented by the functions of phenomenological coefficients as follows putting $\sigma = 1$:

$$\lambda = \frac{t_K + t_A - 1}{F} \tag{5.20}$$

$$\mu = RT\left[(\omega_K + \omega_A) - \left\{L_{P,K}\sigma_K(1 - \sigma_K) + L_{P,A}\sigma_A(1 - \sigma_A)\right\}C_S^*\right] = RT(\omega_K + \omega_A) \tag{5.21}$$

$$\phi = \beta_K + \beta_A \tag{5.22}$$

$$\rho = RT(\sigma_K L_{P,K} + \sigma_A L_{P,A}) = RT\left(L_{P,K} + L_{P,A}\right) \tag{5.23}$$

5.2.3 Concentration Polarization and Limiting Current Density

Under an applied electric current, a boundary layer is formed on the desalting surface of an ion exchange membrane, and the salt concentration in the layer is depleted due to concentration polarization (Fig. 5.3) [15–18].

Fig. 5.3 Concentration polarization

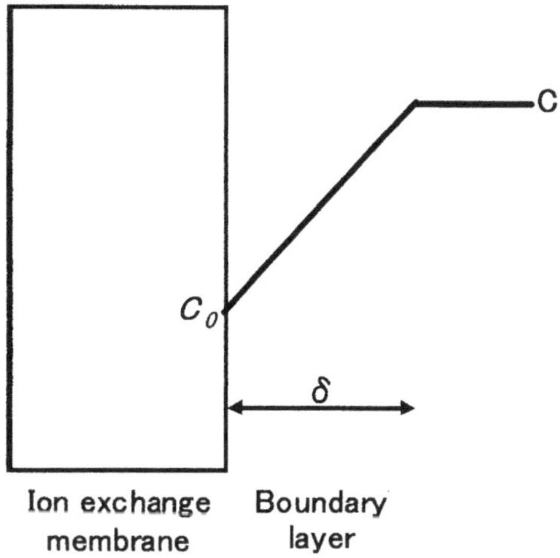

Depletion of the salt at the membrane surface means that an increasing fraction of the voltage drop is dissipated in transporting ions across the boundary layer rather than through the membrane. A point can be reached at which the concentration at the membrane surface is zero. The current through the membrane at this point is the limiting current density of the ion exchange membrane. At over limiting current density, the extra power is dissipated by the side reaction such as water dissociation [19–23]. Concentration polarization is a very important phenomenon in ion exchange membrane electrodialysis because it influences the performance of an electrodialyzer.

We assume here that a cation exchange membrane is immersed in a NaCl solution and an electric current density i is applied across the membrane. In a desalted side of the membrane, Na^+ ion transport number in the solution t_{Na} is 0.4, while that in the membrane $\bar{t}_{Na}$ is 0.95. In this situation, Na^+ ion flux in the solution $J_{Na} = t_{Na}(i/F) = 0.4(i/F)$ is less than Na^+ ion flux in the membrane $\bar{J}_{Na} = \bar{t}_{Na}(i/F) = 0.95(i/F)$. In order to maintain the material balance, NaCl concentration is decreased, and the concentration gradient $(C - C_0)/\delta$ is generated in the boundary layer. The material balance in the boundary layer is consequently given as follows:

$$\bar{t}_{Na}\frac{i}{F} = D\frac{C - C_0}{\delta} + t_{Na}\frac{i}{F} \quad (5.24)$$

in which D is the diffusion constant of NaCl, δ is boundary layer thickness. If current density i is increased in Eq. 5.24, C_0 is decreased and reaches $C_0 = 0$ at $i = i_{lim}$, which is limiting current. Substituting these parameters introduces:

$$i_{\lim} = \frac{FDC}{\left(\bar{t}_{Na} - t_{Na}\right)\delta} \quad (5.25)$$

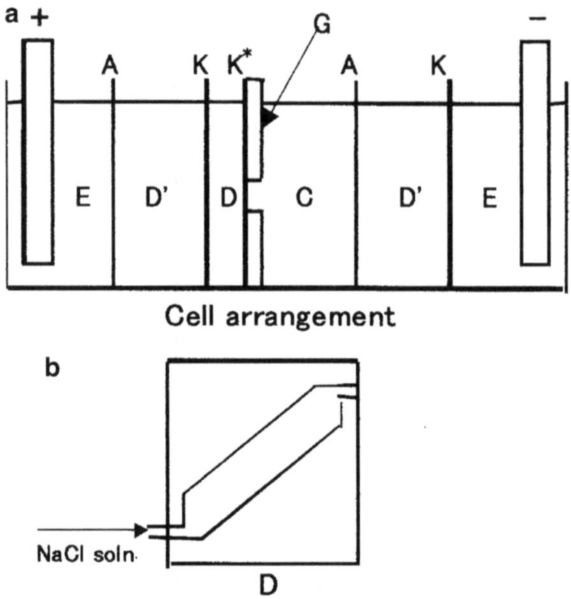

Fig. 5.4 Apparatus for measuring limiting current density (Reproduced from Ref. [27] with kind permission of © Elsevier (2005))

Equation 5.25 is the limiting current density equation which is known widely. In discussing the limiting current density with Eq. 5.25, we cannot determine the boundary layer thickness δ easily. In order to solve such a problem, i_{lim} is usually measured experimentally as follows with current–voltage relationship, *I–V* curves [24–26].

The experimental apparatus (Fig. 5.4a) was assembled with cation exchange membranes (Aciplex K-172, Asahi Chemical Co.) and anion exchange membranes (Aciplex A-172) [27]. The thickness of cell D was adjusted to 0.075 cm. Width and length of the flow pass in cell D were adjusted to 1 and 2 cm, respectively (Fig. 5.4b). A diagonal net spacer was put into cell D. A 25°C NaCl solution was supplied to cell D. Passing an electric current and changing the current density incrementally, the limiting current density i_{lim} of a cation exchange membrane (K* in the figure) was measured from the inflection of V/I versus $1/I$ plot [28]. The experiment was repeated by changing NaCl concentration C and solution velocity u in D step by step. i_{lim} of an anion exchange membrane was measured in the same manner. From the above experimental results, i_{lim} was expressed by the following empirical function of C and u:

$$i_{lim} = \mathrm{m}C^{\mathrm{n}} \tag{5.26}$$

For a cation exchange membrane Aciplex K-172

$$\begin{aligned} \mathrm{m} &= 83.50 + 24.00u \\ \mathrm{n} &= 0.7846 + 8.612 \times 10^{-3}u \end{aligned} \tag{5.27}$$

For an anion exchange membrane Aciplex A-172

$$\begin{aligned} \mathrm{m} &= 66.63 + 14.72u \\ \mathrm{n} &= 0.7404 + 3.585 \times 10^{-3}u \end{aligned} \tag{5.28}$$

5.2.4 Solution Velocity Distribution

In an electrodialyzer, ion exchange membranes, desalting cells, and concentrating cells are arranged alternately, and a solution is supplied into desalting cells. In this flow system, the solution velocity distribution in desalting cells does not become uniform. This phenomenon causes the concentration distribution and influences to the limiting current density of the electrodialyzer (Sect. 5.2.7). The solution velocity distribution is evaluated by measuring the solution velocity ratio ξ defined by the following equation:

$$\xi = \frac{u - \overline{u}}{\overline{u}} \tag{5.29}$$

in which u is solution velocity in each desalting cell, $\overline{u}$ is the average solution velocity in every desalting cell integrated in a stack.

We introduce here a measuring instance of the solution velocity distribution in a practical electrodialysis unit [27]. Seawater was supplied at the average velocity $\overline{u} = 3$ cm/s to desalting cells in a stack of a unit-cell type electrodialyzer incorporated with Selemion CMV/AST membranes (Asahi Glass Co.) and crosspiece spacers. The distance between membranes in a desalting cell was 2 mm. The solution velocity distribution was evaluated by injecting coloring liquid into the inlets of each desalting cell and measuring elapse time until the coloring liquid appears in the outlets of the desalting cells. The frequencies of desalting cells N_{cell} are plotted against the solution velocity ratio ξ (Eq. 5.29) and shown in Fig. 5.5 which is equated with the normal distribution, and the standard deviation of the normal distribution σ is measured as 0.078.

Seawater was supplied to desalting cells in a filter-press type electrodialyzer incorporated with Aciplex K-172/A-172 membrane (Asahi Chemical Co.) and electrodialyzed at the current density of 3 A dm^{-2}. The solution velocity in each desalting cell u was evaluated by measuring the electrolyte concentration at the inlets C_{in} and the outlets C_{out} of the cells. N_{cell} versus $\xi = (u - \overline{u})/\overline{u}$ was confirmed to be equated with the normal distribution. Standard deviation of σ of solution velocity ratio ξ is determined as presented in Table 5.1. σ is distributed in the range of 0.017–0.222.

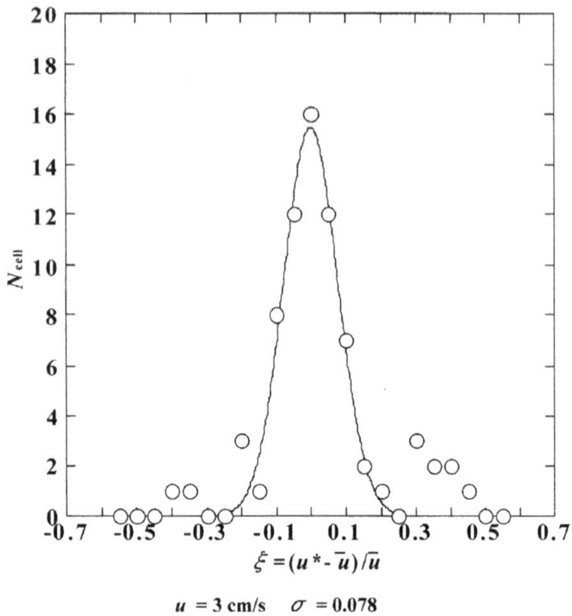

Fig. 5.5 Velocity distribution of desalted solutions between desalting cells (Reproduced from Ref. [27] with kind permission of © Elsevier (2005))

Table 5.1 Standard deviation of solution velocity ratio σ in desalting cells (Reproduced from Ref. [27] with kind permission of © Elsevier (2005))

u cm/s	C'_{in} eq/dm^3	C'_{out} eq/dm^3	σ
1.12	0.590	0.276	0.117
1.58	0.557	0.276	0.222
1.73	0.606	0.392	0.141
2.07	0.603	0.430	0.017
2.24	0.606	0.426	0.134
3.12	0.603	0.488	0.020
3.54	0.606	0.491	0.122
4.93	0.603	0.535	0.102

5.2.5 Current Density Distribution

Ohm's law $V = IR$ is estimated to hold between electrodes in an electrodialyzer. This is because electric resistance R_{in} (current I_{in}) is decreased (increased) at the inlets of desalting cells and electric resistance R_{out} (current I_{out}) is increased (decreased) at the outlets of desalting cells, and the following relationship is realized between electrodes:

$$V_{in} = I_{in}R_{in} = V_{out} = I_{out}R_{out} \tag{5.30}$$

In order to confirm the reasonability of the above assumption, the following experiment was performed [27]. Seawater was supplied to electrodialyzers

Table 5.2 Voltage difference between electrodes in a stack (Reproduced from Ref. [27] with kind permission of © Elsevier (2005))

Experiment 1[a]						
I/S (A dm^{-2})	1	2	3	4	6	8
V_{out} (V)	3.56	5.60	8.69	11.00	15.40	17.90
V_{in}(V)	3.22	5.47	7.82	9.90	13.80	17.00
V_{out}/V_{in}	1.11	1.02	1.11	1.11	1.12	1.05
Experiment 2[b]						
I/S (A dm^{-2})	1.97	2.66	3.38	–	–	–
V_{out} (V)	19.2	23.6	29.7	–	–	–
V_{in}(V)	19.2	23.7	29.6	–	–	–
V_{out}/V_{in}	1.00	1.00	1.00	–	–	
Experiment 3[c]						
I/S (A dm^{-2})	2.50	2.50	–	–	–	–
V_{out} (V)	104.3	104.1	–	–	–	–
V_{in}(V)	103.4	103.6	–	–	–	–
V_{out}/V_{in}	1.01	1.00	–	–	–	

[a]S = 5 dm^2, N = 20 pairs, Selemion CMV/ASV (Asahi Glass Co.), 20 pairs/stack × 1 stack = 20 pairs between electrodes

[b]S = 200 dm^2, N = 87 pairs, Neosepta C66-5T/ACS (Tokuyama Inc.), 87 pairs/stack × 12 syacks = 1,044 pairs between electrodes

[c]S = 140 dm^2, N = 300 pairs, Aciplex K-172/A-172 (Asahi Chemical Co.), 300 pairs/stack × 6 stacks = 1,800 pairs between electrodes

incorporated with diagonal net spacers, and voltage difference at the inlets V_{in} and the outlets V_{out} of desalting cells in a stack was measured as shown in Table 5.2.

Experiment 1 is for the small-scale electrodialyzer in which the number of membrane pairs between electrodes is 20. Current density was changed and V_{out}/V_{in} was in the range of 1.05–1.11. Experiments 2 and 3 are for the large-scale electrodialyzers. The number of membrane pairs between electrodes was, respectively, 1,044 and 1,800, and in both experiments, V_{out}/V_{in} is 1.0. This is because electric resistance and ohmic loss of the electrodes incorporated in the large-scale electrodialyzers are negligible compared to the values between electrodes.

Electrolyte concentrations in desalting cells are decreased along flow-passes and give rise to electrolyte concentration distribution. It causes electric resistance distribution and current density distribution. Current density in an electrodialyzer decreases from i_{in} at the inlet ($x = 0$, x: distant from the inlet of a desalting cell) to $i = I/S$ (average current density of an electrodialyzer, I: electric current, S: membrane area) at $x = pl$ in the desalting cells. It decreases further from i at x to i_{out} at the outlet ($x = l$, l: flow-pass length in a desalting cell). In order to determine current density distribution in an electrodialyzer, we assume that current density i at x is approximated by the following current density distribution equation [27]:

$$i = a_1 + a_2\left(\frac{x}{l}\right) + a_3\left(\frac{x}{l}\right)^2 \tag{5.31}$$

To determine a_1, a_2, and a_3 in Eq. 5.31, the following three simultaneous equations are set up, taking into account the voltage difference measurement between electrodes (cf. Eq. 5.30):

$$V_{in} = V_{out} \tag{5.32}$$

$$V_{in} = V_p \tag{5.33}$$

$$\zeta_{inout} = \zeta_{inp} \tag{5.34}$$

ζ_{inout} is introduced from Eq. 5.32 and expressed by Eq. 5.35, which is equal to ζ_{out} (outlet electric current nonuniformity coefficient):

$$\zeta_{inout} = \frac{\alpha_1 + \alpha_2 p + \alpha_3 p^2}{\beta_1 + \beta_2 p + \beta_3 p^2} = \zeta_{out} = \frac{i_{out}}{I/S} \tag{5.35}$$

ζ_{inp} is introduced from Eq. 5.33 and expressed by Eq. 5.36, which is also equal to ζ_{out}:

$$\zeta_{inp} = \frac{\gamma_1 + \gamma_2 p + \gamma_3 p^2}{(2p - 3p^2)(I/S)} = \zeta_{out} \tag{5.36}$$

We determine a_1, a_2, a_3 with $\zeta_{in} = i_{in}/(I/S)$ (inlet electric current nonuniformity coefficient), ζ_{out}, i_{in}, and i_{out}:

$$\zeta_{in} = \frac{i_{in}}{I/S} = \frac{a_1}{I/S} \tag{5.37}$$

$$\zeta_{out} = \frac{i_{out}}{I/S} = \frac{a_1 + a_2 + a_3}{I/S} \tag{5.38}$$

5.2.6 Cell Voltage and Energy Consumption

Cell voltage V_{cell} is given by the following equation, taking into account the voltage difference measurement between electrodes [29] (Sect. 5.2.5):

$$V_{cell} = V_{\Omega,in} + V_{memb,in} = V_{\Omega,out} + V_{memb,out} \tag{5.39}$$

Ohmic potential $V_{\Omega,in}$ and membrane potential $V_{memb,in}$ at the inlets of desalting cells in Eq. 5.39 are given as:

$$V_{\Omega,in} = \left(r'_{in} + r_{memb,in} + r''\right) i_{in} \tag{5.40}$$

$$V_{memb,in} = 2(t_K + t_A - 1)\left(\frac{RT}{F}\right)\ln\frac{\gamma''C''}{\gamma'_{in}C'_{in}} \tag{5.41}$$

r'_{in} and $r_{memb,in}$ are, respectively, electric resistance of a solution and direct electric resistance of a membrane pair at the inlet of a desalting cell. r'' is electric resistance of a solution in a concentrating cell. r' and r'' include the influence of the electric current screening ratio of a spacer ε (electric current screening area of a spacer/membrane area). C'_{in} and C'' are, respectively, electrolyte concentration at the inlet of desalting cells and in the concentrating cell. t_K and t_A are the transport number of a cation and an anion exchange membrane, respectively. γ'_{in} is the activity coefficient of ions dissolving in a solution at the inlet of a desalting cell. γ'' is the activity coefficient of ions dissolving in a concentrating cell.

Ohmic potential $V_{\Omega,out}$ and membrane potential $V_{memb,out}$ at the outlets of desalting cells in Eq. 5.39 are given by averaging the values for N cells integrated in an electrodialyzer:

$$V_{\Omega,out} = \left(\sum_{j=1}^{j\max} Y_j r'_{out,j} + \sum_{j=1}^{j\max} Y_j r_{memb,out,j} + r''N\right) \times i_{out}\left(\frac{1}{N}\right) \tag{5.42}$$

$$V_{memb,out} = 2(t_K + t_A - 1)\left(\frac{RT}{F}\right)\sum_{j=1}^{j\max}\ln\frac{\gamma''C''}{\gamma'_{out,j}C'_{out,j}} \times \left(\frac{1}{N}\right) \tag{5.43}$$

r'_{out} and $r_{memb,out}$ are electric resistance of a solution and direct electric resistance of a membrane pair at the outlet of a desalting cell, respectively. C'_{out} is electrolyte concentration at the outlet of a desalting cell. γ'_{out} is the activity coefficient of ions dissolving in a solution at the outlet of a desalting cell. Y_j is number of desalting cells in group j, and subscript j denotes group j in the normal distribution within the range of $\xi_j - \Delta\xi_j < \xi_j < \xi_j + \Delta\xi_j$. N is number of desalting cells in a stack.

Energy consumption E is expressed by the following equation:

$$E = \frac{V_{cell}I}{Q_{out}} \tag{5.44}$$

I is an electric current and Q_{out} is solution volume at the outlets of desalting cells, the desalted solution output.

5.2.7 *Limiting Current Density of an Electrodialyzer*

Limiting current density of a cation exchange membrane is less than that of an anion exchange membrane, because the mobility of counter ions in a solution for a cation exchange membrane is less than that for an anion exchange membrane. So the limiting current density of an ion exchange membrane integrated in an electrodialyzer, i_{lim}, is given by the following empirical equation established for a cation exchange membrane (cf. Eqs. 5.26 and 5.27 in Sect. 5.2.3):

$$i_{\lim} = (m_1 + m_2 u_{out}) C'^{\,n_1+n_2u_{out}}_{out} \tag{5.45}$$

When current density reaches the limit of a cation exchange membrane i_{lim} at the outlet of a desalting cell in which linear velocity becomes the least among u_{out}, $u_{out}{}^{\#}$ the average current density applied to an electrodialyzer is defined as its limiting current density $(I/S)_{\lim}$ which is expressed by Eq. 5.46 introduced from Eqs. 5.38 and 5.45 [27]:

$$\left(\frac{I}{S}\right)_{\lim} = \frac{i_{\lim}}{\zeta_{out}} = \frac{\left(m_1 + m_2 u^{\#}_{out}\right)}{\zeta_{out}} \left(C'^{\#}_{out}\right)^{n_1+n_2u^{\#}_{out}} \tag{5.46}$$

in which $C'_{out}{}^{\#}$ is C'_{out} at $u = u_{out}{}^{\#}$.

In Eq. 5.46, $u_{out}{}^{\#}$ is nearly equal to $u_{in}{}^{\#}$. Substituting $u_{out}{}^{\#} = u_{in}{}^{\#}$ in Eq. 5.46 leads to:

$$\left(\frac{I}{S}\right)_{\lim} = \frac{m_1 + m_2 u^{\#}_{in}}{\zeta_{out}} \left(C'^{\#}_{out}\right)^{n_1+n_2u^{\#}_{in}} \tag{5.47}$$

in which $C'_{in}{}^{\#}$ is C'_{in} at $u = u_{in}{}^{\#}$ given by:

$$u_{in}{}^{\#} = u_{in}(1 - 3\sigma) \tag{5.48}$$

in which σ is the standard deviation of the normal distribution of the solution velocity ratio ξ (Eq. 5.29).

On the other hand, the relationship between $(I/S)_{\lim}$ and $C'_{out}{}^{\#}$ is also introduced as follows:

$$\left(\frac{I}{S}\right)_{\lim} = \left(\frac{a}{\lambda l}\right) u^{\#}_{in} \left(C'_{in} - C'^{\#}_{out}\right) \tag{5.49}$$

Putting Eq. 5.47 equal to Eq. 5.49:

$$Z_1 = \frac{\left(C'^{\#}_{out}\right)^{n_1+n_2u^{\#}_{in}}}{C'_{in} - C'^{\#}_{out}} \tag{5.50}$$

$$Z_2 = \left(\frac{a\zeta_{out}}{\lambda l}\right)\left(\frac{u^{\#}_{in}}{m_1 + m_2 u^{\#}_{in}}\right) \tag{5.51}$$

$$Z_1 = Z_2 \tag{5.52}$$

5.3 Electrodialysis Process

5.3.1 Electrodialyzer

The basic structure of an electrodialyzer consists of stacks in which cation exchange membranes, anion exchange membranes, and gaskets (desalting cells and concentrating cells) are arranged alternately (Fig. 5.6) [30].

Fastening frames are put on both outsides of the stack which is fastened up together through crossbar setting in the frames. Inlet manifold slots and outlet manifold slots are prepared at the bottoms and heads of the gaskets, respectively. Spacers are incorporated with the gaskets to prevent the contact of cation exchange membranes with anion exchange membranes and to mix the solution. Many stacks are arranged through the fastening frames, and electrode cells are put on both ends of the electrodialyzer, which are fastened by a press putting on the outsides of electrode cells (Fig. 5.7) [30].

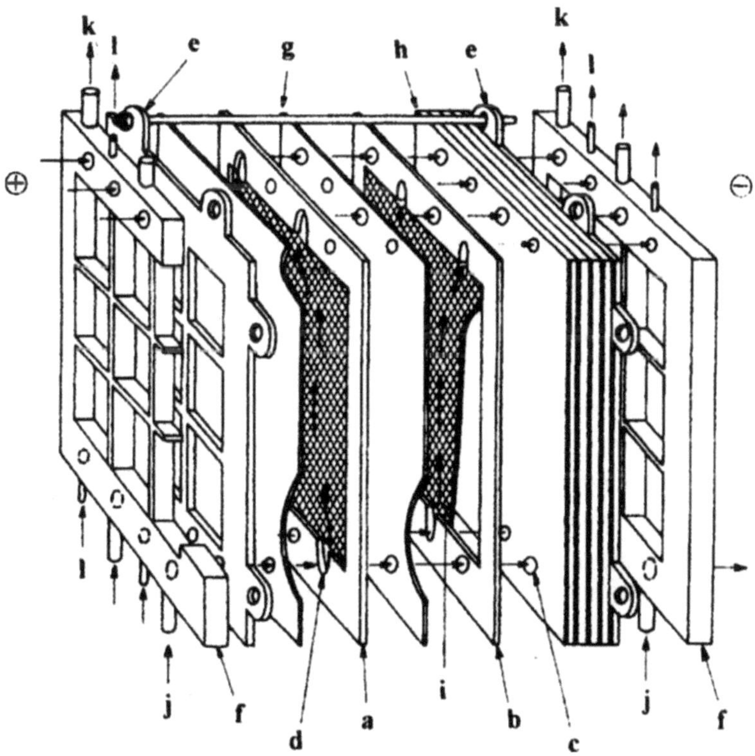

Fig. 5.6 Structure of a stack (Reproduced from Ref. [30] with kind permission of © Sea Water Science, Japan (1980))

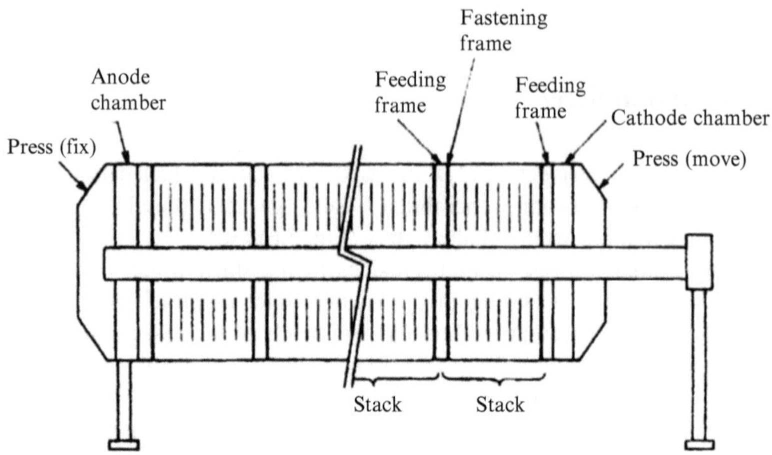

Fig. 5.7 Electrodialyzer (Reproduced from Ref. [30] with kind permission of © Sea Water Science, Japan (1980))

An electrolyte solution to be desalted is supplied from solution feeding frames to entrance manifolds, flows through entrance slots, current passing portions and exit slots, and discharged from exit manifolds to the outside of the stack. A concentrated solution is supplied to concentrating cells with a part of a feeding solution in a circulation flow system and discharged to the outside of the stack through an overflow extracting system.

Electrodialysis process is classified to a continuous (one-pass flow) process (Sect. 5.3.2), a batch process (Sect. 5.3.3), and a feed-and-bleed process (Sect. 5.3.4) [31–33]. The performance of each process is computed by applying the electrodialysis program, including the following steps described in each parenthesized section:

Step 1 Mass transport (cf. Sect. 5.2.2)
Step 2 Current density distribution (cf. Sect. 5.2.5)
Step 3 Cell voltage and energy consumption (cf. Sect. 5.2.6)
Step 4 Limiting current density (cf. Sect. 5.2.7)

The program aims at determining the relationship between the parameters in each process and is developed on the following assumptions:

1. The overall transport number λ, overall solute permeability μ, overall electro-osmotic permeability ϕ, and alternating electric resistance R_{alter} are expressed by the empirical functions of overall hydraulic permeability (leading parameter) ρ (Eqs. 5.12–5.15).
2. Influence of temperature T on the performance of an electrodialyzer is evaluated from the relationship between T and ρ.
3. Solution leakage and electric current leakage in an electrodialyzer are negligible.
4. Direct current electric resistance of a membrane includes the electric resistance of a boundary layer formed on the desalting surface of the membrane due to concentration polarization.

5. Frequency distribution of solution velocity ratio in desalting cells is equated by the normal distribution.
6. Current density i at x distant from the inlets of desalting cells is approximated by the quadratic equation.
7. Voltage difference between the electrodes at the entrance of desalting cells is equal to the value at the exits.
8. Limiting current density of an electrodialyzer is defined as average current density applied to an electrodialyzer when current density reaches the limit of an ion exchange membrane at the outlet of a desalting cell in which linear velocity and electrolyte concentration are the least.
9. Salt concentrations are uniform in concentrating cells, and the concentrated solutions are extracted from the concentrating cells to the outside of the process.

Further, the model equations in the program are developed on the basis of the following experimental measurements:

1. Membrane characteristics (λ, μ, ϕ, ρ)
2. Alternating and direct current electric resistance of membranes
3. Solution velocity distribution between desalting cells and current density distribution in an electrodialyzer
4. Voltage difference between electrodes at the inlets and outlets of desalting cells in an electrodialyzer
5. Physical properties of desalted and concentrated saline water
6. Limiting current density of ion exchange membranes
7. Solution leakage in an electrodialyzer

5.3.2 *Continuous Process*

5.3.2.1 Introduction

The continuous process is applicable to large-scale electrodialysis, and its performance was so far investigated as follows. Belfort and Daly constructed optimization routine for a continuous electrodialysis plant. The algorithm was applied to the Office of Saline Water test bed plant at Webster, South Dakota and compared the actual cost and operating conditions [34]. Avriel and Zeligher developed a mathematical model for preliminary engineering design and economical evaluation of a continuous electrodialysis plant. Detailed cost computations were performed resulting in capital investment and annual operating costs [35]. Lee et al. developed a computer simulation program for describing a continuous flow process and estimated investment and operation costs. Further, electrodialysis plant was designed and optimized in terms of overall costs and the different parameters [36]. Moon et al. investigated ionic transport across membranes based on one- and two-dimensional continuous electrodialysis modeling using the principles of electrochemistry, transport phenomena, and thermodynamics [37]. Fidaleo and Moresi simulated mass transfer, mass balance, potential drop, and limiting current density in a continuous operation based on the Nernst–Planck equation [38]. Sadrzadeh et al. modeled continuous flow desalination starting from a differential equation of steady-state mass balance and give salt concentration in dilute compartments or separation percent for various voltages, flow rates, and feed concentrations [39]. Nikonenko et al. described electrodialysis or electrodeionization characteristics (mass transfer coefficient, Sherwood number, degree of desalination, and others). The reasonability of the functions was discussed with experimental measurements [40]. Brauns et al. developed simulation model through solver software. Experimental verification of the software was performed using industrial type pilot plant. Limiting current density was theoretically evaluated in the model calculations for design purpose and corrected with the experimental results [41]. In the previous investigation, we developed a program for computing the performance of a continuous constant current mode seawater concentrating process. The validity of the program was discussed by comparing with the performance (NaCl concentration in concentrated solutions, energy consumption to obtain 1 t of NaCl) of electrodialyzers operated during 1 year in seven salt-manufacturing plants (output 200,000 t/year for each plant). The differences of data between the computations and the operations are assumed to be due to solution leakages and electric current leakages generated in the electrodialyzers, which are not taken into account in the model. From the above discussion, the validity of the computer simulation model was assumed to be supported by the plant operations [42].

5.3.2.2 Continuous Electrodialysis Process

A continuous electrodialysis process is illustrated in Fig. 5.8, in which an electrodialyzer (effective membrane area S, number of desalting cells N) is operated with constant applied voltage.

A salt solution (concentration C'_{in}) is supplied to the inlets of desalting cells at average linear velocity u_{in}. Salt and solutions transfer from desalting cells to concentrating cells. Average fluxes of ions and solutions transferred through a membrane pair are expressed by J_S and J_V, respectively. In desalting cells, salt concentration is decreased under applied average current density I/S and reaches average salt concentration C'_{out} at the outlets of desalting cells. Salt concentration change in desalting cells causes current density change along the flow-pass from i_{in} at the inlets to i_{out} at the outlets. Figure 5.8 shows that current density becomes i and I/S, respectively, at x and pl distant from the inlets of desalting cells. Further, the figure shows that J_S, J_V, and average salt concentration C' and $C'' = J_S/J_V$ are the values at pl distant from the inlets of desalting cells. Here, l is flow-pass length of a desalting cell, and p is the dimensionless distance from the inlet of a desalting cell at which current density is equal to the average current density I/S of an electrodialyzer. $u = (u_{in} + u_{out})/2$ is the average median velocity in desalting cells. Salt solutions in the concentrating cells are extracted at pl and discharged to the outside of the electrodialyzer. Ionic constituent ratio in the solution supplied is assumed to be the same as that of seawater.

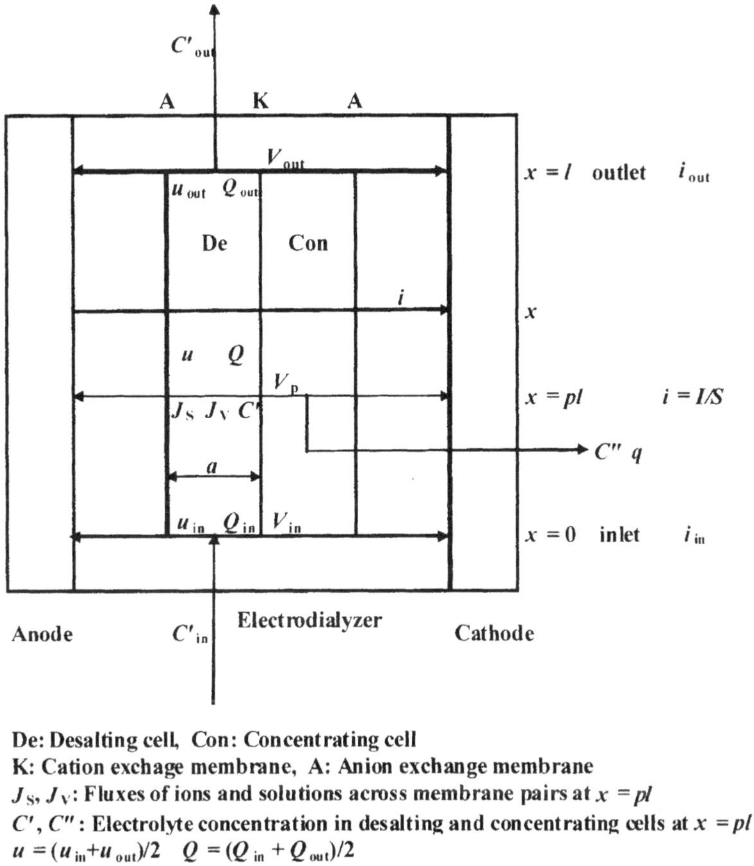

Fig. 5.8 Continuous electrodialysis process (Reproduced from Ref. [43] with kind permission of © Elsevier (2009))

The electrodialysis program [43, 44] is given in Fig. 5.9 (steps 1–3) and Fig. 5.10 (step 4) as described in Sect. 5.3.1.

The equation numbers supplied in the figures correspond to those of equations described in Sects. 5.2 and 5.3. Computation of the electrodialysis program is performed with substituting the following input and control keys in Figs. 5.9 and 5.10.

(a) Input

Overall hydraulic permeability ρ
Salt concentration at the inlets of desalting cells C'_{in}
Average linear velocity at the inlets of desalting cells u_{in}
Flow-pass thickness in a desalting cell a
Flow-pass width in a desalting cell b
Flow-pass length in a desalting cell l

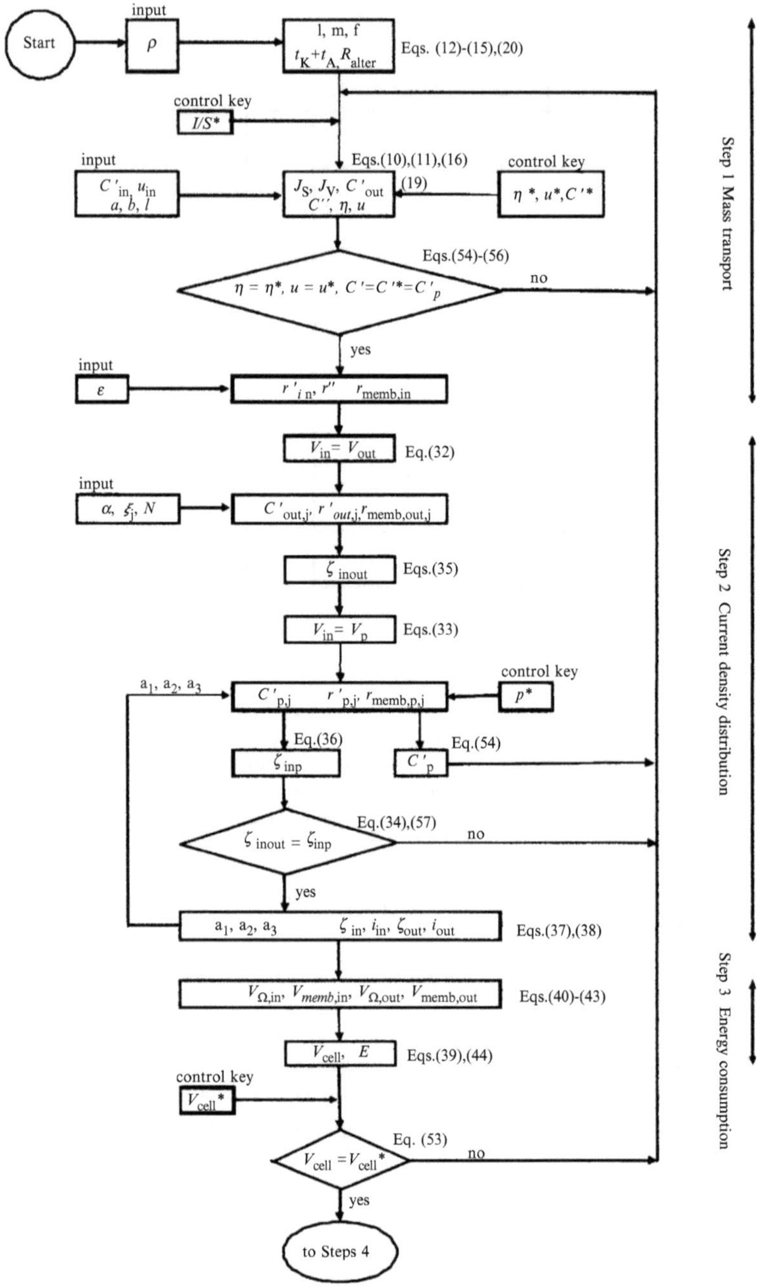

Fig. 5.9 Simulation of mass transport, current density distribution, and energy consumption in a continuous process (steps 1–3) (Reproduced from Ref. [43] with kind permission of © Elsevier (2009))

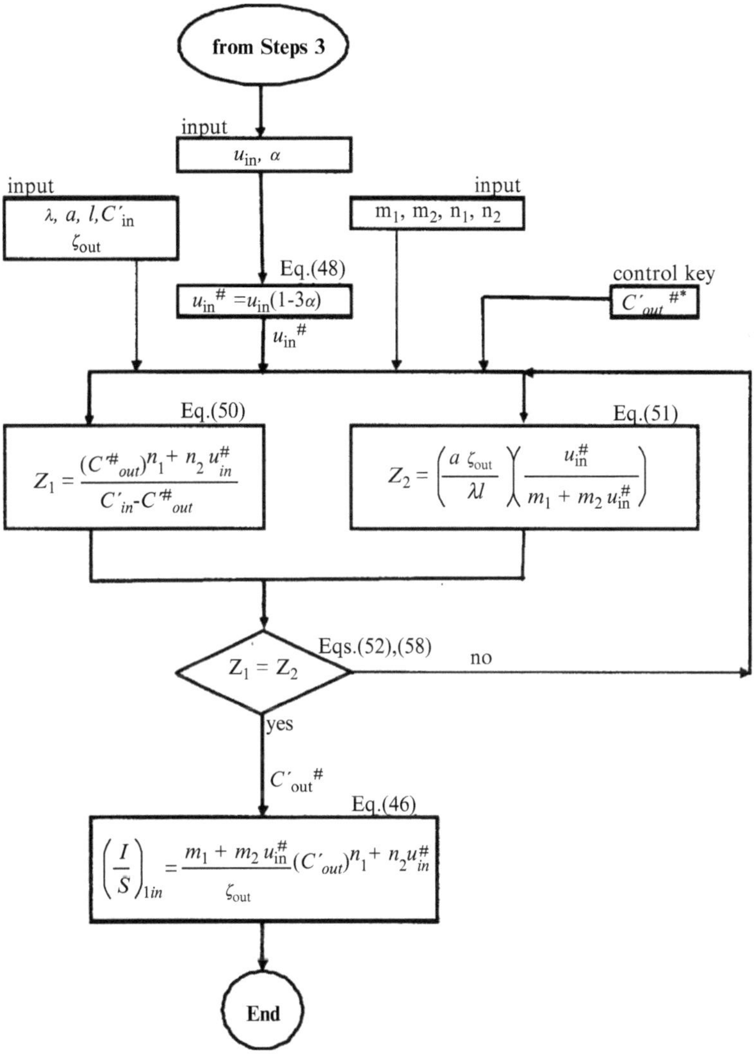

Fig. 5.10 Simulation of limiting current density (step 4) (Reproduced from Ref. [43] with kind permission of © Elsevier (2009))

(b) Control key

Current density I/S^*
Cell voltage $V_{cell}{}^*$
Current efficiency η^*
Average median linear velocity in desalting cells u^*
Average salt concentration in desalting cells at $x = pl$, C'^*
Position at which current density becomes I/S in an electrodialyzer p^*

Computation is carried out using the following trial-and-error calculation:

1. Steps 1–3 (Fig. 5.9)

① I/S^* is adjusted to realize

$$V_{cell} = V_{cell}^* \tag{5.53}$$

② C'^* is adjusted to realize

$$C' = C'_p = \frac{1}{N}\sum_{j=1}^{j\max} Y_j C'_j = C'^* \tag{5.54}$$

③ η^* is adjusted to realize

$$\eta = FJ_S/(I/S) = \eta^* \tag{5.55}$$

④ u is adjusted to realize

$$u = (u_{in} + u_{out})/2 = u^* \tag{5.56}$$

⑤ p^* is adjusted to realize

$$\zeta_{inout} = \zeta_{inp} \tag{5.57}$$

2. Step 4 (Fig. 5.10)
$C'_{out}{}^*$ is adjusted to realize

$$Z_1 = Z_2 \tag{5.58}$$

In the above computation, if an algorithm reaches a decision point (diamond symbol), it is adjusted by changing control keys to realize the equations, Eqs. 5.53–5.58, given in the decision points. Then, it loops back to an earlier portion in the algorithm. The trial-and-error calculation is repeated until all equations are satisfied. The computation is finished within 10 m to obtain one group of plots in the figure shown in Sect. 5.3.2.3.

5.3.2.3 Computation

We try computation assuming that a continuous process is operated applying the following electrodialytic conditions and electrodialyzer specifications and changing the level of C'_{in} and V_{cell}:

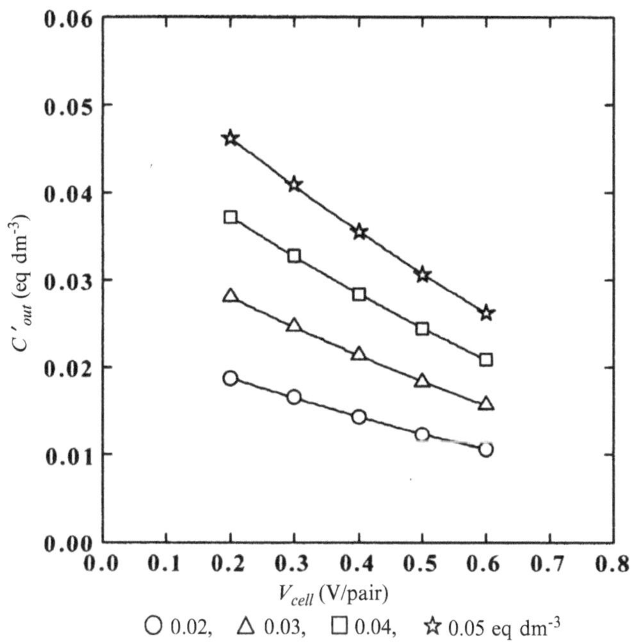

Fig. 5.11 Salt concentration in a desalted solution

Salt concentration at the inlets of desalting cells $C'_{in} = 0.02$ (TDS = 1,157 mg/*l*), 0.03 (1,736), 0.04 (2,315), 0.05 (2,894) eq/dm^3
Cell voltage $V_{cell} = 0.2, 0.3, 0.4, 0.5, 0.6$ V/pair
Linear velocity at the inlets of desalting cells $u_{in} = 10$ cm/s
Standard deviation of normal distribution of solution velocity ratio $\sigma = 0.1$
Flow-pass thickness in a desalting cell $a = 0.05$ cm
Flow-pass width in a desalting cell $b = 100$ cm
Flow-pass length in a desalting cell $l = 100$ cm
Membrane area $= S = bl = 10^4$ cm^2 = 1 m^2
Number of desalting cells integrated in an electrodialyzer $N = 300$ cells
Overall hydraulic conductivity of a membrane pair $\rho = 1 \times 10^{-2}$ cm^4/eq s
Current screening ratio of a spacer $\varepsilon = 0.15$

ρ values of commercially available ion exchange membranes vary to some extent. $\rho = 1 \times 10^{-2}$ cm^4/eq s inputted above is estimated to be a reasonable and convenient value which was decided based on the observed data.

Computed data are explained as follows. Salt concentration at the outlets of desalting cell C'_{out} is decreased with the increase of V_{cell} and the decrease of C'_{in} (Fig. 5.11). Current efficiency η is increased with the increase of V_{cell} and C'_{in} (Fig. 5.12). Desalting ratio $\alpha = (1 - C'_{out}/C'_{in})$ is increased with the increase of V_{cell}; however, it is not influenced by C'_{in} (Fig. 5.13). Water recovery $Re = Q_{out}/Q_{in}$ is higher than 0.9, and it is decreased with the increase of V_{cell} and C'_{in} (Fig. 5.14). Energy consumption E is increased with V_{cell} and C'_{in} (Fig. 5.15).

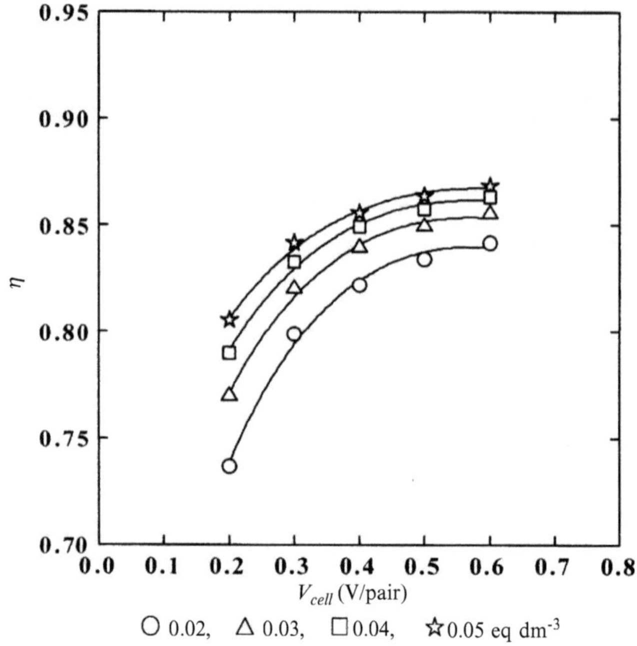

Fig. 5.12 Current efficiency

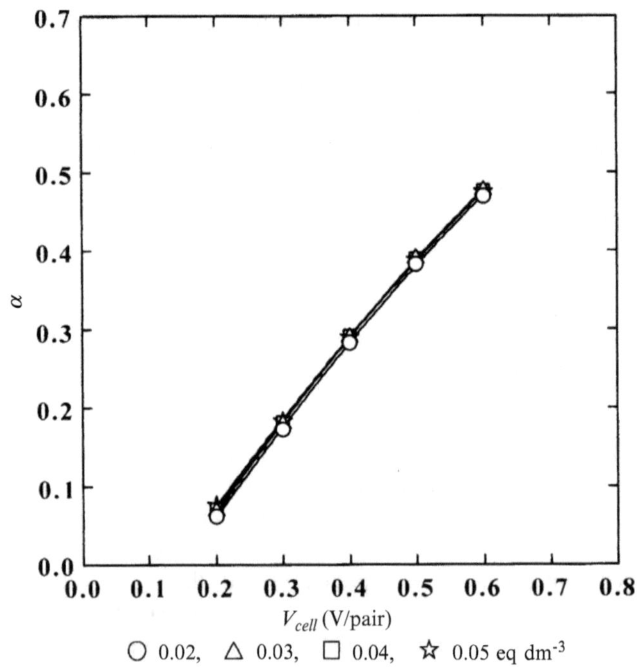

Fig. 5.13 Desalting ratio

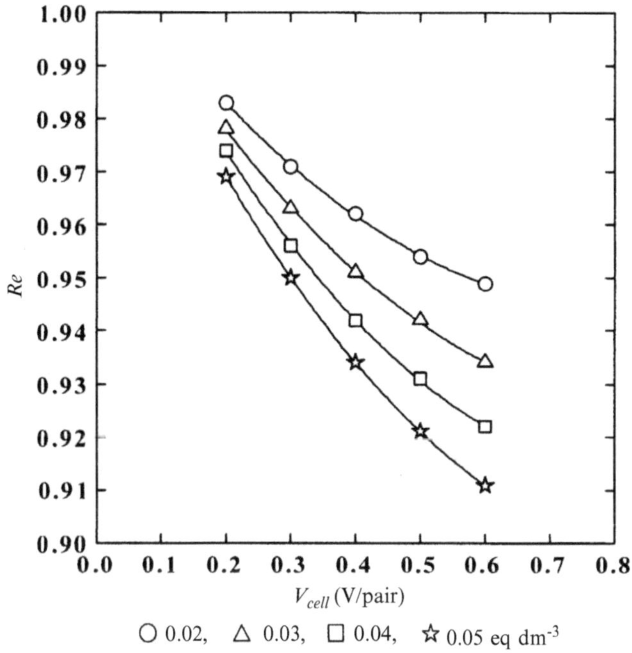

Fig. 5.14 Water recovery

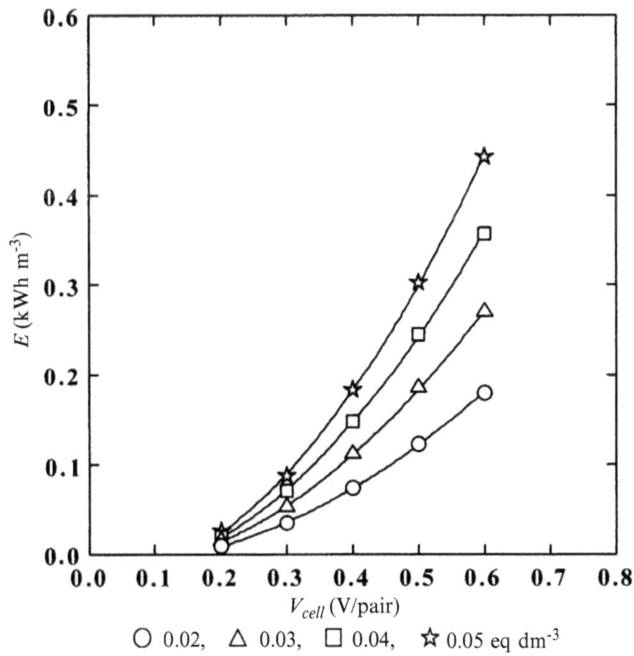

Fig. 5.15 Energy consumption

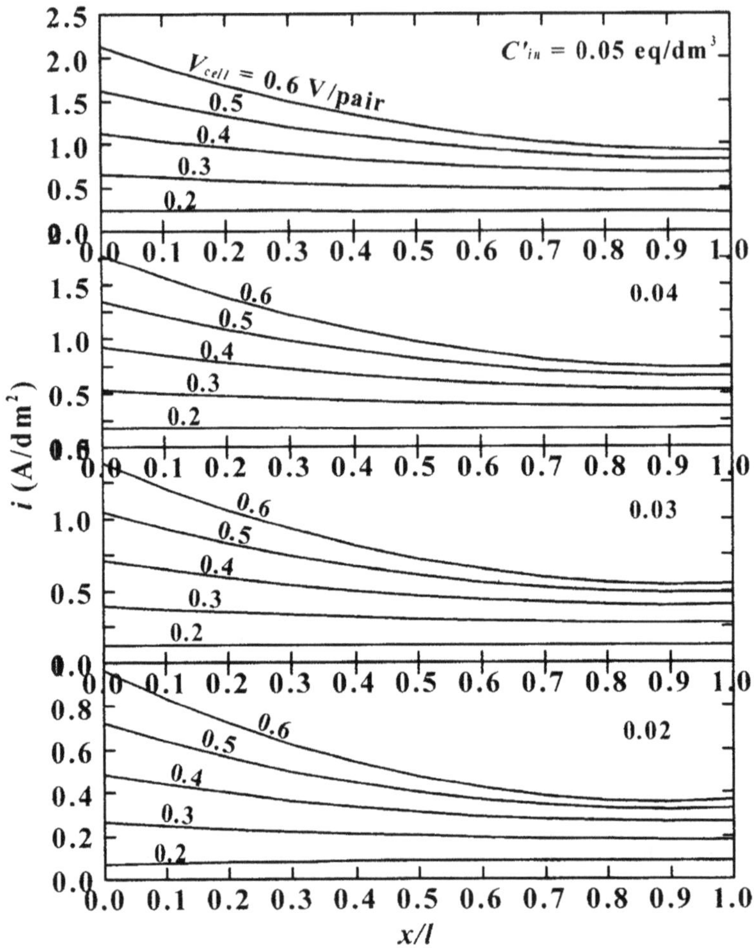

Fig. 5.16 Current density distribution

Figure 5.16 shows current density distribution. Current density i decreases from the inlet toward the outlet of a desalting cell, and it is higher at larger C'_{in}, but its change along the flow-pass is relatively larger at lower C'_{in}. Figure 5.17 shows inlet current density nonuniformity coefficient ζ_{in} and outlet current density nonuniformity coefficient ζ_{out}. ζ_{in} and ζ_{out} coverage 1 at lower V_{cell}. ζ_{in} is increased and ζ_{out} is decreased with the increase of V_{cell} and decrease of C'_{in}. Limiting current density $(I/S)_{lim}$ is plotted against current density I/S taking V_{cell} as a parameter (Fig. 5.18). Intersections between the plots and a $I/S = (I/S)_{lim}$ line show the real limiting current density of the electrodialyzer $(I/S)_{lim}$ operating in this section, which is estimated to be at over 0.6 V/pair.

σ is influenced extremely by precision of part dimensions of a stack and skill of stack assembling work of an electrodialyzer, and it is desirable to keeping σ value as low as possible. An electrodialyzer is operated stably at $\sigma < 0.1$ because the

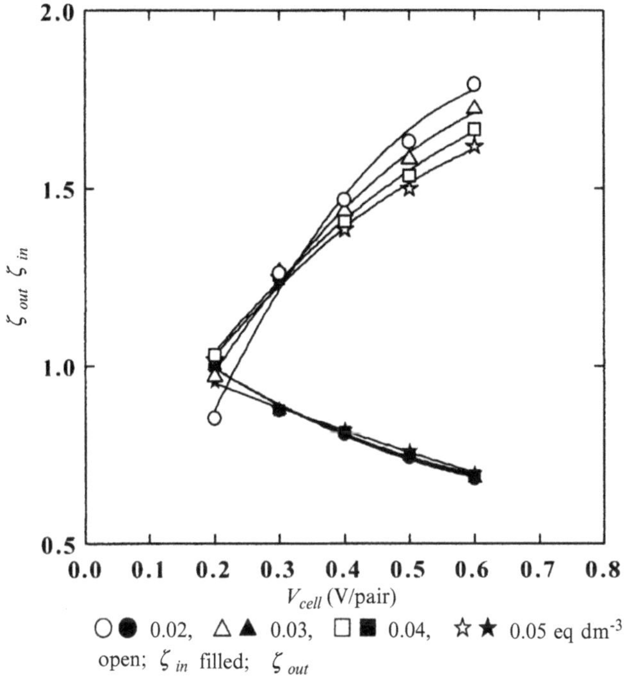

Fig. 5.17 Current density nonuniformity coefficient

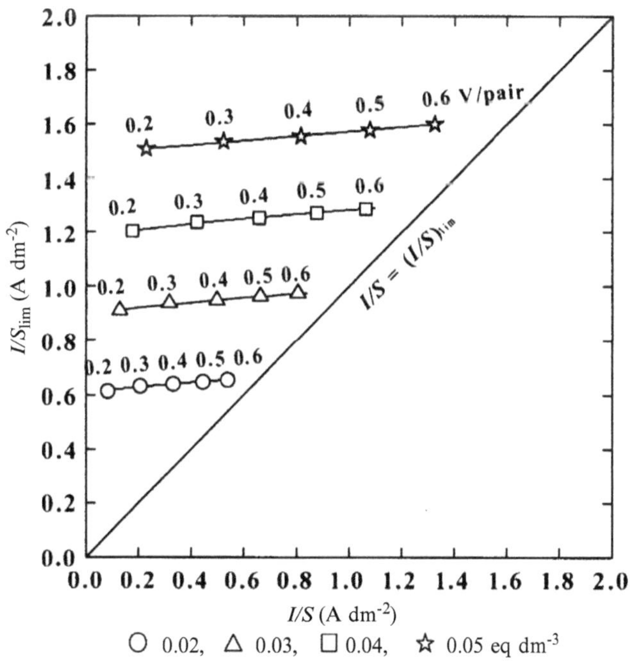

Fig. 5.18 Limiting current density

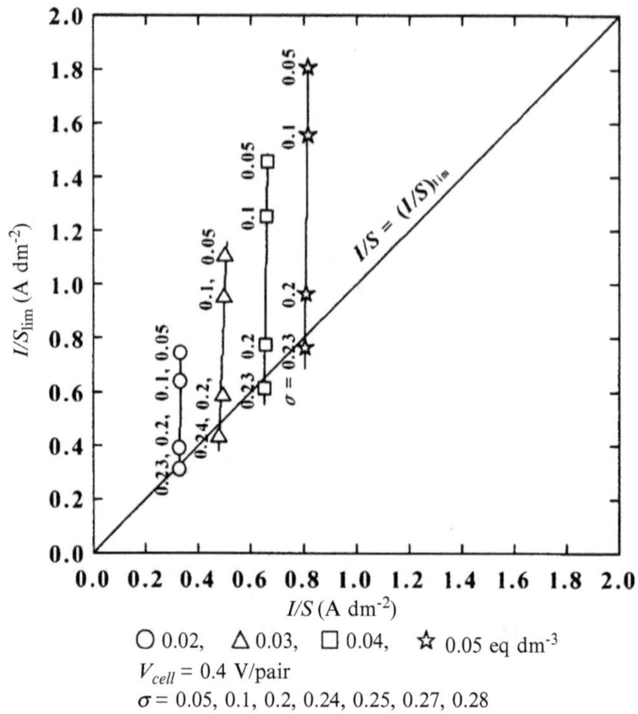

Fig. 5.19 Relationship between σ and $(I/S)_{\text{lim}}$

limiting current density $(I/S)_{\text{lim}}$ is high. However, electrodialyzer operation becomes unstable at $\sigma > 0.15$ because $(I/S)_{\text{lim}}$ is lowered. $\sigma = 0.1$ defined in Sects. 5.3.2.3, 5.3.3.3 and 5.3.4.3 functions as a safty factor for operating an electrodialyzer stably. $(I/S)_{\text{lim}}$ is calculated at $V_{cell} = 0.4$ V/pair taking σ as parameter and keeping the other parameters as described above and shown in Fig. 5.19. The limiting current density of this electrodialyzer is obtained from intersections between $(I/S)_{\text{lim}}$ lines and the I/S-$(I/S)_{\text{lim}}$ line. $(I/S)_{\text{lim}}$ is found at $\sigma = 0.23$ for every C'_{in}. C'_{out}, η, α, Re, and E are not influenced by σ. We conclude that excepting limiting current density, the performance of an electrodialyzer is not influenced by σ.

5.3.3 *Batch Process*

5.3.3.1 Introduction

A batch process is applicable to the operation of every kind of small- or middle-scale electrodialysis operations; Tokuyama Inc. installed a commercial multistage batch system electrodialysis plant for desalination of brackish groundwater at Hatsushima, Atami, Shizuoka Pref., Japan [45]. Tokuyama Inc. developed a seawater desalination batch mode unit operated in a vessel [46]. Rapp and Pfromm removed chloride from the chemical recovery cycle of a kraft pulp mill [47]. Elidaoui et al. demineralized sugar liquor in a beet sugar-manufacturing pilot plant [48]. Banasiak et al. investigated the removal of fluoride and nitrate from brackish groundwater [49]. Walha et al. demineralized brackish groundwater by nanofiltration, reverse osmosis, and electrodialysis in Tunisia and compared their energy consumption [50]. Kabay et al. removed nitrate, fluoride, and boron in a salt solution with a batch electrodialyzer [51–53]. The performance of a batch process has been discussed from various view points; Parulekar investigated energy consumption of batch operation in (1) constant current, (2) constant voltage, (3) constant current followed by constant voltage, (4) constant voltage followed by constant current, and (5) operation with time-variant current and voltage [54]. Demircioglu et al. introduced equations expressing the ionic mass balance around a dilute circulation tank and discussed energy consumption in a batch process [55]. Ahmed et al. developed a mathematical model to predict changes in contaminants with time and to estimate contaminant fluxes of migration, diffusion, and convection in a laboratory-scale batch electrodialysis cell for the regeneration of contaminated hard-chrome plating baths [56]. Moon et al. predicted the performance of a one- and two-dimensional batch electrodialysis process based on the fundamental principles of electrochemistry, transport phenomena, and thermodynamics [37]. Ortiz et al. developed a mathematical model for a batch process and discussed mass balance, ohmic drop, and membrane potential [57].

5.3.3.2 Batch Electrodialysis Process

In Fig. 5.20, a definite volume $Q' = Q'_{start}$ of a salt solution (concentration $C'_{in} = C'_{in,start}$) is prepared in the circulation tank at first. Next, we open valve V_1, close valve V_2, and then the solution is supplied to the inlets of desalting cells in an electrodialyzer setting linear and volume velocity of the solution at u_{in} and q'_{in}, respectively, and applying constant cell voltage V_{cell}. Effective membrane area and number of desalting cells are, respectively, S and N. Flow-pass thickness and length in a desalting cell are a and l, respectively. Current density in an electrodialyzer decreases from i_{in} at the inlets ($x = 0$), via $i = I/S$ at $x = pl$ and i at x, to i_{out} at the outlets ($x = l$) in desalting cells. Voltage difference between a cathode and anode is maintained at V_{in} ($x = 0$) $= V_p$ ($x = pl$) $= V_{out}$ ($x = l$). J_S and J_V are, respectively, fluxes of ions and solutions across membrane pairs from desalting cells to concentrating cells at $x = pl$. C' is salt concentration in desalting cells at $x = pl$. C'_{out}, u_{out}, and q'_{out} are, respectively, salt concentration, linear velocity, and volume velocity of solutions at the outlets of desalting cells. u and q' are, respectively, average median linear and volume velocity in desalting cells ($u = (u_{in} + u_{out})/2$, $q' = (q'_{in} + q'_{iout})/2$). C'' and q'' are, respectively, salt concentration and volume flow velocity of a concentrated solution, which is extracted at $x = pl$ to the outside of the process. Electrodialysis is performed until salt concentration of a desalted solution in the tank C'_{in} changes from $C'_{in,start}$ at operating time $t = 0$ to a definite value $C'_{in,final}$ at $t = t_{ope}$, and then valve V_1 is closed, valve V_2 is opened, and the solution is discharged to the outside of the process to obtain a desalted solution (potable water).

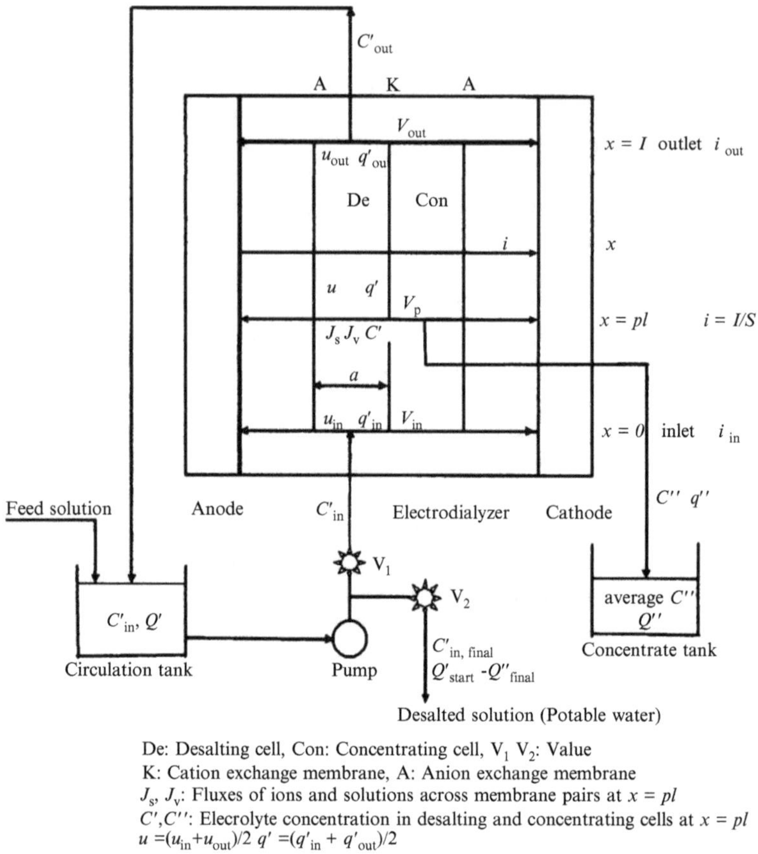

Fig. 5.20 Batch electrodialysis process (Reproduced from Ref. [58] with kind permission of © Elsevier (2009))

The electrodialysis program consists of mass transport (step 1), current density distribution (step 2), and cell voltage (step 3) (Fig. 5.21) [58]. The equation numbers supplied in the figures correspond to those of equations described in Sects. 5.2 and 5.3. Computation is performed with substituting the same input and control keys described in Sect. 5.3.2.2. In this computation, however, it is necessary to determine the relationship between C'_{in} and I/S, C'' and C' which is expressed as follows and exemplified in Fig. 5.22:

$$I/S = \mathrm{X}_1 C'^{\,0.5}_{in} + \mathrm{X}_2 C'_{in} + \mathrm{X}_3 C'^{\,1.5}_{in} \tag{5.59}$$

$$C'' = \mathrm{Y}_1 C'^{\,0.5}_{in} + \mathrm{Y}_2 C'_{in} + \mathrm{Y}_3 C'^{\,1.5}_{in} \tag{5.60}$$

$$C' = \mathrm{Z}_1 C'^{\,0.5}_{in} + \mathrm{Z}_2 C'_{in} \tag{5.61}$$

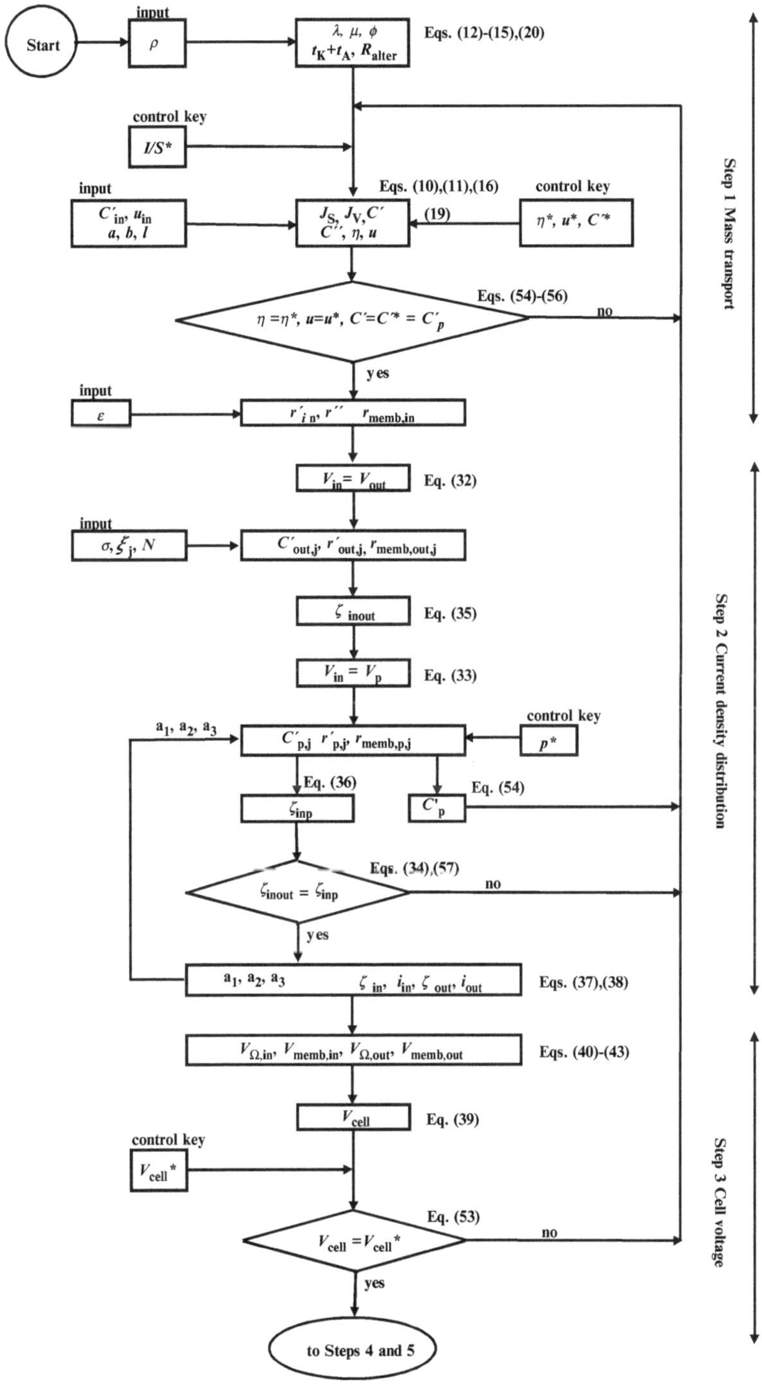

Fig. 5.21 Simulation of performance of a batch electrodialysis process (steps 1–3) (Reproduced from Ref. [58] with kind permission of © Elsevier (2009))

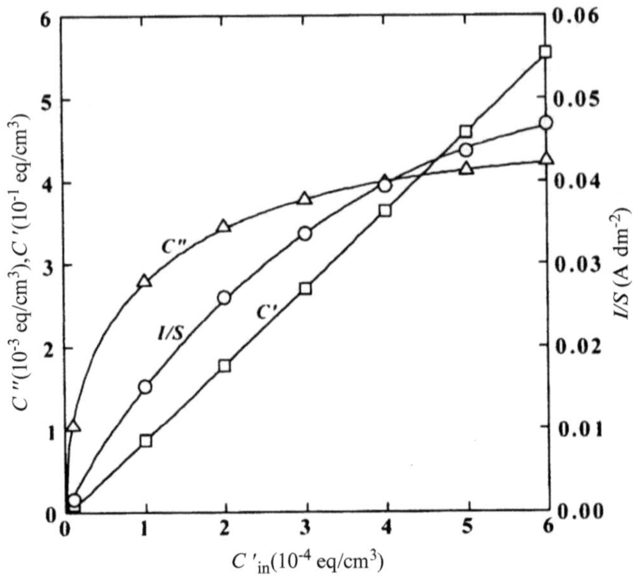

Fig. 5.22 C'_{in} versus I/S, C'', and C' (Reproduced from Ref. [58] with kind permission of © Elsevier (2009))

The coefficients X_1, X_2, X_3, Y_1, Y_2, Y_3, Z_1, and Z_2 are substituted into Eqs. 5.63 and 5.64 in the following step 4 (Fig. 5.23) for computing, respectively, batch duration t_{ope} and the volume of a solution Q'' being discharged from concentrating cells to the outside of the process.

Step 4 Mass balance and energy consumption (Fig. 5.23) in the batch process are expressed as follows, taking into account the overall mass transport equation, Eqs. 5.10 and 5.11:

$$-\left[Q'_{start} - \left\{\phi\frac{I}{S} + \rho(C'' - C')\right\}NS\right]\frac{dC'_{in}}{dt} = \left\{\lambda\frac{I}{S} - \mu(C'' - C')\right\}NS \quad (5.62)$$

Operating time t_{ope} during which C'_{in} decreases from $C'_{in,start}$ to $C'_{in,final}$, i.e., batch duration is introduced from Eq. 5.62 as follows:

$$t_{ope} = \frac{1}{NS}\int_{C'_{in,final}}^{C'_{in,start}} \frac{Q'_{start} - \left(A_1C'^{0.5}_{in} + A_2C'_{in} + A_3C'^{1.5}_{in}\right)NS}{B_1C'^{0.5}_{in} + B_2C'_{in} + B_3C'^{1.5}_{in}} dC'_{in} \quad (5.63)$$

in which

$$A_1 = \phi X_1 + \rho(Y_1 - Z_1)$$

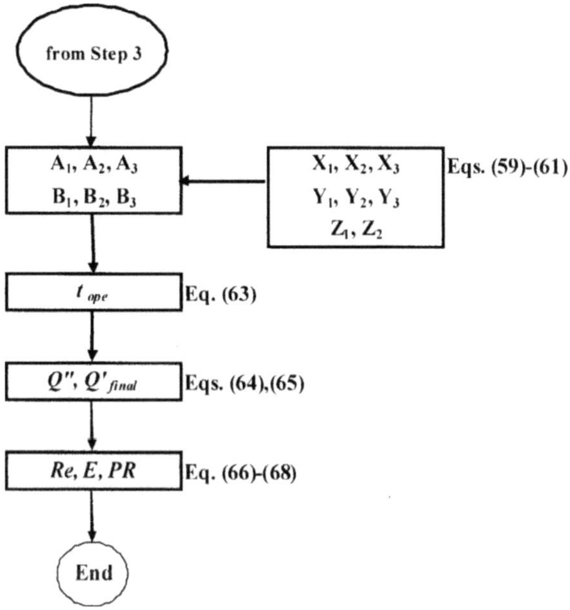

Fig. 5.23 Simulation of mass balance and energy consumption (step 4) (Reproduced from Ref. [58] with kind permission of © Elsevier (2009))

$$A_2 = \phi X_2 + \rho(Y_2 - Z_2)$$

$$A_3 = \phi X_3 + \rho Y_3$$

$$B_1 = \lambda X_1 - \mu(Y_1 - Z_1)$$

$$B_2 = \lambda X_2 - \mu(Y_2 - Z_2)$$

$$B_3 = \lambda X_3 - \mu Y_3$$

The volume of a solution being discharged from concentrating cells to the outside of the process Q'' and output of a desalted solution Q'_{final} is given as follows:

$$Q'' = \int_0^{t_{ope}} \left\{ \phi \frac{I}{S} + \rho(C'' - C') \right\} NSdt = \int_0^{t_{ope}} \left(A_1 C'^{0.5}_{in} + A_2 C'_{in} + A_3 C'^{1.5}_{in} \right) NSdt \tag{5.64}$$

$$Q'_{final} = Q'_{start} - Q'' \tag{5.65}$$

Water recovery Re is

$$\mathrm{Re} = \frac{Q'_{final}}{Q'_{start}} \tag{5.66}$$

Energy consumption to obtain 1 m^3 of a desalted solution (potable water) E (kWh/m^3) and production rate of a desalted solution PR (m^3/h) are, respectively, presented as follows:

$$E = \frac{V_{cell}NS}{Q'_{final}} \int_0^{t_{ope}} \frac{I}{S} dt \tag{5.67}$$

$$PR = \frac{Q'_{final}}{t_{ope}} \tag{5.68}$$

Step 5 Limiting current density is explained in Sect. 5.2.7, and its simulation is given in Fig. 5.10.

5.3.3.3 Computation

A batch process is assumed to be operated applying constant voltage between electrodes for obtaining potable water. We assume the following specifications of an electrodialyzer operating with changing the level of $C'_{in,start}$, V_{cell}, and Q'_{start}.

Salt concentration of a solution supplied to an electrodialyzer at the beginning of operation $C'_{in,start} = 500$, 1,000, 2,000, 3,000, 4,000 mg/l (ionic constituent ratio is the same as that of seawater).
Cell voltage $V_{cell} = 0.2, 0.25, 0.3, 0.4, 0.5, 0.6, 0.7, 0.8$ V/pair
Volume of a solution prepared in a circulation tank $Q'_{start} = 5, 10, 15, 20\ m^3$
Standard deviation of solution velocity ratio in desalting cells $\sigma = 0.1$
Linear velocity at the inlets of desalting cells $u_{in} = 10$ cm/s
Flow-pass thickness in a desalting cell $a = 0.05$ cm
Flow-pass width in a desalting cell $b = 100$ cm
Flow-pass length in a desalting cell $l = 100$ cm
Membrane area $S = bl = 10^4\ cm^2 = 1\ m^2$
Number of desalting cells $N = 300$ cells
Overall hydraulic permeability $\rho = 1 \times 10^{-2}\ cm^4/eq\ s$
Electric current screening ratio $\varepsilon = 0.15$

Figures 5.24 and 5.25 exemplify the change of C'_{in} and I/S with operating time t setting $Q'_{start} = 10\ m^3$ and $V_{cell} = 0.4$ V/pair. Operation time (batch duration) t_{ope} versus V_{cell} computed setting $Q'_{start} = 10\ m^3$ is given in Fig. 5.26. V_{cell} versus potable water production rate PR, water recovery Re, and energy consumption E are computed taking $C'_{in,start}$ as a parameter and shown in Figs. 5.27–5.29 which are not influenced by Q'_{start}. Figure 5.30 shows relationship between Q'_{start} and t_{ope}. Limiting current density $(I/S)_{\lim}$ versus current density I/S is given in Fig. 5.31 setting V_{cell} and C'_{in} as parameters. A $I/S = (I/S)_{\lim}$ line is indicated in the figure showing that I/S approaches $(I/S)_{\lim}$ at increased V_{cell} and decreased C'_{in}.

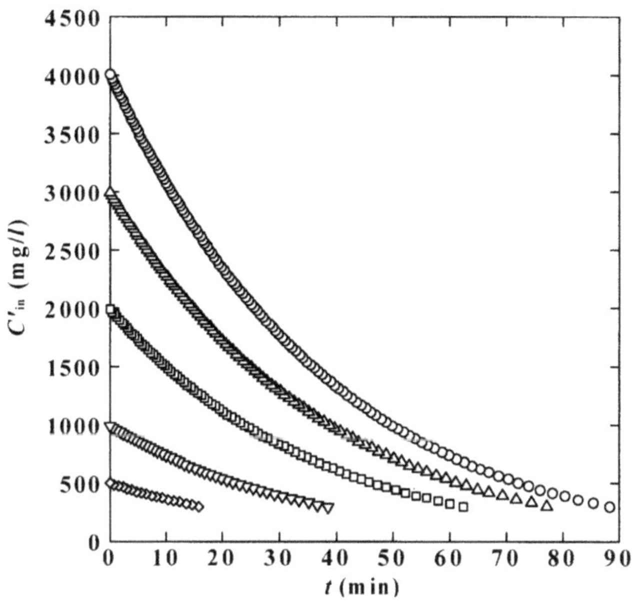

Fig. 5.24 Salt concentration changes in a feeding solution with operating time (Reproduced from Ref. [58] with kind permission of © Elsevier (2009))

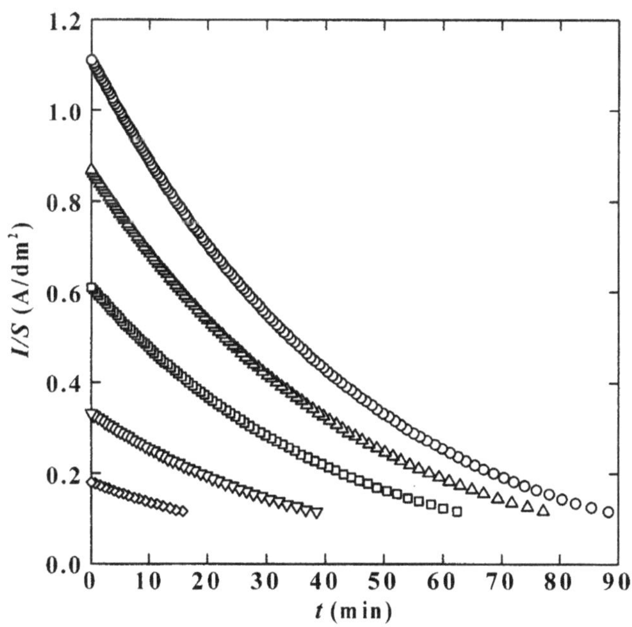

Fig. 5.25 Current density changes in a feeding solution with operating time (Reproduced from Ref. [58] with kind permission of © Elsevier (2009))

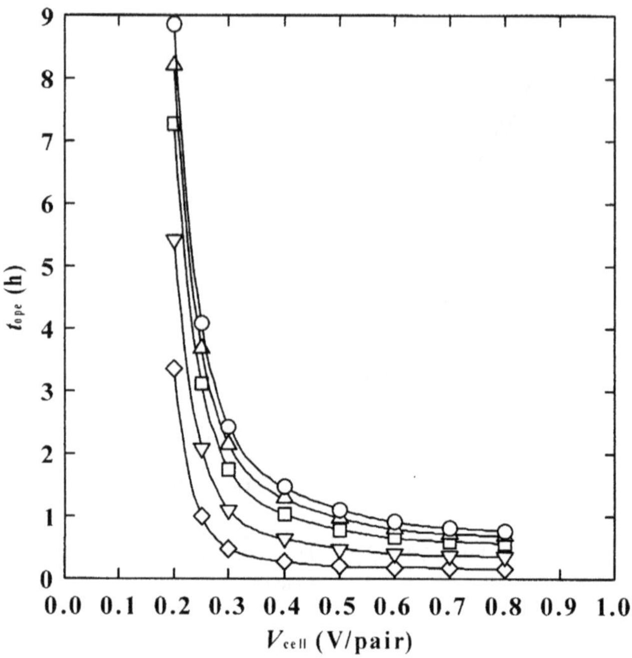

Fig. 5.26 Operating time (Reproduced from Ref. [58] with kind permission of © Elsevier (2009))

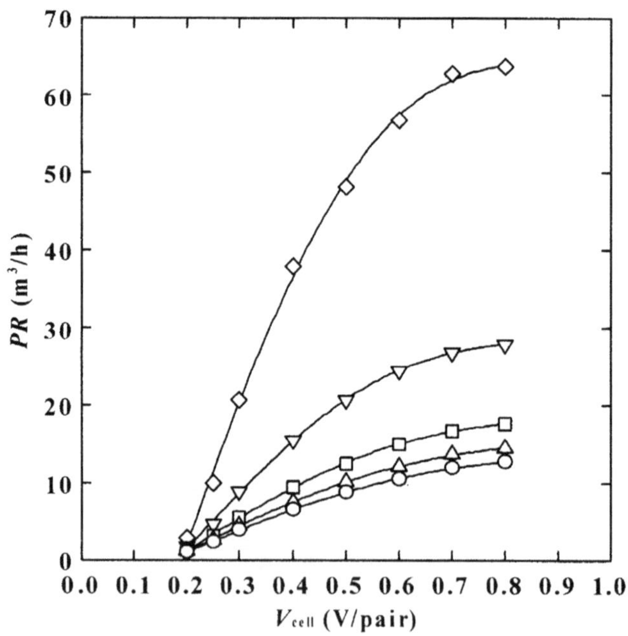

Fig. 5.27 Water production rate (Reproduced from Ref. [58] with kind permission of © Elsevier (2009))

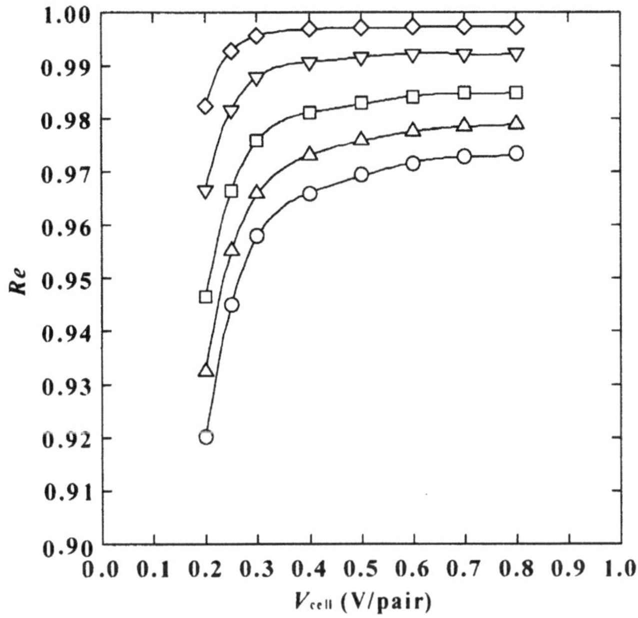

Fig. 5.28 Water recovery (Reproduced from Ref. [58] with kind permission of © Elsevier (2009))

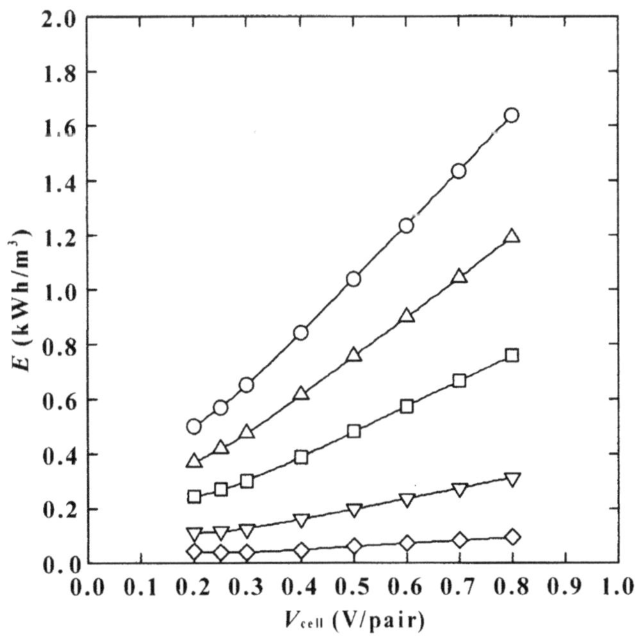

Fig. 5.29 Energy consumption (Reproduced from Ref. [58] with kind permission of © Elsevier (2009))

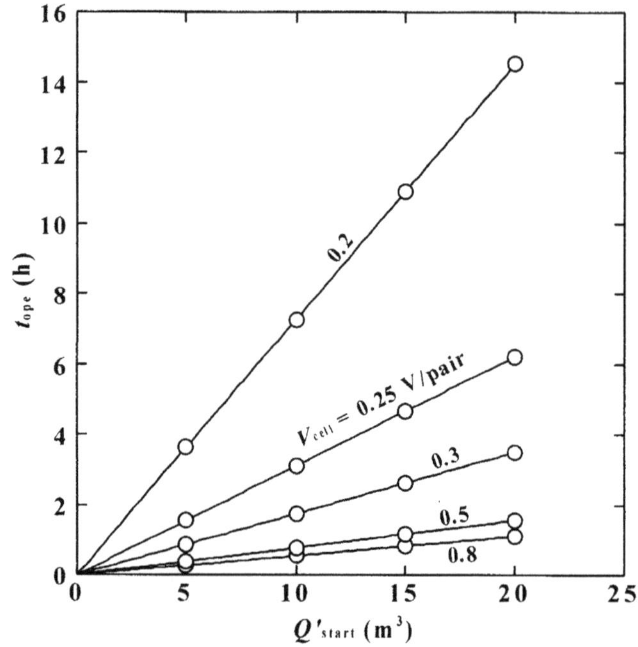

Fig. 5.30 Relationship between volume of a solution prepared and operation time (Reproduced from Ref. [58] with kind permission of © Elsevier (2009))

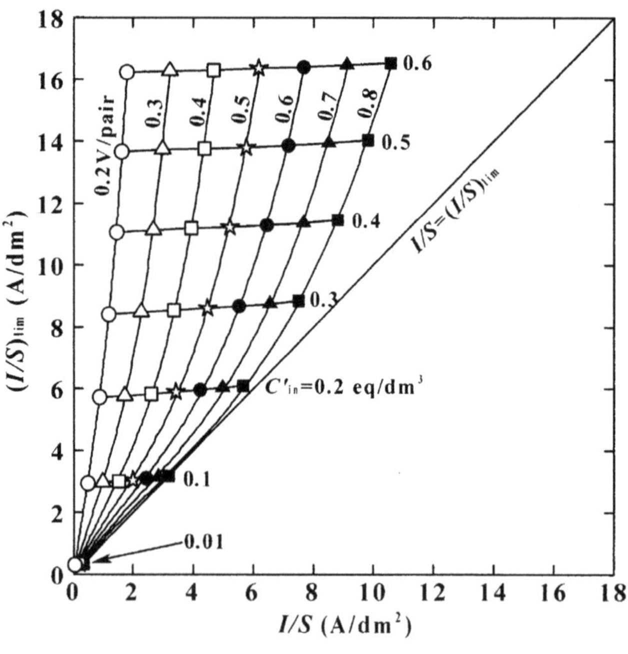

Fig. 5.31 Limiting current density (Reproduced from Ref. [58] with kind permission of © Elsevier (2009))

5.3.4 Feed-and-Bleed Process

5.3.4.1 Introduction

A feed-and-bleed operation is widely performed so far. For instance, Asahi Chemical Co. developed a feed-and-bleed electrodialysis plant (2,500 m^3/day) for producing potable water in Ohoshima island, Tokyo [59]. Asahi Glass Co. constructed large-scale blackish water desalination plants in Kashima Power Station (2,000 m^3/day) and Kashima South Joint Power Station (12,000 m^3/day), Ibaraki Pref., Japan [60]. Morinaga Milk Industry Co. developed a four-stage feed-and-bleed electrodialysis process for demineralizing whey [61]. Tokuyama Inc. developed the technology to establish a closed system by means of electrodialytic reuse of waste water in a plating process [62]. Ryabtsev et al. developed a two-stage feed-and-bleed setup for desalination of underground saline water. The first stage involves desalination of initial water in galvanostatic regime at increased current density, and the second stage involves profound desalination in a potentiostatic regime resulting in desalinated water [63]. Rapp and Pfromm removed chloride in a kraft-pulping process with feed-and-bleed operation incorporated with monovalent-selective anion exchange membranes [64]. Thompson et al. separated sulfide from hydroxide in kraft white liquor with an electrodialysis system operated in both the batch and feed-and-bleed modes, producing sulfide-rich white liquor [65].

5.3.4.2 Feed-and-Bleed Electrodialysis Process

Figure 5.32 shows solution flow in a feed-and-bleed process. Saline water (salt concentration C'_0) is supplied to compartment I in a circulation tank with volume flow rate of Q'_0. Then, it (salt concentration C'_{in}) is supplied (volume flow rate Q'_{in}) to desalting cells in an electrodialyzer in which the salt concentration is decreased to C'_{out} under an applied electric potential field. A concentrated solution (salt concentration C'') is transferred to concentrating cells across the membranes and extracted to the outside of the process. The diluate in the desalting cells is returned (volume flow rate Q'_{out}) to compartment II in the circulating tank, it (salt concentration C'_{out}) flows out (volume flow rate Q''') to the outside of the process to obtain potable water (product), and it is transferred to compartment I (volume flow rate $Q'_{in} - Q'_0$).

The electrodialysis program consists of mass transport (step 1), current density distribution (step 2), and cell voltage (step 3) (Fig. 5.33) [66]. The equation numbers supplied in the figures correspond to those of equations described previously. After step 3, mass balance and energy consumption (step 4) is added.

In Fig. 5.32, the volume of a solution flowing out from desalting cells Q'_{out} and that of a solution flowing out from compartment II in the circulation tank Q''' are:

$$Q'_{out} = Q'_{in} - SNJ_V \tag{5.69}$$

$$Q''' = Q'_0 - SNJ_V \tag{5.70}$$

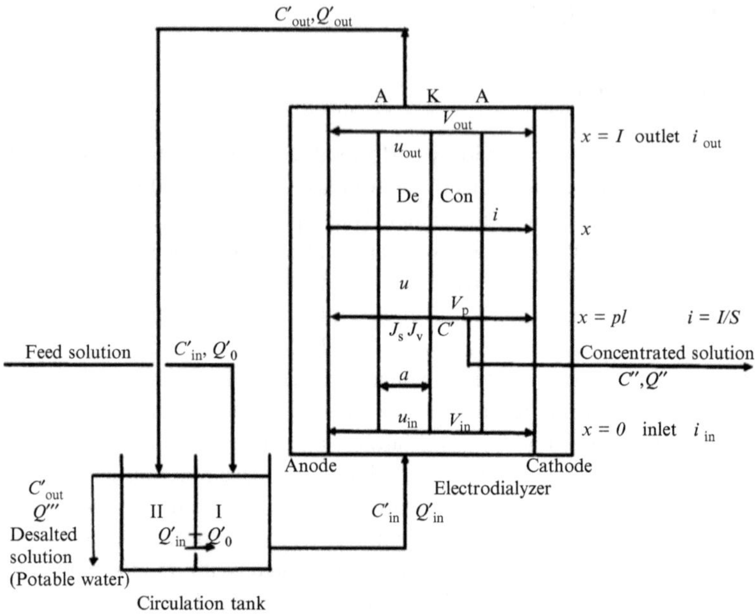

Fig. 5.32 Feed-and-bleed electrodialysis process (Reproduced from Ref. [66] with kind permission of © Elsevier (2010))

Material balance in compartment I is:

$$C'_{in}Q'_{in} = C'_0 Q'_0 + C'_{out}(Q'_{in} - Q'_0) \tag{5.71}$$

Feeding solution volume Q'_0, water recovery Re, energy consumption E, and desalting ratio α in this process are introduced from the above equations as follows:

$$Q'_0 = \frac{C'_{in} - C'_{out}}{C'_0 - C'_{out}} Q'_{in} \tag{5.72}$$

$$\mathrm{Re} = \frac{Q'''}{Q'_0} = \frac{Q'_0 - SNJ_V}{Q'_0} = 1 - \frac{SNJ_V}{Q'_0} \tag{5.73}$$

$$E = \frac{IV_{cell}N}{Q'''} = \left(\frac{1}{Q'''}\right)\left(\frac{I}{S}\right) SV_{cell}N \tag{5.74}$$

$$\alpha = 1 - \frac{C'_{out}}{C'_0} \tag{5.75}$$

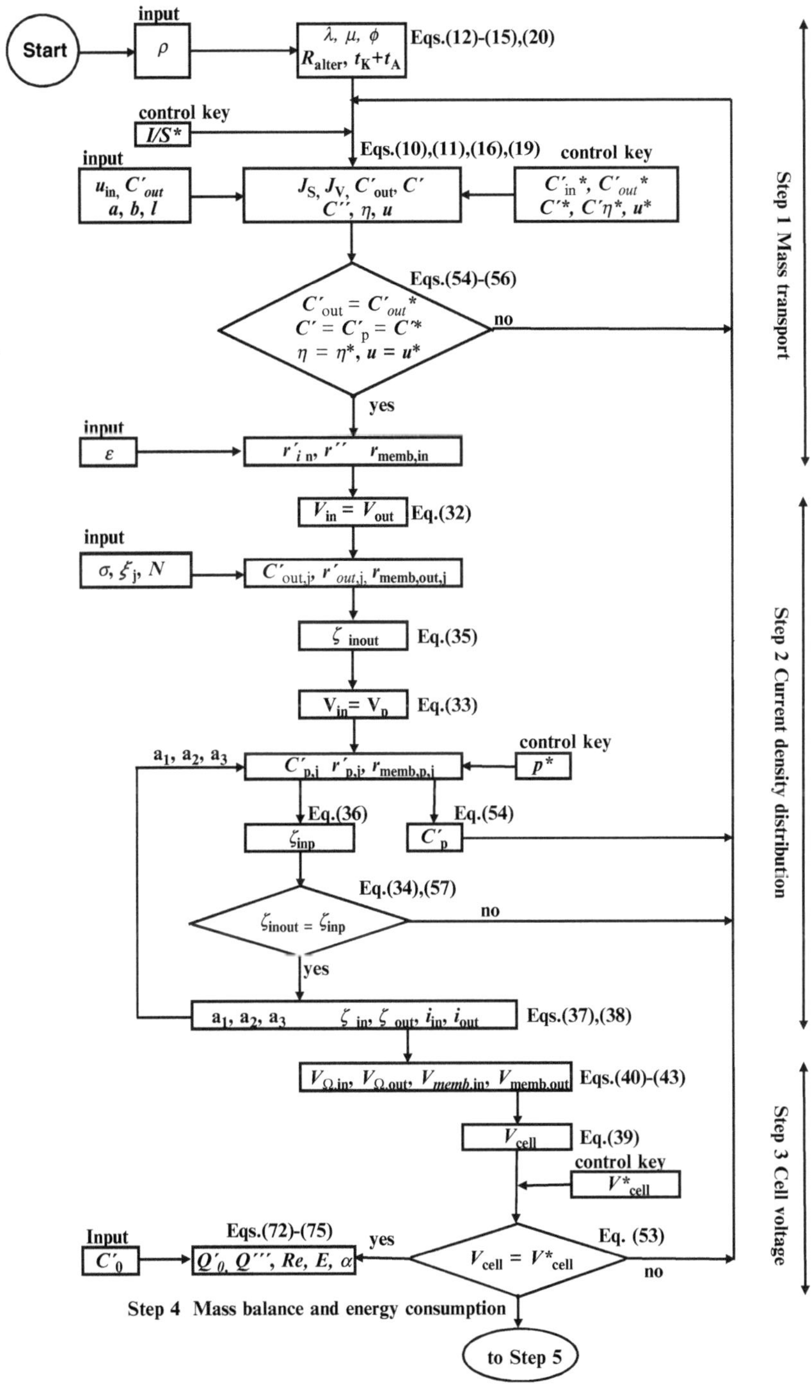

Fig. 5.33 Simulation of performance of a feed-and-bleed electrodialysis process (steps 1–4) (Reproduced from Ref. [66] with kind permission of © Elsevier (2010))

Step 5 Limiting current density is explained in Sect. 5.2.7, and its simulation is given in Fig. 5.10.

5.3.4.3 Computation

A feed-and-bleed process is assumed to be operated applying constant voltage between electrodes for obtaining potable water. We assume the following specifications of an electrodialyzer operating with changing the level of C'_0, and V_{cell}:

Salt concentration of a feeding solution $C'_0 = 500$, 1,000, 2,000, 3,000, 4,000, 5,000 mg/l (ionic constituent ratio is the same as that of seawater)
Cell voltage $V_{cell} = 0.1$, 0.2, 0.25, 0.3, 0.4, 0.5, 0.6, 0.7, 0.8 V/pair
Salt concentration of a desalted solution $C'_{out} = 300$ mg/l (potable water)
Standard deviation of normal distribution of solution velocity ratio $\sigma = 0.1$
Linear velocity at the inlets of desalting cells $u_{in} = 10$ cm/s
Flow-pass thickness of a desalting cell $a = 0.05$ cm
Flow-pass width of a desalting cell $b = 100$ cm
Flow-pass length of a desalting cell $l = 100$ cm
Membrane area $S = bl = 10^4$ cm^2 $= 1$ m^2
Number of desalting cells integrated in an electrodialyzer $N = 300$ cells
Overall hydraulic permeability of a membrane pair $\rho = 1 \times 10^{-2}$ cm^4/eq s
Current screening ratio of a spacer $\varepsilon = 0.15$

Relationships between cell voltage V_{cell} and current density I/S, ionic flux J_S, solution flux J_V, current efficiency η, ohmic potential V_Ω, and membrane potential V_m are shown in Fig. 5.34. These parameters are independent of salt concentration in a feeding solution C'_0. Q''' corresponds to product capacity of the unit, and it increases with the decrease in C'_0 and the increase in V_{cell} (Fig. 5.35). E increases with the increase in C'_0 and V_{cell}; however, minimum E values are found near $V_{cell} = 0.2$ V/pair (Fig. 5.36). The increase in E at $V_{cell} < 0.2$ V/pair is due to decrease in Q''' (Fig. 5.35). Re decreases with the increase in C'_0 and the decrease in V_{cell} (Fig. 5.37) because of decrease in Q'''(Fig. 5.35).

5.4 Conclusion

The main principles being included in the computer simulation program (electrodialysis program) are as follows:

1. Overall mass transport equation, Eqs. 5.10 and 5.11
2. Relationship between the overall hydraulic permeability (leading parameter)ρ and other overall membrane pair characteristics, Eqs. 5.12–5.15
3. Current density distribution, Sect. 5.2.5
4. Limiting current density of an electrodialyzer, Sect. 5.2.7

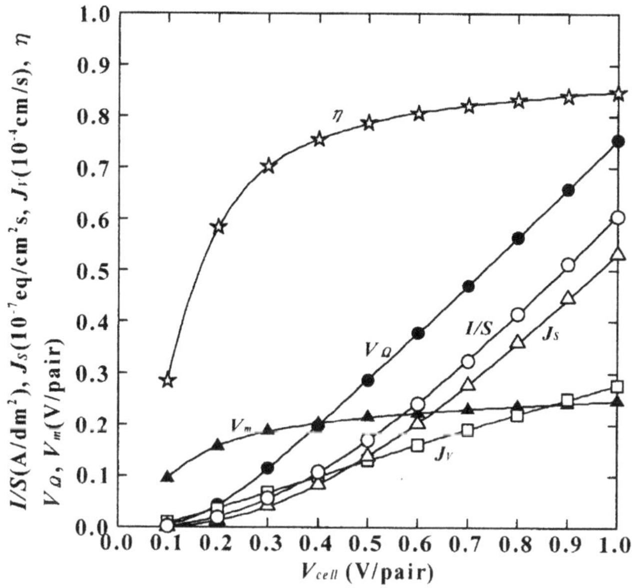

Fig. 5.34 Ionic flux, solution flux, current density, ohmic potential, and membrane potential (Reproduced from Ref. [66] with kind permission of © Elsevier (2010))

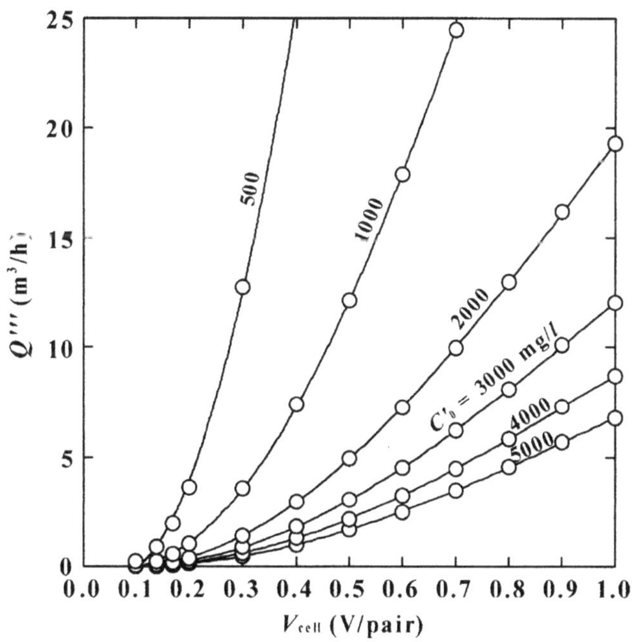

Fig. 5.35 Output of a desalted solution (Reproduced from Ref. [66] with kind permission of © Elsevier (2010))

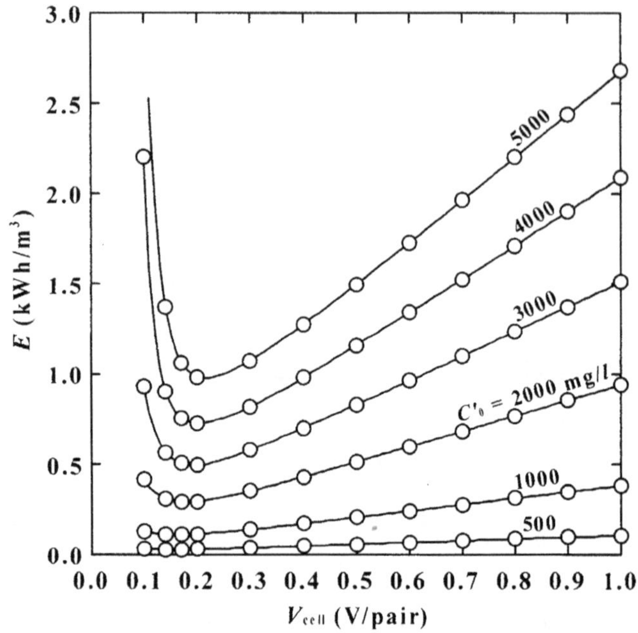

Fig. 5.36 Energy consumption (Reproduced from Ref. [66] with kind permission of © Elsevier (2010))

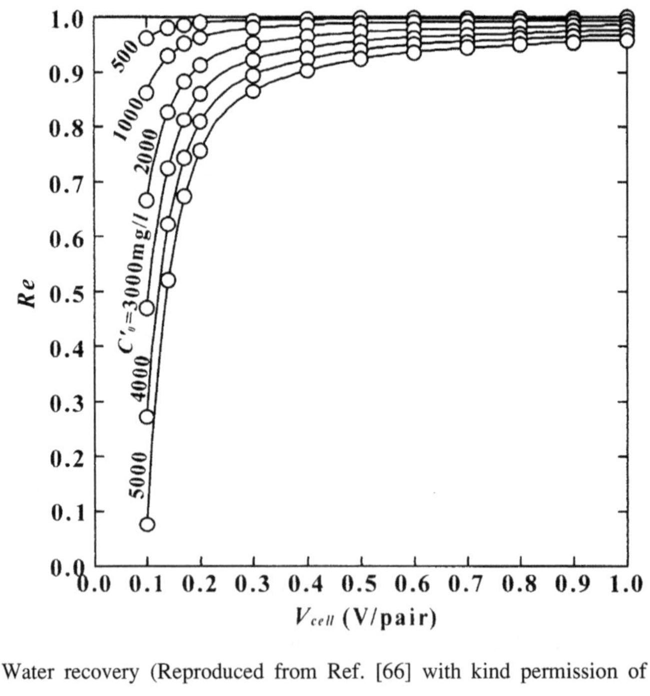

Fig. 5.37 Water recovery (Reproduced from Ref. [66] with kind permission of © Elsevier (2010))

The electrodialysis program is developed based on the above principles and many empirical equations to calculate the performance of a continuous, a batch, and a feed-and-bleed electrodialysis process. The program aims to work as a pilot plant operation, but the ability of the program is assumed to be insufficient at present. In order to improve the precision of the program further, the following functions or factors must be integrated into the program:

1. Influence of temperature on the performance of an electrodialyzer
2. Addition of a feeding solution into concentrating cells for preventing scale precipitation
3. Multiple stage operation
4. Influence of electric current leakage and solution leakage
5. Pressure drop in an electrodialyzer

References

1. Planck M (1890) Ueber die potentialdifferenz zwischen zwei verdunnten losungen binarer elektrolyte. Ann Der Physik u Chemie 40:561–576
2. Donnan FG (1924) The theory of membrane equilibria. Chem Rev 1:73–90
3. Michaelis L, Fujita A (1925) Permselectivity of biological membranes. Biochem Z 158:28, 161:47; 164:23
4. Teorell T (1935) An attempt to formulate a quantitative theory of membrane permeability. Proc Soc Exp Biol Med 33:282–285
5. Meyers KH, Sievers JF (1936) La permeabilite des membranes I, Theorie de la permeabilite ionique. Helv Chim Acta 19:649–664; La permeabilite des membranes II, Essis avec des membranes selectives artifielles. Helv Chim Acta 19:665–677
6. Meyers KH, Straus W (1940) La permeabilite des membranes VI, Sur le passage du courant electrique a travers des membranes selectives. Helv Chim Acta 23:795–800
7. Wyllie R, Patnode HW (1950) The development of membranes from artificial cation-exchange materials with particular reference to the determination of sodium-ion activity. J Phys Chem 54:204
8. Juda W, McRae WA (1950) Coherent ion-exchange gels and membranes. J Am Chem Soc 72:1044; Ion-exchange materials and method of making and using the same. USP at. 2,636,851
9. Kedem O, Katchalsky A (1961) A physical interpretation of the phenomenological coefficients of membrane permeability. J Gen Physiol 45:143–179
10. Onsager L (1931) Reciprocal relations in irreversible process. Phys Rev 37:405–426
11. Kedem O, Katchalsky A (1963) Permeability composite membranes, part 1. Electric current volume flow and flow of solute through membranes. Trans Faraday Soc 59:1918–1930
12. House CR (1974) Water transport in cells and tissues. Edward Arnold, London
13. Schulz SG (1980) Basic principles of membrane transport. Cambridge University Press, Cambridge
14. Tanaka Y (2006) Irreversible thermodynamics and overall mass transport in ion-exchange membrane electrodialysis. J Membr Sci 281:517–531
15. Bird RB, Stewart WE, Lightfoot EN (1965) Transport phenomena. Wiley, New York
16. Taky M, Pourcelly G, Lebon F, Gavach C (1992) Polarization phenomena at the interface between an electrolyte solution and an ion-exchange membrane. J Electroanal Chem 336:171–194
17. Takemoto N (1972) The concentration distribution in the interfacial layer at desalting side in ion exchange membrane electrodialysis. J Chem Soc Jpn 1972:2053–2058

18. Rubinstein I, Shtilman L (1979) Voltage against current voltage of cation-exchange membranes. J Chem Soc, Faraday Trans II 75:231–246
19. Rosenberg NW, Tirrel CE (1957) Limiting currents in membrane cells. Ind Eng Chem 49:780–784
20. Rubinstein I (1977) A diffusion model of "water splitting" in electrodialysis. J Phys Chem 81:1431–1436
21. Simons R (1984) Electric field effects on proton transfer between ionizable groups and water in ion exchange membranes. Electrochim Acta 29:151–158
22. Simons R (1985) Water splitting in ion exchange membranes. Electrochim Acta 30:275–282
23. Tanaka Y (2010) Water dissociation reaction generated in an ion exchange membrane. J Membr Sci 350:347–360
24. Peer AM (1956) Discus Faraday Soc 21:124 (communication in the membrane phenomena special issue)
25. Kooistra W (1967) Characterization of ion exchange membranes by polarization curves. Desalination 2:139–147
26. Spiegler KS (1971) Polarization at ion exchange membrane-solution interface. Desalination 9:367–385
27. Tanaka Y (2005) Limiting current density of an ion-exchange membrane and of an electrodialyzer. J Membr Sci 266:6–17
28. Cowan DA, Brown JH (1959) Effect of turbulence on limiting current in electro- dialysis cells. Ind Eng Chem 51:1445–1448
29. Tanaka Y (2003) Mass transport and energy consumption in ion-exchange membrane electrodialysis of seawater. J Membr Sci 215:265–279
30. Azechi S (1980) Electrodialyzer. Bull Soc Sea Water Sci Jpn 34:77–83
31. Minz MS (1963) Criteria for economic optimization are presented in the form of comparative performance equations for various methods of operation. Ind Eng Chem 55:19–28
32. Itoi S, Komori R, Terada Y, Hazawa Y (1978) Basis of electrodialyzer design and cost estimation. Ind Water 239:29–40
33. Leiz F B (1977) Measurements and control in electrodialysis. Desalination 59:381–401, presented at the international congress on desalination and water re-use, Tokyo
34. Belfort G, Daly JA (1970) Optimization of an electrodialysis plant. Desalination 8:153–166
35. Avriel M, Zeligher N (1972) A computer method for engineering and economic evaluation of electrodialysis plant. Desalination 10:113–146
36. Lee HJ, Sarfert F, Strathmann H, Moon SH (2002) Designing of an electrodialysis desalination plant. Desalination 142:267–286
37. Moon P, Sandi G, Stevens D, Kizilel R (2004) Computational modeling of ionic transport in continuous and batch electrodialysis. Sep Sci Technol 29:2531–2555
38. Fidaleo M, Moresi M (2005) Optimal strategy to model the electrodialytic recovery of a strong electrolyte. J Membr Sci 260:90–111
39. Sadrzadeh M, Kaviani A, Mohammadi T (2007) Mathematical modeling of desalination by electrodialysis. Desalination 206:534–549
40. Nikonenko VV, Pismenskaya ND, Itoshin AG, Zabolotsky VI, Shudrenko AA (2008) Description of mass transfer characteristics of ED and EDI apparatus by using the similarity theory and compartmentation method. Chem Eng Proc 47:1118–1127
41. Brauns E, De Wilde W, Van den Bosch B, Lens P, Pinoy L, Empsten M (2009) On the experimental verification of an electrodialysis simulation model for optimal stack configuration design through solver software. Desalination 249:1030–1038
42. Tanaka Y (2010) A computer simulation of ion exchange membrane electrodialysis for concentration of seawater. Membr Water Treat 1:13–37
43. Tanaka Y (2009) A computer simulation of continuous ion exchange membrane electrodialysis for desalination of saline water. Desalination 249:809–821
44. Tanaka Y (2010) Simulation of an ion exchange membrane electrodialysis process for continuous saline water desalination. Desalin Water Treat 22:271–285
45. Kusakari K, Kawamata F, Matsumoto N, Saeki H, Terada Y (1977) Electrodialysis plant at Hatsushima. Desalination 21:45–50

46. Tani Y, Doi K, Terada Y, Yokota M, Wakayama M (1978) Electrodialysis seawater desalination unit in a vessel. Ind Water 239:86–89
47. Rapp HJ, Pfromm PH (1998) Electrodialysis for chloride removal from the chemical recovery cycle of a kraft pulp mill. J Membr Sci 146:249–261
48. Eimidaoui A, Chay L, Tahaikt M, Menkouchi Sahli MA, Taky M, Tiyal F, Khalidi A, Alaoui Hafidi MyR (2006) Demineralisation for beet sugar solutions using an electrodialysis pilot plant to reduce melassigenic ions. Desalination 189:209–214
49. Banasiak LJ, Kruttschnitt TW, Schafer AI (2007) Desalination using electrodialysis as a function of voltage and salt concentration. Desalination 205:38–46
50. Walha K, Amar RB, Firdaous L, Quemeneur F, Jaouen P (2007) Brackish groundwater treatment by nanofiltration, reverse osmosis and electrodialysis in Tunisia, performance and cost comparison. Desalination 207:95–106
51. Kabay N, Yuksel M, Samata S, Arar O, Yuksel U (2007) Removal of nitrate from ground water by a hybrid process combining electrodialysis and ion exchange process. Sep Sci Technol 42:2615–2627
52. Kabay N, Arar O, Samatya S, Yuksel U, Yuksel M (2008) Separation of fluoride from aqueous solution by electrodialysis. Effect of process parameters and other ionic species. J Hazard Mater 153:107–113
53. Kabay N, Arar O, Acar F, Ghazal A, Yuksel U, Yuksel M (2008) Removal of boron from water by electrodialysis, effect of feed characteristics and interfering ions. Desalination 223:63–72
54. Parulekar SJ (1998) Optimal current and voltage trajectories for minimum energy consumption in batch electrodialysis. J Membr Sci 148:91–103
55. Demircioglu M, Kabay N, Ersoz E, Kurucaovali I, Safak C, Gizli N (2001) Cost comparison and efficiency modeling in the electrodialysis of brine. Desalination 136:317–323
56. Ahmed MI, Chang HT, Selman JR, Holsen TM (2002) Electrochemical chromic acid regeneration process, fitting of membrane transport properties. J Membr Sci 197:63–74
57. Ortiz JM, Sotoca JA, Exposito E, Gallud F, Garcia-Garcia V, Montiel V, Aldaz A (2005) Brackish water desalination by electrodialysis, batch recirculation operating modeling. J Membr Sci 252:65–75
58. Tanaka Y (2009) A computer simulation of batch ion exchange membrane electrodialysis for desalination of saline water. Desalination 249:1039–1047
59. Koga S, Mitsugami Y (1978) Supply of potable water from brackish water by electro-dialysis desalination process at Ohoshima island in Tokyo prefecture. Ind Water 239:41–47
60. Kawahara T (1994) Construction and operation experience of a large-scale electrodialysis water desalination plant. Desalination 96:341–348
61. Okada K, Tomita M, Tamura Y (1975) Electrodialysis in the treatment of dairy products. In: Symposium separation processes by membranes, ion-exchange and freeze-concentration in food industry, Paris, 13–14 Mar 1975
62. Matsunaga Y (1986) Reuse of waste water by electrodialytic treatment. In: Industrial application of ion exchange membranes, vol 1. Research group of electrodialysis and membrane separation technology, Soc Sea Water Sci Jpn 188–196
63. Ryabstev AD, Kotsupalo NP, Titarenko VI, Igumenov IK, Gelfond NV, Fedotova NE, Moroziva NB, Shipachev VA, Tibilov AS (2001) Development of two-stage electrodialysis set-up for economical desalination of artesian and surface waters of sea type. Desalination 137:207–214
64. Rapp HJ, Pfromm PH (1998) Electrodialysis field test for selective chloride removal from the chemical recovery cycle of a kraft pulp mill. Ind Eng Chem Res 37:4761–4767
65. Thompson R, Paleologou M, Wong RY, Berry RM (1997) Separation of sulphide from hydroxide in white liquor by electrodialysis. J Pulp Paper Sci 23:J182–J187
66. Tanaka Y (2010) A computer simulation of feed and bleed ion exchange membrane electrodialysis for desalination of saline water. Desalination 254:99–170

Chapter 6
Structure, Synthesis, and General Properties of Ion Exchangers

Jin-Soo Park

Abstract For decades, industrial applications have demanded removal, recovery, and transport of ionic substances in their processes. The technological demands motivated extensive research activities with great progress in ion exchange science. Great efforts have been and are being made to develop new and improved ion exchangers. Advances in polymer science have been devoted to be able to meet the demand. Polymer tailoring technology has developed novel synthetic methods and varied the structure of ion exchanging materials. A wide spectrum of properties of ion exchangers make it possible to expand various industrial experiences to be involved in desalination, purification, separation, chemical synthesis, etc., addressed in this book. This chapter is devoted to updating the state-of-the-art structure, synthesis, and properties of ion exchangers. Ion exchangers are made of inorganic or organic materials. At present, many ion exchangers used in industrial applications are organic compounds. Thus, this chapter focuses on the structure, synthesis, and properties of organic ion exchangers. Ion exchangers in this chapter contain all types such as resin, membrane, powder, etc.

6.1 Structure of Ion Exchangers

An ion exchanger is an insoluble material which traps ions and simultaneously releases ions. This process is known as ion exchange reactions. The ion exchanger has a functional group. The insoluble host structure allows diffusion of hydrated ions, i.e., a hydrophilic matrix, which must carry a fixed ionic charge, termed the fixed ion. Electrical neutrality of the structure must be established by the presence of a mobile ion of opposite charge to that of the fixed ion [1].

J.-S. Park (✉)
Department of Environmental Engineering, College of Engineering, Sangmyung University, 31 Sangmyungdae-gil, Dongnam-gu, Cheonan, Chungnam 330-720, Republic of Korea
e-mail: energy@smu.ac.kr

Inamuddin and M. Luqman (eds.), *Ion Exchange Technology I: Theory and Materials*, DOI 10.1007/978-94-007-1700-8_6,

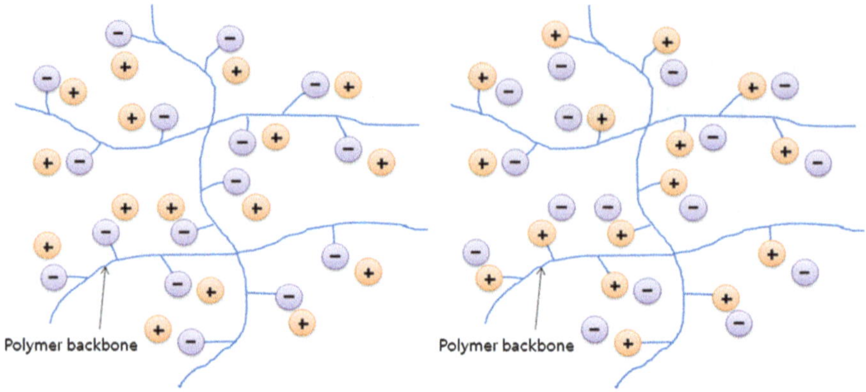

Fig. 6.1 Schematic diagram of chemical structure of cation-exchangers (*left*) and anion-exchangers (*right*)

If it has insoluble fixed anionic/cationic complement, it is a cation-/anion-exchanger, respectively. In a cation-exchanger, the fixed negative charges are in electrical equilibrium with mobile cations in the interstices of the polymer as indicated in Fig. 6.1 (left-hand side), which shows schematically the matrix of a cation-exchanger with fixed anions and mobile cations; the latter are referred to as counter ions [2].

There are four main types differing in their functional groups [3, 4]:

- Strongly acidic (typically sulfonic acid groups, e.g., sodium polystyrene sulfonate)
- Strongly basic (quaternary amino groups, e.g., trimethylammonium groups)
- Weakly acidic (mostly carboxylic acid groups)
- Weakly basic (primary, secondary, and/or tertiary amino groups, e.g., polyethylene amine)

Ion exchangers are characterized by their structures on a microscopic scale. They are represented by a microphase separation between hydrophobic matrix and hydrophilic ionic domains [5]. Ion exchangers based on linear fluorocarbon polymers are quite heterogeneous on a microscopic scale due to a backbone of a highly crystallized polymer and a highly amorphous ionic domain. The amorphous region consists of fixed charges, hydrated counter ions, and neighboring water. The highly hydrophilic ionic domain well separated by the distinct crystalline area can form clusters (water pools embedded in the polymer matrix) with a diameter of ca. 4–6 nm connected by "bottlenecks" between crystalline regions of ca. 1 nm in diameter [2]. The structure of the fluorocarbon-polymer-based membrane has been studied in great detail using transmission electron microscopy, wide and small angle X-ray diffraction [6], differential scanning calorimetry, and infrared and nuclear magnetic resonance spectroscopy [7, 8]. From the experimental evidence, it can be concluded that the fluorocarbon polymer membranes have a heterogeneous two-phase structure composed of crystalline polytetrafluoroethylene and amorphous vinyl ether with fixed charges as depicted in Fig. 6.2.

Fig. 6.2 The structure of the fluorocarbon polymer showing crystalline and fixed charges (the *dotted lined circle* indicates the crystalline of the fluorocarbon polymers) (Reproduced from Ref. [2] with kind permission of © Elsevier (2004))

Most hydrocarbon polymer membranes have a rather regular distribution of cross-linkings and ionic groups attached to a backbone of aromatic rings. As a result, these membranes are quite homogeneous in their structure even on a microscopic scale [2].

6.2 Synthesis of Ion Exchangers

So far, many synthetic methods to prepare ion exchangers have been reported. Although they use different routes to synthesis, all synthesized materials should have [1]:

1. A hydrophilic structure of regular and reproducible form
2. A controlled and effective exchange capacity
3. A reversible and rapid rate of exchange
4. Chemical stability toward electrolyte solutions
5. Physical stability in terms of mechanical strength and resistance to attrition
6. Thermal stability
7. Consistent particle size and effective surface area compatible with the hydraulic design requirements for industrial scale plant
8. An option on the type of exchanger so as to be able to select either cation- or anion-exchange

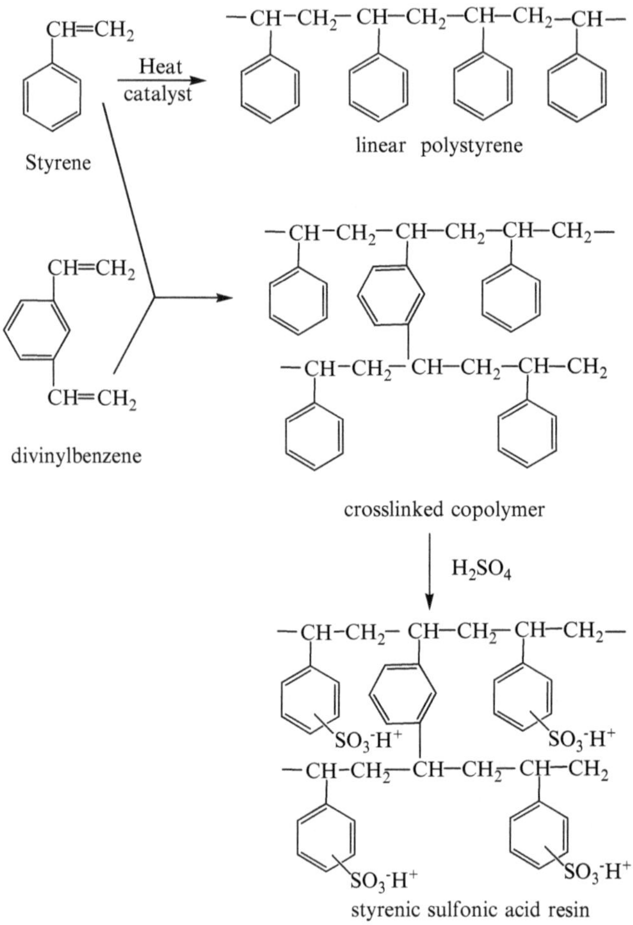

Fig. 6.3 Addition polymerization synthesis of a styrene sulfonic acid cation-exchanger (Reproduced from Ref. [1] with kind permission of © The Royal Society of Chemistry (1994))

The first completely synthetic ion exchangers were prepared by the condensation polymerization of methanol (formaldehyde) with phenol or polysubstituted benzene compounds which give a phenol-methanol polymer chain (a cross-linked copolymer) with the elimination of water. Afterward, the functional groups such as sulfonic acid group or ammonium group are introduced in the polymer [1].

In the recent history of synthesis of ion exchangers, ethenylbenzene (so-called styrene) and diethenylbenzene (divinylbenzene) are most commonly utilized neutral starting material for a traditional hydrocarbon type cation-exchanger for industrial uses, from which a strongly acidic cation-exchanger is usually prepared by sulfonation [9]. These syntheses are based on vinyl polymerization (or addition polymerization) which is carried out by radical reactions between monomers with vinyl (or ethenyl) double bonds ($-CH = CH_2$) (Fig. 6.3).

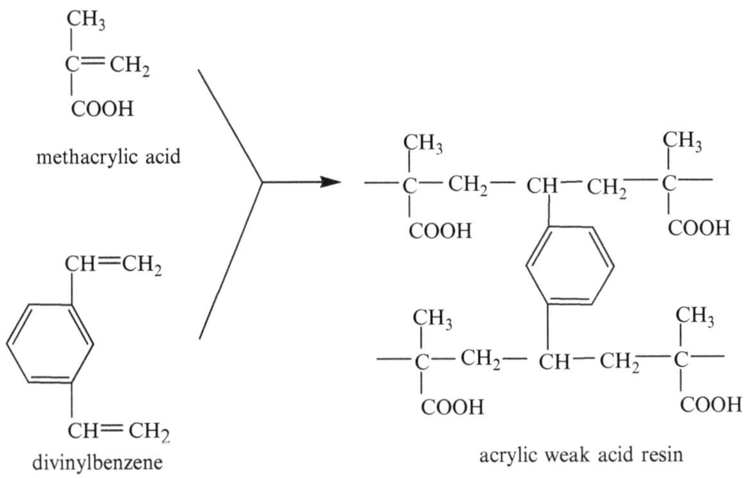

Fig. 6.4 Addition polymerization synthesis of an acrylic carboxylic cation-exchange (Reproduced from Ref. [1] with kind permission of © The Royal Society of Chemistry (1994))

Styrene has been widely used as a grafting vinyl monomer, because it shows favorable radical polymerization kinetics, and the aromatic ring is readily sulfonated to introduce ionic conductivity. For the better performance and durability of styrene-based ion exchangers, divinylbenzene as cross-linking monomer is used. Nevertheless, the shortcoming of sulfonated styrene-based cation-exchangers is that they are intrinsically weak to radical species (HO·, HOO·) [10]. The Achilles' heel appears to be the weak benzylic hydrogen at the α-position of poly(styrene sulfonate), which is easily abstracted during acid-catalyzed HO· radical attack at the aromatic ring, leading eventually to chain scission [11]. To overcome this weakness, substituted styrene monomers, preferably with protected α-position, are potentially offered as alternative monomers such as *m,p*-methylstyrene, *p-tert-* butylstyrene, α-methylstyrene, and α,β,β-trifluorostyrene [10].

There is an alternative monomer to be copolymerized with divinylbenzene, i.e., the propenoic (acrylic) monomers. Various alkyl-substituted propenoic acid monomers may be employed in the manufacture of weakly acidic cation-exchangers, as are propenonitriles (acrylonitriles) and alkyl propenoate (acrylic esters) [1] (Fig. 6.4).

Apart from the ion exchangers based on styrene or acrylic monomers, wholly aromatic polymers are also believed to be one of the promising materials to prepare ion exchangers. Various aromatic polymers or monomers are commercially available, thermally and mechanically stable, cost-effective, etc., specifically poly (arylene ether ether ketone), poly(arylene ether sulfone), polyimide, and their derivatives [12–14] (Fig. 6.5).

Introduction of cation-exchangeable sites to the polymers can be carried out by postsulfonation of a polymer and direct copolymerization of sulfonated monomers. The postsulfonation approach to prepare cation-exchanger is easy to produce, time-saving, and mass productive but hardly controls the degree of

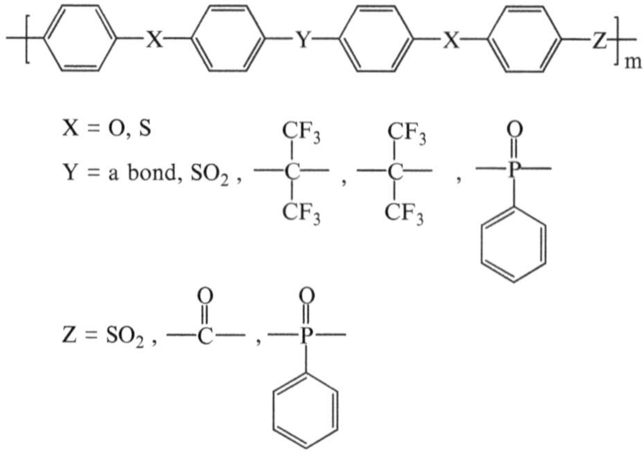

Fig. 6.5 Several possible poly(arylene ether) chemical structures (Reproduced from Ref. [14] with kind permission of © The American Chemical Society (2004))

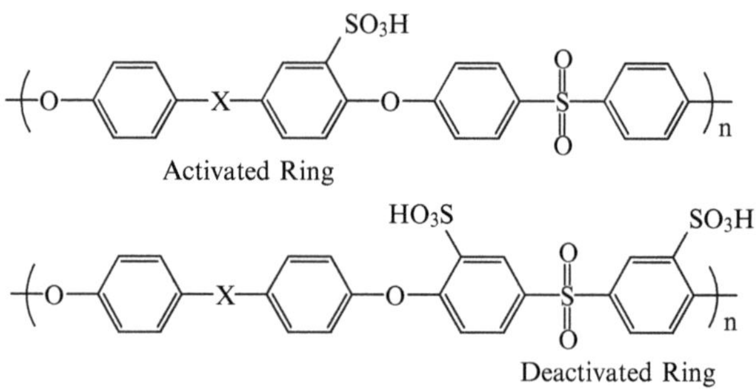

Fig. 6.6 Placement of the sulfonic acid group in postsulfonation (activated ring) versus direct copolymerization (deactivated ring) (Reproduced from Ref. [14] with kind permission of © The American Chemical Society (2004))

sulfonation and location of functionalization because sulfonation depends, to a great extent, on reaction time, temperature, and concentration and type of sulfonating agents. The major drawbacks of postsulfonation approach can be solved by direct copolymerization of sulfonated monomers. As seen in Fig. 6.6, we might expect enhanced stability and higher acidity from two sulfonic acid groups which are attached to an electron-deficient aromatic ring than from one sulfonic acid group bonded to an electron-rich aromatic ring [13, 14].

Robeson and Matzner [15] were the first to have reported the sulfonated monomer. Ueda et al. [16] reported the sulfonation of 4,4′-dichlorodiphyeny sulfone and provided general procedures for its purification and characterization. McGrath's group modified the procedure for disulfonation of the monomer, shown in Fig. 6.7. Sulfonated poly(arylene ether sulfone) copolymers were then synthesized via direct copolymerization in any composition desired as shown in Fig. 6.8 [17]:

Two families of sulfonated polyimides (SPIs) have been reported. The phthalic SPIs are characterized by five-membered imide rings (Fig. 6.9a), while the naphthalenic SPIs contain six-membered imide rings (Fig. 6.9b).

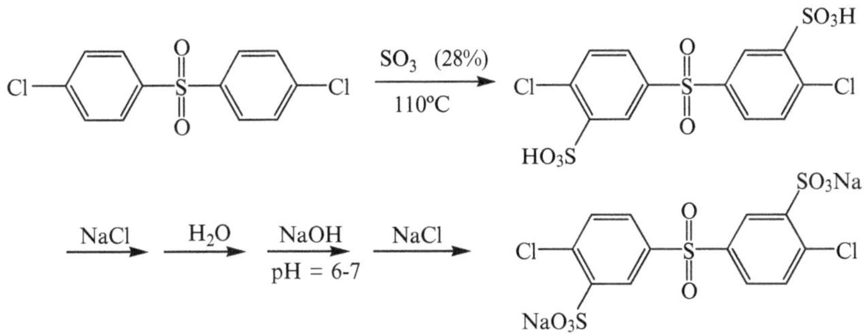

Fig. 6.7 Synthesis of 3,3′-disulfonated 4,4′-dichlorodiphenyl sulfone and its sodium salt (Reproduced from Ref. [14] with kind permission of © The American Chemical Society (2004))

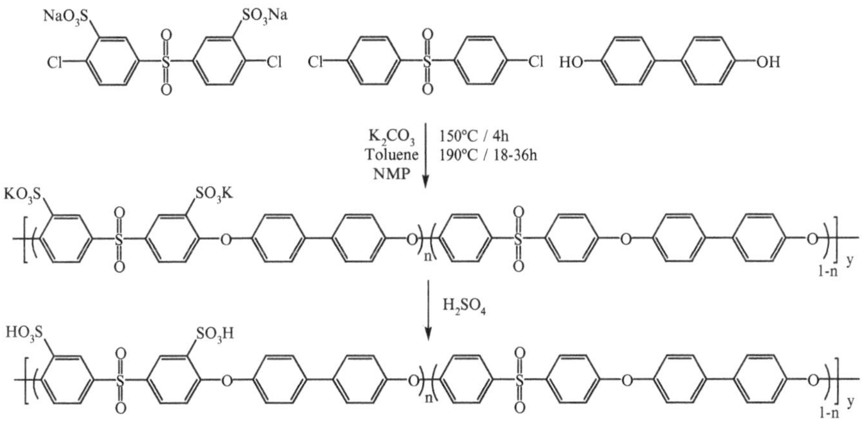

Fig. 6.8 Synthesis of directly copolymerized wholly aromatic sulfonated poly(arylene ether sulfone), BPSH-xx, where xx is the ratio of sulfonated/unsulfonated activated (Reproduced from Ref. [17] with kind permission of © Elsevier (2002))

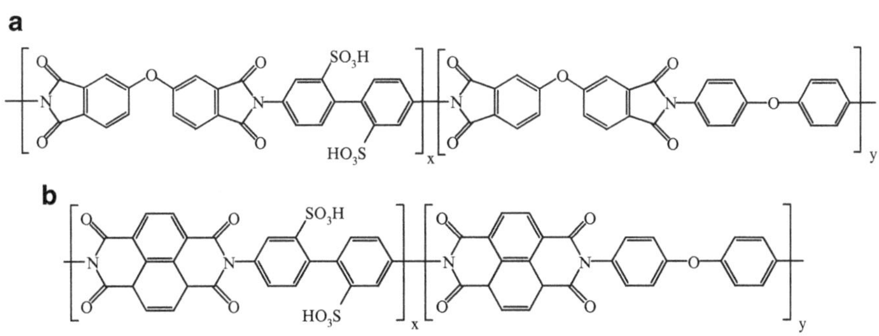

Fig. 6.9 Chemical structure of (**a**) phthalic sulfonated polyimide and (**b**) naphthalenic sulfonated polyimide

Naphthalenic SPIs present a higher stability and a better ionic conductivity since phthalic SPIs are prone to be hydrolyzed, resulting in chain scission. A decrease in molecular weight by the chain scission makes phthalic SPIs brittle.

A strongly basic anion-exchange membrane is usually prepared by two steps, chloromethylation and quaternary amination of various copolymers. Chloromethylation is performed by a Friedel-Crafts reaction which was developed by Charles Friedel and James Crafts in 1877. Amination is the reaction in which an amine group is introduced into the chloromethylated copolymer.

Chloromethylation of the polymerized ethenylbenzene-diethenylbenzene copolymer is carried out by a Friedel-Crafts reaction between the copolymer and chloromethoxymethane with aluminum chloride as the catalyst and introduces chloromethyl groups (-CH_2Cl) into the ethenylbenzene nuclei as shown in Fig. 6.10. The chloromethylated copolymer is aminated by reaction with various alkyl-substituted aliphatic amines. Trimethylamine, $(CH_3)_3$ N, gives the quaternary benzyltrimethylammonium chloride functional group, $RCH_2N(CH_3)_3^+Cl^-$, which is characteristics of most strongly basic anion-exchangers, shown in Fig. 6.10. The equivalent reaction using dimethylethanolamine, $(CH_3)_2(C_2H_4OH)N$, gives the Type II class of strong base anion-exchangers, $RCH_2N(CH_3)_2(C_2H_4OH)^+Cl^-$. If instead of using tertiary trimethylamine, methylation or dimethylation is employed, the resulting anion-exchangers are weakly basic with secondary, $RCH_2NH(CH_3)$, or tertiary, $RCH_2N(CH_3)_2$, functionality [1].

Anion-exchangers can be also prepared by an acrylic matrix which shows the greater hydrophilic nature of the aliphatic skeletal structure of the acrylic matrix. It is attributed to a weaker van der Waals type attraction between the resin matrix and the hydrocarbon structure of an organic counter ion. In general, methyl propenoate (methyl acrylate) is chosen as the monomer for copolymerization with diethenylbenzene to give the host matrix as shown in Fig. 6.11 [1]. Amination with dimethylaminopropylamine introduces a tertiary amine functional group, and a subsequent "quaternization" step employing chloromethane (methyl chloride) converts the weak base product to the strongly basic quaternary ammonium resin as shown in Fig. 6.11.

Recently, many anion-exchangers have also been prepared by aromatic polymers or monomers. A variety of anion-exchangers based on poly(phenylene oxide) [18, 19] and poly(arylene ethers) [20–23] have been reported. Poly (phenylene oxide) has excellent electrical properties, unusual resistance to acids and bases as well as other chemicals, excellent dimensional stability, low moisture absorption, and high mechanical and dielectric strength [24, 25]. Poly(arylene ethers) are high-performance polymers that are thermally and chemically stable [24, 26] (Figs. 6.12 and 6.13).

The stability of anion-exchangers is one of the key issues in the literature. In particular, the stability of anion-exchangers under strong alkaline conditions is of importance.

Sata et al. [27] have studied the effect of trimethyl, triethyl, tri-n-propyl ammonium, and tri-n-butyl ammonium groups on the alkaline stability of anion-exchangers. The comparative analysis showed that as the chain length of alkyl

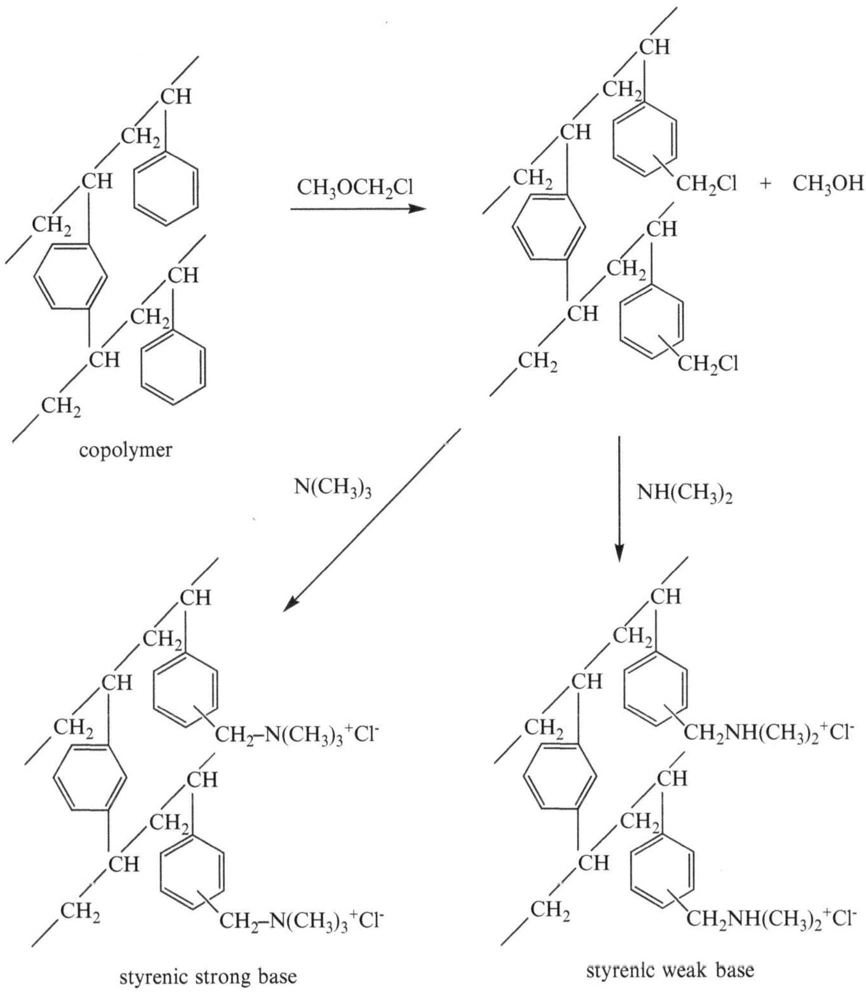

Fig. 6.10 Addition polymerization synthesis of styrenic anion-exchange resins (Reproduced from Ref. [1] with kind permission of © The Royal Society of Chemistry (1994))

groups bonded to ammonium groups increased, the loss of ion exchange capacity was significant. However, anion-exchanger-containing benzyl trimethylammonium groups were stable.

When anion-exchange copolymer cross-linked diethenylbenzene was aminated in the diamine solution, anion-exchanger with high electrical resistance and low water content was obtained due to the double cross-linking of the polymer matrix with diethenylbenzene and diamine [28]. Komkova et al. [29] have investigated a series of anion-exchangers prepared from chloromethylated polysulfone and aliphatic diamine compounds, i.e., *N,N,N′,N′*-tetramethylmetanediamine (TMMDA), *N,N,N′,N′*-tetra-methylethylenediamine (TMEDA), *N,N,N′,N′*-tetramethyl-1,3-propanediamine

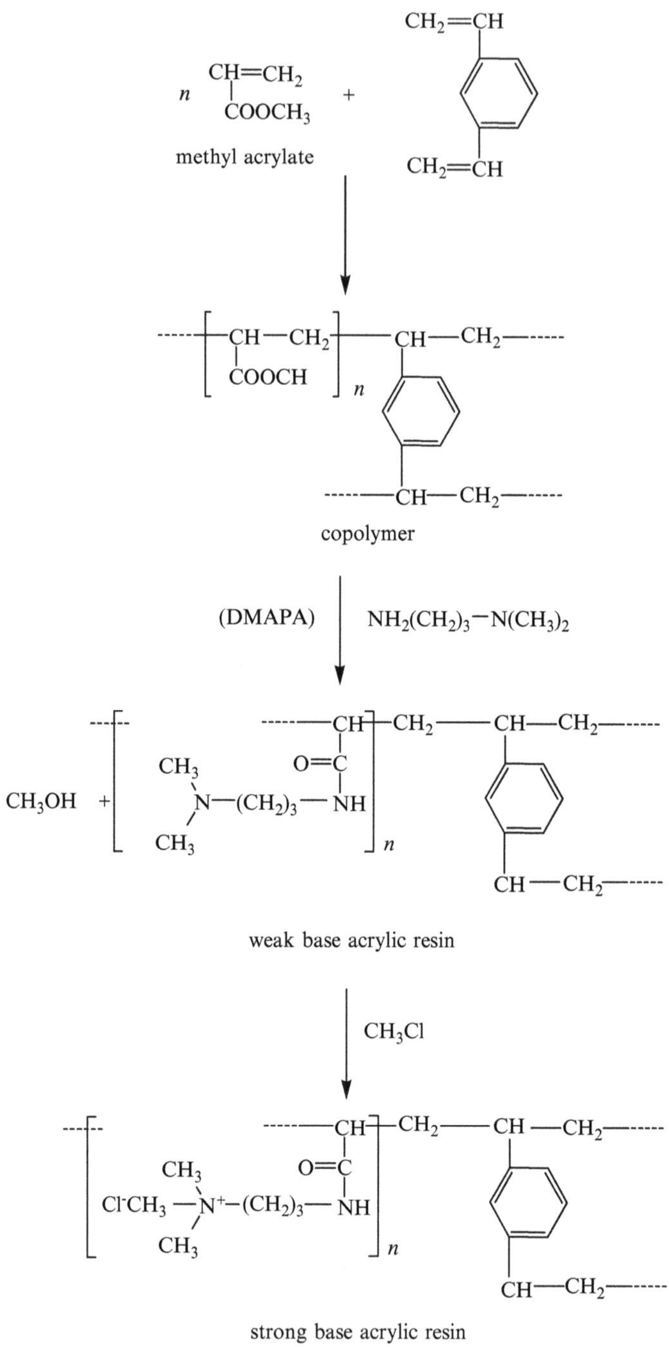

Fig. 6.11 Addition polymerization synthesis of acrylic anion-exchange resins (Reproduced from Ref. [1] with kind permission of © The Royal Society of Chemistry (1994))

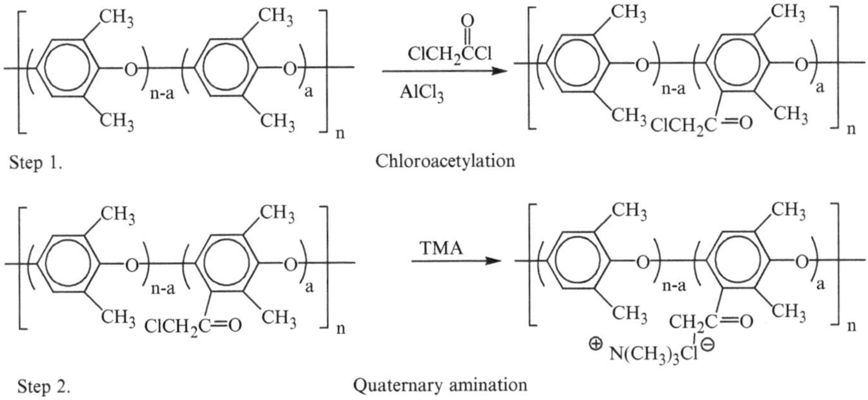

Fig. 6.12 The new synthesis route for anion-exchange membranes based on PPO. Step 1: chloroacetylation and step 2: quaternary amination (Reproduced from Ref. [18] with kind permission of Elsevier (2006))

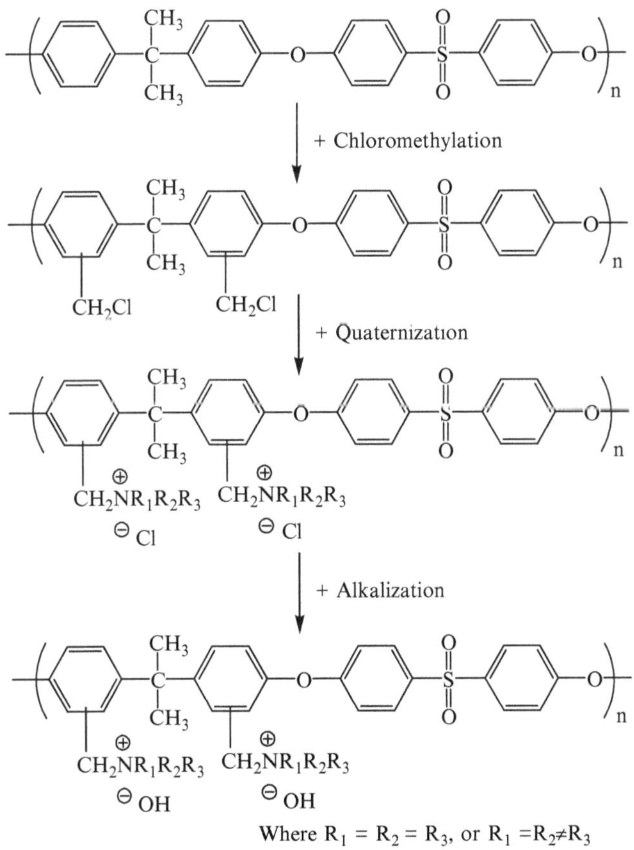

Fig. 6.13 Schematic diagram for synthesis: (**a**) chloromethylation, (**b**) quaternization, and (**c**) alkalization (Reproduced from Ref. [23] with kind permission of © Elsevier (2009))

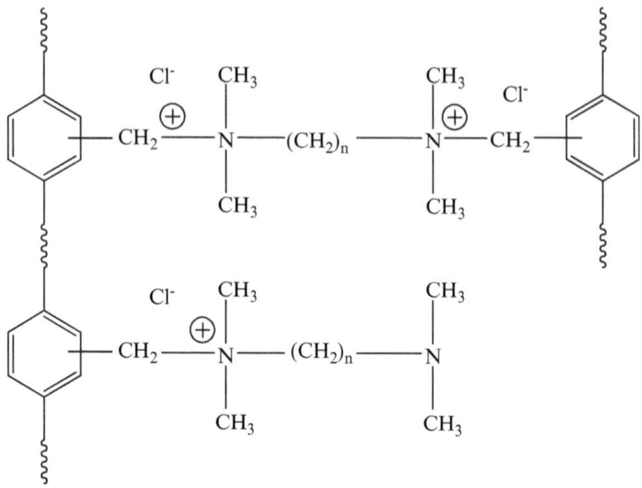

Fig. 6.14 Quaternization with the aliphatic diamines of various lengths of the aliphatic chain (n = 1–6) (Reproduced from Ref. [29] with kind permission of © Elsevier (2004))

(TMPDA), *N,N,N′,N′*-tetramethyl-1,4-butanediamine (TMBDA), and *N,N,N′,N′*-tetramethyl-1,6-hexanediamine (TMHDA). It was generally shown that biquaternization with diamine was more preferable than mono-quaternization. However, an exception was found for the diamine with bulky substitute at the amine nitrogen (Fig. 6.14).

According to the connection way of charge groups to the matrix or their chemical structure, ion exchangers can be further classified into homogenous and heterogeneous resins, in which the charged groups are chemically bonded to or physically mixed with the polymer matrix, respectively. However, most of the practical ion exchangers are rather homogenous and composed of either hydrocarbon or fluorocarbon polymer matrices hosting the ionic groups [30].

Apart from homogeneous polymeric ion exchangers, a heterogeneous ion exchanger can also be prepared from the mixture of polymeric matrix and grinded inorganic material powders made of zeolites, bentonite, or phosphate salts [31–33]. However, these membranes are rather unimportant due to their high cost and other disadvantages, such as relative bad electrochemical properties and too large pores, though they can undergo higher temperatures than organic membranes [34].

6.3 Properties of Ion Exchangers

The development of new ion exchange membranes with better selectivity, lower electrical resistance, and improved thermal, chemical, and mechanical properties and other applications of ion exchange membranes apart from the initial desalination of brackish water have recently gained a broader interest in food, drug, and chemical process industry as well as biotechnology and waste water treatment nowadays [35–42].

Properties of ion exchangers are highly dependent on their chemical structure which can be manipulated by differing the type of polymer matrix, functionalization step (sulfonation or chloromethylation/amination), and posttreatment. The properties can be categorized into two parts: (1) physicochemical properties such as water content, hydraulic permeability, and mechanical strength, and (2) electrochemical properties such as ion exchange capacity, electrical resistance, and permselectivity (or transport number).

6.3.1 Water Content

The water content of an ion exchanger depends on the type of polymer matrix, the nature of the ion-exchangeable groups and their concentration in the ion exchanger, the counter ions, the cross-linking density, and the homogeneity of the ion exchanger. The water content of an ion exchanger depends not only on the properties of the ion exchanger but also on the composition of the solution with which the ion exchanger is in contact. Especially, the concentration of the solution has a significant effect on the water content of the ion exchanger because of osmotic effects that are directly related to the chemical potential difference of the water in the ion exchanger and in the solution [2]. The state of water in an ion exchanger can be categorized into three parts: (1) freezable bound water which is weakly bound to clusters and embedded particles in the polymer, (2) free water which corresponds to water existed in free volume of the polymer, and (3) nonfreezable water which is strongly bound to cluster and embedded particles in the polymer [43]. It can be studied by differential scanning calorimetry, infrared spectroscopy, and nuclear magnetic resonance.

The total water uptake of the ion exchanger in equilibrium with an electrolyte solution can be determined by measuring the weight difference between an ion exchanger in wet and dry state. To determine the water content of an ion exchanger, a sample is equilibrated in any test solution. After removing the surface water from the sample, the wet weight of the swollen ion exchanger is determined. The sample is then dried at elevated temperature over phosphorous pentoxide under reduced pressure until a constant weight is obtained. The water content of an ion exchanger is obtained in weight percent by

$$wt\ \%\ swelling = \frac{W_{wet} - W_{dry}}{W_{wet}} \times 100 \tag{6.1}$$

where W_{wet} and W_{dry} are the weight of an ion exchanger sample in the wet and the dry state [2].

6.3.2 Hydraulic Permeability

Hydraulic permeability measurements provide information on the diffusive or convective transport of components through a membrane under a hydrostatic pressure driving force. The hydraulic permeability of the ion-exchanger is determined at room temperature using deionized water and a hydrostatic pressure driving force in a conventional filtration cell. The permeability can then be calculated from the volumetric flow rate [2].

6.3.3 Ion Exchange Capacity

The ion exchange capacity (IEC) of ion exchangers is an important property and a measure of the number of fixed charges per unit weight of dry polymer. It is usually expressed in milliequivalents per gram of dry ion exchanger. Experimentally, the cationic or anionic exchange capacity of an ion exchanger is readily determined by titration of the fixed ions with NaOH or HCl, respectively. For these measurements, cation- and anion-exchangers are equilibrated for about 24 h in 1 N HCl or 1 N NaOH, respectively, and then rinsed free from chloride or sodium for 24 h with deionized water. The ion exchange capacity of the samples is then determined by back titration with 1 N NaOH or 1 N HCl, respectively. Weak base anion-exchangers are characterized by equilibrium in 1 N sodium chloride and titration with a standardized 0.1 N silver nitrate solution. The samples are then dried, and the ion exchange capacity is calculated for the dry membrane [2]. The ion exchange capacity of the ion exchanger is calculated from the titration data using the following equation:

$$IEC = \frac{V_i\ (mL) \times N_i\ (N)}{W\ (g)} meq\ g^{-1}\ of\ dry\ ion\ exchanger \tag{6.2}$$

where *IEC* is the ion exchange capacity, V_i the volume of consumed NaOH or HCl (*mL*), N_i the molarity of NaOH or HCl, and *W* the weight of a dried ion exchanger.

6.3.4 Electrical Resistance

Electrical resistance of ion exchangers is one of the factors which determine the energy requirements of the processes driven by an electrical field. The specific ion exchanger resistance is usually reported as Ω cm or Ω m. From the engineering point of view, the ion exchanger area resistance in units of Ω cm^2 or Ω m^2 is more useful and generally given in the literature describing commercial products. Alternatively, electrical resistance can be converted into electric conductivity which is an inverse of electrical resistance. The specific electric conductivity of a material is essentially given by the concentrations and mobilities of the charge carriers (electrons or ions) which the material contains. The specific electric conductivity is normally expressed as S cm^{-1} or S m^{-1} [44].

For the measurement of electrical resistance of ion exchange resin beds, a quite good approach has been chosen by the Wyllie group [45]. Here, a simple model is used for calculating electrochemical properties from empirical geometrical parameters. This "porous-plug" model is based on the idea that, in principle, the electric current can take three different paths through the bed: (1) the first path leads through alternating layers of particles and interstitial solution, (2) the second exclusively through particles which are in contact with one another, and (3) the third exclusively through the interstitial solution. In the model, the bed is represented by three conductance elements in parallel which correspond to the three possible paths. The fractional cross section equivalents a, b, and c of the three elements and the contributions d and of the particles and the solution, respectively, to the first element are empirical constants which must be determined experimentally. The specific conductivity of a bed which is filled with, and is in equilibrium with, a solution of given specific conductivity may be written as the sum of the contributions k_1, k_2, and k_3 of the three conductance elements:

$$k_b = k_1 + k_2 + k_3 \tag{6.3}$$

The contributions, when defined according to this relation, are given by

$$k_1 = \frac{ak\bar{k}}{dk + e\bar{k}} \quad k_2 = b\bar{k} \quad k_3 = ck \tag{6.4}$$

With strong-acid, spherical cation-exchange resins of moderate cross-linking (Dowex 50-X8, Amberlite IR 120) and aqueous solutions, the following values for the empirical parameters were obtained:

a = 0.63 b = 0.01 c = 0.34 d = 0.95 e = 0.05

The values depend little on the particle size and the nature of the electrolyte. They are likely to hold rather well for other spherical conductivity of the bed can be calculated from the parameters by use of Eqs. 6.3 and 6.4, provided that the specific conductivities of the resin and the solution are known [44].

Proton conductivity of sheet-like ion exchangers (i.e., ion exchange membranes) is measured using a galvanostatic four-point-probe ac electrochemical impedance spectroscopy (EIS) technique, which is relatively so insensitive to the contact impedance that it could be adequate to accurately test membranes with high conductivity as reported by Cahan and Wainright [46]. A conductivity cell was made up of two platinum foils carrying the current and two platinum wires sensing the potential drop, which were apart a certain distance. The cell configuration was illustrated in Fig. 6.2. Fully hydrated membranes with deionized water (18 MΩ cm)

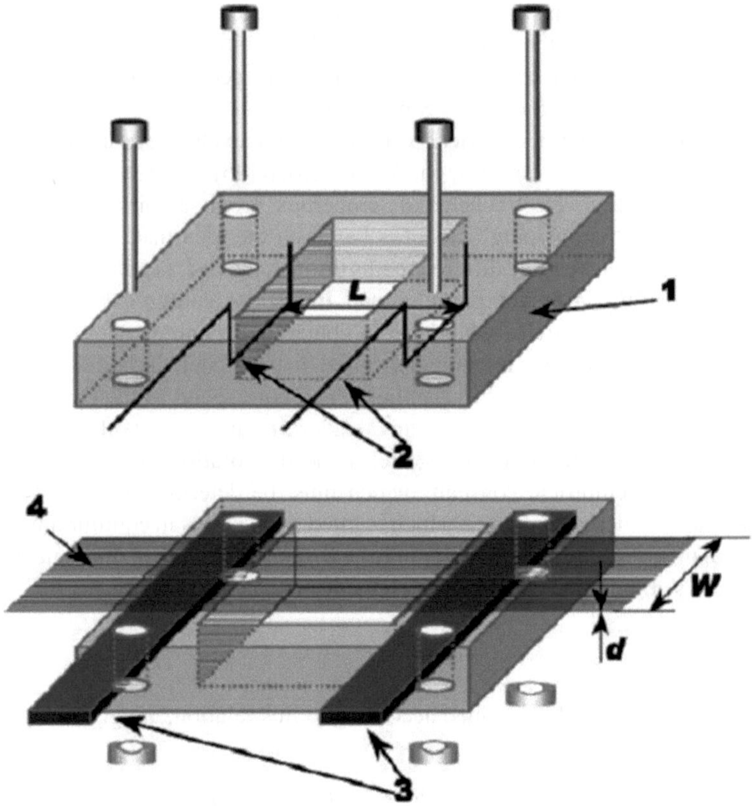

Fig. 6.15 Four-point-probe conductivity cell for measuring ionic conductivities in ion exchange membranes: (1) Teflon block, (2) Pt wires for potential readout, (3) Pt foils for constant current supply, and (4) membrane specimen (Reproduced from Ref. [47] with kind permission of © Elsevier (2003))

during 24 h were cut in a rectangular shape prior to mounting on the cell. After mounting sample onto two platinum foils on the lower compartment, upper compartment was covered, and then the upper and lower compartment were clamped tightly by four bolts and nuts. The impedance analyzer was worked in galvanostatic mode with ac current amplitude of 0.01 mA over frequency range from 1 MHz to 10 mHz by Nyquist method [47] (Fig. 6.15).

6.3.5 Transference Numbers and Transport Numbers

The transference number t_i of a species I is defined as the number of moles of the species transferred by 1 Faraday of electricity through a stationary cross section in the direction of positive current – a complete definition only when the frame of reference is specified. The electric current induces convection in the pores. Thus, it makes a difference whether the matrix, the solvent, or the pore liquid is considered as stationary [48, 49]. A spectator riding on a solvent molecule or floating in the pore liquid will see the matrix being "transferred." In the following, the matrix is chosen as the frame of reference. Transference of the solvent must thus be considered. According to the definition given above, the transference numbers of anions are negative since anions are transferred in the direction opposite to that of positive electric current.

The transport number of a species is defined as the fraction of the electric current which is carried by the species. According to this definition, the transport number is the product of the transference number and the electrochemical valence of the species. The transport numbers are positive for all ions and zero for the (electrically neutral) solvent. By definition, the sum over all transport numbers is equal to unity:

$$\sum_i z_i t_i = 1 \tag{6.5}$$

In homogeneous electrolyte solutions, the concentrations of the cation and the anion are necessarily stoichiometrically equivalent. Hence, the ratio of the transference numbers is determined solely by the ratio of the ionic mobilities. In ion exchangers, the situation is different. Here, the concentrations of the counter ion exceed that of the coion by far, particularly if the solution which is in equilibrium with the ion exchanger is dilute (Donnan exclusion). If an electric field is applied to the ion exchanger, the transport of the electric current thus is accomplished nearly exclusively by the counter ions. The transport number of the counter ion (or the sum over the transport numbers of all counter ion species) is nearly unity. Electric current in an ion exchanger transfers predominantly counter ions. However, when the concentration of the solution is increased, the Donnan exclusion of the coion becomes less efficient. The increasing concentration of the coion in the ion exchanger causes a decrease in the transport number of the counter ion.

6.3.6 Permselectivity

The permselectivity of an ion exchanger is determined by the ratio of the flux of specific components to the total mass flux through the membrane under a given driving force. In ion exchangers, the permselectivity is generally related to the transport of electric charges by the counter ions to the total electrical current through the ion exchanger. However, the transport rate of different components of the same charge in ion exchangers can also be quite different because of kinetic parameters such as the size of the components and the structure of the ion exchangers. The permselectivity of ion exchangers, i.e., its charge selectivity, is determined by the concentrations of counter- and coions in the ion exchanger, which again depends mainly on the ion exchange capacity of the membrane and the ion concentration in the outside solutions because of the Donnan exclusion as discussed earlier. The permselectivity can be calculated from the transport or transference number of the counter- and coions in the ion exchanger and the outside solutions [44].

The permselectivity of an ion exchanger is given by the following equation:

$$\Psi^m = \frac{T^m_{cou} - T_{cou}}{T_{co}} \tag{6.6}$$

The transport numbers are defined by the following equation:

$$T_i = \frac{z_i J_i}{\sum_i z_i J_i} \tag{6.7}$$

Here, Ψ is the permselectivity, T is the transport number, z is the valence, and J is the flux; the subscript i refers to cation or anion, the subscripts *cou* and *co* refer to counter- and coions, and the superscript m refers to ion exchangers.

To determine the transport number of ions in the ion exchanger and the ion exchanger permselectivity, respectively, two methods can be applied. The first method is based on a measurement of the increase in concentration of certain ions in the concentrate and the decrease in the diluate solution during electrodialysis of a test solution and by measuring the amount of current passing through the unit. From the current utilization, the permselectivity of the ion exchanger and the transport number of the counter ion can be calculated. With this "dynamic" measurement, water transport due to electroosmosis and osmosis is taken into account and the "true" ion exchanger permselectivity is obtained. The dynamic method, however, is rather time-consuming and affected by concentration polarization at the membrane surfaces facing the diluate and concentrate solutions [44].

A faster "static" method for the determination of membrane permselectivities, which is not affected by boundary layer transport phenomena, is based on the measurement of the potential gradient across an ion exchanger which separates two electrolyte solutions of different concentrations. The static method, however, does not take the water transport through the ion exchanger into account and is referred to as "apparent" permselectivity.

6.4 Conclusion

Ion exchangers made of organic materials such as various types of polymers have been significantly contributed to recover, remove, or reuse ionic substances in many industrial processes. The state-of-the-art polymer tailoring technology makes it possible to update energy conversion systems. Energy conversion devices such as proton exchange membrane fuel cells and dye sensitized solar cells are being developed to cope with global warming and deficiency in fossil fuels. Those devices accompanied with ionic transport through polymeric ion exchangers are used as electrolytes sandwiched between electrodes. The polymeric electrolytes with functional groups demand more critical ion exchanging properties, and thus further development based on much sophisticated polymer tailoring technology is necessary. Hydrocarbon aliphatic polymers were sufficient to meet the requirements of ion exchangers used in conventional industrial fields, but for the energy conversion devices, perfluorinated aliphatic or hydrocarbon-aromatic-polymer-based ion exchangers are necessary. In addition, the higher mechanical strength of ion exchangers is important due to their thin film casting, and to secure dimensional stability against electrical resistance, the high ratio of ion exchange capacity to water content becomes a critical property to develop ion exchangers. Generally, the development of new synthetic methods is to fulfill the properties of ion exchangers which are required in industrial applications. Thus, understanding the properties of ion exchangers with respect to their structures architected by polymer tailoring technology is of importance.

References

1. Harland CE (1994) Ion exchange: theory and practice. The Royal Society of Chemistry, Cambridge
2. Strathmann H (2004) Ion-exchange membrane separation processes, vol 9, Membrane science and technology. Elsevier, Amsterdam
3. Hideo K, Tsuzura K, Shimizu H (1991) Ion exchange membranes. In: Dorfner K (ed) Ion exchangers. Walter de Gruyter, Berlin
4. Strathmann H (1995) Electrodialysis and related processes. In: Nobe RD, Stern SA (eds) Membrane separation technology–principles and applications. Elsevier Science B.V, Amsterdam
5. Gebel G (2000) Structural evolution of water swollen perfluorosulfonated ionomers from dry membrane to saturation. Polymer 41:5829–5838
6. Gierke TH, Munn GE, Wilson FC (1982) Morphology of perfluorosulfonated membrane products. In: Eisenberg A, Yeager HL (eds) Perfluorinated ionomer membranes, ACS-symposium series 180. American Chemical Society, Washington DC
7. Falk M (1982) Infrared spectra of perfluorosulfonated polymer and of water in perfluorosulfonated polymer. In: Eisenberg A, Yeager HL (eds) Perfluorinated ionomer membranes, ACS-symposium series 180. American Chemical Society, Washington DC
8. Komoroski RA, Mauritz KA (1982) Perfluorosulfonated ionomer membranes. In: Eisenberg A, Yeager HL (eds) Perfluorinated ionomer membranes, ACS-symposium series 180. American Chemical Society, Washington DC
9. Xu T (2005) Ion exchange membranes: state of their development and perspective. J Membr Sci 263:1–29
10. Gubler L, Slaski M, Wokaun A, Scherer GG (2006) Advanced monomer combinations for radiation graft fuel cell membranes. Electrochem Commun 8:1215–1219

11. Hübner G, Roduner E (1999) EPR investigation of HO· radical initiated degradation reactions of sulfonated aromatics as model compounds for fuel cell proton conducting membranes. J Mater Chem 9:409–418
12. Wang S, McGrath JE (2003) Synthesis of poly(arylene ether)s. In: Martin ER, Timothy EL (eds) Synthetic methods in step-growth polymers. Wiley, Hoboken
13. Kim YS, Wang F, Hickner M, McCartney S, Hong YT, Zawodzinski TA, McGrath JE (2003) Effect of acidification treatment and morphological stability of sulfonated poly(arylene ether sulfone) copolymer proton-exchange membranes for fuel-cell use above 100 °C. J Polym Sci Part B: Polym Phys 41:2816–2828
14. Hickner MA, Ghassemi H, Kim YS, Einsla BR, McGrath JE (2004) Alternative polymer systems for proton exchange membranes (PEMs). Chem Rev 104:4587–4612
15. Robeson LM, Matzner M (1983) Flame retardant polyarylate compositions. US Patent 4,380,598
16. Ueda M, Toyota H, Ochi T, Sugiyama J, Yonetake K, Masuko T, Teramoto T (1993) Synthesis and characterization of aromatic poly(ether sulfone)s containing pendant sodium sulfonate groups. J Polym Sci Part A: Polym Chem 31:853–858
17. Wang F, Hickner M, Kim YS, Zawodzinski TA, McGrath JE (2002) Direct polymerization of sulfonated poly(arylene ether sulfone) random (statistical) copolymers: candidates for new proton exchange membranes. J Membr Sci 197:231–242
18. Wu L, Xu T, Yang W (2006) Fundamental studies of a new series of anion exchange membranes: membranes prepared through chloroacetylation of poly(2,6-dimethyl-1,4-pehnylene oxide) followed by quaternary amination. J Membr Sci 286:185–192
19. Wu L, Xu T, Wu D, Zheng X (2008) Preparation and characterization of CPPO/BPPO blend membranes for potential application in alkaline direct methanol fuel cell. J Membr Sci 310:577–585
20. Hibbs MR, Hickner MA, Alam TM, McIntyre SK, Fujimoto CY, Conelius CJ (2008) Transport properties of hydroxide and proton conducting membranes. Chem Mater 20:2566–2573
21. Zhou J, Unlu M, Vega JA, Kohl PA (2009) Anionic polysulfone ionomers and membranes containing fluorenyl groups for anionic fuel cells. J Power Source 190:285–292
22. Xiong Y, Liu QL, Zeng QH (2009) Quaternized cardo polyetherketone anion exchange membrane for direct methanol alkaline fuel cells. J Power Source 193:541–546
23. Wang G, Weng Y, Chu D, Chen R, Xie D (2009) Developing a polysulfone-based alkaline anion exchange membrane for improved ionic conductivity. J Membr Sci 332:63–68
24. Elias HG (2005) Macromolecules, vol 2, Industrial polymers and syntheses. Wiley, New York
25. Fu RQ, Julius D, Hong L, Lee JY (2008) PPO-based acid–base polymer blend membranes for direct methanol fuel cells. J Membr Sci 322:331–338
26. Avram E, Brebu MA, Warshawsky A, Vasile C (2000) Polymers with pendent functional groups. V. Thermooxidative and thermal behavior of chloromethylated polysulfone. Polym Degrad Stab 69:175–181
27. Sata T, Tsujimoto M, Yamaguchi T, Matsusaki K (1996) Change of anion-exchange membranes in an aqueous sodium hydroxide solution at high temperature. J Membr Sci 112:161–170
28. Sata T, Teshima K, Yamaguchi T (1996) Permselectivity between two anions in anion exchange membranes cross-linked with various diamines in electrodialysis. J Polym Sci Part A: Polym Chem 34:1475–1482
29. Komkova EN, Strmatialis DF, Strathmann H, Wessling M (2004) Anion-exchange membranes containing diamines: preparation and stability in alkaline solution. J Membr Sci 244:25–34
30. Risen JW (1996) Applications of ionomers. In: Schuck S (ed) Ionomers–characterization, theory and applications. CRC Press, New Jersey
31. Bishop HK, Bittels JA, Guter GA (1969) Investigation of inorganic ion exchange membranes for electrodialysis. Desalination 6:369–380
32. Bregman JI, Braman RS (1965) Inorganic ion exchange membranes. J Colloid Sci 20:913–922
33. Shrivastava K, Jain AK, Agrawal S, Singh RP (1978) Studies with inorganic ion exchange membranes. Talanta 25:157–165

34. Ohya H, Paterson R, Nomura T, McFadzean S, Suzuki T, Kogure M (1995) Properties of new inorganic membranes prepared by metal alkoxide methods: part I. A new permselective cation exchange membrane based on oxides. J Membr Sci 105:103–112
35. Pourcelly G, Gavach C (2000) Electrodialysis water splitting-application of electrodialysis with bipolar membranes. In: Kemperman AJB (ed) Handbook on bipolar membrane technology. Twente University Press, Enschede
36. Xu TW (2001) Development of bipolar membrane-based process. Desalination 140:247–258
37. Daufin G, Escudier JP, Carrere H, Berot S, Fillaudeau L, Decloux M (2001) Recent emerging applications of membrane processes in the food and dairy industry. Food Bioprod Process 79:89–102
38. Tarvainen T, Svarfvar B, Akerman S, Savolainen J, Karhu M, Paronen P, Jarvinen K (1999) Drug release from a porous ion exchange membrane in vitro. Biomaterials 20:2177–2183
39. Kim YH, Moon SH (2001) Lactic acid recovery from fermentation broth using one-stage electrodialysis. J Chem Technol Biotechnol 76:169–178
40. Saracco G (2003) Ionic membrane technologies for the recovery of valuable chemicals from waste waters. Ann Chim Rome 93:817–826
41. Bazinet L, Lamarche F, Ippersiel D (1998) Bipolar membrane electrodialysis: applications of electrodialysis in the food industry. Trends Food Sci Technol 9:107–113
42. Xu TW (2002) Electrodialysis processes with bipolar membranes (EDBM) in environmental protection – a review. Resour Conserv Recycl 37:1–22
43. Park SH, Park JS, Yim SD, Park SH, Lee YM, Kim CS (2008) Preparation of organic/inorganic composite membranes using two types of polymer matrix via a sol–gel process. J Power Sources 181:259–266
44. Hefferich F (1995) Ion exchange. Dover Publications, New York
45. Sauer MC, Southwick PF, Spiegler KS, Wyllie MRJ (1955) Electrical conductance of porous plugs–ion exchange resin-solution systems. Ind Eng Chem 47:2187–2193
46. Cahan BD, Wainright JS (1993) AC impedance investigations of proton conduction in Nafion^{TM}. J Electrochem Soc 140:L185–L189
47. Woo Y, Oh SY, Kang YS, Jung B (2003) Synthesis and characterization of sulfonated polyimide membranes for direct methanol fuel cell. J Membr Sci 220:31–45
48. Scatchard G (1953) Ion exchanger electrodes. J Am Chem Soc 75:2883–2887
49. Scatchard G (1956) General introduction of discussions of membrane phenomena. Discuss Faraday Soc 21:27–30

Chapter 7
Ion Exchange Membranes: Preparation, Properties, and Applications

Mahadevappa Y. Kariduraganavar, Arjumand A. Kittur, and Srikant S. Kulkarni

Abstract Membrane science is a relatively new area of applied chemistry and chemical engineering and plays a vital role in the field of alternative energy and separation applications. The field of membrane science is therefore an emerging area with enormous industrial and public health significance. Today, various kinds of separation membranes have been studied and utilized industrially in different processes including reverse osmosis, nanofiltration, ultrafiltration, microfiltration, electrodialysis, and pervaporation. Technical feasibility of all these processes largely depends on membrane and its properties. Among the separation membranes, ion exchange membranes are one of the advanced separation membranes. Therefore, this chapter addresses on different types of ion exchange membranes with reference to their preparation, properties, as well as their industrial applications. Although ion exchange membranes are broadly classified into homogeneous and heterogeneous membranes, this chapter also covers other types of ion exchange membranes such as amphoteric, mosaic, bipolar, interpolymer, and graft and block copolymer membranes. At the end, prospectives and conclusions on the ion exchange membranes have also been highlighted. The relevant literature was collected from different sources including Google sites, books, reviews, and research articles.

M.Y. Kariduraganavar (✉)
P. G. Department of Studies in Chemistry, Karnatak University,
Pavate Nagar, Dharwad 580 003, KA, India
e-mail: mahadevappak@yahoo.com

A.A. Kittur
Department of Chemistry, SDM College of Engineering & Technology,
Dhavalgiri, Dharwad 580 002, KA, India
e-mail: aakittur@yahoo.co.in

S.S. Kulkarni
Department of Chemistry, JSS Degree College, Vidyagiri, Dharwad 580 004, KA, India
e-mail: srikantkulkarnidwd@yahoo.com

Inamuddin and M. Luqman (eds.), *Ion Exchange Technology I: Theory and Materials*, DOI 10.1007/978-94-007-1700-8_7,

7.1 Membrane Science

Membrane science is a relatively new area of applied chemistry and chemical engineering. During the 1960s and 1970s, this area was primarily developed for water desalination application for the production of drinking water from sea and brackish water sources. Today, this application is a fast growing one all over the world. During the past two decades, membranes have been developed for a wide variety of chemical separation applications involving the treatment of industrial liquids, gases, and vapors, such as those involved in wastewater treatment, pollution control, wastewater reuse and waste recovery, food processing, gas separations, petroleum engineering, biotechnology, and biomedical devices. Membranes are also now playing a special role in the field of alternative energy as one of the fundamental parts of a fuel cell. In this sense, membrane technology has potential contribution for a green chemistry. The field of membrane science is second to none in terms of economic significance and social relevance as it touches many vital areas of everyday life. Therefore, the field of membrane science is an emerging area with enormous industrial and public health significance. To make the fullest utilization of the potentialities of membranes for practical applications require intensive and extensive R&D work [1].

Despite the apparent proliferation of membrane processes, the science and technology of membrane separations is still in its early stages of development. Tremendous opportunities exist and will continue to exist for the penetration of membrane processes in every facet of chemical process engineering and environmental protection through intensive research and development.

The level of research and development activity in the general field of membrane science and technology has been rising dramatically all over the world during the last two decades. This is happening not only because of the growing recognition of the commercial potentialities of membrane separations but also because of the growing realization of the international scientific and engineering community that synthetic membranes are capable of playing tremendous beneficial roles in society and industry far more than what has hitherto been appreciated. This is particularly a significant and welcoming realization for the health and well-being of all humanity.

Today, various kinds of separation membranes have been studied widely and utilized industrially in various fields, including membranes for reverse osmosis, nanofiltration, ultrafiltration, microfiltration, pervaporation, electrodialysis, and for medical use as an artificial kidney [2]. Among these membranes, ion exchange membranes are one of the most advanced separation membranes.

7.2 Ion Exchange Membranes

Ion exchange membranes are thin sheets or films of ion exchange material, which can be used to separate ions by allowing the preferential transport of either cations (in case of cation-exchange membrane) or anions (in case of anion-exchange

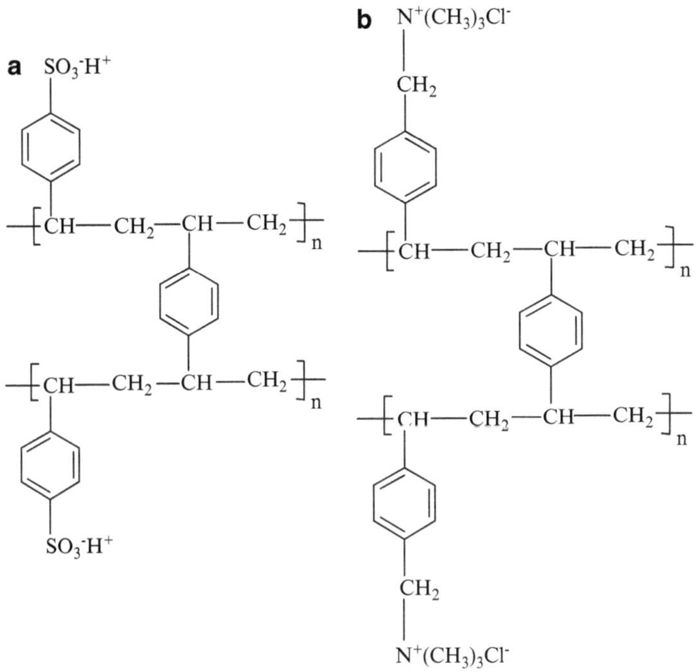

Fig. 7.1 Schematic representation of ion exchange membranes: (**a**) cation-exchange membrane; (**b**) anion-exchange membrane

membrane) [3]. In 1950, ion-selective membranes having high selectivity, low electrical resistance, high mechanical strength, and good chemical stability were described [4]. These were essentially insoluble, synthetic, polymeric ion exchange resins in sheet form. The chemical structures of typical modern membranes of this type are shown in Fig. 7.1. The membrane illustrated consists (cation-exchange membrane) of polystyrene having negatively charged sulfonate groups chemically bonded to most of the phenyl groups in the polystyrene. The negative charges of the sulfonate groups are electrically balanced by positively charged cations (counter ions). Sulfonated polystyrene swells greatly in water. The amount of swelling is typically controlled by introducing cross-linking groups in the polymer backbone (divinylbenzene in Fig. 7.1), by incorporating electrically neutral polymers, or by substituting fluorine groups on the polymer that decrease the affinity for water.

The positively charged counter ions are appreciably dissociated into the imbibed water, and these may be exchanged for other anions from an ambient solution while maintaining the electrical neutrality of the membrane. The high concentration of counter ions in ion exchange membrane is responsible for the low electrical resistance of the membrane. The high concentration of bound negatively charged groups tends to exclude mobile negatively charged ions (co-ions) from an ambient solution and is responsible for high selectivity of the membrane. The anion-selective membrane represented in Fig. 7.1 consists of cross-linked polystyrene having positively charged quaternary ammonium groups chemically bonded to most of the phenyl groups in the polystyrene. In this case, the counter ions are negatively charged and are the principal carriers of the electric current.

7.3 Historical Background

The development of ion exchange membrane–based process began in 1890 with the work of Ostwald [5], who studied the properties of semipermeable membranes and discovered that a membrane can be impermeable for any electrolyte. To illustrate this, the so-called *membrane potential* at the boundary between a membrane and its surrounding solution was postulated as a consequence of the difference in concentration. In 1911, Donnan [6] confirmed the existence of such boundary and developed a mathematical equation describing the concentration equilibrium, which results in the so-called *Donnan exclusion potential*. However, the actual basic studies related to ion exchange membranes were carried out in 1925 by Michaelis and Fujita with the homogeneous, weak acid collodion membranes [7]. In 1932, Sollner presented the idea of a charge-mosaic membrane or amphoteric membrane containing both negatively and positively charged ion exchange groups and showed distinctive ion transport phenomena [8]. Around 1940, interest in industrial applications led to the development of synthetic ion exchange membrane on the basis of phenol–formaldehyde polycondensation [9]. Simultaneously, Meyer and Strauss proposed an electrodialysis process in which anion-exchange and cation-exchange membranes were arranged in alternating series to form many parallel solution compartments between two electrodes [10]. It was hard to go into the industrial implications because commercial ion exchange membranes with excellent properties especially low electric resistance were still not available at that time. With the development of stable, highly selective ion exchange membrane of low electric resistance in 1950 by Juda and McRae of Ionics Inc. [4] and Winger et al. at Rohm in 1953 [11], electrodialysis based on ion exchange membranes rapidly became an industrial process for demineralizing and concentrating electrolyte solutions. Since then, both ion exchange membranes and electrodialysis have been greatly improved and widely used in many fields. For example, in the 1960s, first salt production from seawater was realized by Asahi Co. with monovalent-ion-permselective membranes [12]; in 1969, the invention of electrodialysis reversal (EDR) realized long-term run without salt precipitation or deposition on both membranes and electrodes [13]; in the 1970s, a chemically stable cation-exchange membrane based on sulfonated polytetrafluoroethylene was first developed by DuPont as Nafion, leading to a large scale use of this membrane in the chlor-alkali production industry and energy storage or energy conversion system (fuel cell) [14]; simultaneously, a composition of cation-exchange layer and anion-exchange layer into a bipolar membrane was developed in 1976 by Chlanda et al. [15], which brings many novelty in electrodialysis applications [16]. By the development of new ion

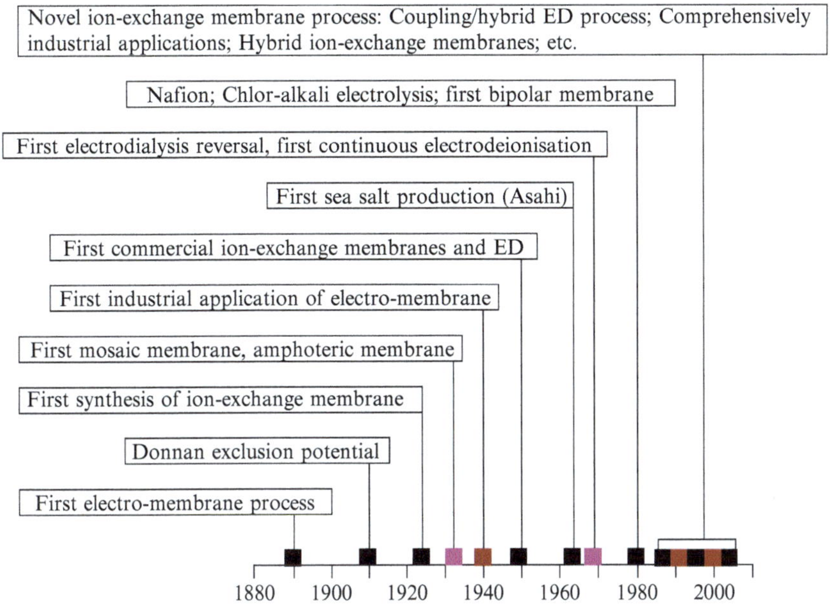

Fig. 7.2 Timeline visualization of ion exchange membrane development and its related processes [30]

exchange membranes with better selectivity, lower electrical resistance, and improved thermal, chemical, and mechanical properties, other applications of ion exchange membranes apart from the initial desalination of brackish water have recently gained a broader interest in food, drug, and chemical process industry including biotechnology and wastewater treatment [16–23]. Apart from polymeric ion exchange membranes, an ion exchange membrane can also be developed from inorganic materials such as zeolites, bentonite, or phosphate salts [24–26]. However, these membranes are rather unimportant due to their high cost and other disadvantages such as relatively bad electrochemical properties and too large pores, though they can undergo higher temperatures than organic membranes [27]. It can be expected that ion exchange membranes prepared from polymers can possess both chemical stability and excellent conductivity, if the membranes were incorporated with inorganic component such as silica. Therefore, inorganic–organic ion exchange membranes were developed in the late 1990s by sol–gel technique for the applications in severe conditions such as higher temperature and strongly oxidizing circumstances [27–29]. Thus, till today, various ion exchange membranes including inorganic–organic (hybrid) ion exchange membranes, amphoteric ion exchange membranes, mosaic ion exchange membranes, and bipolar membranes are available, and their position in the history of ion exchange membrane development as well as the important affairs related to them are schematically shown in Fig. 7.2.

7.4 Preparations and Properties of Ion Exchange Membranes

As far as chemical structure is concerned, ion exchange membranes are very similar to normal ion exchange resins. From a chemical point of view, these resins would make excellent membranes of high selectivity and low resistivity. The difference between membranes and resins arises largely from the mechanical requirement of the membrane process. Unfortunately, ion exchange resins are mechanically weak, cation resins tend to be brittle, and anion resins soft. They are dimensionally unstable due to the variation in the amount of water imbibed into the gel under different conditions. Changes in electrolyte concentration in the ionic form or in temperature may cause major changes in the water uptake and hence in the volume of the resin. These changes can be tolerated in small spherical beads. But, in large sheets that have been cut to fit an apparatus, they are acceptable. Thus, it is generally not possible to use sheets of material that have been prepared in the same way as a bead resin. However, the most common solution to this problem is the preparation of a membrane with a backing of a stable reinforcing material that gives the adequate strength and dimensional stability.

Most commercial ion exchange membranes can be divided into two major categories such as homogeneous and heterogeneous based on their structure and preparation procedure. According to Molau [31], depending on the degree of heterogeneity of the ion exchange, they can be divided into (a) homogeneous ion exchange membrane, (b) microheterogeneous graft and block polymer membrane, (c) snake-in-cage ion exchange membrane, (d) interpolymer membrane, and (e) heterogeneous ion exchange membrane. However, mosaic and bipolar membranes are the recent inventions of ion exchange membranes.

7.4.1 Homogeneous Ion Exchange Membranes

The method of making homogeneous ion exchange membranes can be summarized in three different categories:

1. Polymerization or polycondensation of monomers; at least one of them must contain a moiety that either is or can be made anionic or cationic groups, respectively
2. Introduction of anionic or cationic moieties into a preformed solid film
3. Introduction of anionic or cationic moieties into a polymer such as polysulfone, followed by dissolving the polymer and casting it onto a film

7.4.1.1 Membranes Prepared by Polymerization and Polycondensation of Monomers

Under this category, the membranes prepared from polymerization and polycondensation of monomers, in which at least one of them must contain a moiety that can be made anionic or cationic. Under the polycondensation type of the

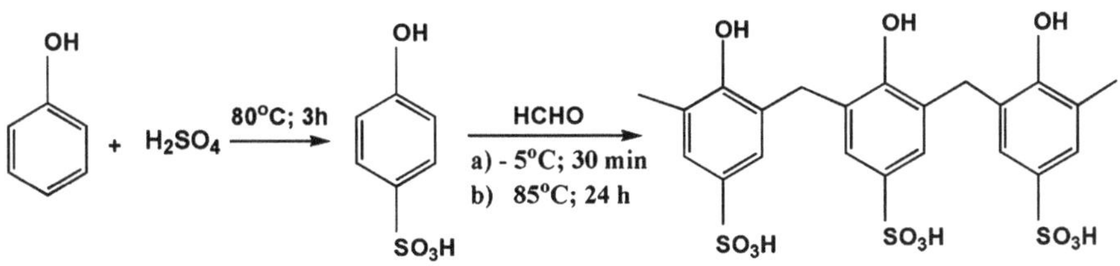

Fig. 7.3 Reaction of phenolsulfonic acid with formaldehyde

membranes, formaldehyde is used as a cross-linking agent to make the membrane water insoluble, while using suitable monomeric/polymeric material endowed with ion exchange properties. The first membrane made by polycondensation of monomer followed by cross-linking with formaldehyde was of phenolsulfonic acid [32]. The preparation scheme is presented in Fig. 7.3.

In the scheme, the phenol was treated with concentrated sulfuric acid at 80°C, which led to phenolsulfonic acid in para form. This was then reacted with 38% solution of formaldehyde in water initially at −5°C for about 30 min and then at 85°C for several hours. The solution was then cast into a film, which forms a cation-exchange membrane. Excess monomer was removed by washing the film in the water. With a slight modification of this procedure, condensation of anisole or sulfonated phenol with formaldehyde and subsequent casting onto a glass plate or mercury followed by curing at 90°C gives the cation-exchange membrane [33]. Similarly, anion-exchange membrane was prepared by condensation of *m*-phenylenediamine or aliphatic diamine compounds such as polyethylenediamine or dicynodiamines with formaldehyde [34]. The resulting membranes showed good electrochemical properties but were lack in mechanical strength. The membranes prepared from dimethyl-2-hydroxy benzyl amine, phenol, and formaldehyde [35]; condensation of $C_6H_4(OH)_2$ or $C_6H_3(OH)_3$ and $NH_2C_6H_4$ COOH with formaldehyde [36]; benzidine–formaldehyde condensate with binder of acrylonitrile–vinyl chloride copolymer [37]; and phenolsulfonic acid cross-linked with formaldehyde supported on orlar cloth [38] were also reported.

Fluorinated Ionomer Membranes

These membranes are prepared by the polymerization of monomers, which contain a moiety that can be made either cationic or anionic by further treatment. These membranes are mainly developed for the applications in the chlor-alkali industries and for the fuel cell applications [39] since most of the conventional hydrocarbon ion exchange membranes degrade in the presence of oxidizing agent especially at elevated temperatures. These are the fluorocarbon-based ion exchange membranes (Nafion) developed by DuPont with good chemical and thermal stability [40]. There is a four-step procedure for the synthesis of Nafion membranes: (a) the reaction of tetrafluoroethylene with SO_3 to form the sulfone cycle, (b) the condensation of these products with sodium carbonate followed by copolymerization with

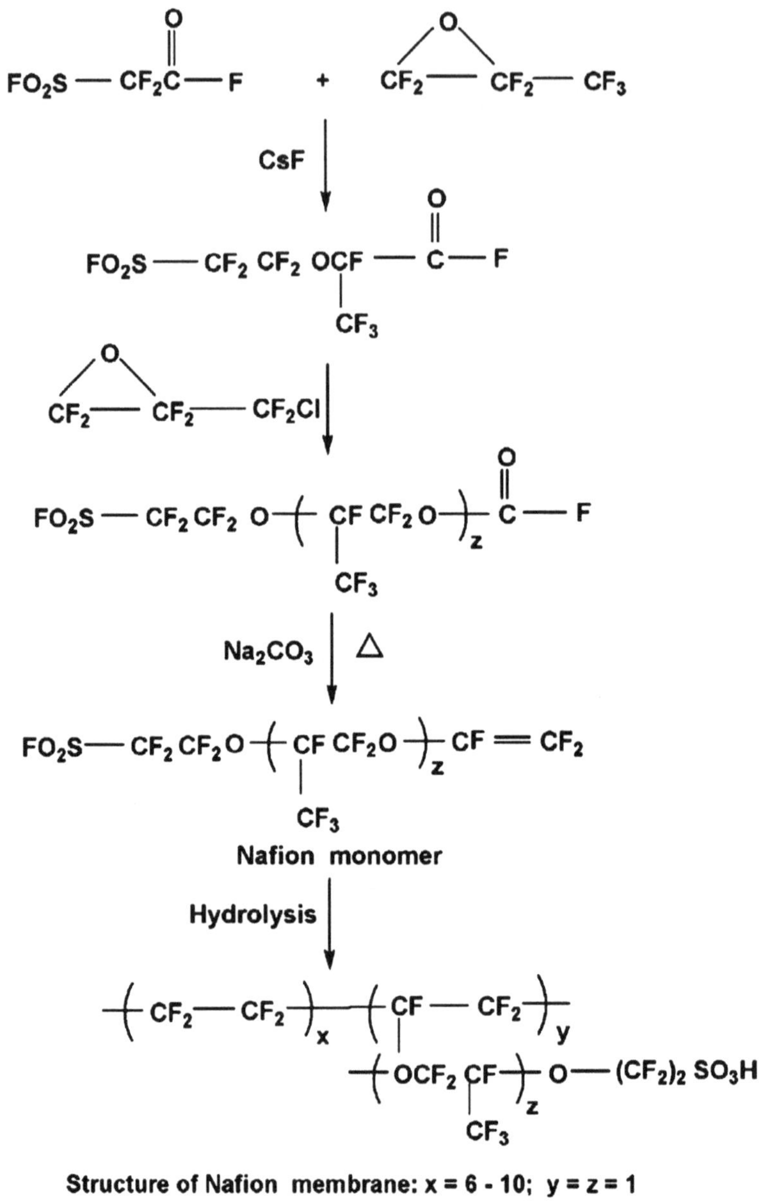

Fig. 7.4 Nafion membrane and its preparation

tetrafluoroethylene to form the insoluble resin, (c) the hydrolysis of this resin to form a perfluorosulfonic polymer, and (d) the chemical exchange of the counter ion Na^+ with the proton in an appropriate electrolyte. DuPont introduced this membrane in 1966 following the reaction route shown in Fig. 7.4.

These are the high equivalent weight (EW) perfluorinated membranes and have limited use in fuel cells because they consume high power density. The Dow Chemical Company has overcome this problem with the development of low EW perfluorinated membranes in 1988. The Dow membrane is prepared by the copolymerization of tetrafluoroethylene with vinyl ether monomer as per the reaction scheme shown in Fig. 7.5.

The polymer has a Teflon-like backbone structure with a side chain attached via ether group. This side chain is characterized by a terminal sulfonate functional group. It must be pointed out that the Dow monomer is more complicated to elaborate than the DuPont monomer. The synthesis of Dow epoxy (Fig. 7.5) is more complicated than that of a Nafion epoxy (Fig. 7.4), which is a commercially available material [41, 42]. The perfluorinated ionomer membrane was also developed by Asahi Glass Company and commercialized as Flemion. The general properties of long-side-chain perfluorinated ionomer membranes (e.g., Nafion/Flemion/Aciplex) and short-side-chain perfluorinated ionomer membranes (e.g., Dow) are as follows: (a) EW range = 800–1,500 and (b) conductivity = 0.20–0.05 S/cm^2.

7.4.1.2 Introduction of Anionic or Cationic Moieties into a Preformed Solid Film

Styrene–Divinylbenzene-Based Membranes

Styrene–divinylbenzene-based membranes are the best-known membranes in the literature for the application of electrodialysis [43–46]. Copolymerization of styrene and divinylbenzene followed by sulfonation and amination in solution or in bulk gives the cation- and anion-exchange membranes, respectively. The cation-exchange membrane was prepared by the sulfonation of polymer with chlorosulfonic acid or with concentrated sulfuric acid in dichloroethane using the silver sulfate as catalyst. The anion-exchange membrane was prepared by chloromethylation of the polymer followed by amination. The reaction scheme for these membranes is shown in Fig. 7.6. It is known [45] that the selectivity of pure styrene–divinylbenzene cation-exchange membrane is poor for monovalent ion. It is also known that there is no remarkable effect of cross-linking agent on the selectivity of particular cations, though the mobility of multivalent cations decreases with increasing the cross-linking. This was due to a compact structure of the membrane.

From the point of commercial applications, special types of ion-selective membranes have been prepared using these monomers. For example, membranes for monovalent ion selective for the production of table salt, the membranes for high hydrogen ion retention to recover acids from the wastewater, etc. The hydrogen ion can pass selectively through the membrane against all other ions in electrodialysis, and therefore, the development of membranes with high acid retention was really a challenging task for the applications of acid recovery. Sata [45–47] reported on the monovalent-ion-selective membranes for the production of table salt from seawater. They reported the modification of permselectivity between ions with the same charge through ion exchange membranes in electrodialysis, and

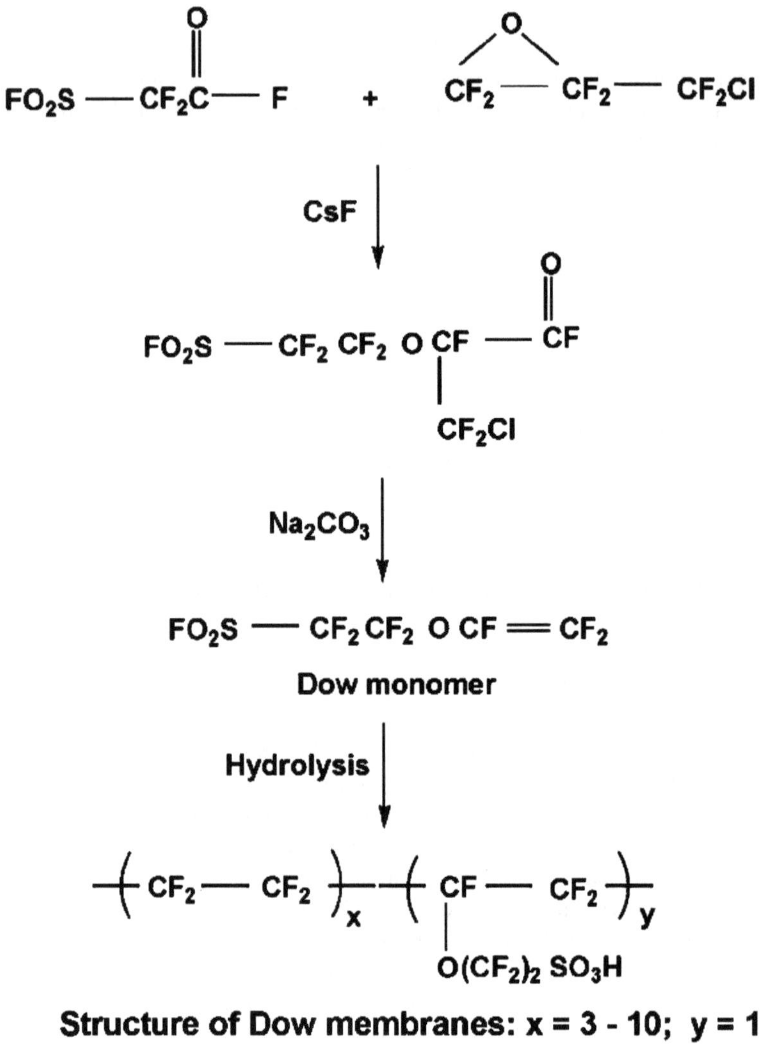

Fig. 7.5 Dow membrane and its preparation

they have been classified according to its mechanism. The first is the sieving of ions with changing cross-linkage of the membrane; the second is the effect of charge and electrostatic repulsion on permselectivity of ions; and the third is the specific interaction between the specific ion exchange groups, the membrane matrix, and the ions. In case of anion-exchange membrane, sieving of ions by highly cross-linked layer was effective on changing the permselectivity between chloride and sulfate ions. Sieving can also be done effectively by introducing the dense polyelectrolyte layer on the membrane surface, like polymerization of pyrrole on the surface of cation- and anion-exchange membranes to change the permselectivity of the larger

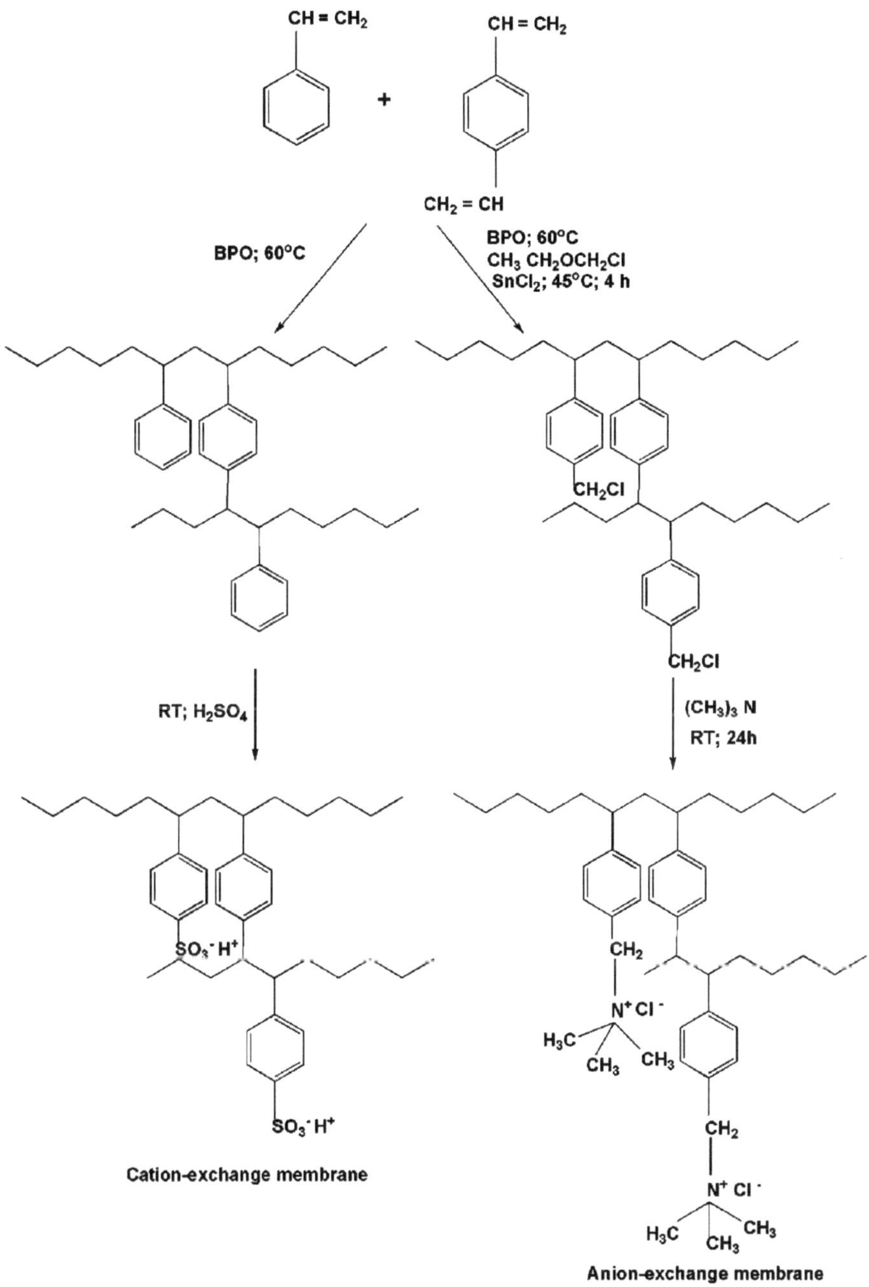

Fig. 7.6 Reaction scheme for styrene–divinylbenzene-based ion exchange membranes

ion, such as calcium and sulfate ion with respect to sodium and chloride ions, respectively. Selective permeation of specific ions is mainly dependent on the degree of affinity of specific anions with the anion-exchange membrane and partially on the change in mobility through the membrane. The affinity of the specific ions to the membrane depends on the balance of hydrophilicity of the membrane as done by introducing the polysoap layer made up of poly(n-dodecyl-4-vinylpyridine bromide) on the anion-exchange membrane. Utilization of electrostatic repulsion force to change the permselectivity was studied from various aspects, and this was effective on achieving monovalent cation and monovalent anion permselectivity [46–48]. It is also reported that a bipolar ion exchange membrane is selectively permeated to lower valent cations compared to higher ones. In a special type of membrane [48], synergistic effect of the change in ion exchange equilibrium constants of various anions to chloride ions and the change in the ratio of mobility between anions have been studied by developing membranes from poly(vinyl alcohol), β-cyclodextrin, and poly(n-ethyl-4-vinylpyridinium bromide) by the casting method, followed by cross-linking with formaldehyde. Cyclodextrin is a naturally occurring cyclic oligosaccharide, mostly consisting of six, seven, and eight D-glucose units for α-, β-, and γ-cyclodextrin, respectively. The doughnut-shaped cyclodextrin has an interior cavity with a diameter of 4.7–5.3 Å (α-CD), 6.0–6.5 Å (β-CD), or 7.5–8.3 Å (γ-CD) [48]. They have a unique configuration, which makes the outer surface hydrophilic and the inner cavity hydrophobic in nature. Due to the presence of hydrophilic outer surface and hydrophobic inner cavity, it was expected that the permeation of different anions through anion-exchange membrane would be different. On the other hand, it was reported that the permselectivity of anions through anion-exchange membranes in electrodialysis is mainly governed by a balance of the hydrophilicity of anion-exchange membranes with the hydration energy of anions in permeating through the membrane [48]. β-cyclodextrin makes the membrane hydrophilic, in which water content of the membranes was increased with increasing the content of β-cyclodextrin. Consequently, transport numbers of the less hydrated anions (nitrate and bromide ions) relative to chloride ions decrease and those of strongly hydrated anions like sulfate and fluoride ions increase with increasing the content of β-cyclodextrin.

The anion-exchange membrane prepared from the copolymerization of 4-vinylpyridine and divinylbenzene followed by quaternization with methyl iodide gives the membrane with good electrochemical properties but lacks in chemical stability. Very recently, Kariduraganavar et al. [49] developed the novel homogeneous anion-exchange membranes by polymerization of 4-vinylpyridine and simultaneously cross-linking with epichlorohydrin and aniline on an oven support cloth at different ratios. The reaction was carried out at 80°C in the presence of benzoyl peroxide as an initiator. The resulting membranes were subsequently quaternized with methyl iodide using hexane as a solvent. The scheme of the reaction is shown in Fig. 7.7. These membranes exhibited good electrochemical properties and mechanical strength and can be used for the conversion of sea and brackish water into potable water, production of table salt from seawater, and production of chlor-alkali.

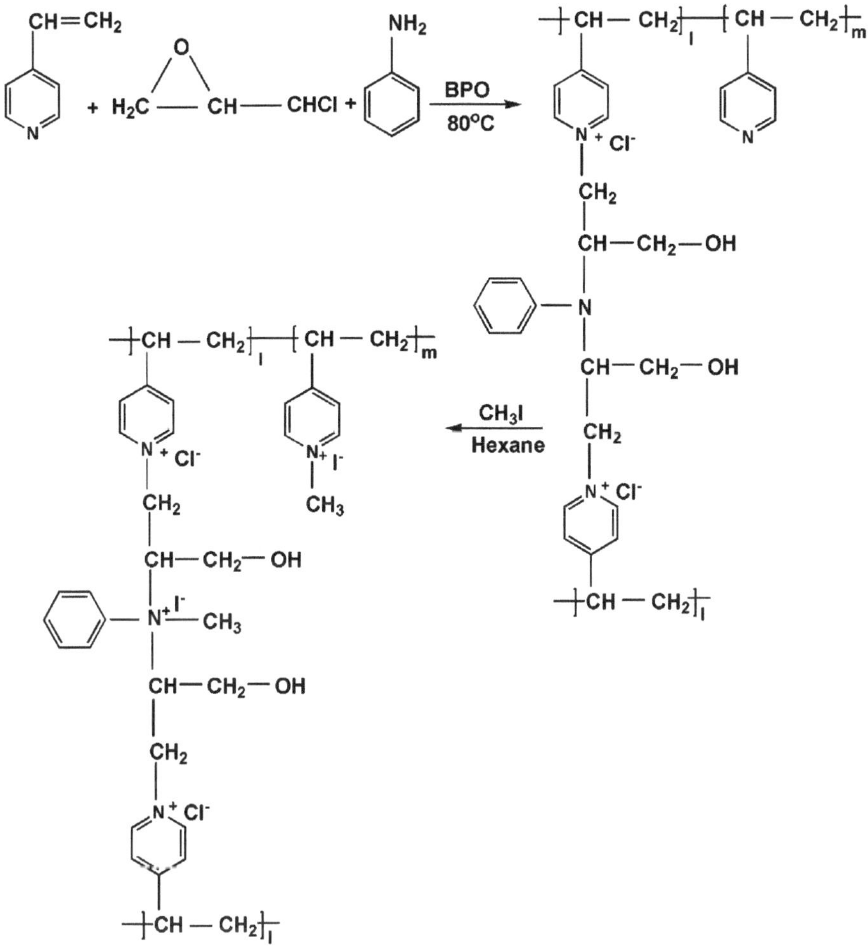

Fig. 7.7 Poly(4-vinylpyridine) based anion-exchange membranes

Partially Fluorinated Ionomer Membranes

The preparation of monomer α,β,β-trifluorostyrene was carried out by Cohen et al. [50], even though the poly-α,β,β-trifluorostyrene was first synthesized by Prober [51]. The structure of the polymer is shown in Fig. 7.8.

Hodgdon [52] investigated the sulfonation of α,β,β-trifluorostyrene seeking their applications in fuel cells. From the optimization of the reaction conditions, he has prepared a multiplicity of equivalent weight or different ion exchange capacity of both linear and cross-linked poly-α,β,β-trifluorostyrene sulfonic acids. He has shown that the extreme difficulty in sulfonation of poly-α,β,β-trifluorostyrene was caused by the beta-directing influence of the perfluorinated polyalkyl group attached to the aromatic ring [51]. The degree of sulfonation was

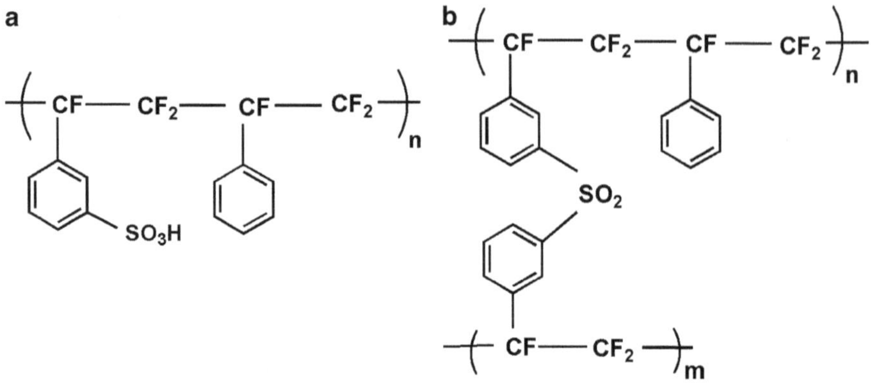

Fig. 7.8 Poly(trifluorostyrene) based ion exchange membranes: (**a**) linear; (**b**) cross-linked

very much dependent on the temperature as well as on the concentration of chlorosulfonic acid. The linear poly-α,β,β-trifluorostyrene sulfonic acid was prepared by a direct combination of chlorosulfonic acid with trifluorostyrene. Sulfone cross-linked poly-α,β,β-trifluorostyrene sulfonic acid was prepared and was found to be totally resistant to oxidative mode of degradation, which causes drastic oxidative depolymerization of polystyrene and polystyrene–divinylbenzene sulfonic acids. This is attributed to the substitution of fluorine in place of benzylic hydrogen atom of polystyrene, leading to the added C–F bond strength required to resist the oxidation environment. Thus, it was claimed that the high stability of fluorine atoms attached to alkyl carbon atoms imparts the oxidative and thermal stability much superior to that of conventional ion exchange membranes such as styrene sulfonic acid.

7.4.1.3 Introduction of Anionic or Cationic Moieties into the Polymer

Polysulfone-Based Ion Exchange Membranes

For the purpose of applying electrodialysis process under severe conditions such as high temperature and strongly oxidizing conditions, a more stable Nafion and Dow ion exchange membranes have been developed and successfully applied under those conditions. It was found that only a few major industrial applications were sought other than the chlor-alkali industry, primarily because of its high cost. Thus, there has been a challenge to develop new types of ion exchange membranes, which would be low cost but also have good electrochemical properties and excellent resistance to degradation by heat and chemical attack. Engineering plastics such as polysulfone and polyethersulfone have been widely used as a base polymer for ultrafiltration and gas separation because of their excellent workability and mechanical strength [53–56]. In particular, a polysulfone membrane having excellent chemical resistance has been studied for its application as an

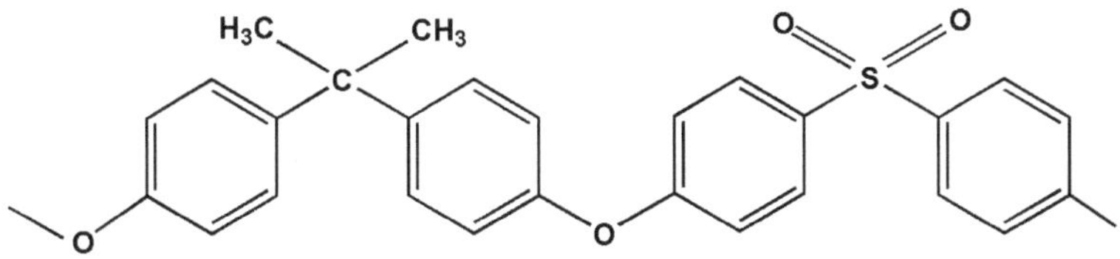

Fig.7.9 Structure of Udel polysulfone

ion exchange membrane by improving the permeability for ultrafiltration and reverse osmosis or imparting ion permselectivity by introducing the ion exchange groups into the membrane [57–59]. Several researchers [60–62] have prepared the ion exchange membranes using polysulfone as a base polymer of the membrane. The sulfonation was achieved either by solution method [63] or by slurry method [64], and their properties have also been described [61, 65, 66]. The sulfonation procedure for poly(arylene ether sulfone) Udel P-1700 (PSU) and Vitrex PES 5200P (PES) was described in the patent literature [64]. A new sulfonation process for polysulfone ionomers has also been proposed [61, 67–71]. It was reported that these polymers have been cross-linked into membranes by a new cross-linking procedure. The basic working hypothesis for the new type of sulfonation method was that polysulfone Udel as shown in Fig. 7.9 is usually sulfonated at the *ortho* position to the ether bridge of the bisphenol-A position by electrophilic substitution [72] because this part of the molecule has high electron density in contrast to the diarylsulfone portion of the repeating monomer unit, which has a low electron density due to the electron-withdrawing SO_2 group.

One disadvantage of electrophilically sulfonated polysulfone is that the SO_3H group at this position can undergo *ipso*-substitution with H^+, which lowers the hydrolytic stability of the sulfonated polysulfone in strong acids. This limits the application of the commonly sulfonated polysulfone as a cation exchange polymer in electro-membrane process. The new method based on the sulfonation of polysulfone in the diarylsulfone part should show increased hydrolytic stability against *ipso*-substitution by H^+ compared to electrophilically sulfonated polysulfone, due to the electron-deficient of the diarylsulfone position of the polymer.

Based on the literature [71], it was found that polysulfone Udel can be substituted in the electron deficient of the monomer by first deprotonating the aromatic ring position ortho to the sulfone bridge with metal-organic reagents and subsequent reaction of the metalated site with the electrophiles. Deprotonation reactions by attack of metal organics at the polymer backbones are also known from poly (2,6-dimethyl-para-phenylene ether) (PPO), polystyrene, poly(vinyl-thiophene), and poly(methylphenylphosphazene). To convert the metalated site of the deprotonated polymer into sulfonic acid group, the metalated polymer is quenched with the electrophile SO_2, and the metal sulfinate derivative of the respective polymer is formed. Sulfinates and particularly the corresponding sulfinic acids are known to be chemically unstable and can be easily oxidized to sulfonic acids.

Based on the aforementioned facts, a three-stage process has been developed for introduction of sulfonic acid group in diarylsulfone part of the polysulfone Udel. The scheme for this process is presented in Fig. 7.10.

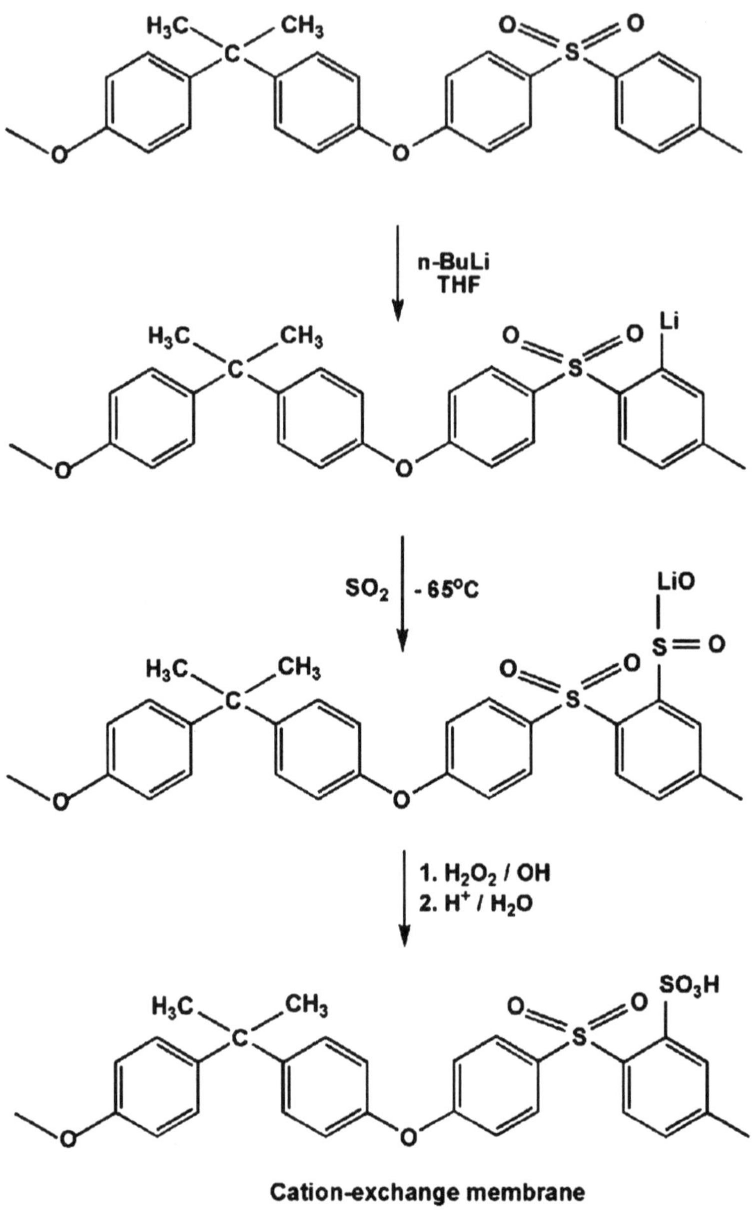

Fig. 7.10 Schematic presentation of sulfonation of Udel polysulfone membrane

The sulfonation process consists of the following steps:

(a) Lithiation of the polymer in a 2.4 wt% THF solution with 10 N n-butyl lithium at temperature ranging from −50 to −80°C under argon atmosphere. The lithiation proceeds very fast (0.5 h) and is nearly quantitative, as it was proved by NMR;
(b) Introduction of SO_2 into the reaction vessel containing the lithiated polymer at temperature range from −65 to −20°C. The reaction of the lithiated polymer

with the electrophile, SO_2 is also nearly quantitative as it was proved by NMR. The sulfonation reaction was finished after 5–10 min of SO_2 introduction;

(c) Oxidation of the formed sulfinate. Alkaline or neutral H_2O_2, sodium hypochlorite, or $KMnO_4$ was used for oxidation in an aqueous solution or suspension after isolating the polymeric sulfinate.

(d) Conversion of lithium sulfonate to the corresponding sulfonic acid by aqueous HCl (water-insoluble polymeric sulfonic acids) or by stirring of the sulfonate with a cation-exchange resin (water-soluble polymeric sulfonic acid).

The ion exchange capacity (IEC) of the resulting polymer is largely dependent upon the oxidation step. A loss in sulfinate groups can partially be caused by splitting off of the sulfinate group during the oxidation process and subsequent substitution by hydrogen (ipso reaction). However, by selecting the suitable oxidation conditions, the loss in IEC can be minimized as can be seen at the oxidation of poly (sulfone)-SO_2-25 with $KMnO_4$, where 98.9% yield of the oxidation products are obtained. At higher to medium degree of sulfination, $KMnO_4$ and NaOCl are the most suitable oxidants (3.3–1.9 mequiv/g of IEC). For the low degree of sulfination, nonionic oxidant H_2O_2 should be used. The membrane swelling largely depends upon the IEC. At IECs of $\geq$2.4 mequiv SO_3H/g dry polymer, the polymer becomes water-soluble. The specific conductance of the membranes is dependent on the IEC and is lower than that of Nafion membrane. This can be explained by stronger clustering of SO_3H in Nafion due to its extremely hydrophobic backbone [72].

It is reported [61, 73, 74] that the ion exchange membranes prepared using polysulfone as a base polymer are likely to undergo a dimensional change during flocculation and tend to have defects. As the ion exchange capacity (IEC) increases, the affinity toward water molecules increases, whereby it tends to hardly flocculate. Hence, it is difficult to obtain a membrane having sufficient mechanical strength since dimensional stability tends to be poor [62]. The water tends to concentrate as molecular clusters around the ionic groups, and therefore, the affinity of the membrane toward water molecules increases with IEC. The high affinity toward water leads to low cohesive forces, and thereby, membrane swells greatly in water [73].

Terada et al. [63, 75] have reported a new type of ion exchange membrane. Their study demonstrated that anion-exchange membrane formed using a block copolymer having segments, to which ion exchange groups can be introduced readily, had a higher mechanical strength as compared with the type of ion exchange membranes formed by segments, to which ion exchange groups can hardly be introduced. The advantage of the block copolymer is that the cohesive force of the part of polymer without ionic groups controls the membrane swelling. The anion-exchange membranes have been prepared by block copolymers of polysulfone and poly-phenylenesulfidesulfone in two steps. In the first step, polymer undergoes chloromethylation and in the second step, amination of chloromethylated copolymers takes place [73]. These membranes have shown good dimensional stability with poor electrochemical properties. Sulfonation of the copolymer with triethylphosphate and sulfuric acid yields cation-exchange membranes with poor electrochemical properties [76].

Partially Sulfonated Poly(Ether Ether Ketone) Membranes

The electrophilic aromatic sulfonation of the poly(ether ether ketone) (PEEK) was reported in the literature [77]. The general concept of the sulfonation is to dissolve the dried PEEK in concentrated sulfuric acid. The desired level of sulfonation depends on the time and temperature of the process. For instance, Bailey et al. [78] have sulfonated the PEEK by dissolving 10 g of the oven dried PEEK in 100 ml of 96% H_2SO_4. The time and temperature of the deep-red viscous solution were varied in order to achieve the desired level of sulfonation. The reaction was quenched by slowly pouring the acidic solution into 1 L of distilled water. The precipitation of the polymer was appeared instantly. In the last step, the polymer was extensively washed to remove the excess acid and dried in a vacuum oven at 100°C.

Unfortunately, these membranes have swollen in a gel-like state, making them mechanically fragile. Due to this, the lifetime of the membrane in electrolysis system will be limited. There are different strategies for the reduction of swelling of the ion exchange membranes: cross-linking the polymer with suitable cross-linking agent and blending the polymer, which are capable of formation of hydrogen bonds. The formation of hydrogen bonds leads to compatibilization of the blend polymers. Many examples of this type of blend are mentioned in the literature [79]. Kerees et al. [80] developed the ion exchange membranes by blending of sulfonated poly (ether ether ketone) with the aminated polysulfone Udel (PSU-NH_2), polyamide, PA 6-3-T (Trogamid P) (Huls), and poly(ether imine) PEI Ultem (General Electrical). In the blend membrane, swelling was reduced by specific interaction in case of blend components PA and PEI hydrogen bonds and in the case of the blend component PSU-NH_2 (partial) polysalt formation, leading to electrostatic interaction between the blend component macromolecules and hydrogen bonds.

The acid–base interactions also led to a decrease of ionic conductivity by partial blocking of SO_3 groups for cation transport, when compared with the ionic conductivity of the hydrogen bond containing blends, the acid–base blends showed better ion permselectivities than the hydrogen bond containing blends, even at higher electrolyte concentrations, and thus, they have a better performance in electrodialysis. The thermal stability of the blends is good, and in case of acid–base blend, it is still better. The investigated blend membranes showed similar performance as indicated from the commercial cation-exchange membrane CMX in electrodialysis (ED) application. The performance of the acid–base blend membrane is better than the performance of the hydrogen-bonded PEEK-PA blend, especially in ED experiment applying the higher NaCl concentration. This may be due to lower swelling and thus, better ion permselectivity of the acid–base blend membranes when compared to the PEEK-PA blend.

Polyphosphazene Ion Exchange Membranes

Polyphosphazenes are the new class of base polymers for ion exchange membranes because of their reported thermal and chemical stability and the ease of chemically

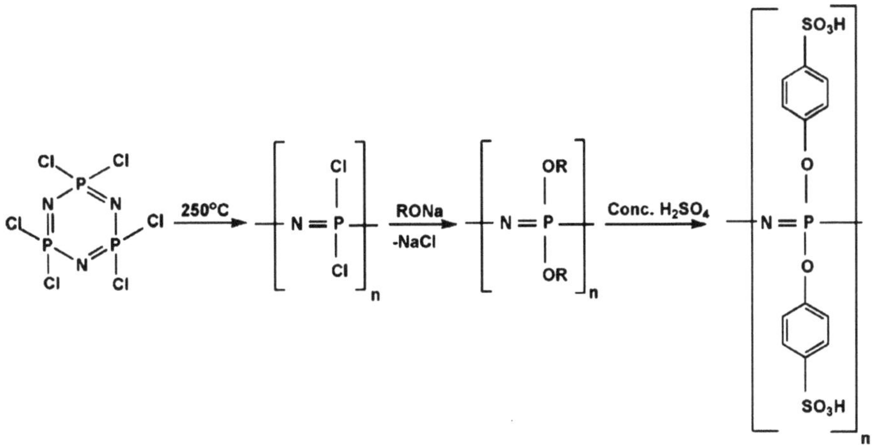

Fig. 7.11 Reaction scheme for the preparation of sulfonated polyphosphazene cation-exchange membrane

altering the polymer by adding various side chains onto the –P = N– backbone. Sulfonated polyphosphazene cation-exchange membranes are the attractive alternative to perfluorosulfonic acid and polystyrene sulfonate membranes. However, there are few reports in the literature on the fabrication and properties of ion exchange membranes from the sulfonated polyphosphazenes. The difficulty associated with producing such membranes lies in preparing the sulfonated polyphosphazene and, more importantly, in balancing the hydrophilicity of the resulting polymer to prevent dissolution in aqueous solution [81]. One of the main methods for the synthesis of these polymers is illustrated in Fig. 7.11.

Poly(dichlorophosphazene) reacts with a wide variety of nucleophiles to yield high-molecular-weight organic-derivatized polymers with properties that vary widely with the side-group structure. The stability of the phosphorus–nitrogen backbone makes this class of macromolecules particularly suitable for side-group and surface modification. Allcock et al. [82] reported that the molecular level and surface sulfonation of aryloxy and arylamino phosphazene were accomplished through the use of concentrated sulfuric acid. Allcock et al. also reported [83] the sulfonation of aminophosphazene with 1,3-propanesulfone, but the yield was very low. Montoneri et al. [84, 85] showed that aryloxy polyphosphazene can also be sulfonated via the use of sulfur trioxide. In the sulfonation of poly[bis(phenoxy) phosphazene] in dichloromethane with SO_3, it was found that when the mole ratio of SO_3 to phosphazene was less than one, there was no C-sulfonation and up to 50% of the skeleton nitrogens were attacked to form $\equiv N \rightarrow SO_3$ complex. When the molar ratio of SO_3/polyphosphazene is greater than one, the C-sulfonation can be observed mainly at the *meta*-position in the phenoxy side groups. Usually, sulfonation of polyphosphazene yields water-soluble polymers [81]. So, in practice, water-insoluble sulfonic acid membranes from polyphosphazene polymers can be made which involve the cross-linking polyphosphazene membranes followed by heterogeneous sulfonation [85]. Alternatively, homogeneous or heterogeneous polymer sulfonation is carried out followed by film casting and cross-linking. These two methods require that the polyphosphazene contains side groups that can be used for cross-linking, or in the second case, sulfonate groups can serve this purpose [86]. The water insoluble polyphosphazene can be made by introducing the alkyl groups on the phenoxy ring followed by controlling mole ratio of sulfonating agent to polyphosphagene mole ratios [87].

7.4.2 Styrene/Ethylene–Butadiene/Styrene Triblock Copolymers

Styrene/ethylene–butadiene/styrene triblock polymer is a commercially available product (Kraton G1650) containing a saturated carbon center block, which should be inert to the sulfonation reaction. This polymer exhibits good mechanical and chemical stability [88, 89]. The typical casting solvent of the sulfonation process can be reached in n-propanol, which should preferentially solvate the sulfonated styrene blocks. It has been claimed that membranes with reproducible properties at a constant sulfonation were obtained. It was also indicated that the conductivities at sulfonation levels above 50 mol% of styrene units exceed that of Nafion under similar measurement conditions. In particular, a sulfonation level of 60% was found to be a good balance of electrical and mechanical properties.

7.4.2.1 Poly(Benzylsulfonic Acid) Siloxane (PBSS)-Based Membrane

The synthesis of this membrane was obtained by sol–gel process with different steps. It was based on the hydrolysis condensation of the benzyl-triethoxylane precursor in a methanol solution containing water and triflic acid (F_3SO_3H). The steps of preparation are indicated elsewhere [90], and the structure is as shown in Fig. 7.12.

The polymer was sulfonated in dichloromethane solvent using chlorosulfonic acid. Several copolymer compositions prepared from different proportions of the mixture of benzyl-triethoxysilane, n-hexyltrimethoxysilane, and ethoxysilane (e.g., different organically modified alkoxy silanes). Cross-linking was also performed in THF by hydrosilylation of silane groups with divinylbenzene using divinyltetramethyldisiloxane platinum complex as catalyst. These membranes are thermally stable up to 250°C and have shown conductivity of $1.6 \times 10^{-2}/\Omega/\mathrm{cm}$ at room temperature [90].

7.4.2.2 Membranes Prepared by the Grafting Method

Radiation grafting of polymers is a well-known technique for modification of the physical and chemical properties of the polymer [91–96]. Irradiation of polymers

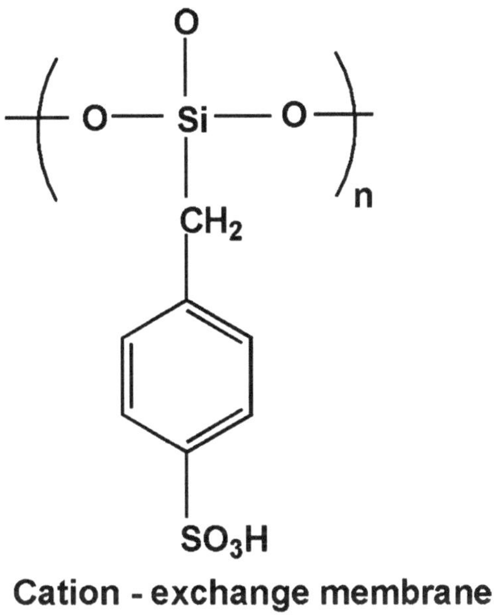

Fig. 7.12 Structure of poly(benzylsulfonic acid) siloxane–based cation-exchange membrane

by means of ionizing radiation such as γ-rays induces the formation of radicals on the polymers, polymer main chain from which polymerization of another monomer can be initiated. Therefore, from this technique, it is possible to bring two highly incompatible polymers in one material having combined properties. In recent years, the amount of efforts devoted to use this technique for the preparation of various types of membranes increased significantly because it provides the ability to closely control the membrane composition and properties by proper selection of the degree of grafting. Moreover, it overcomes the membrane shaping problem where grafting can be started with a film already in a membrane form. The ion exchange membranes made using polyethylene sheets by the grafting method followed by sulfochlorination and amination have exhibited low electrical resistance combined with high permselectivity and excellent mechanical strength. The reaction scheme for the preparation of these membranes is shown in Fig. 7.13.

A large number of papers were published on radiation grafting of monomers onto polymer films to obtain ionomers for versatile applications; they are used as anion- and cation-exchange membranes in electrodialysis processes, water desalination, carrier for immobilization of medical products, separators in alkaline batteries, etc.

Nasef et al. [95] prepared the proton exchange membrane by simultaneous irradiation of γ-rays from a ^{60}Co source at a dose rate of 1.32–15 kGy/h for styrene grafting onto tetrafluoroethylene-co-hexafluoropropylene films at room temperature and sulfonic acid cation-exchange membranes by radiation grafting of styrene onto poly(tetrafluoroethylene-co-perfluorovinyl ether) films followed by a sulfonation reaction. The content of the polystyrene in the membrane was found to be strongly dependent upon the grafting conditions like nature of solvent, dose rate,

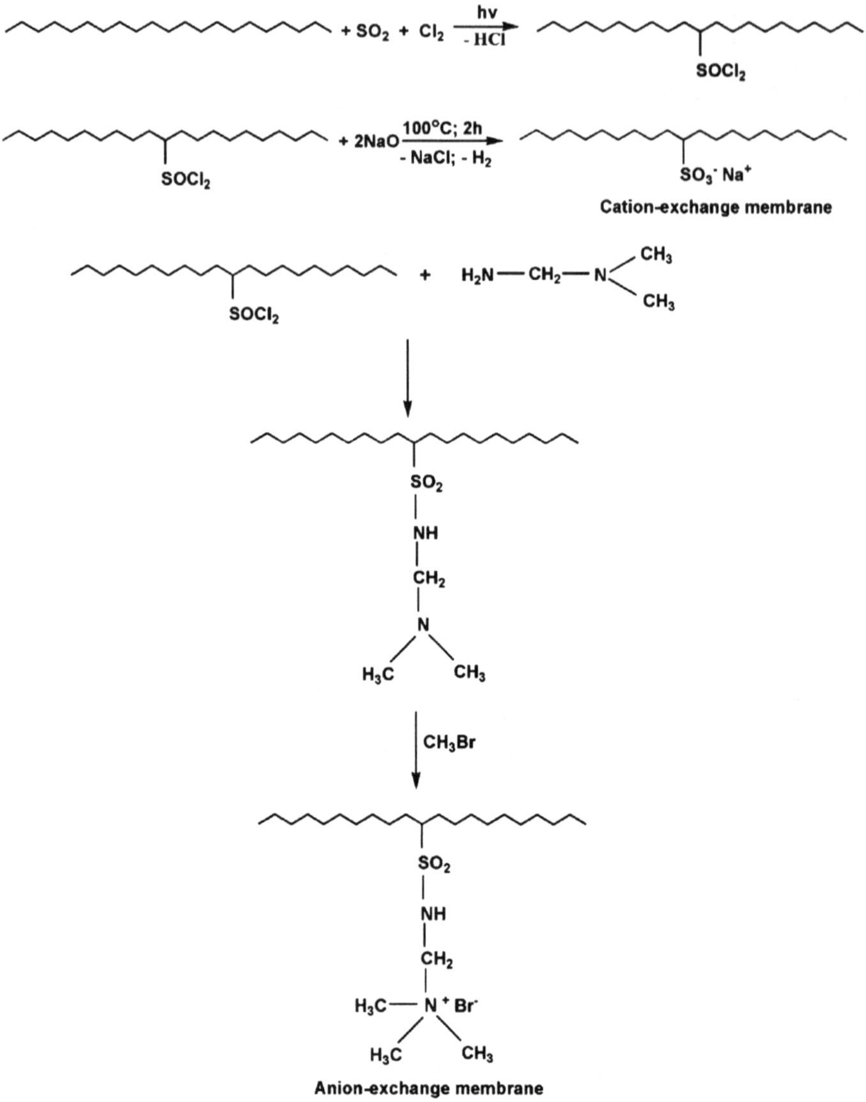

Fig. 7.13 Schematic presentation of grafting of polyethylene membrane

irradiation time, monomer concentration, etc. Similar behavior was also obtained upon grafting of other monomers such as acrylic acid onto poly(ethyleneterphthalate) (PET) films and for grafting of styrene onto normal PET films by pre-irradiation. It was found that the grafted PET films possess good mechanical stability, suggesting their use for the preparation of proton exchange membranes.

Kostov et al. [91–93] showed that the similar effect of irradiation dose on the anion-exchange membrane prepared by grafting of 4-vinylpyridine (4-VP) onto low-density polyethylene (LDPE) and polytetrafluoroethylene (PTFE) films by irradiation from ^{60}Co γ-source or high-energy electron beam. The initial properties of the polymer such as mechanical strength, thermal and chemical resistance combined with ion exchange properties of grafted polymer give valuable properties of the polymer. By varying the reaction conditions of the radiation copolymerization, grafted copolymers were prepared with degree of grafting from 17.1% to 74.5% for LDPE and from 0.7 to 13.4 for PTFE. The method of multiple grafting was used involving post-polymerization effect, which provides possibility for the monomer to penetrate into the polymer matrix together with the stepwise generation of free radicals. It is observed that the linear dependence of the irradiation dose on grafting degree was noticed for both the films. Irradiation exerts stronger effect on the grafting onto PTFE compared to LDPE, which is connected with the lower radiation resistance of C–F compared with C–H bonds. The water content of the membrane increases linearly with the grafting degree of 4-VP. The thickness of the grafted layer of the 4-VP affects the value of the specific resistances, which decreases with increase in the amount of grafted 4-VP.

7.4.3 Amphoteric Ion Exchange Membranes

Amphoteric ion exchange membranes contain both acidic (negative charge) and basic (positive charge) groups that are randomly distributed within the membrane matrix [97–101]. The sign of the charge groups in these membranes exhibits a pH response to an external solution. Generally, amphoteric membranes are classified into four types: (1) membranes in which both anion and cation-exchange groups are distributed randomly during the preparation, (2) snake-in-the-cage-type amphoteric membranes, (3) polysalt membrane, and (4) membrane prepared by layer-by-layer method. In the first case, styrene, 2-methyl-5-vinylpyridine, and divinylbenzene are copolymerized in presence of an inert polymer, and the membrane thus obtained was sulfonated and then quaternized [102]. Amphoteric polyelectrolytes such as copolymer of n-diethyl amino methacrylate and methacrylic acid are dissolved in a suitable solvent together with an inert polymer, and then, it was cast onto a flat surface [103].

Snake-in-the-cage ion exchange (retardion) membranes can be synthesized by saturation of a suitable ion exchange resin (the cage) with an organic counter ion that can be polymerized to form a linear chain of oppositely charged polymer (the snake) [104, 105]. The polymer chain cannot be removed from the host resin due to strong electrostatic interaction and the restricted mobility of the linear polymer chains. The ionic sights of the resin cage are neutralized in the presence of oppositely charged polymer chains. This provides mechanical stability leading to decreased swelling and shrinking properties. The main retardion resins have a low affinity for nonelectrolyte, coupled with an ability to selectively retard ionic substances. They can therefore be used not only for salting nonionic solutions but also for separating unwanted ions from ionic substances. In addition to retention of ions, retardion resins absorb acids and can be used for such purposes as removing excess acid following acid hydrolysis of proteins. The relative mobility of snake polymer chains provides the possibility of very close contact between oppositely charged groups on retardion resins and is a prime reason for the characteristics of these materials.

For polysalt amphoteric membrane, a cationic polyelectrolyte solution such as poly(benzyl trimethyl ammonium salt) and an anionic polyelectrolyte solution such as polystyrene sulfonate were mixed. The resulting precipitated polymer salt was then dissolved in a suitable solvent and formed into a membrane by casting technique [106]. Recently, membranes prepared by a "layer-by-layer" technique have been actively studied. For example, a cationic polyelectrolyte can be absorbed on a flat plate as a monolayer and then anionic polyelectrolyte layer was added to it. The same procedure can be repeated so as to obtain ultrathin, multilayer, amphoteric ion exchange membrane [107]. The multilayers formed on a porous membrane have been used as separation membrane [108], and further, such membranes have also been used as new functional material for sensors, devices for photoelectronics [109], etc.

7.4.4 Mosaic Ion Exchange Membranes

In 1932, Sollner reported ion transport through a mosaic ion exchange membrane [8]. Such membranes contain a set of anion- and cation-exchange elements arranged in parallel, each elements providing a continuous pathway from one bathing solution to the other [8, 97]. To prepare such mosaic membranes, various methods have been employed: (1) distributing particulate cation- and anion-exchange resins in an inert polymer film so as to have each resin particle penetrate both membrane surfaces; (2) cutting a laminated block of alternating cation- and anion-exchange membranes into films perpendicular to the membrane surface; (3) forming a film using two different copolymers (copolymers of styrene–butadiene and vinylpyridine–butadiene) and introducing anion- and cation-exchange groups into respective copolymer domains together with a cross-linking reaction [110]; (4) forming a film by a casting method using a block copolymer composed of a part in which cation-exchange groups can be introduced and a part for insulation, and then introducing the respective ion exchange groups into the domain; and (5) forming a film by the casting method using a dispersion that contains cationic microsphere gel, anionic microsphere gel, and matrix polymer and introducing a charge group into the gels if necessary [111].

After the preparation of a film from a block copolymer composed of five components (isoprene-4-vinylbenzenedimethylamine-isoprene-styrene-isoprene) by the casting method, cation-exchange groups (sulfonic acid) and quaternary ammonium groups were introduced into the respective domains after a suitable cross-linking. Also, after preparing a film from a block copolymer composed of styrene-butadiene-4-vinyl pyridine, cross-linking was carried out by γ-ray irradiation, and then the respective ion exchange groups were introduced [112].

7.4.5 Bipolar Membranes

Bipolar membranes are a special type of layered ion exchange membranes. They consist of two polymer layers carrying fixed charges, one is only permeable for the anions and the other only for cations [113–117]. Actually, unlike with membranes used for separation purposes, nothing should be transported from one side to the other. The desired function is a reaction in the bipolar junction of the membrane where the anion and the cation permeable layers are in direct contact: water is split into hydroxide ions and protons by a disproportionation reaction. The produced hydroxide ion and proton are separated by migration in the respective membrane layer out of the membrane. Unlike water splitting at electrodes during electrolysis, no gases are formed as a side product nor gases are used up [30].

Bipolar membrane efficiently converts an aqueous salt solution such as NaCl into the base NaOH and the acid HCl. Similarly, other salts such KF, Na_2SO_4, NH_4Cl, KCl, etc., as well as the salts of organic acids and bases can be converted. It is an electrodialysis process since ion exchange membranes are used to separate ionic species in solution with the driving force of an electrical field, but it is different by the unique water splitting capability of the bipolar membrane. In addition, the process offers unique opportunities to directly acidify or basify process streams without adding chemicals, avoiding by-products or waste streams and costly downstream purification steps. For the preparation of bipolar membranes, various methods have been initiated, such as preparing directly from commercial cation- and anion-exchange membranes by adhering with heat and pressure or with an adhesive paste [118], preparing by casting a cation-exchange polyelectrolyte solution (or an anion-exchange polyelectrolyte solution) on a commercial anion-exchange membrane (or on a cation-exchange membrane), respectively [119, 120], or preparing from the same base membrane by simultaneous functionalizing at the two membrane sides [121–124] or selectively functionalizing on one side to give cation selectivity and on the other side to give anion selectivity [125]. Among these, the casting method seems to be the most attractive one for preparing such membranes as it is simple, low cost, and allows a bipolar membrane with desired properties for commercial use such as good mechanical strength, ability to operate at high current density, high permselectivity, low potential drop, etc. [126]. Using this method, novel bipolar membranes have been prepared by casting the sulfonated poly(phenylene oxide) (PPO) solution on a series of anion-exchange membranes as shown in Fig. 7.14 [127].

The two ion exchange layers in a bipolar membrane allow for the selective transport of the water dissociation products – protons and hydroxyl ions and block co-ions in the electrolyte. To facilitate the water-splitting effect, a bipolar membrane also includes a contact region, also referred to the interfacial layer, where the desired water dissociation reaction occurs. The charged groups and structure of this region are of great significance for water dissociation and thus generally are modified elaborately to improve the bipolar membrane's performance. Now, it is well concluded that materials with the best catalytic activity are sufficient amounts of weak acids (and the corresponding bases) such as amino groups, pyridine, carboxylic acid, and phenolic and phosphoric acid groups [128–131] as well as

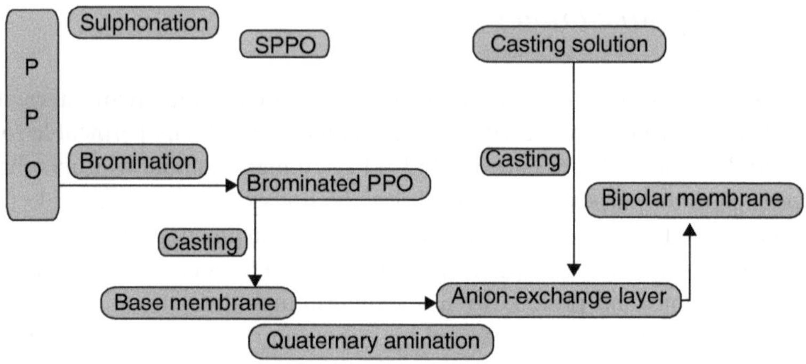

Fig. 7.14 Schematic diagram of bipolar membranes prepared from material poly(2,6-dimethyl-1,4-phenylene oxide) (PPO)

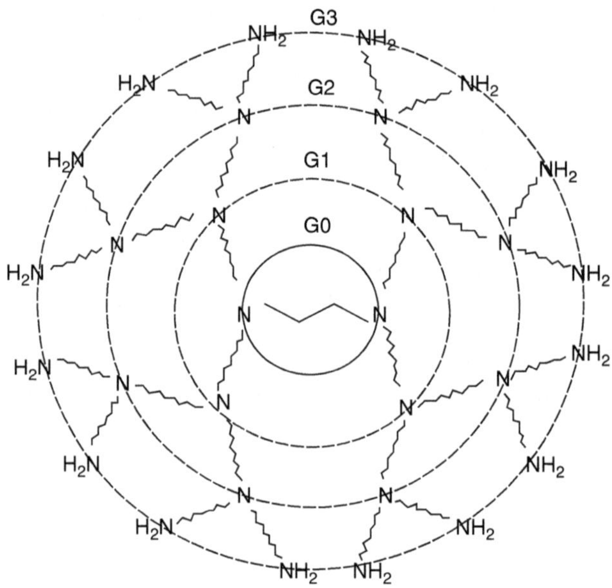

Fig. 7.15 Schematic diagram of PAMAM G3: the initiate core is ethylenediamine and the repeating unit is $-CH_2CH_2CONHCH_2CH_2N-$

heavy metal ion complexes such as ruthenium trichloride, chromic nitrate, indium sulfate, hydrated zirconium oxide, etc. [132–135]. The catalytic mechanism is underlined by chemical reaction model of water dissociation, i.e., the water splitting could be considered as some type of proton-transfer reaction between water molecules and the functional groups or chemicals [130, 136].

Another interesting approach reported recently is the modification of a bipolar membrane with the starburst dendrimer polyamidoamine (PAMAM) [137], which possesses much higher amino groups than conventional macromolecule as shown in Fig. 7.15. As expected, these amino groups have appreciable catalytic function when they are incorporated into the middle layer of a bipolar membrane [137]. However, due to the PAMAM steric effect, the catalytic function of amino groups varies with both PAMAM concentrations and generations.

7.4.6 Interpolymer Ion Exchange Membranes

The interpolymer ion exchange membranes developed in the literature [138–140] possess an excellent combination of both electrochemical and mechanical properties. For the preparation of interpolymer membrane, linear polyethylene was used as a binder. It was made into an organosol with suitable solvating monomers like styrene–divinylbenzene. This blend on polymerization of monomers under a free radical mechanism yields a chemical polyblend of two interpenetrating networks of linear and cross-linked polymer molecules [139, 140]. In addition, due to grafting, there may be an inter cross-linking between polyethylene and polystyrene phase. Such a chemical polyblend behaves like a homogeneous type having less probability of microvoids compared to heterogeneous-type membranes. For lack of such available space, when there is solvation of ionic groups, there is an expansion of matrix with the dimensional changes.

Recently [141], ion exchange membranes, by the semi-interpenetrating polymer networks (s-IPNs), have been prepared by mixing of two polymers such as PVA and polyelectrolytes, followed by gaseous cross-linking with dibromoethane. The polyelectrolytes used to develop tailor-made ion exchange membranes are poly (sodium styrenesulfonate) (PSSNa), poly(styrene sulfonic acid) (PSSH), and poly (acrylic acid) (PAA) for cation-exchange membranes and polyethyleneimine (PEI), poly(diallyldimethylammonium chloride) (PDDMACl), and poly(1,1-dimethyl-3,5-dimethylenepiperidinium chloride) (PDDPCl) for anion-exchange membranes. It was found that the homogeneous film with suitable mechanical property was obtained for the 60/40 mass ratio of poly(vinyl alcohol) to polyelectrolyte. The membranes PVA/PAA show less ion exchange capacity and swelling compared to PVA/PSSNa and PVA/PSSH membranes. The anion-exchange membranes prepared using the above polyelectrolytes show weak ion exchange capacity.

7.4.7 Heterogeneous Ion Exchange Membranes

Ion exchange membranes, both the homogeneous and heterogeneous, being unique in their nature overcome each other in one way or another. Homogeneous membranes having good electrochemical properties lack in their mechanical strength, whereas heterogeneous membranes having very good mechanical strength are comparatively poor in their electrochemical performance [142]. However, by choosing a suitable binder to make non-reinforced membranes or by choosing a suitable reinforcing fabric, it is possible to have good ion exchange membranes by an optimum combination of electrochemical properties and mechanical strength with heterogeneous method. Heterogeneous ion exchange membranes can be made [143] by the incorporation of powdered ion exchange resin into sheets of rubber, PVC, acrylonitrile copolymers, or some other extrudable or moldable matrix. Such membranes can be prepared [144] either (1) by calendaring ion exchange particles into an inert plastic film, (2) by dry molding of inert film forming polymers and ion exchange particles and then milling the mold stock, or (3) resin particles can be dispersed in a solution containing a film-forming binder, and then the solvent is evaporated to give ion exchange membrane. Such heterogeneous membranes may also be reinforced with a chemically resistance fabric [145]. Kariduraganavar et al. [146] and Adhikari et al. [138, 139] found that the particle size distribution of the resin particles of different mesh affected on the electrochemical and mechanical properties of the membranes. It is observed that with increase in resin loading, the membrane becomes more and more brittle, and at the same time, it is also observed that the finer the resin particles are, the more flexible is the membrane. For PVC binder, with resin particles of −300 + 400 mesh (39 μm), it is possible to obtain flexible membrane up to 60% resin loading, whereas the membrane becomes brittle even with 40% loading for −100 + 200-mesh-sized (80 μm) resin particles. The brittleness of the membrane may be due to increased resin loading, which increases after a certain point when the phase inversion takes place. The cross-linked resin particles tend to form the continuous phase, whereas binders tend to form the discrete phase. Actually, resin particles being more brittle in nature compared to binder fail to act as impact modifier, and crack propagation becomes facile, resulting in a brittle membrane. At any particular blend ratio, the finer the resin particles, the more homogeneous the blend is, resulting in a more flexible membrane. As the resin load increases, the burst strength of the membrane decreases. This may be due to the formation of continuous phase by the resin particles, which has a lesser impact on the strength. However, the burst strength can be improved by decreasing the particle size.

For practical applications, the dimensional stability in different ionic forms is the most desirable criteria for any commercially successful ion exchange membrane. The heterogeneous ion exchange membranes show the good dimensional stability compare to the homogeneous membranes. This may be due to the absence of microvoids in the homogeneous ion exchange membranes since it forms the continuous phase during the polymerization. In case of heterogeneous membranes, the loss of solvent due to evaporation introduces microvoids between the resin particles and binder regions. These microvoids are sufficient to accommodate solvent molecules for the solvation of ionic species in the resin, so the salvation does not manifest the dimensional changes of the membranes. The particle size of the resin also affects the ion exchange capacity and areal resistance of the membrane. At a definite resin loading and thickness of the membrane, the electrical resistance decreases and ion exchange capacity increases with decreasing the particle size. Decreasing the particle size results in the higher surface area of the resin particles. With increase in the surface area, the number of functional groups which may actively participate in the transport of counter ions through the membrane results to lower the resistance and higher the ion exchange capacity of the membranes.

7.5 Applications of Ion Exchange Membranes

Ion exchange membranes are widely used in various separation processes like electrodialysis, electrolysis, diffusion dialysis, Donnan dialysis, piezodialysis, reverse osmosis, etc. The driving force for permeation of ions and water through the membrane at initial stage was thought primarily because of electrochemical and chemical potential. But in the recent papers [147, 148], it is reported that hydraulic pressure, temperature difference, difference of proton concentration, and so on are also responsible for the driving force. This is especially observed in tube and hollow fiber type ion exchange membranes wherein driving force is not electrochemical potential. Therefore, the recent applications of ion exchange membranes are due to the development of an anisotropic structure or other new types of membranes: charged reverse osmosis membrane, mosaic ion exchange membrane, gas separation membranes having amino groups, and so on. As it is impossible to explain all the applications reported in the literature, only the important applications of ion exchange membranes are summarized in Table 7.1.

Table 7.1 Applications of ion exchange membranes [149]

Methods	Types of ion exchange membranes	Purposes
Electrodialysis	Cation-, anion-, and bipolar-exchange membranes	Concentration or desalination of electrolyte solution, separation of ionic materials from nonionic materials, production of acid and base by water splitting, double decomposition of salt, sterilization of water, etc.
Electrolysis	Cation- and anion-exchange membranes	Separator for electrolytes (chlor-alkali, organic synthesis, water electrolysis, etc.)
Diffusion dialysis	Anion-exchange membrane	Recovery of acid from waste acid solution, etc.
Donnan dialysis	Cation-exchange membrane	Recovery of trace amount of ions, ion exchange, etc.
Piezodialysis	Mosaic, amphoteric exchange membranes	Concentration, desalination, and separation of electrolytes
Reverse osmosis	Anisotropic cation- and anion-exchange membranes	Desalination, concentration, and separation of electrolytes
Facilitated transport	Cation- and anion-exchange membranes	Separation between ions with same charge
Cell and battery	Cation- and anion-exchange membranes	Electrolyte of fuel cell, reverse electrodialysis, separator of redox-flow cell and others
Others	Cation- and anion-exchange membranes	Sensor for humidity, proton, etc. Pervaporation, gas separation, etc.

7.5.1 Ion Exchange Membranes in Electrodialysis

Initially, the purpose of developing ion exchange membrane was to produce sodium chloride from seawater by electrodialysis. Later, it was spread to demineralization of cheese whey, deacidification of fruit juice, removal of potassium tartarate from wine, demineralization of amino acid solution in food industry, desalting of sugar and molasses, protein recovery from blood plasma, etc.

Electrodialysis (ED) was commercially introduced during the 1960s well before the reverse osmosis for water desalination. The most important large-scale application of ED is the production of portable water from brackish water [150–156], in which ED competing directly with reverse osmosis (RO) and multistage flash evaporation. ED is generally the most economical process for water with relatively low salt concentration (less than 5,000 ppm). One significant feature of ED is that the salt can be concentrated to comparatively high value (in excess of 18–20% by wt.) without affecting the economics of the process. The most modern ED units operate with the so-called reverse polarity (electrodialysis reversal, EDR), which are effective on protection of membrane from organic fouling as well as scaling of low-solubility salt such as $CaSO_4$ [157].

Desalination by ED process had been commercialized in the USA during the 1970s, and several ED/EDR plants were installed. For desalination of brackish water, more than 2,000 plants with a total capacity of more than 1,000,000 m^3 of product water were installed [158]. In India, Central Salt & Marine Chemicals Research Institute (CSMCRI) installed several brackish water desalination plants of desired capacity in rural area during 1985–1995. A substantial number of installations of ED/EDR units can also be found in Russia and China for the production of potable water [158, 159]. ED units powered by photovoltaic cell were also developed and installed for providing potable water from brackish water [160, 161].

Figure 7.16 shows the applications of membrane technology which covers not only ED but also applications of RO and UF in diary industry. Other applications to food industries are the mineralization of molasses in order to increase the sugar recovery [162], the acid removal from fruit juice, the separation of amino acids and minerals in amino acid production process [163, 164], and so on. It is also reported that acid removal by ED has made grape juice stable against precipitation of potassium tartarate [165]. In these cases, organic fouling of anion-exchange membrane is serious, and therefore in several cases, neutral membranes are being used instead of anion-exchange membranes.

Treatment of industrial wastewater by electrodialysis using NEOSEPTA membranes has contributed to pollution abatement and recovery of valuable minerals, for example, the concentrating recovery of nickel galvanization effluent as shown in Fig. 7.17. The treatment of copper-, zinc-, and tin galvanizing liquors and the recovery of noble metals are made efficiently. Similar applications are developer regeneration by electrodialysis for color paper processing [166] and electrolytic recovery of silver from photographic fixing solutions and the reuse of the solution by simultaneous treatment of the waste photographic fixing and bleaching solution in a cell [167].

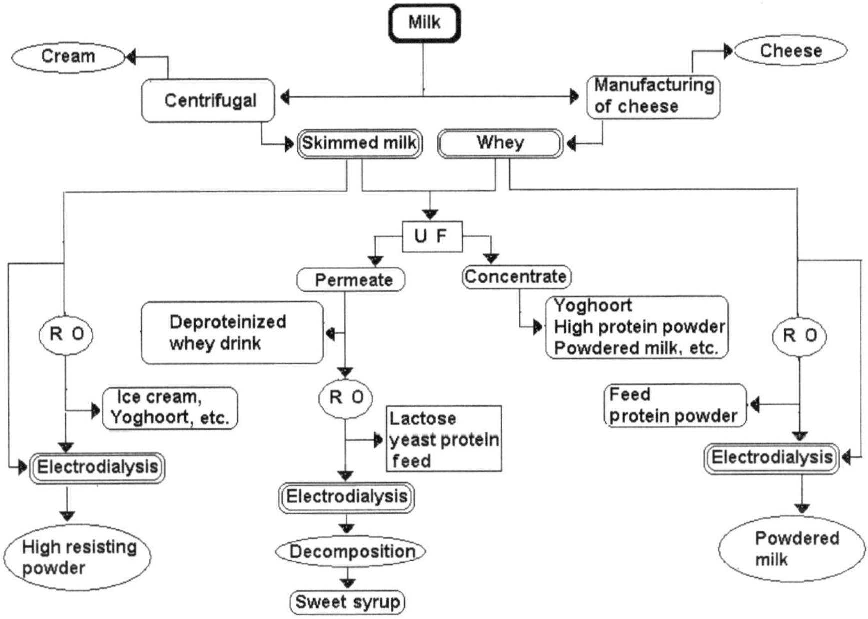

Fig. 7.16 Application to dairy industry

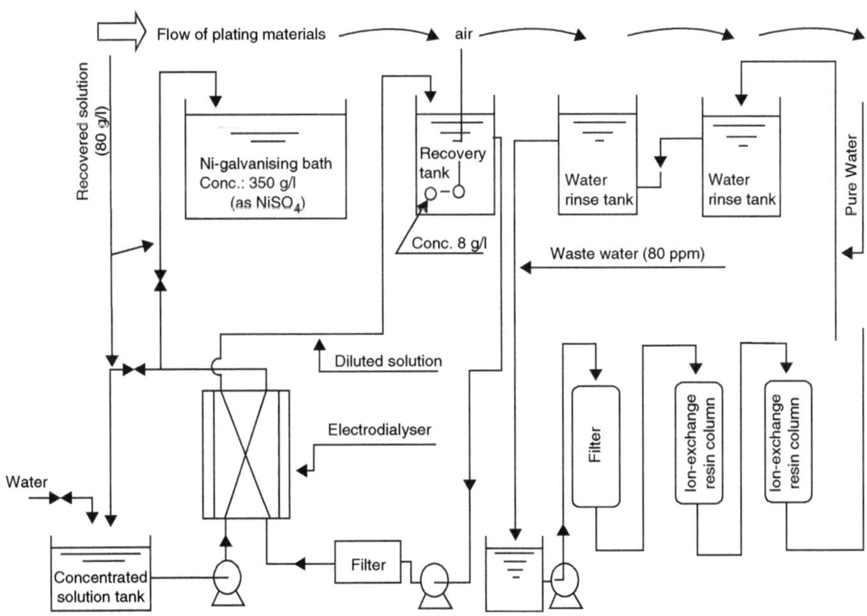

Fig. 7.17 Flow diagram of treatment of Ni-galvanization effect

Other interesting applications of ion exchange membranes are bactericidal effect of electrodialysis on *E. coli* and accelerated fermentation to produce glutamic acid and lactic acid. Bactericidal effect of electrodialysis on *E. coli* is observed when the electrodialysis is carried out over the limiting current density and is dependent on cation and anion species contained in the solution. This effect is connected with pH change of the solution and observed remarkably in the presence of ionic species, which permeate easily through the membrane [168]. Accelerated fermentation is made by removing glutamic acid and lactic acid from culture fluid by electrodialysis continuously in order to maintain the low concentration in products [169]. Applications of the electrodialysis to dilute solution are enrichment of trace level ions (10^{-6}–10^{-4} mol/l; ionic strength is adjusted by KC1) [170] and removal of radioactive ions by multistep electrodialysis [171].

7.5.2 Ion Exchange Membranes in Electrolysis

Typical example of this case is electrolysis of sodium chloride to produce chlorine, hydrogen, and sodium hydroxide (chlor-alkali process). With the development of chemically and thermally stable perfluorocarbon ion exchange membranes, the chlor-alkali industry is nowadays one of the largest industries producing annually more than 48 million tons of sodium hydroxide and more than 42 million tons of chlorine [172]. This has been attained by the improvement of membrane, electrolyzer, and electrolysis methods as well as development of new cathode of low hydrogen overvoltage. Especially, perfluorocarbon carboxylic acid group membranes gave high current efficiency in caustic soda production, and surface treatment of the membrane served to release bubble from the membrane surface, thereby reducing the cell voltage significantly.

Ion exchange membrane is used for electrolysis of water to produce hydrogen and oxygen gas as for utilization of hydrogen energy. In this case, the usage of perfluorocarbon cation-exchange membranes and inorganic ion exchange membranes [173] is being made. Especially, it is reported that solid polymer electrolyte (SPE) method produces hydrogen gas at high efficiency, where perfluorocarbon cation-exchange membranes are used with catalyst coated as electrodes [174]. Ion exchange membranes are also used as separator in organic synthesis by electrolysis. Typical example is hydrodimerization of acrylonitrile to produce adiponitrile [175]. Other various applications to electrolytic organic synthesis have been reported [176]. Tokuyama Soda Co. Ltd. has made significant contribution to this area. Application of SPE electrolysis method to organic synthesis is also reported [177]. There are many other applications of ion exchange membranes to the electrolysis as separator [164]: the recovery of precious metal in catholyte by deposition on the cathode, the reduction of uranyl ions to uranous ions by the cathode reaction, and so on.

Table 7.2 Examples of diffusion dialysis applied to various acids recovery [149]

Process	Acid recovery
Acid recovery from pickling solution or pickling waste solution of steel	H_2SO_4, HCl, HNO_3, HF
Purification of battery waste acid	H_2SO_4
Acid recovery from alumite process waste acid solution	H_2SO_4, HNO_3
Acid recovery from metal refining process	H_2SO_4, HCl, H_3PO_4
Treatment of etching waste solution (Al, Ti)	HCl
Acid recovery from surface treatment acid of electroplating line	H_2SO_4, HCl, HNO_3, HF
Purification and acid removal from organic synthetic process	H_2SO_4, HCl

7.5.3 Ion Exchange Membranes in Diffusion Dialysis

Diffusion dialysis is one of the oldest membrane processes that has been established for many years where ion exchange membranes are used instead of uncharged membranes. Acid recovery from waste acid solutions is made by this process using anion-exchange membranes. The examples of diffusion dialysis applied to various acids recovery are given in Table 7.2. Flow rate of the solution in dialysis is about 1–5 cm/min, and therefore, this is an energy saving process.

The anion-exchange membranes for this purpose were developed by Tokuyama Soda Co. Ltd., i.e., NEOSEPTA APN [178]. This process requires membranes with high permeation of acid, less leakage of metal ions, good mechanical strength, and so on. Industrial recovery of HNO_3 and HF acids has been retarded because of oxidative deterioration of anion-exchange membranes.

Donnan dialysis is also one kind of diffusion dialysis. This process has been used for water softening using cation-exchange membrane and sodium chloride solution [179], the removal of potassium hydroxide from gold plating bath [180], the removal of hazardous metal ions, the pretreatment of ion chromatography in order to concentrate ions [181], and so on. There is a report on the utilization of pH difference as the driving force to transport ions through the cation-exchange membrane against the concentration gradient [182].

7.5.4 Ion Exchange Membranes in Fuel Cells and Battery

Recently, proton exchange membrane (PEM) gained broader attention for fuel cells and electrochemical energy conversion devices [183–189]. The usage of perfluorocarbon sulfonic acid to hydrogen–oxygen fuel cell as an electrolyte is well known. Other cases are the usage of membranes of low electrical resistance to alkali battery as a separator where the membrane is made by grafting of acrylic acid to polyethylene, separator for zinc–bromine battery [190], Zn–Cl_2, and redox-flow battery [191]. Recently, many promising polymers based on aromatic thermoplastics such as poly(ether ether ketone) (PEEK), poly(ether sulfone) (PES), poly(benzimidazole) (PBI), and polyimides (PI) have been studied, which have demonstrated excellent chemical resistance, high thermo-oxidative stability, good mechanical property, and low cost [192–196]. By introducing sulfonic acid groups to the polymer chains, these sulfonated PEEK, PES, PI, and PBIs were applied as a polymer electrolyte membrane [197–199].

7.5.5 *Ion Exchange Membranes in Other Applications*

Ion exchange membranes are interesting materials as they are polymeric in nature containing ion exchange groups, i.e., hydrophilic groups. In view of this, ion exchange membranes for pervaporation (PV) applications are now actively pursued [200]. Especially, cation- and anion-exchange membranes of various ionic forms are examined in case of separation of ethanol–water mixtures [201]. Further attempts have also been made to increase the flux and selectivity by suitably modifying the membranes [202]. As olefinic compounds react reversibly with silver ions, cation-exchange membrane, i.e., ion exchanged with silver ion, is expected to separate olefinic compound from alkane compound. Similarly, CO_2 is also separable by the cation-exchange membrane, i.e., ion exchange with ethylenediamine [203]. For the purpose of gas separation, anion-exchange membrane derived from poly (2,6-dimethyl-p-phenelyne oxide) shows the oxygen-enriching properties [204].

Water content of the ion exchange membrane varies with humidity. Therefore, ion exchange membrane is being used as a hydrometer. When water content of the membrane increases with increasing the humidity, the increase in the current or ionic conductivity between the electrodes on both sides of the membrane is detectable [205]. Membrane potential is also used to determine the concentration of ions [206]. Ion exchange membranes are used as carriers for photosensitive materials as they have ion exchange groups: photovoltaic behavior of bacteriorhodopsin-loaded cation-exchange membranes [207], photophysical properties of cadmium sulfide loaded on cation-exchange membrane [208], and others [209].

7.6 Conclusions

From the literature, it is concluded that ion exchange membranes play a vital role in solving different problems. For any purpose, preparation of ion exchange membrane is the most crucial. During the last two decades, development of new ion exchange membranes has gained the advantage of an interdisciplinary approach integrating recent advances in the field of material science. This chapter demonstrates the interest of using innovative methods for material processing with several examples on the preparation of ion exchange membranes. As shown in Fig. 7.18, conventional mode starts from a specific application. To satisfy with a specific requirement, one has to choose a proper membrane from the sea of membranes, optimize the operation condition then design the process. Such application

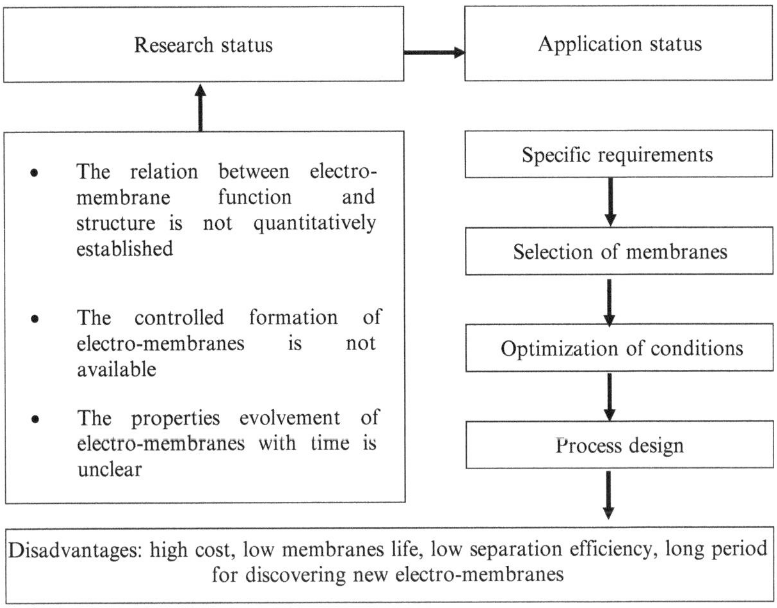

Fig. 7.18 Current status in research and application of electro-membrane fields

status is limited by the current research status, the relation between electro-membrane function and structure, the controlled formation of electro-membrane and the properties evolution of electro-membranes with time leave to be quantitatively determined at this time. Therefore, it is necessary to discard this old mode and establish a new one for synthesis and applications of ion-exchange membranes.

As shown in Fig. 7.19, the new mode is expected to go to the reverse route to the conventional one: designing a specific membrane for a specific purpose. The new mode starts from required membrane properties for a specific purpose, determination of membrane properties, membrane structure, and finally, design of the membrane from molecular level, i.e., such mode goes to form mole to process.

Apart from the preparation of ion exchange membranes, the technical and commercial relevance of the ion exchange membranes based processes should also be considered. As analyzed by Strathmann in his recent work [97], some of the applications can be considered as state-of-the-art technology, such as the applications using conventional electrodialysis, production of pure water using continuous electro-deionization, and some specific applications using bipolar membranes such as producing organic acid from the fermentation broth or recovering HF and HNO_3 from a waste stream generated by neutralization of a steel pickling bath. Often, ion exchange membrane separation processes are in competition with other mass separation techniques and their application is determined based on economic considerations. Some of the ion exchange membranes have better quality and eco-friendly in nature, and therefore, they are being used in spite of their high cost. In view of the above, the development of ion exchange

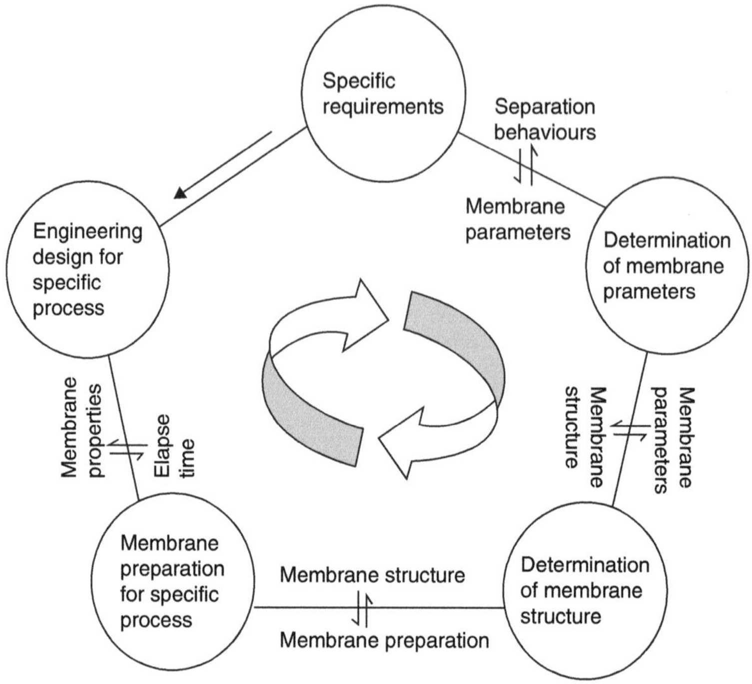

Fig. 7.19 New mode for an electro-membrane designed for a specific process

membranes has gained the importance of an interdisciplinary approach, integrating recent advances in every field of science and technology. Therefore, technologists and researchers should give prime consideration to interdisciplinary knowledge such as material, chemistry, polymer science and technology, mathematics, as well as engineering to solve some challenging issues.

References

1. Sourirajan S (1991) Lectures on membrane separations. The Indian Membrane Society, Baroda
2. Seno M, Takagi M, Takeda K et al (1993) Handbook of separation science. Kyoritsu Shuppan Co, Tokyo
3. (1993) IUPAC compendium of chemical technology, 2nd edn, vol 65, p 856
4. Juda W, McRae WA (1950) Coherent ion-exchange gels and membranes. J Am Chem Soc 72:1044
5. Ostwald W (1890) Elektrische eigenschaften halbdurchlässiger scheidewände. J Phys Chem 6:71–82
6. Donnan FG (1911) The theory of membrane equilibrium in presence of a non-dialyzable electrolyte. Z Elektrochem 17:572–581
7. Michaelis L, Fujita A (1925) The electric phenomena and ion permeability of membranes. II. Permeability of apple peel. Biochem Z 158:28–37

8. Sollner K (1932) Uber mosaikmembranen. Biochem Z 244:390
9. Wassenegger H, Jaeger K (1940) Effecting cation-exchange in removing calcium from hard waters. US Patent 2,204,539
10. Meyer KH, Strauss H (1940) La perm éabilité des membranes VI, sur le passage du courant électrique a travers des membranes sélective. Helv Chim Acta 23:795–800
11. Winger AG, Bodamer GW, Kunin R (1953) Some electrochemical properties of new synthetic ion exchange membranes. J Electrochem Soc 100:178–184
12. Nishiwaki T (1972) Concentration of electrolytes prior to evaporation with an electromembrane process. In: Lacey RE, Loeb S (eds) Industrial processing with membranes. Wiley, New York
13. Mihara K, Kato M (1969) Polarity reversing electrode units and electrical switching means thereof. US Patent 3,453,201
14. Grot WG (1973) Laminate of support material and fluorinated polymer containing pendant side chains containing sulfonyl groups. US Patent 3,770,567
15. Chlanda FP, Lee LTC, Liu KJ (1976) Bipolar membranes and method of making same. US Patent 4,116,889
16. Pourcelly G, Gavach C (2000) Electrodialysis water splitting-application of electrodialysis with bipolar membranes. In: Kemperman AJB (ed) Handbook on bipolar membrane technology. Twente University Press, Enschede
17. Xu T (2001) Development of bipolar membrane-based process. Desalination 140:247–258
18. Daufin G, Escudier JP, Carrere H et al (2001) Recent emerging applications of membrane processes in the food and dairy industry. Food Bioprod Process 79:89–93
19. Tarvainen T, Svarfvar B, Akerman S et al (1999) Drug release from a porous ion exchange membrane in vitro. Biomaterials 20:2177–2183
20. Kim YH, Moon SH (2001) Lactic acid recovery from fermentation broth using one-stage electrodialysis. J Chem Technol Biotechnol 76:169–178
21. Saracco G (2003) Ionic membrane technologies for the recovery of valuable chemicals from waste waters. Ann Chim Rome 93:817–826
22. Bazinet L, Lamarche F, Ippersiel D (1998) Bipolar membrane electrodialysis: applications of electrodialysis in the food industry. Trends Food Sci Technol 9:107–113
23. Xu T (2002) Electrodialysis processes with bipolar membranes (EDMB) in environmental protection – a review. Resour Conserv Recy 37:1–22
24. Bishop HK, Bittels JA, Guter GA (1969) Investigation of inorganic ion-exchange membranes for electrodialysis. Desalination 6:369–380
25. Bregman JI, Braman RS (1965) Inorganic ion exchange membranes. J Colloid Sci 20:913–922
26. Shrivastava K, Jain AK, Agrawal S et al (1978) Studies with inorganic ion exchange membranes. Talanta 25:157–159
27. Ohya H, Paterson R, Nomura T et al (1995) Properties of new inorganic membranes prepared by metal alkoxide methods. Part I. A new permselective cation exchange membrane based on Si/Ta oxides. J Membr Sci 105:103–112
28. Kogure M, Ohya H, Paterson R et al (1997) Properties of new inorganic membranes prepared by metal alkoxide methods. Part II. New inorganic–organic anion exchange membranes prepared by the modified metal alkoxide methods with silane coupling agents. J Membr Sci 126:161–169
29. Ohya H, Masaoka K, Aihara M et al (1998) Properties of new inorganic membranes prepared by metal alkoxide methods. Part III. New inorganic lithium permselective ion exchange membrane. J Membr Sci 146:9–13
30. Xu T (2005) Ion exchange membranes: State of their development and perspective. J Membr Sci 263:1–29
31. Molau GE (1981) Heterogeneous ion-exchange membranes. J Membr Sci 8:309–330
32. Helfferich F (1962) Ion-exchange. McGraw-Hill, New York
33. Nichuus H, Stephanm R, Wolf F et al (1960) Anion-exchange membranes by condensation of m-phenylene diamine and poly(ethylenediamine) with formaldehyde. Germany Patents 1,041,486 and 1,056,103

34. Haagen K, Helfferich F (1961) Ion-exchange membrane. Ger Patent 971,729
35. Onoue Y, Mizutani Y, Izumi Y (1961) Amine exchange membranes from dimethyl (2-hydroxy-benzyl)amine. Denki Kagaku 29:653
36. Geyer R, Erxleben E, Syring W et al (1964) Resin membranes as indicators in titrimetric reactions. Tech Hochsch Chem Leuna-Merseburg 6:163
37. Gregor HP, Bruins PF, Kramer RM (1966) Chloride-sulfate equilibria and transport processes in benzidine-formaldehyde and other anion-permeable membranes. J Appl Polym Sci 10:807–815
38. Oster EA, Fickett AP (1965) Ion-exchange membrane. US Patent 3,207,708
39. Savadogo O (1998) Emerging membranes for electro-chemical systems: (I) solid polymer electrolyte membranes for fuel cell systems. J New Mat Elect Syst 1:47–66
40. Grot W (1978) Use of Nafion perfluorosulfonic acid products as separators in electrolytic cells. Chem Ing Tech 50:299–301
41. Ezzell BR, Carl WP, Mod WA (1982) High performance polymer films and membranes. US Patent 4,358,412
42. Steck A (1995) In: Savadogo O, Roberge PR, Veziroglu TN (ed) Proceedings of the first international symposium on new materials for fuel cell systems, Montreal
43. Murphy EA, Paton FJ, Ansell J (1943) Apparatus for the electrical treatment of colloidal dispersion. US Patent 2,331,494; McRae WA, Alexander SS (1960) Sulfonating reagent and its use in preparing cation-exchange membranes. US Patent 2,962,454
44. Kuzumoto K, Sata T, Mizutani Y (1976) Modification of anion-exchange membranes with polystyrenesulfonic acid. Polym J 8:225–226
45. Sata T (1978) Modification of properties of ion-exchange membranes. IV. Change of transport properties of ion-exchange membranes by various polyelectrolytes. J Polym Sci Polym Chem Ed 16:1063–1080
46. Sata T (1994) Studies on ion exchange membranes with permselectivity for specific ions in electrodialysis. J Membr Sci 93:117–135
47. Sata T (2000) Studies on anion-exchange membranes having permselectivity for specific anions in electrodialysis-effect of hydrophilicity of anion-exchange membranes on permselectivity of anions. J Membr Sci 167:1–31
48. Sata T, Kawamura K, Matsusaki K (2001) Electrodialytic transport properties of anion-exchange membranes prepared from poly(vinyl alcohol), poly(N-ethyl 4-vinylpyridinium salt) and β–cyclodextrin. J Membr Sci 181:167–178
49. Aminabhavi TM, Kulkarni PV, Kariduraganavar MY (2004) Ion exchange membranes, methods and processes for production thereof and uses in specific applications. US Patent 6,814,865
50. Cohen SG, Wolosinski HT, Scheur PJ (1949) α, β, β–Trifluorostyrene and α-chloro-β, β difluorostyrene. J Am Chem Soc 71:3439–3440
51. Prober M (1952) The synthesis and polymerization of some fluorinated styrenes. J Am Chem Soc 75:968–973
52. Hodgdon RB Jr (1968) Polyelectrolytes prepared from perfluoroalkylaryl macromolecules. J Polym Sci 6:171–191
53. Sara M, Wolf G, Sleytr UB (1988) Structure and characterization of ultrafiltration membranes from crystalline bacterial envelope layer. Sen-I Gakkaishi 44:4–10
54. Tam CM, Dal-Cin M, Guiver MD (1993) Polysulfone membrane. IV. Performance evaluated of Radel A/PVP membranes. J Membr Sci 78:123–134
55. Popescu G, Nechifor G, Albu B et al (1989) Microporous polyethersulfone membrane. Rev Roum Chim 34:577–580
56. Pusch W (1988) Synthetic membranes for separation processes. Sen-I Gakkaishi 44:20–35
57. Swinyard BT, Sagoo PS, Barrie JA et al (1990) The transport and sorption of water in polyethersulfone, polysulfone and polyethersulfone/phenoxyl blends. J Appl Polym Sci 41:2479–2485
58. Inoue H, Kaibara K, Ohta N et al (1988) Ion transport across charged ultra-filtration membrane. Mem Fac Sci Kyushu Univ Ser C 16:137–148

59. Breitbach L, Hinke E, Staude E (1991) Heterogeneous functionizing of polysulfone membranes. Die Angew Makromol Chem 184:183–196
60. Kinzer KE, Lloyd DR, Wlghtman JP et al (1983) Asymmetric membrane preparation from nonsolvent casting systems. Desalination 46:327–334
61. Zschocke P, Quellmalz D (1985) Novel ion-exchange membrane based on an aromatic polyethersulfone. J Membr Sci 22:325–332
62. Kedem O, Warshawsky A (1988) Composite ion-exchange membranes. Sen-I Gakkaishi 44:11–19
63. Terada I, Horie H, Sugaya Y et al (1991) Sankaishyuyouno sinkina ionkoukanmaku. Kobunshi Kakou 40:38–41
64. Wainright JS, Savinell RF, Litt MH (1997) In: Savadogo O, Roberge PR (ed) Proceedings of the second international symposium on new materials for fuel cell and modern battery systems, Montreal
65. Coplan MJ, Gootz G (1983) Heterogeneous sulfonation process for difficultly sulfonatable poly(ether sulfone). US Patent 4,413,106
66. Arnold C Jr, Assink RA (1988) Development of sulfonated polysulfone membranes for redox flow batteries. J Membr Sci 38:71–83
67. Chao HS, Kelsey DR (1986) Process for preparing sulfonated poly(aryl ether) resins. US Patent 4,625,000
68. Kerres J, Cui W, Eigenberger G et al (1996) In: Veziroglu TN, Winter CJ, Baselt JP et al (ed) Proceedings of the 11th hydrogen conference, Stuttgart
69. Kerres J, Cui W, Reichle S (1996) New sulfonated engineering polymers via the metalation route. I. Sulfonated poly(ether solfone) PSU Udel via methalation- sulfination-oxidation. J Polym Sci Pol Chem 34:2421–2438
70. Kerres J, Cui W, Reichle S et al (1994) Lecture (Speaker J. Kerres), Jahres-kolloquium 1994 des Sonderforschungsbereichs 270 Energietrager Wasserstoff, VDI Verlag, S
71. Kerres J, Cui W, Neubrand W et al (1995) In: Euromembrane 92 congress, Bath
72. Nolte R, Ledjeff K, Bauer M et al (1993) Partially sulfonated poly(arylene ether sulfone) – a versatile proton conducting membrane material for modern energy conversion technologies. J Membr Sci 83:211–220
73. Rieke PC, Vanderborgh NE (1987) Temperature dependence of water content and proton conductivity in polyperfluorosulfonic acid membranes. J Membr Sci 32:313–328
74. Hwang GJ, Ohya H (1998) Preparation of anion-exchange membrane based on block polymers, Part I. Amination of the chloromethylated co-polymers. J Membr Sci 140:195–203
75. Terada I, Horie H, Sugaya Y et al (1992) Kakusan tousekiyou anion koukanmaku no sintenkai, sanwo kaisyusurutameno atarashi ionkoukanm-ku. Hyoumen 30:133
76. Hwang G, Ohya H, Nagai T (1999) Ion-exchange membrane based on block copolymers. Part III. Preparation of cation-exchange membrane. J Membr Sci 156:61–65
77. Linkous CA, Anderson HR, Kopitzke RW et al (1996) In: Veziroglu TN, Winter CJ, Baselt JP et al (ed) Proceedings of the 11th hydrogen conference, Stuttgart
78. Bailey C, Williams DJ, Karasz FE et al (1987) The sodium salts of sulfonated poly(aryether-ketone) (PEEK): preparation and characterization. Polymer 28:1009–1016
79. Katime IA, Iturbe CC (1996) In: Salomon JC (ed) Polymeric materials encyclopedia. CRC Press, Boca Raton
80. Cui W, Kerres J, Eigenberger G (1998) Development and characterization of ion-exchange polymer blend membranes. Sep Purif Technol 14:145–154
81. Wycisk R, Pintauro PN (1996) Sulfonated polyphosphazene ion-exchange membranes. J Membr Sci 119:155–160
82. Allcock HR, Fitzpatrick RJ, Salvati L (1991) Sulfonation of (aryloxy)-and (arylamino) phosphazenes: small-molecule compounds, polymers and surfaces. Chem Mater 3:1120–1132
83. Allcock HR, Klingenberg EH, Welker MF (1993) Alkanesulfonation of cyclic and high polymeric phosphazenes. Macromolecules 26:5512–5519
84. Montoneri E, Gleria M, Ricca G et al (1989) New acid polyfunctional water soluble phosphazenes: synthesis and spectroscopic characterization. J Macromol Sci Chem 26:645–661

85. Montoneri E, Gleria M, Ricca G et al (1989) On the reaction catenapoly(diphenoxy-λ^5-phosphazene) with sulfur trioxide. Die Makromol Chem 190:191
86. Wycisk R, Pintaura PN, Wang W et al (1995) Polyphosphazene ion-exchange membranes. In: Proceedings of the North American membrane society, 7th annual meeting, Portland
87. Wycisk R, Pintaura PN (1996) Sulfonated polyphosphazene ion-exchange membranes. J Membr Sci 119:155–160
88. Ehrenberg SG, Serpico JM, Wnek GE et al (1995) Fuel cell incorporating novel ion-conducting membrane. US Patent 5,468,574
89. Wnek GE, Rider JN, Serpico JM et al (1995) In: Gottesfeld S, Halpert G, Landgrebe A (ed) Proceedings of the first international symposium on proton conducting membrane fuel cells. Electrochem Soc Proc 247:95–123
90. Gautier-Luneau J, Denoyelle A, Sanchez JY et al (1992) Organic-inorganic protonic polymer electrolytes as membrane for low-temperature fuel cell. Electrochim Acta 37:1615–1618
91. Kostov GK, Matsuda O, Machi S et al (1992) Radiation synthesis of ion exchange carboxylic fluorine containing membranes. J Membr Sci 68:133–140
92. Kostov GK, Atanassov AN (1993) Properties of cation-exchange membranes prepared by radiation grafting of acrylic acid onto tetrafluoroethylene-ethylene copolymers. J Appl Polym Sci 47:1269–1276
93. Kostov GK, Turmanova SC (1997) Radiation-initiated graft copolymerization of 4-vinylpyridine onto polyethylene and polytetrafluoroethylene films and anion-exchange membranes therefrom. J Appl Polym Sci 64:1469–1475
94. Tsuneda S, Saito K, Mitsuhara H et al (1995) Novel ion-exchange membranes for electrodialysis prepared by radiation-induced graft polymerization. J Electrochem Sci 142:3659–3663
95. Nasef MM, Saidi H, Nor HM (2000) Proton exchange membranes prepared by simultaneous radiation grafting of styrene onto poly(tetrafluoroethylene-co-hexafluoropropylene) films. I. Effect of grafting conditions. J Appl Polym Sci 76:220–227
96. Brack HP, Buhrer HG, Bonorand L et al (2000) Grafting of pre-irradiated poly(ethylene-alt-tetrafluoroethylene) films with styrene: influence of base polymer film properties and processing parameters. J Mater Chem 10:1795–1803
97. Strathmann H (2004) Ion exchange membrane separation processes. Elsevier, Amsterdam
98. Burns NL, Holmberg K, Brink C (1996) Influence of surface charge on protein adsorption at an amphoteric surface: effects of varying acid to base ratio. J Colloid Interface Sci 178:116–122
99. Jimbo T, Higa M, Minoura N et al (1998) Surface characterization of poly(acrylonitrile) membrane graft-polymerized with ionic monomers as revealed by zeta potential measurement. Macromolecules 31:1277–1284
100. Jimbo T, Tanioka A, Minoura N (1998) Characterization of amphoteric-charged layer grafted to pore surface of porous membrane. Langmuir 14:7112–7118
101. Jimbo T, Tanioka A, Minoura N (1999) Fourier transform infrared spectroscopic study of flat surfaces of amphoteric-charged poly(acrylonitrile) membranes: attenuated total reflection mode. Langmuir 15:1829–1832
102. Izeo YR, Mizutani Y (1965) Permselectivity of the amphoteric ion-exchange membrane. Denkin Kagaku J Electrochem Soc 33:589–593
103. Jacobson H (1962) Preparation and properties of a variable charge ion-exchange membrane. J Phys Chem 66:570–572
104. Theodoropoulos AG, Tsakalos VT, Valkanes GN (1993) Suplhone-type crosslinks in sulphonation of macronet polystyrene backbone. Polymer 34:3905–3910
105. Gerome R, Desreux V (1971) The preparation and stability of snake cage membranes. J Appl Polym Sci 15:199–207; Gerome R, Desreux V (1972) Electrochemical behaviour of snake cage cation exchange membrane. J Appl Polym Sci 16:2739–2749
106. Michaels AS, Miekka RG (1961) Polycation-polyanion complexes-preparation and properties of poly(vinylbenzyltrimethylammonium) and polystyrene sulphonate. J Phys Chem 65:1765–1773; Michaels AS, Mir L, Sahneider NS (1965) A conductometric study of polyanion-polycation reactions in dilute aqueous solutions. J Phys Chem 69:1447–1455

107. Schlenoff JB, Ly H, Li M (1998) Charge and mass balance in polyelectrolyte multilayers. J Am Chem Soc 120:7626–7634; Ichinose I, Mizuki S, Ohno S et al (1999) Preparation of crosslinked ultrathin film based on layer-by-layer assembly polymers. Polym J 31:1065–1070
108. Levasalmi JM, McCarthy TT (1997) Poly(4-methyl-1-pentene) supported polyelectrolyte multilayer films: preparation and gas permeability macromolecoles 30:1752–1757; Harrish JJ, Steir JL, Bruening ML (2000) Layered polyelectrolyte films as selective, ultrathin barrier for anion transport. Chem Mater 12:1941–1946
109. Wu A, Yoo D, Lee JK et al (1999) Solid state light emitting devices based on the tris chelated ruthenium (II) complex: high efficiency devices via layer-by-layer molecular-level bending approach. J Am Chem Soc 121:4883–4891
110. Paltt LK, Schindler A (1971) Ionic membranes for water desalination. I. Charge mosaic membranes from blends of random copolymers. Die Angew Makromol Chem 19:135–155
111. Ma GH, Fukutomi T (1992) Preparation and chemical fixation of poly(4-vinyl pyridine) microgel film with ordered structure. Macromolecules 25:1870–1875; Takizawa M, Sujimoto Y, Doi S et al (2001) Fabrication of novel charge-mosaic membranes using microsphere and their application. Curr Trends Polym Sci 6:59–66
112. Liang L, Ying S (1993) Charge mosaic membranes from gamma-irradiatted poly(styrene-butadiene-4-vinylpyridene) triblock copolymer. J Polym Sci Polym Phys 31:1075–1081
113. Bauer B, Gerner FJ, Strathmann H (1988) Development of bipolar membranes. Desalination 68:279–292
114. Mueller H, Puetter H (1987) Production of bipolar membrane. US Patent 4,670,125
115. Francesco P, Rosignano SL (1998) Method for making a bipolar membrane. US Patent 5,849,167
116. Simons RG, Bay R (1993) High performance bipolar membranes. US Patent 5,227,040
117. Lee LTC, Dege GJ, Liu KJ (1997) High performance, quality controlled bipolar membrane. US Patent 4,057,48
118. Frilette V (1956) Preparation and characterization of bipolar ion exchange membranes. J Phys Chem 60:435–439
119. Wilhelm F, Vegt NVD, Wessling M et al (2000) Bipolar membrane preparation. In: Kemperman AJB (ed) Handbook bipolar membrane technology. Twente University Press, Enschede
120. Hao JH, Chen C, Li L et al (2001) Preparation of bipolar membranes. J Appl Polym Sci 80:1658–1663
121. Fu RQ, Xu TW, Yang WH, Pan ZX (2003) Preparation of a mono-sheet bipolar membrane by simultaneous irradiation grafting polymerization of acrylic acid and chloromethylstyrene. J Appl Polym Sci 90:572–576
122. Jendrychowska-Bonamour AM (1973) Semipermeable membranes obtained by grafting polytetrafluorethylene films. Manufacturing process and properties of these films. 1. Anionic and cationic mono-grafted membranes. J Chim Phys 70:12–19; Jendrychowska-Bonamour AM (1973) Semipermeable membranes obtained by grafting polytetrafluorethylene films. Manufacturing process and properties of these films. 2. Anionic and cationic membranes. J Chim Phys70:20–26
123. Xu Z, Gao H, Qian M et al (1993) Preparation of bipolar membranes via radiation peroxidation grafting. Radiat Phys Chem 42:963–966
124. El-Sayed A, Hegazy H, Kamal N et al (1999) Membranes prepared by radiation grafting of binary monomers for adsorption of heavy metals from industrial wastes. Nucl Instrum Meth B 151:386–392
125. Moussaoui RE, Hurwitz H (1998) Single-film membrane-process for obtaining it and use thereof. US Patent 5,840,192
126. Hao JH, Yu L, Chen C, Li L et al (2001) Preparation of bipolar membranes (II). J Appl Polym Sci 82:1733–1738
127. Xu TW, Yang W (2004) Fundamental studies on a novel series of bipolar membranes prepared from poly(2,6-dimethyl-1,4-phenylene oxide) (PPO): 1. Effect of anion exchange layers on I-V curves of bipolar membranes. J Membr Sci 238:123–129

128. Sheldeshov NV, Gnusin NP, Zabolotskii VI et al (1987) Chronopotentiometric study of electrolyte transport in commercial bipolar membranes. Sov Electrochem 22:742
129. Simons R (1979) Strong electric field effects on proton transfer between membrane-bound amines and water. Nature 280:824–826
130. Simons R (1984) Electric field effects on proton transfer between ionizable groups and water in ion exchange membranes. Electrochim Acta 29:151–158
131. Simons R (1985) Water splitting in ion exchange membranes. Electrochim Acta 30:275–282
132. Hanada F, Hirayama K, Ohmura N et al (1993) Bipolar membrane and method for its production. US Patent 5,221,455
133. Simons R (1986) A novel method for preparing bipolar membranes. Electrochim Acta 31:1175–1177
134. Simons R (1993) Preparation of a high performance bipolar membrane. J Membr Sci 78:13–23
135. Umemura K, Naganuma T, Miyake H (1995) Bipolar membrane. US Patent 5,401,408
136. Simons R (1979) The origin and elimination of water splitting in ion exchange membranes during water demineralization by electrodialysis. Desalination 28:41–42
137. Fu RQ, Xu TW, Cheng YY et al (2004) Fundamental studies on the intermediate layer of a bipolar membrane: Part III. Effect of starburst dendrimer PAMAM on water dissociation at the interface of a bipolar membrane. J Membr Sci 240:141–147
138. Narayanan PK, Adhikary SK, Harkare WP et al (1987) Indian Patent 160,880
139. Adhikary SK, Dave NJ, Narayanan PK et al (1983) Studies on interpolymer membranes. Part III. Cation exchange membranes. React Polym 1:197–206
140. Govindan KP, Narayanan PK (1969) Indian Patent 124,573
141. Lebrun L, Da Silva E, Metayer M (2002) Elaboration of ion-exchange membranes with semi-interpenetrating polymer networks containing poly(vinyl alcohol) as polymer matrix. J Appl Polym Sci 84:1572–1580
142. Vyas PV, Shah BG, Trivedi GS et al (2001) Characterization of heterogeneous anion-exchange membrane. J Membr Sci 187:39–46
143. Bodamer GW (Rohm and Hass) (1954) US Patents 2,681,319 (1954), 2,681,320 2,737,486 (1956).
144. Winston WS, Sarkar KK (1992) Membrane handbook. Van Nostrand Reinhold, New York
145. Mark HF, Gaylord NG (1968) Encyclopedia of polymer science and technology. Wiley, New York
146. Kariduraganavar MY, Nagarale RK, Kulkarni SS (2006) Electrodialytic transport properties of heterogeneous cation-exchange membranes prepared by gelation and solvent evaporation methods. J Appl Polym Sci 100:198–207
147. Ceynowa J (1984) Osmotic flow across ion-exchange membranes in H_2SO_4 solution. Angew Makromol Chem 127:187–202
148. Tasaka M, Kishi K, Okita M (1984) Thermoosmosis of various electrolyte solutions through anion-exchange membranes. J Membr Sci 17:149–160
149. Sata T (1989) Recent trends in ion-exchange membrane research. Pure Appl Chem 58:1613–1626
150. Lee H-J, Sarfert F, Strathmann H et al (2002) Designing of an electrodialysis desalination plant. Desalination 142:267–286
151. Demircioglu M, Kabay N, Kurucaovali I et al (2002) Demineralization by electrodialysis (ED)-separation performance and cost comparison for monovalent salts. Desalination 153:329–347
152. Tanaka Y (2003) Mass transport and energy consumption in ion-exchange membrane electrodialysis of seawater. J Membr Sci 215:265–279
153. Narayanan PK, Harkare WP, Adhikary SK et al (1985) Performance of an electrodialysis desalination plant in rural area. Desalination 54:145–150
154. Goldstein AL (1979) Electrodialysis on the American continent. Desalination 30:49–58
155. Harkare WP, Adhikary SK, Narayanan PK et al (1982) Desalination of brackish water by electrodialysis. Desalination 42:97–105

156. Mehta DJ, Rao AV, Govindan KP (1979) Water Desalination a perspective for membrane processes in India. Desalination 30:325–335
157. Strauss SD (1985) Power 129:72
158. Veza JM (2001) Desalination in the Canary Islands: an update. Desalination 133:259–270
159. Yuan Z, Richard S, Tol J (2004) Implications of desalination for water resources in China – an economic perspective. Desalination 164:225–240
160. Adiga MR, Adhikary SK, Narayanan PK et al (1987) Performance analysis of photovoltaic electrodialysis desalination plant at Tanote in Thar desert. Desalination 67:59–66
161. Kuroda O, Takahashi S, Kubota S et al (1987) An electrodialysis sea water desalination system powered by photovoltaic cells. Desalination 67:33–41
162. Nakasone S, Miyagi S, Aragaki T et al (1985) Process for clarifying and desalinating sugar cane syrup or molasses. US Patent 4,492,601
163. Odagiri H, Kogyo S (1983) 19 83:20
164. Sata T (1984) Nenryo oyobi Nensho (Fuel Combust) 51:166
165. Niketi-Aleksic G, Jaskoic MM (1980) Application of selective membranes in grape juice stabilization. Desalination 35:317–327
166. Yamada M, Matsushita S, Hirai H, Tsuyuki I, Ogawa Y (1981) J Appl Photogr Eng 7:53
167. Tasaka A, Yasuzawa T, Nagafuji Y, Yamano A, Miyazaki H, 0-oka H, Sugimoto H (1985) Denki Kagaku, 53: 466; ibid. 53 (1985) 400; ibid. 53 (1985) 404
168. Sato T, Tanaka T, Suzuki T (1985) Denki Kagaku 53:706
169. Iwahara M, Takamoto H, Nomura Y, Motoe M (1981) Presented at meeting of Nishinihon Shibu of Agricutural Society of Japan, Shimane University, 3 Oct 1981
170. Cox JA, Carlson H (1981) Quantitative enrichment of trace levels of ions by electrodialysis. Anal Chim Acta 130:313–321
171. Sugimoto S, Aikawa H (1980) Nihon Genshiryoku Gakkaishi 22:813–820
172. Balster J, Stamatialis DF, Wessling M (2004) Electrocatalytic membrane reactors and the development of bipolar membrane. Chem Eng Process 43:1115–1127
173. Vandenbarre H, Leysen R, Nackaerts H (1983) mt. J Hydrogen Energ 8:81
174. Torikai E, Takenaka K (1982) Soda to Enso 33:51; (1983) Denki Kagaku 51:633
175. Seko M, Ogawa S, Yoshida M, Shiroki H (1984) Dechema-Monographien Band 97:27
176. Kranpelt M (1979) Electro organic synthesis technology. AICHE Symp Ser 185:25
177. Ogumi Z, Ohashi S, Takehara Z (1985) Application of the SPE method to organic electrochemisty-VI. Oxidation of cyclohexanol to cyclohexanone on Pt-SPE in the presence of iodine and iodide. Electrochim Acta 30:121–124
178. Kobuchi Y, Motomura H, Noma Y et al (1985) Application of ion exchange membranes to the recovery of acids by diffusion dialysis. J Membrane Sci 27:173–179
179. Joshi BS, Govindan KP (1981) Water softening by diffusion using cation exchange membranes. Desalination 38:509–516
180. Law HH (1983) Plat Surf Finish 70:72
181. DiNunzio JE, Jubara M (1983) Donnan dialysis preconcentration for ion chromatography. Anal Chem 55:1013–1016
182. Oaki H, Ishida M, Shirai T (1980) J Chem Eng Japan 13:251–252
183. Winter M, Brodd RJ (2004) What are batteries, fuel cells, and superconductors? Chem Rev 104:4245–4270
184. St-Pierre J, Wilkinson DP (2001) Fuel cells: a new, efficient and cleaner power source. AIChE 47:1482–1486
185. Hirschenhofer JH, Stauffer DB, Engleman RR, Klett MG (1998) Fuel cell handbook, 4th edn. Parsons Corporation, Reading
186. Scholta J, Berg N, Wilde P et al (2004) Development and performance of a 10 kW PEMFC stack. J Power Sources 127:206–212
187. Wang C, Zhou S, Hong X, Qiu T, Wang S (2005) A comprehensive comparison of fuel options for fuelcell vehicles in China. Fuel Process Technol 86:831–845
188. Yeom J, Mozsgai GZ, Flachsbart BR et al (2005) Microfabrication and characterization of a silicon-based millimeter scale, PEM fuel cell operating with hydrogen, methanol, or formic acid. Sens Actuators B Chem 107:882–891

189. Yamaguchi T, Kuroki H, Miyata F (2005) DMFC performances using a pore-filling polymer electrolyte membrane for potable usages. Electrochem Commun 7:730–734
190. Lim HS, Lackner AM, Knechtli HC (1977) Zinc-bromine secondary battery. J Electrochem Soc 124:1154–1157; Will FG (1979) Bromine diffusion through nafion perfluorinated ion exchange membranes. J Electrochem Soc 126:36–43; Donepudi VS, Conway BE (1984) Electrochemical calorimetry of the zinc and bromine electrodes in zinc-bromine and zinc-air batteries. J Electrochem Soc 131:1477–1485
191. Oei D (1985) Permeation of vanadium cations through anionic and cationic membranes. J Appl Electrochem 15:231–235; Fedkiw PS, Watts RW (1984) A mathematical model for the iron/chromium redox battery. J Electrochem Soc 131:701–709
192. Savadogo O (2004) Emerging membranes for electrochemical systems. Part II. High temperature composite membranes for polymer electrolyte fuel cell (PEFC) applications. J Power Sources 127:135–161
193. Kerres J, Ullrich A, Haring Th et al (2000) Preparation, characterization and fuel cell application of new acid-base blend membranes. J New Mater Elect Syst 3:229–239
194. Kerres J, Zhang W, Jorissen L et al (2002) Application of different types of polyaryl-blend-membranes in DMFC. J New Mat Elect Syst 5:97–107
195. Nunes SP, Ruffmann B, Rikowski E et al (2002) Inorganic modification of proton conductive polymer membranes for direct methanol fuel cells. J Membr Sci 203:215–225
196. Kim J, Kim B, Jung B (2002) Proton conductivities and methanol permeabilities of membranes made from partially sulfonated polystyrene-block-poly(ethylene-ran-butylene)-block-polystyrene copolymers. J Membr Sci 207:129–137
197. Kreuer KD (2001) On the development of proton conducting polymer membranes for hydrogen and methanol fuel cells. J Membr Sci 185:29–39
198. Kerres J, Ullrich A, Meier F et al (1999) Synthesis and characterization of novel acid-base polymer blends for application in membrane fuel cell. Solid State Ionics 125:243–249
199. Guo X, Fang J, Watari T et al (2002) Novel sulfonated polyimides as polyelectrolytes for fuel cell application. 2. Synthesis and proton conductivity of polyimides from 9,9-Bis (4-aminophenyl)fluorine-2,7-disulfonic acid. Macromolecules 35:6707–6713
200. Tealdo G, Castello G, D'Amato G et al (1982) Water-glycerol permeation through styrene-grafted and sulfonated PTFE membranes. J Membr Sci 11:3–9
201. Wenzlaff A, Boddeker KW, Hattenbach K (1985) Pervaporation of water-ethanol through ion exchange membranes. J Membr Sci 22:333–344
202. Ukihashi H, Asawa T, Yamabe M, Miyake H Japan Kokai Tokkyo Koho (unexamined application) JP 58—921405, 58—921407, 58—921108, 58—921113 etc,.
203. Le Blanc OH, Jr WWJ, Matson SL et al (1980) Facilitated transport in ion-exchange membranes. J Membr Sci 6:339–343
204. Yamada T, Sugie K, Yamaji T Japan Tokkyo Kokai Koho (unexamined Application) JP 57—1 17321
205. Kinjo N, Ohara S, Sugawara T, Tsuchitani S (1983) Changes in electrical resistance of ionic copolymers caused by moisture sorption and desorption. Polym J 15:621–623
206. Eriksson T, Johansson G (1973) Determination of the concentration of strong acids with a permaplex membrane electrode in the presence of hydrofluoric acid and various cations. Anal Chim Acta 63:445–453
207. Caplan SR, Fischer G (1983) Photovolataic behavior of bacteriorhodospin-loaded cation-exchange membranes. J Membr Sci 16:391–405
208. Kuczynski JP, Milosavljevic BH, Thomas JK (1984) Photophysical properties of cadmium sulfide in Nafion film. J Phys Chem 88:980–984
209. Lee PC, Meisel D (1980) Luminescence quenching in the cluster network of perfluoro-sulfonate membrane. J Am Chem Soc 102: 5477–5481; Lee PC, Rodgers MAJ (1984) Kinetic properties of singlet oxygen in a polymeric microheterogeneous system. J Phys Chem 88:4385–4389

Chapter 8
Synthetic Inorganic Ion Exchange Materials

Wolfgang Schmidt

Abstract Synthetic ion exchangers are used for a wide range of different applications, ranging from water softening, water remediation, catalysis, molecular sieving, and selective adsorption to medical applications. Their well-defined compositions and tunable structural features allow minute tailoring of their specific properties. A selection of the most prominent ion exchangers for aqueous ion exchange is introduced in this chapter, and the ion exchange process is illustrated for selected materials. Furthermore, structural features of individual materials are discussed and typical preparation methods and applications are addressed.

8.1 Introduction

In this chapter, synthetic ion exchangers will be introduced. The focus is on solid exchangers for ion exchange in aqueous solutions and at temperatures up to about 100°C. A selection of materials has been made which is by far not comprehensive but illustrates features and properties of the most abundantly used synthetic inorganic ion exchangers. The survey will start with a class of materials denoted as zeolites. They are the most commonly used synthetic inorganic ion exchangers and therefore will be discussed in some more detail. The concept and theory of ion exchange on synthetic solids will be illustrated on zeolites; however, they can be also transferred directly to the other materials that will be described in the following sections.

W. Schmidt (✉)
Max-Planck-Institut für Kohlenfoschung, Kaiser-Wilhelm-Platz 1,
Mülheim an der Ruhr 45470, Germany
e-mail: schmidt@kofo.mpg.de

Inamuddin and M. Luqman (eds.), *Ion Exchange Technology I: Theory and Materials*, DOI 10.1007/978-94-007-1700-8_8,

8.2 Benefits of Synthetic Ion Exchangers

Ion exchange on solids is a type of reaction that is often thought to proceed mainly on organic ion exchangers since ion exchange resins are nowadays commodities that are used for many application in industry and private households. However, ion exchange is also a feature that is quite commonly observed on inorganic solids. Basically, any polar surface is able to coordinate ionic species and thus can act as ion exchanger. The ion exchange capacity then simply scales with the specific surface area of that material. Accordingly, a broad variety of materials with ion exchange properties can be found in nature, be it clay minerals, natural zeolites, metal oxides, or others. Thus, one could ask why one should want to generate synthetic ion exchangers. There are two main reasons to do so. One problem of natural products is their inhomogeneity with respect to chemical and phase composition as well as to morphological features, such as particle sizes, particle size distributions, and particle morphologies. Synthetic products can be synthesized with well-defined phase and chemical composition. Furthermore, a high degree of control on morphological properties is allowed for. The second reason why synthetic materials have advantages is that they can be produced in a bottom-up approach. This allows tailoring of properties already at the stage of the particle formation which is hardly possible for solids that already exist as macroscopic particles. Disadvantage of synthetic materials can be higher production costs in comparison to most natural products. Ion exchange materials on the basis of minerals can be found in large natural deposits. Production thus involves mining, crushing, sieving, and may be crude phase separation, all of which are usually not cost intensive. Therefore, natural ion exchangers usually have a much lower price than synthetic ones. For any practical application, expenses and advantages have to be balanced thoroughly.

8.3 Zeolites

Synthetic zeolites are the most abundant inorganic ion exchange materials for technical applications. Very unique structural features make zeolites excellent ion exchangers. They are typically crystalline aluminosilicates with structures that form pore networks throughout the crystallites. These intrinsic pores accommodate exchangeable cations which compensate negative charges of the aluminosilicate framework. About 200 different structures are known for zeolites, from which approximately 20% can be found as natural minerals. The others exist only as synthetic materials. Due to their specific structural features, basically, all zeolites can serve as ion exchangers, but for practical application, only a little number is actually used. Since its ion exchange capacity is directly related to the aluminum content of the zeolite, high-alumina zeolites are preferred for ion exchange applications. Furthermore, a well accessible pore system is advantageous for a rapid ion exchange. If such zeolites then can be

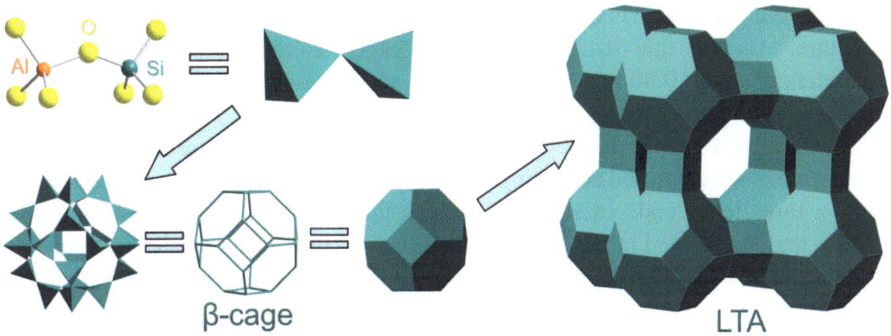

Fig. 8.1 Schematic construction of zeolites from SiO_4 and AlO_4 tetrahedra and simplified polyhedra representation of a zeolite structure

synthesized with high yields at moderate temperatures, this is of course highly welcome for large-scale applications. For those reasons, high-alumina zeolites with easily accessible pore systems, such as the zeolites Linde type A (LTA), Linde type X (FAU), or zeolite P (GME), are the most commonly used synthetic zeolites for ion exchange applications.

8.3.1 Zeolite Structures

Zeolite structures are formed by SiO_4 and AlO_4 tetrahedra, the corners of which are connected via shared oxygen anions, as shown in Fig 8.1. Since either silicon or aluminum atoms occupy the centers of the tetrahedra, they are denoted as T-atoms. The tetrahedra are connected to each other in a way that larger channels and/or cages are formed. Such cages, formed by tetrahedra, can be represented in a simplified manner as polyhedra that form the final zeolite structure, as shown in the Fig. 8.1. In this representation, the oxygen atoms are omitted and straight lines connect the silicon and aluminum sites of the zeolite framework. The polyhedra representation allows an easy visualization of the larger cages and windows of the open zeolite pore system.

Often, zeolite structures are described with the help of structural subunits, such as rings and cages. As for example, the β-cages of the Linde type A (LTA) structure consist of rings that are formed by either six or four tetrahedra, denoted as 6MR (6-membered ring) or 4MR (4-membered ring). The LTA structure, as shown in Fig. 8.1, consists of β-cages that are connected to each other via double 4-membered rings (D4R). In that way, a larger α-cage with an approximate diameter of 1.18 nm is formed which is accessible only via 8MR windows. The diameter of this 8MR window is about 0.41 nm, and only cations or molecules that are smaller can pass through these windows. The smaller β-cages are only accessible via the very narrow 6MR windows with free diameters of only 0.21 nm.

Connecting β-cages via double 6-membered rings (D6R) results in the FAU structure. Also in this structure, large cages (so-called Faujasite cages) with diameters of about 0.90 nm are formed that are accessible via 12MR with a free diameter of about 0.70 nm.

The topologies of the frameworks are characteristic for each zeolite, and therefore, zeolites are classified according to their framework types. For each framework type, a three-letter code exists that specifies the zeolite structure. Exchangeable cations that are located in the channels and cages of the zeolite are generally rather mobile and not considered for this kind of classification. As for some examples, the three-letter code *LTA* is assigned to the structure of synthetic *L*inde *t*ype *A* zeolites and the code *FAU* to structures of *Fau*jasite-type zeolites. Faujasites are naturally occurring minerals, but the structure-type FAU is also found in the synthetic zeolites Linde type X and Linde type Y. Natural zeolites can be easily recognized by their names, but for synthetic zeolites, the nomenclature is more or less arbitrary and many zeolite producers have their own names for a given zeolite material. Thus, many different names may exist for the same type of zeolite material. In order to avoid confusion, in the following, the three-letter codes will be used for the assignment of zeolites. A prefix will be used to specify exchangeable cations, for example, Na-LTA specifies a Linde type A zeolite, the framework charges of which are compensated by Na^+ cations. Formally, the composition of the zeolite can be represented by a stoichiometric formula, such as:

$$Na_xAl_xSi_{y-x}O_{2y} \cdot n\ H_2O \text{ or } Na_x(AlO_2)_x(SiO_2)_{y-x} \cdot n\ H_2O,$$

with n the number of water molecules adsorbed within the pores of the zeolite. This formula reflects that, for each aluminum in the structure, one exchangeable cation, also called extra-framework cation, is present (here Na^+). Al-O-Al bonds are generally not observed for aluminum in tetrahedral coordination. Even though exceptions to this so-called Löwenstein rule have been reported, it applies to zeolites in general. Thus, the minimum Si/Al ratio achievable is 1.

For zeolites, their stoichiometric formula is often reported as the number of elements in one unit cell of the respective structure. Thus, for an LTA-type zeolite with Si/Al = 1, the unit cell content may be given as $Na_{96}Al_{96}Si_{96}O_{384}$ 216 H_2O, and for a FAU-type zeolite with Si/Al = 2, as $Na_{64}Al_{64}Si_{128}O_{384}$ 240 H_2O.

It should be noted that zeolites not necessarily have to be aluminosilicates. Substitution of Si or Al in the tetrahedral by other elements also results in zeolitic structures. Some typical elements for partial or full substitution are Ti, Ge, Ga, or P.

8.3.2 Ion Exchange Properties of Zeolites

As explained above, zeolite frameworks are formed by corner-connected SiO_4 and AlO_4 tetrahedra. Each oxygen anion coordinates either two silicon cations or one

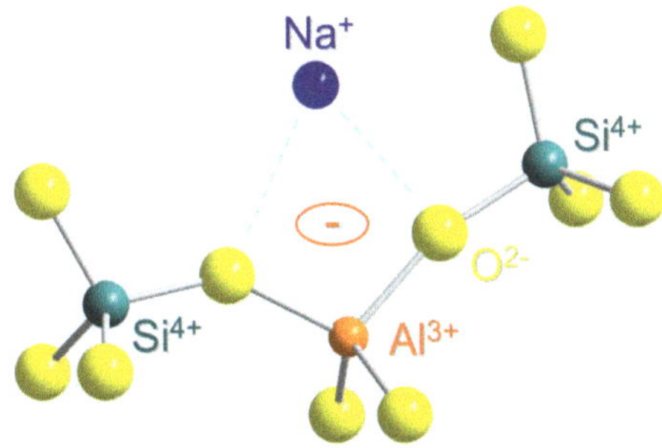

Fig. 8.2 Compensation of negative zeolite framework charge by sodium cation

silicon and one aluminum cation. Thus, it formally contributes one negative charge to each T-atom. Each four-valent silicon cation is coordinated by four oxygen anions, and the charges are balanced. However, also each three-valent aluminum cation is coordinated by four oxygen anions, and thus, one negative charge from the oxygen atoms remains unbalanced. This creates one negative framework charge per aluminum site in the structure, as illustrated in Fig 8.2.

These negative charges are compensated by cations that are located in the channels and cavities of the respective zeolite. Since such cations are only weakly coordinated by the zeolite framework, they can be easily exchanged against other cations, provided the latter are small enough to fit into the pores of the zeolite. If water is present in the pores of the zeolite, the zeolite framework and the exchangeable cations are usually also coordinated by water molecules. Small cations with high positive charge can be coordinated by several shells of water molecules. This can cause problems for the ion exchange since for ion exchange not only the cation must enter the zeolite pores but also its hydration shell which is sometimes bigger than the diameter of the pores. In such cases, ion exchange is either limited or stripping of the hydration shell takes place, especially if the ion exchange is performed at elevated temperature. Upon drying of the zeolite, for example, at high temperature and/or under vacuum, water molecules are removed from the pores and from the hydration shells. As a result, the cations move closer to the oxygen anions of the zeolite framework. Since different cations have different sizes, pore openings can be modified by exchanging a given zeolite with different cations.

Such cations are often found on well-defined positions in the cages and/or pores of zeolites. Some typical preferred cation sites for LTA- and FAU-type zeolites are shown in Fig. 8.3. For LTA zeolites, cations are often found facing 6MR (S1) or 4MR (S3) units in the larger cages of the zeolite. Others occupy sites in the 8MR window opening (S2) of the LTA structure. However, for steric reasons, not all of these four cation sites are occupied simultaneously. In addition, cations are also located inside of the β-cages. The occupation of these sites depends on the types, charges, and sizes of the respective cations. Similarly as LTA zeolites, FAU-type zeolites also are formed by β-cages (see Fig. 8.1) that are differently connected to each other by these two zeolites. As for zeolite LTA, cations are also found within the β-cages (S1′ and S2′) and in D6R units, connecting the β-cages (S1) for FAU-type zeolites. In the large cages of the FAU zeolite, further cations can be located at sites facing 6R units (S2), 4R units (S3), and in the center of the large 12MR pore window (S4).

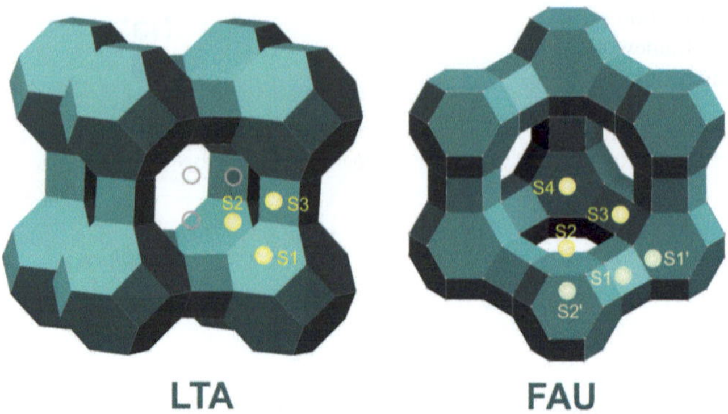

Fig. 8.3 Typical cation sites in LTA- and FAU-type zeolites

The change of free energy associated with the ion exchange is

$$\Delta G^o = \Delta H - T\Delta S. \tag{8.1}$$

As for any other chemical reaction, ion exchange is favored if the change in free energy of ion exchange ΔG^o is < 0. Thus, changes of enthalpy, ΔH, as well as of entropy, ΔS, at a given temperature, T, have to be considered. Entropy is lower for cations that are bound to a zeolite framework compared to those in solution. Furthermore, entropy changes occur during release of water molecules from a coordination shell of cations and/or from the zeolite pores. As a result, ion exchange is a complex interaction of changes of enthalpy and entropy. Knowing the thermodynamic equilibrium constant K_a, ΔG^o may be calculated as

$$\Delta G^o = -\frac{RT}{ab}\ln K_a, \tag{8.2}$$

with R being the ideal gas constant, T the temperature, and a and b the valences of the cations A^{a+} and B^{b+}. As for every chemical reaction, the equilibrium constant is characteristic for a given reaction, and in the case of ion exchange in zeolites,

$$\mathrm{b\ A_s{}^{a+} + a\ B_z{}^{b+} \rightleftharpoons b\ A_z{}^{a+} + a\ B_s{}^{b+}},$$

where z and s assign ions in the zeolite (z) or in the solution (s). The equilibrium constant is defined as [1–5]

$$K_a = \frac{a^b_{A_Z} a^a_{B_S}}{a^b_{A_S} a^a_{B_Z}}, \tag{8.3}$$

with $a^b{}_{Az}$, $a^a{}_{Bs}$, being the ion activities in the solution and the zeolite. Ion exchange of a given cation A^{a+} proceeds until equilibrium is reached, i.e., until the ratio described in the equation meets K_a. More comprehensive thermodynamic descriptions of ion exchange processes in zeolites can be found in literature, for example, by Breck [4] or Dyer [6]. One should note that strict thermodynamic control of the ion exchange process is only coming into effect for cases where other factors, such as ion size exclusion or space restrictions, can be excluded.

For cations A^{a+} and B^{b+} in the zeolite, equivalent fractions, Z_A and Z_B, can be calculated as

$$Z_A = \frac{a \cdot m_{A_Z}}{a \cdot m_{A_Z} + b \cdot m_{B_Z}} \text{ and } Z_B = \frac{b \cdot m_{B_Z}}{a \cdot m_{A_Z} + b \cdot m_{B_Z}}, \tag{8.4}$$

with m_{A_Z} and m_{B_Z} being the molalities of the cations in the zeolite (moles of respective cation in 1 kg of zeolite (mol/kg)). The denominator in these equations is the total ion exchange capacity, Q, of the zeolite, i.e.,

$$Q = a \cdot m_A + b \cdot m_B. \tag{8.5}$$

Likewise, equivalent fractions of the cations in solution can be calculated as

$$S_A = \frac{a \cdot M_{A_S}}{a \cdot M_{A_S} + b \cdot M_{B_S}} \text{ and } S_B = \frac{b \cdot M_{B_S}}{a \cdot M_{A_S} + b \cdot M_{B_S}} \tag{8.6}$$

with M_{A_S} and M_{B_S} being the molarities of the cations (moles of respective cations in 1 L of solution, (mol/l)). Thus, S_A and S_B represent the fractions of the negative charges due to anions in solution, compensated by the cations A^{a+} and B^{b+}. The total number of cations in the system is supposed to be constant. Thus, the system is sufficiently described by two equivalent fractions. This allows describing the complete ion exchange process at a given temperature by plotting the equivalent fraction of one cation M^{m+} in the zeolite, Z_M, against the equivalent fraction of the same cation in the solution, S_M. The total normality of all cations remains constant. Such a plot is denoted as ion exchange isotherm. What information does it provide? Let us consider four hypothetical isotherms, such as shown in Fig. 8.4 [4].

For isotherm (a), the cations A^{a+} from solution readily replace the cation B^{b+} in the zeolite. Such isotherms are observed for zeolites that have a preference for the entering cation A^{a+}. Isotherm (b) shows the exact opposite situation. It is found for zeolites that prefer the cation B^{b+} that has to be exchanged. Isotherm (c) has a sigmoidal shape. It is typical for zeolites with two different exchange sites that have opposite preferences for cations A^{a+} and B^{b+}. At low degree of exchange, cations from sites that prefer the entering cation are exchanged, whereas at higher exchange degree, also cations from sites that prefer the leaving cation are replaced (= selectivity reversal). If the exchange cannot proceed to completeness, for example, for steric reasons, isotherm (d) is observed.

The preference of a zeolite for one specific cation (A) over another one (B) may be expressed by the separation factor α_B^A:

$$\alpha_B^A = \frac{Z_A S_B}{S_A Z_B}, \tag{8.7}$$

where $\alpha_B^A > 1$ if A^{a+} is preferred by the zeolite over B^{b+} [4].

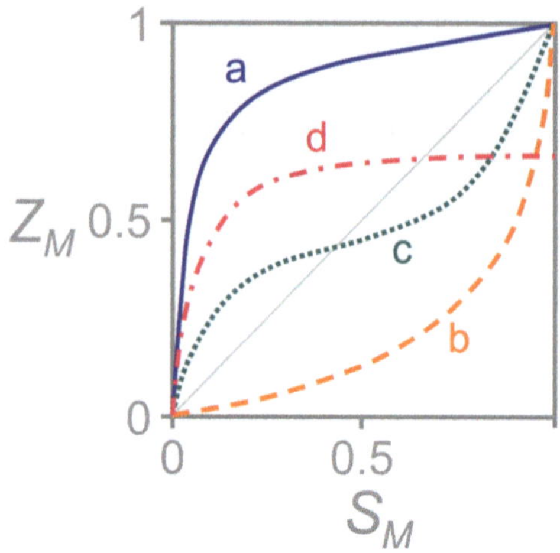

Fig. 8.4 Typical ion exchange isotherms for binary exchange with (**a**) selectivity (preference) of the zeolite for the entering cation over the entire range, (**b**) selectivity for the leaving cation over the entire range, (**c**) selectivity reversal with increasing fraction of the entering cation in the zeolite, and (**d**) incomplete exchange of an initially preferred cation (Reproduced from Ref. [4] with kind permission of © John Wiley & Sons (1974))

The different preferences of zeolites for given cations will be illustrated in the following for the zeolites Na-A (LTA) and Na-Y (FAU). Zeolite Na-A with composition $Na_{96}Al_{96}Si_{96}O_{384}$ 216 H_2O has a cation-exchange capacity (CEC) of 5.48 meq g^{-1}. As can be easily seen from Fig. 8.5a [7], the affinity of alkali cations for zeolite LTA generally decreases with increasing size of the cations, Li^+ being an exception from this trend. ΔG is always positive for exchange of Na^+ by any of the other alkali cations, i.e., Na^+ has the highest affinity to zeolite LTA. The overall affinities of the cations for the zeolite may be summarized as $Na^+ > K^+ > Rb^+ > Li^+ > Cs^+$ [7]. Due to space limitations, the large Cs^+ cations cannot replace all Na^+ in Na-A. All other alkali cations can fully replace Na^+ in that zeolite.

The bivalent alkaline earth cations Ca^{2+}, Sr^{2+}, and Ba^{2+} readily replace Na^+ in Na-A as illustrated in Fig. 8.5b [1, 8]. Their affinities to zeolite LTA is much higher than that of the smaller Mg^{2+} cation. More energy is required to strip the hydration shell from the smaller Mg^{2+} cation which is a prerequisite for entering the narrow LTA pores. Exchange of Na^+ by Mg^{2+} is thus less favored in comparison to the other alkaline earth cations. Affinities of alkaline earth cations to zeolite LTA can be summarized as $Sr^{2+} > Ba^{2+} > Ca^{2+} > Mg^{2+}$ [4].

The major use of Na-A is in detergents for replacing of Ca^{2+} and Mg^{2+} by Na^+ in aqueous washing liquor (water softening), but also, other transition metal cations can be removed from aqueous solutions as illustrated in Fig. 8.5c [9]. In contrast to Zn^{2+} and Cd^{2+}, which may fully replace Na^+ in Na-A, exchange of Na^+ by Co^{2+} and Ni^{2+} is not complete at room temperature. Insufficient water stripping from their hydration shells prevents full exchange of these cations at 298 K. Increasing the temperature of the exchange solution results in easier water stripping and, as the consequence, in enhanced replacement of Na^+ by Co^{2+} and Ni^{2+} in zeolite LTA [9].

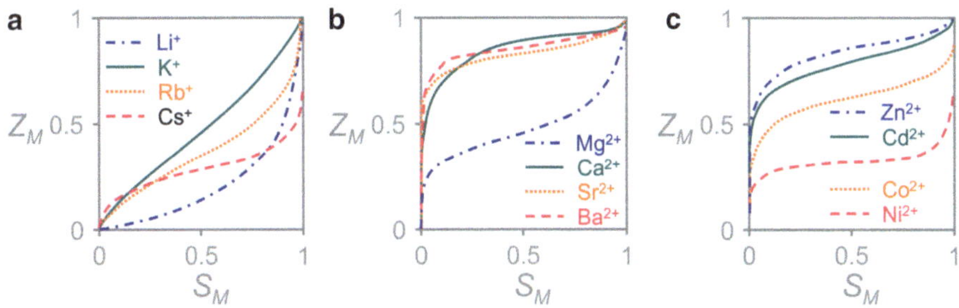

Fig. 8.5 Ion exchange isotherms of Na-A (LTA) with (**a**) Li^+, K^+, Rb^+, and Cs^+ [7]; (**b**) Ca^{2+}, Sr^{2+}, Ba^{2+} [1], and Mg^{2+} [8]; and (**c**) Zn^{2+}, Cd^{2+}, Co^{2+}, and Ni^{2+} [9] (total normalities 0.1 N, 298 K) (Reproduced from Ref. [1] with kind permission of © The American Chemical Society from Ref. [7] with kind permission of the Royal Society (1963), Ref. [8] with kind permission of © Elsevier (1989) and from Ref. [9] with kind permission of © The Royal Society of Chemistry (1971))

A relatively low pH of more concentrated transition metal salt solutions, as the result of autoprotolysis of water, may cause severe damage to the aluminosilicate framework of low-silica zeolites, such as LTA zeolites. In acidic solutions, aluminum is easily extracted from the structure of low-silica zeolites which may result in complete disintegration of the zeolite structure [10–12].

For the same reason, low-silica FAU-type zeolite X suffers of dealumination in acidic solutions. Its somewhat more siliceous counterpart zeolite Y is more stable toward acidic solutions, but nevertheless, its framework is also susceptible to dealumination in solutions with lower pH [12, 13]. Zeolite Na-Y with composition $Na_{54}Al_{54}Si_{138}O_{384}$ 241 H_2O has a CEC of 3.17 meq g^{-1}. However, ion exchange in zeolite Y often proceeds not to full exchange. This is illustrated for alkali cations in Fig. 8.6a [14]. The ion exchange isotherms show that K^+ and Li^+ can fully replace Na^+ in zeolite Y, whereas exchange with Rb^+ and Cs^+ proceeds to only about 68%. These larger cations cannot pass the 6MR windows of the β-cages and thus cannot replace Na^+ from sites within these cages and from D6R units. In such cases, ion exchange is limited to cations that are located within the large Faujasite cages. From the shape of the isotherms for these cages, a preference for cations in the order of $Cs^+ > Rb^+ > K^+ > Na^+ > Li^+$ can be deduced. Similar observations have been made for alkaline earth and transition metal cations. Even though they are obviously preferred by sites within the large cages, as can be seen from Fig. 8.6b [15], only about 68% of Na^+ can be replaced by Ca^{2+}, Sr^{2+}, and Ba^{2+}. The smaller β-cages cannot be populated by these cations. Water molecules in their hydration shells are strongly bound to the bivalent alkaline earth cations and thus prevent their access to the β-cages. Applying elevated temperatures of 323 K does not change this exclusion of Ca^{2+}, Sr^{2+}, and Ba^{2+} from the smaller cages of that Y zeolite.

More or less the same behavior has been reported for exchange of Na^+ from Na-Y with bivalent transition metal cations [16]. Figure 8.6c shows that Zn^{2+}, Cu^{2+}, Co^{2+}, and Ni^{2+} are preferred over Na^+ in the larger cages and they completely replace Na^+, but they cannot fully replace the monovalent sodium cation in Na-Y. Ni^{2+} is completely excluded from its β-cages, and the three other bivalent transition metals only partially replace Na^+ in these cages.

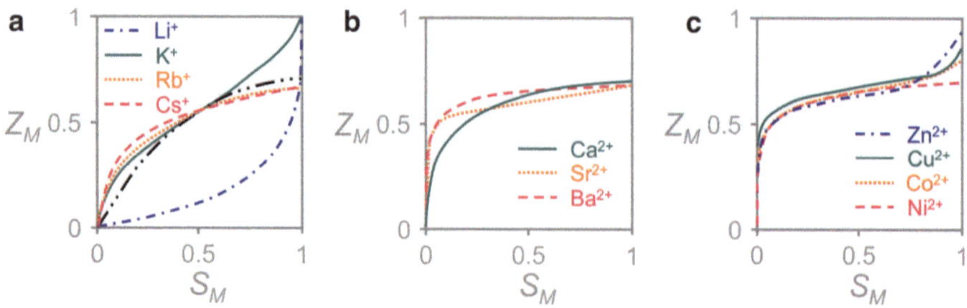

Fig. 8.6 Ion exchange isotherms of Na-Y (FAU) with (**a**) Li^+, K^+, Rb^+, and Cs^+ (anhydrous Na-Y: $Na_{50}Al_{50}Si_{142}O_{384}$) [14]; (**b**) Ca^{2+}, Sr^{2+}, and Ba^{2+} (anhydrous Na-Y: $Na_{51}Al_{51}Si_{141}O_{384}$) [15]; and (**c**) Co^{2+}, Ni^{2+}, Cu^{2+}, and Zn^{2+} (Na-Y: $Na_{54}Al_{54}Si_{138}O_{384}$ 241 H_2O) [16] (298 K, (**a**) and (**b**) total normalities 0.1 N, (**c**) 0.01 N) (Reproduced from Ref. [14] with kind permission of © The American Chemical Society (1966), from Ref. [15] with kind permission of © The American Chemical Society (1968), and from Ref. [16] with kind permission of © The Royal Society of Chemistry (1975))

It should be mentioned at this point that these preferences are not specific for all FAU-type zeolites. Zeolite X has somewhat larger 6MR windows and therefore allows access to β-cages for cations that might be excluded from those in zeolite Y.

These examples show that ion exchange in zeolites is very specific and based on a very delicate balance between space limitations within cages and channels, window size, cation size, and sizes of rigidity of hydration shells. Furthermore, thermodynamic parameters such as enthalpies of exchange and entropy changes are factors that determine ion exchange processes in zeolites.

The kinetics of ion exchange in zeolites is generally diffusion controlled resulting in slower exchange rates for larger crystallites. This is shown in Fig. 8.7 for the exchange of Na^+ by Ca^{2+} in aqueous solution [17]. The ion exchange proceeds faster with the smaller crystallites, whereas the ion exchange equilibrium is identical as expected. With the smaller crystallites, a quite rapid exchange is observed and the solution depletes substantially of Ca^{2+} within less than a minute.

8.3.3 Application of Synthetic Zeolites as Ion Exchangers

Ion exchange of Ca^{2+} and Mg^{2+} ions proceeds very rapidly in synthetic LTA-type zeolite Na-A or GIS-type zeolite Na-P. This feature makes them valuable water-softening agents that are used to large extents in detergents. At room temperature, Na-A and Na-P are able to remove up to 165 mg CaO (NaP) or 175 mg CaO (Na-A) and up to 40 mg MgO per gram of zeolite from aqueous solution [18, 19]. Modern laundry detergents may contain up to 25% of such zeolites [18]. The annual consumption of synthetic zeolites in detergents worldwide is about 1.3 million metric tons, which makes their application in detergents by far the largest application of synthetic zeolites with respect to production capacity. This enormous production volume is only exceeded by the use of natural zeolites that are used in annual

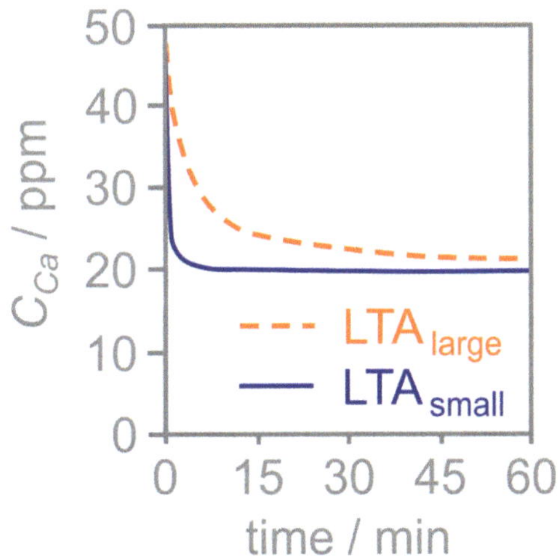

Fig. 8.7 Kinetics of calcium removal from aqueous Ca^{2+} solution (0.0025 N) by either 2.5–3 μm (LTA_{large}) or 1.0–1.5 μm (LTA_{small}) zeolite A crystallites at 303 K (Reproduced from Ref. [17] with kind permission of © Elsevier (1997))

amounts of more than 2.5 million metric tons as cements additives, nutrient release agents in agriculture, as odor control agents in animal husbandry, as pH control agents in aquaria, as pet litter, as soil conditioner, and as ion exchangers for the removal of radioactive isotopes from ground water and reactor effluents [20]. For most of these applications, synthetic zeolites are too cost intensive and therefore replaced by the low-priced natural zeolites.

Another important market for synthetic zeolites is their use as acidic catalysts. Cations in synthetic zeolites can be also exchanged by protons. Direct protonation with mineral acids is possible but often results in severe damage of the zeolite framework. Therefore, usually ammonium cations are used for the ion exchange:

$$\mathrm{Z-Na + NH_4^+ \rightarrow Z-NH_4 + Na^+}.$$

The resulting ammonium-exchanged zeolite is subjected to calcination at temperatures of about 450–600°C. During that procedure, the ammonium cations decompose, releasing ammonia and protons. The ammonia is released from the zeolite pores, and the protons remain in the zeolite:

$$\mathrm{Z-NH_4^+ \rightarrow Z-H + NH_3} \uparrow$$

The protons are located within the micropores of the zeolite and loosely bound to the framework of the zeolite. They are mobile within the pores and can be used for the protonation of guest molecules. In that way, zeolites act as proton donators, i.e., they are solid acids. The acidity of zeolites is very high, and acidic zeolite catalysts are indispensible for various processes in oil refinery, petrochemistry [21–24], and/or production of fine chemicals [21, 25, 26]. Even though about 200 zeolite structure types are known, only a few of them are actually used in catalytic processes. The most abundant zeolitic catalysts consist of FAU-type zeolite

(>90%), followed by MFI-, BEA-, MOR-, and MWW-type zeolites. The world-wide production capacity of zeolite catalysts of about 230,000 t is significantly smaller than that of zeolites for detergents, but the highest market value is in catalysis. A major application of zeolite-containing catalysts is in processes, such as fluid catalytic cracking (FCC) for gasoline production, in crude oil refinery. Zeolite catalysts are highly selective since only molecules that are small enough to enter the pore systems of the respective zeolite get transformed and only product molecules that are small enough to diffuse through the pores of the same zeolite are released to the product stream.

Another application of ion-exchanged zeolites is in adsorption and separation technology. The well-defined pore entrances of zeolites make them not only highly selective catalysts, as pointed out above, but also very selective adsorbents. Zeolites specifically adsorb smaller molecules and are therefore often called molecular sieves. Using appropriate zeolite adsorbents, separation of specific components from gas mixtures is possible. Zeolites are therefore used in air separation and the removal of contaminants from gas streams. In liquids, zeolites are used as molecular sieves for smaller molecules, for example, for removal of trace amounts of water from organic solvents. As illustrated above, specific cations occupy defined sites in the pore openings of given zeolites. Since different cations have different cation radii and/or charges, the free diameter of the pores of a zeolite changes if one type of cation is replaced by another one. Thus, a zeolite adsorbent can be adjusted to the sizes of specific molecules. Commercialized molecular sieves are, for example, zeolites Na-A (molecular sieve 4A), K-A (molecular sieve 3A), and Ca-A (molecular sieve 5A) with pore openings of 0.30, 0.38, and 0.43 nm, respectively.

8.3.4 Production of Zeolites for Ion Exchange Applications

Zeolites for ion exchange applications, such as zeolite Na-A, are usually synthesized in stirred tank reactors at temperatures of 80–100°C. For a typical synthesis, two premixed aqueous solutions, one containing a silica source, such as sodium silicate, and the other one containing an alumina source, such as sodium aluminate, are admixed in the reactor. The reaction mixture, containing defined amounts of the components SiO_2, Al_2O_3, Na_2O, and H_2O, is then reacted at 80–100°C. An aging step at lower temperature may precede the crystallization at that temperature. The crystallization time for the formation of the respective zeolite ranges from a couple of hours to some days. Reaction temperature, composition of the reaction gel, as well as the choice of the educts have a strong influence on the type of zeolite formed and the morphology of the individual crystallites. Zeolites are thermodynamically metastable intermediates that are transformed into dense silica and/or silicate phases at prolonged reaction time and/or at higher reaction temperature. Also, transformations of one zeolite phase into another one might occur. It is thus important for the crystallization of a well-defined product to precisely control the gel composition and reaction parameters.

After termination of the reaction, the crystalline zeolite product is filtered and dried at elevated temperature. The obtained zeolite powder can be used directly for some ion exchange processes, such as in laundry detergents, and for others, pellets or extrudates might be advantageous. If so, the zeolite powder can be mixed with water and a binder, such as clays, silica sols, or alumina, and then granulated or alternatively extruded. Calcination at 300–600°C results in mechanically stable granules or extrudates.

8.4 Pillared Clays

The structures of clay minerals consist of layers that are formed by octahedra and tetrahedra as basic building blocks. The tetrahedral are typically $Si(O,OH)_4$ units, and the octahedral, $Me(O,OH)_6$ units with Me = Al^{3+}, Mg^{2+}, Fe^{3+}, or Fe^{2+}. These building blocks form two-dimensional sheets consisting of either one tetrahedra and one octahedra monolayer (1:1 minerals) or one octahedral monolayer between two tetrahedral layers (2:1 minerals) [27]. Substitution of Si^{4+} by Al^{3+} in the tetrahedral layer or Al^{3+} by divalent cations, such as Mg^{2+}, in the octahedral layer results in the formation of negative charges in the sheets, as illustrated in Fig. 8.8. Between the layers, cations are located that compensate the negative charges. These cations can be exchanged by others, which makes these clays natural ion exchange materials.

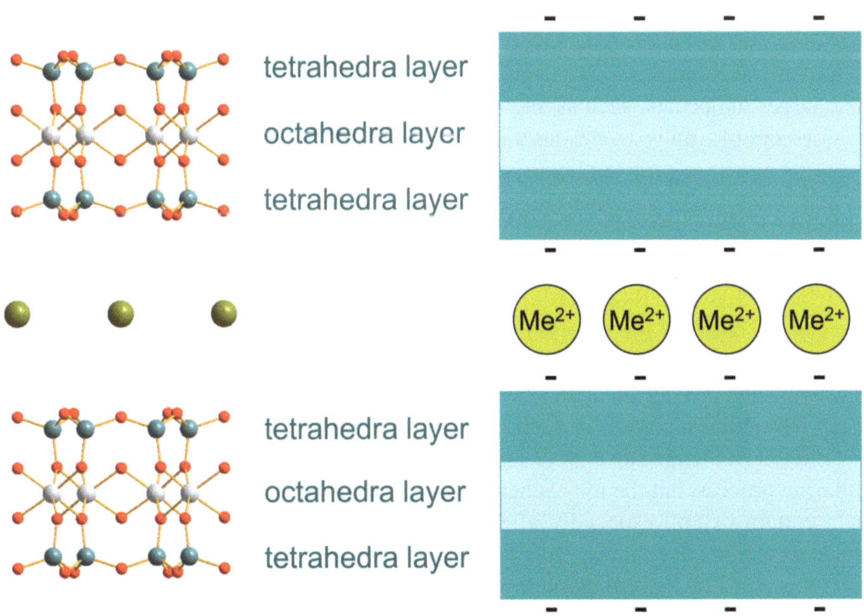

Fig. 8.8 Structure of the clay mineral montmorillonite with composition $(Ca)_{x/2}[(Al_{2-x},Mg_x)Si_4O_{10}(OH)_2]$. The Ca^{2+} cations are located between $[(Al_{2-x},Mg_x)Si_4O_{10}]^{x-}$ layers

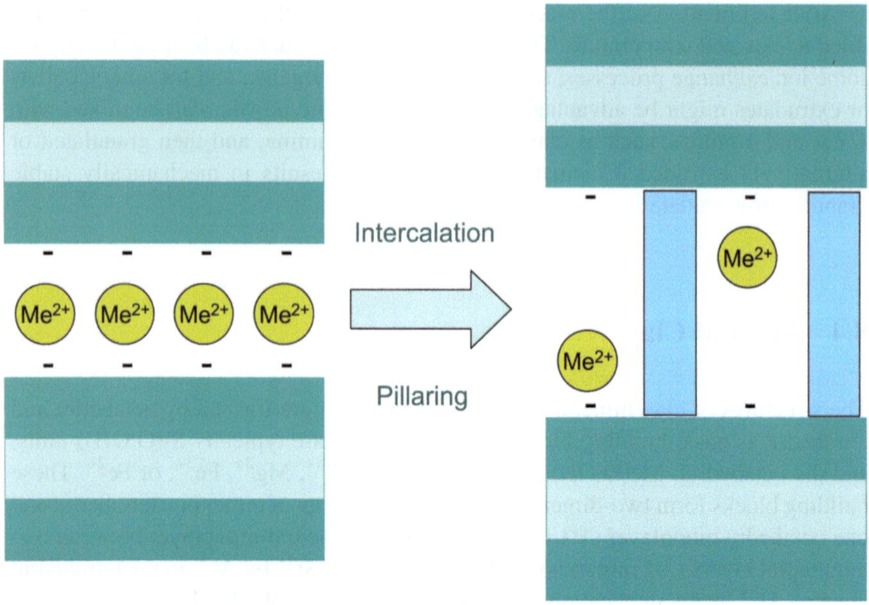

Fig. 8.9 Generation of pillared clays from clay minerals using pillaring agents

Water molecules that coordinate the cations are also located between the layers. The layer to layer distance depends on the size of the extra-layer cations and the amount of water between the layers. A specific features of such clays is their ability to not only accommodate large amounts of water between their layers (swelling) but also larger molecules, such as alkyl ammonium cations, and charged species, such as, for example, polyoxo cations, metal clusters, or metal oxides. Using appropriate pillaring agents to prop the layers, micropore systems with pore dimensions similar as in zeolites can be generated in the clay materials. The products of that process are pillared clay (PILC) materials.

Prominent pillaring agents are Al, Zr, or Ti-based Keggin-type polyoxo cations, such as $[Al_{13}O_4(OH)_{24}(H_2O)_{12}]^{7+}$. The polyoxo cations replace a fraction of the cations between the layers (Fig. 8.9). By this ion exchange, the layer to layer distance becomes larger due to the size of the Keggin cations. The pillars cross-link the layers and provide an open-pore system. Free interlayer distances of up to about 1 nm have been reported for such PILCs [27, 28].

The ion exchange capacity of the clay layers can be significantly reduced by the pillaring process, but the ion exchange capability of the material can be maintained to a certain extent, especially after calcination [29–31]. During calcination, the polyoxo cations decompose into metal oxide pillars and the micropore diameters are reduced [32]. Nevertheless, the open pores allow a fast exchange kinetics and exchange and have been tested for various applications, for example, for the removal of uranyl cations from nuclear waste water. Clays pillared with copper polyoxo cations has been reported to be efficient in removal of UO^{2+} from aqueous solutions with a depollution efficiency of 99.99% [31].

8.5 Hydrotalcite

In brucite, a layered magnesium hydroxide, Mg^{2+}, is octahedrally coordinated by OH^- anions. The $Mg(OH)_2$ octahedra form layers in which all positive charges of the Mg^{2+} cations are balanced by negative charges of OH^- anions. The layers are held together by hydrogen bonds. Similar layers are found in hydrotalcite, as illustrated in Fig. 8.10. Hydrotalcite is a mixed magnesium and aluminum hydroxycarbonate. Substitution of Mg^{2+} by Al^{3+} in the brucite layer creates positive excess charges that are compensated by $CO_3{}^{2-}$ anions. The carbonate anions are located between the $[Mg_6Al_2(OH)_{16}]^{2+}$ layers.

In addition to carbonate anions, water molecules are located in the interlayer. Carbonate in the interlayers can be replaced via ion exchange by other anions. Hydrotalcite with the general composition $Mg_6Al_2(CO_3)(OH)_{16} \cdot 4\ H_2O$ is thus an inorganic anion-exchange material. It can be synthesized from aqueous solutions/slurries containing magnesium hydroxide, carbonates, and aluminum hydroxide under hydrothermal conditions with the precursors being either in solution or finely dispersed solids [33–35]. Hydrotalcites can be also achieved with other anionic species than carbonates in the interlayer as well as with different bi- and trivalent cations, such as Fe^{3+}, Fe^{2+}, Ni^{2+}, Zn^{2+}, Cu^{2+}, etc., in the brucite-like layers [e.g., 36–40].

Anion exchange proceeds readily in hydrotalcites that are thus potent agents for the removal of anionic species. Depending on the size of cations within the interlayer, the basal spacing between the brucite-like sheets may change significantly [39, 40]. Typical anion-exchange isotherms of some hydrotalcite materials are shown in Fig. 8.11 [40]. S_A and H_A have the same meaning, now for anions in solution and in hydrotalcite, as S_M and Z_M have for metal cations in solution and a zeolite, as discussed above.

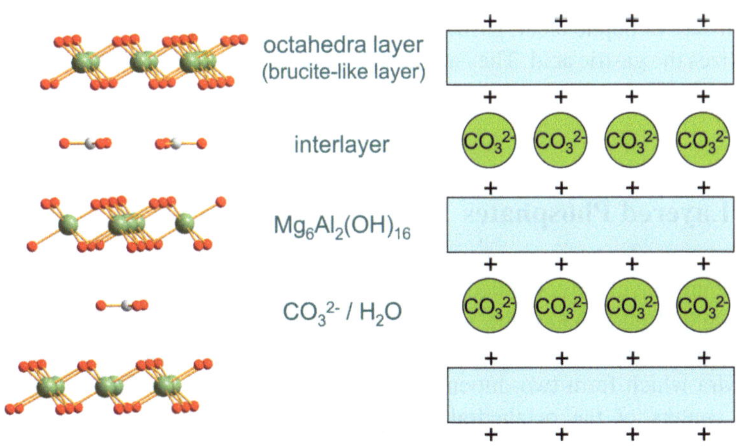

Fig. 8.10 Structure of hydrotalcite of composition $Mg_6Al_2(CO_3)(OH)_{16} \cdot 4\ H_2O$

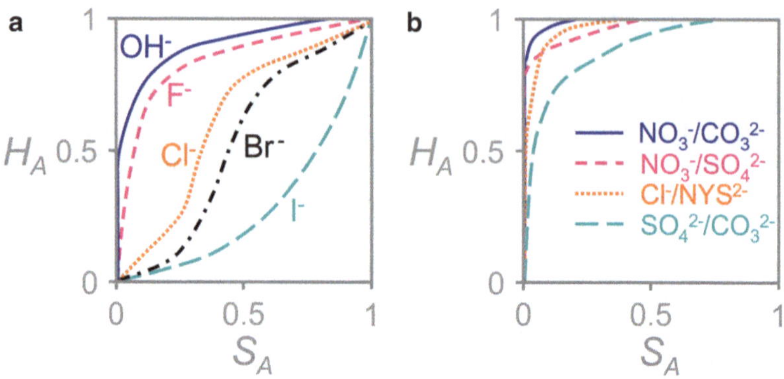

Fig. 8.11 Anion-exchange isotherms for exchange of (**a**) CO_3^{2-} in carbonate-containing hydrotalcite (HT-CO_3) by OH^-, F^-, Cl^-, Br^-, and I^- and (**b**) HT – NO_3, HT – Cl, and HT – SO_4 by either CO_3^{2-}, SO_4^{2-}, or Naphthol Yellow S, an organic anionic molecule (total normalities 0.1, 298 K) (Reproduced from Ref. [40] with kind permission of © Clay Minerals Society (1983))

As can be seen from Fig. 8.11a, the preference of the Mg/Al-containing hydrotalcite for monovalent anions changes as follows: $OH^- > F^- > Cl^- > Br^- > I^-$. For the exchange with bivalent anions as shown in Fig. 8.11b, preferences can be indicated as follows: $CO_3^{2-} > NYS^{2-} > SO_4^{2-}$. There exists a high preference for CO_3^{2-} and OH^- in hydrotalcite. Their selectivity and rather high anion-exchange capacity (AEC) of about 3 meq g^{-1} makes them valuable agents for the removal of anionic pollutants, such as phosphates (HPO_4^{2-}), cyanides (CN^-, $Fe(CN)_4^{2-}$), chromates (CrO_4^{2-}), arsenates (AsO_4^{3-}), or dye molecules, from waste waters [40]. They are also used as additives for the production of halogen-containing polymers in which they act as stabilizers, for example, for neutralization of hydrochloric acid that is generated upon age- or temperature-induced decomposition of polyvinyl chloride [40, 41].

In medical applications, hydrotalcites are widely used as antacid for symptomatic relief in case of peptic ulcer, gastritis, and heartburn. The carbonate in the interlayer neutralizes the gastric acid. They are furthermore under investigation as drug release agents [42–45].

8.6 Layered Phosphates

Another intensively investigated class of materials with ion exchange properties are layered metal phosphates, the most prominent ones being zirconium phosphates, often denoted as ZrP [46]. The structure of α-ZrP consist of octahedra and tetrahedra which form two-dimensional $Zr(HPO_4)_2$ layers, the Zr^{4+} being located in the centers of the octahedral and coordinated by oxygen atoms from PO_4 tetrahedra. One hydrogen atom per phosphate group compensates negative charges. Water molecules are present in the interlayer, and the $Zr(HPO_4)_2$ layers are held together by hydrogen bonds. The structural features of α-ZrP are illustrated in

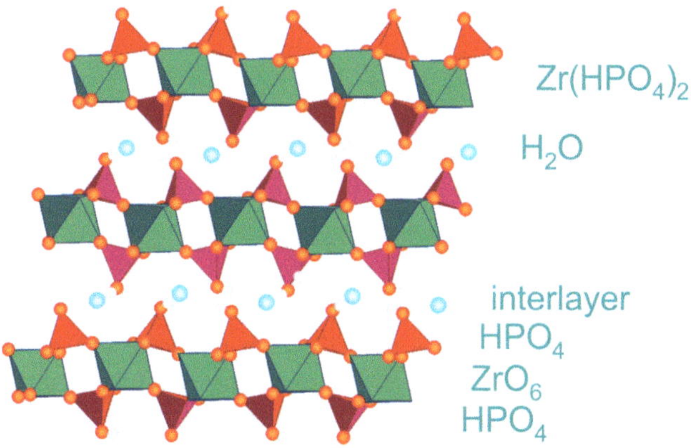

Fig. 8.12 Structure of layered zirconium phosphate (α-ZrP) of composition $Zr(HPO_4)_2\ H_2O$ 47 [47, 48]

Fig. 8.12 [47–49]. Even though well-crystallized materials can be synthesized, ZrP is often used as a material with little structural order and or very small particles. XRD patterns of ZrP often consist of rather broad and ill-defined signals indicating low crystallinity of the products. Nevertheless, the zirconium phosphates possess high ion exchange capacities of up to 6.64 meq g^{-1} [46, 49]. The protons are exchangeable and can be replaced by cations.

Ion exchange proceeds readily for smaller cations, such as Li^+, Na^+, or K^+. The voids between the layers are large enough to accommodate such cations. They migrate into the interlayer and exchange the protons in the ZrP. The exchange proceeds in a stepwise manner, for example, for Na^+ exchange, in a first step a $ZrNaH(PO_4)2$ 5 H_2O phase is formed, which then is transformed into $Zr(NaPO_4)_2$ 3 H_2O with further exchange. Similar observations are made for the two other smaller alkali cations; for K^+, also two phases, and for Li^+, three individual phases have been identified [46, 47, 50]. For the exchange of Rb^+ and Cs^+, the interlayer has to expand, a process that requires additional energy. In exchange experiments, the latter can be provided by bases that neutralize the released protons [47]. The exchange of these cations, however, proceeds only to 75% in alkaline solution. Exchange of larger cations is thus not favored in α-ZrP, but for more hydrated ZrP with larger interlayers, such as, for example, θ-ZrP, exchange of Cs^+, Sr^{2+}, or Ba^{2+} proceeds readily [51]. Layer expansion is thus an option to modify the ion exchange properties of ZrP. There also exist reports on other ZrP phases, such as $Zr_2O_3(HPO_4)$ n H_2O ($n = 0.5, 1.5$) [52], that have specific ion exchange properties.

In addition to surveys on water remediation using ZrP phases [53–55], layered zirconium phosphates have been studied as inorganic proton-conducting component of mixed inorganic/organic membranes for fuel cells. Admixing ZrP to the polymers resulted in membranes with higher stiffness and more stable conductivity at high temperatures [56–58].

It should be noted that high CECs are reported not only for ZrP but also for other layered phosphates, such as $Ti(HPO_4)_2\ H_2O$ (7.76 meq g^{-1}), $Ge(HPO_4)_2\ H_2O$ (7.08 meq g^{-1}), $Sn(HPO_4)_2\ H_2O$ (6.08 meq g^{-1}), $Hf(HPO_4)_2\ H_2O$ (4.17 meq g^{-1}), or $Pb(HPO_4)_2\ H_2O$ (4.79 meq g^{-1}) [46, 59].

8.7 Synthetic Silicates

There exist a number of silicates that have very good ion exchange properties, such as the sodium disilicates that is commercially used for water softening in laundry detergents. It is a layered silicate that is observed in different phases, assigned as α-, β-, γ-, or δ-$Na_2Si_2O_5$ [18, 60, 61]. The exchangeable sodium cations are located between two-dimensional layers of SiO_4 tetrahedra. The major differences of these phases are modifications of the curvatures of layers. The ion exchange properties of these materials do not depend strongly on the phase, but for commercial applications, mainly δ-$Na_2Si_2O_5$ is used. Sodium disilicate can be produced by spray drying of sodium water glass. The resulting amorphous silicate is then transformed into the layered δ-$Na_2Si_2O_5$ by calcination at about 600–800°C. Alternatively, β-$Na_2Si_2O_5$ can be synthesized from sodium water glass via a hydrothermal reaction at 250°C [18, 62, 63].

Amorphous aluminosilicates of the approximate composition $Na_2O\ Al_2O_3 \times SiO_2\ y\ H_2O$ are usually referred to as permutites. They can be obtained either from melts of quartz, kaolin, and sodium/potassium carbonate or from via a gel process from solutions of sodium silicate and sodium aluminate (gel-permutite) [64, 65]. At the beginning of the last century, the terms zeolites and permutites had been often used synonymously, which caused some confusion in the literature. Here, the term permutite will be used only for amorphous silicates. Permutites have been the first commercially available ion exchangers. Due to their low stability toward temperature and acidic environment, permutites tended to show variability of behavior. Even though they were able to remove a variety of cations from solutions [66], they have been widely replaced by other ion exchanges since then [67]. However, they are still in use and considered for applications, such as water remediation [68, 69].

8.8 Others

Ion exchange reactions are quite common on surfaces of polar solids or as bulk reactions at elevated temperature. It would exceed the scope of the present chapter by far to deal with all synthetic materials that show ion exchange properties to a certain extent. There are excellent reviews to be found in literature that deal with those, concededly less commonly used, materials [64, 67, 70–73]. A broad variety of materials are described, ranging from gel particles, hydrated oxide, hydroxides, to heteropoly acids, layered antimonates and arsenates and salt compounds, all with very specific ion exchange properties. However, even though being of scientific interest, most of these materials are of only limited importance for applications on a larger scale. Readers that are interested in these materials are encouraged to refer to the references given above.

8.9 Conclusions

The majority of inorganic ion-exchanger materials consist of crystalline phases that possess a negatively or positively charged framework, the charges of which are compensated by exchangeable cations. The latter are often located either within defined pores or within interlayers between sheet-like structural units. Synthetic inorganic ion exchangers can be synthesized in a wide range of structural and compositional variation which allows a minute tuning of their properties. This allows the synthesis of well-defined materials which makes synthetic inorganic ion exchangers advantageous for many applications. The materials with the widest range of applications are zeolites. The largest production capacities of inorganic ion exchangers will be found in the field of water softening agents, for example, for laundry detergents. However, as pointed out above, there are many other fields in which specific inorganic materials are of importance for certain applications in fields such as water softening, water remediation, catalysis, adsorption, and medicine.

References

1. Sherry HS, Walton HF (1967) Ion-exchange properties of zeolites. 2. Ion-exchange in synthetic Linde 4A. J Phys Chem 71:1457–1465
2. Barrer RM, Munday BM (1971) Cation exchange reactions of sedimentary phillipsite. J Chem Soc A:2904–2909
3. Barrer RM, Munday BM (1971) Cation exchange reactions of zeolite Na-P. J Chem Soc A:2909–2914
4. Breck DW (1974) Ion exchange reactions in zeolites. In: Breck DW (ed) Zeolite molecular sieves. Wiley, New York
5. Schmidt W (2002) Application of microporous materials as ion exchangers. In: Schüth F, Sing KSW, Weitkamp J (eds) Handbook of porous solids. Wiley-VCH, Heidelberg
6. Dyer A (2007) Cation exchange properties of individual zeolites. In: Čejka J, van Bekkum H, Corma A, Schüth F (eds) Introduction to zeolite science and practice, vol 168, 3 Revisedth edn, Studies in surface science and catalysis. Elsevier, Amsterdam
7. Barrer RM, Rees LVC, Ward DJ (1963) Thermochemistry and thermodynamics of ion exchange in crystalline exchange medium. Proc Roy Soc A 273:180–197
8. Rees LVC (1989) Calcium and magnesium exchange in Na-A, Na-X and their precursor gels. In: Karge H, Weitkamp J (eds) Zeolites as catalysts, sorbents and detergent builders, vol 46, Studies in surface science and catalysis. Elsevier, Amsterdam
9. Gal IJ, Jankovic O, Malcic S, Radovanov P, Todorovic M (1971) Ion-exchange equilibria of synthetic 4A zeolite with Ni^{2+}, Co^{2+}, Cd^{2+}, and Zn^{2+} ions. Trans Faraday Soc 67:999–1008
10. Breck DB, Eversole WG, Milton RM, Reed TB, Thomas TL (1956) Crystalline zeolites. 1. The properties of a new synthetic zeolite, Type A. J Am Chem Soc 78:5963–5971
11. Bhat R, Babu GP, Bhat AN (1995) Nickel ion-exchanged zeolite 4A – Reduction behavior of nickel and structure stability. J Chem Soc Faraday Trans 91:3983–3986
12. Weidenthaler C, Schmidt W (2000) Thermal stability and thermal transformations of Co^{2+}- or Ni^{2+}-exchanged zeolites A, X, and Y. Chem Mater 12:3811–3820

13. Weidenthaler C, Zibrowius B, Schimanke J, Mao Y, Mienert B, Bill E, Schmidt W (2005) Oxidation behavior of ferrous cations during ion exchange into zeolites under atmospheric conditions. Microp Mesop Mater 84:302–317
14. Sherry H (1966) Ion-exchange properties of zeolites. I. Univalent ion exchange in synthetic Faujasite. J Phys Chem 70:1158–1168
15. Sherry H (1968) Ion-exchange properties of zeolites. 4. Alkaline earth ion exchange in synthetic zeolites Linde X and Y. J Phys Chem 72:4086–4094
16. Maes A, Cremers A (1975) Ion-exchange of synthetic zeolite X and Y with Co2+, Ni2+, Cu2+, and Zn2+ ions. J Chem Soc Faraday Trans 1 71:265–277
17. Borgstedt E, Sherry H, Slobogin JP (1997) Ion-exchange behavior of zeolite NaA and maximum aluminum zeolite NaP. In: Chon H, Ihm SK, Uh YS (eds) Progress in zeolite and microporous materials, vol 105, Studies in surface science and catalysis. Elsevier, Amsterdam
18. Upadek H, Kottwitz B, Schreck B (1996) Zeolites and novel silicates as raw materials for detergents. Tenside Surfactant Deterg 33:385–392
19. Adams CJ, Araya A, Carr SW, Chapple AP, Franklin KR, Graham P, Minihan AR, Osinga TJ, Stuart JA (1997) Zeolite MAP: the new detergent zeolite. In: Chon H, Ihm SK, Uh YS (eds) Progress in zeolite and microporous materials, vol 105, Studies in surface science and catalysis. Elsevier, Amsterdam
20. Maesen T (2007) The zeolite scene – an overview. In: Čejka J, van Bekkum H, Corma A, Schüth F (eds) Introduction to zeolite science and practice, vol 168, 3 Revisedth edn, Studies in surface science and catalysis. Elsevier, Amsterdam
21. Weitkamp J, Puppe L (1999) Catalysis and zeolites, fundamentals and applications. Springer, Berlin/Heidelberg
22. Scherzer J (1990) Octane-enhancing zeolite FCC catalysis. Marcel Dekker, New York
23. von Ballmoos R, Harris DH, Magee JS (1997) Catalytic cracking. In: Ertl G, Knözinger H, Weitkamp J (eds) Handbook of heterogeneous catalysis. Wiley-VCH, Weinheim
24. Maxwell IE, Stork WJH (1991) Hydrocarbon processing with zeolites. In: van Bekkum H, Flanigen EM, Jansen JC (eds) Introduction to zeolite science and practice, vol 58, Studies in surface science and catalysis. Elsevier, Amsterdam
25. Hölderich WF, van Bekkum H (1991) Zeolites in organic synthesis. In: van Bekkum H, Flanigen EM, Jansen JC (eds) Introduction to zeolite science and practice, vol 58, Studies in surface science and catalysis. Elsevier, Amsterdam
26. van Bekkum H, Kouwenhoven HW (2007) Progress in the use of zeolites in organic synthesis. In: Čejka J, Bekkum H, Corma A, Schüth F (eds) Introduction to zeolite science and practice, vol 168, Studies in surface science and catalysis. Elsevier, Amsterdam
27. Schoonheydt RA (1991) Clays: from two to three dimensions. In: van Bekkum H, Flanigen EM, Jansen JC (eds) Introduction to zeolite science and practice, vol 58, Studies in surface science and catalysis. Elsevier, Amsterdam
28. Pinnavaia TJ, Tzou MS, Landau SD, Raythatha RH (1984) On the pillaring and delamination of smectite clay catalysts by polyoxo cations of aluminium. J Mol Catal 27:195–212
29. Schoonheydt RA, van den Eyden J, Tubbax H, Leeman H, Stuyckens M, Lenotte I, Stone WEE (1993) The Al pillaring of clays. 1. Pillaring with dilute and concentrated Al solutions. Clays Clay Miner 41:598–607
30. Suzuki K, Mori T (1994) US Patent 5,369,069
31. Nistor ID, Miron ND (2007) Depollution of uranyl polluted waters using pillared clays. J Therm Anal Calorim 89:977–981
32. Hutson ND, Hoekstra MJ, Yang RT (1999) Control of microporosity of Al_2O_3-pillared clays: effect of pH, calcination temperature and clay cation exchange capacity. Microp Mesop Mater 28:447–459
33. Cavani F, Tifirò F, Vaccari A (1991) Hydrotalcite-type anionic clays: preparation, properties and applications. Catal Today 11:173–301
34. Misra C, Perrotta AJ (1992) Composition and properties of synthetic hydrotalcites. Clays Clay Miner 40:45–150

35. Eisgruber M, Ladebeck J, Koy J, Schiessling H, Buckl W, Ebert H (2011) US Patent 7,897,136
36. Ookubo A, Ooi K, Hayashi H (1993) Preparation and phosphate ion-exchange properties of a hydrotalcite-like compound. Langmuir 9:1418–1422
37. Miyata S (1975) Synthesis of hydrotalcite-like compounds and their structural and physico-chemical properties. Clays Clay Miner 23:369–375
38. Miyata S, Okada A (1977) Synthesis of hydrotalcite-like compounds and their physicochemical properties. Clays Clay Miner 25:14–18
39. Kapustin AE (1991) Inorganic anion exchangers. Russ Chem Rev 60:1398–1416
40. Miyata S (1983) Anion-exchange properties of hydrotalcite-like compounds. Clays Clay Miner 31:305–311
41. Wypych G (2008) PVC, degradation & stabilization. ChemTex Publishing, Toronto
42. Costantino U, Ambrogi V, Nocchetti M, Perioli L (2008) Hydrotalcite-like compounds: versatile layered hosts of molecular anions with biological activity. Microp Mesop Mater 107:149–160
43. Xu ZP, Lu GQ (2006) Layered double hydroxide nanomaterials as potential cellular drug delivery agents. Pure Appl Chem 78:1771–1779
44. Nalawade P, Aware B, Kadam VJ, Hirlekar RS (2009) Layered double hydroxides: a review. J Sci Ind Res 68:267–272
45. Choy JH, Park M, Oh JM (2008) Gene and drug delivery system with soluble inorganic carriers. In: Shoseyov O, Levy I (eds) Nanobiotechonology: bioinspired devices and materials of the future. Humana Press, Totowa
46. Clearfield A (1984) Inorganic ion exchangers with layered structures. Ann Rev Mater Sci 14:205–229
47. Albertsson J, Oskarsson Å, Tellgren R, Thomas JO (1977) Inorganic ion exchangers. 10. A neutron powder diffraction study of the hydrogen bond geometry in alpha-zirconium bis (monohydrogen orthophosphate) monohydrate. A model for the ion exchange. J Phys Chem 81:1574–1578
48. Čapková P, Beneš L, Melánová K, Schenk H (1998) Structure analysis of intercalated zirconium phosphate using molecular simulation. J Appl Cryst 31:845–850
49. Clearfield A (1995) Inorganic ion exchangers: a technology ripe for development. Ind Eng Chem Res 34:2865–2872
50. Kullberg L, Clearfield A (1981) Mechanism of ion exchange in zirconium phosphates. 32. Thermodynamics of alkali metal ion exchange on crystalline alpha-zirconium phosphate. J Phys Chem 85:1585–1589
51. Alberti G, Costantino U, Gill JS (1976) Crystalline insoluble acid salts of tetravalent metals – XXIII: preparation and main ion exchange properties of highly hydrated zirconium bis monohydrogen orthophosphates. J Inorg Nucl Chem 38:1733–1738
52. Bortun AI, Bortun LN, Clearfield A (1997) A novel layered zirconium phosphate $Zr_2O_3(HPO_4)$. Synthesis and characterization of properties. Solvent Extract Ion Exch 15:305–328
53. Zhang QR, Du W, Pan BC, Pan BJ, Zhang WM, Zhang QJ, Xu ZW, Zhang QX (2008) A comparative study on Pb^{2+}, Zn^{2+} and Cd^{2+} sorption onto zirconium phosphate supported by a cation exchanger. J Hazard Mater 152:469–475
54. Pan B, Zhang Q, Du W, Zhang W, Pan B, Zhang Q, Xu Z, Zhang Q (2007) Selective heavy metals removal from waters by amorphous zirconium phosphate: Behavior and mechanism. Water Res 41:3103–3111
55. Zhang Q, Pan B, Zhang W, Pan B, Lv L, Wang X, Wu J, Tao X (2009) Selective removal of Pb (II), Cd(II), and Zn(II) ions from waters by an inorganic exchanger $Zr(HPO_3S)_2$. J Hazard Mater 170:824–828
56. He R, Li Q, Xiao G, Bjerrum NJ (2003) Proton conductivity of phosphoric acid doped polybenzimidazole and its composites with inorganic proton conductors. J Membr Sci 226:169–184

57. Alberti G, Casciola M, Capitani D, Donnadio A, Narducci R, Pica M, Sganappa M (2007) Novel Nafion-zirconium phosphate nanocomposite membranes with enhanced stability of proton conductivity at medium temperature and high relative humidity. Electrochim Acta 52:8125–8132
58. Truffier-Boutry D, De Geyer A, Guetaz L, Diat O, Gebel G (2007) Structural study of zirconium phosphate-Nafion hybrid membranes for high-temperature proton exchange membrane fuel cell applications. macromolecules 40:8259–8264
59. de Farias RF, Nunes LM, Airoldi C (2000) The ion-exchange capacity of Ti and Zr lamellar hydrogenphosphates – TG measurements. J Therm Anal Calorim 60:517–521
60. Coker WN, Rees LVC (1993) Solubility and water-softening properties of a crystalline layered sodium silicate, SKS-6. J Mater Chem 3:523–529
61. Wilkens J (1995) Structure – property relationship of sodium disilicates. Tenside Surfactant Deterg 32:476–481
62. Just G (1990) Verfahren zur hydrothermalen Herstellung von kristallinem Natriumdisilikat. DE Patent 4,038,388
63. Rieck HP (1987) Use of crystalline layered sodium silicates for softening water and a process for softening water. US Patent 4,664,839
64. Naushad M (2009) Inorganic and composite ion exchange materials and their applications. Ion Exch Lett 2:1–14
65. Derleth H, Hansen M, Lange I, Kurs A (1978) Water softening materials. US Patent 4,098,714
66. Schultze G (1921) Kationenvolumina im Permutit. Mitteilung der Physikalisch-Technischen Reichsanstalt 5:324–330
67. Clearfield A (1988) Role of ion exchange in solid-state chemistry. Chem Rev 88:125–148
68. Evans LR, Miller JE (2004) Sandia report, SAND2004-5461, US Department of Commerce, National Technical Information Service, Springfield
69. Pless JD, Maxwell RS, Philips F, Helean KB, Axness MY, Nenoff TM (2005) Structure-property relationship of permutite-like amorphous silicates, $Na_{x+2y}M_x^{3+}Si_{1-x}O_{2+y}$ (M^{3+} = Al, Mn, Fe, Y), for ion-exchange reactions. Chem Mater 17:5101–5108
70. Yaroslavtsev AB (1997) Ion exchange on inorganic sorbents. Russ Chem Rev 66:579–596
71. Yaroslavtsev AB, Nikonenko VV, Zabolotsky VI (2003) Ion transfer in ion-exchange and membrane materials. Russ Chem Rev 72:393–421
72. Vesselý V, Pekárek V (1972) Synthetic inorganic ion-exchangers – I: hydrous oxides and acidic salts of multivalent metals. Talanta 19:219–262
73. Belinskaya FA, Militsina EA (1980) Inorganic ion-exchange materials based on insoluble antimony(V) compounds. Russ Chem Rev 49:933–952

Chapter 9
Fibrous Ion Exchangers

E.G. Kosandrovich and V.S. Soldatov

Abstract The chapter is a review covering the most important aspects of fibrous ion exchangers: syntheses, physical chemical properties, equilibria and kinetics of sorption processes, and possible and real fields of their applications. Their properties, methods of preparation, and applications are in many ways different from those of conventional ion exchange resins. Advantages and problems connected with fibrous ion exchangers are considered in the chapter. Sorption of different substances from air and applications of fibrous ion exchangers for air purification from substances of different natures (acid, base, neutral substances, water vapors) is a most important field for their practical use. Fibrous catalysts, color-changing sorbents, and hybrid fibrous sorbents impregnated with nanoparticles of inorganic substances are also described.

9.1 Introduction

Each new physical form of ion exchangers requires a specific technology and a niche for their practical application. We know how different are technologies of water treatment with ion exchange beads, membranes, powders, or liquid ion exchange extractants. Fibrous ion exchangers also require specific technologies and apparatuses for their efficient applications. In some cases they can be used in place of other forms of ion exchangers, but there are also specific cases in which only fibrous

E.G. Kosandrovich
Institute of Physical Organic Chemistry of the National Academy of Sciences of Belarus,
13, Surganov street, Minsk 220072, Republic of Belarus

V.S. Soldatov (✉)
Institute of Physical Organic Chemistry of the National Academy of Sciences of Belarus,
13, Surganov street, Minsk 220072, Republic of Belarus

Lublin University of Technology, 40, Nadbystrzycka, 20-618 Lublin, Poland
e-mail: soldatov@ifoch.bas-net.by

Inamuddin and M. Luqman (eds.), *Ion Exchange Technology I: Theory and Materials*, DOI 10.1007/978-94-007-1700-8_9,

ion exchangers can be used. For several reasons considered in this chapter, fibrous ion exchangers cannot be used in conventional industrial ion exchange columns.

Many natural and synthetic fibers have small amounts of ionogenic groups in their structure, improving their dyeability, wetting, or reducing accumulation of an electrostatic charge. In this chapter we consider only synthetic fibers with properties and applications similar to those of ion exchange resins. Fibrous ion exchangers on the base of cellulose and its derivatives, carbonized fiber, and biological fibers will not be considered.

Syntheses of fibrous ion exchangers are described in monographs and review articles [1–16]. The data on their properties and applications are scattered in numerous original publications, many of which are not readily available to Western readers because they are published in Japanese and Russian. The terms *fibrous ion exchangers* and *ion exchange fibers* used in this chapter are identical.

9.2 Common Properties of Fibrous Ion Exchangers: Advantages and Problems

9.2.1 Some Specific Terms and Definitions for Fibrous Materials

Fibrous ion exchangers can be produced in any textile form, but the main forms used in practice are staple fibers, nonwoven canvases, and short-cut fibers (Fig. 9.1).

The staple of chemical fibers is obtained by cutting or breaking the bunch of parallel endless filaments into pieces with a length of 40–70 mm (Fig. 9.1a). They can be used in the sorption application as such or as starting materials for production of nonwoven canvases, threads, felts, etc.

The nonwoven materials under consideration are canvases manufactured from the staple fibers without application of spinning or weaving (Fig. 9.1b). They are characterized by the surface density, measured by the mass per square meter and thickness (mm), measured at standardized conditions.

The short-cut fibers usually have a filament length of 0.25–1.5 mm.

There is also a great variety of other textile forms from which fibrous ion exchangers are obtained. The most important of them are threads, cloth, belts, paper, and some others for example: nonwoven fabric, sew-knit fabric.

The filaments of ion exchange fibers can have different cross-sectional shapes, e.g., circular, rectangular, or more complicated figures. Therefore, characterizing their properties by "diameter" is not always correct. The term *effective diameter* can be used as a more customary one for chemists than the terms used by professionals in the textile field. This quantity usually has a value of 10–40 μ.

An important set of properties of fibrous ion exchangers are their mechanical characteristics, on which their processibility into textile goods strongly depends. The most important of them are force at rupture of a single filament, tensile strength, elongation at rupture, and modulus of elasticity.

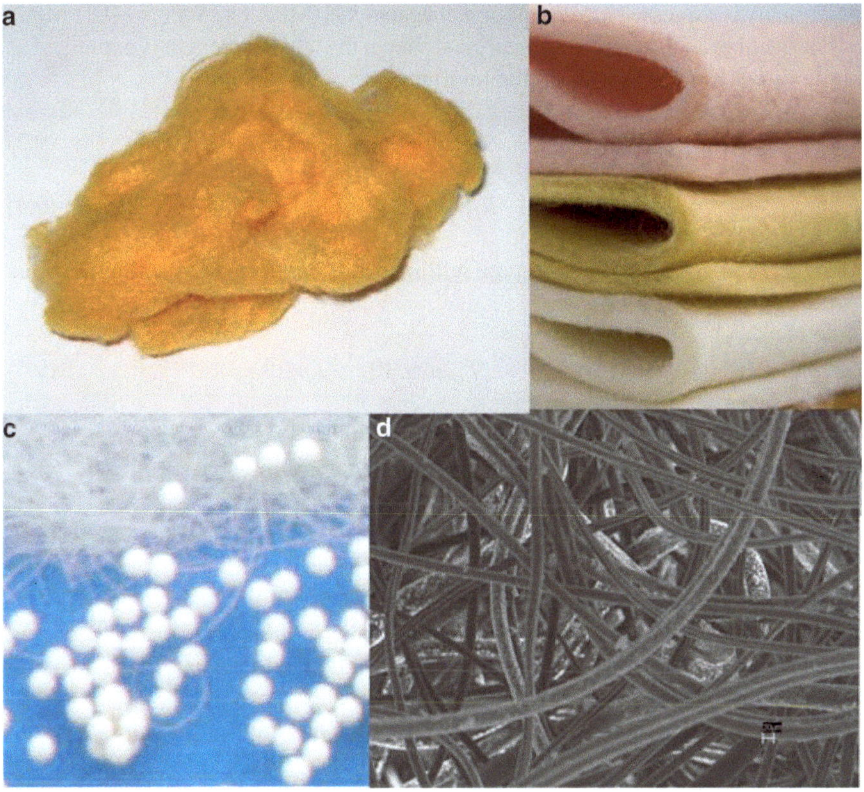

Fig. 9.1 Forms of fibrous ion exchangers: (**a**) Staple fiber of anion exchanger FIBAN A-6. (**b**) Nonwoven fabrics of different FIBAN materials. (**c**) Comparison of granular ion exchanger (bead diameter 0.5 mm) and fibrous ion excanger (fiber diameter 22 μm). (**d**) Microphotograph of staple fiber of ion exchanger FIBAN A-6 with 33% Nitron fiber (magnification = ×600)

For filaments with an arbitrary sectional shape, the thickness is characterized indirectly by the mass of a fiber of a certain length (linear density). *Denier* is an old international unit used to denote the fitness or coarseness of yarn designated by the weight in grams of 9,000 m of the yarn. At present, the most often used unit is named *tex*. It is defined as a weight in grams of 1,000 m of a fiber. Linear density of a fiber expressed in denier is called *titer*. Obviously, titer in denier, T_D, and titer in tex, T_t, relate as

$$T_D = 9 \cdot T_t \tag{9.1}$$

Force at rupture, r, related to a single filament is measured in grams, kilograms, or Newtons.

Tensile strength, R, is a ratio of the force at rupture to the area of cross-section of the fiber expressed in square millimeters. If the fiber section is round:

$$R = 4 \cdot r/(\pi \cdot d^2) \tag{9.2}$$

R usually is expressed in kg/mm^2 or in megapascals (MPa; 1 kg/mm^2 = 9.81 MPa).

For the fiber with an arbitrary sectional shape, the tensile strength is expressed as a ratio of the force at rupture to the titer number:

$$R_t = r/T_t \tag{9.3}$$

$$R_D = r/T_D \tag{9.4}$$

The area of cross-section in square millimeters is connected to the titer numbers by the equation

$$S[\text{mm}^2] = 10^{-3} \cdot T_t/\rho = 10^{-3} \cdot T_D/(9 \cdot \rho), \tag{9.5}$$

if density ρ is expressed in gram per cubic centimeter.

Tensile strength expressed in kilogram per square millimeter is

$$R[\text{kg/mm}^2] = 10^3 \cdot \rho \cdot r/T_t = 10^3 \cdot 9 \cdot \rho \cdot r/T_D, \tag{9.6}$$

where ρ is density in gram per cubic centimeter.

Elongation at rupture is recorded at the moment of rupture of the specimen, often expressed as a percentage of the original length. It corresponds to the breaking or maximum load.

Modulus of elasticity is the ratio of stress, which has units of pressure, to strain, which is dimensionless and can be calculated as

$$\begin{aligned} E &= (tensilestress)/(tensilestrain) = (F/A_0)/(\Delta L/L_0) \\ &= (F \cdot L_0)/(A_0 \cdot \Delta L), \end{aligned} \tag{9.7}$$

where E is the Young's modulus (modulus of elasticity), Pascal; F is the force applied to the object, Newtons; A_0 is the original cross-sectional area through which the force is applied, square meter; ΔL is the amount by which the length of the object changes, meter; and L_0 is the original length of the object, meter.

The approximate limiting values of these quantities sufficient for reprocessing the staple fibers into nonwoven canvases are: R = 7–10 cN/tex.

9.2.2 Advantages and Problems Connected with Fibrous Ion Exchangers

In this section we consider some common properties of fibrous sorbents irrespective of their chemical nature and structure, and originating exclusively from their physical form.

Fibrous ion exchangers would be identical in chemical properties to granular ion exchange resins of the same chemical composition and molecular structure.

However, granular and fibrous ion exchangers are not exactly the same. This makes it necessary to study a full set of properties needed for successful application of each particular ion exchange fiber. Such properties are, e.g., selectivity and kinetics of ion exchange, acid–base strength, structural characteristics, and effect of different destructive factors on the properties of ion exchange fibers.

It is often stated that the main advantage of fibrous ion exchangers compared with their granular analogues is the faster processes of sorption and regeneration ("better kinetics," in the customary slang). That is not true because in many publications comparing the rate of sorption by fibrous and granular exchangers, the fibrous sorbents with effective filament diameters of 20–40 μ are compared with resins that have globular particles with diameters of 300–600 μ. As far as we know the rate of sorption of fibers and beads with equal diameters of the particles has never been studied. The diffusion processes in fibers most often are slower than in the similar ion exchange resins. Moreover, the spherical form of the resin beads is more favorable for faster sorption than the cylindrical form of the filaments with equal diameter. The high rate of sorption on fibrous ion exchangers is caused exclusively by the shortness of the diffusion path from the surface to the middle of the fiber, which overcompensates for the slower internal diffusion.

A common statement, that the faster sorption process is due to the higher surface area of the fibers compared with that of granules of conventional ion exchangers, is also incorrect. The ion exchange fibers are not porous materials and their specific surface in the absence of uncontrolled defects is equal to the geometric area of the fibers with a mass of 1 g. It is easy to calculate that the value for filaments with a diameter of 20 μ is only about 0.16 m^2/g. Nevertheless, the rate of sorption of such a fiber is about a 100 times faster than that of the macroporous resins with a specific surface area of 10^2 m^2/g. It proves that the shortness of the diffusion path, not the surface area, is the real reason for the high rate of sorption.

The real advantage of a fibrous ion exchanger is the possibility of making a packed bed of a sorbent with thin filaments that have a high hydrodynamic permeability combined with the fast sorption similar to that of the microspheric resins. The density of the filtering layer formed by fibrous materials is not directly connected with the filament diameter and can be varied in the range 0.1–0.5 g/cm^3 by compressing it to the desired density in the filtering units or in the layers of textile materials using a wide range of textile processing methods. That allows choosing the sorption layer density according to the admissible pressure loss on the sorption unit.

To understand the real potential possibilities of fibrous ion exchangers in specific applications, one has to consider the properties of their packed beds compared with traditional sorbents used for the same purposes. In any application in which the ion exchanger is used in dynamic conditions (extraction of target substances from liquid or gas, chromatographic separations, catalysis, etc.), the properties of filtering layers that are of primary interest for the user are working capacity of the bed and the pressure loss on the filtering layer, depending on the specific conditions of the process. In turn, they are determined by the exchange capacity of the ion exchanger, the fraction of void volume, the rate of the sorption process, and the external conditions of the dynamic process: the liquid or gas flow velocity and the contact time of the medium and the sorbent.

The volume sorption capacity, E_v, depends on the full capacity of the sorbent related to its mass unit, E, and the density of package of the sorbent in the bed, expressed by the mass of the sorbent in the volume unit of filtering bed, m:

$$E_v = E \cdot m \tag{9.8}$$

The full capacity of a fibrous ion exchanger is always lower than that of its granular analogue. Typical values for the strong base or strong acid fibers are about 2.5–3.0 meq/g, whereas the same values for the same type of granular ion exchange resins are 4.5–5.0 meq/g.

The package density of globular particles in the column is a fixed value (~70% of the volume). The typical volume capacity in the working state is 1.5–2.0 meq/cm^3. The package density of the fibrous ion exchangers in the filtering beds can be varied in the range 0.05–0.5 g of the dry material per cubic centimeter of the filtering bed. A value of 0.25 g/cm^3 can be taken as the one used in common practice. That means that the volume capacity of the filtering beds with fibrous ion exchangers is usually 0.7–0.8 meq/cm^3, which is 2.5 times lower than the value for the granular resins.

Increasing the package density of the fibers in the bed causes an increase in the pressure drop on the filtering layer. An empirical relation between the amount of the dry fiber in the volume unit of the filtering layer, its water uptake, and the flow rate of water was derived in Ref. [17].

It has to be taken into account that loose fibrous filtering layers can be irreversibly compressed by the flow of the eluent, and their compressibility significantly changes with the temperature. So, the pressure drop may not be stable in cyclic processes. Nevertheless, the pressure drop on the filtering layers recommended for practice (thickness <20 cm) never exceeds 20 kPa.

This consideration does not deny the possibility of applying ion exchange fibrous materials in widely used, large-scale water treatment processes. It only proves that special processes and apparatuses accounting for the features of these materials have to be developed. At the same time, fibrous ion exchangers easily can be used in small water treatment devices such as home drinking water purification filters and softeners, in which the thickness of the filtering layer is usually not larger than 20 cm.

9.2.2.1 Kinetics of Sorption

As has already been mentioned, the faster rate of sorption of fibrous ion exchangers compared with "conventional" ion exchange resins (i.e., with particles 300–1,000 μ in diameter) is exclusively due to the shortness of the diffusion path of the sorbate to the middle of the filament. Nevertheless, it does not mean that the rate of the sorption process is determined by the particle diffusion. It will be shown later that, in the case of simple ion exchange of inorganic ions, the rate-determining step of the sorption process is the diffusion of ions through the film of the solution adhered to the particle because the time required for particle transport is shorter than that for the transport across the film. In cases of slower processes such as sorption of organic ions or formation of chelate structures in the sorbent, the process rate is controlled by the rate of transport of the sorbate ions inside the sorbent particle. Intermediate cases probably also occur.

In the literature [18–23], kinetic equations for the sorption processes controlled by film and particle diffusion into particles of spherical shape (resins) and endless cylinders are often used. The most obvious kinetic parameter characterizing the rate of the process is the time of half process, $t_{1/2}$.

In the case of ideal particle diffusion for the spherical particles under a condition of infinite volume solution,

$$t_{1/2} = 0.030 \cdot {r_0}^2/D, \tag{9.9}$$

where r_0 is a radius of the spherical particle and D is the diffusion coefficient in the particle.

The equation for the endless cylinder (the model of fiber) is:

$$t_{1/2} = 0.065 \cdot {r_f}^2/D, \tag{9.10}$$

where r_f is a radius of the fiber.

In the case of film diffusion in the absence of selectivity, the similar dependences are as follows:the sphere:

$$t_{1/2} = 0.231 \cdot \frac{r_0 \cdot \delta \cdot \bar{C}}{D' \cdot C} \tag{9.11}$$

the cylinder:

$$t_{1/2} = 0.345 \cdot \frac{r_f \cdot \delta \cdot \bar{C}}{D' \cdot C} \tag{9.12}$$

In these equations, C and $\bar{C}$ are concentrations of exchanged ion at the interface of film-solution and ion exchanger-film, respectively; δ is a film thickness and D' is the diffusion coefficient in the film.

It follows from Eqs. 9.9–9.12 that if the diffusion coefficients are equal, the process on the fibers should be slower than on the spherical beads.

The shapes of kinetic curves for the film and particle diffusion processes are not so greatly different that they can be used for distinguishing between the two mechanisms without a thorough mathematical analysis. There are simple ways to see whether the process includes a substantial fraction of the film or particle diffusion mechanism: the method of interruption, dependence of the process rate on the intensity of mixing, etc.

It has to be noted that the ion exchange fibers in many cases (probably even most often) do not have a cylindrical shape. The ion exchanger fibers of the Nichibi company [24] have rectangular cross-section with the sides 10–20 and 30–60 μ (the authors name it "small diameter and "large diameter"); FIBAN and VION fibers on the base of polyacrylonitrile (PAN) have a kidney-shaped cross-section [25]; IONEX fibers are a combination of different shapes, and FIBAN fibers on a base of polypropylene (PP) have a cylindrical shape [20]. Kinetic equations can be derived for the fibers of any specific shape [19, 26]. The shortest distance between the periphery and the center of the kidney shaped fiber is much shorter than that for the cylindrical cross-section with the same area. The diffusion path in these cases is shorter and the time of particle transport also is shorter.

Independent of the mechanism, the rate of sorption on the fibrous ion exchangers whose effective diameter is 10–40 μ is always faster by of one or two order of magnitude than that on the granular resins with particles diameter larger than 200 μ used in common practice. Typical half process times ($t_{1/2}$) for simple ion exchange such as Na^+–Ca^{2+}, Cl^-–SO_4^{2-} under conditions of intensive stirring is 2–3 s; for sorption of the cations by chelating ion exchangers $t_{1/2}$ is 20–30 s, whereas these values for the granular resins are 3–5 min and 40–60 min respectively [27–29]. That means that fibrous ion exchangers can be used under conditions of much shorter contact time between the sorbent and solutions, i.e., thinner filtering layers and/or higher flow rate of the mobile phase. It follows from the previous consideration that fibrous ion exchangers can be advantageous compared with the granular ones if they are applied in shallow filtering beds. The experimental confirmation of this issue was obtained in Ref. [30]. In the example presented in Fig. 9.2 the breakthrough curves of Ca^{2+} through the shallow beds (2 cm) of strong acid granular and fibrous ion exchangers are compared.

It is shown that selected conditions are favorable for the fibrous ion exchanger. The efficiency of the filtering layer is approximately 70% if the breakthrough point is selected as $C_{Ca}/C_{0Ca} = 0.05$ (absolute concentration in the effluent is 0.68 meq/l). The immediate breakthrough of Ca^{2+} occurs on the granular resin. In our specific example working capacity of the fibrous ion exchanger was 47 bed volumes (BV), whereas it was zero for the granular resin. Nevertheless the dynamic activity of the resin at $C_{Ca}/C_{0Ca} = 0.5$ is higher for the granular resin because of the higher capacity of the filtering bed. The volumes of the product water are 50 and 100 BV respectively. This example shows that the expediency of using granular or fibrous ion exchanger in the same system depends on the specific requirements of the technological parameters of the process (the process time, pressure drop, frequency of regenerations, etc.) and quality of the final product.

These considerations can be used in designing the contacting apparatuses on fibrous ion exchange materials [31]. It has to be taken into account that their most suitable textile form is nonwoven fabric with a thickness of 2–10 mm and a density of 0.2–1.0 kg/m^2. Accounting for that, the filtering bed can be imagined as several layers of such materials with a total thickness of 20 mm compressed to the surface density of 0.25 kg/m^2. The specific constructions of ion exchange contactors on fibrous ion exchangers can be different: frame types similar to those used in membrane technologies, cylindrical cartridges with radial water flow, moving

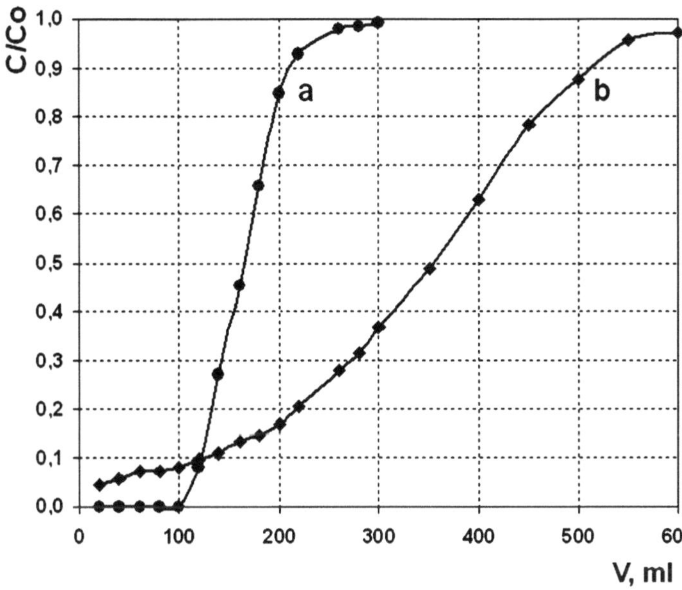

Fig. 9.2 Breakthrough curves of Ca^{2+} for sulfonic-type fibrous FIBAN K-1 (*curve* a, $E_v = 0.83$ meq/cm^3) and granular KU-2 (*curve* b, $E_v = 1.90$ meq/cm^3) ion exchangers. Conditions of the experiment: flow rate = 16.4 cm/min; thickness of the filtering layer = 20 mm; initial concentration of Ca^{2+} (as chloride) = 13.7 meq/dm^3

bands, etc. The first of the mentioned types is the simplest for evaluation of the main parameters of the apparatus and its comparison to the conventional ion exchange column. To be certain, we evaluated parameters of the contactor with an operational volume of 1 m^3.

The conventional column of such volume has the full ion exchange capacity of about 2,000 equivalents. If a rectangular chamber of the same volume is filled with layers of the fibrous ion exchanger 2 cm thick and the distance between them is 1 cm, the number of such layers is 33 and the volume occupied by the fibrous ion exchanger is 660 dm^3. The mass of ion exchanger in the unit is 165 kg and the total capacity is 165×3 eq/kg = 495 ~ 500 equivalents (that is four times less than the capacity of the column with conventional resin). The filtration area of such a unit is 33 m^2 compared with 1 m^2 of the column. The filtration of purifying water goes simultaneously through all layers as shown in Fig. 9.3.

Assume that the purifying water contains 10^{-3} equivalent/l of the component to be removed, the efficiency of the filtering beds is 70%, and the linear flow rate is 20 m/h per square meter, we obtain the following comparative parameters for the contactors on the granules and fibers (Table 9.1).

There are a number of other possibilities for rational use of fibrous ion exchangers in water treatment reviewed in Refs. [32, 33]. Some of them, like using moving bands of ion exchange fabrics, sound quite realistic but they have not yet been developed into real technologies.

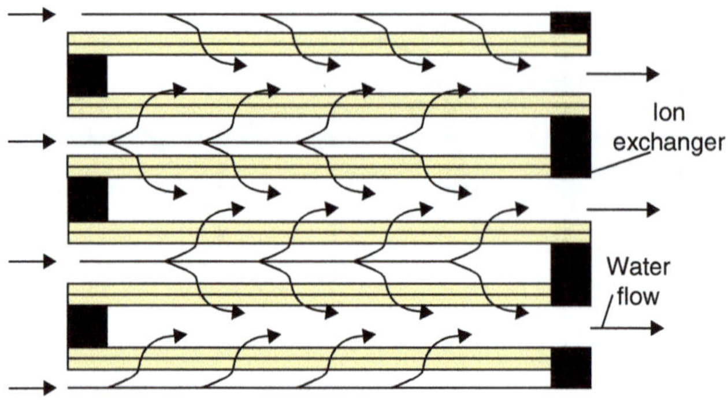

Fig. 9.3 Scheme of water filtration through the absorber

Table 9.1 Comparative parameters for the contactors with granular and fibrous ion exchangers

Parameter	Granules	Fibers
Flow through the absorber, m^3/h	20	660
Purified water volume, m^3	1,400	350
Time before regeneration, h	70	0.53
Number of cycles for treatment of 10,000 m^3 water	5	29
Time for treatment of 10,000 m^3 water, h	350	18.3
Pressure loss on the filtering bed, kPa	20	1

Summarizing potential applications of fibrous ion exchangers in water treatment we conclude that they can be rationally used in the constructions of thin filtering layers used for fast treatment of large volumes of dilute solutions.

This conclusion is also applicable for air purification processes from ionizable impurities. In these cases the rate of sorption is controlled by the particle diffusion, and the length of the diffusion path of the sorbate in the particle plays a decisive role in the kinetics of sorption. Practical large-scale application of granular ion exchangers in this case is problematic (if at all possible) because of slowness of the sorption process and difficulties with regeneration. In spite of numerous research works in this field done during more than 50 years, granular ion exchangers did not find significant application in practice.

In contrast with that, fibrous ion exchange materials have good prospects for massive application in air purification, especially in view of increasing requirements to the quality of breathing and technological air (removal of toxic and badly smelling substances; trace amounts of aggressive substances for the clean rooms of semiconductive, precise machinery, pharmaceutical industries). They are convenient for the air filter physical form, fast sorption rate, low aerodynamic resistance, and a high osmotic stability allowing their plentiful drying and wetting (in the regeneration process) without significant mechanical destruction.

One more problem connected with fibrous ion exchangers should be noted. That is their relatively high cost. On one hand it is caused by the low scale of their production at present. On the other hand the cost of their production seems to be always higher than their granular analogues because it includes more operations than the synthesis of resins. Therefore the most prospective area of their application is air purification and individual means for human's protection from harmful impurities, where application of the granular resins is less convenient and effective. There are also examples of economically justified applications of ion exchange fibers in low-scale water treatment processes.

9.3 Preparation of Fibrous Ion Exchangers

Conventional ion exchangers are not fiber forming materials. Their spatial heavily cross-linked and electrically charged network is not suitable for preparing fibers using traditional wet or melt spinning technologies. Increasing exchange capacity of the fiber always leads to the deterioration of its mechanical properties down to complete destruction of the fiber. Therefore special technologies developed for their preparation always compromise the exchange capacity, water uptake, and mechanical properties of the ion exchange fiber (specifically tensile strength and elasticity). The latter should satisfy severe requirements to the fibers undergone textile processing. A brief review of the existing methods for preparation of fibrous ion exchangers is given in this section.

9.3.1 Preparation of Fibrous Ion Exchangers from Special Polymers

One of the ways for preparing ion exchange fibers is synthesis of special copolymers from monomers containing ionizable groups and nonionizable monomers, and spinning solutions or melts of such copolymers with cross agents. At the following heating the spatial network is formed in the fibers. A number of monomers were used in such syntheses: acrylic and methacrylic acids, maleic anhydride, 5-vinyl pyridine, 5-vinyl-2-methylpyridine [3], sodium methallylsulfonate [34, 35], monomers with quaternary ammonium groups [36, 37], aminoalkylacrylates, N-vinylphthalimide [38], diallylaminoacrylamide monomers [39], and monomers containing carboxylic acid and sulphite groups [40]. Acrylonitrile and (more seldom) styrene were used as fiber-forming inert monomers. Hydrazine [41, 42] or epoxy compounds [43, 44] were most often used as cross agents.

Such methods for preparation of ion exchange fibers are rather expensive and did not find a large application. As we know, the only industrial fiber produced in this way is weak base ion exchanger VION AN-1(Russia) obtained by spinning of copolymer of 5-vinyl-2-methylpyridine and cross-linked with an epoxide oligomer. These fibers have good textile properties and were used for preparing nonwoven canvases.

9.3.2 Composite Fibers

Ion exchange fibers can be prepared by mechanical mixing of inert fiber-forming polymer solutions or melts with finely dispersed ion exchange resins or inorganic ion exchangers with their following spinning into fibers. Fine dispersed particles of granular ion exchangers with the size 3–5 μm [45] in the solution or melt of the fiber forming polymer can be spun to form the fibers containing up to 35% of the ion exchange material. Such technology for obtaining ion exchange fibers is universal and it was described in a number of papers. Another variant of this technique is mixing the fiber-forming polymer with the polymer that can be easily converted to the ion exchanger after formation of the fibers [46–58]. Ion exchange fibers can also be obtained by polymerization, copolymerization, or polycondensation of monomeric carriers of functional groups inside or on the surface of the parent fiber [59–64]. The ion exchange fibers prepared in this way did not have good properties. Their mechanical properties and osmotic stability were too low for textile processing or prolonged use in technological processes.

A method for preparation of ion exchange fibers containing polymeric reinforcement in the volume of the fiber was developed by Yoshioka and Shimamura [20, 65, 66] (Toray Company, Trade name IONEX). The composite parent fiber consists of a mixture of polystyrene and polyethylene (1:5) and strings of polyethylene (16 strings with a diameter of 4.2 μ each per filament). The polystyrene component was cross-linked by formaldehyde and functionalized, forming composite ion exchange fibers of a different type. A great variety of ion exchange fibers have been obtained on the basis of the composite fibers described. They contain sulfonic, quaternary ammonium, primary, secondary, and ternary amino groups; quaternary phosphonium groups; crown ether, and different chelating groups. The structure of these fibers obtained the name "islands–in-the-sea" because their cross-section resembles a map of an archipelago with 16 islands represented by the polyethylene rods. IONEX ion exchangers are produced in forms of short-cut fibers (0.5 mm long). There is no information on the textile materials on their base or of their industrial applications.

9.3.3 Fibrous Ion Exchangers on the Base of Graft Copolymers

Graft copolymerization is a universal technique for syntheses of a great variety of ion exchange fibers. In such fibers the side chains carrying ionogenic groups are grafted on the chain of the main fiber-forming polymer. The latter should be sufficiently chemically inert to withstand active chemicals used in the functionalization of the side chains and further use of the ready ion exchange materials. Therefore, sufficiently chemically inert chemical fibers are used as parent materials for grafting, such as polyolefins, polyethyleneterephtalate, etc. The ionogenic grafts can be formed by monomers containing ionizable groups (e.g., acrylic acid, vinylpyridine, and their derivatives) or by functionalization of nonionogenic grafts, such as side chains formed by polymerization of styrene, glycydyl methacrylate, or acrylonitrile.

The grafting requires generation of free radicals or ions on the main polymer chain. It can be caused by heat, ionizing radiation, or a chemical reaction. There are numerous publications in the literature on all aspects of graft polymerization and properties of graft copolymers (e.g., [67–75]). Both radiation and chemical grafting were used for syntheses of ion exchange fibers.

9.3.3.1 Direct Radiochemical Grafting

In this method the main polymer fibers immersed in the solution of the grafting monomer are placed in the field of penetrating radiation [76–78], usually in the ^{60}Co γ radiation or accelerated electrons. The amount of the grafted polymer increases with the absorbed dose of radiation and depends on the nature and concentration of the monomer, solvent, additives controlling the grafting process (usually initiators or inhibitors), and temperature. The grafted monomer can contain the ionizable group (e.g., vinylpyridine or acrylic acid) or form a polymer chain that can be converted into polyelectrolyte in the following chemical modification (e.g., styrene or glycydyl methacrylate). By its nature the radiation activation of polymerization is not selective. It causes formation of the polymer in all parts of the irradiated system. Therefore, along with grafting, formation of the homopolymer occurs in the solution and on the surface of the fiber. That causes a loss of the monomer and deterioration of the properties of the fiber. Formation of the homopolymer is a serious technological problem because its washing out from the product fiber is a prolonged and expensive process. Formation of the homopolymer of styrene in the process of its grafting into polypropylene fiber was minimized in the process of its grafting in the field of ^{60}Co γ radiation by careful choice of the process parameter. A family of ion exchangers, FIBAN, obtained on the base of polypropylene fiber with grafted polystyrene with divinylbenzene (PP-ST-DVB) is produced at the experimental production plant of the Institute of Physical Organic Chemistry [79] (Minsk, Belarus). Using conventional methods for synthesis of granular ion exchangers on the base styrene–divinylbenzene copolymers, their fibrous analogues have been synthesized. Some reactions used in the syntheses of FIBAN ion exchangers on the polypropylene fiber base are summarized in Fig. 9.4.

The problems connected with formation of homopolymer in the solution can be avoided by carrying out the grafting process from the vapor phase of the monomers. Such a technique was used for grafting of acrylic acid and acrylonitrile [80] into polypropylene fibers. The rate of grafting rapidly increases with temperature and decreasing diameter of the fiber and is strongly inhibited by the residual oxygen in the system. The carboxylic acid fiber obtained by this method had excellent physico-chemical properties. It was produced under the trade name KATIOLAN and used for production of protective cloth. Along with this advantage this technique has serious drawbacks. The process is rather slow (the longevity of the

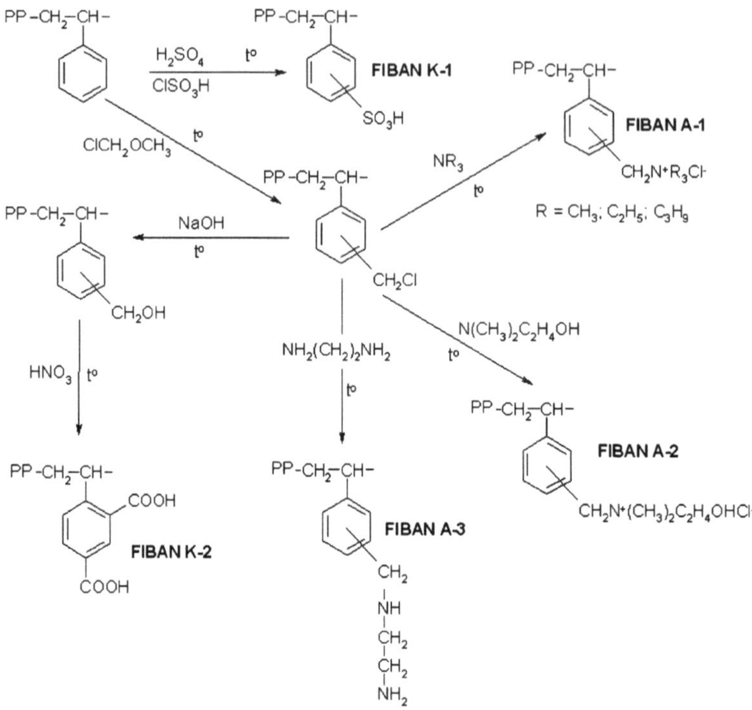

Fig. 9.4 The reactions used in the syntheses of ion exchangers on the PP-ST-DVB fiber base

working cycle is as long as several tens of hours), sensitive to occasional disturbances, and requires complicated equipment working in the radiation field. Polymerization from the gas phase is mainly located in the surface layer of the fiber.

9.3.3.2 Post Irradiation Grafting

Polyolefin fibers irradiated in the air are oxidized by oxygen. In the process of oxidation peroxide and hydroperoxide groups fixed on the polymer chain are accumulated in the volume of the fiber. They are rather stable under ambient conditions and stay in the irradiated polymers for several days. Upon contact with a monomer solution, the grafting can proceed spontaneously or be accelerated by increasing temperature. Grafting of acrylic and methacrylic acids and the other monomers, which can easily form ionizable groups on polyolefin fibers, was described in a number of works [3, 81–88]. This method has been used in production of commercially available fibrous ion exchanger FIBAN K-4, a graft copolymer of polypropylene and acrylic acid.

The homopolymerization is usually suppressed by salts of divalent iron added to the liquid phase. It converts the free radicals formed in the solution into OH^- anions:

$$HO^\bullet + Fe^{2+} \rightarrow OH^- + Fe^{3+} \quad (9.13)$$

The inhibition of graft polymerization proceeds in a much lesser degree because the scavenger (Fe^{2+}) is present in the fiber in much lower concentration than in the solution.

9.3.3.3 Chemical Grafting

Most of the research on chemically initiated grafting was done on polyamide fibers. In this case functional groups able to form fixed radicals or ions have to be introduced first into the structure of the polymer. They are complex-forming groups able to fix ions with variable valences, peroxide or hydroperoxide groups and quaternary ammonium ions. The graft polymerization can be initiated by thermal decomposition of these groups or by redox reactions. In some cases the centers of graft polymerization can be obtained owing to the chain-transfer reaction because of localization of one of the components of the redox system on the main polymer.

The redox systems are formed using Cu^{2+}/Cu^{+} or Fe^{3+}/Fe^{2+} couples or their complexes. A great variety of syntheses of fibrous ion exchangers (mainly weak acid and weak base) utilizing this type of redox systems have been described [89–99]. Stable complexes of these metals in the fibers can be activated by treatment them with oxidizing agents such as hydrogen peroxide.

In Refs. [90, 91] stable copper complexes were incorporated into the fibers from the melts. The subsequent treatment of such fibers with hydrogen peroxide in the presence of vinyl monomers results in graft polymerization of side chains with different functional groups: dimethylaminoethylmethacrylate [92, 93], diethylaminoethylmethacrylate [94], and glycidylmethacrylate [95]. Interesting variants of this method were suggested in Refs. [100–109]. A technology of industrial production of ion exchange fibers by grafting of methacrylic acid (fiber trademark KM-K-1) or dimethylaminoethylmethacrylate (fiber trademark KM-A-1) has been developed. The fibers had a sufficiently high exchange capacity and excellent mechanical properties. Unfortunately we could not find information of the commercial products with such names and of the fate of this development.

It is of note that the chemical stability of ion exchangers on the base of polyamides is not high and they probably can be used as disposable materials.

Another method of initiation of graft polymerization is oxidizing of the main polymer, causing formation in its structure of peroxide and hydroperoxide groups [110–113], followed by their thermal decay in the media containing vinyl monomers (e.g., acrylic or methacrylic acids) with formation of OH• radicals.

Polycaproamide fibers pre-treated with a Cu^{2+}–H_2O_2 solution and placed in solutions of monomers after several hours of storage in air were able to cause graft polymerization at low temperatures (20–30°C) without additional introduction of the metal of variable valence to the solution [100]. Efficiency of grafting approaching 100% was possible at temperatures below or equal to 50°C. The theory of this process is given in reference [101], and the industrial process is described in Refs. [102–105]. In this way, the weak-base ion exchange fiber KM-A1 was produced (polycaproamide with graft dimethylaminoethylmethacrylate). Its exchange capacity was 2.8 meq/g and it had excellent mechanical properties.

Examples of application of redox systems with fixed quaternary ammonium groups in the form of persulfate ion in contact with aqueous solutions of vinyl monomers are given in Refs. [114–117]. In these cases efficient grafting occurs without formation of homopolymers. It is supposed that in this case the polymerization is initiated by $\bullet SO_4^-$ ion-radical joined to the quaternary ammonium group.

9.3.3.4 Polymer Analogues Conversion of Existing Polymeric Fibers

Fibers modified by grafting of polystyrene, poly(glycidyl methacrylate), polyesters of acrylic acid and some other nonionogenic polymers can be converted to ion exchangers by polymer analogues conversion reactions such as sulfonation, chloromethylation with subsequent amination, saponification, etc. These conversions are well described in the literature on the synthesis of conventional ion exchangers and do not need to be repeated in the present paper (see for example review [18]).

Here we consider polymer analogues conversions of some industrial fibers, which led to synthesis of practically important fibrous ion exchangers. The parent fibers should contain some reactive groups that can be converted to ionizable groups without destruction of the fibrous structure. The most important among such fibers are polyacrylonitrile (PAN), modacrylic and polyvynylalcohol (PVA) fibers. A number of commercial products were obtained on the base of these fibers.

There are two important side effects at the syntheses of fibrous ion exchangers by this technique: the fibers lose their mechanical properties with increasing exchange capacity; the swelling of the fibers sharply increases with the capacity increase. The conditions of synthesis should be chosen so that the product fibers would have a sufficiently high ion exchange capacity, the water uptake is suitable for ion exchange technological processes, and mechanical properties maintained, allowing their reprocessing in the textile goods. That means that the conversion of the functional groups present in the parent fiber is usually incomplete. In some cases the reactions of conversion are accompanied by cross-linkage of the fiber-forming polymer, preventing its dissolution in the reaction mixtures, or exceedingly high swelling. In another cases additional cross linkage of the fiber is necessary.

The PAN fibers. The most important reactions leading to obtaining fibrous ion exchangers are alkaline hydrolysis of nitrile groups into carboxylic acid groups; their aminolysis by aqueous solutions of polyamines with formation of amido-amino groups, and amination in nonaqueous solutions or pure polyamines with formation of imidazoline structures. A special and very important reaction is cross-linkage of the PAN fibers by hydrazine [118–121] and polyamines accompanied by accumulation of weak base groups. Another important reaction is formation of amidoxime and hydroxamic acid groups in the fibers treated with hydroxylamine [122–124].

The reactions used in the syntheses of ion exchangers on the PAN fiber base are summarized in Fig. 9.5.

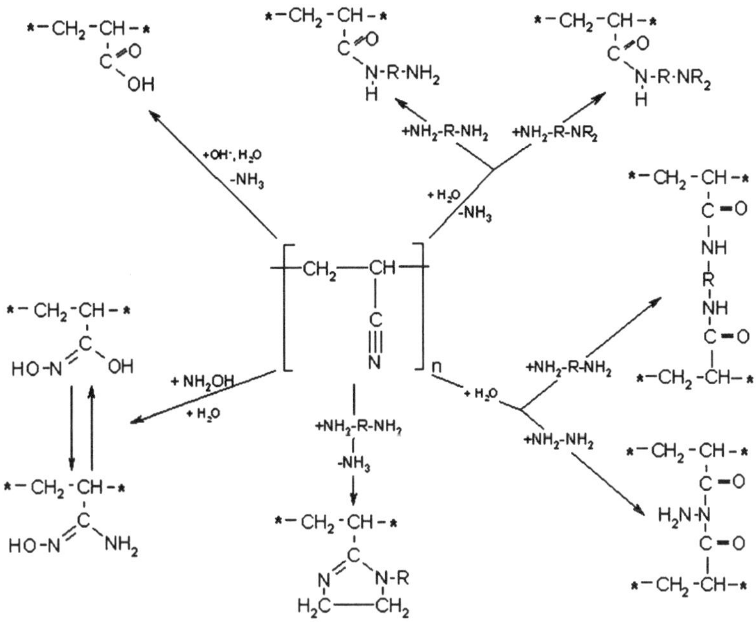

Fig. 9.5 The reactions used in the syntheses of ion exchangers on the PAN fiber base

It has to be taken into account that reaction of hydrolysis proceeds in parallel with some other reactions. The final products of these reactions always contain carboxylic acid groups. The ratio of carboxylic and weak base groups can be varied in a wide range and, to some extent, can be controlled by the synthesis conditions. The weak base anion exchangers with ternary amino groups can be quaternized with different alkylating agents forming strong base groups [125].

Different types of commercially available fibrous exchangers have been synthesized on the base PAN fibers having the family names VION (Moscow, Russia) and FIBAN (Minsk, Belarus).

Polyvinylalcohol fibers have been used for preparation of several practically important fibrous ion exchangers. Reactions of conversion of the hydroxyl groups were used in syntheses of ion exchange fibers [126–137] but they did not find a significant application in practice. Most often, thermal treatment of PVA was used as the first stage of its chemical modification. It leads to cross-linkage and formation of polyvinylene structures (Fig. 9.6). The dehydrated polymer has a high reactivity and can be used for preparation of a large variety of ion exchange fibers using Diels-Alder reaction with vinyl derivatives containing carboxylic acid groups. Brominated polyenic fibers treated with ammonia, amines, or pyridine can be used for introduction into the fibers anion exchange groups [132]. The reactions used in the syntheses of ion exchangers on the PVA fiber base are presented in Fig. 9.6.

The Nichibi Company (Japan) has developed technology for synthesis of sulfonic acid ion exchange fibers by direct sulfonation of thermally dehydrated PVA. The trademark of this product is "Nichibi IEF-SC3050." Reactions of diene

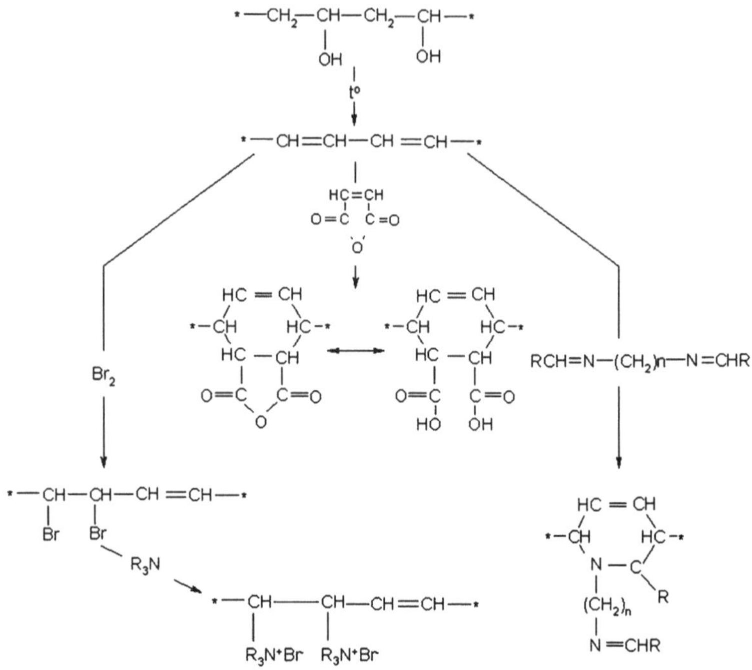

Fig. 9.6 The reactions used in the syntheses of ion exchangers on the PVA fiber base

condensation between the polyene part of the dehydrated PVA and azomethyne derivatives can also be a universal method for preparation of PVA anion exchange fibers [2].

Polyvinylchloride fibers after dehydrochloration also form polyvinylene structures, and the same reactions as those described above can be used for their functionalization. Emission of HCl in these syntheses is a serious technological drawback of these processes.

Reactions of fibrous copolymers of vinylchloride and acrylonitile with polyamines were used by Egava et al. [138] for preparation of ion exchange fibers.

9.4 Physical, Chemical, and Sorption Properties

9.4.1 Strong Acid Fibrous Ion Exchangers

Properties and applications of strong acid fibrous ion exchangers are predetermined by the nature of their predominant functional groups and are similar to those for sulfonic type ion exchange resins. Nevertheless, as described in the review of methods for their preparation, they have specific features.

Toray fibers (IONEX) [139] have sulfonic groups in polystyrene cross-linked with methylene bridges and mechanically mixed with polyolefin.

Ebara fibers [140] contain linear sulfonated polystyrene grafted on the polypropylene chains. The ion exchange component is evenly spread in the volume of polypropylene fiber.

FIBAN fibers [141] contain domains of sulfonated copolymer of styrene and divinylbenzene graft on the polypropylene chains.

Nichibi fibers [142–144] contain sulfonic groups in the structure of partially carbonized fibers of polyvinylalcohol.

In many applications, such as water deionizing or softening, the difference in structure of these materials is not very important. In the other cases, such as catalysis, chromatographic separation of ions, sorption of organic substances, or bacteria fixation, the structure of the sorbent can be of great importance.

Exchange capacity of different fibers usually is in the range of 2–3 meq/g (Table 9.7).

Ion exchanger FIBAN K-1 is strictly monofunctional, with a capacity of 2.9–3.3 meq/g. As the computer analysis of its potentiometric titration shows [145] the amount of weak acid groups is less than 0.15 meq/g, i.e., the same as in the styrene–DVB resins. The Nichibi material contains a significant number of weak acid groups: its salt splitting capacity is 2.4 meq/g and the total capacity is 3.1 meq/g. The IONEX fiber has strong acid capacity of 2.6 meq/g; the total capacity is not given in the company information. The Ebara fibers (similar to FIBAN) are monofunctional by the method of their synthesis. Their capacity can be easily controlled by the amount of grafted polystyrene. The capacity of about 3 meq/g is probably optimal. Its further increase using the irradiation-induced processes is possible, but the mechanical and osmotic properties of the ion exchange fibers sharply deteriorate.

In our experiments we observed that ion exchangers obtained from the graft copolymers of styrene without cross-agent (DVB) lose their capacity in the sorption-regeneration cycles in the process of their prolong exploitation. Therefore, in production of FIBAN K-1 addition of 0.5–2% of DVB is used for stabilizing its capacity in cyclic work.

Chemical and thermal stability of all sulfonic fibrous ion exchangers is similar to that of their bead analogues.

Chemical, thermal, and osmotic stability of FIBAN PP-ST-DVB- based ion exchange fibers have been described in refs. [146–150]. Stability of these fibers in aggressive media was similar to that for the conventional ion exchange resins on ST-DVB matrix. The main properties of the ion exchange fibers FIBAN and effect of different destructive factors on them is illustrated by Table 9.2.

The basic properties of strong acid ion exchange fibers of different producers are summarized as follow: *Toray (TIN-100)* – E = 3.0 meq/g, working temperature (max) = 120°C, diameter of the fiber = 40 μ, fiber length = 500 μ; *Nichibi (IEF-SC3050)* – E = 2.4–3.1 meq/g, density = 1.3 g/cm^3, water uptake = 45%, large diameter of fiber = 30 μ, small diameter of fiber = 10 μ, fiber length = 500 μ. Separate data on mechanical characteristics of the other ion exchange fibers of sulfonic type are similar to those given in the Table 9.2 and can be found in the cited references.

Table 9.2 Effect of different destructive factors on the properties of polypropylene with grafted polystyrene with divinilbenzene (PP-ST-DVB)-based ion exchangers

FIBAN type	Initial property	0.5 h 5 M NaOH 100°C	5 h 5 M NaOH 100°C	0.5 h 5 M HCl 100°C	5 h 5 M HCl 100°C	5 h 5 M HNO_3 100°C	48 h 10% H_2O_2 20°C
Exchange capacity E, meq/g							
K-1, H^+	3.63	3.69	3.37	3.58	3.58	3.43	3.90
A-1, Cl^-	3.22	3.17	2.87	3.17	3.10	2.11	3.22
Water uptake W, g H_2O/meq							
K-1, H^+	0.340	0.344	0.283	0.322	0.293	0.288	0.349
A-1, Cl^-	0.229	0.221	0.221	0.227	0.203	0.299	0.216
Force at rupture, r (percent of initial)							
K-1, H^+	8.1, 100%	102	102	105	98	77	107
A-1, Cl^-	12.2,100%	93	72	93	73	44	89
Tensile strength, R Kg/mm^2 (percent of initial)							
K-1, H^+	6.7, 100%	99	99	99	97	94	102
A-1, Cl^-	8.7, 100%	99	94	97	90	59	98
Elongation at rupture, L% (percent of initial)							
K-1, H^+	11, 100%	157	52	166	43	28	138
A-1, Cl^-	30, 100%	93	77	87	50	40	93
Elasticity modulus, F, Kg/mm^2 (percent of initial)							
K-1, H^+	282, 100%	82	94	84	95	107	82
A-1, Cl^-	248, 100%	85	98	106	104	95	95

All these ion exchangers are (or can be) produced in form of different textile goods (nonwoven fabrics, felt, threads, cloths etc.) and as standard staple fibers or short cut fibers (0.25–1 mm long). It seems that the latter is the main form of Toray and Nichibi fibers, whereas FIBAN is mainly produced as staple fiber and nonwoven canvases with the width up to 140 cm and thickness 2–15 mm (0.2–1.5 kg/m^2).

Observation of the graft matrix fiber and ion exchange filaments in the scanning electron and optical transmission microscope [151] gave compatible results evidencing axial macroscopic nonregularities in their structure. Starting from $\Delta P = 120\%$ the filaments have clearly observed corrugate-like shape with the period of thickening 3–5 μ (Fig. 9.7).

The nature of these structures is not known. It seems most probable that we see the appearance of internal cracks or zones of mechanical destruction of the fibers caused by filament shrinkage and the strength evolved with increasing the filament volume in the process of grafting.

The supermolecular structure of this type of fibrous ion exchangers was studied by X-ray analysis in wide and low angles and IR spectroscopy [152, 153]. The difractograms were used to compute the mean size of the crystallites in the direction perpendicular to the main molecular chains and mean orientation angle of the crystals relative to the direction of the fibers stretching. Data on the low angle X-ray scattering were used to determine the value of the large period of the polypropylene (PP) and the dimensions of the scattering structures in parallel and perpendicular direction to the axis of stretching.

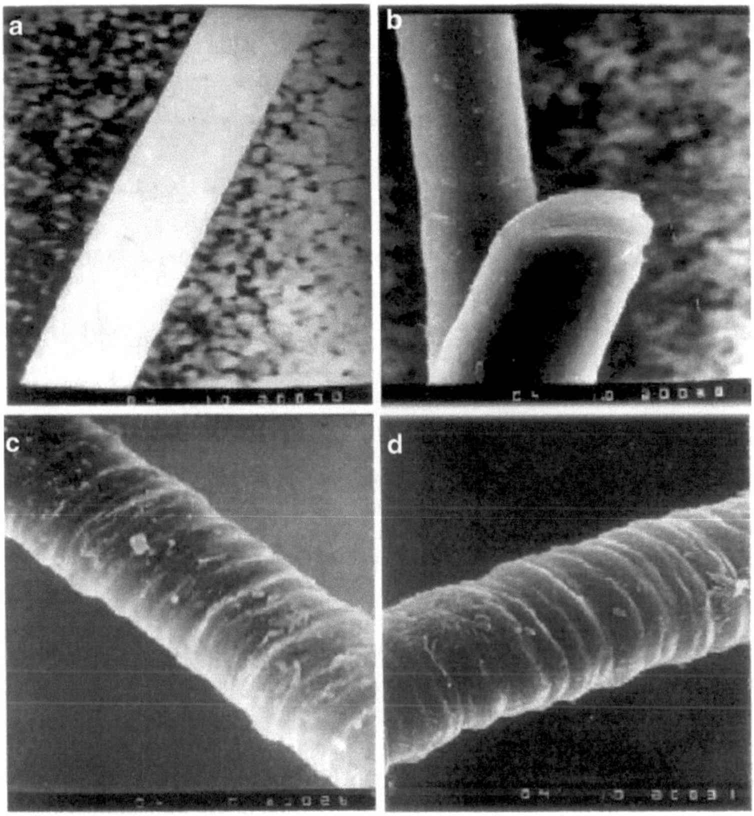

Fig. 9.7 Electron microscopic photographs (X1000) of the PP-ST-DVB fibers with 2% DVB and different ΔP values (%). (**a**) 0. (**b**) 75. (**c**) 100. (**d**) 151

The X-ray analysis has shown that the initial PP fibers have crystalline monoclynic oriented structure.

The most important simple deduction from the X-ray studies is that ion exchange fibers of the considered type contain regularly distributed domains of PP crystallites and functionalized ST-DVB copolymer with the size 10–14 nm (Fig. 9.8).

Investigation of ion exchange equilibrium for H^+–K^+ [148], Na^+–Ca^{2+} [28], Cl^-–NO_3^-, Cl^-–SO_4^{2-}, NO_3^-–SO_4^{2-} [29] and K^+–Cs^+ [154] have shown that selectivity of sorption on the fibrous ion exchanger and its dependence on the degree of exchange is the same as that on the granular ion exchanger with the same water uptake.

Kinetics of ion exchange on sulfonic acid fibers were not deeply studied. In the cited works, mostly comparisons of the fibers with "conventional" resins with unknown bead sizes are presented to demonstrate advantages of the fibers. There is no convincing data either on the mechanism of sorption (interparticle diffusion or film diffusion) or on the kinetic parameters (diffusion coefficient, thickness of the diffusion film etc.). In reference [20] the kinetics of ion exchange for Na^+-Ca^{2+}

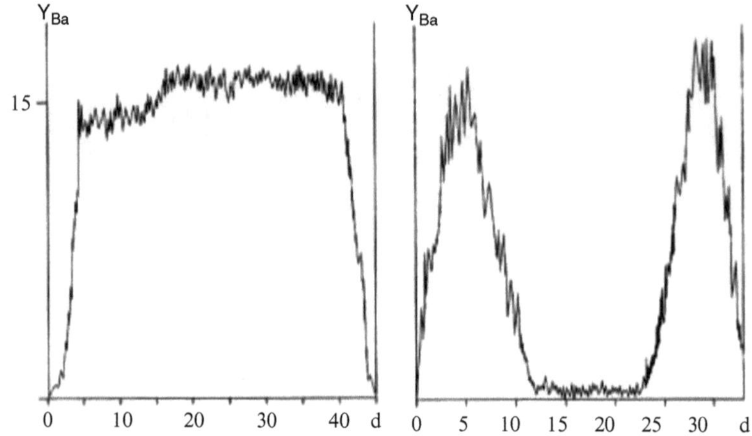

Fig. 9.8 Electron probe microanalysis diagram, resolution 1.5 μ. Y_{Ba} – concentration of Ba^{2+} in conventional units along the diameter of the fiber; d – diameter of the fiber. Cameca instruments (France). Sulfonic type PP-ST –2% DVB bases ion exchangers. $\Delta P = 120\%$. Left figure: E = 3.1 meq/g; Ba^{2+} form; fiber diameter 45 μ. Right figure: E = 2.2 meq/g; Ba^{2+} form; fiber diameter 35 μ

was studied. It appeared that the process is controlled by the film diffusion at a concentration of the solution less than 0.1 mol/l, whereas at the higher concentration the particle diffusion controls the process rate. In publication [23] extensive studies of gadolinium ion (Gd^{3+}) from acid solution were performed. The ion exchanger was sulfonated polyvinyl alcohol fiber with ion exchange shell 4.6 μ. The author presents the particle diffusion coefficient of the order 10^{-8} cm^2/s, i.e., the same as that for the granular ion exchangers. On the other hand the experimental results of this work clearly show that under conditions of the experiment the rate of the process was controlled by the film diffusion because it was strongly dependent on the stirring rate and the solution concentration.

Having excellent kinetic characteristics and uniform thickness of the filaments, ion exchange fibers appeared to be prospective chromatographic material both for preparative and analytical applications. In Ref. [6] it was shown that sulfonic ion exchanger FIBAN K-1 in different textile forms in the process of separation of K^+ and Cs^+ is approximately equivalent to Dowex 50X2 resin with the size of particles 100–200 mesh. Probably lower volume capacity and a large void volume fraction in the filtering bed of the column filled with the fibrous ion exchanger did not allow revealing better kinetic properties of the fibrous packed bed. At the same time, separation of different pairs of amino acids (threonine-valine, alanine-leucine, lysine-histidine) was much more efficient on the sulfonic ion exchanger FIBAN K-1 fiber than on Dowex 50X2 (Fig. 9.9).

High efficiency of Nichibi Company short cut fibrous ion exchangers has been demonstrated in experiments on separation of mixtures of lysine-histidine-arginine and ions of Gd-Sm.

Water treatment is a traditional field of application of strong acid ion exchangers, often in combination with anion exchangers. Extensive research in

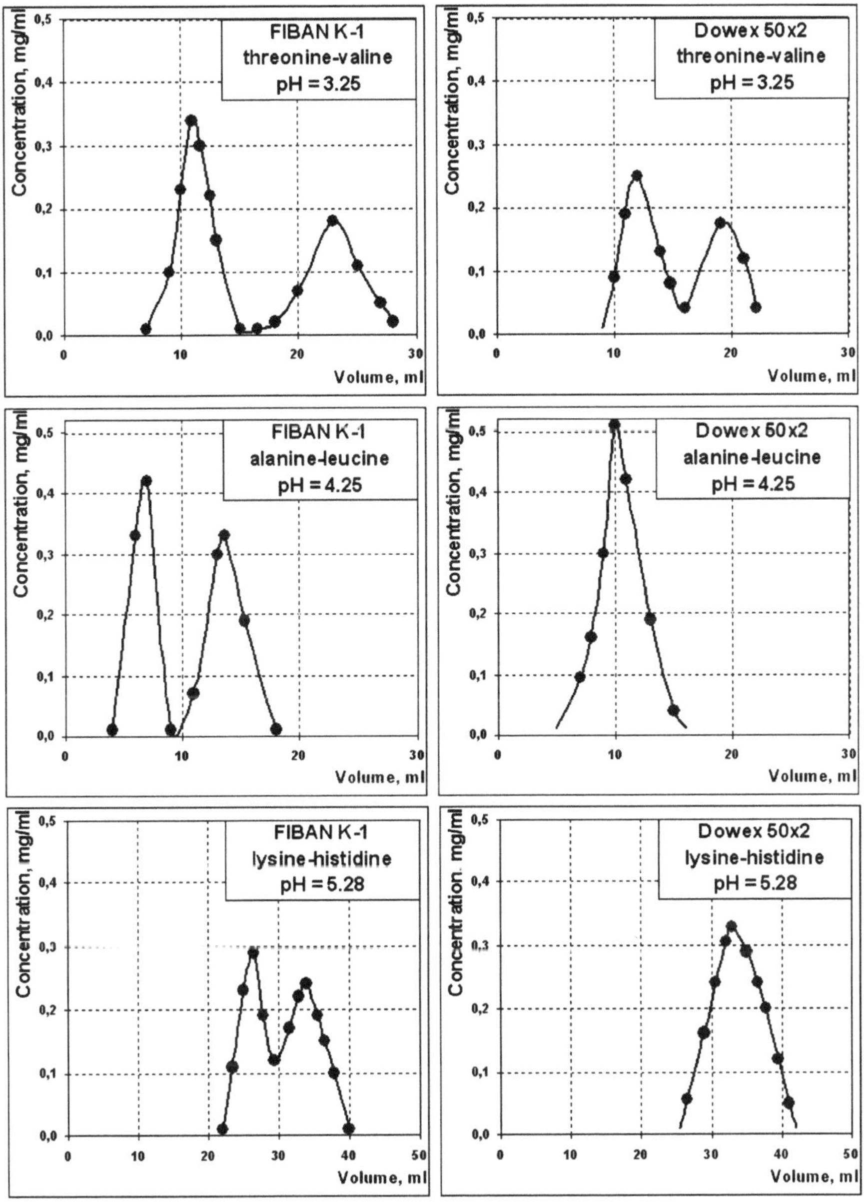

Fig. 9.9 Separation of amino acid mixture on fibrous FIBAN K-1 and granular Dowex 50X2 sulfonic types in exchangers (eluent: citrate buffer). Column parameters: d = 0.8 cm; h = 18 cm; C = 0.92 meq/cm^3; W = 0.5 ml/min; L = 1.74 meq/eq [6] (Reproduced from Ref. [6] with kind permission of © Elsevier (1988))

this area was done by all research teams developing fibrous ion exchangers. Here we briefly review actual and potential applications of these materials in different processes of water treatment.

The filtering layers of fibrous ion exchangers combine two properties oppositely affecting its efficiency in dynamic processes in comparison to granular resins. A higher rate of ion exchange is a factor favoring high performance of fibrous ion exchangers in dynamic processes of water treatment such as softening. A lower volume capacity and large void volume of the filtering layer deteriorate the process. Studies on the effect of package density, thickness of the filtering layer, and flow rate on the performance of Ca^{2+} removal from the water in column conditions allowed outlining the rational conditions for application of fibrous ion exchangers in water treatment. The largest density of the filtering layer in practice is about 0.27 g/cm^3, which corresponds to ion exchange capacity of 0.83 meq/cm^3 for ion exchanger FIBAN K-1 (the full capacity is 3.10 meq/g). The experiments have shown that the filtering layer with a thickness of 7 mm already allows water softening with 34% efficiency of the filtration layer at a linear high flow rate (16.4 cm/min). This corresponds to the solution residence time in the filtering layer 2.56 s. An acceptable column efficiency of 70% occurs when the thickness of the bed is 30 mm. Such characteristics are unachievable for the granular analogue.

Toray ion exchangers found practical application in purification of low radioactivity waters of atomic power stations as an addition of short cut fiber IONEX to the powder of ion exchangers circulating in cooling water. Addition of the fibers sharply increases permeability of the precoat ion exchange filters and increases their exploitation period several times.

The fibers of the Nichibi Company were tested for purification of radioactive wastes in combination with anion exchange resins. The fibers and resin beads form fluffy flakes leading to formation of filtering layers with low hydrodynamic resistance, which allows increasing the operation time of the purifying systems.

The Toray Company has developed several apparatuses and technologies for preparation of high purity water using a combinations of reverse osmosis and ion exchange on IONEX fibers. The apparatuses of small and medium size for production of ultrapure water have become commercial products (Toraypure LV-10T) [24]. Good perspectives have electrodeionizers with fibrous strong-acid and strong base ion exchange fibers in the deionizing chamber. Nonwoven fibrous ion exchange materials are extremely convenient for this purpose. The information of the construction of such devices is scarce. It is only known that such apparatuses are produced and their efficiency is higher than those of granular resins.

Ion exchange paper has been developed on the base of Toray and Nichibi fibers. It found application for analytical purposes.

All developers and producers of ion exchange fibers note their applicability to air purification from impurities of basic nature (ammonia and amines). This is the field of application where fibrous materials have great advantages over the ion exchange resins and it will be discussed in detail in Sect. 9.5 of this paper.

IONEX sulfonic ion exchangers reveal a high performance in sorption of organic substances and bacteria and have good prospects in separation of biologically active substances from cultural liquids and other biological media.

A potential special use of fibrous cation exchangers is in solid acid catalysts. This question will be considered in Sect. 9.7 for all fibrous catalysts on the base of ion exchangers.

9.4.2 Strong Base Fibrous Ion Exchangers

In contrast with strong acid fibers, represented by only one type of functional group and two types of matrix polymer (polystyrene and polyvinylalcohol), there are many strong base fibers differing by the type of the functional group and the polymeric matrix.

Strong base ion exchangers with different functional groups on the polystyrene-containing matrix (IONEX, FIBAN A-1), polyninyl alcohol (NICHIBI IEF-SA), and polyacrylamide (FIBAN A-6–A-10) have become commercial products.

Ion exchange fibers on the base of polystyrene containing copolymers are similar in properties to granular ion exchangers of the same type. Their syntheses are following the same procedure as those for the bead ion exchangers that is chloromethylation of the polystyrene containing fiber followed by quaternization, most often with trimethylamine. Such fibers have become commercial products IONEX TIN-200 and FIBAN A-1.

The IONEX anion exchanger revealed excellent chromatographic efficiency in separation of niobium and tantalum anions – an extremely difficult mixture for separation. Its efficiency in sorption of caramel pigment from sucrose solutions was higher than that of the granular resin in spite of the higher exchange capacity of the latter.

The main properties and its stability to different destructive factors are given in Table 9.2. Their selectivity to simple inorganic ions is similar to that of the bead styrene divinylbenzene resins with identical swelling [147]. The selectivity of ion exchange of inorganic ions of the fiber with 2% DVB and 104% degree of grafting was practically the same as that for Dowex 1X2 resin. Extensive studies of chloride-sulfate exchange on FIBAN A-1 and some other fibrous anion exchangers [21] showed convincingly that the rate of the process is controlled by the film diffusion at the studied ionic concentration (up to 1.8 equiv/l). The process is extremely fast and the half process time ranges within 3–10 s depending on the stirring rate and the solution concentration.

Cases of true particle diffusion control of the process rate were observed in the experiments on sorption of Na^+ salts organic sulfonic acids on Cl^- forms of strong base fibrous ion exchanger with different percentage of DVB in comparison with granular resins Dowex 1 [155]. Sorption rates of benzene, naphthalene, and benzpenicilline sulfonic acids have been studied. The effective diameters of these anions were respectively 0.85, 1.0, and 1.5 nm. The rate of these processes was lower than that in the cases of sorption of inorganic anions, what allowed more precise evaluation of the kinetic parameters. It appeared that the diffusion coefficient in the fibers having the same swelling in water was the same as that in the resins if the ion exchanger contained less than 10 water molecules per sulfonic group. At higher swelling the diffusion coefficient was higher for the resins.

Strong base FIBAN fibers with groups triethyl-, tri-n-propyl, and tri-n-buthylbenzyl ammonium revealed extremely high selectivity to nitrate ions compared with sulfate and chloride and can be used as alternative to nitrate selective anion exchange resins with much better sorption kinetics [29].

The polyfunctional anion exchangers with quaternary ammonium groups will be considered in the following section of the paper.

9.4.3 *Weak Acid Fibrous Ion Exchangers*

9.4.3.1 Carboxylic Acid Ion Exchangers on Polypropylene Base

Many authors described syntheses and properties of fibrous ion exchangers obtained by radiochemical grafting of carboxyl containing monomers into polyolefin fibers, but only few of these products found practical applications.

Ion exchanger FIBAN K-4 produced at the experimental plant of the Institute of Physical Organic Chemistry (Minsk, Belarus) is a graft copolymer of acrylic acid to industrial polypropylene fiber. Its synthesis is described in the previous section and, in more detail, in Ref. [156]. It has an exchange capacity of about 5 meq/g, good textile properties, and low hydro and aero-dynamic resistance in filtering layers. Its drawbacks are relatively short shelf life (about 2 years) and steady destruction at prolonged work in aqueous media, especially at elevated temperatures, with loss of exchange capacity and mechanical properties. Nevertheless, it can be used as a disposable material for air purification from basic impurities and water purification from some heavy metals, specifically led and copper. Its instability is caused by destruction of polypropylene matrix in the course of its γ irradiation in the synthesis process and hydrolysis of the ether bonds between the polypropylene trunk and the polyacrylic acid chain.

In an important work [157] carboxylic acid ion exchangers FIBAN K-4 and granular resin Purolite C-104 (the bead diameter 500–1,200 μ) have been compared in the process of water softening and regeneration with the pressurized aqueous solution of carbon dioxide. In experiments carried out under identical conditions, it was established that although the total Ca^{2+} dynamic capacity was greater for the resin, the Ca^{2+} breakthrough from the fiber column was much sharper and, as a result, the working capacity of FIBAN K-4 was about twice that for the resin. The authors proved that the rate of sorption of Ca^{2+} by the fiber is controlled by the film diffusion, compared with the beads wherein the particle diffusion is predominant. This is similar to the results obtained at comparison of strong acid resin and fiber [30] described in Sect. 9.4.1. Especially interesting are results on regeneration of the ion exchangers by CO_2 pressurized aqueous solutions. The degree of regeneration of the fiber from Ca^{2+} reached about 90% at the pressure 6.8 atm, whereas

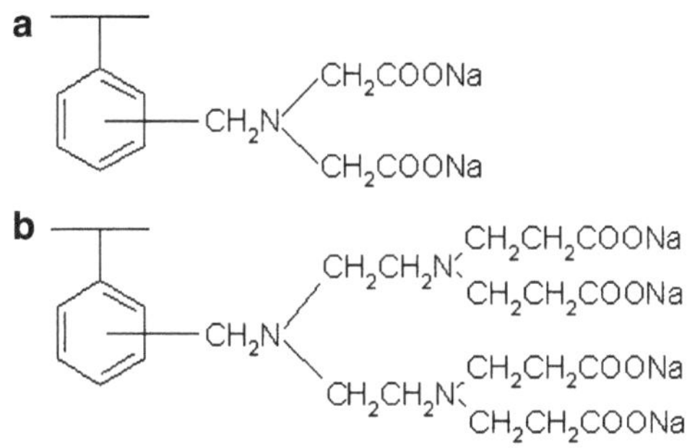

Fig. 9.10 Structure of chelating functional groups of fibrous ion exchangers IONEX: a – functional groups of iminodiacetic acid; b – functional groups of imidodipropionic acid

this value for the resin was less than 10%. The authors explained this striking difference by kinetic hindrances at regeneration of the beads with CO_2 caused by formation of a low swelling shell of the H^+ form.

Work on improvement of chemical stability of FIBAN K-4 is in progress [158, 159]. We do not have information on the other ion exchangers of that type produced as commercial products.

In Ref. [27] some properties of chelation fibers on the base of the "island-in the sea" matrix have been described. The fibers with the following functional groups were obtained (Fig. 9.10).

These fibers revealed a high selectivity to Cu (II) and Mn (II). The sorption processes in this case appeared not as fast as that in the case of simple ion exchange; the half time of the process was 1–7 min, depending on the water uptake by the fibers. This allowed obtaining reliable data whose careful analysis confirmed the particle diffusion mechanism of sorption. The analysis of the kinetic curves indicates that the process is controlled by particle diffusion with $D = 0.33–1.78 \cdot 10^{-9}$ cm^2/s for the fibers and swelling increasing from 1.0 to 2.6 g H_2O/g fiber. Comparison of the fibrous ion exchangers with their granular analogue Diaion CR-10 showed that the half process time is 10–50 times shorter for the fibers than for the resin.

9.4.3.2 Amino-Carboxylic Acid Ion Exchangers on the Base of PAN Fibers

Ion exchange fibers of this family are separated in a special group because their classification into cation or anion exchangers is conventional. They are polyampholytes containing amino and carboxylic groups in different proportions. Conventionally they are named cation or anion exchangers in the literature, depending on the predominant functional groups. They deserve great attention because some of them found significant practical applications and are produced in ton amounts.

Syntheses of ion exchangers on the PAN base are complicated processes including simultaneous reactions of several types. In order to avoid dissolution or exceedingly high swelling of the final products, preliminary cross linkage of the PAN fibers has to be done.

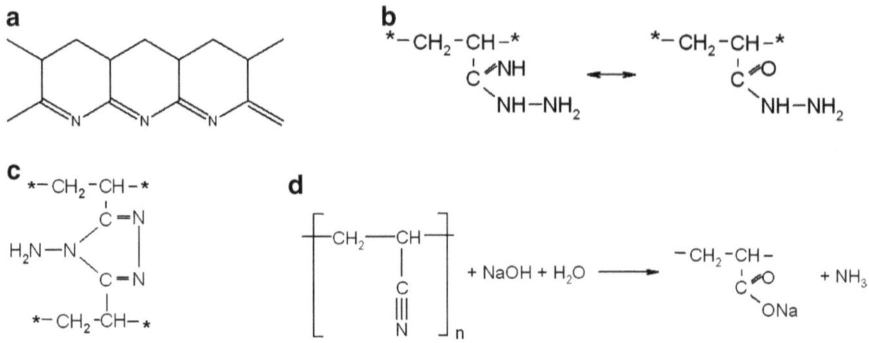

Fig. 9.11 Most probable structures of hydrazinated PAN fiber: a – nanapthyridine; b – hydrazide; c – triazole derivatives; d – carboxylic acid groups

Table 9.3 Properties of carboxylic acid fibrous cation exchangers on the base of PAN

Fiber	E_{COOH}, meq/g	E_{amine}, meq/g	Water uptake, g H_2O/g ion exchanger: H_2O	NaOH 5%	H_2SO_4 5%	Linear density, tex	Strength sN/tex	Elongation, %
Fiban K-5	4.0–8.0	0.5–2.0	0.4–0.5	1.3–1.9	0.4–0.5	0.5–1.4	5–10	15–25
Vion KN-1	5.0–7.0	0.2–1.0	0.4–0.5	1.5–2.0	0.4–0.5	0.2–1.0	7–10	20–30

The most efficient cross linking agent for PAN fibers is hydrazine. It reacts with polyacrylonitrile, forming insoluble polymer owing to formation of interlinking units and different nitrogen containing groups having anion exchanging properties. The structure of the final product can be realized in many ways [118]. The most probable structures in the final product are nanapthyridine (Fig. 9.11a) (can appear at heating of polyacrylonitrile polymers to temperatures above 100°C), hydrazide (Fig. 9.11b), triazole derivatives (Fig. 9.11c), and carboxylic acid groups formed at hydrolysis of nitrile groups (Fig. 9.11d).

4-N-amino-1,2,4-triazole units (Fig. 9.11c) are the most probable interchain bridges in the resulting polymer. The triazole nucleuses are very stable to hydrolysis by mineral acids.

Hydrolysis in alkali and acid media is used for introduction of carboxylic acid groups into the fiber. The process proceeds through a number of stages and ends up with formation of carboxylic acid groups and ammonia. Treatment of PAN fibers with hydrazine has been used in preparation of carboxylic acid cation exchangers by the following or simultaneous saponification of the nitrile groups with alkali [41, 118, 160–170].

Fibrous ion exchangers of this type are produced under trademarks FIBAN K-5 and VION KN-1. Their main properties are presented in Table 9.3.

These ion exchangers have found significant practical application as components of the cartridges for drinking water purification in small-size kitchen filters, light chemical respirators [171], and air purification from ammonia (after additional treatment) [14, 172].

Reactions of modacrylic and polyacrylonitrile fibers with polyamines were used for syntheses of fibrous anion exchangers:

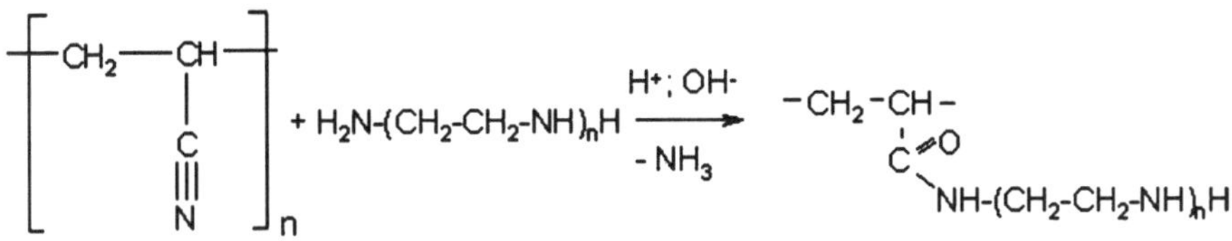

(9.14)

This reaction has been used by Trochimczuk and Kolarz for preparation of granular ion exchangers from copolymers of acrylonitrile and divinylbenzene [173, 174]. The mechanism of this type of reaction was studied by Dragan et al. [175, 176].

In the absence of water amination of PAN leads to formation of imidazoline groups:

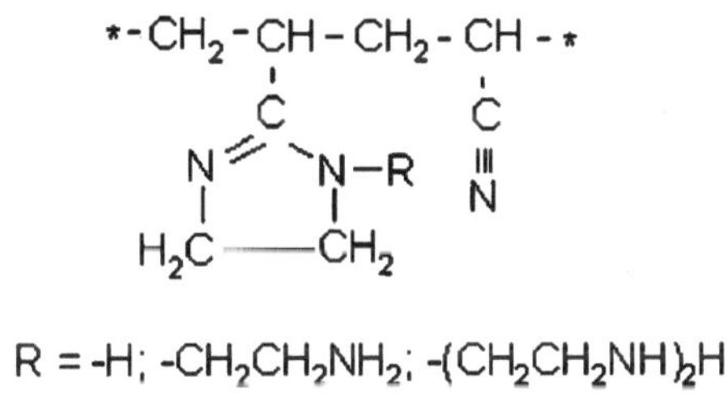

They steadily hydrolyze in aqueous (especially acid) media with opening the circle and of formation amidoamine groups [177]. These ion exchangers reveal excellent chromatographic properties in separation of the ions of complex forming metals. In its turn the materials with residual nitrile groups in aqueous media, especially in alkalis, hydrolyze accumulating carboxylic acid groups. It was observed [178–180] that in ion exchanger FIBAN AK-22 with initial anionic and cationic capacities of 6.3 and 0.5 meq/g, respectively, after keeping during 90 days at ambient temperature in 1.0 N solution of Na_2CO_3, 0.2, 0.5, and 1.0 N solutions of NaOH changed their capacities to the following values, respectively: *1.0 N* Na_2CO_3 – $E_{amine} = 5.6$ meq/g, $E_{COOH} = 1.0$ meq/g; *0.2 N NaOH* – $E_{amine} = 6.1$ meq/g, $E_{COOH} = 2.5$ meq/g; *0.5 N NaOH* – $E_{amine} = 6.1$ meq/g, $E_{COOH} = 3.1$ meq/g; *1.0 N NaOH* – $E_{amine} = 6.0$ meq/g, $E_{COOH} = 3.4$ meq/g. Its mechanical properties remained sufficient.

Table 9.4 Properties of ion exchange fibers FIBAN AK-22

				Water uptake, g H_2O/g ion exchanger	
Fiber	Anion exchange group	E_{COOH}, meq/g	E_{amine}, meq/g	Cl^- form	Free amine form
AK–22(1)	$-(CH_2)_2NH_2$	0.4	2.7	0.52	0.72
AK–22(2)	$-[(CH_2)_2NH]_2H$	0.5	6.3	0.85	0.92
AK–22(3)	$-[(CH_2)_2NH]_3H$	1.1	7.0	0.68	0.80

The functionalized polymers are insoluble, which proves that in parallel with the main reactions the reaction of cross-linkage proceeds:

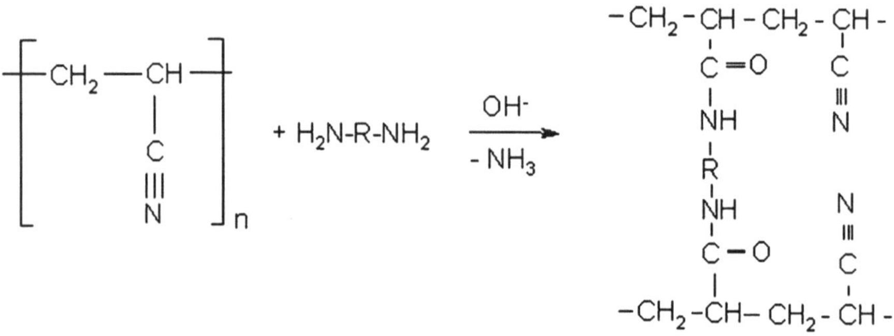

(9.15)

This reaction was used instead of hydrazination in synthesis of carboxylic acid ion exchanger FIBAN K-3.

A family of ion exchangers FIBAN AK-22(n) was developed and found practical applications. The n in parenthesis denotes the number of amino groups in the structure of the predominant functional group. Ethylenediamine (EDA), diethylenetriamine (DETA), and triethylenetetramine (TETA) have been used to obtain fibrous ion exchangers FIBAN AK-22 (1)–(3). Their main properties are summarized in Table 9.4.

These ion exchangers reveal a high selectivity to complex-forming cations increasing with the number of amino groups in the radical of the functional group. They showed promising results in extraction of color and noble metals from technological solutions and waste waters of hydrometallurgical and metal finishing production (Fig. 9.12).

Ion exchangers FIBAN AK-22 are selective to oxy-anions of heavy metals such as chromate, molybdate, and tungstenate. The sorption of these anions passes through maxima at pH = 3–5, reaching values of 300–800 mg of the metal per gram of sorbent. They also selectively and reversibly absorb chloro complexes of platinum and palladium [181].

By variation of conditions of the syntheses a number of aminocarboxylic polyampholytes with controlled total amounts and ratio of the amino and carboxylic acid groups were obtained. By alkylation of fibers FIBAN AK-22(n) with

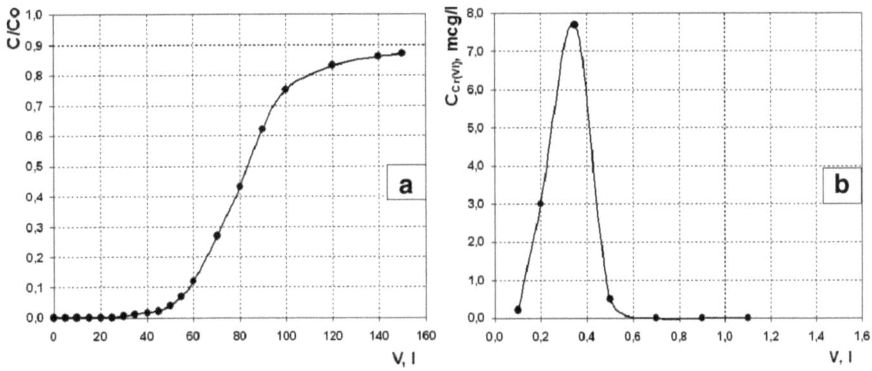

Fig. 9.12 Sorption curve (**a**) of Cr(VI) by amino carboxylic ion exchanger FIBAN AK-22 from hydrogen chloride solution (thickness of filtering layer – 7 mm; mass of ion exchanger = 20.7 g; pH = 2.5; velocity of water filtration = 0.093 m/h; initial concentration of Cr(VI) = 20 mg/dm^3), and desorption curve (**b**) of Cr(VI) (regeneration solution – 1 N NaOH; rate of regenerant filtration = 2.5 ml/min)

mono-chloroacetic acid chelating fibers, FIBAN X-(1)–(3) have been obtained [182, 183]:

$$\begin{array}{c} R - CO - NH - (CH_2CH_2NH)_nH + ClCH_2COOH \rightarrow \\ R - CO - NH - (CH_2CH_2N(COOH))_{n-1})_nCH\ (COOH)_2 \end{array} \tag{9.16}$$

Because of its high selectivity to the ions of heavy metal ions and a high rate of sorption, FIBAN X-1 found application in fine water purification from these ions in drinking water home filters. Its comparison with the granular analogue and the other carboxylic acid fibrous ion exchangers showed its superiority [30, 184]. The usage of granular resins under conditions of the free flow filters was inefficient. Ion exchangers FIBAN K-4 and K-5 have good rates of sorption but their selectivity was not sufficient for efficient use of these fibers in removal of Co, Ni, and Zn.

FIBAN X-1 found application in analytical chemistry for pre-concentration of the heavy metal ions present in concentrations below that required for their accurate and reliable determination from the aqueous media containing Ca^{2+} and the other usual cations in a large excess [185–188]. The pre-concentration process is very fast – practically all ions (Cu, Pb, Cd, Ni, and Mn) are absorbed by 1 g of the fiber by passing 1 l of the solution through the miniature column with this fiber during 10 min (for comparison: this process would take at least 100 min with conventional chelating ion exchange resin). This method is extremely convenient for collecting probes under field conditions. A complete extraction of the absorbed ions from the fiber required not more than 10 ml of nitric acid. The lowest selectivity of FIBAN X-1 was to Mn^{2+}. The selectivity to this ion appeared much higher for ion exchange FIBAN X-2 [182]. FIBAN X-1 passed all strict requirements for officially accepted method analysis and was included into the list of official recommended Methods of Determination of Concentrations of the heavy metal ions in drinking water [189, 190]

A fibrous ion exchanger with ternary amino groups FIBAN A-5 was obtained by reaction of the PAN fiber with dimethylaminopropylamine (DMAPA) (as well as its

Table 9.5 Properties of some strong base fibrous ion exchangers FIBAN

FIBAN type	Alkylation agent of ternary ammonium group	$E_{\text{strong base}}$, meq/g OH^- form	pK_b	Water uptake, g H_2O/g OH^- form
A-6	$(C_2H_3O)CH_2Cl$	2.4	2.2	0.8
A-7	$ClCH_2CH_2Cl$	2.5	2.3	2.1
A-8	CH_3I	2.6	3.0	0.5
A-9	$C_6H_5CH_2Cl$	2.6	1.5	1.6

analogue with diethylaminopropylamine (DEAPA)). Since these amines are not able to form cross-linking units, preliminary cross-linking agents or those in the process of aminations were used.

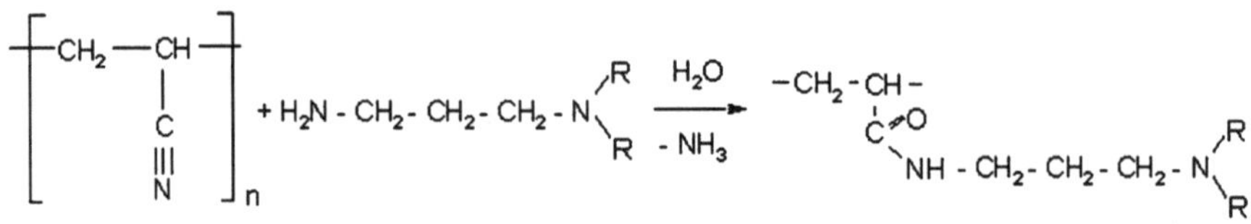

$R = CH_3$ or C_2H_5

(9.17)

As it will be shown in Sect. 9.5, ion exchanger FIBAN A-5 reveals a high efficiency in removal of acid gases from the air. The ion exchangers of this type also appeared to be good parent materials for preparation of composite fibers containing immobilized amorphous iron (III) oxide, having a good sorption ability to arsenic anions [25, 191–195].

A family of strong base fibrous ion exchangers was synthesized by quaternization of this fiber with different quaternizing agents. These fibers also contain residual ternary amino groups (about 30% of their capacity). The largest application has found ion exchanger FIBAN A-6, the product of interaction of the FIBAN A-5 with epichlorohydrine. It is used in fine air purification from SO_2 [196]. It also revealed good sorption ability to viruses and bacteria as well as natural colorizing agents present in the surface waters [197]. The main properties of ion exchangers of this group are presented in Table 9.5.

Except for quaternary ammonium groups, these ion exchangers contain weak base groups (1.0–1.5 meq/g) and 0.3–0.8 meq/g of carboxylic acid groups. These fibers are described in more detail in Ref. [198]. It is seen that the basic strength of these ion exchangers is lower than that for the ion exchanger with benzyltrimethylammonium fixed anion and strongly depends on the structure of N-alkyl radicals.

9.4.4 *Acid–Base Properties of Ion Exchangers FIBAN*

Acid base properties of most of the fibrous ion exchangers were not quantitatively characterized with the exception of FIBAN fibers. Potentiometric titration curves, needed for calculating the acidity parameters of ion exchangers, in the latter case were obtained in a wide range of pH and, in some cases, at different concentration of the supporting electrolyte (KCl or NaCl). They were treated using the approach and computer program developed in refs. [199–203]. This approach allows analyzing the titration curves of polyfunctional ion exchangers with unresolved inflection points. The following quantities characterizing the ion exchanger can be obtained by the computer analysis of the titration curves: the number of types of the functional groups ($i = 1, 2,..n$); the exchange capacity of each type group E_i; and the parameters of their acidity pK_i and Δpk_i.

The k_i is an equilibrium coefficient of ion exchange of H^+ with the cation of titrant (in our case K^+)

$$K_i = x_{Ki} \cdot C_H / (x_{Hi} \cdot C_K) \tag{9.18}$$

where the x_{Ki} and x_{Hi} are the equivalent fractions of K^+ and H^+ on the i type of the functional groups and C_H and C_K are the concentrations of the relative cations in the solution.

K_i is an equilibrium constant [199, 200]:

$$\log K_i = \int_0^1 \log k_i dx_{Ki} \tag{9.19}$$

In approximation of linear dependence of $\log k_i = f(x_{Ki})$ accepted in the model

$$\mathrm{p}k_i = \mathrm{p}K_i + \Delta \mathrm{p}k_i \cdot (1/2 - x_{Ki}) \tag{9.20}$$

$$\mathrm{p}K_i = \mathrm{p}k_{i(x=1/2)} \tag{9.21}$$

$$\Delta \mathrm{p}k_i - \mathrm{p}k_{i,x=1} - \mathrm{p}k_{i,x=0} \tag{9.22}$$

The overall titration curve is described by a set of n Eqs. 9.20–9.22. The computer program fits this set to the experimental data with minimal n and predetermined precision and in this way returns values E_i, pK_i, and Δpk_i.

Anion exchangers are described in the same way with OH^- in place of H^+ and Cl^- (as a standard anion of the supporting electrolyte) in place of K^+.

The examples of titration curves of several fibrous ion exchangers are given in Fig. 9.13; the acidity parameters are presented in Table 9.6; the main properties of fibrous ion exchangers are given in Table 9.7.

There are numerous publications on different specific fibrous ion exchangers and their applications that were not considered in the present section, whose significance and potential practical extraction of uranium from sea water is an attractive goal for many researchers. The most promising sorbent for this purpose is probably fibrous ion exchangers with amidoxime and hydroxamic acid functional group [122–124].

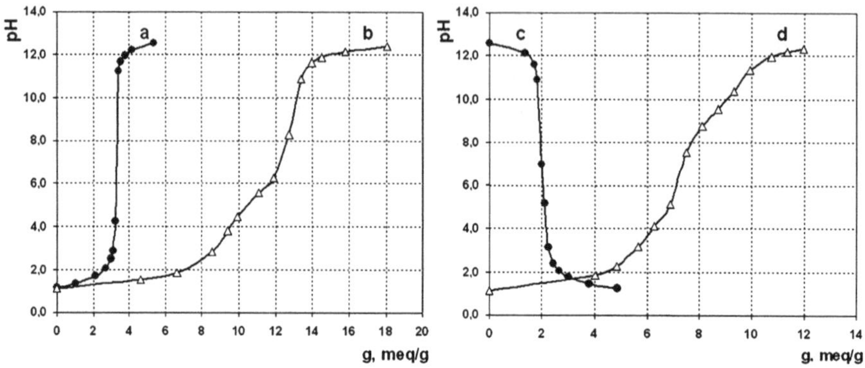

Fig. 9.13 Potentiometric titration curves of fibrous ion exchangers FIBAN: *a* – K-1; *b* – K-5; *c* – A-1; *d* – X-1. In the abscissa the amount of the KOH or HCl per gram of the ion exchanger are given. The points are experimental data; the lines are calculated from the model with the parameters in Table 9.6

Table 9.6 Acidity parameters of FIBAN fibrous ion exchangers

Ion exchanger FIBAN	Functional groups	pK_a	Δpk	E, meq/g
K-1	$-SO_3H$	−0.3	0.3	3.1
K-4	–COOH	5.3	2.0	4.2
K-5	–COOH	8.0	1.0	1.0
		5.0	0	2.8
	$-NH\text{-}NH_2$, =NH, ≡N	3.0	0	1.5
X-1	$-N(CH_2COOH)_2$	9.3	1.0	2.4
		4.6	0.5	1.4
	$=NH, -NH_2$	2.9	1.0	1.4
AK-22(2)	$=NH, -NH_2$	2.0	0	0.5
		4.6	1.5	1.3
		7.8	1.5	2.0
	–COOH	10.0	0	0.9
AK-22(3)	$=NH, -NH_2$	2.0	0	0.3
		4.6	1.5	2.4
		8.9	1.0	3.2
	–COOH	11.0	0	1.1
AK-22 G	$=NH, -NH_2$	5.0	1.5	1.5
		8.0	1.0	1.1
		10.0	0	1.3
	-COOH	11.0	0	0.5
A-1	$-N^+(R)_3$	3.0	0	0.5
		15.0	0	2.1
A-5	$-N(R)_2$	9.3	0	3.3
	–COOH	11.0	0	1.0
A-6	$-N^+(R)_3$	11.8	0	1.9
	$-N(R)_2$	9.0	0	0.7

Table 9.7 The main properties of fibrous ion exchangers

Trademark of the fiber	Functional groups	Exchange capacity, meq/g	Water uptake, g H_2O/g	Max temperature, °C
FIBAN A-1	$–N^+(CH_3)_3Cl^-$	2.7	0.8	50 (OH^-)
				100(Cl^-)
FIBAN A-5	$–N(CH_3)_2$; =NH	4.2	1.4	80
	–COOH	0.5		
FIBAN A-6	$–[N(CH_3)_2C_3H_5O]^+Cl^-$	2.0	1.2	80(Cl^-)
	$–N(CH_3)_2$	0.8		
FIBAN A-7	$–[N(CH_3)_2C_2H_4OH]^+Cl^-$	2.1	1.6	80(Cl^-)
	$–N(CH_3)_2$	1.0		
FIBAN K-1	$–SO_3H$	3.0	1.0	100(H^+)
FIBAN K-1-1	$–SO_3^-(K^+, Co^{2+})$	3.0	0.7	100
	$K_xCo_y[Fe(CN)_6]$	10 mass.%		
FIBAN K-3	–COOH	5.0	0.5	100
	=NH, ≡N	2.0		
FIBAN K-4	–COOH	5.0	1.1	80
FIBAN K-5	–COOH	5.0	0.5	80
	$–NH–NH_2$, =NH, ≡N	1.2		
FIBAN X-1	$–N(CH_2COOH)_2$	3.5	0.6	80
	$–NH_2$	0.5		
FIBAN X-2	$–N(CH_2COOH)_2$	5.2	0.5	80
	$–NH_2$	0.8		
FIBAN AK-22	$–NH_2$, =NH, ≡N	4.5	0.8	80
	–COOH	0.5		
FIBAN AK-22-1	=NH, ≡N	4.5	0.7	80
	–COOH	1.0		
FIBAN AK-22-B	$–NH_2$, =NH	2.0	–	80
	–COOH	2.0		
Nichibi IEF-SC3050	$–SO_3Na$	2.4–3.1	0.5	–
Nichibi IEF-SA	$–CH_2–N(CH_3)_3{}^+Cl^-$	1.8–2.1	0.6	–

(continued)

Table 9.7 (continued)

Trademark of the fiber	Functional groups	Exchange capacity, meq/g	Water uptake, g H_2O/g	Max temperature, °C
Nichibi chelating	$–N(CH_2COONa)_2$	2.0	0.6	–
TIN-100	$–SO_3Na$	3.0	–	120
TIN-200	$–CH_2–N(CH_3)_3{}^+Cl^-$	2.0	–	60(OH^-) 80(Cl^-)
TIN-300	No	–	–	–
TIN-400	$–CH_2Cl$	2.5	–	–
TIN-500	$–CH_2NH_2$	2.5	–	100
TIN-600	$–CH_2–N(CH_2COONa)_2$	1.5	–	80(H^+) 120(Na^+)
VION KN-1	–COOH	5.0–7.0	0.4–0.5	–
VION KC-2	$–SO_3H$	0.8–1.1	0.3–0.4	–
VION AN-1	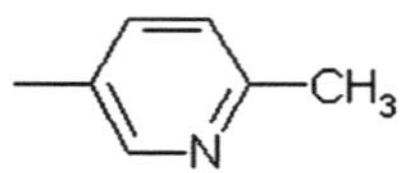	2.0–2.5	0.1	–
VION AN-3	$=NH, –NH_2$	3.0–3.5	0.1–0.2	–
VION AC-1	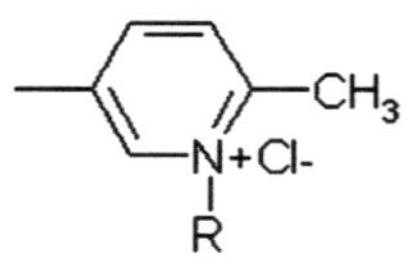	0.8–1.2	0.2–0.3	–
VION AC-2	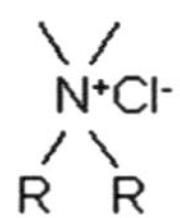	0.7–1.1	0.2–0.3	–

Ion exchange fibers are suitable for immobilization of bacteria and purification of water and air from viruses and bacteria [197].

Fibrous ion exchangers are already used in analytical chemistry for preconcentration of trace amounts of different ions [189, 190] in place of microspheric resins and have good prospects for future.

9.5 Removal of Volatile Substances from Gaseous Phase

The first publication on application of ion exchange resins to treatment of gas media was done in 1955 [204].

Since then an immense amount of scientific work on studies of sorption of gases by ion exchangers has been done. Their final aim was development of new technologies for separation of gases and their drying; purification of exhaust and ventilation air of different industries and agricultural productions; gas analysis; and deep purification and deodoration of the air. The research done on the ion exchange resins has shown that in spite of principle possibility of their using in the mentioned processes they have drawbacks preventing from their wide industrial applications. They are slow sorption of substances from gases, exceedingly high aerodynamic resistance of their filtering layers and poor osmotic stability in wetting–drying cycles. Fibrous ion exchange materials are free of these drawbacks, and air purification seems to be the most promising field for their large-scale practical application.

There are several ways for using ion exchangers in air purification from impurities of different chemical nature.

The most straightforward process is neutralization of H^+ or OH^- forms of ion exchangers with impurities of the base or acid nature. The examples of such processes are sorption of ammonia by the cation exchangers

$$RH + NH_3(\text{gas}) \leftrightarrow RNH_4 \tag{9.23}$$

and hydrogen chloride by the anion exchanger

$$ROH + HCl(\text{gas}) \leftrightarrow RCl + H_2O \tag{9.24}$$

Ion exchangers can also function as carriers of acid and base concentrated solutions, absorbing the base and acid gases by the same mechanism. Air purification from trace amounts of weak acids such as HCN or H_2S is retarded by the carbon dioxide–stronger acid present in the natural air in a higher concentration. In such cases the other chemical mechanisms for removal of such substances can be utilized, for example catalytic conversion or specific interaction of the impurities with the functional groups or impregnating agents in the ion exchanger. It is also of note that the simple main processes like the two mentioned almost always are accompanied by different side processes, making the overall process much more complicated than it looks at first sight.

On one hand gaseous sorption systems are simpler than the liquid ones because the gas phase is much closer to the ideal systems than the liquid one. On the other hand they are more complicated in theoretical description because they include one more degree of freedom compared with the systems with aqueous solutions: that is the activity of water vapor always present in natural air. Typical specific examples of practically important processes will be considered in this section of the paper.

In the aqueous systems the water activity is nearly constant and close to the activity of pure water; in gaseous systems the water activity can change from zero to unity, being one of the major factors controlling the equilibrium and rate of the sorption process in the system. Therefore we start consideration of the gaseous sorption systems with a description of water sorption dependent on its activity.

9.5.1 Water Sorption

The role of water in the sorption of substances from the gaseous phase is manifold: it may be a participant or product of the main reaction; a diffusion medium in which transport of the sorbate molecules to the sorption centers occurs; an agent causing ionization of the functional groups of ion exchanger and the sorbate molecules; or a solvation agent affecting the effective concentration of the interacting ions. Therefore the water activity (relative water humidity) is one of the key factors in processes of air purification. The water sorption isotherms $W = f(\alpha)$ (isopiestic curves, where $\alpha = P/P_0$) give valuable information of the state of water and ions in the phase of ion exchanger and have been extensively studied for both ion exchange resins and fibers [205–221]. The results of these studies were used for calculating thermodynamic functions of the sorption processes and different parameters characterizing intermolecular and interionic interactions in ion exchangers [214, 222–225].

Nevertheless none of the existing models allow expressing the water sorption isotherm in terms suitable for its explicit application in prediction of the influence of water activity on the sorption of a target substance. Such predictions have a great practical importance for calculation of the working resources of gas filters for obtaining ultrapure air in electronic and similar industries. Direct experimental determination of the working resource is difficult because the concentration of the target substance in the product air is low (order of 1 ppb). For a filter with filtering layer thickness of 3 mm, the period of work before regeneration is about 1 year; the sorption strongly depends on the relative air humidity.

We have developed a mathematical model allowing expression of the total water sorption in a whole interval of water activity via the amount of water in the sorbent, in which the sorbate can be dissolved (free water) and the hydrate water, in which the sorbate cannot be dissolved [226, 227].

Knowledge of these values is necessary to calculate the amount of sorbate that can be extracted from the air before its saturation and in so doing correctly predict longevity of its working time.

The main statements of the model are as follows:

1. The total amount of water in the ion exchanger is expressed by the sum of the water bound to hydrates of the ionic pairs counter ion-fixed ion with the number of water molecules q_i and constant of their formation K_i, and free water in which the sorbate can be dissolved.
2. The concentration of sorbate in the free water is determined by Henry's law with the same distribution constant as that for the system water-gas.
3. The internal solution is an ideal mixture of the free water molecules, hydrated and nonhydrated functional groups in H form, and obeys Rault's law; that means that hydration is assumed the only reason for nonideality of the system.

From these statements the following equation for the water sorption isotherm was derived:

$$W = \sum_i \frac{q_i \cdot K_{hi} \cdot \alpha^{q_i}}{1 + K_{hi} \cdot \alpha^{q_i}} + \frac{\alpha \cdot \left(W_0 - \sum_i q_i \cdot n_{h,i}\right) \cdot f_W}{1 + W_0 - \sum_i q_i \cdot n_{h,i} - \alpha \cdot \left(W_0 - \sum_i q_i \cdot n_{h,i}\right)} \tag{9.25}$$

where $\alpha = P/P_0$ is relative humidity of air; W_0 is total water uptake ($\alpha = 1$), mol/eq; f_W is activity coefficient of the free water; $n_{h,i}$ and n_W are numbers of moles in the hydrate i containing q_i water molecules and free water respectively; and K_{hi} is the equilibrium constant of hydrate formation, determined from the equation of the hydrate formation:

$$K_{hi} = \frac{n_{h,i}}{n_{R,i} \cdot \alpha^{q_i}} \tag{9.26}$$

The quantities q_i and K_{hi} are fitting parameters.

Application of this equation to different ion exchangers showed that in most cases assumption of one hydrate and $f_W = 1$ is sufficient for accurate description of the water sorption as a function of α. In more complicated cases (H^+, OH^-, F^- forms of ion exchangers) assumption of the presence of two hydrates significantly increases the accuracy of description [226, 227].

The presence of higher hydrates can be accounted in the inexplicit form by the dependence of the free water activity coefficient of α in the form

$$f_W = \alpha^{\lambda} \tag{9.27}$$

where constant λ is a fitting parameter.

This equation has been applied to hydrogen forms of strong and weak acid fibrous cation exchangers that were systematically investigated for sorption of

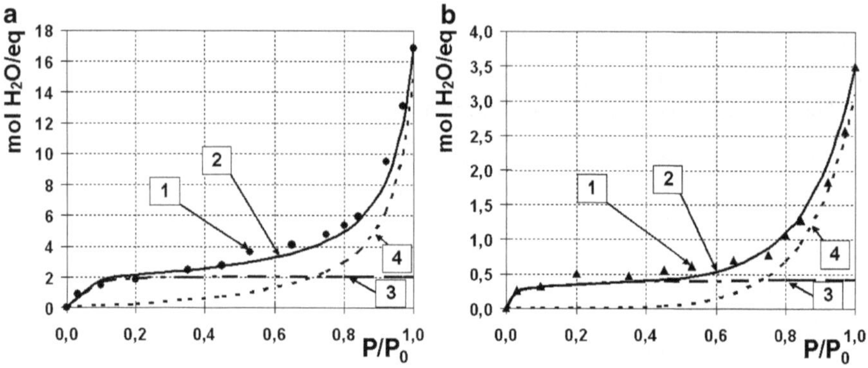

Fig. 9.14 Isopiestic curves of ion exchangers FIBAN K-1 (**a**) and FIBAN K-4 (**b**) (the symbols calculated from experimental data; 1 – experimental points; 2 – total amount of water; 3 – amount of hydrate water; 4 – amount of free water; 2, 3, 4 – theoretical curves computed from Eq. 9.25)

ammonia from the air under dynamic conditions dependent on the relative air humidity and external conditions of the process (Fig. 9.14). The amounts of free water at different relative humidities needed for the calculation of the ammonia sorption have been found.

Ion exchangers also can be used as drying agents for gases [228–232] at a high relative humidity ($P/P_0 > 0.2$) because they reveal a higher drying efficiency than zeolites or silica gel.

9.5.2 *Sorption of Bases*

Air purification by both granular and fibrous ion exchangers from the compounds of basic nature (mainly ammonia) has been described in plentiful publications, e.g., [14, 153, 172, 233–254]. Ion exchangers used had different functional groups (sulfonic, carboxylic, or phosphonic acid) and different polymeric matrixes (polypropylene with grafted sulfonated polystyrene, cellulose, polyacrylonitrile). The filtering layers of fibrous ion exchange materials had much higher dynamic sorption activity compared with the fixed beds of their granular analogues as well as more suitable physical form for the gaseous filters. The sorption efficiency of these materials increases with the increasing acid strength of the functional groups, relative air humidity, and concentration of ammonia in the air.

The authors of works [172, 236, 248, 249] presented results of comparative studies of different filtering plants loaded with granular and fibrous ion exchangers. It has been shown that because of a low aerodynamic resistance and small size of the filters on fibrous ion exchangers, they can be easily built into existing ventilation systems and provides efficient air purification from ammonia and the other plentiful alkaline contaminants to the level below their maximal admissible concentrations. In paper [238] sorption of ammonia from the air by cellulose phosphate has been studied at different relative humidity and thickness of the filtering layers. These materials were recommended for practical applications. However in Ref. [234] it was noted that such ion exchangers have rather low exchange capacity (<2.5 meq/g) and their chemical stability in the acid and alkaline media is low. Reference [250] presents the data on ammonia sorption from the air with its very low concentration (the starting concentration of NH_3 was 5 ppm) by the fibrous sulfonated graft copolymer of polystyrene and polypropylene. The sorption was studied under dynamic conditions by H^+, Cu^{2+}, and Ni^{2+} ionic forms. It was observed that the amount of absorbed ammonia depends on the exchange capacity of the fiber and does not depend on its concentration in the air. The ion exchangers in Cu and Ni forms absorb more ammonia than in H form because NH_3 absorbs because of its coordination by the metal ions and coordination numbers for Cu and Ni are 4 and 6, respectively. The use of a similar ion exchanger for air purification from the basic impurities was described [251]. Air purification from ammonia by carboxylic acid ion exchange fiber VION KN-1 in the form of nonwoven fabric was studied in a wide range of concentrations of NH_3 (5–1,000 mg/m^3) [252]. The conditions of sorption and regeneration of the ion exchanger were established.

Systematic studies of ammonia sorption from the air flows by two commercially available fibrous cation exchangers FIBAN K-1 and K-4 in the form of nonwoven fabric were performed in Ref. [253]. Their application to air purification from ammonia was described in detail in review [254]. The influence of the following factors on the efficiency of sorption has been studied: concentration of ammonia (2.5–20 mg/m^3), relative air humidity (0.075–0.85), the filtering layer thickness (3–12 mm). The air flow rate was 0.081 m/s; temperature was 25, in some cases 15°C.

The simplest behavior was observed for the sulfonic ion exchanger FIBAN K-1: the sorption is practically unaffected either by ammonia concentration (in the breakthrough points at different concentration of ammonia the sorption was practically equal: ~2.4 mmol NH_3/g) or by the air relative humidity within the studied range of these parameters.

Markedly different regularities were observed for the carboxylic acid fiber FIBAN K-4. In this case the sorption strongly depends on the ammonia concentration and relative air humidity. At a concentration of ammonia in the air of 18 mg/m^3, efficient air purification is possible only if the relative air humidity is above 0.52. At this condition the ion exchanger in equilibrium with the initial gas solution absorbs 2 mmol NH_3/g (~40% of its full capacity). The time before NH_3 breakthrough (the protection action time) increases with α in parallel with increasing full dynamic capacity of the ion exchanger reaching about 80% of the total capacity at $\alpha = 0.94$. At fixed air humidity $\alpha = 0.55$ the protection action time decreases with increasing concentration of NH_3 but the full dynamic capacity increases. This means that weak acid ion exchangers are inefficient for removal of trace amounts of ammonia from the gas phase. A moderate variation of temperature at a constant relative humidity does not affect the breakthrough curve. In our experiments at 15 and 25°C at the same relative humidity ($\alpha = 0.56$) we observed that the sorption curves coincide in spite of a large difference in concentration of the water vapors (7.18 g/m^3 at 15°C and 12.90 g/m^3 at 25°C).

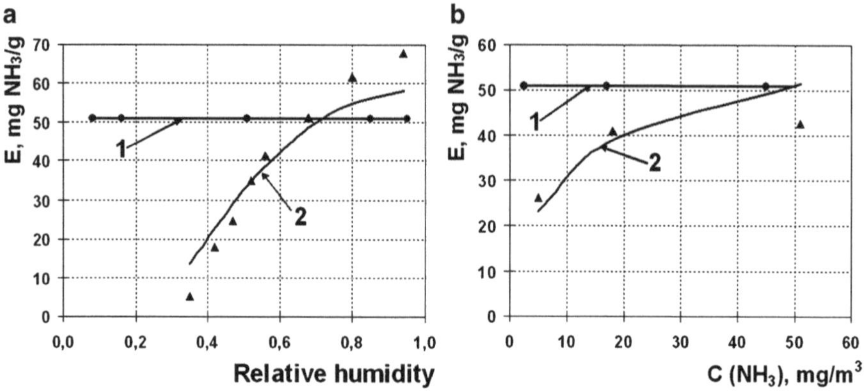

Fig. 9.15 Equilibrium ammonia sorption on ion exchangers FIBAN K-1 (*curve* 1) and FIBAN K-4 (*curve* 2) depending on relative air humidity at $[NH_3]_G = 17.5$ mg/m^3 (**a**), and concentration of ammonia at $\alpha = 0.48$ for K-1 and $\alpha = 0.55$ for K-4 (**b**). Points on the figures are experimental data; the curves are computed from Eq. 9.28 with the following parameters: (FIBAN K-1) $q = 2$, $K_h = 500$, $\lambda = 0$, p$K = 0$, Δp$k = 0$, $E = 3.0$ meq/g; (FIBAN K-4) $q = ½$, $K_h = 5$, $\lambda = 3.5$, p$K = 4.93$, Δp$k = 2.0$, $E = 4.2$ m-eq/g

The length of mass transfer zone calculated from the breakthrough curves appeared equal to 1.7 mm for FIBAN K-1 and 3.6 mm for FIBAN K-4.

The experimental data obtained in these studies have been used for testing the theoretical model allowing aprioristic calculation of equilibrium sorption of gaseous bases and acids by ion exchangers, depending on their concentration and the air humidity [227]. The model accounts for the acidic strength of the functional group in the ion exchanger and its variation with degree of sorption via the acidity parameters of the ion exchanger pK and Δpk. It also accounts for the ionization constant of the sorbate via its ionization constant K_D. With knowledge of the mole fraction of free water in the ion exchanger as a function of the water activity in the gas phase $X_W = X_W(\alpha)$, the degree of loading of the ion exchanger with the sorbate X can be found from Eq. 9.28 (Fig. 9.15). All parameters in this equation are accessible from independent experiment. Details of derivation and calculations on the model are given in refs. [227, 255].

$$X = K_H \cdot X_W \cdot [NH_3]_G \cdot K_D / (k_{hyd} + K_D \cdot K_H \cdot X_W \cdot [NH_3]_G) \qquad (9.28)$$

where K_H is Henry's constant; X_W is mole fraction of "free" water in ion exchanger; $[NH_3]_G$ is concentration of ammonia in gas phase; K_D is ammonia dissociation constant; and k_{hyd} is equilibrium coefficient of hydrolysis reaction of NH_4^+-form of ion exchanger.

The model suggested allowed correct prediction of the interval of relative air humidity and ammonia concentration in which these materials can be used. Especially important was calculation of the working range of the filters for removal of traces of ammonia from outdoor air, containing usually less than 100 μg/m^3 of

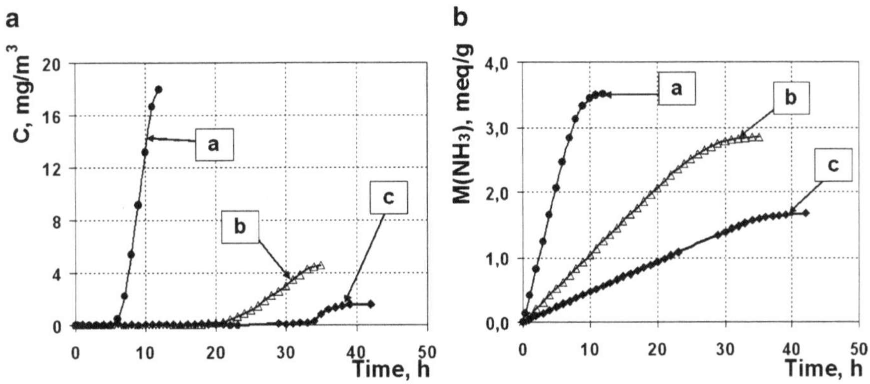

Fig. 9.16 The breakthrough and sorption curves of ammonia on ion exchanger FIBAN AK-22B impregnated with phosphoric acid: a – C (NH_3) = 18.6 mg/m^3; b – C (NH_3) = 4.7 mg/m^3; c – C (NH_3) = 2.0 mg/m^3

NH_3, down to the level suitable for high performance clean rooms (1–5 μg/m^3). Direct experimental determination of these values is extremely difficult because their working period usually lasts 1–2 years.

Efficient sorbents for air purification from alkaline contaminants were obtained by impregnation of fibrous ion exchangers with concentrated strong nonvolatile acids, specifically phosphoric and sulfuric acid. Aminocarboxylic fibers FIBAN AK-22 and FIBAN K-5 can retain up to 4–5 mmole of these acids per gram and remain surface dry at the ambient air humidity. They reveal excellent sorption properties toward ammonia (Fig. 9.16) and found practical application in purification of the air in the clean rooms of electronic production [256].

The sorption of amines from the air was studied in much less detail than that of ammonia. At the same time this problem is receiving increasing attention in connection with the removal of bad smells from the air. Separate studies devoted to this topic show that, in principle, the regularities of amines sorption are similar to that of ammonia [257–259]. Nevertheless existing data are not sufficient for estimating efficiency of fibrous ion exchangers in the process of extracting amines from the air.

Recently we studied sorption of a typical representative of this class of substances, diethylamin, on different fibrous ion exchangers. These data allow choosing the optimal ion exchanger and external conditions for air purification from this and similar substances. Diethylamine (DEA) is a volatile substance with a boiling temperature of 56.3°C and $pK_b = 3.07$ (compared with 4.75 for NH_3). It has a strong unpleasant smell. The conditions of the experiment were as follows: initial concentration of DEA 35–42 mg/m^3; temperature 19 ± 0.5°C, linear flow rate of the air 0.08 ± 0.005 m/s; and the filtering layer thickness 6 mm.

The data obtained compared with those for ammonia are present in Table 9.8.

The regularities of diethylamine (DEA) sorption can be summarized as follows: The breakthrough curves on the strongest acid ion exchanger FIBAN K-1 are

Table 9.8 Comparison of sorption parameters of ammonia and diethylamine

			Breakthrough capacity, mmol/g; NH_3/DEA		Capacity at $C/C_0 = 0.5$, mmol/g; NH_3/DEA	
FIBAN type	E, meq/g	Min α NH_3/DEA	$\alpha = 0.55$	$\alpha = 0.80$	$\alpha = 0.55$	$\alpha = 0.80$
K-1	3.0	<0.07/<0.15	2.4/2.1	2.4/2.1	2.8/2.7	2.8/2.7
AK-22M	3.7	–/0.08	1.7/1.6	–/1.6	2.5/1.9	–/1.9
K-4	4.9	0.45/0.50	1.7/0.3	2.9/1.0	2.3/0.9	3.5/1.9
K-5	7.8	–/0.55	–/0.1	–/1.5	–/0.9	–/2.1

practically independent of the relative air humidity, its working capacity to the breakthrough point is 2.0–2.1 mmol/g, the full capacity 2.8–3.1 mmol/g independent of the thickness of the filtering layer (3, 6, and 9 mm) and relative air humidity (18–85%).

A similar behavior was observed for FIBAN AK-22 M (amino-carboxylic polyelectrolyte impregtated with phosphoric acid): the breakthrough capacity was 1.5–1.7 mmole/g, the full capacity 1.8–2.1 mmole/g; the layer thickness and relative air humidity does not affect the sorption.

Different regularities were observed for the carboxylic acid fiber FIBAN K-4 and FIBAN K-5. In this case the sorption strongly depends on the DEA concentration and relative air humidity. At a concentration of DEA in the air of 40 mg/m^3, efficient air purification is possible only if the relative air humidity is greater than 0.50 (for K-4) and 0.55 (for K-5).

Dependencies of the protection time, τ, corresponding to the breakthrough capacities of the filtering layer thickness, L, have been used for calculation of the most important technical characteristic of the filtering layer, that is the length of the mass transfer unit L_0 [260]: for K-1 $L_0 = 1.7$ mm; for K-5–4.8 mm; K-4–4.5 mm; AK-22M–2.3 mm.

9.5.3 *Sorption of Acids*

9.5.3.1 Removal of Acids and Anhydrides through Ion Exchange and Molecular Sorption in the Absence of Side Processes

The simplest case of acid sorption by alkaline form of anion exchanger is expressed by equations:for strong base groups

$$R^+OH^- + H^+ + A^- \leftrightarrow R^+A^- + H_2O \tag{9.29}$$

for weak base groups

$$R + H^+ + A^- \leftrightarrow RH^+A^- \tag{9.30}$$

These processes are symmetric to the sorption of ammonia by cation exchangers and can be theoretically described in the same way as it was done in Sect. 9.5.2. Important examples of such processes are sorption of monobasic carboxylic acids and hydrohaloid (most often acetic and the other fatty acids, hydrogen chloride and fluoride).

More complicated is the mechanism of sorption of dibasic acids. In this case the process includes two steps:

$$2R^{+}OH^{-} + 2H^{+} + A^{2-} \leftrightarrow R^{+}{}_{2}A^{-} + 2H_{2}O \tag{9.31}$$

$$R^{+}{}_{2}A^{-} + H^{+} + HA^{-} \leftrightarrow 2R^{+}HA^{-} \tag{9.32}$$

and

$$2R + 2H^{+} + A^{2-} \leftrightarrow RH^{+}{}_{2}A^{2-} \tag{9.33}$$

Practically important cases of such processes are sorption from the air CO_2 and SO_3.

In air purification it is especially important to account for the presence in the air of carbon dioxide, whose concentration in atmospheric air is about 300 mg/m^3. This greatly exceeds the concentration of many other acid impurities intended for removal from the air and present in concentrations of few tens milligrams per cubic meter (e.g., HCN, H_2S). Being weaker acids than carbonic acid they practically will not be absorbed by the anion exchanger in the presence of CO_2. In the process of air purification with ion exchanger the latter is converted into the carbonate-bicarbonate form soon after the beginning of the process and the sorption of acid proceeds owing to substitution of carbonate (bicarbonate) with the anion of the absorbing acid:

$$R^{+}\left(HCO_{3}{}^{-}, CO_{3}{}^{2-}\right) + H^{+} + A^{-} \rightarrow R^{+}A^{-} + CO_{2} + H_{2}O \tag{9.34}$$

Another feature of the acid sorption is the ability to form in the phase of ion exchanger associates such as

$$R^{+}A^{-} + nH^{+} + nA^{-} \leftrightarrow R^{+}A^{-}(HA)_{n} \tag{9.35}$$

where $n = 1, 2, \ldots$

This causes super-equivalent sorption of the acids even at their low concentrations in the air.

An example of such processes is air purification from HCl. This process has been studied mainly on granular [261–264] and some fibrous ion exchangers [12, 265–267], but systematic data on its regularities at low concentrations of HCl in the air have not been obtained. In our work sorption of HCl has been systematically studied in dynamic conditions on different fibrous ion exchangers depending on the

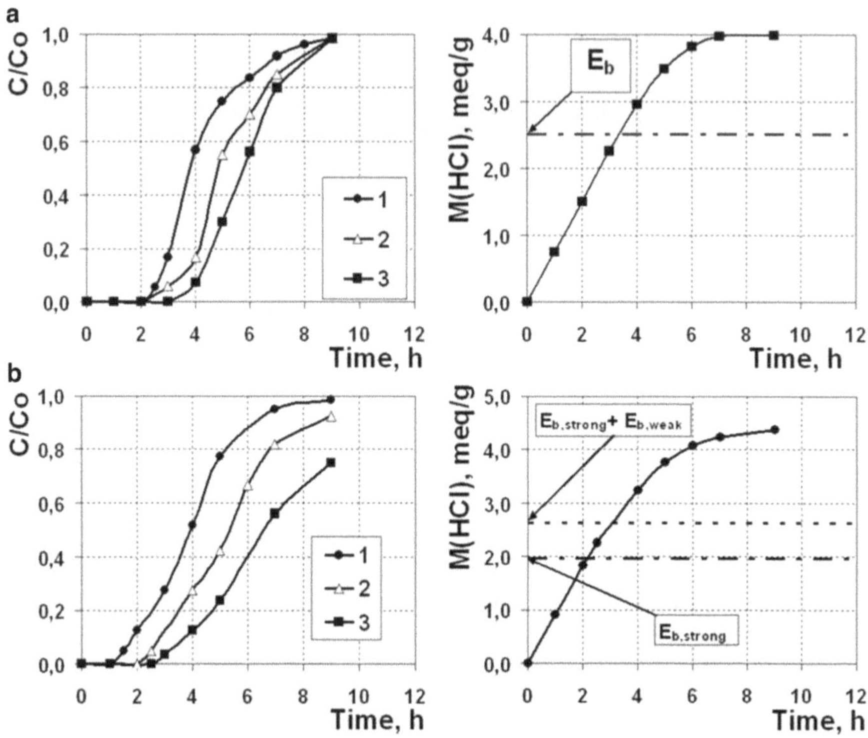

Fig. 9.17 The breakthrough and sorption curves of HCl on strong base FIBAN anion exchangers: (**a**) A-1, (**b**) A-6 (α: 1–0.35; 2–0.50; 3–0.60). The conditions of the experiment: initial concentration of HCl = 80 mg/m^3, thickness of the filtering layer = 9.5 ± 0.5 mm, the air flow velocity = 0.11 m/s, temperature = 18°C

air humidity at concentrations of HCl 50–100 mg/m^3. The ion exchangers studied differed in the type of functional groups, their basic strength, polymeric matrix, and hydrophilicity. The initial ionic form of the fibers was obtained by coating them with a solution of 0.1 M sodium carbonate, which corresponds to conditions of their regeneration in multicyclic work. The main properties of studied ion exchangers are presented in Table 9.6. The experiments were carried out in the same way as those described in Sect. 9.5.2 for sorption of ammonia.

The experimental breakthrough and sorption curves of HCl on different ion exchangers are presented in Figs. 9.17–9.18. The main regularities of these processes can be briefly summarized as follows.

In all the cases super-equivalent sorption is observed. No peculiarities in dependencies g = g(τ) occur at the point corresponding ion exchange capacity of the fiber. This confirms the opinion that the sorption HCl on the ion exchange group and formation of the associates proceed in parallel [268].

A large super-equivalent sorption of HCl (parts of curves in Figs. 9.17–9.18 above the horizontal line) even from its low concentrated mixtures with air is a good pre-condition for the easy regeneration of the ion exchangers with water or

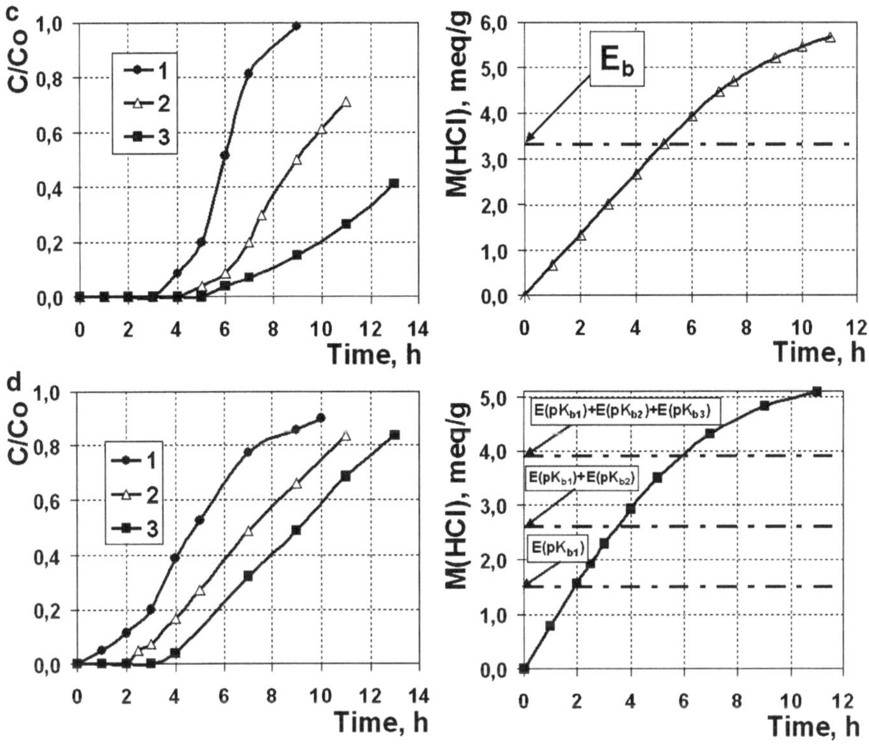

Fig. 9.18 The breakthrough and sorption curves of HCl on weak base FIBAN anion exchangers: (**c**) A-5, (**d**) AK-22G. See captions for Fig 9.17

low concentrated solutions of HCl. In Ref. [12] it was shown that by recycling regenerating water in a regeneration bath it was possible to obtain 6–7% HCl solution from the air containing 100 mg/m^3 of HCl.

In contrast with the sorption of ammonia on the strong acid cation exchanger, sorption of HCl on the strong base anion exchanger depends on the relative air humidity (Fig. 9.17a).

All studied ion exchangers can be used in air purification from HCl at the relative air humidity of 0.35 and higher with the exception of FIBAN AK-22G, for which the lowest α is 0.50. Ion exchanger FIBAN A-5 has the highest working capacity in the studied humidity range. The decrease in this value with decreasing humidity is the lowest for the strong base ion exchanger FIBAN A-1 and the largest for the weak base FIBAN AK-22G (Fig. 9.19).

It is interesting to note that breakthrough capacity of hydrofluoric acid (HF) on this ion exchanger does not depend on the relative humidity (α = 0.05–0.95) and its concentration in the air (10–1,000 mg/m^3) [12]. This peculiarity is a result of the specific features of HF, which can play a role of solvating agent in the phase of ion exchanger similar to water molecules causing protonization of the functional groups.

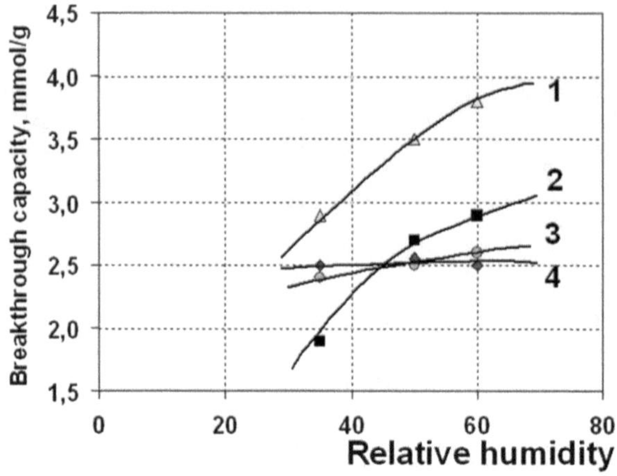

Fig. 9.19 Breakthrough capacity on HCl of different fibrous ion exchangers FIBAN as a function of relative air humidity: 1 – A-5; 2 – AK-22G; 3 – A-6; 4 – A-1

The dynamics of sorption of hydrofluoric acid (HF) by fibrous ion exchangers VION AN-1 in the concentration range 10–30 mg/m^3 at temperatures from −20°C to +24°C was studied. It was observed that the breakthrough capacity increases 2–3 times at the decreasing the temperature to −20°C.

Sorption of acetic acid from diluted solutions in the air strongly depends on the relative air humidity (Fig. 9.20).

At the studied concentration (23–27 mg/m^3) the super-equivalent sorption is not observed. At an acetic acid concentration of 2 g/m^3 a large superequivalent sorption on weak base anion exchanger VION AN-1 was observed in Ref. [269]. Ion exchanger FIBAN A-6, having quaternary ammonium groups, provides deeper air purification than the fiber FIBAN A-5 with ternary amino groups but its capacity at 10% breakthrough is substantially higher than that for the FIBAN A-6, (1.4 and 1.0 mmol/g). The critical humidity at which the ion exchangers can be used for air purification is lower for the FIBAN A-5. Finally we can conclude that both ion exchangers can be efficiently used in air purification from the vapors of acetic acid at relative air humidity above 0.5.

9.5.3.2 Sorption Processes Accompanied by the Catalytic Conversion of the Sorbates

Among such processes the most practically important is air purification from sulfur dioxide. Sorption of SO_2 by anion exchangers is accompanied by its oxidation by the atmospheric oxygen at room temperature in the presence of the water vapor [270, 271]. The process is catalyzed by the anion exchange because of strong binding of the product of oxidation that is mainly sulfate ion. The latter is a bivalent anion of a strong acid and its binding by the ion exchanger is much stronger than

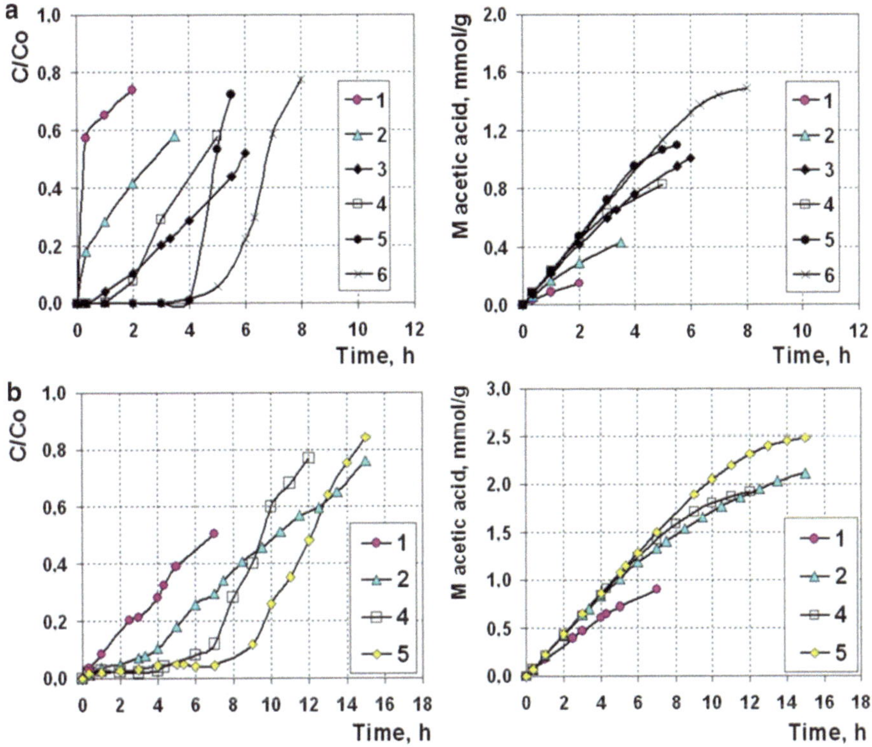

Fig. 9.20 The breakthrough and sorption curves of CH_3COOH on different FIBAN anion exchangers: (**a**) A-6, (**b**) A-5 (α: 1–0.30; 2–0.40; 3–0.50; 4–0.60; 5–0.70; 6–0.80). The conditions of experiment: initial concentration of acetic acid = 23–27 mg/m^3, thickness of the filtering layer = 6.0 ± 0.5 mm, the air flow velocity = 0.1 m/s, temperature = 19–21°C

that of the predominant anion HSO_3^- forming in the ion exchanger upon sorption of SO_2 and whose concentration in the internal solution of the ion exchanger is low owing to incomplete dissociation of H_2SO_3. The higher basicity of the ion exchanger and decrease of air humidity increase the degree of oxidation [271].

The air purification from SO_2 is a problem of great practical importance and a large number of scientific works have been devoted to it. The majority of them concerns purification of industrial wastes to the sanitary admissible level (from 10^4 to 10^0 mg/m^3). Purification to the level of 10^2 mg/m^3 can be efficiently achieved by wet scrubber technology. The removal of residual SO_2 is economically more expedient by using sorption technologies. The new problem that appeared owing to development of production of microcircuits of high integration, precise machinery, and optics, is air purification in the clean rooms of these productions to the level 10^{-3} mg/m^3. The two problems tasks most efficiently can be solved by application of filtering systems equipped with fibrous anion exchangers.

Regularities of SO_2 removal from the air by granular ion exchangers have been studied in numerous papers (e.g., [272–293]) but systematic data on the fibrous ion exchangers are presented only in a few works [2, 77, 196, 266, 267, 294–304].

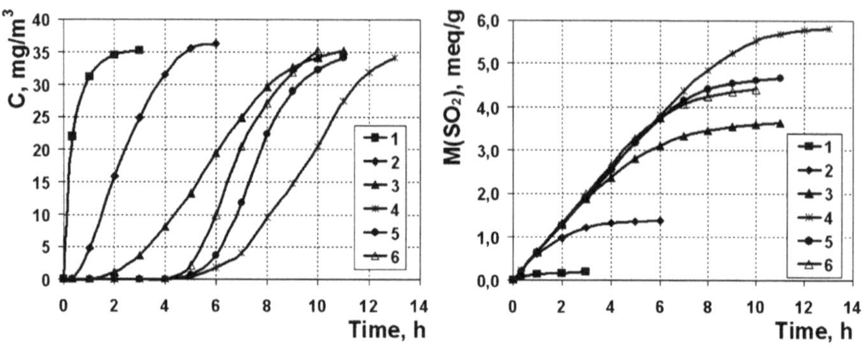

Fig. 9.21 The breakthrough and sorption curves of sulfur dioxide on anion exchanger FIBAN A-5 (HCO_3^- form) at various relative air humidity. The experiment conditions: temperature – 25°C; velocity of the air filtration – 0.08 m/s; SO_2 concentration – 37 mg/m^3; thickness of the filtering layer – 6 mm. α: 1–0.15; 2–0.22; 3–0.28; 4–0.49; 5–0.68; 6–0.91

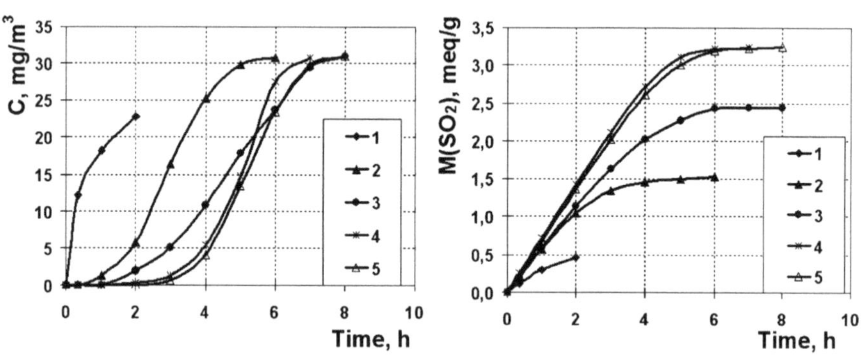

Fig. 9.22 The breakthrough and sorption *curves* of sulfur dioxide on anion exchanger FIBAN A-6 (HCO_3^- form) at various relative air humidity. The experiment conditions: temperature – 25°C; velocity of the air filtration – 0.08 m/s; SO_2 concentration – 32 mg/m^3; thickness of the filtering layer – 6 mm. α: 1–0.28; 2–0.33; 3–0.33; 4–0.48; 5–0.75

Equilibrium sorption in static conditions of SO_2 and water vapors by fibrous anion exchangers on the base of PAN fibers VION with different weak and strong base groups have been studied [304]. It has been established that all studied ion exchangers (in different degree) can be recommended for air purification but the lowest relative air humidity should be above 0.4.

In Ref. [196] the sorption of sulfur dioxide was studied in dynamic conditions on two fibrous ion exchangers, FIBAN A-5 weak base fiber and FIBAN A-6 strong base fiber in HCO_3^- ionic form. This ionic form has been chosen because in practical air purification OH^- form rapidly converts to mixed carbonate-bicarbonate form because of its reaction with atmospheric CO_2, whose concentration is about 300 mg/m^3.

The influence of relative humidity on the breakthrough curves of SO_2 is illustrated by Figs. 9.21 and 9.22.

It is rather surprising that weak base anion exchanger FIBAN A-5 with ternary amino groups reveals better ability for removal SO_2 from the air at equal relative humidity than strong base FIBAN A-6 at concentration of 32–37 mg/m^3.

One possible reason for a higher sorption ability of FIBAN A-5 may be catalytic oxidation of SO_2 to SO_3 with the following formation of H_2SO_4 in the phase of ion exchanger. If this conversion on the ternary amino groups of FIBAN A-5 is faster than on quaternary ammonium groups of FIBAN A-6, the rate of the overall SO_2 sorption process is controlled by this process and the difference between weak and strong base ion exchanger is not significant at the studied concentration of SO_2 in the air. That is because a difference in degree of neutralization of the weak and strong base groups by strong sulfuric acid can be noticed only at very low concentrations of SO_2.

Ion exchangers FIBAN A-5 and A-6 have found significant practical application in industrial exhaust air cleaning and the air of clean rooms of semiconductor industries. In the latter case, it is critically important to remove SO_2 from the air with relative humidity about 0.4.

Another practically important case of application of catalytic conversion of the sorbate on the fibrous ion exchanger is removal of hydrogen sulfide from the air. Plentiful attempts to apply an alkaline form of anion exchangers in air purification from low concentration of H_2S were doomed to failure because of strong competition with CO_2 present in the air in higher concentration and forming much stronger than H_2S carbonic acid. In refs. [305–308] it was suggested to remove H_2S from the air flow by its absorption with the alkaline form of the ion exchanger with a following catalytic oxidation to elemental sulfur on a suitable catalyst fixed in the fiber. Fe(III) complex with etylenediaminetetraacetate (EDTA) in alkaline solutions was applied as a catalyst. The main advantage of this catalytic system is selective oxidation of H_2S to S^0 without formation in noticeable amounts of side products, such as thiosulfates, sulfates, or sulfites.

The scheme of this process can be formulated as follows:

$$H_2S_{gas} \rightarrow H_2S_{solution} \tag{9.36}$$

$$H_2S + OH^- = HS^- + H_2O \tag{9.37}$$

$$2Fe^{3+} + EDTA + HS^- = S^0 + H^+ + 2Fe^{2+} + EDTA \tag{9.38}$$

$$2Fe^{2+} + EDTA + H_2O + 1/2O_2 = 2Fe^{3+} + EDTA + 2OH^- \tag{9.39}$$

The dry catalyst in these processes is inactive. In many cases it was required to remove H_2S from the air with relative humidity of about 50%, which was achieved by adding different humidifying agent to the fiber. The performance of H_2S removal from air with the optimal composition of the catalytic fiber is illustrated by Fig. 9.23.

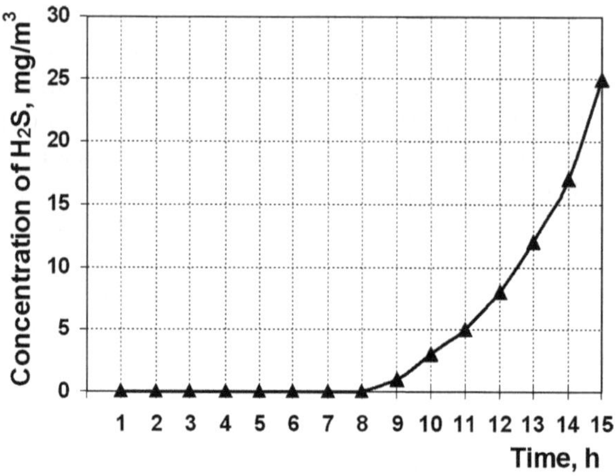

Fig. 9.23 Breakthrough curve of H_2S on fibrous catalyst (the parent material FIBAN AK-22; C_0 $_{(H2S)}$ = 60 mg/m^3; thickness of catalyst layer = 3 mm (1 layer, mass 13.0 g); linear air flow = 0.033 m/s; relative air humidity = 50%; the catalyst contain 0.18 mmol Fe/g)

The working capacity of such catalysts is higher than that for pure ion exchangers and restricted only by accumulation of elemental sulfur in the filtering bed. At the same time this process requires longer contact time of the air and the fiber, which can be achieved by reducing the rate of air filtration or application of thicker filtration layers. The air purifying devices on such catalytic fibers can be disposable or continuously operating units with a regenerable catalyst.

Fibrous ion exchangers can be used for removal from the air of many other bases and acids not described in this section, as well as neutral substances. In the literature applications of fibrous ion exchangers to removal of the following substances have been described: nitrogen oxides [309–311], hydrazine [312], elemental halogens [313–315], and ozone [316, 317].

This allows hope that fibrous ion exchangers will find their niche in practice in a not too distant future.

9.5.4 Color Changing Materials

The gas filtering materials can be impregnated with different color changing indicators showing presence in the outlet of the filter of a target impurity by changing the color of filtering materials. The simplest case is indication of acids or bases. In this case the ion exchange filtering fabric is impregnated with one of the acid–base indicators. Such materials are convenient to use in industrial air purifying filtering plants, chemical gas masks, protective clothing, and wiping materials for immediate indication of their exhaust.

Most of ion exchange fibers can be obtained in indicator forms. The important requirement to the initial material is absence of its own intensive color. In some cases

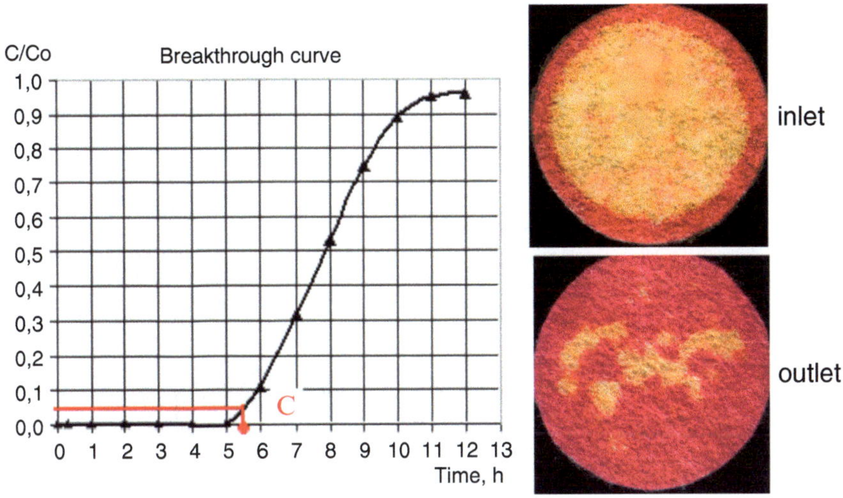

Fig. 9.24 Breakthrough curve of NH_3 and change color of indicating material FIBAN AK-22 M with methyl orange

the indicator is not absorbed or deactivated by the ion exchangers. We have prepared a number of color changing materials and in this section present some examples.

Ion exchange materials used for deep air purification from ammonia are polyampholyte FIBAN AK-22 with almost balanced capacities on amino and carboxylic acid groups (2.1 and 2.8 meq/g, respectively) and impregnated by sulfuric or ortho-phosphoric acid to have about 3 mmole of free acid per gram of the sorbent (FIBAN AK-22M). They are white-yellow and after the impregnation receive the indicator color in the strong acid medium. In the experiment the air flow, with concentration of ammonia 13 mg/m^3 and relative humidity 0.5, was passed through the layer of the nonwoven material 6 mm thick at a rate of 0.081 m/s at temperature 25°C. The color of the filtering material on the outlet was observed at the breakthrough point (0.6–0.7 mg NH_3/m^3).

A similar experiment was done with anion exchanger FIBAN A-5 (carbonate form) in sorption of SO_2 under the same conditions except for the concentration of SO_2 which was 32 mg/m^3 and the breakthrough point 1.5–1.7 mg/m^3 SO_2.

Examples given in Figs. 9.24 and 9.25 show a good correlation between the position of the breakthrough point and the moment of the color change of the material.

The color indication can be done at different concentrations of the acid or alkaline sorbate by choosing an appropriate indicator. For example, if concentration of the ammonia at the outlet should be lower than that indicated with methylorange (pH color change 3.0–4.4) then the indicator with lower transition pH should be used, for example methyl violet, having two distinct color transition: at pH = 0.13–1.5–3.2 (yellow-blue and blue-violet respectively).

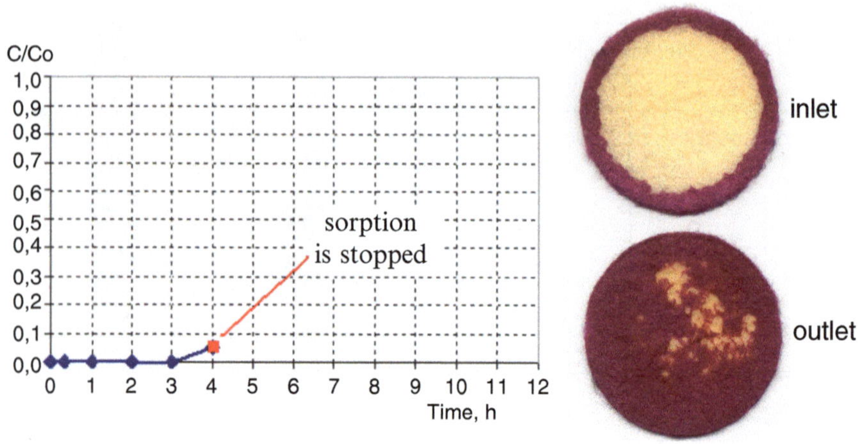

Fig. 9.25 Breakthrough curve of SO_2 and change color of indicating material FIBAN A-5 with phenol red

9.6 Hybride Fibrous Sorbents

Recently ion exchange resins [318–327] and fibers [328–332] were obtained containing in the polymer body finely dispersed inorganic precipitates with sorption properties [Cs, Mn, Ni, F, Fe, Ti, Zr, etc.]. Such materials allow combining a high selectivity to specific ions of inorganic ion exchangers with better mechanical properties and a higher rate of sorption on the polymeric ion exchangers.

Composite ion sorbents are obtained by precipitation of a desired inorganic compound in the phase of parent polymeric sorbent. For example fibrous composition sorbents for selective removal of arsenic anions from water were obtained by precipitation of iron (III) hydroxy-oxide in different ion exchange fibers.

Iron hydroxy-oxide FeOOH in granular form is used as sorbent of As anions and produced under trademarks Bayoxide E33 (Bayer), GFH (USFilter Co.), AD-33L (Adedge Technologies, Inc.). It has a high selectivity to these anions but its drawbacks are slowness of the sorption process, low mechanical strength, and spontaneous peptization with release of the particles of sorbent absorbed As to purifying water. The low rate of sorption allows its use only in large filters of centralized drinking water purification with the depth of the filtering layer more than 1 m.

The composition of sorbents contained in the phase of polymeric ion exchanger FeOOH was obtained in several varieties. The precipitate can be formed both in cation and anion exchangers. The method of precipitate formation can be different:

1. The cation exchanger saturated with Fe^{3+} ions put in contact with the alkaline solution
2. The anion exchanger in OH^- form put in contact with Fe^{3+} solution

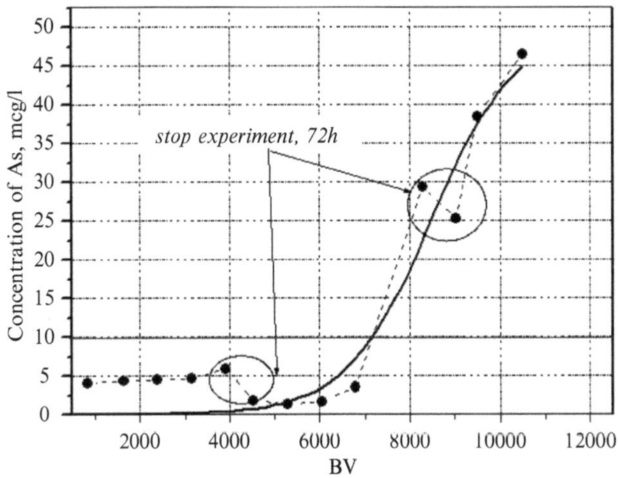

Fig. 9.26 Retention of As(V) by the layer of FIBAN-As 4 cm. Conditions of the process: initial $C_{As} = 50$ μg/dm^3, sorbent-solution contact time 40 s, thickness of filtering layer = 35 mm (Reproduced from Ref. [25] with kind permission of © Elsevier (2006))

In both cases at least a part of the precipitate remains in the ion exchanger.

It appeared that the sorption properties of such sorbents strongly depend on the nature of the parent ion exchanger and details of the procedure of precipitation. An efficient fibrous As sorbent (FIBAN-As) was obtained on the base of anion exchanger with ternary amino groups [25]. It contains 1.2 mmol/g of FeOOH and absorbs both As(III) and As(V) anions and can be produced in the form of nonwoven fabric. The process of sorption is very fast and controlled by the particle diffusion, with the diffusion coefficient 1.8–10^{-8} cm^2/s; the process half time is 23 s. This allows using it in thin filtering beds with the high flow rates usual for point-of-use filters such as small kitchen portable water filters. Even at a filtering layer thickness of 4 cm it allows purifying of 5,500 bed volumes of tap water containing 100 μg/l of arsenate down to the level below sanitary norm 10 μg/l (Fig. 9.26). Recently, the method of water purification from As(III) anions was improved by introducing in the purifying system preliminary step of oxidizing of As(III) to As(V) by the composition fiber impregnated with Mn(III) oxide [333].

In a similar way some other fibrous composite sorbents have been obtained: for sorption Cs^+ – cation exchanger impregnated by cyanofferrate K-Co ($K_xCo_y[Fe(CN)_6]$) [334]; for sorption F^- – anion exchanger impregnated by TiO_2.

9.7 Fibrous Ion Exchange Catalysts

Since the sorption of interacting molecules is the first stage of the reaction in the heterogeneous catalysis, the high rate of sorption processes on the fibrous ion exchangers is a good precondition for their good catalytic properties. Not much

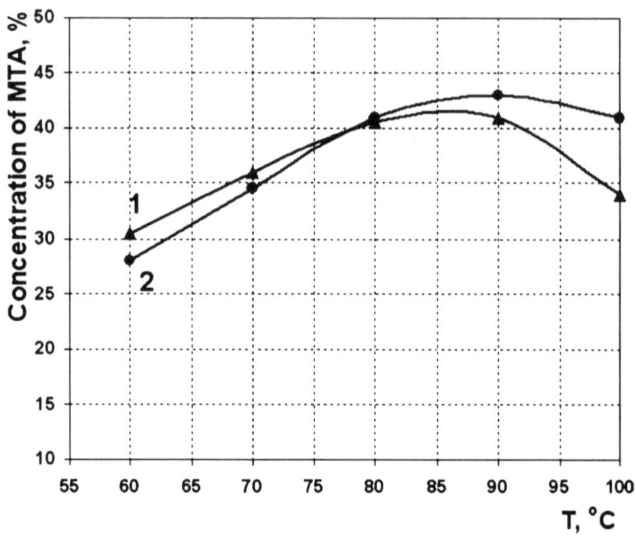

Fig. 9.27 Catalysis of reaction MTA ether formation by granular Dowex M-31 (*curve* 1; E = 3.7 meq/g) and fibrous FIBAN K-1 (*curve* 2; E = 4.4 m-eq/g) sulfonic types ion exchangers. Conditions of the experiment: pressure = 0.8 MPa; CH_3OH : i-C_5H_{10} = 1:1 (Reproduced from Ref. [336] with kind permission of © The Publishing House of Siberian Department of Russian Academy of Sciences (2001))

research on catalytic properties of fibrous ion exchangers has been done until now, but the information collected up to date shows that this area of catalysis is extremely promising.

The catalytic properties of fibrous ion exchangers have been tested in processes of different types. The most common case is acid catalysis. In Ref. [335] it was shown that fibrous ion exchanger IONEX is an efficient catalyst of reaction of sucrose inversion and methyl acetate hydrolysis in aqueous media.

Ion exchanger FIBAN K-1 revealed good catalytic properties in reaction of synthesis of methylthributhyl and methylthriamyl ether from methanol and butenes mixture (MTB and MTA, the modifier of gasoline increasing its octane number). Its catalytic activity and selectivity were comparable to that of Dowex M-31 used in industrial process [336] (Fig. 9.27).

It is of note that Dowex M-31 is a macroporous sulfostyrene ion exchanger with specific surface area 30 m^2/g, whereas FIBAN K-1 is not a porous material. Its high catalytic activity may not be explained by the small diameter of the fibers. In comparative experiment with finely ground sulfonic gel type ion exchange beads (to the size of the fiber diameter) it was observed that its catalytic activity is very low. The reason for the high catalytic ability of the fibrous ion exchanger of this type can be a close alliance of strongly polar domains of sulfonated polystyrene and nonpolar domains of the polypropylene. The interface is the location of the catalytic reaction. In favor of this assumption is the fact that FIBAN K-1 can absorb some amount of butenes (~5% mass), whereas neither cross-linked sulfostyrene nor polypropylene absorb butenes in significant amounts.

The acid catalysis of isomerisation reaction was successfully used for synthesis of isobornyl acetate, valuable aromatic substance, by isomerization of camphene [337].

In the previous examples the fibrous ion exchanger itself was used as a catalyst. Another group of catalysts was obtained on the base of ion exchange fibers where they are used as a carrier of the catalytic agent.

Important examples of such catalysts are fibrous ion exchangers impregnated by metallic palladium. The palladium is introduced into the ion exchanger as a cation or as an anion by ion exchange and then reduced to metal palladium with controlled size of the particles. Sulfonic ion exchanger containing 0.1–1.0% palladium appeared to be an effective catalyst of reaction low temperature oxidation of hydrogen with oxygen [338, 339]. The further development of this research has led to obtaining extremely active catalysts of this type on the base of weak base anion exchanger and found application in deoxygenation of boiler water of power stations.

A catalyst of reaction of oxidation of hydrogen sulfide to elemental sulfur was obtained by impregnation of fibrous ion exchangers with concentrated solutions of Fe-EDTA complexes in strongly alkaline solutions. This catalyst has found an application in air purification from H_2S and is described in more detail in Sect. 9.5.

References

1. Economy J, Dominguez L, Mangun C (2002) Polymeric ion exchange fibers. Ind Eng Chem Res 41:6436–6442
2. Volf LA (1980) Fibers with special properties (Volokna s osobymi svojstvami (Russ)). Khimia, Moscow
3. Zverev MP (1981) Chemosorption fibers (Khemosorbtsionnye volokna (Russ)). Khimia, Moscow
4. Soldatov V, Pawlowski L, Shunkevich A et al (2004) New materials and technologies for environmental engineering. Part I. Syntheses and structure of ion exchange fibers. Monographs of Polish Academy of Sciences, Lublin
5. Perepelkin KE (2001) Chemical fibers with specific properties for industrial application and personnel protection. J Ind Text 31:87–102
6. Soldatov VS, Shunkevich AA, Sergeev GI (1988) Synthesis, structure and properties of new fibrous ion exchangers. React Polym Ion Exch Sorbents 7(2–3):159–172
7. Shcherbinina NI, Myasoedova GV, Savvin SB (1988) Fibrous chelating sorbents in inorganic analysis. Russ J Anal Chem (Zhurnal analiticheskoy khimii (in Russian)) 43(12):2117–2131
8. Shimamura M, Teramoto K, Yoshioka T et al (1989) Polystyrene based functional fibers. In: Lewin M, Preston J (eds) Handbook of fiber science and technology. Marcel Dekker, New York
9. Kotze MH (1992) The status of ion-exchange fibers for metallurgical applications, overview. JOM 44:46–50
10. Shimizu H (1993) Development of ion-exchange technology in Japan, part III. J Ion Exch 4 (2):2–19
11. Jaskari T, Vuorio M, Kontturi K (2001) Ion exchange fibers and drugs: an equilibrium study. J Control Rel 70:219–229
12. Soldatov VS, Elinson IS, Shunkevich AA (1994) Application of fibrous ion exchangers in air purification from acidic impurities. In: Proceedings of the international symposium on hydrometallurgy'94. Chapman & Hall, London/Glasgow/Weinhem/New York/Tokyo/ Melbourne/Madras, pp 837–855

13. Dominguez L, Benak KR, Economy L (2001) Design of high efficiency polymeric cation exchange fibers. Polym Adv Technol 12:197–205
14. Soldatov VS, Elinson IS, Shunkevich AA et al (1996) Air pollution control with fibrous ion exchangers. In: Pawlowski L, Lacy WJ, Uchrin C et al (eds) Proceedings of the 10th international conference. Chemistry for the protection of the environment, vol 2. Plenum, New York/London, pp 55–67
15. Jyo A, Kugara J, Trobradovic H et al (2004) Fibrous iminodiacetic acid chelating cation exchangers with a rapid adsorption rate. Ind Eng Chem Res 43:1599–1607
16. Soldatov VS, Pawlowski L, Wasag H et al (1996) Prospects of fibrous ion exchangers in water pollution control. In: Pawlowski L, Lacy WJ, Uchrin C et al (eds) Proceedings of the 10th international conference. Chemistry for the protection of the environment, vol 2. Plenum, New York/London, pp 107–121
17. Ogorodnikov VA, Soldatov VS, Shunkevich AA (2003) Hydrodynamic resistance of the filtration layers of fibrous carboxylic acid cation exchanger FIBAN K-4. Proc Natl Acad Sci Belarus: Chem Ser (Vesti NAN Belarusi: seriya khimicheskikh nauk (Russ)) 2:83–87
18. Helferich F (1962) Ion exchange. McGraw-Hill, New York
19. Kokotov YuA, Zolotarev PP, El'kin GE (1986) Theoretical foundations of ion exchange (Teoreticheskie osnovy ionnogo obmena (in Russian)). Khimia, Leningrad
20. Yoshioka T, Shimamure M (1983) Studies of polystyrene-based ion exchanger fibers. I. The preparation and fundamental characteristics of polystyrene-based ion exchange fiber. Bull Chem Soc Jpn 56(12):3726–3729
21. Petruzzelli D, Kalinichev A, Soldatov V et al (1995) Chloride/sulfate ion exchange kinetics on fibrous resins. Two independent models for film diffusion control. Ind Eng Chem Res 34:2618–2624
22. Petruzzelli D, Tiravanti G, Liberti L et al (1993) Chloride – sulfate exchange kinetics on fibrous resins. A complete study with spherural exchangers. In: Duer A, Hudson MJ, Williams PA (eds) Ion exchange processes: advances and applications. Royal Society of Chemistry, Cambridge
23. Chen L, Yang G, Zhang J (1996) A study of the exchange kinetics of ion exchange fibers. React Funct Polym 29:139–144
24. Lewin M, Preston J (eds) (1989) Handbook of fiber science and technology, vol 2, High technology fibers. CRC Press, Boca Raton
25. Vatutsina OM, Soldatov VS, Sokolova VI et al (2006) A new hybrid (polymer/inorganic) fibrous sorbent for arsenic removal from drinking water. React Funct Polym 67(1):90–108
26. Kokotov YuA, Pasechnik VA (1970) Equilibrium and kinetics of ion exchange (Ravnovesie i kinetika ionnogo obmena (in Russian)). Khimia, Leningrad
27. Yoshioka T (1985) Studies of polystyrene-based ion exchanger fiber. III. A novel fiber-form chelating exchanger and its adsorption properties for heavy-metal ions. Bull Chem Soc Jpn 58 (9):2618–2625
28. Hautari HH, Mironova TV, Sosinovich ZI et al (2002) Selectivity of sodium-calcium exchange on a fibrous sulfonic ion exchanger. Proc Natl Acad Sci Belarus: Chem Ser (Vesti NAN Belarusi: seriya khimicheskikh nauk (in Russian)) 4:46–49
29. Soldatov VS, Sokolova VI, Medyak GV et al (2007) Binary ion exchange equlibria in systems containing NO_3^-, Cl^- and SO_4^{2-} on fibrous anion exchangers with tetraalkylammonium groups. React Funct Polym 67:1530–1539
30. Soldatov VS, Shunkevich AA, Elinson IS et al (1999) Chemically active textile materials as efficient means for water purification. Desalination 124:181–192
31. Soldatov V, Shunkevich A, Wasag H et al (2002) Prospects of fibrous ion exchanger in technology of water purification. In: Pawlowski L, Dudzinska MR, Pawlowski A (eds) Environal engineering studies. Kluwer/Plenum, New York/Boston/Dordrecht/London/Moscow, pp 153–165
32. Soldatov V (1991) Fibrous ion exchangers: new materials for hydrometallurgy. In: Abe M, Kataoka T, Suzuki T (eds) Proceedings of the international conference on ion exchange. New

developments in ion exchange: materials, fundamentals and applications (ICIE '91), Tokyo, 2–4 Oct 1991. Amsterdam, Oxford/New York/Tokyo, pp 511–516
33. Zagorodni AA (2007) Ion exchange materials: properties and applications. Elsevier, St. Louis
34. Barash AN, Kalyanova NF, Litovchenko GD et al (1977) The effect of heating on the properties of anion exchange fiber VION AN. Fibre Chem (Khimicheskie volokna (in Russian)) 6:44–45
35. Zakharova NN, Zashchepkina ES, Volkov LA et al (1990) Synthesis of copolymer of acrylonitrile and sodium metallyl sulfonate in dimethylsulfoxide and preparation chemosorption fibers on its base. Fibre Chem (Khimicheskie volokna (in Russian)) 4:16–19
36. Zverev MP (1989) Chemosorption fibers VION: materials for protection of the environment from harmful substances. Fibre Chem (Khimicheskie volokna (in Russian)) 3:32–37
37. Khamrakulov G, Kamilova M, Ismailov I et al (1992) On the properties of fiber-forming copolymers of acrylonitrile with quaternary ammonium salt of N, N-dimethylaminoethylmethacrylate. Russ J Appl Chem (Zhurnal prikladnoy khimii (in Russian)) 65(5):1193–1196
38. Pulatova FA, Valieva GA, Ergasheva VA (1992) New fiber-forming copolymers of acrylonitrile. In: Proceedings of the conference, Chemical technologies of textile materials, Tashkent, pp 33–37
39. US Patent 5009951 (1991)
40. US Patent 4988364 (1991)
41. Shoushtari AM, Zargaran M, Abdouss M (2006) Preparation and characterization of high efficiency ion-exchange crosslinked acrylic fibers. J Appl Polym Sci 101:2202–2209
42. Myasoedova GV, Nikashina VA, Molochnikova NP et al (2000) Properties of new types of fibrous sorbents with amidoxime and hydrazidine groups. J Anal Chem (Zhurnal analiticheskoy khimii (in Russian)) 55(6):611–615
43. Zverev MP, Barash AN, Kalyanova NF et al (1978) Synthesis of fiber-forming carbo-chain chemosorption copolymers and studies of reaction of formation of three-dimensional polymer net. In: Proceedings of the international symposium, Tashkent, USSR 17–21 Oct 1978. Macromolecular chemistry. USSR, Nauka, Moscow, pp 113–116
44. Pakshver EA (1973) Carbochain synthetic fibers. Khimia, Moscow
45. Kuleznev VN (1980) Polymer mixtures: structure and properties. Khimia, Moscow
46. FR Patent 2237953 (1974)
47. German Patent DE 2323656 (1973)
48. US Patent 2980635 (1961)
49. Kobuke Y, Tabushi I, Aoki T et al (1988) Composite fiber adsorbent for rapid uptake of uranyl from sea water. Ind Eng Chem Res 27:1461–1466
50. Mezhirov MS, Chegolya AS, Idiatulov RK et al (1981) New fibrous materials with adsorption properties. In: Proceedings of third international conference. Chemical fibers, vol 5. Kalinin, USSR, pp 13–18
51. FR Patent 2197938 (1973)
52. Mazovetskaya VP, Danilova EYa, Volf LA (1976) Investigation of bicomponent mixtures of some fiber-forming polymers with polyethyleneimines. In: Methods of syntheses and applications of polyethyleneimine in practice. Nauka, Moscow, pp 231–239
53. Emets LV, Danilova EYa, Ivanova GV et al (1981) Modification of carbochain fibers by polyethyleneimine aiming preparation of the ion exchangers. In: Proceedings of third international conference. Chemical fibers, vol 5. Kalinin, USSR pp 288–294
54. Ratushnyak IB, Vasiljeva OO, Danilova EYA (1978) The compositions for preparation of ion exchange fibers. Fibre Chem (Khimicheskie volokna (in Russian)) 2:17–19
55. Zverev MP (1975) Research of VNIIV in the field of preparation of ion exchange fibers. Fibre Chem (Khimicheskie volokna (in Russian)) 5:3–6

56. Barash AN, Zverev MP, Kalyanova NF (1981) Preparation of the chemosorption fiber from poly-2-vinylpyridine and vinylidene fluoride. Fibre Chem (Khimicheskie volokna (in Russian)) 4:20–21
57. US Patent 30386797 (1968)
58. UK Patent 968328 (1964)
59. Japanese Patent 55110125 (1980)
60. Japanese Patent 1272863 (1989)
61. Author's certificate USSR 1616930 (1990)
62. Artemenko SE, Kardash MM, Svekol'nikova OYu (1992) The effect of fiber additives on the structure formation of cation exchange membranes. Fibre Chem (Khimicheskie volokna (in Russian)) 5:29–32
63. Ratushnyak IB, Danilova EYa, Emets LV et al (1978) Ion exchange polyacrylonitrile fibers containing polyethyleneimine. Fibre Chem (Khimicheskie volokna (in Russian)) 6:33–35
64. Kazakevich YuT, Danilova EYa, Emets LV et al (1992) Synthesis and studies of fibrous anion exchangers on the base of polyamines. Fibre Chem (Khimicheskie volokna (in Russian)) 5:12–14
65. Shimizu H (ed) (1993) Handbook of adsorption techniques (Kunchaku Gijutsu Hando Bukku). NTS, Tokyo
66. Yoshioka T, Shimamura M (1986) Studies of polystyrene-based ion exchange fiber. IV. A novel fiber-form material for adsorption and immobilization of biologically-active proteins. Bull Chem Soc Jpn 59(3):399–403
67. Mark HF, Norman GG, Norbert MB (1965) Encyclopedia of polymer science and technology: plastics, resins, rubbers, fibers. Interscience, New York - London
68. Ponomarev AN (1980) Functionalization of polymer supports for polymerization catalysts by graft polymerization methods. J Appl Polym Sci 25(2):349–357
69. Manson JA, Sperling LH (1976) Polymer blends and composites. Plenum, New York
70. Tsetlin BL, Vlasov AV, Babkin IYu (1973) Radiation graft polymerization. In: Kargin VD (ed) Radiation chemistry of polymers. Nauka, Moscow
71. Kabanov VJa, Sidorova LP, Spitsin VI (1973) Kinetics and mechanism of radiation ionic graft polymerization. Proc Acad Sci USSR (Doklady akademii nauk SSSR (in Russian)) 211 (4):889–892
72. Huglin MB (1976) Radiation-induced graft copolymerization. Proc R Austral Chem Inst 43 (2):43–50
73. Kabanov VJa, Sidorova LP, Spitsin VI (1974) Radiation-induced graft polymerization by ionic mechanism. Eur Polym J 10(12):1153–1158
74. Lawler JP, Charlesby A (1980) Grafting of acrylic acid onto polyethylene using radiation as initiator. Radiat Phys Chem 15(5):595–602
75. Kabanov VJa, Aliyev RE, Chasovnikov IA (1982) Radiation graft polymerization of styrene and acrylonitrile at a high dose rate. High Mol Compd ser B (Vysokomoleculyarnye soedineniya ser B (in Russian)) 24(2):134–136
76. Kapustina IB, Kurilenko AI, Khoroshko RP (1972) Preparation of combined ion exchange fibers by liquid-phase copolymerization of styrene and divinylbenzene on polypropylene fibers. Proc Acad Sci Belarusian SSR Phys-Energy Ser (Vesti academii nauk Belaruskay SSR seriya fiziko-energeticheskikh nauk (in Russian)) 2:29–32
77. Soldatov VS, Elinson IS, Shunkevich AA (1986) Purification of air from acid gases (SO_2) by non-woven strong-base filtering materials. In: Pawlowski L, Alaerts G, Lacy W (eds) Proceedings of the fifth international conference. Chemistry for protection of the environment, Leuven 9–13 Sept 1985. Elsevier, Amsterdam/Oxford/New York/Tokyo, pp 369–386
78. Hideaki S (2000) FAPIG: First Atom Power Ind Group 155:20–26
79. http://ifoch.bas-net.by. Accessed 01 Aug 2010
80. Malakhova LI, Vlasov AV, Mikhailov NV et al (1972) Kinetics of radiation graft polymerization of acrylic acid from the gas phase on polypropylene fibers. High Mol Compd Ser A (Vysokomoleculyarnye soedineniya ser A (in Russian)) 14(4):751–755
81. Junjuj U, Rogovin ZA, Konkin AA (1962) Grafting of polyacrylonitrile and polyvinylacetate to polypropylene fibers. Fibre Chem (Khimicheskie volokna (in Russian)) 6:11–14

82. Kostrov YuA, Konkin AA, Kostrova KA et al (1966) Electron microscopic studies of the fibers made of the graft copolymer of polyethylene and polyacrylic acid. Fibre Chem (Khimicheskie volokna (in Russian)) 5:15–19
83. Druzhinina TV, Andrichenko YuA, Konkin AA et al (1962) Studies of the process of formation of the polyethylene fiber. Fibre Chem (Khimicheskie volokna (in Russian)) 2:17–20
84. Druzhinina TV, Konkin AA, Stasyuk HA et al (1967) Synthesis of graft copolymers of polyethylene amine with polyacrylic acid. Fibre Chem (Khimicheskie volokna (in Russian)) 3:16–19
85. Chapiro A (1958) Synthesis of grafted copolymers having undergone ionizing radiation. J Polym Sci 29:321–342
86. Chapiro A (ed) (1962) Radiation chemistry of polymeric systems. Interscience, New York
87. Rao MH, Rao KN (1980) Radiation initiated grafting. In: Proceedings of the symposium. Industrial polymers and radiation, vol 1. Vallabh Vidyanagar, Gujarat, 12–14 Feb 1979 pp 275–286
88. Gupta BD, Shapiro A (1989) Preparation of ion exchange membranes by grafting acrylic acid into pre-irradiated polymer films. Eur Polym J 25(11):1145–1148
89. Muchin BA, Andrichenko YuD, Druzhinina TV et al (1976) Synteze von pfropf copolymeren des polycaproamids durch radikalishe polymerisation. Fazers Textil 27(6):277–285
90. Kislyuk MS, Gabrieljan GA, Andrichenko YuD et al (1982) Synthesis of the graft copolymers of polyacrylamide and polydimethylaminoethylmethacrylate using reversible reductive-oxidative systems. High Mol Compd Ser A (Vysokomoleculyarnye soedineniya ser A (in Russian)) 24(11):2321–2325
91. Kislyuk MS, Gabrieljan GA, Galbraich LS et al (1983) Studies kinetics and mechanism of reactions of graft polymerization of dimethylaminoethylmethacrylate. High Mol Compd Ser A (Vysokomoleculyarnye soedineniya ser A (in Russian)) 25(5):1095–1101
92. Zheltobryukhov VF, Efros AV, Andrichenko YuD et al (1978) Properties of polycaproamide fibers modified by quaternary ammonium salts of dimethylaminoethylmethacrylate. Fibre Chem (Khimicheskie volokna (in Russian)) 6:47–48
93. Author's certificate USSR 1650215 (1991)
94. Druzhinina TV, Chelysheva LV, Galbraich LS (1987) Optimizing of the preparation method for graft copolymers of polycaproamide and polydiethylaminoethylemethacrylate. Fibre Chem (Khimicheskie volokna (in Russian)) 2:20–22
95. Mosina NYu, Druzhinina TV, Galbraich LS (1992) Features of hetero-phase emulsion graft polymerization of glycidylmethacrylate to polycaproamide fibers. Fibre Chem (Khimicheskie volokna (in Russian)) 5:14–16
96. Alexandrijskij AS, Tsukanova NP, Druzhinina TV et al (1991) Preparation of fibrous anion exchangers on the base of graft copolymer of polycaproamidoglycidylmethacrylate and polyethylenepolyamine. Fibre Chem (Khimicheskie volokna (in Russian)) 5:34–35
97. Alexandrijskij AS, Druzhinina TV, Gembitskij LA et al (1991) Preparation of fibrous anion exchangers containing guanidine groups. Fibre Chem (Khimicheskie volokna (in Russian)) 1:29–31
98. Andrichenko YuD, Druzhinina TV (1992) Initiation of the graft polymerization of potassium parastyrenesulfonate at the expense of oxidation of polycaproamide by hydrogen peroxide. Russ J Appl Chem (Zhurnal prikladnoy khimii (in Russian)) 65(11):2633–2637
99. Andrichenko YuD, Druzhinina TV (1993) Preparation of cation exchange polyacrylamide fibers. Fibre Chem (Khimicheskie volokna (in Russian)) 2:12–14
100. Hardin AP, Zheltobryukhov VF, Gulbina TI et al (1983) Synthesis of graft copolymers of polyacryloamide by polymerization of vinyl monomers, initiated by the occluded macroradicals. High Mol Compd Ser B (Vysokomoleculyarnye soedineniya ser B (in Russian)) 25 (8):554–557
101. Zheltobryukhov VF, Tatarnikov MK (1984) Foundations of technology of production of fibers from the graft copolymers of polyamide I. Izvestiya vysshikh uchebnykh zavedeniy Ser technologii legkoy promyshlennosti 27(6):35–38

102. Zheltobryukhov VF, Tatarnikov MK (1985) Foundations of technology of production of fibers from the graft copolymers of polyamide II. Izvestiya vysshikh uchebnykh zavedeniy Ser technologii legkoy promyshlennosti (in Russian) 28(1):45–47
103. Tatarnikov MK, Mcrtychev KN, Zheltobryukhov VF et al (1986) Experimental industrial realization of the process for preparation of chemosorption polyamide fiber. Fibre Chem (Khimicheskie volokna (in Russian)) 6:35–36
104. Valdman DI, Zheltobryukhov VF, Mcrtychev KN et al (1985) Determination of the amount of graft copolymer at preparation of modified fibers. Fibre Chem (Khimicheskie volokna (in Russian)) 1:59–60
105. Tatarnikov MK, Zvezdin VI, Zheltobryukhov VF et al (1988) The computer control of production of fibers from graft copolymers of polycaproamide. Fibre Chem (Khimicheskie volokna (in Russian)) 2:56–57
106. Hardin AP, Zheltobryukhov VF, Tatarnikov MK (1984) Low temperature synthesis of graft copolymers of polycaproamide with polydimethylaminoethylmethacrylate. Fibre Chem (Khimicheskie volokna (in Russian)) 3:33–34
107. Kuzmin VS, Morozenko TF, Zheltobryukhov VF (1991) Development of a method for preparation of fibrous sorbent KM-K1 for purification of air from basic gases and vapors. In: Proceedings of all-union conference. Concept of ecologically pure regions, Volgograd, USSR, pp 18–22
108. Perevalova EA, Korsunskij SN, Morozenko TF et al (1991) Development of a method for preparation of fibrous sorbent KM-A1 for purification of air from acid gases and vapors. In: Proceedings of all-union conference. Concept of ecologically pure regions, Volgograd, USSR, pp 22–27
109. Godenko AE, Korsunskij SN, Perevalova EA et al (1991) Mathematical modeling of the process of preparation of graft copolymers of polyacryloamide and polydimethylaminoethylmethacrylate. Referative Chem J (Referativnyi zhurnal khimia (in Russian)), 24C576
110. Battaerd J, Tregear DW (1967) Graft copolymers. Wiley, New York
111. Korshak VV, Mozgova KK, Shkolina MA et al (1963) Preparation of graft copolymers. High Mol Compd (Vysokomoleculyarnye soedineniya (in Russian)) 5(3):338–341
112. Buchenska J, Skwarski T (1991) Grafting of vinyl monomers into caproamide-6 fibers. Polim Tworz Wielkoc Zastcezk 36(1):23–27
113. Yurkevich VV, Konkin AA (1972) Investigation of physical mechanical properties of grafted polyacrylamide fibers. Fibre Chem (Khimicheskie volokna (in Russian)) 4:49–50
114. Author's Certificate USSR 840053 (1981)
115. Shalabi SE, Gabrielijan GA, Rogovin ZA (1982) Synthesis of the "sandwich" polymers on the base of polycaproamide. High Mol Compd Ser B (Vysokomoleculyarnye soedineniya ser B (in Russian)) 24(3):222–225
116. Shalabi SE, Afanasijeva IS, Gabrielijan GA et al (1984) Properties of polycaproamide fibers modified by grafting of polyacrylonitrile. Fibre Chem (Khimicheskie volokna (in Russian)) 3:32–33
117. Author's Certificate USSR 1599454 (1990)
118. Sonnerskog S (1958) On the reaction between polyacrylonitrile and hydrazine. Acta Chem Scandinavica 12:1241–1246
119. Kudryavtsev GI, Zharkova MA (1956) Saponification of the polyacrylonitrile fiber: acid hydrolysis of copolymers on the base of polyacrylonitrile. Russ J Appl Chem (Zhurnal prikladnoy khimii (in Russian)) 7:1103–1108
120. Kudryavtsev GI, Matiash TA, Zharkova MA et al (1961) Hydrazination of polyacrylonitrile fibers. Fibre Chem (Khimicheskie volokna (in Russian)) 4:13–19
121. Romanova TA, Zharkova MA, Kudryavtsev GI et al (1968) Modification of polyacrylonitrile fibers by hydrazine hydrate. Fibre Chem (Khimicheskie volokna (in Russian)) 5:23–25
122. Kulinsky DA, Emets LV, Kostetsky VV (1976) Modification of polyacrylonitrile and the fibers on its base by hydroxylamine. Fibre Chem (Khimicheskie volokna (in Russian)) 6:21–22
123. Kato T, Kago T, Kusakake K et al (1990) Preparation of amidoxime fibers for recovery of uranium from sea water. J Chem Eng Jpn 23(6):744–750

124. Omichi H, Katakai A, Sugo T et al (1986) A new type of amidoxime group containing adsorbent for the recovery of uranium from sea water: effect of grafting of hydrophilic monomers. Sep Sci Technol 21(3):299–313
125. Akulich ZI, Sokolova VI, Medyak GV et al (2001) New fibrous polyacrylonitrile ion exchangers FIBAN with strong base functional groups. In: Proceedings of II Belarus science conference. Scientific and technical problems in production of chemical fibers in Belarus, Mogilev, pp 267–271
126. Volf LA, Kirilenko YuK, Urban ZA et al (1969) Thermostabilization of the polyvinylalcohol fibers. Fibre Chem (Khimicheskie volokna (in Russian)) 3:15–17
127. Sugasaka K, Katoh S, Takai N et al (1981) Recovery of uranium from sea water. Sep Sci Tech 16(9):271–313
128. Author's certificate USSR 183375 (1966)
129. Volf LA, Meos AI, Inkina SA (1962) Sulfonation of the polyvinylalcohol fibers. Russ J Appl Chem (Zhurnal prikladnoy khimii (in Russian)) 35:2047–2050
130. Author's certificate USSR 191043 (1969)
131. Bekshieva EV, Besprozvannyh AV, Volf LA et al (1970) On sulfonation of dehydrochlorinated polyvinylchloride fibers. Russ J Appl Chem (Zhurnal prikladnoy khimii (in Russian)) 43(10):2367–2368
132. Sergeeva LN, Rogovin ZA (1964) Carbon chain fibers: modified polyvinylalcohol fibers with ion exchange properties. Fibre Chem (Khimicheskie volokna (in Russian)) 2:27–30
133. Tsetlina LA, Meos AI, Volf LA (1961) Preparation of fire-resistant polyvinylalcohol fibers and cloths. Fibre Chem (Khimicheskie volokna (in Russian)) 6:22–24
134. Tsetlina LA, Volf LA, Meos AI (1963) Composition and structure of phosphoric acid ethers of polyvinylalcohol fibers. Fibre Chem (Khimicheskie volokna (in Russian)) 5:23–25
135. Tsetlina LA, Yanovskaya NV, Volf LA (1965) Phosphorilation of polyvinylalcohol fibers vinol in the presence of ternary bases. Fibre Chem (Khimicheskie volokna (in Russian)) 4:16–19
136. US Patent 3275575 (1966)
137. Messalem R, Forgacs C, Michael J et al (1977) Ion exchange fibers: preparation and application. J Appl Polym Sci 31:383–388
138. Egawa H (1965) Preparation of carboxylic acid type cation exchange fibers. J Chem Soc Jpn 68(7):1304–1306
139. Henmi M, Yoshioka T (1993) Studies of ion exchange fiber "IONEX" for precoating material. Desalination 91(3):319–332
140. Fujiwara K (2007) Preparation of functional fibers by radiation induced graft polymerization and application. Nucl Instrum Methods Phys Res Sect B Beam Interact Mater Atoms 265 (1):150–155
141. Medyak GV, Shunkevich AA, Soldatov VS (1989) Physical chemical properties of fibrous ion exchangers FIBAN on the base of polypropylene. Proc Natl Acad Sci Repub Belarus Chem Ser (Vesti NAN Belarusi seriya khimicheskikh nauk (in Russian)) 1:69–74
142. Japanese Patent 8012774 (1996)
143. Japanese Patent 62164734 (1987)
144. Japanese Patent 62237924 (1987)
145. Mironova TV (2005) Methods of fast potentiometric titration and their applications to characterization of ion exchangers. Diss, Minsk
146. Medyak GV, Shunkevich AA, Soldatov VS (1989) Chemical and mechanical stability of FIBAN polypropylene-based fibrous ion exchangers. Proc Natl Acad Sci Belarus Chem Ser (Vesti NAN Belarusi seriya khimicheskikh nauk (in Russian)) 2:74–78
147. Soldatov VS, Pokrovskaya AI, Martsinkevich RV (1984) Polypropylene-based fibrous anion exchangers. Russ J Appl Chem (Zhurnal prikladnoy khimii (in Russian)) 57(6):1410–1413
148. Soldatov VS, Pokrovskaya AI, Martsinkevich RV (1984) Fibrous polypropylene-based sulfonic cation exchangers. Russ J Appl Chem (Zhurnal prikladnoy khimii (in Russian)) 57 (9):2030–2034
149. Soldatov VS, Tsygankova AV, Elinson IS et al (1987) Osmotic stability and freeze resistance of strongly basic fibrous anion exchange resins based on polypropylene. Proc Acad Sci

Belorussian SSR Chem Ser (Vesti Academii nauk Belorusskoy SSR seriya khimicheskikh nauk (in Russian)) 2:78–84
150. Soldatov VS, Tsygankova AV, Elinson IS et al (1988) Chemical, osmotic and freeze resistance of FIBAN A-1 strongly basic fibrous anion exchangers. Russ J Appl (Zhurnal prikladnoy khimii (in Russian)) 61(11):2465–2472
151. Soldatov VS (2008) Syntheses and the main properties of FIBAN fibrous ion exchangers. Solv Extract Ion Exch 26(5):457–513
152. Yegiazarov YuG, Soldatov VS (2001) Catalytic systems on the base of fibrous ion exchangers. Sorpt Chromatogr Proc (Sorbtsionnie i khromatograficheskie protsessy (in Russian)) 1(4):591–600
153. Soldatov VS (1984) New fibrous ion exchangers for purification of liquids and gases. In: Pawlowski L, Verdier AJ, Lacy WJ (eds) Proceedings of international conference chemistry for protection environment, Tolouse, 19–25 Sept 1983. Studies in environmental science, vol 3. Elsevier, Amsterdam, pp 353–364
154. Soldatov VS, Martzinkevich RV, Shunkevich AA et al (1998) Selectivity of cesium sorption by sulfonic ion exchangers present in aqueous solutions in trace quantities. Russ J Phys Chem (Zhurnal fizicheskoy khimii (in Russian)) 72(9):1686–1689
155. Soldatov VS, Popova OP, Shunkevich AA (1994) Kinetics of exchange of chloride with organic anions on fibrous strong base ion exchange FIBAN A-1. Russ J Phys Chem (Zhurnal fizicheskoy khimii (in Russian)) 68(4):763–765
156. Medyak GV, Shunkevich AA, Polikarpov AP et al (2001) Preperation and properties of fibrous ion exchanger FIBAN K-4. Russ J Appl Chem (Zhurnal prikladnoy khimii (in Russian)) 74(10):1608–1613
157. John E, Greenleaf AK, Sengupta N (2006) Environmentally benign hardness removal using ion exchange fibers and snowmelt. Environ Sci Technol 40:370–376
158. Prigozhaeva LM, Polikarpov AP, Shunkevich AA et al (2005) Influence of addition of bifunctional co-monomers on the grafting of acrylic acid into polypropylene fibers. Proc Natl Acad Sci Belarus Chem Ser (Vesti NAN Belarusi seriya khimicheskikh nauk (in Russian)) 3:46–49
159. Prigozhaeva LM, Polikarpov AP, Shunkevich AA (2009) Influence of addition of bifunctional co-monomers on the chemical stability and of oxydability of the aqueous extracts from the fibrous ion exchanger FIBAN K-4. Proc Natl Acad Sci Belarus Chem Ser (Vesti NAN Belarusi seriya khimicheskikh nauk (in Russian)) 2:87–90
160. Russ Patent 1051989 (1995)
161. Khobotova EB, Zarechensky VM (1997) Regeneration of ammonium copper-bearing washing solutions. Calvanothech Surf Treat (Galvanotechnica y obrabotka poverkhnosty (in Russian)) 5(3):43–49
162. Russ Patent 2044748 (1995)
163. Barash AN, Kostina TF, Egorov KK et al (1988) Alkaline hydrolysis of nitrile groups in the hydrazinated Nitron fibers. Fibre Chem (Khimicheskie volokna (in Russian)) 3:7–8
164. Myasoedova GV, Nikashina VA, Molochnikova NP et al (2000) Properties of new types of fibrous sorbents with amidoxyme and hydrazine groups. Russ J Anal Chem (Zhurnal analiticheskoy khimii (in Russian)) 55(6):611–615
165. Barash AN, Zverev MP, Litovchenko GD et al (1984) The influence of nature of the second component on the process of hydrazination of polyacrylonitrile copolymers. High Mol Comp (Vysokomolekulyarnye soedineniya (in Russian)) 26(9):687–691
166. Romanova TA, Zharkova MA, Kudryavtsev GI et al (1968) Modification of polyacrylonittryle fiber by hydrazine. Fibre Chem (Khimicheskie volokna (in Russian)) 5:23–24
167. Dorokhina IS, Zharkova MA (1974) Investigation of hydrolysis of Nitron fiber structurized by hydrazine. Fibre Chem (Khimicheskie volokna (in Russian)) 3:50–51
168. Kudryavtsev GI, Matyash TA, Zharkova MA et al (1961) Hydrazination of polyacrylonitrile fibers. Fibre Chem (Khimicheskie volokna (in Russian)) 4:13–19
169. Russ Patent 2262557 (2004)

170. Elinson IS, Martinovich VI, Titova LI (2004) Physical chemical properties of of a fibrous ion exchanger on the base of hydrazinated Nitron. Sorp Chromatogr Processes (Sorbtsionnie i khromatograficheskie protsessy (in Russian)) 4:482–489
171. Martinovich VI, Elinson IS, Kashinsky AV et al (2004) Light dust-gas respirators with ion exchange fibers. Labor Soc Protect 10:59–61
172. Nizovtseva OP, Shunkevich AA (1978) Ammonia sorption by fibrous cation exchangers in H^+-form. Proc Acad Sci Belorussian SSR Chem Ser (Vesti Academii nauk Belorusskoy SSR seriya khimicheskikh nauk (in Russian)) 5:62–65
173. Trochimczuk A, Kolarz BN, Wojaczynska M (1988) Acrylic anion exchangers and their sorption properties toward copper (II) and cobalt (II). React Polym Ion Exch Sorbents 7(2/3):197–202
174. Kolarz BN, Jezierska J, Bartkowiak D et al (1994) Acrylic resins with complexes of guanidyl groups and copper (II). React Polym Ion Exch Sorbents 23(2/3):53–61
175. Romanian Patent 70119 (1980)
176. Dragan S, Barboiu V, Petrariu I et al (1981) Cationic polyelectrolytes. III. Nitrile group reaction of macromolecular compounds with n, n-dialkylaminoalkylamines. J Polym Sci Polym Chem Ed 19:2869–2894
177. Kasperchik VP, Sergeev GI, Soldatov VS (1988) Kinetics of alkali hydrolysis of poly (2-vinyl-2-imidazoline). Proc Acad Sci Belorussian SSR (Doklady Academii Nauk Belorusskoy SSR (in Russian)) 32(7):621–623
178. Soldatov VS, Martsinkevich RV, Sergeev GI et al (1988) Effect of alkaline treatment on physicochemical properties of a FIBAN AK-22 fibrous ion exchanger. Russ J Appl Chem (Zhurnal prikladnoy khimii (in Russian)) 61(10):2271–2275
179. Soldatov VS, Martsinkevich RV, Pokrovskaya AI et al (1994) Hydrolytic stability, swelling and potentiometric titration of FIBAN AK-22-1 amino carboxylic ion exchanger. Russ J Appl Chem (Zhurnal prikladnoy khimii (in Russian)) 67(10):1644–1647
180. Soldatov VS, Martsinkevich RV, Sergeev GI et al (1989) Physical chemical properties of FIBAN-type new fibrous ion exchanger. Russ J Appl Chem (Zhurnal prikladnoy khimii (in Russian)) 62(7):1676–1679
181. Soldatov VS, Sergeev GV (1990) Fibrous ion exchangers: perspective sorbents for extraction of the heavy metals ions from aqueous solutions. Zhurnal vsesoyuznogo khimicheskogo obschestva Mendeleeva (in Russian) 35(1):101–106
182. Grachek VI, Shunkevich AA, Martsinkevich RV et al (2005) Chelating sorbents for water purification. Ecol Ind Russ (Ecologiya i promyshlennost Rossii (in Russian)) 1:25–27
183. Grachek VI, Lysenko GN, Akulich ZI et al (2005) Study of the structure of chelate fibrous ion exchangers by IR spectroscopy. Russ J Gen Chem (Zhurnal obschey khimii (in Russian)) 79 (3):350–355
184. Shunkevich AA, Martsinkevich RV, Medyak GV et al (2004) Comparative evaluation of fibrous carboxylic acid cation exchangers as the means for the water purification from the heavy metal ions. Russ J Appl Chem (Zhurnal prikladnoy khimii (in Russian)) 77(2):253–258
185. Orlovskaya LA, Soldatov VS, Shunkevich AA (2004) Analytical characteristics of sorbent FIBAN X-1 and its application for sorption concentrating of transition metal ions in the tap water. Proc Natl Acad Sci Belarus Chem Ser (Vesti NAN Belarusi seriya khimicheskikh nauk (in Russian)) 4:100–104
186. Orlovskaya LA (2003) Methods for concentration of the heavy metal ions in analysis of the natural waters. Proc Natl Acad Sci Belarus Chem Ser (Vesti NAN Belarusi seriya khimicheskikh nauk (in Russian)) 2:42–44
187. Orlovskaya LA (2008) Evaluation of reliability of the method for preliminary sorption concentration of the heavy metal ions on FIBAN X-1 sorbent. Proc Natl Acad Sci Belarus Chem Ser (Vesti NAN Belarusi seriya khimicheskikh nauk (in Russian)) 1:32–37
188. Orlovskaya LA, Kremko LM, Shunkevich AA et al (2002) The sorption concentration of the heavy metal ions in distilled water by FIBAN X-1 sorbent. Proc Natl Acad Sci Belarus Chem Ser (Vesti NAN Belarusi seriya khimicheskikh nauk (in Russian)) 4:22–26
189. MDC.MINSK (МВИ.МН) 2256-2005
190. Republic of Belarus Patent 9746 (2005)

191. Vatutsina OM (2005) The sorption of arsenic anions in dependence of the solution pH. Proc Natl Acad Sci Belarus Chem Ser (Vesti NAN Belarusi seriya khimicheskikh nauk (in Russian)) 5:19–22
192. Vatutsina OM, Sokolova VI, Sokol VP et al (2006) A new composition sorbent for purification of drinking water from arsenic. Proc Natl Acad Sci Belarus Chem Ser (Vesti NAN Belarusi seriya khimicheskikh nauk (in Russian)) 1:58–61
193. Vatutsina OM, Sokol VP, Soldatov VS (2006) The pH effect on the arsenic anions sorption by a fibrous iron containing sorbent. Proc Natl Acad Sci Belarus Chem Ser (Vesti NAN Belarusi seriya khimicheskikh nauk (in Russian)) 3:83–86
194. Vatutsina OM (2006) Preperation of the iron containing fibrous composition sorbent. Proc Natl Acad Sci Belarus Chem Ser (Vesti NAN Belarusi seriya khimicheskikh nauk (in Russian)) 5:19–22
195. Egorova OM, Zelenkovsky VM, Soldatov VS (2007) The structure of sorption complexes of Fe (III) in a weak base anion exchanger with amido amine groups. Proc Natl Acad Sci Belarus (Doklady Natsionalnoy Akademii Nauk Belarusi (in Russian)) 51(1):83–88
196. Soldatov VS, Kosandrovich EG (2004) Sorption of sulfur dioxide by fibrous anion exchange materials. Proc Natl Acad Sci Belarus (Doklady Natsionalnoy Akademii Nauk Belarusi (in Russian)) 48(5):62–64
197. Amvrosieva TV, Diakonova OV, Poklonskaya NV et al (2000) Fibrous ion anion exchangers and poly ampholytes FIBAN as perspective sorbents for the removal of viral agents from water. In: Proceedings of 4th international congress. Water: ecology and technology, Moscow, 30 May–2 June 2000, p 816
198. Shunkevich AA, Akulich ZI, Medyak GV et al (2005) Acid-base properties of ion exchangers. III. Anion exchangers on the basis of polyacrylonitrile fiber. React Funct Polym 63:27–34
199. Soldatov VS (1998) Potentiometric titration of ion exchangers. React Funct Polym 38 (2–3):73–112
200. Soldatov VS (1995) Quantitative presentation of potentiometric titration curves of ion exchangers. Ind Eng Chem Res 34:2605–2611
201. Soldatov VS (2000) Potentiometric titration of polyfunctional ion exchangers. In: Greig JA (ed) Ion Exchange at the Millennium. Imperial College, London, pp 193–200
202. Soldatov VS, Sosinovich ZI, Korshunova TA et al (2004) Acid-base properties of ion exchangers. I. Optimizing of potentiometric titration of ion exchangers exemplified by carboxylic acid resins. React Funct Polym 58:3–12
203. Soldatov VS, Sosinovich ZI, Mironova TV (2004) Acid-base properties of ion exchangers. II. Determination of the acidity parameters of ion exchangers with arbitrary functionality. React Funct Polym 58:13–26
204. IEC report on the Cincinnati ACS meeting (1955) Ind Eng Chem 47(5):7A–16A
205. Ashirov A (1983) Ion exchange purification of waste waters, solutions and gases (Ionoobmennaya ochistka stochnykh vod, rastvorov i gazov (in Russian)). Khimia, Leningrad
206. Soldatov VS, Tsigankov VI, Elinson IS et al (1990) Sorption of water vapor by salt forms of fibrous anion exchanger FIBAN A-1. Russ J Appl Chem (Zhurnal prikladnoy khimii (in Russian)) 63(10):2285–2291
207. Glueckauf E, Kitt GP (1955) A theoretical treatment of cation exchangers. III. The hydration of cations in polystyrene sulphonates. Proc Roy Soc 228:322
208. Gregor HP, Sundheim BR, Held KM et al (1952) Studies on ion-exchange resins. V. Water vapor sorption. J Colloid Sci 7(5):511–534
209. Boyd GE, Soldano BA (1953) Osmotic free energies of ion exchangers. Z Electrochem 57:162–175
210. Gregor HP, Frederick M (1953) Thermodynamic properties of ion exchange resins: free energy of swelling as related to ion selectivity. Ann NY Acad Sci 57(3):87–115
211. Van Krevelen DW (1976) Properties of polymers correlated with chemical structure. Russian translation In: Malkin AYa (ed) Khimia, Moscow
212. Chalych AE (1987) Diffusion in polymeric systems (Diffuzia v polimernykh sistemakh (in Russian)). Khimia, Moscow

213. Soldatov VS, Tsigankov VI, Elinson IS et al (1990) Quantitative description of water sorption by salts forms of fibrous anion exchanger FIBAN A-1. Russ J Appl Chem (Zhurnal prikladnoy khimii (in Russian)) 63(10):2291–2296
214. Gantman AI, Veshev SA (1985) Water uptake and swelling of ion exchangers. Russ J Phys Chem (Zhurnal fizicheskoy khimii (in Russian)) 59(10):2615–2618
215. Sosinovich ZI, Hogfeldt E, Novitskaya LV et al (1978) Investigation of the hydration of strong anion exchangers using the model of stepwise hydration. Proc Natl Acad Sci Belarussian SSR (Doklady Akademii Nauk BSSR (in Russian)) 22(10):920–923
216. Katz BM, Kutarov VV, Kutovaja LM (1991) Kinetics of water vapor sorption by anion exchange fiber on the base of cellulose and polyacrylonitrile. Russ J Appl Chem (Zhurnal prikladnoy khimii (in Russian)) 64(8):1713–1717
217. Katz BM, Kutovaja LM, Kutarov VV (1991) Water vapor sorption by carboxylic chemosorption fiber in different ionic forms. Russ J Appl Chem (Zhurnal prikladnoy khimii (in Russian)) 64(4):846–849
218. Grebennikov S, Serpinski V, Kunin A et al (1983) Thermodynamics of solvent vapor sorption by synthetic fibers. Khim Ind (Sofia) 7:305–308
219. Chiyomi M, Yoshihiko O, Hiroshi I et al (1992) Adsorption of water by a weak acid carboxylated cotton fiber. Sen'I Gakkaishi Fiber 48(12):677–681
220. Razumovsky LP, Zaikov VG, Druzhinina TV et al (1990) Water sorption by the graft copolymere of polyamide and poly-N-N-dimethylaminoethylmethacrylate. Russ J Appl Chem (Zhurnal prikladnoy khimii (in Russian)) 63(10):2296–2301
221. Rabinovich IB, Krylov EA (1999) Separate determination of the “bound” and “free” water in organic ion exchangers. Russ J Phys Chem (Zhurnal fizicheskoy khimii (in Russian)) 73 (5):924–927
222. Brunauer S (1943) The adsorption of gases and vapors: V.1. Physical adsorption. Humprey Milford, London
223. Jovanovich DS (1969) Physical adsorption of gases. Kolloid Zeitschrift und Zeitschrift fur Polymere 235(1):1203–1225
224. White HJ, Eyring H (1947) The adsorption of water by swelling high polymeric materials. Text Res J 17(10):523–553
225. Sosinovich Z, Novitskaya L, Soldatov V et al (1985) Thermodynamics of water sorption on Dowex 1 of different cross-linking and ionic forms. In: Marinsky J, Markus Y (eds) Ion exch and solvent extraction, vol 9. Marcel Dekker, New York, pp 303–339
226. Soldatov VS, Kosandrovich EG (2005) A new equation of isopiestic curve for polyelecrolytes. Proc Natl Acad Sci Belarus (Doklady NAN Belarusi (in Russian)) 49 (4):66–69
227. Soldatov VS, Kosandrovich EG (2008) Chemical equilibria between the ion exchanger and gas phase. In: Recent advances in ion exchange theory and practice (Proceedings of IEX 2008), Fitzwilliam College, Cambridge, pp 103–110
228. US Patent 3275549 (1966)
229. Radl V, Krejkar E (1962) Cation exchangers as drying agents for gases and liquids. Chem Prumysl (Poll) 12(10):579–582
230. Wymore CE (1962) Sulfonic type cation exchange resins as desiccants. Ind Eng Chem Prod Dev 1(3):173–178
231. Shamilov TO, Zhirova LF, Kadyrova MB et al (1980) The gas drying by fibrous anion exchangers. Chem Ind (Khimicheskaya promyshlennost (in Russian)) 3:181
232. Miagkoj ON, Krutskikh AS, Astakhova EV et al (1981) The gas drying by fibrous anion exchangers. In: Application of ion exchange materials (Primenenie ionoobmennykh materialov (in Russian)). Voronezh University, Voronezh, pp 50
233. Zverev MP (1982) Chemosorption fibers Vion: a perspective material for the environment protection. Ind Energetic (Promyshlennaya energetika (in Russian)) 2:12–14
234. Kurilenko OD, Ennan AA, Nekryach EF et al (1975) Ion exchange fibrous materials on the base of cellulose in gas purification. Proc Acad Sci Ukranian SSR (Vestnik AN USSR (in Russian)) 7:37–45

235. Emets LV, Danilova EYa, Strukova IM (1979) Purification of the gas-air mixtures by fibrous ion exchangers. Proc Leningrad Technol Inst (Mezhvuz sbornik nauchnyh trudov (in Russian)) 2:96–99
236. Vulikh AI, Aloviajnikov AA, Varlamova LV (1981) Purification of ventilation gases by fibrous ion exchange sorbents. Color Metals (Tsvetnye metally (in Russian)) 4:38–41
237. Vulikh AI, Zagorskaya MK, Aloviajnikov AA (1970) Sorption of gases and vapors by ion exchange resins. Proc Res Inst Color Metals (Sbornik nauchnyh trudov NII tsvethoj metallurgii (in Russian)) 31:120–133
238. Ermolenko IN, Liubliner IP (1972) Ammonia sorption by fibrous ion exchangers in dynamic conditions. Russ J Appl Chem (Zhurnal prikladnoy khimii (in Russian)) 45(4):748–751
239. Malinovky EK, Osokov VK, Zverev MP et al (1990) The influence of exchange capacity of carboxylate fiber VION KN-1 on its sorption and physical mechanical characteristics. Russ J Appl Chem (Zhurnal prikladnoy khimii (in Russian)) 63(1):64–67
240. Soldatov VS, Zenon P, Malgorzata P et al (2004) A strong acid nonwoven filtering medium for deep air purification. Fibres Text East Eur 12(4(48)):56–61
241. Kosandrovich EG, Soldatov VS (2004) Sorption of ammonia from air by fibrous sulfonic cation exchanger FIBAN K-1. Proc Natl Acad Sci Belarus Chem Ser (Vesti NAN Belarusi seriya khimicheskikh nauk (in Russian)) 3:95–98
242. Soldatov VS, Kosandrovich EG, Wasag H (2005) Fibrous polymeric sorbents for air purification. In: Modyfikacja Polimerov. Ofycina Wydawnicza Politechniki Wroclawskiej, Wroclaw, pp 423–427
243. Soldatov VS, Wasag H, Kosandrovich EG et al (2008) Odour control by fibrous ion exchangers. In: Proceedings of NOSE2008 1st international conference on environ odour monitoring and control. Chemical engineering transactions, vol 15. Rome, 6–8 July 2008, pp 387–395
244. Wasag H, Pawlowski L, Soldatov V et al (2009) Removal of ammonia from air by fibrous ion exchangers. Environ Prot Eng 35(3):293–304
245. Hwang TS, Park LW, Kim SM et al (2004) Synthesis of PU-g-AAc ion exchangers by UV radiation-induced graft copolymerization and adsorption of ammonia. J Korean Ind Eng Chem 10(3):409–415
246. US Patent 5783608 (1998)
247. Soldatov VS (1996) Chemically active textile materials for filtration and purification of gases and liquids. In: Proceedings of 7th world filtration congress, vol 1. Budapest, pp 213–217
248. Vulikh AI, Aloviajnikov AA, Nikandrov GA (1979) The gas purification by ion exchangers. Color Metals (Tsvetnye metally (in Russian)) 7:48–52
249. Vulikh AI, Nikolaev AV, Zagorskaya MK et al (1965) Sorption of ammonia and chlorine by ion exchange resins in dynamic conditions. Proc Acad Sci USSR (Doklady Akademii Nauk SSSR (in Russian)) 160(5):1072–1074
250. Okamoto J, Sugo T, Fujiwara K et al (1990) The synthesis of a new type adsorbent for the removal of toxic gas by radiation-induced graft polymerization. Int J Radiat Appl Instr Part C Radiation Phys Chem 35(1–3):113–116
251. Takanobu S (2003) Application of graft polymerization technique to air purification. Kuki Seijo to Kontamineshyon Kontororu Kenkyu Taikai Yokoshu 21:23–26
252. Baskin ZL, Utkin VV, Abdulkhakova ZZ et al (1987) Dynamics of the sorption of hydrogen chloride and ammonia on the carboxyl group containing fibre VION KN-1. Fibre Chem (Khimicheskie volokna (in Russian)) 19(6):48–50
253. Kosandrovich EG (2005) Sorption of ammonia and sulfur dioxide by fibrous ion exchangers. Diss, Minsk
254. Soldatov VS, Kosandrovich EG (2011) Ion exchange in air purification. In: Sengupta Arup K (ed) Ion exchange solvent extraction, vol 20. CRC Press, Boca Raton, pp 45–117
255. Soldatov VS, Kosandrovich EG (2006) Theoretical description of the sorption equilibria in the gas-ion exchanger system. In: Chemical technology of new substances and materials. Belorusskaya Nauka, Minsk, pp 206–228
256. www.mwgroup.net. Accessed 01 August 2010
257. Hashida Isuo (1974) Development of deodorant fibers and filters. Sci Ind 66(7):256–264

258. Wasag H, Guz L, Sobczuk H et al (2009) Removal of methylamine from air by means of fibrous ion exchangers. In: Ozonek J, Pawlowski A (eds) Monografie Komitetu Inzynierii Srodowiska Polskiej Akademii Nauk, Lublin, Poland 59 (2):247–253
259. Wasag H (2009) Application of fibrous ion exchanger in filters for air deodorization. In: Ozonek J, Pawlowski A (eds) Monografie Komitetu Inzynierii Srodowiska Polskiej Akademii Nauk, Lublin, Poland, 59 (2):255–262
260. Harjuda R, Lehto J (1995) Memorandum of the international workshop on uniform and reliable nomenclature, formulations and experimentation for ion exchange. React Funct Polym 27:147–153
261. Kats BM, Lazarev MYu, Malinovsky EK (1980) Influence of the capacity and porosity of anion exchanger AN-221 on its sorption ability toward SiF_4 and HCl. Russ J Appl Chem (Zhurnal prikladnoy khimii (in Russian)) 53(5):1175–1178
262. Kats BM, My L (1982) Swelling of the weak base macroporous ion exchangers in the process of sorption HCl gas and water vapor. Russ J Appl Chem (Zhurnal prikladnoy khimii (in Russian)) 55(9):1971–1974
263. Kats BM, Lazarev MYu, Artyushin GA et al (1982) Sorption of HCl by weak base macroporous anion exchanger AN-511. Russ J Appl Chem (Zhurnal prikladnoy khimii (in Russian)) 55(9):2093–2095
264. Hashida I, Nishimura M (1975) Adsorption of hydrogen chloride on porous resins with different functional groups. J Chem Soc Jap Chem Ind Chem 3:569–571
265. Kutovaya LM, Bartkovskaya YuF, Zverev MP (1990) Application of new anion exchange fibers to sulfur dioxide, hydrogen sulfide and hydrogenchloride sorption. In: Abstracts of presentations of the all-union meeting. The modern aspects of synthesis and production of ion exchange materials (Tez. Dokl. Vsesojuznjgo soveschania “Sovremennie acpekty sinteza I proizvodsyva ionoobmennyh materialov (in Russian)), Cherkassy, pp 133–134
266. Barash AN, Zverev MP, Kalianova NF (1987) Exploitation properties of chemosorption fiber Vion AN-1. Fibre Chem (Khimicheskie volokna (in Russian)) 3:37–38
267. Zverev MP, Barash AN, Grebennikov SF (1988) The processes of mass exchange at the gas sorption by fibers Vion AN-1 and Vion AS-1. Fibre Chem (Khimicheskie volokna (in Russian)) 6:9–10
268. Soldatov VS, Polhovski EM, Sosinovich ZI (2004) Non-exchange sorption of electrolytes by ion exchangers. I. Sorption of hydrochloric and perchloric acids and their sodium salts by Dowex 1x8 resin. React Funct Polym 60:41–48
269. Subbotin AI, Tkachenko VI (1975) The air purification from vapor of acetic acid by ion exchangers. Plastics (Plasticheskije massy (in Russian)) 5:38–39
270. Kelman BYa, Tereschenko VN, Novikov PD (1968) Ion exchange resins (Ionoobmennye smoly (in Russian)). NII Plastmass, Moscow
271. Smola VI, Keltsev NV (1976) Atmosphere protection from sulfur dioxide (Zaschita atmocfery ot dvuokisi sery (in Russian)). Metallurgia, Moscow
272. Cole R, Shulman HL (1960) Adsorbing sulfur dioxide on dry ion exchange resins for reducing air pollution. Ind Eng Chem 52(10):859–860
273. Vulikh AI, Bogatyrev VA, Aloviajnikov AA (1970) Application of ion exchange resins for sorption gases. Zhurnal vsesoyuznogo khimicheskogo obschestva Mendeleeva (in Russian) 15(4):425–429
274. Hashida I, Nishimura M (1976) Adsorption and desorption of sulfur dioxide by macroreticular strong base anion exchangers. J Chem Soc Jpn Chem Ind Chem 4:131–135
275. Ksendzenko VI, Zil’berg GA, Gorilovskaya NB (1976) Adsorption of sulfur compounds containing gases. In: Ion exchange chromatography (Ionnyj obmen i khromatografiya (in Russian)), vol 20. Voronezh University, Voronezh, p 257
276. Vulikh AI, Zagorskaya MK, Varlamova LV et al (1982) Sorption of sulfur dioxide by stationar layers of the granular ion exchangers from humid gas mixtures. Russ J Appl Chem (Zhurnal prikladnoy khimii (in Russian)) 55(6):1297–1302
277. Krejcar E (1965) Sorption of sulfur dioxide by anion exchangers. Chem Prum 15(2):77–79

278. Gloviak B, Gostomczyk A (1973) Investigation of sorption of sulfur dioxide by ion exchangers. Staub-Reinhalt Luft 33(10):387–390
279. Khr B, Pantofchieva L, Khistov Y (2000) Adsorption of sulfur dioxide by fixed bed anion exchange resin. Theor Technol (Teoreticheskie osnovy khimicheskoy tekhnologii (in Russian)) 34(2):160–164
280. Pinaev VA, Muromtseva LS (1968) The sorption of sulfur dioxide by the synthetic resins. Russ J Appl Chem (Zhurnal prikladnoy khimii (in Russian)) 41(9):2092–2094
281. Ksendzenko VI, Zilberg TA, Gorilovskaya NB et al (1974) The sorption of sulfur containing gases on anion exchangers. Proc Moscow Inst Fine Chem Technol 4(2):68–74
282. Buzanova GP, Volf IF, Onokhin SA (1976) The gas purification from sulfur dioxide by ion exchange sorption. Theory Pract Ion Exch Processes (Teoria i praktika sorbtsionnykh protsessov (in Russian)) 11:102–106
283. Nikandrov GA, Vulikh AI, Zagorskaya MK (1971) The sorption of sulfur dioxide by anion exchangers. In: Ion Exch and Chromatography (Ionnyj obmen i khromatografiya (in Russian)), vol 15. Voronezh University, Voronezh, pp 218–220
284. Nikandrov GA (1978) Gas purification from sulfur dioxide by ion exchange sorbents. Ind Sanitary Gas Purif (Promyshlennaya i sanitarnaya ochistka gazov (in Russian)) 4:14
285. Youngquist GR, Garg SK (1972) Sorption of sulfur dioxide by macroreticular ion exchange resins. Ind Eng Chem Proc Des Dev 11(2):259–261
286. Hashida I, Nishimura M (1972) Adsorption of sulfur dioxide from gas on porous resins. J Chem Soc Jpn Chem Ind Chem 1(1):179–184
287. Layton L, Youngquist GR (1969) Sorption of sulfur dioxide by ion exchange resins. Ind Eng Chem Proc Des Develop 8(3):317–324
288. Hashida I, Nishimura M (1973) Adsorption of sulfur dioxide by porous ion exchangers. J Chem Soc Jpn Chem Ind Chem 6:1195–1200
289. Avgul NN, Belyakova LD, Dadugina NG (1976) The adsorption heat of sulfur dioxide on macroporous weak base anion exchangers. Colloid J (Kolloidnyj zhurnal (in Russian)) 38 (1):129–132
290. Belyakova LD, Galitskaya NB, Kiselev AV (1976) Adsorption of sulfur dioxide by macroporous vinylpyridine anion exchangers. Colloid J (Kolloidnyj zhurnal (in Russian)) 38(6):1060–1064
291. Belyakova LD, Galitskaya NB, Kiselev AV (1977) Adsorption of sulfur dioxide by macroporous vinylpyridine anion exchangers. Colloid J (Kolloidnyj zhurnal (in Russian)) 39(2):339–343
292. Belyakova LD, Galitskaya NB, Kiselev AV (1979) Adsorption of sulfur dioxide by macroporous vinyl pyridine anion exchangers. Colloid J (Kolloidnyj zhurnal (in Russian)) 41(4):627–631
293. Belyakova LD, Galitskaya NB, Kiselev AV (1981) Adsorption of sulfur dioxide by macroporous vinyl pyridine anion exchangers. Colloid J (Kolloidnyj zhurnal (in Russian)) 43(6):1027–1033
294. Varlamova LV, Ksenzenko VI, Aloviajnikov AA (1984) Dynamics of sulfur dioxide sorption on fibrous cation exchanger Vion KN-1. Ind Sanitary Gas Purif (Promyshlennaya i sanitarnaya ochistka gazov (in Russian)) 6:11
295. Bergen RL (1971) Removal of SO2 from the gas mixture by the fibers containing polymeric amines. In: Special center symposium. Proceedings of symposium on technology for the future to control industrial and urban wastes. University of Missouri-Rolla, pp 73–75
296. Shunkevich AA, Sergeev GI, Elinson IS (1990) Fibrous ion exchangers in protection of the atmosphere. Zhurnal vsesoyuznogo khimicheskogo obschestva Mendeleeva (in Russian) 35 (1):64–72
297. Kosandrovich EG, Soldatov VS (2006) Sorption of sulfur dioxide from the air by fibrous polyampholyte. In: Proceedings of the III Belarus conference. Scientific technical problems of production of chemical fibers in Belarus, Mogilev, pp 28–33

298. Hwang TS, Kim YS, Lee HK et al (2004) Removal of toxic gases on strong- and weak-base anion exchange fibers. J Korean Ind Eng Chem 10(4):504–510
299. Hwang TS, Kim YS, Park JW et al (2004) Adsorption properties of SO_2 on PAN-based fibrous ion exchanger and its potential for air purification. J Korean Ind Eng Chem 10(1):139
300. Hwang TS, Choi JE, Kang KS (2002) Adsorption properties of SO_2 using fibrous strong-base anion ion exchange scrubber. Polymer (Korea) 26(5):661
301. Aldabas I, Bogoeva G, Stojanovski G et al (1990) Cation exchangers on the base of PAN with capability to sorb SO_2. Hem Ind 44(4):129–136
302. Druzhinina TV, Lishevskaya MO, Lazarev My et al (1990) Influence of the functional groups nature of the fibrous anion exchangers on the polycaproamide base on the sorption of sulfur and nitrogen oxides and hydrogen chloride. In: Abstracts of presentations of the all-union meeting. The modern aspects of synthesis and production of ion exchange materials (Tez. Dokl. Vsesojuznjgo soveschania Sovremennie acpekty sinteza i proizvodsyva ionoobmennyh materialov (in Russian)), Cherkassy, pp 48–49
303. Varlamova LV, Larina TV, Sudareva VE (1985) The influence of low temperatures on sorption of sulfur dioxide on ion exchangers. Color Metals (Tsvetnye metally (in Russian)) 11:31–32
304. Katz BM, Kutovaya LM, Lazarev MYu et al (1989) Sorption of sulfur dioxide and water vapor by chemosorption fibers on the base of polyacrylonitrile. Russ J Appl Chem (Zhurnal prikladnoy khimii (in Russian)) 62(2):289–293
305. Neyaglov AA, Digurov NG, Bukharkina TV et al (1991) Kinetics and mechanism of the liquid phase oxidation of hydrogen sulfide by chelate complex of tri-valent iron (Fe^{3+}-EDTA). Kinet Catal (Kinetika i kataliz (in Russian)) 32(3):548–552
306. Potapova LL, Yegiazarov YuG, Soldatov VS et al (1998) The oxidation of hydrogen sulfide on fibrous anion exchanger FIBAN A-1 with coordination saturated complex Fe-EDTA. Proc Acad Sci Belarus (Doklady Akademii Nauk Belarusi (in Russian)) 42(3):54–57
307. Potapova LL, Shunkevich AA, Akulich ZI et al (2000) The oxidation of hydrogen sulfide on fibrous anion exchangers with fixed Fe-EDTA complexes. Russ J Appl Chem (Zhurnal prikladnoy khimii (in Russian)) 73(5):780–784
308. Soldatov VS, Kashinsky AV, Martinovich VI (2009) Catalytic air purification from hydrogen sulfide by ion exchange fibers FIBAN. Chem Technol (Khimicheskaya tekhnologia (in Russian)) 6:364–369
309. Nabieva ZM, Merenkov KV, Yusupov MM (1968) Nitrogen oxides sorption from the exhaust by ion exchange adsorbents. In: Proceedings of conference. Ion exchange materials and their applications (Ionoobmennye materially i ikh primenenie (in Russian)), Alma-Ata, SSSR, pp 232–235
310. Ganzha GF, Filipov VI, Mazur NA et al (1978) Sanitary air purification from nitrogen oxides by fibrous ion exchange materials. Ind San Gas Purif (Promyshlennaya i sanitarnaya ochistka gazov (in Russian)) 3:14–15
311. Soldatov VS, Kashinsky AV, Korshunova TA et al (2006) The air purification from nitrogen dioxide by aqueous solution of carbamide on ion exchange fibrous carrier. Chem Technol (Khimicheskaya tekhnologia (in Russian)) 5:34–38
312. Miagkoy ON, Serdiukova MI, Perunova NA (1982) Sorption of the hydrazine vapor by ion exchange materials. In: Theory and practice of sorption processes (Teoria i praktika sorbtsionnykh protsessov (in Russian)), vol 15. Voronezh University, Voronezh, pp 73–77
313. Asaulova TA, Kats BM (1976) Air purification from chlorine and chlorine compounds by ion exchange fibrous materials. In: Ion exchange and chromatography (Ionnij obmen i khromatografia (in Russian)). Voronezh University, Voronezh, pp 253–254
314. Serdiukova MI, Petrunin AN, Miagkoj ON et al (1976) Sorption of iodine on fibrous polyacrylonitrile anion exchangers. In: Ion exchange and chromatography (Ionnij obmen i khromatografia (in Russian)). Voronezh University, Voronezh, pp 200–201

315. Khamizov RKh, Tikhonov NA (2002) On the possibility of purification of anodic gases of aluminum production from fluoride and sulfur compounds by the wet filtration on anion exchange fibrous materials. Sorption and Chromatographic 3(2):331–339
316. Miagkoj ON, Ivanova EV, Serdiukova MI et al (1985) Hygienic properties of ion exchangers AV-17-8 and VION AN-1 for sanitary air purification from ozone In: Theory and practice of sorption processes (Teoria i praktika sorbtsionnykh protsessov (in Russian)). Voronezh University, Voronezh, pp 84–89
317. Serdiukova MI, Lovchinovskaya TA, Krutskikh AS et al (1986) Sanitary chemical properties of filtering ion exchange material VION AN-1 for air purification from ozone. Higiene Sanit (Gigiena i sanitariya (in Russian)) 5:76–79
318. Mohan D, Charles U, Jr P (2007) Arsenic removal from water/wastewater using adsorbents – a critical review. J Hazard Mater 142(1–2):1–53
319. DeMarco MJ, SenGupta AK, Greenleaf JE (2003) Arsenic removal using a polymeric/inorganic hybrid sorbent. Water Res 37:164–176
320. Cumbal L, SenGupta AK (2005) Arsenic removal using polymer-supported hydrated iron (III) oxide nanoparticles: role of Donnan membrane effect. Env Sci Tech 39:6508–6515
321. Lin J-C, SenGupta AK (2009) Hybrid anion exchange fibers with dual binding sites: simultaneous and reversible sorption of perchlorate and arsenates. Environ Eng Sci 11:1673–1683
322. Sarkar S, Blaney LM, Gupta A et al (2008) Arsenic removal from ground water and its safe containment in a rural environment: validation of a sustainable approach. Environ Sci Technol 42:4268–4273
323. Blaney LM, Cinar S, SenGupta AK (2007) Hybrid anion exchanger for trace phosphate removal from water and wastewater. Water Res 41:1603–1613
324. Sarkar S, Blaney LM, Gupta A et al (2007) Use of ArsenXnp, a hybrid anion exchanger, for arsenic removal in remote villages in the Indian subcontinent. React Funct Polym 67:1599–1611
325. Puttamaraju P, SenGupta AK (2006) Evidence of tunable on-off sorption behaviors of metal oxide nanoparticles: role of ion exchanger support. Ind Eng Chem Res 45:7737–7742
326. Cumbal L, Greenleaf J, Leun D et al (2003) Polymer supported inorganic nanoparticles: characterization and environmental applications. React Polym 54:167–180
327. Zhao D, SenGupta AK (2000) Ligand separation with a copper(II) -loaded polymeric ligand exchanger. Ind Eng Chem Res 39(2):455–462
328. Solntseva DP, Krasnov MS, Kalinina RN et al (2003) The modified ion exchange for fluorination of the drinking water. Plastics (Plasticheskie massy (in Russian)) 11:43–45
329. Emel'yanova GI, Gorlenko LE, Voronova LV et al (1999) Low-temperature catalytic oxidation of hydrogen sulfide into sulfur on VION metal-containing filament. Kinet Catal 40 (1):15–17
330. Seko N, Basuki F, Tamada M et al (2004) Rapid removal of arsenic (V) by zirconium (IV) loaded phosphoric chelate adsorbent synthesized by radiation induced graft polymerization. React Funct Polym 59(3):235–241
331. Rajakovic LV, Mitrovic MM (1992) Arsenic removal from water by chemisorption filters. Environ Pollut 75(3):279–287
332. Lenoble V, Chabroullet C, Al Shukry R et al (2004) Dynamic arsenic removal on MnO_2-loaded resin. J Colloid Interface Sci 280(1):62–67
333. Mitskevich DE, Soldatov VS, Sokol VP et al (2010) A system oxidant: sorbent for the purification of drinking water from oxyanions of As(III) и As(V). Russ J Appl Chem (Zhurnal prikladnoy khimii (in Russian)) 83(3):415–420
334. Soldatov V (2003) Selective fibrous ion exchangers and sorbents. In: Science papers of the institute of organic and polymer technology of the Wroclaw University of Technology, Wroclaw. Proceedings of XVI science conference, Wroclaw. Oficyna wydawnicza Polytechniki Wroclawskiej, pp 110–113

335. Yoshioka T, Shimamure M (1984) Studies of polystyrene-based ion exchange fiber. II. A novel fiber-form catalyst for sucrose inversion and methyl acetate hydrolysis. Bull Chem Soc Jpn 52(2):334–337
336. Yegiazarov YuG, Potapova LL, Radkevich VZ et al (2001) New catalytic systems on the base of fibrous ion exchangers. Chem Sustain Dev (Khimia v interesakh ustoichivogo razvitia (in Russian)) 9:417–431
337. Author's certificate USSR SU 1768580 A1 (1990)
338. Yegiazarov YuG, Radkevich VZ, Johann J et al (2000) Fibrous catalyst for oxidation of hydrogen with oxygen. React Funct Polym 44:145–152
339. Radkevich VZ, Shunkevich AA, Kistanova IE et al (2000) The activity of palladium catalysts on the base of fibrous sulfonic cation exchanger FIBAN K-1 in reaction of hydrogen oxidation. Russ J Appl Chem (Zhurnal prikladnoy khimii (in Russian)) 73(11):1861–1864

Chapter 10
Chelating Ion Exchangers: Theory and Applications

Dhiraj Sud

Abstract The lack of selectivity, sensitivity, and capacity of the conventional ion exchange resins particularly for trace heavy metal ions had led to the development of metal-ion-specific exchange resins known as chelating ion exchangers or chelating ion exchange resins. The chelating ion exchangers are the polymers covalently bonded to ligands forming complexes with metal ions through functional groups. The chelating ion exchangers consist essentially of two components – polymeric matrix and chelating ligands. A variety of polymeric matrices, namely, inorganic (silica) and organic – both natural and synthetic [polystyrene divinyl benzene (PS-DVB), polymethacrylate (PMA)] – have been employed for the synthesis of chelating exchangers. Most of the commercially available chelating exchangers are silica- or PS-DVB-based and have diverse applications. A large number of diversified chelating ligands may be employed for the synthesis of chelating exchangers, such as the carbamates, β-diketones, diamine, iminodiacetic acid and amino acids, aldoxime, aminophosphonic acids, various azo-triphenylmethane dyes, and 8-hydroxyl quinolinol. These chelating ligands are incorporated into a polymeric matrix by different methods. The chelation exchange mechanisms are found to be slower than ion exchange, and efficient separations are possible only by the choice of the correct chelating functional group. Furthermore, the chelating ligand should have a broad spectrum of chelating action and have no special selectivity for one or two metal ions. The structure, coordination chemistry, and applications of most commonly employed chelating ion exchangers have been discussed. Chelating ion exchangers containing iminodiacetic acid (IDA) chelating group have been the most studied for metal separations. The efficiency and versatility of another chelating exchanger containing the chelating ligand aminophosphonate have been addressed. The development of new chelating ion exchange materials with special

D. Sud (✉)
Department of Chemistry, Sant Longowal Institute of Engineering and Technology, (Deemed to be University), Longowal 148106, India
e-mail: suddhiraj@yahoo.com

Inamuddin and M. Luqman (eds.), *Ion Exchange Technology I: Theory and Materials*, DOI 10.1007/978-94-007-1700-8_10,

chelating properties can provide better kinetics of interaction between metal ion and chelating groups, and better understanding of their coordination chemistry for surfaces can revolutionized the domain of chromatographic separations and analysis of trace metal ions particularly from complex matrices.

10.1 Introduction

Ion chromatography using highly efficient ion exchange substrates is now a well-established technique for the separation of metal cations. This technique is preferred for routine analysis because of its relative cheapness, ease of automation, and online capability. However, a limitation of the technique is the selectivity of metal separation, particularly of trace metals in the presence of other massive-amount metals. The high concentration of alkali metal salts could swamp the column and degrade or destroy the separation of metal cations, thereby lowering the column capacity. These limitations of the selectivity and capacity of the conventional ion exchange resins have led to the development of a new class of specific and selective ion exchange resins known as chelating ion exchange resins. A chelating ion exchange resin consists of polymeric complexing and chelating compounds showing selectivity and specificity toward the particular ion. These chelating ion exchangers can be used in analytical separation columns in an ion chromatograph system just like ion exchange columns and have added advantage of insensitivity to changes to ionic strength. Chelating ion exchange resins/materials have increasing application in trace metals as well as alkali and alkaline earth metal ion separations, preconcentration, and recovery of trace metals in pollution control and industrial process, hydrometallurgy, and various miscellaneous applications [1–3].

The first chelating resin, a polystyrene derivative containing a dipicrylamine group, was originally proposed in 1940 by Skogseid [4] for the selective isolation of K^+. The earliest covalently bonded commercially available chelating exchange substrate was Chelex 100 (or Dowex A-1), a low cross-linked polystyrene substrate containing bonded IDA. The first notable analytical achievement with this resin was the determination of trace metals in seawater by Riley and Taylor [5] in 1968. There onward, a considerable number of chelating substrates other than Chelex 100 are being developed and studied. The development of these high-performance chelating substrates for trace metal separations and determinations, opened vistas for their use in the IC system and led to a new important technique called high-performance chelation ion chromatography (HPCIC).

10.2 Chelation

The term chelation is used for a special type of metal complexation having ring structures with two or more bonds between binding sites. A chelating ligand possesses more than one coordination position, and binding takes place through

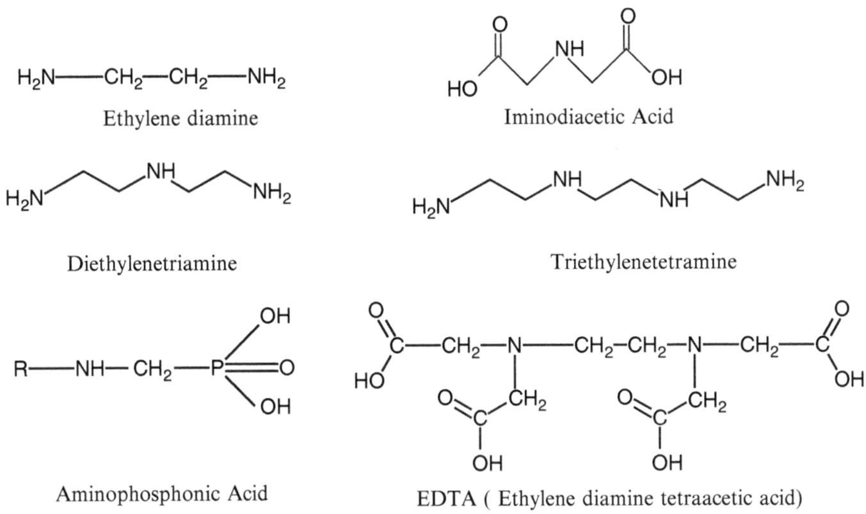

Fig. 10.1 Examples of chelating ligands

more than one donor atom present in the ligand. Depending upon the donor atom – two, three, four, and six – the chelating ligands are termed as bidentate, tridentate, tetradentate, and hexadentate ligands. Examples of some common polydentate ligands are given in Fig. 10.1.

Chelating or coordinating resins are polymers with covalently bonded functional group atoms containing one or more donor atoms that are capable of forming complexes directly with metal ions. In these resins, the most frequently used chelating ligand contains functional group atoms – nitrogen (e.g., N present in amines, azo groups, amides, nitrilcs), oxygen (e.g., O present in carboxylic, hydroxyl, phenolic, ether, carbonyl), sulfur (e.g., S present in thiols, thiocarbamates, thioethers), the ligands containing N and O as donor atoms (hydroxylamine, oxine), S and N as donor atoms (dithiocarbamates), and other related ligand atoms.

A chelating ion exchanger consists essentially of two components, that is, a chelating ligand and a polymeric matrix or the support.

10.3 Polymeric Matrix

Fundamentally, the two types of polymeric matrices used for the synthesis of chelating resins are inorganic and organic.

10.3.1 Inorganic Polymeric Matrix

Silica is one of the most extensively used inorganic supports and has diverse applications. The advantages of inorganic supports are high mechanical strength, high thermal stability, ability to withstand high pressure, and stability toward organic solvents and mineral acids. However, these supports lack high degree of functionalization and have relatively low ion exchange capacity. A number of silica-based chelating exchangers are commercially available. Few examples of silica-based chelating exchangers are given in Table 10.1.

Although silica-bonded chelating exchangers exhibited favorable high efficiency, they are unstable at high pH value, the use of silica support for the attachment of chelating ligands working in alkaline solution is not possible [7].

10.3.2 Organic Polymeric Matrices

Organic matrices used in chelating ion exchangers may be natural or synthetic polymers. Among the naturally occurring polymers used as support, cellulose has been the most studied. Other naturally occurring polymers are chitin-poly (N-acetyl-D-glucosamine) and its deacylated derivative, chitosan.

Synthetic organic polymeric matrices used as supports or matrices for chelating ion exchangers include polymers of esters, amides, and alkyl halides, and styrene divinylbenzene copolymers synthesized by addition or condensation polymerization processes. The most frequently used substrate, styrene divinylbenzene copolymers, is synthesized via addition polymerization. The structure of a styrene divinylbenzene (ST-DVB) is shown in Fig. 10.2.

This additional resin contains primarily polystyrene, and a small amount of divinylbenzene is added during the polymerization to cross-link the resin. These cross-linked polymers are mechanically stable, resist hydrolytic cleavage and pH changes, and are stable at relatively high temperature. Typically, 2–25% weight of the cross-linking compound is used for microporous resins and upto 55% weight cross-linking for macroporous resins. Table 10.2 includes some of the commonly used ST-DVB-based chelating ion exchange resins.

10.3.3 Microporous and Macroporous Resins

Polymeric matrices can be further classified as microporous or macroporous resins. Microporous resins are produced by suspension polymerization in which styrene and divinylbenzene are suspended in water droplets and benzoyl peroxide is added to initiate the polymerization process. The monomers are kept in suspension in a reaction vessel through rapid uniform stirring. The resulting uniform beads are microporous. The size distribution of the beads depends on the stirring rate, that is, faster stirring produces smaller beads. However, the microporous resin undergoes a large change in volume upon conversion from one ionic form to another. Such volume changes can disrupt the uniform packing in a column. Even a low degree of swelling is undesirable for chromatography.

Table 10.1 Commercially available chelating ion exchanger (Adapted from Ref. [6] with kind permission of © Elsevier (1984))

Chelating ligand	Manufacturer	Commercial name	Polymeric matrix
Iminodiacetic acid	Diamond Shamrock	IMAC SYN 101	Polystyrene X% DVB
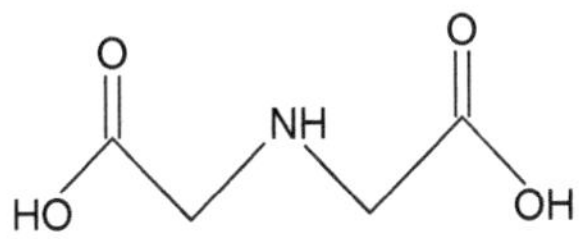	Diamond Shamrock	Duolite ES 466	Polystyrene X% DVB
Coordinating atoms: N,O,O	DOW chemical	Dowex A-1	Polystyrene X% DVB
	Bio-rad	Chelax-100	Polystyrene X% DVB
	Mitsubishi	Diaion CR-10	Polystyrene X% DVB
	Reanal	Ligandex E	Polystyrene X% DVB
	Rohm & Hass	IRC-718, Wofatet MC50	Polystyrene X% DVB
	Permutit Co.Ltd.	Permutit S-1005	Polystyrene X% DVB
	Bayer A.G.	TP-2007	Polystyrene X% DVB
Aminophosphonic acid	Diamond Shamrock	Duolite ES-467	Polystyrene
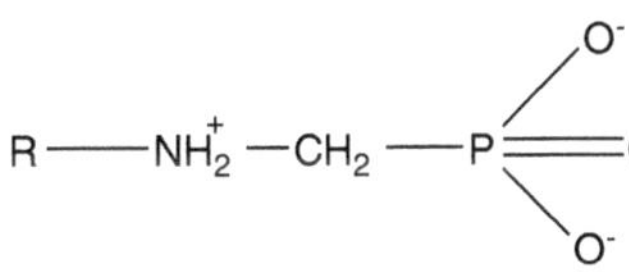	Purolite	Purolite S950	Polystyrene-divinyl benzene copolymer
Coordinating atoms: N, O, O			
Dithiocarbamate	Nippon soda	Misso ALM-525	Polystyrene
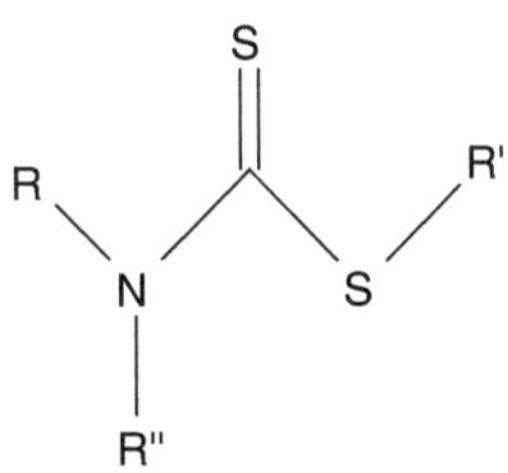	Sumitomo	Sumichelate Q-10R	Polyacryclic
Coordinating atoms: N, S, S			

(continued)

Table 10.1 (continued)

Chelating ligand	Manufacturer	Commercial name	Polymeric matrix
8-Hydroxyquinoline 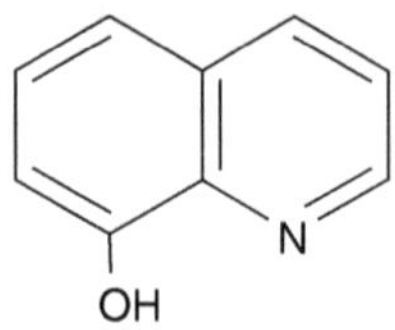	Lachema	Spheron oxime 1000	Poly(hydroxyl ethyl methacrylate co-ethylene diamine metha acrylate)
Coordinating atoms: N, O	Iontosorb	Iontosorb oxin (IO)	–

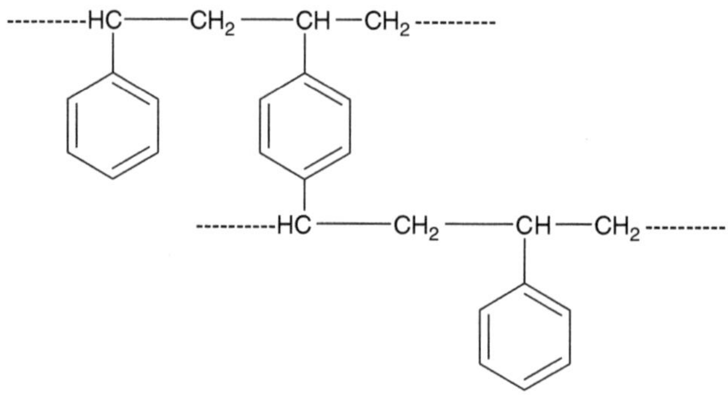

Fig. 10.2 Structure of styrene –divinyl benzene (SDVB) copolymer

Another type of copolymer are macroporous resins, sometimes called macroreticular resins, which are prepared by a special suspension polymerization process. During polymerization, the monomers are kept as a suspension of a polar solvent. The suspended monomer droplets also contain an inert diluent that is a good solvent for monomers but not for polymerized beads. After polymerization is complete, the diluents are washed out of the beads to get rigid, spherical macroporous beads having high surface area. Macroreticular resins have a permanent macroscopic structure regardless of the ionic form. The surface area of macroporous resins is much higher (25–800 m^2/g) as compared with microporous resin surface area (1 m^2/g). The macroporous resins have an extensive network of large pores up to 50 nm in diameter and even larger throughout the interior of the resin beads, which allows exchange reactions to proceed rapidly.

Traditionally, most of the polymeric resins contain microporous beads, but now both microporous and macroporous resin beads are extensively employed for synthesis of chelating ion exchangers (Table 10.3).

Table 10.2 Chelating exchangers containing silica as polymeric matrix

S. No.	Structure of chelating ligand	Matrix(type. pore size, surface area, capacity)	Column and particle size	Separation	Ref.
1	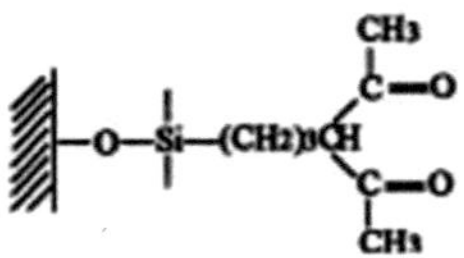	Nucleosil Si, 100–5, 10 nm, 350 m^2/g, 835 μmol ligand/g	135 × 4.6 mm, 5 μm	Co(II), Cd(II), Cu(II)	[8]
2	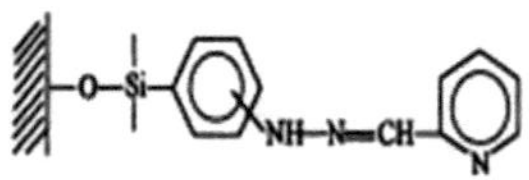	Pierce CPG, 24 nm,130 m^2/g, 25 μmol metal/g	250 × 4.6 mm, 37–74 μm	Fe(II), Co(II), Ni(II), Cu(II)	[9]
		Chromosorb LC-6, 12 nm, 400 m^2/g, 18 μmol ligand/g	250 × 4.6 mm, 5 μm	Mn(II), Fe(II), Cd(II), Zn(II), Co(II), Pb(II), Cu(II)	[10]
3	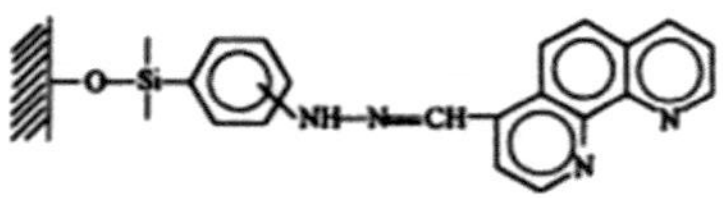	Chromosorb LC-6, 12 nm, 400 m^2/g, 39 μmol ligand/g	250 × 4.6 mm, 5 μm	Mn(II), Pb(II), Fe(II), Cd(II), Zn(II), Co(II)	[11]
4		Silasorb Si 300, 10 nm, 3000 m^2/g, 38 μmole metal/g	100 × 4.6 mm	Mg(II), Ca(II), Mn(II), Co(II), Cd(II), Zn(II), Pb(II), Cu(II)	[12]
5	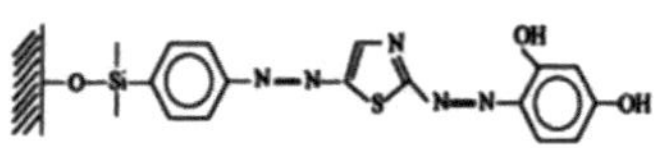	Silasorb Si 600, 6 nm, 570 m^2/g, 30 μmole metal/g	250 × 4.6 mm, 10 μm	Mn(II), Cd(II), Pb(II), Zn(II), Co(II), Cu(II)	[13]

Table 10.3 Chelating exchangers containing styrene-divinyl benzene (ST-DVB) as polymeric matrix

S. No.	Functional group	Type of resin and capacity	Column and particle size	Separated ions	Ref.
1	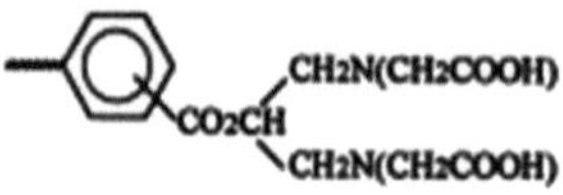	ST-DVB, XAD-4	28×6 mm, 45–75 μm	Cu(II), U(VI), Th (IV), Zr(IV)	[62]
2	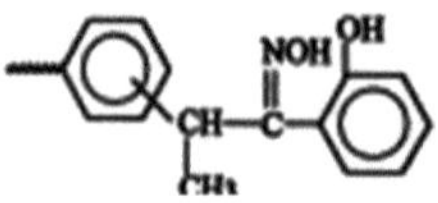	ST-DVB, XAD-4		Cu(II), Mo (VI)	[63]
3	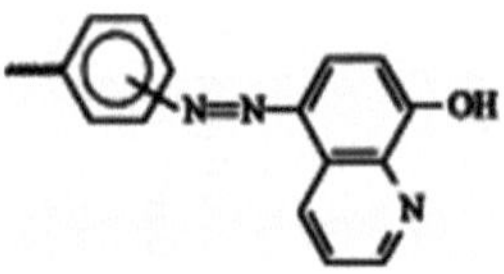	ST–4% DVB, BN-4, gel type, 0.042–1.33 mequiv. /g	250 × 3 mm, 7–10 μm	Mn(II), Fe (II), Co (II), Zn (II), Ni (II), Cu (II)	[64]
4	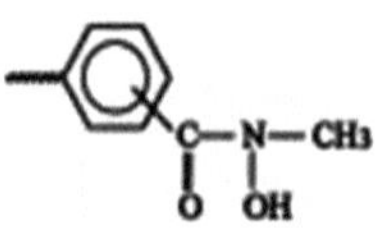	ST–DVB, XAD-4	–	Ti(IV), Zr (IV), Mo (VI), Fe(III), Th (IV), etc.	[65]
5	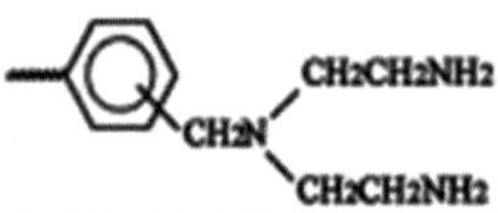	ST–2% DVB	45–75 μm	Zn(II), Ni (II), Cu (II)	[66]

10.4 Chelating Ligand

The chelating ligand is the most critical factor for the development of chelating ion exchangers, which can result in efficient and selective separations. Criteria for the selection of the chelating ligand to anchor onto the polymeric matrices are of paramount importance.

Gregor et al. [14] postulated the properties a suitable chelating ligand should have:

1. It should be capable of resin formation or substitution in the resin matrix.
2. It should be sufficiently stable to withstand the polymerization process.

3. It should be compact so as not to be sterically hindered by the dense resin matrix.
4. Both arms of the chelate structure should be on the same monomer in proper spatial configuration.

The chelating ligands which do not form 1:1 complexes, e.g., anthranilic acid, are not found suitable. A number of chelating ligands containing different functional groups such as iminodiacetic acid, hydroxylamine, aminophosphonic acid, oxines, thiols, dithiocarbamates, schiff bases, crown ethers, etc., have been incorporated into a variety of polymeric matrices.

Sahni and Reedijk [6] pointed that the essential requirements necessary for the synthesis of chelating ion exchange resins are:

1. The chelating ligand attached to a solid matrix should possess strong metal-binding properties and selectivity toward certain metal ions. According to the hard soft acid base (HASB) theory by Pearson [15], soft metal ions, for instance, gold(III), silver(I), and palladium(II) ions, show affinity to soft bases with donor atoms as O< N < S. On the other hand, hard metal ions, for instance, copper (II), iron(III), and aluminum(III) ions, show affinity to hard bases with donor atoms as O> N > S.
2. The chelating group should be capable of undergoing incorporation into a polymeric network and allows the process of polymerization or resinification. The chelating polymer should be resistant to strong acids and alkalies at elevated temperatures.
3. The chelating ligand should preferably be multidentate, permitting the formation of 1:1 chelate with metal ion. Chelating ion exchangers should be monofunctional as far as possible.
4. The chelating ion exchanger should possess good swelling properties, and compatibility between the polymer and the medium (generally aqueous) is essential; this can be regulated by the presence of hydrophilic groups on the polymer and the extent of cross-linking. A high degree of cross-linking results in increased mechanical strength, but the swelling properties of the reagents are adversely affected.

Based on the experience of the researchers working on the use of chelating ion exchangers on columns for the separation of trace metal ions, the following points are highlighted:

1. The choice of the correct chelating functional group is a critical factor for the efficient functioning of the separation system.
2. The efficiency of the separation system depends on the homogeneity of the bonded functional group.
3. Generally, bi- or tridentate ligands are preferred over those with higher denticity. The ligands with higher denticity have slower kinetics of dissociation because of the number of bonds involved with the metal.
4. The chelating groups should have a broad spectrum of chelating action and have no special selectivity for one or two separate metals.

10.4.1 Chemical Functionalization: Incorporation of Chelating Ligands into Polymeric Matrices

Chemical functionalization means incorporation of chelating ligands into the polymeric matrices. The chelating ion exchange resin is developed by fictionalization of the polymeric matrices. Some of the most frequently used chelating ligands are the carbamates, β-diketones, diamine, aminodiacetate and imino acids, various azo-triphenylmethane dyes, and 8-hydroxyl quinolinol.

Akelah and Sherrengton [16] have discussed three methods which have been employed to incorporate active functional groups into polymer chain:

(a) Direct polymerization and copolymerization of monomers containing the desired functional groups
(b) Chemical modification of preformed polymer
(c) Both (a) and (b)

The method involving the incorporation of chelating ligands into preformed polymers or matrices to synthesize chelating ion exchange resins is preferred over other methods. This method offers to control the degree of functionalization by varying the amount of the cross-linking agent and the extent of modification in preformed matrices. A large number of chelating ion exchangers has been prepared by attaching a chelating ligand to a preformed polymer, and the most extensively used preformed polymers are the copolymers of styrene divinylbenzene.

Apart from covalently bound and permanent impregnation techniques of forming a chelating surface, there is another approach that, as yet, has been with little study. This is the dynamic modification of the substrate where the chelating compound is added to the mobile phase.

10.5 Exchange Equilibrium Involving Chelating Ion Exchangers

The chelation exchange process involves the formation of a covalent coordinate bond between a metal cation and the functional groups present on the surface of the chelating exchange resin. On the other hand, ion exchange mechanism involves the equivalent exchange of ions between two or more ionized species located in different phases, and one of which is an ion exchanger, without the formation of a covalent bond. Thus, in chelation exchangers, the thermodynamics and kinetics of metal complex formation and dissociation control the process.

For efficient chromatographic separations of metal ions on columns containing chelating exchange resins, the chelating groups should be readily accessible so that the chelation of metal ions takes place readily without steric hindrance. The equilibrium between a divalent metal ion (M^{2+}) and a chelating resin ($RL^{-}H^{+}$) may be written:

$$M^{2+} + 2RL^{-}H^{+} \rightleftarrows (RL)_2M + 2H^{+}.$$

Mostly, the chelating ligands are conjugate bases of weak-acid groups and accordingly have strong affinity for hydrogen ions. Increasing the H^+ concentration in the eluent weakens the chelates and speeds up the elution. Therefore, an acidic eluent is used to control this equilibrium so that the retention factor of the sample metal ion is in the desired range. Separations of different metal ions will occur due to differences in their metal stability constants. Ligands being charged species, cation exchange due to electrostatic interactions may also take place, and generally the dominance of the chelation mechanism is achieved by ionic strength regulations. Other dominant factors affecting the separation are pH of the eluent and temperature. A strong correlation between the retention of metal ions on IDA-silica [17] and the corresponding stability constants has been demonstrated at high ionic strength of the eluent and increased temperature.

The separation of metal ions on a chelating resin column can also be achieved by using a complexing eluent (E^-), such as oxalate and tartrate at a fixed pH.:

$$(RL)_2M + 2E^- \rightleftarrows 2RL^- + ME_2.$$

The presence of a complexing agent in the mobile phase results in a competition between the chelating group on the surface of the substrate and the chelating group in the mobile phase. Thus, the retention of metal ions will be influenced by the type and concentration of L^- in the chelating resin, the pH of the eluent, and the type and concentration of E^- in the eluent. Because the formation and dissociation of metal chelates are slower than a simple ion exchange equilibrium, it is essential to select chelating resins with fast kinetics. When a complexing eluent is involved, the kinetic situation may become more difficult, and slower equilibrium between the metal chelate and the eluent chelate also exists.

10.6 Chelating Ion Exchangers Containing Chelating Ligands

Chelating ion exchangers containing different chelating ligands are extensively used for chromatographic separations. The chelating ligands include the iminodiacetic acid, dithiocarbamates, β-diketones, diamine, aminodiacetate and imino acids, various azo-triphenylmethane dyes, and 8-hydroxyl quinolinol. The chelating exchangers containing commonly used ligands along with their applications are discussed in this section.

10.6.1 Chelating Ion Exchangers Containing Iminodiacetic Acid Functional Group

Iminodiacetic acid and its derivatives have been incorporated into a variety of polymeric matrices using addition and condensation polymerization. The chelating

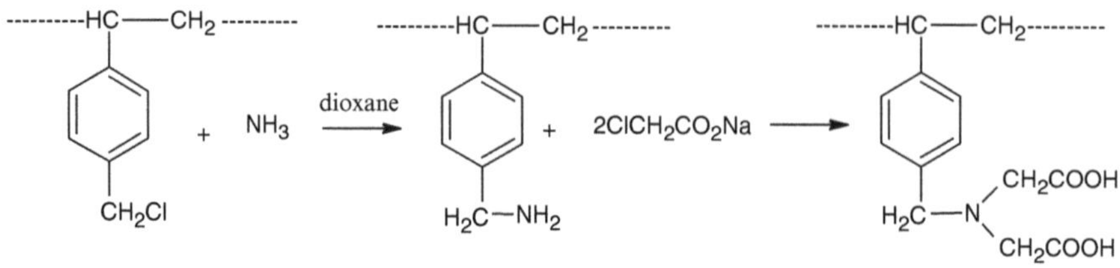

Fig. 10.3 Synthesis of polystyrene-iminodiacetic acid [18]

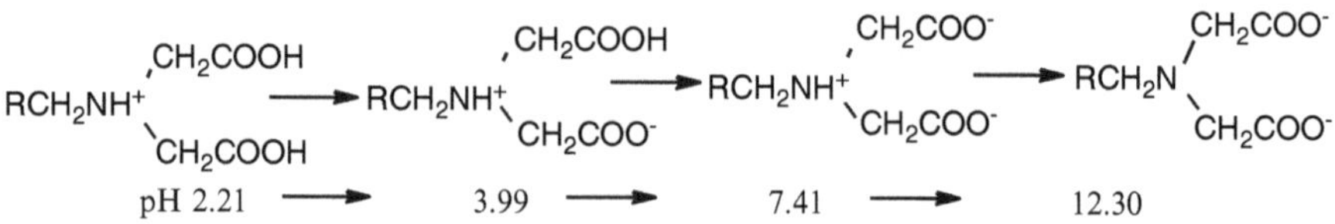

Fig. 10.4 Structure of iminoacetic acid ligand as function of pH

ion exchangers are synthesized by the modification of a polymer through a sequence of reactions on the polymer itself. The best route for the synthesis of iminodiacetic acid resin on a polystyrene matrix consists in the treatment of cross-linked polystyrene with iminodiacete, and its subsequent hydrolysis [18] is depicted in Fig. 10.3. The chelating resin was first manufactured by the Dow Chemical Company and was marketed as Dowex A-1. The iminodiacetic acid resin contains average out of every 8 rings in the resin 5 is substituted with iminodiacetic acid.

The resins containing iminodiacetic functional groups that are now commercially available under different names are given in Table 10.1. The free acid form of an IDA resin has a symmetric structure, and various forms of the ligand as a function of pH are given in Fig. 10.4. In alkaline solutions, the protons can be neutralized in two consecutive steps, leading to a deprotonated form which exists only at $pH > 12$. Furthermore, at high proton concentrations in the liquid phase, the nitrogen atom is protonated, and thus, the resin can absorb anions [19].

Dowex A-1 acts as a tridentate ligand bonding through the imino nitrogen atom and through two carboxylic oxygen atoms involving deprotonation (Fig. 10.5). These chelating ligand forms mainly 1:1 complexes with different metal ions, though the possibilities of 1:2 metal to ligand complexes may occur.

Most of the IDA metal complexes are being kinetically labile and moderately soluble. These properties ensure the reversible sorption of metal ions. Thus, the IDA function is one of the most favorable from the view of kinetics and specificity of chelate formation, and IDA emerged as the most promising ligand suitable for chromatographic separations. Among the commercially available resins, Dowex A-1 and Chelex-100 have been used for the selective separation of various metal ions from alkali and alkaline earth metal ions as a function of pH, even from higher ionic strength. The selectivity coefficients of resin relative to calcium, K^M Ca, are 4.9 for Mn^{2+}, 15.2 for Co^{2+}, 19.8 for Zn^{2+}, 52 for Ni^{2+}, and 500 for Cu^{2+} [20]. The affinity of IDA stationary phase toward transition metal ion is of great

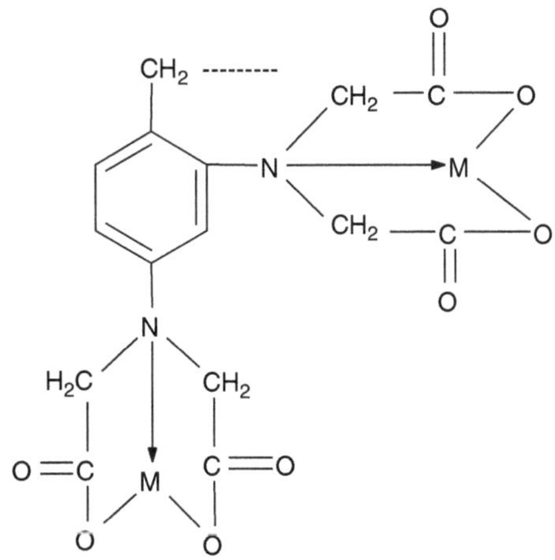

Fig. 10.5 Possible metal coordination of chelating ion exchangers

advantage for the determination of trace metals in samples containing high levels of alkali and alkaline earth metal ions.

A comparative study between macroreticular and microreticular chelating resins containing iminodiacetate groups has been preformed. Macroreticular chelating ion exchangers containing iminodiacetate groups prepared from XAD-1, macroreticular styrene-divinyl copolymers, show selectivity among metal ions similar to those of the commercially available microreticular chelating resins [21]. The selectivity order for the macroreticular iminodiacetate resin is $K^+ < Mg^{2+} < Ca^{2+} < Mn^{2+} < Ni^{2+} < Co^{2+} < Zn^{2+} << Cu^{2+}$. Hering [22] determined the following order for a gel-type iminodiacetate resin: $Na^+ < Mg^{2+} < Ca^{2+} < Co^{2+} < Zn^{2+} < Mn^{2+} < Ni^{2+} << Cu^{2+}$, while Rossets [20] determined the selectivity of Dowex A-l as $Mg^{2+} < Mn^{2+} < Co^{2+} < Zn^{2+} < Ni^{2+} << Cu^{2+}$. The elution order with the macroreticular chelating resin is therefore similar to that previously observed with the microreticular chelating resins. However, the chromatographic separation on the macroreticular chelating resin is rapid and sharp (Fig. 10.6).

A silica-based (Nucleosil 300-7) IDA substrate was synthesized for the determination of trace metal ions in seawater [23]. The effect of various complexing eluents (ethylenediamine, PDCA, tartrate) on the speed of separation was studied. A tartrate eluent (pH 2.5) was found suitable for the determination of Co(II) < Zn (II) < Cd(II) in seawater, the analysis time was ~20 min. The factors affecting the elution parameters to optimize separation on a commercial IDA-silica packing (Biochem Mach, Ruslar) are ionic strength and pH. Increasing the ionic strength negates the ion exchange interactions and ensures that chelation is the dominant mechanism. With pH being the fundamental parameter to manipulate retention, chelating groups are conjugate bases of weak acids and correspondingly have high affinity for protons. Increasing the temperature increases the retention of metal ions due to the "chelate effect" and changes in entropy. Under optimized conditions, 35 mM potassium chloride, 65 mM potassium nitrate, pH 2.5, 50°C, and Cd(II) < Co(II) < Zn(II) were determined in freshwater samples, using PAR

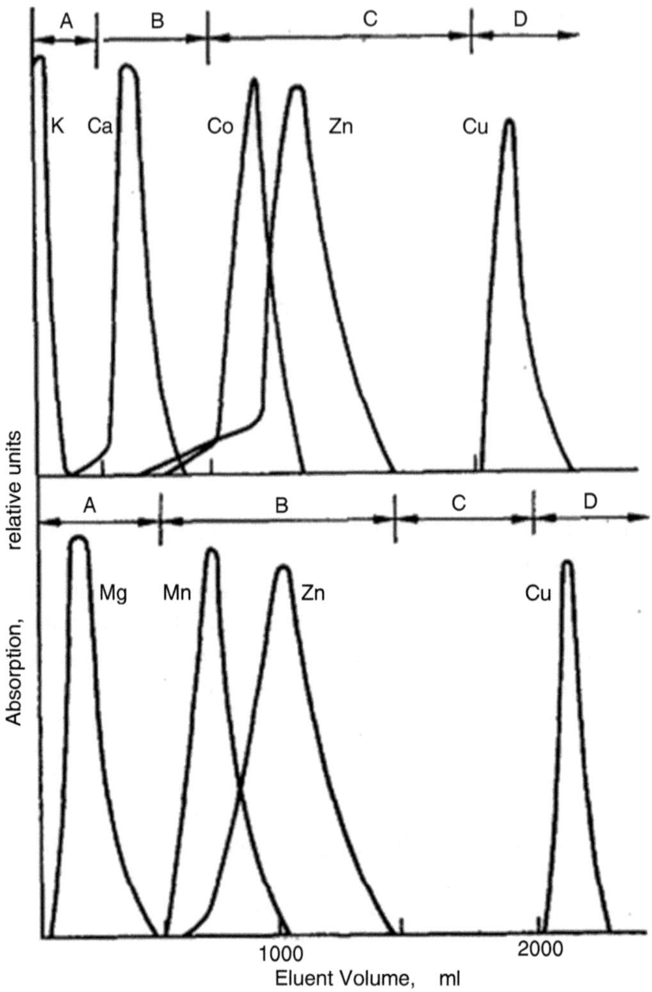

Fig. 10.6 Chromatographic separations with the macro reticular iminodiacetate ion exchange resin. Buffer solutions: A-pH 4.3, B-pH 3.1, C-pH 2.0, D-pH 1.2 (Reprinted from Ref. [21] with kind permission of © Elsevier (1970))

(495 nm). Mn(II) coeluted with the alkaline earths on the solvent front at lower temperatures [24].

The silica-gel-bonded IDA substrate-based chromatography IC column was employed for chromatographic separation of 14 lanthanide ions and yttrium [25]. The influence of nitric acid concentration, ionic strength of the eluent, and temperature as well as complexing agents (diglycolic, maleic acid) was investigated. It was shown that with nitric acid as the eluent, in the presence of 0.5–1.0 M potassium nitrate, the retention of the lanthanides under increased temperature is defined mainly by the stability of the corresponding surface complexes with

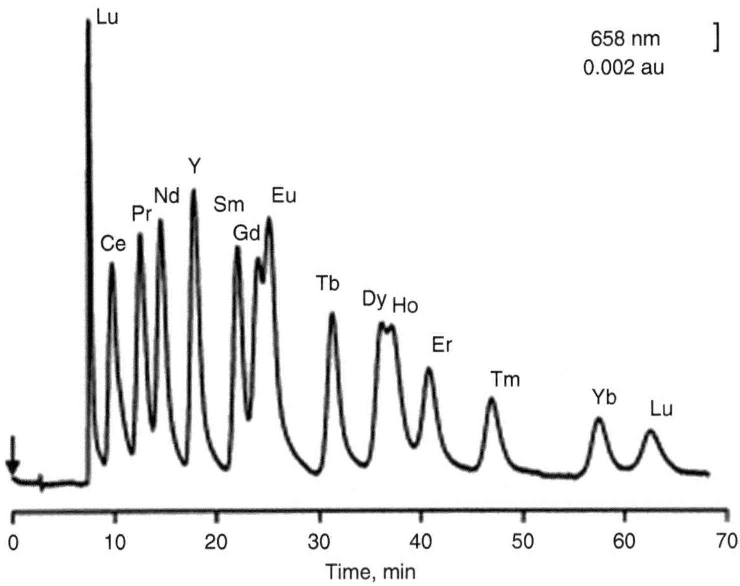

Fig. 10.7 Isocratic separation of standard mixture of 14 Lanthanides and Yttrium. Eluent: 1.6×10^{-2} M HNO_3 with 0.5 M KNO_3; flow-rate 1.0 ml /min; column temperature 65°c; sample volume 20 μl; sample concentration of each metal was 4 ppm in 0.2% HNO_3 (Reprinted from Ref. [25] with kind permission of © Elsevier (1997))

iminodiacetic acid functional groups (Fig. 10.7). The combination of increased ionic strength of the eluent (0.5 M potassium nitrate) and increased temperature of the chromatographic column (65°C), under an optimal concentration of nitric acid (1.6 × 10 M), allowed the isocratic separation of 14 lanthanide ions and yttrium in 65 min. This work also suggests that high-performance chelation of complex matrixes is definitely possible.

An other application of the IDA-silica-based exchanger is in the determination of Fe(II) and Fe(III) in many sample types. One important example is in the power industry where the measurement of the two forms in boiler water can give important information about potential corrosion problems. IDA-silica is ideal for this analysis as the eluent can be optimized to give very short retention times. Figure 10.8 shows a chromatogram of Fe(II) and Fe(III) separations at two different concentration levels. The high sensitivity and low LODs are due to the short retention times [17].

An iminodiacetic resin, amberlite IR-718, is employed for the removal of Hg(II) selectively from solutions containing iron, zinc, cadmium, and lead in sulfuric acid at pH 1.5. The resin is highly selective for Hg(II). Hg capacity at pH 1.5 is about 3.3 meq/g, although the full capacity will not be realized in applications requiring very low Hg leakage [26]. The effectiveness of IRC-718 is very dependent on pH, and regeneration with 3 M NaCl at near-neutral pH yields a solution from which mercury can be recovered by reduction to an insoluble and commercially valuable form.

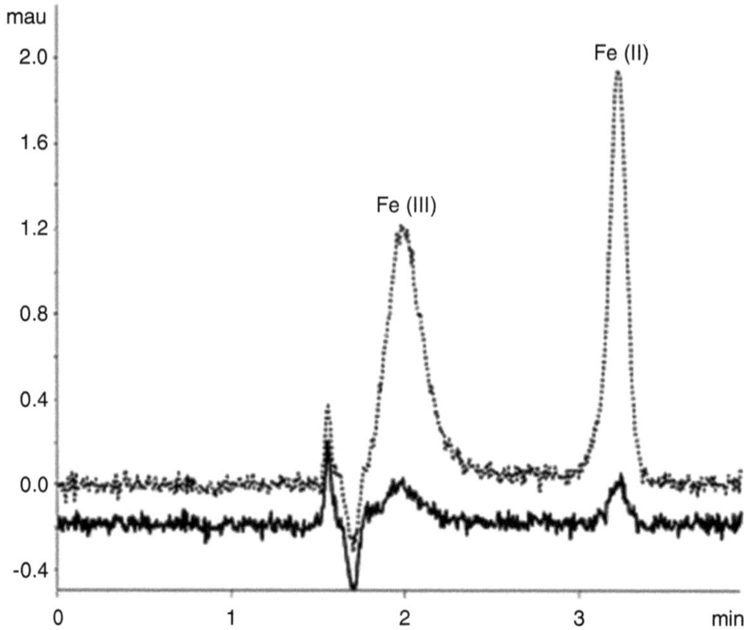

Fig. 10.8 Separation of Fe(III)/Fe(II) at two different concentration levels using a 100 × 4.0 mm ID-silica, 5 lm column. Injection volume, 100 μL, 20 μg/L of each species (*dashed line*) and 2 μg/L of each species (*solid line*) (Reprinted from Ref. [17] with kind permission of © Elsevier (2007))

10.6.2 Chelating Ion Exchangers Containing Amino Phosphonic Acid Functional Group

Resins containing aminophosphonic acid functional groups show increased selectivity toward toxic heavy metals and are commercially available under the trade names Lewatit OC1060 (Bayer), Duolite ES-467 (Rohm & Hass), Purolite S940 and 950 (Purolite Intl.), and Chelate P (SERVA). The kinetics of the complexation of Aminophosphonic acid silica (APAS) is as good as with IDA, and a more closely spaced selectivity coefficient under acidic conditions allows the high-efficiency isocratic separation of a great number of metal ions. These reagents have been used for preconcentration of uranium and thorium and other trace metals from natural and industrial water [27–30].

APAS was synthesized [31] by the reaction of aminopropylsilica with hypophosphorous acid and formaldehyde (Fig. 10.9). APAS contains 0.35 mmol/g unreacted aminopropyl groups and 0.1 mmol/g aminophosphonic acid functional groups at the surface.

Aminophosphonate functionality can be represented as shown in Fig. 10.10, depending on the pH of the eluent. APAS can exhibit properties of acids due to the dissociation of the phosphonic acid functional groups and base properties due to the

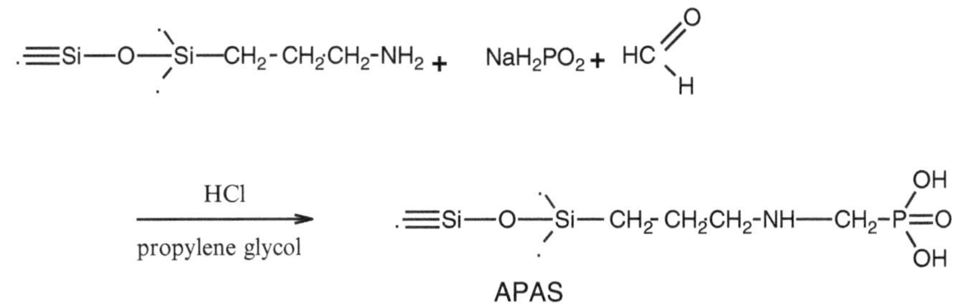

Fig. 10.9 Synthesis of aminophosphonic acid silica (Reprinted from Ref. [31] with kind permission of © Elsevier (1998))

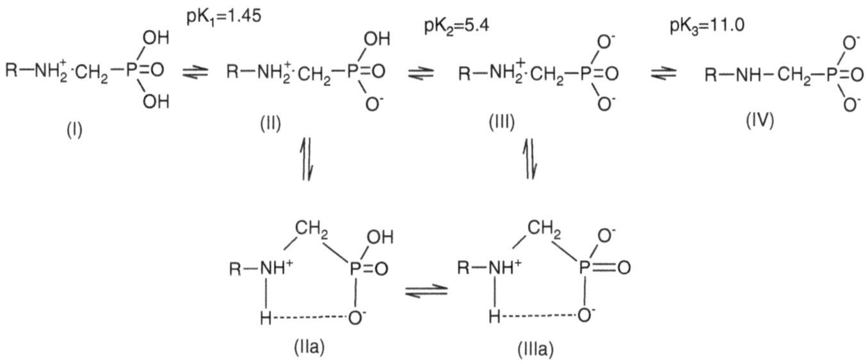

Fig. 10.10 Structures of aminophosphonate at various pH

presence of a secondary amine group. The pKa values for the corresponding equilibrium between different forms of aminophosphonic acid are 1.45, 5.4, and 11.0, respectively [32].

The amino methyl phosphonic acid chelating group has three ligating atoms and so is potentially a tridentate ligand (Fig. 10.11), having two bonding sites at phosphonic acid group and one coordination site yielding secondary nitrogen atom, respectively, N, O, and O chelation system. Thus, aminomethylphosphonic acid can bind metal ions via oxygen atoms from the phosphoric group normally under acidic conditions and nitrogen from secondary amino groups. As pH increases, the formation of complexes involving N, O, and O cannot be ruled out depending on the magnitude of individual metal stability constant. This type of ligand shows 1:1 M-L stoichiometry, and mostly the formation of a four-member ring occurs through deprotonation of the P–OH group. Normally, amine nitrogen does participate under acidic conditions, and chelation occurs through the O–O chelating system. However, the formation of N, O, and O chelation cannot be ruled out.

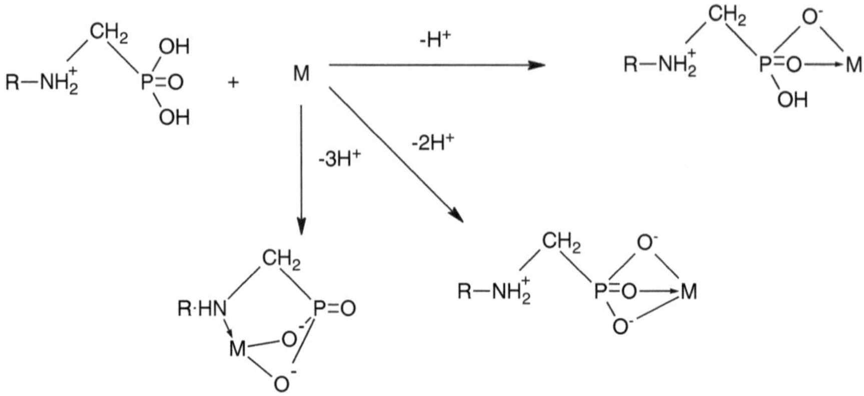

Fig. 10.11 Possible metal coordination of APAS

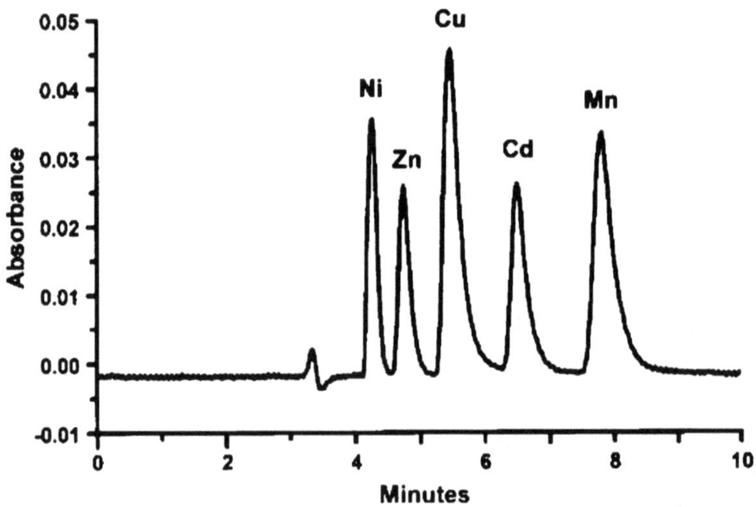

Fig. 10.12 Separation Of Ni(II) 1 ppm, Zn(II) 2 ppm, Cu(II) 1 ppm, Cd(II) 1 ppm, And Mn(II) 1 ppm on the aminophosphonic acid-functionalized silica column, 250 × 4.6 mm. Eluant:1MKNO$_3$, 100 mm HNO$_3$ (Reprinted from Ref.[32] with kind permission of © Elsevier (1999))

Aminophosphonate bonded to silica is reported to be more efficient and versatile than the IDA-based substrate. An aminomethylphosphonic acid-functionalized silica column was employed to study the chromatographic behavior of alkaline earth metals, Cu(II), Mn(II), Co(II), Zn(II), Ni(II), and Pb(II). The observed retention order of transition and heavy metal ions is Ni(II) < Mg(II) < Co(II) < Zn (II) < Cu(II) < Pb(II) < Cd(II) < Mn(II). This order agrees very closely with the order of stability constants for metal proton ligand [MHL] complexes of aminomethylphosphonate measured by Mohn and Abbot [33]. The chromatogram (Fig. 10.12) illustrates the high selectivity for Mn(II) at pH 1, eluting last after Cu

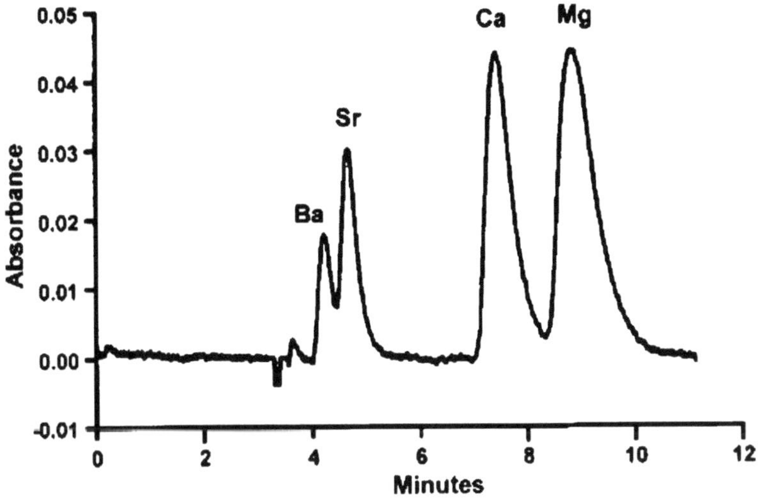

Fig. 10.13 Separation Of Ba(II) 1 ppm, Sr(II) 1 ppm, Ca(II) 1 ppm, And Mg(II) 1 ppm on the aminophosphonic acid-functionalized silica column, 250 × 4.6 Mm. Eluent:1MKNO$_3$ at pH 5 (Reprinted from Ref.[32] with kind permission of © Elsevier (1999))

(II). The high-efficiency isocratic separation of a greater number of metal ions in a reasonable time takes place.

APAS also shows a very high selectivity between the alkaline earth group and 2^+ transition and heavy metal ions at high pH as well as good resolution between the individual members of each group at selected pH values as shown in Fig. 10.13. These characteristics make aminophosphonate chelating substrates a useful and more versatile phase for high-performance chelation ion chromatography than iminodiacetate phases.

Purolite S950, a microporous styrene divinylbenzene chelating resin containing an aminophosphonic acid functional group, has been shown to have a high affinity for various heavy metals, and its successful application in metallurgical and waste water treatment processes has been encouraging [28, 34]. The aminophosphonate chelating reagents have greater success in the recovery of metals from complex solutions in comparison to cation exchange resins.

Purolite S950 was investigated for the adsorption of Ni and Co present as metal organic complexes during the biological leaching of low grade nickel laterite ores [35]. The adsorption of the various Ni and Co complexes on Purolite was quite low, 16–18 and 5.4–9 mg/g of resin, respectively, in comparison to the smaller nickel ions and nickel sulfate. This was attributed to the bulky organic ligands which promoted a crowding effect or steric hindrance. The adsorption of these complexes was further hampered by the strong affinity of the resin to H^+ ions under acidic conditions. The elution efficiency of both Ni and Co complexes from Purolite S950 resin was high (82.5%).

In another study, resins with typical chelating functional groups Purolite's iminodiacetate S930, aminophosphonate S940 and S950, thiouronium S920 resins,

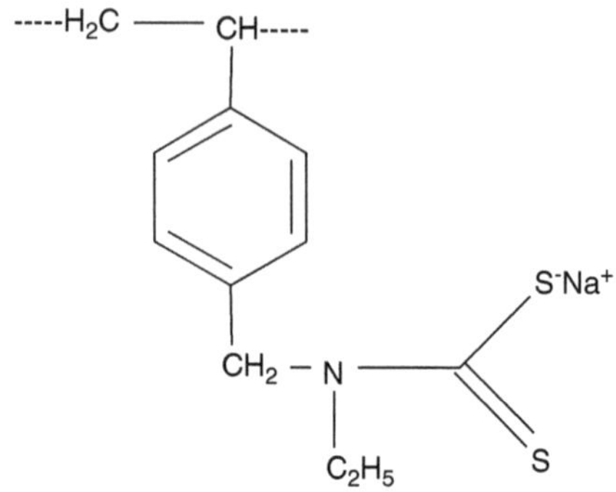

Fig. 10.14 Structure of chelating exchanger containing ligand dithiacarbamate

and salicylic acid (Spheron Salicylide) were tested for their abilities to remove Zn, Ni, Cu, and Cd from solutions simulating waste effluents from the metal-plating industry [36]. From the pure-metal chloride solutions, the aminophosphonate resin showed the best performance, and it behaved in the same way in most of the studied solutions. Capacities between 0.85 and 1.4 meq/mL at 5% breakthrough (BT) point were obtained in the series of Ni–Cd–Zn which is well in line with literature [6]. The decontamination factors (DFs) were good at around 250. In copper sulfate solutions, the uptake and the DFs were considerably lower, being 0.35 meq/mL and 100, respectively. Complexing agents, especially cyanide, considerably reduced the performance of the resins with only a few exceptions. Ammonium seemed to improve the ion exchange performance of some chelating resins, and capacities higher than the theoretical values, given by the manufacturer, were measured. Comparative experiments between chelating-strong acid and weak-acid ion exchange resins showed that the advantage of chelating exchangers over strong- and weak-acid exchangers is the very low metal BT level, even as low as 2 μg/L, which is very important, especially in the end-of-pipe polishing.

10.6.3 Chelating Ion Exchanger Containing Dithiocarbamate Functional Group

Dithiocarbamates form exceptionally strong insoluble complexes with a large number of metal ions except alkali and alkaline earth metals. The chelating ability of dithiocarbamate has been explored by anchoring onto a variety of polymeric supports.

Several dithiocarbamate resins are synthesized by treating polyamine-polymer urea resins containing some available secondary amine groups with carbon disulfide [37]. These dithiocarbamate chelating resins form mainly (1:1) metal to ligand complexes, and the capacity of resins for various metal ions tested appears to follow the order $Ag^+ > Hg^{2+} > Cu^{2+} > Sb^{2+} > Pb^{2+} > Cd^{2+} > Ni^{2+} > Co^{2+}$, and Ca^{2+} was not taken up. Poly(dithiocarbamate) has been used for concentration and separation of trace elements from complex matrixes. The structure of the styrene vinylbenzene-based dithiocarbamate ligand is given in Fig. 10.14.

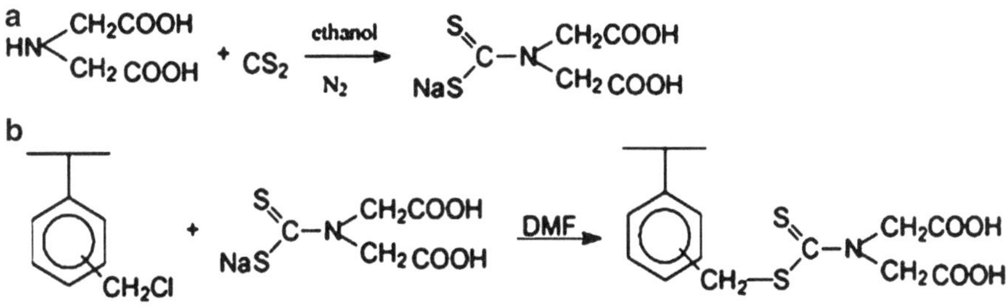

Fig. 10.15 (**a**) Preparation of N-DC (**b**) Preparation of PSDC

According to the principle of HSAB, sulfur-containing groups are, namely, soft bases, and therefore, it is reasonable to use sulfur-containing functional group materials to remove with much efficiency such soft acids like Cu(II), Pb(II), Ni (II), etc. [15]. Various types of dithiocarbamate (DTC) which were widely used as sulfur-bearing complexing agents were anchored to diverse polymeric supports to remove and separate heavy metals from different systems [38–40]. Nevertheless, those synthesized chelating resins possessing DTC as the sole functional group still suffer from disadvantages including the low-adsorption selectivity and capacity toward divalent metals. Recently, a new PS-DVB copolymer (PSDC), consisting of DTC and IDA moiety, with the chemical functionalization with sodium N,N′-di (carboxymethyl) dithiocarbamate, has been synthesized [41]. The typical characters and the special performances of such typical divalent heavy metal ions like Cu(II), Pb(II), and Ni(II) are comprehensively investigated. The synthesis of the PS-DVB copolymer (PSDC) was done as follows: N, N′ di(carboxymethyl) dithiocarbamate (N-DC) acid was prepared by the reaction of carbon disulfide with iminodiacetic acid. The chloromethylated PS-DVB copolymer was reacted with N-DC to form the N,N′-di(carboxymethyl) dithiocarbamate resin (Fig. 10.15).

The adsorption behavior of Pb(II), Cu(II), and Ni(II) by PSDC was pH dependent, with the most effective uptake at pH of 5–6. The strong affinity of PSDC toward these target soft acids could be well demonstrated with the electrostatic attraction and chelating interaction caused by IDA moiety and sulfur. This new material shows the characteristics of a chelating exchanger and can therefore be used as an alternative adsorbent in selective separation processes.

10.6.4 Chelating Ion Exchanger Containing Phenolic and Carboxylic Acid Groups

Compounds containing phenolic groups can easily be converted into polymeric compounds by condensing them with HCHO or any other aldehyde, with or without phenol, in the presence or absence of a cross-linking agent to get gel-type ions.

The salicylic acid exchangers exhibited selectivity for copper, aluminum, and uranyl ions. These resins were used successfully in column operations to determine the iron and copper contents of the brine. A chelating resin containing salicylic acid, chemically bound via the side chain azo group to hydrophilic glycol methacrylate gel, has been reported by Jing et al. [42] and is commercially available under the trade name Spheonsalicyl-1000. Table 10.1 shows high selectivity for iron(III) and aluminum(III) above pH 2.5.

Salicylate-containing ion exchange resins exhibit the selectivity order as per the stability constant of metal salicylate complexes. The order of the stability constants of metal salicylate complexes, is $Fe^{3+} > Al^{3+} > Cu^{2+} > Ni^{2+} = Co^{2+} = Zn^{2+}$ [43]. This is the controlling factor in the metal take-up by the resin, the effect of hydrated ionic radius playing at most a minor role. However, a lightly cross-linked, polystyrene-based exchanger demonstrated higher selectivities than a highly cross-linked dense condensation resin. A salicylate resin showed better selectivity for iron (III) than an equivalent p-hydroxybenzoate resin, and it is possible to produce salicylic acid condensation resins suitable for column operations [44]. The resin used for column work showed selectivity for iron(III) and copper(II) ions over manganese and sodium ions. The results with the polystyrene-azo-salicylate resin confirmed the findings of Vernon and Eccles [45] that the functional group shows selectivity for uranyl ions in the pH range 3–4.

Iontosorb oxin (IO) and iontosorb salicyl (IS), cellulose resins containing covalently bonded 8-hydroxyquinoline, and salicylic acid functional group were used for the determination of speciation of aluminum in water samples. The determined ratio of reactive Al in the water samples was relatively high – from 64.2% to 89.9% of total Al content. This method offers a relatively simple technique, without a need for high-cost instrumentation, and can also be performed in the field [46].

10.6.5 Chelating Ion Exchange Resins Containing Hydroxamic Acid and Amidoxime Functional Group

Chelating ion exchange resins containing hydroxamic acid groups have been synthesized from cross-linked poly(carboxylic acid) resins by the conversion of the carboxyl groups to the acid chloride [47, 48] or to an ester [49], followed by treatment with hydroxylamine.

Alternatively, the poly(hydroxamic acids) prepared from the hydrolysis of cross-linked poly(acrylonitriles) and subsequent condensation with hydroxylamine show high stability toward concentrated acids and bases during regeneration and, in turn, retain their high capacity for iron(III) ions (3.5 mmol/g) [50]. This type of chelating ion exchanger is cheaper and easier to prepare than Dowex A-l, and the swelling, equilibration rates, and total capacities are comparable. They possess higher selectivity, which makes them more versatile ion exchangers for analytical purposes.

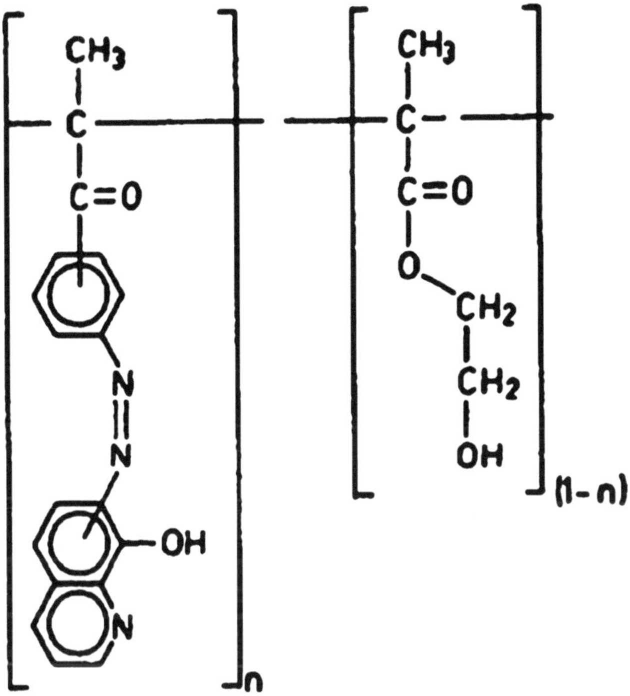

Fig. 10.16 The proposed structure of poly(acrylamidooxime) resin

The synthesis of a macroporous cross-linked poly(hydroxamic acid) has been achieved through the hydrolysis of a cross-linked poly(acrylonitrile)-DVB copolymer to an amide upon treatment with hydroxylamine [51]. This polymeric ligand has been reported to be superior to those previously reported and was found selective for iron(III), vanadium(V), and mercury(II). These types of resins are completely stable toward 5 M HCl or H_2SO_4 and also to 1 M NaOH. The main advantage of such types of resin appears to be their ability to separate multivalent ions, e.g., iron(III) from Cu(II) at pH 1.5 and Cu(II) from Co(II) at pH 3.5, whereas iminodiacetic acid resin is most useful for blanket extraction of these metal ions at pH > 5. The application of these resins for the recovery and separation of iron and uranium from sea water samples is studied. Sorption and desorption of gold and silver by this exchanger and separation from each other are reported [52].

Another class of chelating ligands, amidoxime, are being increasingly studied for their incorporation into chelating exchangers because of their ability to selectively bind to several metal ions such as iron(III), copper(II), and uranyl(II) at very low pH range. A poly(acrylamidoxime) metal chelating resin is cross-linked with divinylbenzene to improve its mechanical stability for use in columns [53]. The proposed structure of a poly(acrylamidoxime) resin is depicted in Fig. 10.16.

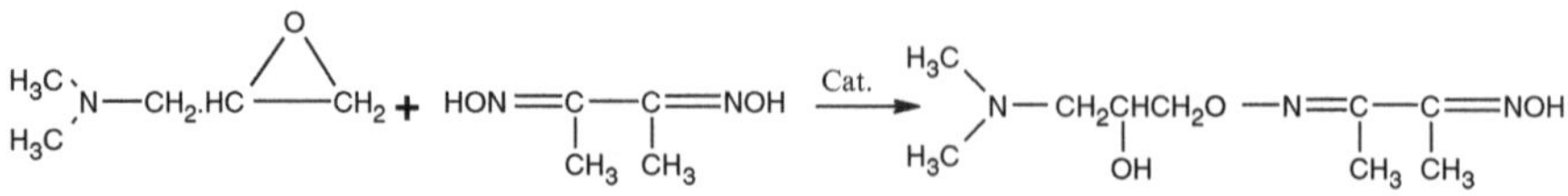

Fig. 10.17 Synthesis of epoxidized polybenzimidazole (EPBI) (Reprinted from Ref. [56] with kind permission of © Elsevier (1989))

In comparison with Chelex-100 and a poly(dithiocarbamate) resin [54], the poly (acrylamidoxime) resin (CII) has been shown to possess a distinct advantage in the full recovery of sequestered metals. A commercial resin, Duolite CS-346, cross-linked to a polyacrylic matrix containing an amidoxime functional group, strongly binds iron(III), uranium(VI), vanadium(II), and copper(II) and binds moderately with cobalt(II) and nickel(II), whereas metals like chromium(III) also form relatively stable complexes with this resin at relatively low pH [52]. Using this chelating resin, the separation of iron(III), copper(II), and uranium(VI) ions, using pH control, was studied by Vernon and Kyffin [55].

A new resin, epoxidized polybenzimidazole (Dimethylglyoxime) [EPBI (DMG)], was prepared from microporous polybenzimidazole (PBI) of 250–500-μ m spherical bead size from Celanese reaction with epichlorohydrin and sodium hydroxide, and the resulting epoxidized product (EPBI) (Fig. 10.17) further reacted with DMG using a quaternary ammonium salt as the phase transfer catalyst [56]. The resin has high sorption capacity for UO_2^{2+} and Fe^{3+}. The sorption of UO^{2+} is the maximum in weakly acidic or neutral solution pH (5–7) while the uptake of Fe^{3+} is not affected by pH. The sorbed UO_2^{2+} is readily stripped by 1N HCl, with nearly 90% stripping taking place in 1 min under vigorous agitation, and thus provides a method for selective stripping and recovery.

10.6.6 Chelating Ion Exchanger Containing 8-Hydroxyquinoline

The chelating ion exchange resins incorporating 8-hydroxyquinoline (oxine or 8-quinolinol) form an important and extensively studied class of selective ion exchangers. A large number of resins with oxine as a functional group having different types of matrices have been synthesized. The oxine resins studied so far can be divided into four groups: (1) cross-linked polystyrene resins with oxine groups attached to a styrene ring via an azo linkage; (2) condensation resins derived from formaldehyde, phenol, and oxine; (3) resins derived from poly(5-vinyl-oxine); and (4) oxine attached to silica gel and porous glass beads. Parrish synthesized [57] an oxine resin using cross-linked polystyrene as a matrix in which the chelating group is linked through an azo group, and the resin, polystyrene-(4-azo-5)-8-hydroxyquinoline, was synthesized (Fig. 10.18).

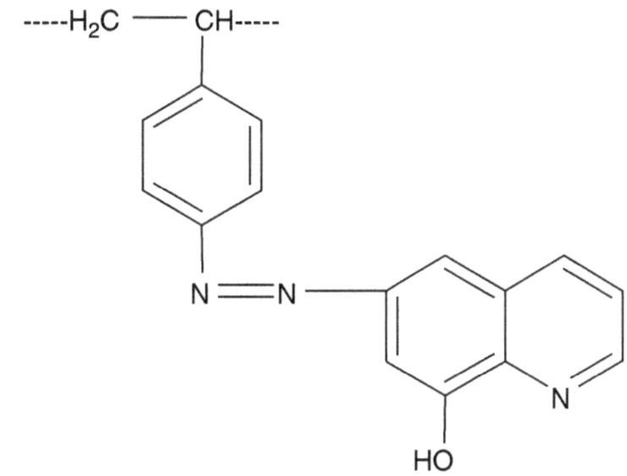

Fig. 10.18 Structure of poly (acrylamidoxime) resin

A commercial resin containing 8-quinolinol as a chelating group is now available commercially as Spheron Oxine-1000 (Table 10.1). This resin consists of a matrix of a copolymer of hydroxyethyl methacrylate, and ethylenedimethyl acrylate, and the functional group is bound to the polymeric matrix via an azo group [58]. Spheron Oxine-1000 does not interact with ammonium and alkali metal ions or (in acidic media) with ions of alkaline earths. This resin thus differs from classical chelating resins containing iminodiacetic acid groups, which also possess the nonselective properties of weak carboxylic cation exchangers. In contrast, the affinity of Spheron Oxine is remarkably high for heavy metal ions and is only slightly influenced by ionic strength. The exchange capacity of this resin for most heavy metal ions varies from 0.2 to 0.3 mmol/g at pH 5. Spheron Oxine-1000 is useful for group-selective separation of metal ions like Mn^{2+}, $A1^{3+}$, Pb^{2+}, Zn^{2+}, Cd^{2+}, Co^{2+}, Ni^{2+}, Fe^{3+}, Cu^{2+}, and $UO_2{}^{2+}$ [59].

8-Hydroxyquinoline immobilized on silica gel has been prepared and used for the preconcentration of Cd, Pb, Zn, Cu, Fe, Mn, Ni, and Co from seawater prior to their determination by graphite furnace atomic absorption spectrometry [60].

10.7 Chelating Ion Exchangers for Separation of Anions

The chelating resins can also be employed for the separation of anions by loading metal ions onto a weak base chelating resin. Recently, six new chelating resins were prepared by functionalizing three commercially available XAD resins of various matrix properties, including XAD1180, XAD16, and XAD7HP for selective separation of strong ligands such as arsenate [61]. Two types of pyridinyl functional groups were loaded on these sorbents, one with two nitrogen donor atoms (2 N) per functional group and the other with three N-donor (3N) atoms per functional group. Consequently, six polymeric ligand exchangers (PLEs) were prepared by immobilizing Cu(II) ions onto these chelating resins, where Cu(II) serves as the surface central metal (Fig. 10.19). The highest Cu(II) loading was determined to be 44 mg/g. The PLEs displayed greater affinity toward arsenate than for sulfate. The

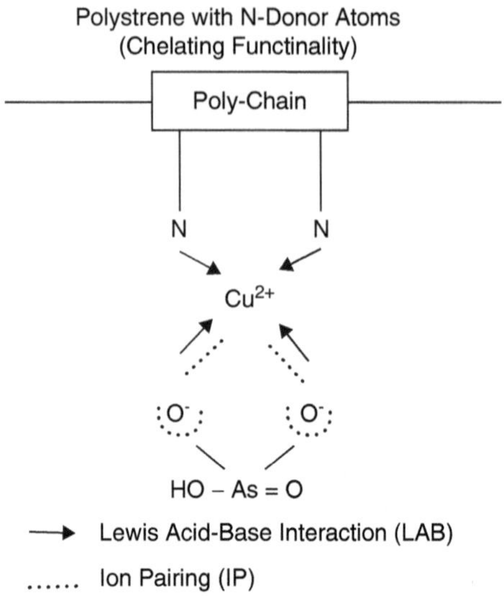

Fig. 10.19 Structure of a PLE and binding mechanisms of Arsenic onto an XAD-based PLE (Reprinted from Ref. [58] with kind permission of © Springer (1978))

binary arsenate-sulfate separation factor ($\alpha_{As/S}$) ranged from 5.1 to 10. Bench-scale column breakthrough tests confirmed the greater selectivity for arsenate over other common anions (sulfate, bicarbonate, and chloride). The breakthrough of arsenate occurred last at ~920 BV for XAD1180-3N–Cu. The PLEs can be effectively regenerated using 8% NaCl at pH 10. The arsenate capacity can be completely recovered using the brine. Despite the high surface area, the PLE's sorption capacity was limited by the hyper-cross-linkage and the degree of methylation/amination of the template XAD resins.

10.8 Concluding Remark

The development of new chelating ion exchange materials with special chelating properties can provide better kinetics of interaction between metal ion and chelating groups, and better understanding of their coordination chemistry at surfaces can revolutionize the domain of chromatographic separation of trace metal ions particularly from complex matrices.

References

1. Schmuckler G (1965) Chelating resins-their analytical properties and applications. Talanta 12:281–290
2. Iseid SS, Kuehn CG, Lyon JM, Merrifield RB (1982) Specific peptide sequences for metal ion coordination. 1. Solid-phase synthesis of cyclo-(gly-his). J Am Chem Soc 104:2632–2634

3. Inczedy J (1966) Analytical applications of ion exchangers. Pergamon Press, Oxford
4. Skogseid A (1947) Norway Patent 72,583; (1952) US Patent 2,592,350
5. Riley JP, Taylor D (1968) Chelating resins for the concentration of trace elements from sea water and their analytical use in conjunction with atomic absorption spectrophotometry. Anal Chim Acta 40:479–485
6. Sahni SK, Reedijk J (1984) Coordination chemistry of chelating resins and ion exchangers. Coord Chem Rev 59:1–139
7. Nesterenko PN, Jones P (1997) First isocratic separation of fourteen Lanthanides and Yttrium by high-performance chelation ion chromatography. Anal Commun 34:7
8. Slebioda M, Wodecki Z, Kolodziejczyk AM, Nowicki W (1994) Chem Anal (Warsaw) 39:149
9. Watanesk S, Schilt AA (1986) Separation of some transition-metal ions on silica-immobilized 2-pyridinecarboxaldehyde phenylhydrazone. Talanta 33:895–899
10. Simonzadeh N, Schilt AA (1988) Metal-ion chromatography on silica-immobilized 2-pyridinecarboxyaldehyde phenylhydrazone. Talanta 35:187
11. Simonzadeh N, Schilt AA (1989) J Coord Chem 20:117
12. Nesterenko PN, Smirnov IP, Brykina GD, Bolshova TA (1991) Vestnik Mosk Univ Khimia 32:358
13. Elefterov AI, Kolpachnikova MG, Nesterenko PN, Shpigun OA (1997) Ion-exchange properties of glutamic acid-bonded silica. J Chromatogr A 769:179–188
14. Gregor HP, Taifer M, Citarel L, Becker EI (1952) Ind Eng Chem 44:2834
15. Pearson GR (1963) Hard and soft acids and bases. J Am Chem Soc 85:3533–3539
16. Akelah A, Sherrington DC (1981) Application of functionalized polymers in organic synthesis. Chem Rev 81:557–587
17. Nesterenko PN, Jones P (2007) Review: recent developments in the high-performance chelation ion chromatography of trace metals. J Sep Sci 30:1773–1793
18. Okawara M, Komeda Y, Imoto E (1960) Chem High Polym (Jpn) 17:30
19. Loewenschuss H, Schmuckler G (1964) Chelating properties of the chelating ion exchanger Dowex A-l. Talanta 11:1399–1408
20. Rosset R (1966) Bull Soc Chim Fr 59
21. Hirsh RF, Gancher E, Russo FR (1970) Macroreticular chelating ion-exchangers. Talanta 17:483–489
22. Hering R (1967) Chelatbildende Ionenaustauscher. Akademie, Berlin
23. Bonn G, Reiffenstuhl S, Jandik P (1990) Ion chromatography of transition metals on an iminodiacetic acid bounded stationary phase. J Chromatogr 76:499–669
24. Bashir W, Paull B (2001) Ionic strength, pH and temperature effects upon selectivity for transition and heavy metal ions when using chelation ion chromatography with an iminodiacetic acid bonded silica gel column and simple inorganic eluents. J Chromatogr A 942:73–82
25. Nesterenko PN, Jones P (1998) Isocratic separation of lanthanides and yttrium by high-performance chelation ion chromatography on iminodiacetic acid bonded to silica. J Chromatogr A 804:223–231
26. Becker NSC, Eldridge RJ (1993) Selective recovery of mercury(II) from industrial wastewaters I. Use of a chelating ion exchanger regenerated with brine. React Polym 21:5–14
27. Lehto J, Vaaramaa K (1997) H^+/Na^+ exchange in an aminophosphonate-chelating resin. React Funct Polym 33:19–24
28. Leinonen H, Lehto J, Makela A (1994) Purification of nickel and zinc from waste waters of metal-plating plants by ion exchange. React Polym 23:221–228
29. Yebra-Biurrun MC, Bermejo-Barrera A, Bermejo-Barrera MP (1992) Synthesis and characterization of a poly (aminophosphonic acid) chelating resin. Anal Chim Acta 264:53–58
30. Yebra-Biurrun MC, Bermejo-Barrera A, Bermejo-Barrera MP, Barciela-Alonso MC (1995) Determination of trace metals in natural waters by flame atomic absorption spectrometry following on-line ion-exchange preconcentration. Anal Chim Acta 303:341–345
31. Nesterenko PN, Zhukova OS, Shpigun OA, Jones P (1998) Synthesis and ion-exchange properties of silica chemically modified with aminophosphonic acid. J Chromatogr A 813:47–53

32. Nesterenko PN, Shaw MJ, Hill SJ, Jones P (1999) Aminophosphonate-functionalized silica: a versatile chromatographic stationary phase for high-performance chelation ion chromatography. Microchem J 62:58–69
33. Mohan M, Abbott E (1978) J Coord Chem 8:175
34. Lehto J, Vaaramaa K, Leinonen H (1997) Ion exchange of zinc on an aminophosphonate-chelating resin. React Funct Polym 33:13–18
35. Deepatana A, Valix M (2006) Recovery of nickel and cobalt from organic acid complexes: adsorption mechanisms of metal-organic complexes onto aminophosphonate chelating resin. J Hazard Mater B 137:925–933
36. Koivula R, Lehto J, Pajo L, Gale T, Leinonen H (2000) Purification of metal plating rinse waters with chelating ion exchangers. Hydrometallurgy 56:93–108
37. Dingman J Jr, Siggia S, Barton C, Hiscock KB (1972) Concentration and separation of trace metal cations by complexation on polyamine-polyurea resins. Anal Chem 44:1351–1357
38. Kantipuly G, Katragadda S, Chow A, Gesser HD (1990) Chelating polymers and related supports for separation and preconcentration of trace metals. Talanta 37:491–498
39. Mathew B, Rajasekharan Pillai VN (1994) N, N′-2-methylene bisacrylamide-crosslinked polyacrylamides as supports for dithiocarbamate ligands formetalion complexation. Polym Int 28:201–208
40. Lezzi A, Cobianco S (1994) Chelating resins supporting dithiocarbamate and methylthiourea groups in adsorption of heavy metal ions. J Appl Polym Sci 54:889–897
41. Roy PK, Rawat AS, Rai PK (2003) Synthesis, characterisation and evaluation of polydithiocarbamate resin supported on macroreticular styrene–divinylbenzenecopolymer for the removal of trace and heavy metal ions. Talanta 59:239–246
42. Jing X, Liu F, Yang X, Ling P, Li L, Long C, Li A (2009) Adsorption performances and mechanisms of the newly synthesized N, N′-di(carboxymethyl) dithiocarbamate chelating resin toward divalent heavy metal ions from aqueous media. J Hazard Mater 167:589–596
43. Slovak Z, Slovakova S, SmIz M (1976) Analytische eigenschaften von hydrophilen glykolmethacrylat-gelen mit chemisch gebundener salicylsäure. Anal Chim Acta 87:149
44. Hanclkova L, Bartusek M (1969) Collect Czech Chem Comm 34:3722
45. Vernon F, Eccles H (1974) Chelating ion-exchangers containing salicylic acid. Anal Chem Acta 72:331–338
46. Degeiso RC, Donaruma LG, Tomic EA (1962) Polymeric ligands. I. Some salicylic acid derivatives. J Org Chem 27:1424
47. Matus P, Kubova J, Steresko V (2003) Utilization of chelating ion exchange for aluminium. Speciation Chem Pap 57:176–178
48. Comaz JP, Hutschneker K, Deuel H (1957) Helv Chim Acta 40:2015
49. Vrancken M, Smets G (1954) J Polym Sci 14:521
50. Petrie G, Locke D, Meloan CE (1965) Hydroxamic acid chelate ion exchange resin [2]. Anal Chem 37:919–920
51. Vernon F, Eccles H (1976) Chelating ion-exchangers containing n-substituted hydroxylamine functional groups: part IV. Column separations on a hydroxamic acid resin. Anal Chim Acta 83:187–193
52. Vernon F, Zin WMd (1981) Chelating ion-exchangers containing n-substituted hydroxylamine functional groups. Part 6. Sorption and separation of gold and silver by a polyhydroxamic acid. Anal Chim Acta 123:309–313
53. Vernon F, Eccles H (1975) Chelating ion-exchangers containing n-substituted hydroxylamine functional groups: part II. N-acylphenylhydroxylamines original research article. Anal Chim Acta 79:229–236
54. Hackett DS, Siggia S (1977) In: Ewing G (ed) Environmental analysis. Academic, New York, p 2
55. Vernon F, Kyffin TW (1977) Chelating ion-exchangers containing n-substituted hydroxylamine functional groups. Part V. Iron, copper, and uranium separations on Duolite CS-346 resin. Anal Chim Acta 94:317–322

56. Chanda M, Rempel GL (1989) Polybenzimidazole resin-based new chelating agents. Uranyl and ferric ion selectivity of resins with anchored dimethylglyoxime. React Polym 11:165–176
57. Parrish JR (1955) Chem Ind 386
58. Slovak Z, Slovakova S (1978) Use of hydrophilic glycolmethacrylate gels with bound 8-hydroxyquinoline in trace analysis – Part II. Batch sorption methods [Verwendung von hydrophilen Glykolmethacrylat-Gelen mit chemisch gebundenem 8-Hydroxychinolin bei Spurenanalysen – II. Statische sorptionen]. Fresenius Z Anal Chem 292:213–215
59. Slovak Z, Toman J (1976) Use of hydrophilic glycolmethacrylate gels with bound 8-hydroxyquinoline in trace analysis|[Verwendung von hydrophilen Glykolmethacrylat-Gelen mit chemisch gebundenem 8-Hydroxychinolin bei Spurenanalysen]. Fresenius Z Anal Chem 278:115–120
60. Sturgeon RE, Berman SS, Willie SN, Desauiniers JAH (1981) Preconcentration of trace elements from seawater with silica-immobilized 8-hydroxyquinoline. Anal Chem 53:2337–2340
61. An B, Fu Z, Xiong Z, Zhao D, SenGupta AK (2010) Synthesis and characterization of a new class of polymeric ligand exchangers for selective removal of arsenate from drinking water. React Funct Polym 70:497–507
62. Moyers EM, Fritz JS (1977) Preparation and analytical applications of a propylenediaminete-traacetic acid resin. Anal Chem 49:418–423
63. King JN, Fritz JS (1978) Separation of metal ions using an aromatic o-hydroxy-oxime resin. J Chromatogr A 153:507–516
64. Jonas PMM, Eve DJ, Parrish JR (1989) Preparation, characterization and performance of surface-loaded chelating resins for ion-chromatography. Talanta 36:1021–1026
65. Phillips RJ, Fritz JS (1980) Synthesis and analytical properties of an n-phenyl-hydroxamic acid resin. Anal Chim Acta 121:225–232
66. Suzuki TM, Yokoyama T (1984) Preparation and complexation properties of polystyrene resins containing diethylenetriamine derivatives. Polyhedron 3:939–945

Chapter 11
Ion Exchange Voltammetry

Paolo Ugo and Ligia M. Moretto

Abstract This chapter focuses on recent advances in the field of ion-exchanger-based voltammetric sensors, whose widespread use has given rise to a new electroanalytical technique named ion exchange voltammetry (IEV). The chapter starts with a brief historical overview and a short introduction to voltammetric methods of analysis, followed by the presentation of the fundamentals of ion exchange voltammetry. Analytical advantages and limits up to the most recent developments in the technique are then presented and discussed. Typical examples of application to determinations of trace concentrations of inorganic and organic electroactive molecules of interest for environmental and biomedical analysis are finally given along with future prospects.

11.1 Introduction

11.1.1 Ion Exchange Voltammetry: A Brief History

Ion exchange voltammetry (IEV) is a new analytical technique which conjugates recent advances in the field of ion exchange in thin films with detection capabilities of modern dynamic electrochemical methods of analysis. The idea as well as fundamentals of IEV date back to the beginning of the 1980s, when pioneering studies carried out in Anson's, Bard's, Murray's, and other laboratories [1–4] showed the possibility and advantages of modifying the surface of electrodes by simply depositing a thin layer of an ion exchange polymer onto the electrode surface. The term ion exchange voltammetry was firstly used by Chuck Martin and coworkers in a paper

P. Ugo (✉) • L.M. Moretto
Department of Molecular Sciences and Nanosystems,
University Ca' Foscari of Venice, Santa Marta 2137, 30123 Venice, Italy
e-mail: ugo@unive.it

Inamuddin and M. Luqman (eds.), *Ion Exchange Technology I: Theory and Materials*, DOI 10.1007/978-94-007-1700-8_11,

published in 1986 [5]. These and following studies opened the way to extended research efforts which brought rapidly to the development of a large arsenal of polymer-modified electrodes whose surface properties were tailored for specific analytical purposes [6–9]. Side by side with continuously growing application for analytical purposes, further recent advances in the field of IEV came from improvements of the techniques used for controlling the molecular structure of ultrathin ion-exchanger layers, thanks to the application to this aim of Langmuir-Blodgett techniques, of self-assembled monolayers and layer-by-layer techniques, whose principles can be found in references [10, 11]. This allowed one to obtain nanostructured electrode interfaces which represent, indeed, successful examples of the bottom-up nanotech approach, by which the careful and controlled assembly of different molecular bricks allows the fulfillment of specialized functions in an analytical device, in this case, an electroanalytical sensor.

The modification of electrode surfaces is presently finding a large impulse, thanks to the widespread interest for electrochemical sensors to be applied in different fields, for instance, in advanced biomedical devices [12, 13] or for environmental monitoring [7]. It can be noted that many of the early studies on polymer-coated electrodes employed a variety of Ru, Ir, Os, and Fe complexes as electroactive reversible probes which were used for understanding the fundamentals of charge transfer at electrode/polymer interfaces. At that time, the electrochemistry of such metal complexes was considered of almost no interest as far as practical applications were concerned; nowadays, many of such molecules are widely used as redox mediators in a variety of biosensors, including some of the most advanced DNA chips [14–16] and glucose sensors [17].

The present chapter deals mainly with electrodes functionalized by thin films of ion exchangers, deposited onto the surface of metal electrodes. It is divided into two parts. In the first, the principles of IEV as well as its analytical advantages and limits are presented and discussed. In the second, recent examples of application to determinations of inorganic and organic electroactive molecules of interest for environmental and biomedical analysis are given along with prospects for future developments. For the case of electrodes in which the ion exchanger is incorporated into the body of the electrode, such as carbon paste or screen-printed electrodes, the reader is invited to refer to the review article published recently by Svancara et al. on this specific topic [18].

11.1.2 An Introduction to the Detection Technique: Voltammetry

This section presents some short basic notions on electrochemical methods of analysis, useful for understanding the functioning and better appreciating most recent advancements in IEV, which constitutes the core of this chapter. Readers interested in a deeper insight on fundamentals and applications of electroanalytical techniques are addressed to specialized textbooks [19, 20].

The analytes detected by electrochemical methods are redox species, that is, molecules or ions which can release (or acquire) electrons to (from) proper reactants (molecules or metal surfaces) characterized by oxidizing (or reducing) properties.

From a classification viewpoint, electrochemical methods of analysis are divided in equilibrium methods (potentiometry) and dynamic ones (polarography, amperometry, voltammetry, coulometry, and others) [19, 20].

In the former, no current circulate through the electrochemical cell and the tendency of a certain redox reaction to occur is obtained by measuring the open-circuit electrochemical potential (E(OC)) of the cell; it is linked to the concentration of redox species in the electrolyte solution by the Nernst equation.

In dynamic (or transient) electrochemical techniques, a net current (or charge) is obtained as the result of an external excitation (generally, a voltage waveform) applied to the electrochemical cell; since such a current signal obeys Faraday's law, it is named *Faradaic current*. Experimental conditions are chosen so that the system operates under concentration polarization conditions and the Faradaic current (at proper applied potentials) is a function of the concentration in the electrolyte solution of the redox species (analyte) being reduced or oxidized at the electrode.

In voltammetry, the quantity being measured is the intensity of the electrical current which circulates through the circuit composed of the so-called working electrode (which is the "sensing" electrode), the electrolyte solution, and the counter electrode (a large area electrode, with polarity opposite to the working electrode). The excitation waveform is given as an electrical potential applied (and changed in a known manner) to the working electrode with respect to a third electrode, which acts as an unpolarized reference electrode (generally, a second species electrode, e.g., an Ag/AgCl electrode).

In linear sweep and in cyclic voltammetry (CV), the potential (E_{appl}) applied to the working electrode changes linearly with time (t) according to Eq. 11.1:

$$E_{appl} = E_i + vt \tag{11.1}$$

where E_i is the initial potential which is lower than the Nernst potential (E^0) of the redox couple (i.e., $|E_i| < |E^0|$); v is the scan rate (in V s^{-1} or mV s^{-1})

In CV, the potential is scanned forward and backward (once or many times, depending on the measurement mode used) from E_i (where no Faradaic current is observed since $|E_i|<|E^0|$) to move (with a certain scan rate) toward a vertex potential (E_v chosen so that $|E_v|>|E^0|$) where a net current corresponding to the Faradaic reduction (or oxidation) of the analyte is observed. In the backward scan (from E_v to E_i), the opposite electrode process occurs, which corresponds roughly to the return to the initial redox state of the product generated; if the forward process is a reduction, the process expected in the backward scan is the reoxidation of reduction products.

Transport of the analyte from the bulk solution to the electrode/solution interface can take place only by diffusion since convection and migration are inhibited by keeping the solution quiet, by avoiding thermal gradients (eventually by a thermostat), and by making negligible migration by annulling the effect of the applied electric field using an excess supporting electrolyte (electrochemically inert).

With planar electrodes of dimensions larger than the thickness of the diffusion layer (see below), that is, in the millimeter size range, diffusion follows the so-called semi-infinite planar diffusion model. The CV obtained under this diffusion regime is characterized by two associated voltammetric peaks, one recorded during the forward scan and one in the backward scan. The two peaks are located at potential values symmetric with respect to a characteristic potential value named half-wave potential ($E_{1/2}$). For reversible redox couples, $E_{1/2} = (Ep_f + Ep_b)/2$, where Ep_f and Ep_b are the potentials of the forward and backward peak, respectively; $E_{1/2}$ is strictly related to E^0 [19, 20]. If the analyte is an oxidized species, then the forward peak corresponds to the direct process:

$$\mathrm{Ox} + \mathrm{ne} \rightarrow \mathrm{Red} \tag{11.2}$$

and the backward peak to the reverse process (from electrogenerated Red back again to Ox).

It would be worthy to note that in voltammetry, an electrochemical process is considered reversible when the charge-transfer process between the analyte and the electrode is faster than mass transport (diffusion). Therefore, under diffusion control and when $|E_{appl}| >> |E^0|$, a concentration gradient is generated between the electrode surface (where the concentration of analyte is zero since all redox molecules which come or are on the electrode surface are reduced) and the bulk of the solution (where the concentration is the initial analyte concentration). The region where a concentration gradient exists is called diffusion layer. Under these conditions, the Faradaic current relevant to reaction 11.2 depends on the flux of the analyte which diffuses from the bulk solution to the electrode surface, this being inversely proportional to the thickness of the diffusion layer. This points out that voltammetric responses are time dependent since the thickness of the diffusion layer, δ, depends on time according to Eq. 11.3:

$$\delta = (\mathrm{D\ t})^{1/2} \tag{11.3}$$

where D is the diffusion coefficient of the analyte (in $cm^2\ s^{-1}$) and t is the timescale of the experiment. Note that in voltammetry, t depends on the scan rate according to:

$$\mathrm{t} = \mathrm{RT/Fv} \tag{11.4}$$

Scan rates normally used in voltammetry are between 0.020 and 1.000 V s^{-1}, and diffusion coefficients for solution species are of the order of 10^{-5}–$10^{-6}\ cm^2\ s^{-1}$, so that it is easily estimated that the thicknesses of diffusion layers are in the micrometer scale. This will be relevant when using electrodes with some peculiar dimensions (electrode diameter or thickness of electrode coatings) in such or even smaller dimension range.

The Faradaic current signal (S) is not the only current being measured by voltammetric electrodes; an electrode dipped in an electrolyte solution behaves, in fact, also as a capacitor, and a certain amount of current (charging current) is consumed to build up the so-called electrical double layer [19, 20]. The double-layer charging current increases with the electrode area and the scan rate. Other small currents, named residual currents and related to the electroactivity of trace concentrations of redox impurities, can interfere with the recording of the Faradaic current of the analyte. In voltammetry, double-layer charging currents and residual currents constitute the main component of the noise (N). The voltammetric S/N ratio can be improved by using the pulsed voltammetric techniques, such as differential pulse voltammetry (DPV) or Osteryoung's square wave voltammetry (SWV) [19, 20]. These electroanalytical techniques use excitation waveforms which are the combination of linear (or staircase) potential ramps with superimposed sequences of short potential pulses (10–50 mV for 10–50 ms). Proper sampling and combination of current values measured at fixed time lag with respect to the pulses application allow significant improvements in S/N ratios and consequent lowering of detection limit (DL) values. In fact, while with classical cyclic voltammetry (where no noise reduction is applied), DLs are typically in the 10^{-4} M range, with pulsed techniques, DLs go down to the 10^{-6}–10^{-7} M range. Newly advanced pulsed techniques such as, for instance, multiple square wave voltammetry (MSWV) [21, 22] and double-differential MSWV [23, 24], allowed even further lowering of voltammetric DLs.

The already low DLs achievable by using these pulsed techniques can be improved by introducing a preconcentration step before the voltammetric detection. The preconcentration of the analyte is performed directly at the transducer (electrode) surface via Faradaic (for anodic, ASV, or cathodic, CSV, stripping voltammetry) [25] and/or non-Faradaic (adsorption, ion exchange, complexation, etc.) processes [26]; this allows one to reach DLs as low as 10^{-9}–10^{-11} M.

11.1.3 Classical Electroanalysis with Mercury and Related Electrodes

Transient electrochemical techniques are very sensitive, and they allow one to work rather easily at very low detection limits. In addition, they can give information on the redox state and complexation of the analyte in the sample [27]. They can be used in turbid or colored samples and do not suffer for interference from ambient light. However, in some cases, dissolved oxygen (which is electroactive) can constitute a problem and should be eliminated; indeed, some experimental tricks can allow one to shorten or even avoid the necessity to eliminate dissolved oxygen [28, 29]. All these characteristics make, in principle, electrochemical methods and sensors attractive for determining redox analytes at trace (micromolar) and ultratrace (submicromolar) concentration levels in complex samples such as those of interest, for instance, in environmental monitoring or for biomedical analyses. However, up to the recent time, some limits to the widespread use of dynamic electrochemical methods came from the fact that a great part of these methods were based on the use of mercury as electrode material. Mercury has some advantages in electroanalysis, and the very first dynamic electroanalytical method, polarography, for which Jaroslav Heyrowski was prized with the Nobel Prize in 1959, was based on the use of the dropping mercury electrode. However, because of the high toxicity of mercury, nowadays, there is an increasing interest in developing methods which employ different electrode materials. Note that, in addition to toxicity-related problems, the use of mercury electrodes suffers also from other limitations. Even the most recent mercury drop electrode systems are still rather expensive and cumbersome; they use relatively high amounts of mercury and are rather complicate for set up of unmanned automatic analyses. Recently, silver-mercury amalgam [30] and bismuth [31] have been proposed as alternatives for mercury. However, mercury as well as these alternative materials cannot be used at positive potential values because of their easy oxidation, so that all the world of oxidation processes is precluded to this kind of electrodes.

In order to widen the application field of electroanalytical methods, it is important to develop novel electrode systems characterized by high sensitivity, good reproducibility, wide usable potential window, and low cost and made of nontoxic materials. Moreover, they should be characterized by intrinsic molecular recognition properties so that they can give specific responses in complex real samples and can be used even for automatic continuous monitoring. Obviously, this is not an easy task, also because interferences and artifacts are always a possible important source of mistake; however, some significant steps in this direction have been taken recently, also thanks to the development of new electrode systems and IEV.

11.2 Films of Ion Exchangers on Metal Electrodes

11.2.1 Principles of Ion Exchange Voltammetry

For simplicity, principles of IEV are discussed here for the case of the preconcentration and electrochemical one-electron reduction of a cationic-oxidized analyte, M^{n+}, at an electrode coated with a polyanionic film containing, for instance, sulfonic groups as ion exchange sites. Analogue considerations can be obtained, with the proper easy corrections, for IEV of anionic oxidized species on a polycationic-coated electrode and for reduced cationic or anionic analytes as well.

The preconcentration capability of the coating is ruled by the ion exchange equilibrium between the electroactive counter ion (analyte) and non-electroactive counter ions present in the solution or already incorporated into the coating, according to the following reaction [32]:

$$M^{n+} + n(SO_3^{-}X^{+}) \leftrightharpoons n(SO_3^{-})M^{n+} + X^{+} \qquad (11.5)$$

where SO_3^- are the ion exchange sites in the polymer, X^+ is an electrochemically inert counter ion (for simplicity, a cation of charge $z = 1$, as generally is the supporting electrolyte cation, e.g., Na^+), and M^{n+} is the multiply charged electroactive analyte.

The extent of such equilibrium reaction is expressed quantitatively by the selectivity coefficient K_X^M:

$$K_X^M = \frac{[M^{n+}]_p [X^+]_s^n}{[M^{n+}]_s [X^+]_p^n} \tag{11.6}$$

where subscripts p and s indicate concentrations in the polymer and solution phases, respectively, and square bracket indicates molar concentrations. With proper adaptations, ionic equivalent fractions can be used instead of molar concentrations. In some cases, a corrected selectivity coefficient is used, where activity coefficients of the solution species are used instead of concentrations. It is worth stressing that the selectivity coefficient is not strictly constant but can depend on experimental conditions [32].

When the ion-exchanged analyte is a trace species, the changes in $[X^+]_s$ and $[X^+]_p$ can be negligible and the selectivity coefficient can be conveniently substituted by the distribution coefficient, k_D, given as:

$$k_D(M^{n+}) = \frac{[M^{n+}]_p}{[M^{n+}]_s} \tag{11.7}$$

It is obvious that the analyte will be preconcentrated by coatings characterized by $k_D > 1$.

When a reducing electrochemical potential is applied (i.e., $E_{appl} < E^0$), M^{n+} is reduced at the polymer/electrode interface and a voltammetric peak is correspondingly recorded. If, as stated above, the reduction is a one-electron process (i.e., number of electrons exchanged, $n = 1$), reaction 11.8 holds [33, 34]:

$$n(SO_3^-)M^{n+} + e^- + X^+ \rightleftharpoons (n-1)(SO_3^-)M^{(n-1)+} + SO_3^-X^+ \tag{11.8}$$

Under proper experimental conditions (three-electrode electrochemical cell; mass transport controlled by semi-infinite planar diffusion), the voltammetric reduction current is a function of the concentration of electroactive species incorporated into the coating. It is interesting to note that the open circuit electrochemical potential at the coated electrode ($E(OC)_{CE}$), obtained by applying the Nernst equation to reaction (11.8), is given by:

$$E(OC)_{CE} = E^{0'} + \frac{RT}{F} \ln \frac{[X^+]_s}{[X^+]_p} + \frac{RT}{F} \ln \frac{[M^{n+}]_p}{[M^{(n-1)+}]_p} \tag{11.9}$$

where $E^{0\prime}$ is the formal potential of the incorporated redox couple at the ionic strength of the coating; ionic activities have been approximated by their concentrations. Equation 11.9 can be easily rearranged for cases where n or z $\neq$ 1.

It was shown [33, 35, 36] that, when differences in diffusion coefficients of oxidized and reduced species are negligible, the half-wave potential obtained by IEV at the coated electrode, $(E_{1/2})_{IEV}$, exhibited by the redox couple within the coating is given by:

$$(E_{1/2})_{IEV} = E^{0\prime} + \frac{RT}{F} \ln \frac{[X^+]_s}{[X^+]_p} \tag{11.10}$$

The partitioning of the oxidized and reduced species between the solution and the coating is ruled by relevant distribution coefficients; Eq. 11.7 for the oxidized analyte is complemented by Eq. 11.11 relevant to the reduced partner:

$$k_D(M^{(n-1)+}) = \frac{[M^{(n-1)+}]_p}{[M^{(n-1)+}]_s} \tag{11.11}$$

By proper substitution, one gets:

$$E(OC)_{CE} = E^{o\prime} + \frac{RT}{F}\left\{ \ln\frac{[X^+]_s}{[X^+]_p} + \ln\frac{k_{D(M^{n+})}}{k_{D(M^{(n-1)+})}} + \ln\frac{[M^{n+}]_s}{[M^{(n-1)+}]_s} \right\} \tag{11.12}$$

This equation is important for cases where both redox partners M^{n+} and $M^{(n-1)+}$ are present in the sample. In particular, when the concentration of electroactive analytes M^{n+} and $M^{(n-1)+}$ is low, almost all the ion exchange sites are saturated by X^+ so that changes in $[X^+]_p$ are negligible [37], and Eq. 11.10 can be rearranged as:

$$E(OC)_{CE} = K + \frac{RT}{nF} \ln \frac{[M^{n+}]_s}{[M^{(n-1)+}]_s} \tag{11.13}$$

where

$$K = E^{0\prime} + \frac{RT}{nF} \ln \frac{k_{D(M^{n+})}}{k_{D(M^{(n-1)+})}} + \frac{RF}{F} \ln \frac{[X^+]_s}{[X^+]_p} \tag{11.14}$$

This means that open-circuit potential values, $E(OC)_{CE}$, measured by potentiometry [37] at the polymer-coated electrode, change linearly with ln $([M^{n+}]_s/[M^{(n-1)+}]_s)$ with a slope equal to 59/n mV (at 25°C) and with an intercept given by Eq. 11.14. If all the other terms are independently known, $E(OC)_{CE}$ values allow one to calculate the concentration ratio in the sample solution between the redox species M^{n+} and $M^{(n-1)+}$. Equation 11.13 is rather similar to the usual Nernst equation for a redox indicator electrode [20] with the advantage that the coated electrode can preconcentrate the redox ions so, in principle, it can be used for trace analysis; moreover, the coating rejects interfering species of opposite ionic charge than the analyte.

These principles are the basis for performing redox state speciation of trace analytes of environmental interest, as for instance, Fe^{2+} and Fe^{3+} cations [37], by using ionomer-coated electrodes. By combining ion exchange voltammetry and potentiometry (with both data sets obtained at the same polymer-coated electrode), it is possible to obtain quantitative information on speciation. IEV is used for quantifying total concentrations and for obtaining relevant k_D values (see below); potentiometry at the coated electrode is employed for obtaining the concentration ratio between chemical species characterized by different redox states.

11.2.2 Current Signals at Polymer-Coated Electrodes

Many research papers have been devoted to study the mechanisms which rule mass and charge transport phenomena in polymeric matrices (see Ref. [38–40] for reviews). Depending on the experimental conditions involved (mainly, type of polymer, redox properties, and concentration of the ion-exchanged analyte), the overall transport process can be controlled by a variety of phenomena such as physical diffusion of redox species, electron hopping, segmental motion of polymer chains, polymer diffusion, counter ion migration, and ion pairing [38, 39]. Regardless of the microscopic mechanism, such transport phenomena obeys Fick's law of diffusion, and the rate of transport depends on an observable parameter characteristic for the analyte (and polymer layer) which is named apparent diffusion coefficient, D_{app}.

Analogously with voltammetric experiments performed at the electrode/solution interface, also at polymer-coated electrodes, peak currents are determined by the dimensions and time dependence of the diffusion layer which (for the polymer case) is the region of the electrode/polymer interface where an analyte concentration gradient is generated by the applied potential. Note that, at polymer-coated electrodes, the rate-limiting step which determines current responses is diffusion through the electrode/polymer interface. It is therefore important to distinguish whether the measurements are carried out in a timescale, t, for which the thickness of the diffusion layer, δ, is smaller or larger than the thickness of the polymeric coating, Φ. This determines the existence of two different kinds of voltammetric responses, typical of polymer-coated electrodes, which are [4]:

(a) A linear diffusion controlled behavior, when $\delta << \Phi$
(b) A thin-layer-like behavior, when $\delta >> \Phi$

The passage from situation (a) to situation (b) depends on relevant values of the coating thickness and voltammetric scan rate. Under conditions (a), there is a pool outside the diffusion layer which furnishes fresh analyte able to diffuse to the electrode in the time course of the scan; for case (b), the analyte incorporated in

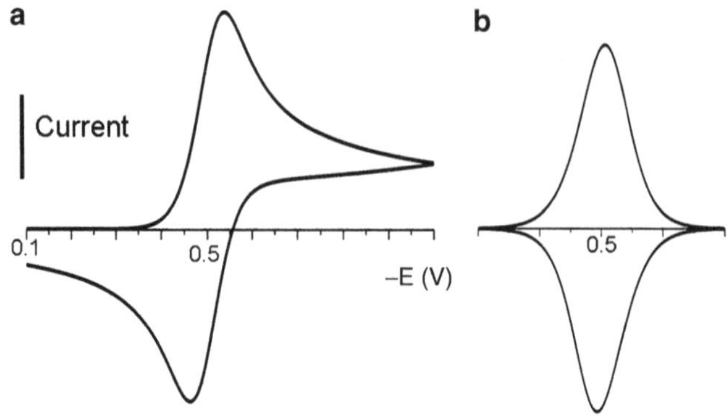

Fig. 11.1 Typical CV patterns observed at electrodes coated with thin films of ion exchangers: (**a**) under semi-infinite linear diffusion (film thickness > diffusion layer thickness; Ip scales with $v^{1/2}$); (**b**) under thin-layer-like conditions (film thickness $\leq$ diffusion layer thickness; Ip scales with v)

the coating is reduced (or oxidized) exhaustively during the voltammetric scan. As illustrated in Fig. 11.1, the characteristics of the voltammograms recorded under the two conditions are very different:

- Under linear diffusion conditions (case a), the voltammetric peak shows the classical diffusive tail due to the Cottrellian current decay observed at potential values cathodic enough with respect to the peak potential (see Fig. 11.1a); the peak current depends on the square root of v, according to easy rearrangement [5] of the classical Randles-Sevcik equation, where D_{app} substitutes D and $[M^{n+}]_p = k_D[M^{n+}]_s$ substitutes the bulk solution concentration.
- For a thin-layer-like behavior (case b), the voltammetric peak shows a characteristic symmetric Gaussian shape (see Fig. 11.1b); the peak current Ip depends linearly on the scan rate v, according to Eq. 11.15 [4]:

$$(Ip)_p = 9.39 \times 10^5 n^2 \text{ v V } [M^{n+}]_p \tag{11.15}$$

where V is the volume of the polymer thin layer (electrode area × film thickness).

Signals given by Eq. 11.15 are independent of D_{app} but depend on the film volume V. This means that the peak current in thin-layer-like conditions increases with the film volume, that is, film thickness.

Measurements under thin-layer-like conditions are useful for measuring voltammetrically the amount of analyte incorporated in the polymer layer after equilibration with a certain concentration of analyte in the sample solution; this quantity is obtained via coulometric integration of the voltammetric peak recorded in thin-layer-like conditions (low scan rate, generally 2–5 mV/s). The plots $[M^{n+}]_p$ versus $[M^{n+}]_s$ at constant temperature give the ion exchange distribution isotherms from which one can calculate relevant K_X^M or k_D values (see Eqs. 11.6, 11.7) [41, 42].

Ion exchange voltammetric signals recorded under semi-infinite linear diffusion conditions are strongly influenced by D_{app} values. The ratio between peak currents recorded in the same experimental conditions at unmodified, $(Ip)_s$, and at polymer-coated electrodes, $(Ip)_p$, (of the same surface area A and with both electrodes operating under semi-infinite planar diffusion control) is given by Eq. 11.16 [5]:

$$\frac{(Ip)_p}{(Ip)_s} = \left(\frac{D_{app}}{D_s}\right)^{1/2} k_D \tag{11.16}$$

It shows that the factor $(D_{app}/D_s)^{1/2}$ k_D plays the role of an amplification factor which quantifies the increase of voltammetric signals when polymer-coated electrodes are used instead of bare ones. D_{app} values are typically in the 10^{-9}–10^{-12} $cm^2\ s^{-1}$ range [38–40]; they are two to three orders of magnitude lower than D_s values, usually in the 10^{-5}–10^{-6} $cm^2\ s^{-1}$ range. Because of such a large difference in diffusion coefficient values, the ion exchange preconcentration results effective in increasing voltammetric signals for those analytes who are characterized by ion exchange distribution coefficient values $k_D > 100$–1,000. In such cases, the increase in sensitivity can then be from one to several orders of magnitude.

Without entering into a detailed analysis of the dynamics of electron transport in ion-exchanger coatings, it is worth noting that the operativity of different microscopic charge-transfer mechanisms can have consequences also for the analytical application of IEV since it can cause the appearance of a dependence of D_{app} (and therefore of $(Ip)_p$) on $[M^{n+}]_p$. Although such complications can be considered more exceptions than the rule, the possibility that D_{app} changes with $[M^{n+}]_p$ cannot be neglected. If electron hopping between the redox centers incorporated into the polymeric coating is the rate-determining step [3, 43, 44], then D_{app} can increase with $[M^{n+}]_p$ [45, 46]. According to the Dahms-Ruff electron hopping model [39, 47, 48], D_{app} is given as:

$$D_{app} = \frac{1}{6} k_{ex} d^2 [M^{n+}]_p \tag{11.17}$$

where k_{ex} = site to site electron exchange rate constant and d = distance of closest approach between two neighboring sites.

The incorporation of a multiply-charged analyte which can interact electrostatically with more than one polymeric chain can cause a sort of electrostatic cross-linking effect which reflects in a decrease of D_{app} with increasing $[M^{n+}]_p$ [49–54]. With electrodes modified with perfluorosulfonate ionomers, the decrease of diffusion coefficients with increasing concentration was explained by a bottleneck effect caused by the narrow channels that interconnect ionic clusters in these polymers [49]. It was shown that changes in the microenvironment within the ionomeric coating can influence such a behavior [55].

11.3 Electrode Modifiers for Ion Exchange Voltammetry

11.3.1 Ion Exchangers and Coatings

The electrode coating most widely used in IEV of cationic analytes is probably Nafion®, followed by the sulfonated polyesters such those belonging to the Eastman AQ™ series. For anionic analytes, poly(4-vinylpyridine) and Tosflex® IE-SA 48 (produced by Tosoh Co., Japan) are the most used polymeric coatings. Nafion®, Tosflex®, and sulfonated polyesters (see Scheme 11.1) belong to the ionomer group that are ionic polymers characterized by a rather low content of ionic groups [56].

These ionic groups interact or associate to form ion-rich hydrophilic regions surrounded by hydrophobic domains [57], such a clustering being the basis of the high stability of ionomer films.

The easiness of modification of the electrode surface and mechanical and chemical stability of ionomer coatings explain their successful use in IEV. The physical properties of Nafion® such as density, ion exchange capacity, etc., which are relevant to the successful preparation and analytical application of coated electrodes have been the subject of careful studies by Leddy's group [58–60].

Apart from ionomers, also other ion exchangers were applied for ion exchange voltammetry such as, for instance, poly(4-vinylpyridine) [1], poly(2-vinylpyrazine) [35], and poly-L-lysine [61, 62]. However, because of the higher water solubility

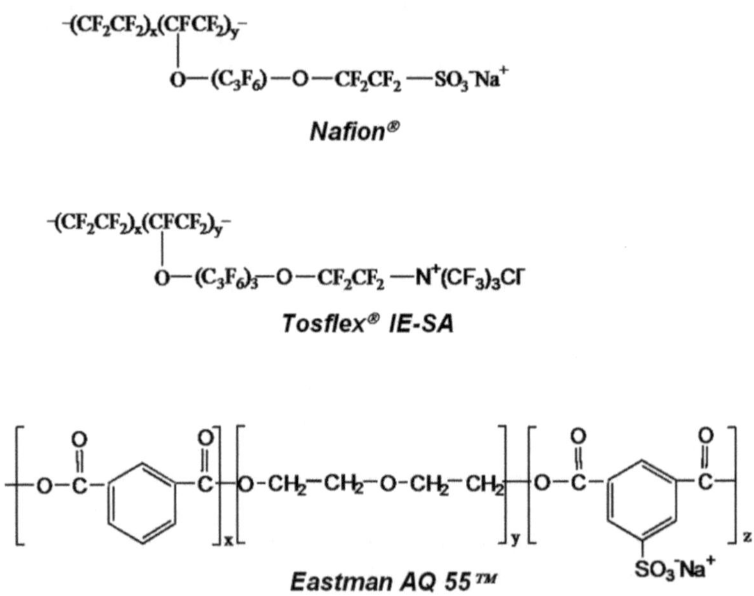

Scheme 11.1 Structural formulas of some ionomers used in IEV

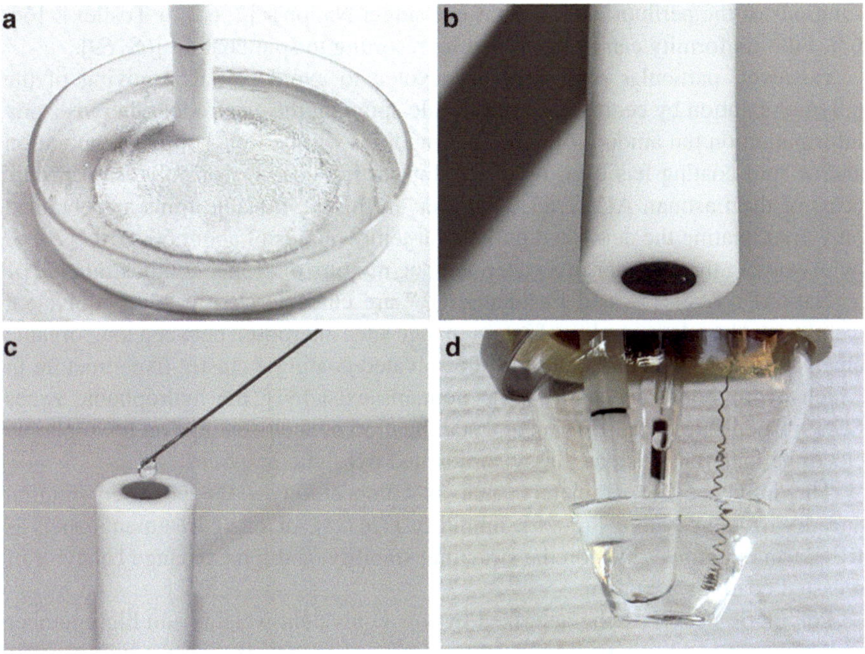

Fig. 11.2 Sequence of the main steps used in the deposition of an ionomer by microvolume evaporation: (**a**) mirror polishing of a glassy carbon disk electrode on graded alumina powder (0.3 and, then, 0.05 μm particle diameter), followed by rinsing in deionized water and ultrasonication; (**b**) a mirror-polished glassy carbon electrode with a disk diameter of 5 mm; (**c**) application of a microvolume (typically 3 μl) of 0.05% (w/v) Nafion solution in water/methanol/isopropanol (1/1/1); (**d**) after drying at room temperature for about 30 min, the modified electrode is dipped in the electrolyte in a three-electrode electrochemical cell; on the *left*: modified working electrode; in the *center*: Ag/AgCl reference electrode; on the *right*: Pt spiral counter electrode

of these polyelectrolytes which contain a large number of ion exchange sites, the preparation of stable coatings requires a chemical cross-linking stabilization procedure. Another class of electrode coatings of inorganic nature is based on the use of clays. Applications and recent trends on the promising use of these inorganic modifiers for IEV has been nicely reviewed by Walcarius [63, 64].

11.3.2 Classical Methods of Coating Depositions

Different ways of film deposition can be used, the simplest and more popular procedure being casting a microvolume of polymer solution onto the electrode surface, then allowing the solvent to evaporate slowly. The steps used for the modification of the electrode surface are summarized in Fig. 11.2. This procedure has been used successfully for modifying electrodes starting from alcohol-water solutions of the perfluorinated cation-exchanger Nafion® [2, 65] or Tosflex® [66, 67]. Film uniformity can be improved by resorting to spin-coating [68, 69].

However, particular care must be devoted to avoid partial removing of the polymer solution by centrifugal force while spinning the electrode; otherwise, any information on the amount of deposited modifier will be lost. Droplet evaporation and/or spin coating has been used also for casting coatings of poly(estersulfonic acid) of the Eastman AQ series. However, in this case, stable films are obtained only after heating the deposited polymer at temperatures of about 60–80°C [70] or by dissolving the ionomer in a water-acetone mixture of proper composition [71].

Nafion®, Tosflex®, and Eastman AQ™ are characterized by high equivalent weights and by a low number of ion exchange sites interposed between long organic chains. This makes the structure of the solvated coating a cluster-like structure in which hydrophobic and hydrophilic domains exist [57], the hydrophobic zones being responsible for adsorption and stabilization of such coatings on hydrophobic electrode materials such as glassy carbon and pyrolytic graphite.

The influence of parameters such as composition of the polymer solution and drying temperature [72–75], humidity [76, 77], or other treatments such as RF-plasma treatment [78] on the structure, stability, and ion exchange behavior of recasted coatings is known.

Polymeric materials characterized by low equivalent weights and high number of ion exchange sites have been used to recast polymer films on electrode surfaces by microvolume evaporation, followed by stabilization via chemical cross-linking. A typical example is given by poly(4-vinylpyridine), which gives more stable and reproducible coatings when the polymer backbone is cross-linked by reacting with alkyl dihalides [79, 80]. This procedure was used also for other polymers containing heterocyclic nitrogen, such as poly(2-vinylpyrazine) [35]. Note that poly (4-vinylpyridine) behaves as an anion exchanger when the pyridinic nitrogen is in the pyridinium form, e.g., protonated by dipping the modified electrode in solution at $pH < 5.2$ [1], or quaternized by reaction with alkyl halides [81]. Cross-linking of polyelectrolytes can be obtained also by irradiation methods, as for the case of poly(dimethyldiallylammonium chloride) [82].

Another popular and effective way of preparing ion exchange coatings is electropolymerization [4], this procedure being used for depositing films of electroactive polymers prepared by polymerization of electroactive monomeric precursors. They can be redox polymers (see, e.g., [83]), conducting polymers [84], or functionalized conducting polymers [85–89]. However, in principle, performing ion exchange voltammetry with electrodes coated with polymers which are electroactive themselves can suffer some limitations since the electroactivity window relevant to the polymer can overlap with the detection window of some analytes. On the other hand, the signal relevant to the electroactive moiety of the polymer can be used to good advantage as an internal standard to evaluate signals quantitatively [90].

As above mentioned, electrochemical oxidation is used to electropolymerize anionic or cationic derivatives of pyrrole [87–89], so obtaining coatings in which the potential dependent ion exchange properties typical of conducting polymers are enhanced and become independent on the applied potential. Moreover, functionalization of polypyrrole with fixed ion exchange sites increases its ion exchange capacity about four times [88], compared with plain polypyrrole in its fully oxidized form. Electrochemical polymerization is used for the preparation of coatings which incorporate the electroactive mediators used in electrochemical biosensors. This approach has been introduced some years ago by Heller and coworkers [91], originating a class of electroactive polymeric materials tailored for the electrochemical "wiring" of redox enzymes and other biomacromolecules, successfully used for glucose sensors [92].

11.3.3 Molecular Engineering of Ionomer Deposition

Advances in the reliable use of polymer-coated electrodes for analytical purposes depend on improvements in the control and reproducibility of the deposition of the polymer layer. This allows the lowering of the film thickness and shortening of the response time. Moreover, molecular control of the coating structures allows the deliberate control of the hydrophobic/hydrophilic properties as well as clustering and surface exposures of the ion exchange sites.

Hoyer et al. [93] presented an electrostatic spray deposition method which allows the preparation of very uniform thin films (ca 0.3 μm), keeping unaltered the peculiar features of Nafion® as far as permselectivity and rejection of interferents are concerned.

New possibilities to built ordered structures of ionomer films on a smaller scale (nanometer) have been demonstrated by the application of layer-by-layer deposition techniques [12]. The layer-by-layer strategy allows the construction of supramolecular sandwiches, assembled by exploiting electrostatic interactions. It is based on the alternate deposition of molecular layers of cationic and anionic (poly)ions arranged in sandwiched ordered structures, thanks to favorable electrostatic interactions between the layers of opposite charge. Typical examples of application of the layer-by-layer strategy to the preparation of modified electrodes and sensors are found in the biosensors literature [94–101]; for instance, a thin film consisting of alternating layers of Nafion® and ferric ions was used to develop calcification-resistant implantable biosensors [102] or other nanolayered materials [103–105].

Other possibilities in the molecular engineering of ionomer films on electrodes surfaces were opened recently, by studying the preparation and characteristics of ionomer coatings prepared by using the Langmuir-Blodgett (LB) technique [10]. It is used to build up monolayers or multilayers of organic amphiphilic molecules in which the order and 2-D structure are controlled at the molecular level by controlling the compression extent in the LB trough of a monomolecular layer of amphiphilic molecules spread at the air/water interface. By this way, it is possible to obtain ultrathin films characterized by long-range order, which can provide new insights on electron transfer processes at molecularly ordered interfaces. The 2-D-ordered interfacial layer can be transferred on solid surfaces by controlled dipping through the interface. There are two ways of dipping the substrate: the classical

Langmuir-Blodgett technique (vertical transfer) and the Langmuir-Schaefer (LS) technique (horizontal transfer). The LS technique is usually employed for the transfer of rigid films of materials such as polymers. Although the LS technique showed to work well for Nafion® [106], it has the limitation to rely upon the manual ability and experience of the experimentalist. Recently, it was shown that successful transfer of Langmuir layers of ionomers [107] can be achieved also by optimizing the conditions for the classical LB technique which is a vertical dipping-lift deposition technique for which the rate and extent of the movement of the substrate through the interfacial film is controlled automatically by a mechanical dipper driven by suitable software.

It was shown that if proper electrolyte is dissolved in an aqueous subphase, Nafion® forms an interfacial film at the air/water interface, which can be compressed in a typical Langmuir-Blodgett experiment, giving a stable film transferable on the surface of a variety of materials, including electrode surfaces. The thickness of the Nafion film can be as thin as few nm and scales with the number of layers deposited, keeping its typical permselectivity [106].

Very recent advances showed the possibility to characterize the structure of these ultrathin coatings by using electroactive photoluminescent probes such as $Ru(bpy)_3^{2+}$ complexes [108] as well as to exploit electrochemically induced luminescence to detect suitable co-reactants (analytes) such as tertiary amines or oxalate [109, 110].

11.3.4 Coating Regeneration and Electrochemical Control of Ion Exchange Voltammetry

When considering the analytical application of IEV, it is important to assess the reusability of the modified surface. If the ion exchange selectivity coefficient for the analyte is not too large, regeneration can be achieved by simply exposing the film to a high concentration of non-electroactive salt solution so obtaining expulsion of the analyte from the ion-exchanger film. However, such a simple regeneration procedure cannot be successful for analytes characterized by very large ion exchange selectivity coefficients. For this reason, a number of studies were devoted to two alternative approaches which can overcome this problem; they are:

- Preparing single-use electrodes based on the preparation of cheap and disposable electrode systems such as screen-printed electrodes [111, 112]
- Developing coatings which can facilitate the regeneration of the ion exchange membranes

The modifiers used for the preparation of screen-printed electrodes are typically different from the polymeric membrane systems which are the subject of the present chapter, so we will not go into many details of this approach.

The regeneration of the coatings can be improved by developing polymer films whose net charge can be switched chemically and/or electrochemically from positive to neutral to negative. Pioneering studies in this direction were outlined already in the 1980s by researches performed in Anson's group where poly(4-vinylpyridine) [113] and poly(2-vinylpyrazine) [35], containing electroactive pendant groups such as $[Fe(CN)_5{}^{3-/2-}]$, were used to this aim.

A similar approach, in which the ion exchange characteristics were modulated by electrochemical switching, implied terpolymers based on styrene (backbone), styrene sulfonate (unswitchable cation exchanger), and vinylferrocene (electrochemically switchable from neutral to anion exchanger) [114]. When the Fc/Fc^+ couple in the film was reduced, the terpolymer acted as a cation exchanger. When Fc was oxidized, the electrogenerated Fc^+ cation became the counter ion for the $SO_3{}^-$ groups, thus allowing the release of the incorporated cations (analyte). A similar approach was extended recently to add electrochemical control to ion-permeation capabilities in templated nanotubes functionalized with 11-ferrocenyl-1-undecanethiol [115]. The switching from cation exchanger to neutral polymer was achieved also by using electrodes coated with conducting polymer and polyelectrolytes composites [116–125]. Typical examples of such an approach were the electrochemically controlled binding and release of cations at poly(N-methyl-pyrrole)/polystyrene sulfonate [123] or at polypyrrole/polyester sulfonate [125]. When the conducting polymer was reduced, cations such as protonated amines, $Ru(bpy)_3{}^{2+}$, or methylviologen were incorporated into the film, while the release of these compounds was accomplished by simple reoxidation [113].

More recently, these principles were applied for developing solid-phase microextraction (SPME) electrodes in which the redox switching of the conducting polymer was exploited for performing the electrochemically controlled uptake and release of ionic analytes [126–129].

By taking advantage of the anion preconcentration capabilities of oxidized conducting polymers, Mark and coworkers [126] used a Pt microfiber coated with poly (3-methylthiophene) (PMeT) to perform the electrochemically controlled extraction and desorption of arsenate anions. The extension of this principle to cationic analytes was presented by Nyholm and coworkers [129], who used polypyrrole films doped with anions of low mobility in order to obtain a film which in the reduced state acts as a cation exchanger.

All these studies were performed using aqueous solution as the sample. However, both from a SPME viewpoint and when thinking to chromatographic applications of electrochemically switchable ion exchange devices [130], the development of electrochemically switchable polymer films usable in organic solvents is rather attractive; recently, a composite between poly(3-methylthiophene) and polyester sulfonate was developed to this aim [131]. The microscopic structure of this composite is shown in Fig. 11.3.

Note that, among the ionomers used as coatings on electrode surfaces, Eastman AQ55™ is the only one which is both stable and shows very good cation-exchange properties also in acetonitrile solutions [70, 71, 132].

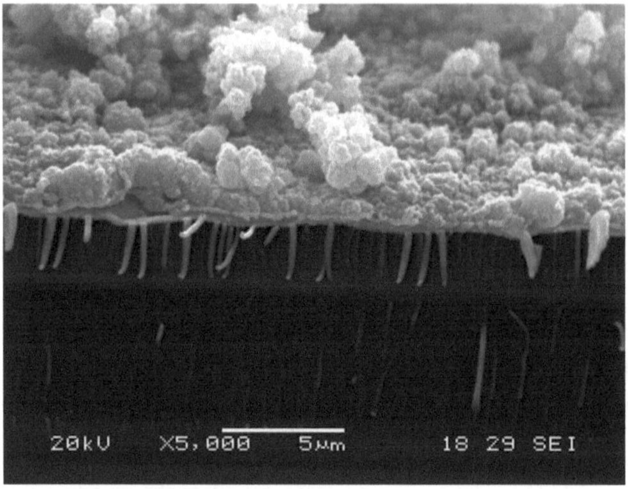

Fig. 11.3 SEM image of a cross-section of the bilayer composite Eastman AQ55™ (*lower* layer) and poly(3-methylthiophene) (*upper* granulated layer) for achieving electrochemical control of the ion exchange (Reproduced from Ref. [131] with kind permission of © Elsevier (2006))

The development of ionomer-based SPME devices can open the way to extended application of IEV for environmental electroanalysis and in electrochemically controllable separation methods.

11.4 Recent Analytical Applications of Ion Exchange Voltammetry

Some of the earlier analytical applications of IEV have been reviewed previously [6–9]. In the following, Tables 11.1–11.3 are listed and briefly commented those examples which can be considered as typical for IEV, together with some very recent updates. These examples of analytical applications of IEV methods demonstrate the excellent capabilities of the technique; note that many of the methods listed are characterized by very low detection limits, typically in the 10^{-9} M range or lower. In many cases, the use of ion-exchanger-coated electrodes can reduce drastically the effects of interferences and electrode poisoning and allows one to perform trace speciation analysis, that is, to identify and to measure the concentration of one or more individual chemical species of the same element. In the same cases, suitable electrocatalysts incorporated in the ion-exchanger film are used to sense a soluble analyte (substrate), generating an electrocatalytic current which scales with the analyte concentration. In Tables 11.1–11.3, we included only those examples of electrocatalysis at polymer-coated electrodes in which the ion exchange process plays a key role.

Table 11.1 Organic cation exchangers

Exchanger	Analytes	Notes	Reference
Nafion Nafion + cellulose acetate	Dopa, epinephrine, norepinephrine	Amperometric detection in flow systems. Advantages of bifunctional coatings are described	[133, 134]
Nafion	Dopamine, serotonin	Carbon fiber electrodes for in vivo detection of the analytes in the nM range	[135–137]
Nafion	Yb^{3+}, Eu^{3+}	Determination of trace Yb^{3+} and Eu^{3+}. DL = 30 nM for Eu^{3+} and 20 nM for Yb^{3+}	[138, 139]
Nafion	Promethazine	Good exclusion of interfering anions and satisfactory results in flow analysis	[140]
Nafion	UO_2^{2+}	Effect of oxidation state on the partitioning of UO_2^{2+} is examined. Influence of local proton activity inside Nafion is studied	[42]
Nafion	4-Nitroso-N,N-diethylaniline	Study of the redox mechanism at the Nafion GC modified electrode. DL 3 nM	[141]
Nafion	Salbutamol, fenoterol, metaproterenol	Nafion is deposited on the surface of a CPE. DL 2.5×10^{-8} M	[142]
Nafion	Pb(II)	Nafion/copper-mercury film electrode. DL 80 ng L^{-1}	[143]
Nafion/1-(2-pyridylazo)-2-naphthol	Pb(II)	The chelating agent in Nafion increases sensitivity and selectivity	[144]
Nafion and Nafion/poly (vinylsulfonic acid)	Cu(II),Cd(II), Pb(II), Zn(II)	Different kinds of coatings are compared for batch analysis of environmental samples such as river and run-off waters	[145]
Nafion	CH_3Hg^+	DPV and multiple square wave voltammetry (MSWV) allows ultratrace determination of CH_3Hg^+ with Nafion	[23]
Nafion	Tl(I)	Square wave stripping, in the presence of EDTA. DL 0.01 ppb	[146]
Nafion	Parathion	DL 50 nM, linear range up to 15 μM	[147]
Nafion	Selenium(IV)	Nafion-coated mercury film electrode modified with 3,3′-diaminobenzidine. CSV, DL 0.48 μg L^{-1}	[148]
Polyester sulfonate	Cu(II), Cd(II), Pb(II), Zn(II)	Electrodes coated with ionomer/mercury thin film for ASV of heavy metals	[149]
Nafion	Apomorphine	Ion exchange preconcentration, detection by SWV. DL 3 nM	[150]

(continued)

Table 11.1 (continued)

Exchanger	Analytes	Notes	Reference
Nafion	Eu^{3+}	MSWV for ultratrace determination, DL 10 pM, application to nuclear plants cooling waters	[24]
Nafion	Phenol	Adsorptive stripping in the presence of ethyltrimethylammonium which favors the analysis. DL 1 nM	[151]
Sulfopolyester (Eastek)	N-nitrosamines	Study on the effect of the degree of sulfonation; comparison with Nafion, cellulose acetate, and butyrate. DL 0.1 nM	[152]
Eastman AQ55™	Cyt c	GCE modified with AQ55™. Incorporation of cyt c; electrocatalytic activity is observed	[153]
Eastman AQ55™/silica composite	Oxalate, TPA, chlorpromazine	$Ru(bpy)_3^{2+}$ immobilized in AQ55™/silica film. ECL detection	[154]
Polydiphenylamine/Na-dodecylsulfate	Electro-inactive cations	Gold electrode modified with ion exchanger Amperometric detector for IEC	[155]
PDMDAAC	Cations	Sol–gel preparation of CPE with ion exchanger incorporated. Preconcentration by IE, detection by SWV	[156]
Poly(sodium 4-styrene sulfonate)	Trace heavy metals	Application for heavy metal analysis in estuarine waters	[157]
Nafion Eastman AQ55™	Cytochrome *c*/ ascorbate	Nafion or AQ55™ film deposited by LB technique incorporating cyt c. Electrocatalytic detection of ascorbate	[158]
Eastman AQ55™	Ethyl 4-iodobenzoate Di-bromocyclohexane	Incorporation of nickel complexes by ion exchange, as detecting electrocatalyst	[159]
Polylysine/polystyrene sulfonate	Pb(II), Cu(II), Cd(II)	GC/TMFE, preconcentration by IEV and detection by SW-ASV	[160]
2,5-Bis(2-thienyl)-*N*-(3-phosphorylpropyl) pyrrole	DNA	Electropolymerization of PPy, target 27- and 18-mer DNA incorporated in the film by IE. Detection of 0.16-fmol DNA	[161]
Hexacyanoferrate nanocomposite	Radioactive caesium	Incorporation of Cs by ion exchange in the reduction of nickel hexacyanoferrate	[162]
Nafion	TPA	Immobilization of $Ru(bpy)_3^{2+}$ in Nafion LB film for ECL, detection of tertiary amines	[109, 163]

(continued)

Table 11.1 (continued)

Exchanger	Analytes	Notes	Reference
Nafion/CNT	NADH	Ion exchange immobilization of thionine, electrocatalyst for NADH detection	[164]
Nafion/PSS	TPA	Immobilization of $Ru(bpy)_3^{2+}$ on ITO/Nafion/PSS electrode ECL detection. DL 3 nM	[165]
Poly (sodium-4-styrene sulfonate) (PSS)	Pb(II)	GC/TMFE, preconcentration by IEV and detection by SW-ASV	[166]
Polyvinyl ferrocenium	Hg^{2+}	DL 5×10^{-10} M	[167]
Nafion/nano-Au/thionine multilayer	Thrombin (TB)	Au electrode coated LBL by ion exchange with Nafion, thionine, and nano-Au. DL 40 pM	[168]
Nafion	Cr(VI)	Preparation of silver nanoparticles in GCE/Nafion film. Electrocatalysis of Cr (VI)	[169]
Nafion	Phenolic compounds	Electrospun $Ru(bpy)_3^{2+}$-doped Nafion nanofibers. ECL detection. DL 1.0 nM	[170]
Phosphonic acids bearing pyridinium groups	Cations	Phosphonic acids bearing pyridinium groups grafted on mesoporous TiO_2. Cations are incorporated by ion exchange	[171]
Nafion	NO_2^-	$Fe(bpy)_3^{2+}$ immobilized in Nafion by IE. Amperometric sensor	[172]

Table 11.2 Organic anion exchangers

Exchanger	Analytes	Notes	Reference
Protonated poly (4-vinylpyridine)	Cr(VI)	Detection by reduction to Cr (III). Linear range: 10^{-6}–10^{-8} M	[173]
Quaternized poly (4-vinylpyridine)	Cu/Cl^-	Copper preconcentrated as electrogenerated anionic cuprous complexes	[174]
Tosflex	Cu/Cl^-	Copper preconcentrated as electrogenerated anionic cuprous complexes. DL 3×10^{-6} M. Influence of amino acids is examined	[66]
Tosflex	Hg/Cl^-	Trace determination of Hg(II) in chloride-containing media. DL 4×10^{-11} M. Application to seawater analysis	[175]

(continued)

Table 11.2 (continued)

Exchanger	Analytes	Notes	Reference
Poly(4-vinylpyridine)	Bi(III)	Square wave stripping of anionic complexes of Bi (III) with chloride	[176]
Tosflex	$HgCl_4^{2-}$	The mechanism of IEV preconcentration/detection is examined	[177]
Tosflex/8-quinolinol	Te(IV)	Cathodic stripping of anion $TeCl_6^{2-}$. DL 0.1 μg/L	[178]
Quaternized poly (4-vinylpyridine) (QPVP)	Pb(II)	Anionic complexes generated in the coating are preconcentrated; application to groundwater analyses	[179]
Tosflex	Bi(III)	Preconcentration of the analyte by IE. DL 0.58 ppb	[180]
PDMDAAC	Anions	Cross-linking of the modifier by μ-irradiation. Application to $Fe(CN)_6^{3-}$ analysis	[181]
Tosflex	Zn^{2+}	TMFE modified with Tosflex DL 0.1 ng/mL	[182]
Tosflex	Diethyldithiophosphoric acid	DL 0.4 ppb	[183]
Poly(N-chloranil N,N,N′, N′-tetramethylethylene diammonium dichloride)	β-Lactam antibiotics	DL 2.12 nM	[184]
Poly(vinylsulfonic acid) (PVSA)	Anions $Fe(CN)_6^{4-}$	Sol–gel preparation of the modified electrode. SWV detection	[185]
Poly-L-lysine	Iodide Idoxuridine	DL 0.017 μg/mL	[62]
Tosflex	S^{2-}	SPE modified with Tosflex/Fe $(CN)_6^{3-}$. Electrocatalytic detection. DL 8.9 nM	[186]
Cationic poly(allylamine hydrochloride) (PAH)	Pb(II)	Anionic complexes of Pb are detected DL 6 nM	[187]
Tosflex	2-Naphthol	DL 0.2×10^{-7} M	[188]
Poly(allylamine) + poly (acrylic acid)	$PdCl_4^{2-}$	Ion exchange loading of $PdCl_4^{2-}$ for application as nanocatalyst	[189]

Table 11.3 Clays, zeolites, and silica and organosilica modifiers

Exchanger	Analyte	Notes	Reference
Zeolite-Y	MV^{2+}	Study of ion exchange/charge transfer mechanism	[190]
Zeolite-Y	Cl^-, Br^-	Ag-modified zeolite	[191]
Zeolite-Y + anionic surfactant	MV cation radical	Preconcentration by ion exchange, detection by double potential step chronocoulometry	[192]
Montmorillonite/cellulose acetate (NCA)	Cu(II)	NCA-modified GCE; preconcentration by IE. DL 1.73 ppb	[193]
Zeolite (nano)	Ag^+, Cd^{2+}, Pb^{2+}	GCE modified by nanoscale zeolites building blocks via layer-by-layer technology	[194]
Clays and organoclays	Cations or anions	Grafting of amine groups and protonating with HCl. It is not better than Nafion	[195]
Na-montmorillonite nanoparticles + anthraquinone	Cd(II), Pb(II)	Preconcentration by ion exchange, detection by DP-ASV DL 3-nM Cd^{2+}; 1-nM Pb^{2+}	[196]
Clay	TPA, oxalate	$Ru(bpy)_3{}^{2+}$ immobilized in clay multilayer films. Detection by ECL	[197]
Silica	Anions	Sol–gel preparation of quaternary amine functionalized silicon alkoxide	[198]
Polysiloxane-immobilized amine or thiol ligands	Hg(II)	Selective accumulation of anionic chloro-complexes of Hg(II). Speciation analysis	[199]
Porous silica sphere functionalized with ammonium quaternary groups	Anions	Silica microspheres obtained by the surfactant template route	[200]
Kaolinite/dimethylsulfoxide triethanolamine	CN^-	GCE modified with triethanolamine grafted kaolinite. Electrochemical preconcentration of cyanide	[201]
Mesoporous silica functionalized	Cations or anions	CPE modified with solid ion exchanger	[202]
Nanomontmorillonite	Methyl jasmonate	DL 5×10^{-7} M	[203]
Clay + p-phenylene dimethylene bis-dodecyl N,N-dimethylammonium	Methyl parathion	DL 70 nM	[204]

11.5 Conclusion and Prospects

The functionalization of voltammetric electrodes by depositing on their surface thin films of ion exchangers has contributed significantly to develop a new class of tailored electrochemical sensors characterized by very high sensitivity and very low detection limit, both characteristics being improved dramatically with respect to conventional electrodes.

Ion exchange voltammetry at ionomer-coated electrodes has grown progressively up to reaching nowadays the rank of a widespread powerful electroanalytical technique. It is suitable for trace and ultratrace analyses and can be used as a valuable tool in speciation analysis for environmental studies and biomedicine. The ion-exchanger coating gives to the electrode high sensitivity and selectivity and protects it from surfactant and organic interferences. Further efforts should be devoted in the near future to continue improving the reproducibility and control of the coating deposition procedure and to shorten the analysis time. The application of molecular engineering procedures such as LB, electrostatic spray, layer-by-layer deposition, or other bottom-up approaches can play a crucial role in this issue.

Acknowledgments We wish to thank all collaborators of the Laboratory for Electrochemical Sensors of the University Ca' Foscari of Venice (their names can be found as coauthors of many papers). Partial financial support by MIUR (Rome, Project PRIN 2008MWHCP2_001) is acknowledged.

References

1. Oyama N, Anson FC (1980) Electrostatic binding of metal complexes to electrode surfaces coated with highly charged polymeric films. J Electrochem Soc 127:247–250
2. Rubinstein I, Bard AJ (1980) Polymer films on electrodes. 4. Nafion-coated electrodes and electrogenerated chemiluminescence of surface-attached tris(2,2′-bipyridine)ruthenium(2+). J Am Chem Soc 102:6641–6642
3. Murray RW (1981) Modified electrodes. Chemically modified electrodes for electrocatalysis. Philos Trans R Soc Lond A 302:253–265
4. Murray RW (1984) Chemically modified electrodes. In: Bard AJ (ed) Electroanalytical chemistry, vol 13. Marcel Dekker, New York, pp 191–368
5. Espenscheid MW, Ghatak-Roy AR et al (1986) Sensors from polymer modified electrodes. J Chem Soc Faraday Trans I 82:1051–1070
6. Ugo P, Moretto LM (1995) Ion-exchange voltammetry at polymer modified electrodes: principles and analytical prospects. Electroanalysis 7:1105–1113
7. (a) Ugo P, Moretto LM, Vezzà F (2002) Ionomer-coated electrodes and nanoelectrode ensembles as electrochemical environmental sensors: recent advances and prospects. Chem Phys Chem 3:917–925; (b) Ugo P, Moretto LM, Vezzà F (2003) Ionomer-coated electrodes and nanoelectrode ensembles as electrochemical environmental sensors: recent advances and prospects. In: Baltes H, Fedder GK, Korvink JG (eds) Sensors update, vol 12. Wiley-VCH, Weinheim, pp 121–140
8. Zen JM, Kumar AS, Tsai DM (2003) Recent updates of chemically modified electrodes in analytical chemistry. Electroanalysis 15:1073–1087

9. Arrigan DWM (1994) Voltammetric determination of trace metals and organics after accumulation at modified electrodes. Analyst 119:1953–1966
10. Ulman A (1991) Ultrathin organic films. Academic, San Diego
11. Decher G (1997) Toward layered polymeric multicomposites. Science 277:1232–1237
12. Marquette CA, Bouteille F, Corgier BP et al (2009) Disposable screen-printed chemiluminescent biochips for the simultaneous determination of four point-of-care relevant proteins. Anal Bioanal Chem 393:1191–1198
13. Willner I, Katz E (eds) (2005) Bioelectronics. Wiley VCH, Weinheim
14. Thorp HH, Loomis CR, Napier ME (1999) Polymer-electrodes for detecting nucleic acid hybridization and method of use thereof. US Patent 5,968,745
15. Drummond TG, Hill MG, Barton JK (2003) Electrochemical DNA sensors. Nat Biotechnol 21:1192–1199
16. Dill K, Montgomery DD, Ghindilis AL et al (2004) Immunoassays and sequence-specific DNA detection on a microchip using enzyme amplified electrochemical detection. J Biochem Biophys Methods 59:181–187
17. Heller A, Feldman B (2010) Electrochemistry in diabetes management. Acc Chem Res 43:963–973
18. Svancara I, Vytras K, Kalcher K et al (2009) Carbon paste electrodes in facts, numbers, and notes: a review on the occasion of the 50-years jubilee of carbon paste in electrochemistry and electroanalysis. Electroanalysis 21:7–28
19. Bard AJ, Faulkner L (2000) Electrochemical methods, 2nd edn. Wiley, New York
20. Wang J (2000) Analytical electrochemistry, 2nd edn. VCH, Weinheim
21. Fatouros N, Simonin JP, Chevalet J et al (1986) Theory of multiple square wave voltammetry. J Electroanal Chem 213:1–16
22. Krulic D, Fatouros N, Chevalet J (1990) Multiple square wave voltammetry: experimental verification of the theory. J Electroanal Chem 287:215–227
23. Moretto LM, Ugo P, Lacasse R et al (1999) Determination of methylmercury at Nafion® coated electrodes by single and multiple pulse voltammetric techniques. J Electroanal Chem 467:193–202
24. Moretto LM, Chevalet J, Mazzocchin GA et al (2001) Advances in multiple square wave techniques for ion-exchange voltammetry at ultratrace levels: the europium(III) case. J Electroanal Chem 498:117–126
25. Brainina Kh, Neyman E (1993) Electroanalytical stripping methods in chemical analysis. Wiley, New York
26. Wang J (1989) Voltammetry following nonelectrolytic preconcentration. In: Bard AJ (ed) Electroanalytical chemistry, vol 16. Marcel Dekker, New York, pp 1–88
27. Buffle J (1988) Complexation reactions in aquatic systems – an analytical approach. Ellis Horwood, Chichester
28. Wojclechowski M, Go W, Osteryoung J (1985) Square-wave anodic stripping analysis in the presence of dissolved oxygen. Anal Chem 57:155–158
29. Colombo C, van den Berg C (1988) In-line deoxygenation for flow analysis with voltammetric detection. Anal Chim Acta 377:229–240
30. Mikkelsen Ǿ, Schrøder KH (2000) Dental amalgam in voltammetry – some preliminary results. Anal Lett 33:3253–3269
31. Wang J, Lu J, Kirgoz UA et al (2001) Insights into the anodic stripping voltammetric behavior of bismuth film electrodes. Anal Chim Acta 434:29–34
32. Helfferich F (1962) Ion-exchange. McGraw-Hill, New York
33. Naegeli R, Redepenning J, Anson FC (1986) Influence of supporting electrolyte concentration and composition on formal potentials and entropies of redox couples incorporated in Nafion coatings on electrodes. J Phys Chem 90:6227–6232
34. Redepenning J, Anson FC (1987) Permselectivities of polyelectrolyte electrode coatings as inferred from measurements with incorporated redox probes or concentration cells. J Phys Chem 91:4549–4553

35. Ugo P, Anson FC (1989) Poly(2-vinylpyrazine) as a soluble polymeric ligand and as an electrode coating. Reactions with pentacyanoferrate(II). Anal Chem 61:1799–1805
36. Toniolo R, Comisso N, Bontempelli G et al (1994) Potential shifts at electrodes coated with ion-exchange polymeric films. Talanta 41:473–478
37. Ugo P, Moretto LM, De Boni A et al (2002) Iron(II) and iron(III) determination by potentiometry and ion-exchange voltammetry at ionomer-coated electrodes. Anal Chim Acta 474:147–160
38. Hillman AR (1987) Reactions and applications of polymer modified electrodes. In: Linford RG (ed) Electrochemical science and technology of polymers, vol 1. Elsevier Applied Science, London, Chap 5
39. Majda M (1992) Dynamics of electron transport in polymeric assemblies of redox centers. In: Murray R (ed) Molecular design of electrode surfaces. Wiley, New York, pp 159–206
40. Oyama N, Ohsaka T (1992) Voltammetric diagnosis of charge transport on polymer coated electrodes. In: Murray R (ed) Molecular design of electrode surfaces. Wiley, New York
41. Szentirmay MN, Martin CR (1984) Ion-exchange selectivity of Nafion films on electrode surfaces. Anal Chem 56:1898–1902
42. Ugo P, Ballarin B, Daniele S et al (1992) Electrochemical behaviour and preconcentration of uranyl(VI) at Nafion-coated glassy carbon electrodes. J Electroanal Chem 324:145–159
43. Kaufman FB, Engler EM (1979) Solid-state spectroelectrochemistry of crosslinked donor bound polymer films. J Am Chem Soc 101:547–549
44. Facci J, Murray RW (1981) Charge transport by electron exchange cross reaction in cyclic voltammetry of $IrCl_6^{3-}$- $Fe(CN)_6^{3-}$ mixtures trapped in polycationic films on electrodes. J Phys Chem 85:2870–2873
45. Buttry DA, Anson FC (1983) Effects of electron exchange and single-file diffusion on charge propagation in Nafion films containing redox couples. J Am Chem Soc 105:685–689
46. Anson FC, Blauch DN, Saveant JM et al (1991) Ion association and electric field effects on electron hopping in redox polymers. Application to the tris(2,2′-bipyridine)osmium(3+)/tris (2,2′-bipyridine)osmium(2+) couple in Nafion. J Am Chem Soc 113:1922–1932
47. Dahms HI (1968) Electronic conduction in aqueous solution. J Phys Chem 72:362–364
48. Ruff I, Friedrich VJ (1971) Transfer diffusion. I. Theoretical. J Phys Chem 75:3297–3302
49. Whiteley LD, Martin CR (1989) Fresh look at transport in perfluorosulfonate ionomers: ultramicroelectrode investigations of Nafion and the Dow ionomers. J Phys Chem 93:4650–4658
50. Kuo K-N, Murray RW (1982) Electrocatalysis with ferrocyanide electrostatically trapped in an alkylaminesiloxane polymer film on a Pt electrode. J Electroanal Chem 131:37–60
51. Majda M, Faulkner LR (1984) Electrochemical behaviour of tris(2,2′-bipyridine)ruthenium complexes in films of poly(styrenesulfonate) on electrodes. J Electroanal Chem 169:77–95
52. Martin CR, Rubinstein I, Bard AJ (1982) Polymer films on electrodes. 9. Electron and mass transfer in Nafion films containing tris(2,2′-bipyridine)ruthenium(2+). J Am Chem Soc 104:4817–4824
53. Shi M, Anson FC (1997) Some consequences of the significantly different mobilities of hydrophilic and hydrophobic metal complexes in perfluorosulfonated ionomer coatings on electrodes. Anal Chem 69:2653–2660
54. Daniele S, Ugo P, Bragato C et al (1996) Use of Nafion® coated carbon disk microelectrodes in solution without and with different concentrations of supporting electrolyte. J Electroanal Chem 418:29–34
55. Amatore C, Sella C, Thouin L (2003) Effects of chemical environment on diffusivities within thin Nafion® films as monitored from chronoamperometric responses of generator–collector double microband assemblies. J Electroanal Chem 547:151–161
56. Kroschwitz JI (1990) Concise encyclopedia of polymer science and engineering. Wiley, New York
57. Yeo RS, Yeager HL (1985) Structural and transport properties of perfluorinated ion-exchange membranes. In: Conway BE, White RE, Bockris JO'M (eds) Modern aspects of electrochemistry, vol 16. Plenum, New York, pp 437–504

58. Zook LA, Leddy J (1996) Density and solubility of Nafion: recast, annealed, and commercial films. Anal Chem 68:3793–3796
59. Chen T-Y, Leddy J (2000) Ion exchange capacity of Nafion and Nafion composites. Langmuir 16:2866–2871
60. Oberbroeckling KJ, Dunwoody DC, Minteer SD et al (2002) Density of Nafion exchanged with transition metal complexes and tetramethyl ammonium, ferrous, and hydrogen ions: commercial and recast films. Anal Chem 74:4794–4799
61. Pereira FC, Fogg AG, Ugo P et al (2005) Determination of iodide and idoxuridine at a glutaraldehyde-cross-linked poly-L-lysine modified glassy carbon electrode. Electroanalysis 17:1309–1316
62. Anson FC, Saveant JM, Shigehara K (1983) New model for the interior of polyelectrolyte coatings on electrode surfaces. Mechanisms of charge transport through protonated poly (L-lysine) films containing FeIII(edta)$^{-}$ and FeII(edta)$^{2-}$ as counter ions. J Am Chem Soc 105:1096–1106
63. Walcarius A (2008) Electroanalytical applications of microporous zeolites and mesoporous (organo)silicas: recent trends. Electroanalysis 20:711–738
64. Walcarius A (2001) Electroanalysis with pure, chemically modified and sol–gel-derived silica-based materials. Electroanalysis 13:701–718
65. Buttry DA, Anson FC (1981) Electron hopping vs. molecular diffusion as charge transfer mechanism in redox polymer films. J Electroanal Chem 130:333–338
66. Ugo P, Moretto LM, Mazzocchin GA (1993) Voltammetric determination of trace mercury in chloride media at glassy carbon electrodes modified with polycationic ionomers. Anal Chim Acta 273:229–236
67. Dunsch L, Kavan L, Weber J (1990) Perfluoro anion-exchange polymeric films on glassy carbon electrodes. J Electroanal Chem 280:313–325
68. Kaufman FB, Schroeder AH, Engler EM et al (1980) Ion and electron transport in stable, electroactive tetrathiafulvalene polymer coated electrodes. J Am Chem Soc 102:483–488
69. Schroeder AH, Kaufman FB (1980) The influence of polymer morphology on polymer film electrochemistry. J Electroanal Chem 113:209–224
70. Gennett T, Purdy WC (1990) Voltammetric determination of the ion-exchange behaviour of AQ polymers [poly(ester sulfonic acid) anionomers] in acetonitrile. Anal Chem 62:2155–2158
71. Brunetti B, Ugo P (1999) Factors influencing the ion-exchange preconcentration and voltammetric behaviour of redox cations at polyestersulfonated ionomer coated electrodes in acetonitrile solutions. J Electroanal Chem 460:38–45
72. Weber J, Janda P, Kavan L et al (1986) Study of Nafion film on electrodes prepared from dimethylacetamide solution. J Electroanal Chem 200:379–381
73. Moore RB III, Martin CR (1988) Chemical and morphological properties of solution-cast perfluorosulfonate ionomers. Macromolecules 21:1334–1339
74. Goebel G, Aldebert P, Pineri M (1987) Structure and related properties of solution-cast perfluorosulfonated ionomer films. Macromolecules 20:1425–1428
75. Hoyer B, Jensen N (1994) Signal stability of Nafion-coated thin mercury film electrodes for stripping voltammetry. Talanta 41:449–453
76. Stribel KA, Scherer GG, Haas O (1991) Effect of curing-humidity on recast-Nafion films. J Electroanal Chem 304:289–296
77. Shi M, Anson FC (1996) Effects of hydration on the resistances and electrochemical responses of nafion coatings on electrodes. J Electroanal Chem 415:41–46
78. Shimazu K, Kita H, Kuwana T (1989) RF-plasma treatment of Nafion film electrodes. J Electroanal Chem 258:49–59
79. Nishide H, Deguchi J, Tsuchida E (1977) Adsorption of metal ions on cross-linked poly (4-vinylpyridine) resins prepared with a metal ion as template. J Polym Sci 15:3023–3029
80. Lindholm B, Sharp M (1986) Simple preparation and characteristics of stabilized, protonated poly-4-vinylpyridine films containing electroactive anions on glassy carbon electrode surfaces. J Electroanal Chem 198:37–52

81. Oyama N, Shimomura T, Shigehara K et al (1980) Electrochemical responses of multiply-charged transition metal complexes bound electrostatically to graphite electrode surfaces coated with polyelectrolytes. J Electroanal Chem 112:271–280
82. De Castro ES, Huber EW, Villaroel D et al (1987) Electrodes with polymer network films formed by g-irradiation cross-linking. Anal Chem 59:134–139
83. Denisevich P, Abruna HD, Leidner CR et al (1982) Electropolymerization of vinylpyridine and vinylbipyridine complexes of iron and ruthenium: homopolymers, copolymers, reactive polymers. Inorg Chem 21:2153–2161
84. Kaner RB (1990) Preparations and properties of electrochemically synthesized polymers. In: Linford RG (ed) Electrochemical science and technology of polymers, vol 2. Elsevier, London, pp 97–148
85. Ugo P, Sperni L, Moretto LM (1997) Ion-exchange voltammetry of trace mercury(II) at glassy carbon electrodes coated with a cationic polypyrrole derivative. Application to pore-waters analysis. Electroanalysis 9:1153–1158
86. Mao H, Pickup PG (1989) Electronically conductive anion exchange polymers based on polypyrrole. Preparation, characterization, electrostatic binding of ferrocyanide and electrocatalysis of ascorbic acid oxidation. J Electroanal Chem 265:127–142
87. Pickup PG (1987) Poly-(3-methylpyrrole-4-carboxylic acid): an electronically conducting ion-exchange polymer. J Electroanal Chem 225:273–280
88. Basak S, Rajeshwar K, Kaneko M (1990) Ion binding by poly{pyrrole-co-[3-(pyrrol-1-yl) propanesulfonate]} thin films. Anal Chem 62:1407–1413
89. Cosnier S, Deronzier A, Moutet J-C et al (1989) Alkylammonium and pyridinium group-containing polypyrroles, a new class of electronically conducting anion-exchange polymers. J Electroanal Chem 271:69–81
90. Guadalupe AR, Abruna HD (1985) Electroanalysis with chemically modified electrodes. Anal Chem 57:142–149
91. Degani Y, Heller A (1989) Electrical communication between redox centres of glucose oxidase and electrodes via electrostatically and covalently bound redox polymers. J Am Chem Soc 111:2357–2358
92. Heller A, Feldman B (2008) Electrochemical glucose sensors and their applications in diabetes management. Chem Rev 108:2482–2505
93. Hoyer B, Jensen N, Busch LP (2001) Effect of the pretreatment of recast Nafion membranes on their rejection of the albumin interference in anodic stripping voltammetry. Electroanalysis 13:843–848
94. Wilner I, Katz E (2000) Integration of layered redox proteins and conductive supports for bioelectronic applications. Angew Chem Int Ed 39:1180–1218
95. Mano N, Kuhn A (2002) Affinity assembled multilayers for new dehydrogenase biosensors. Bioelectrochemistry 56:123–126
96. Mano N, Kuhn A (2001) Electrodes modified with nitrofluorenone derivatives as a basis for new biosensors. Biosens Bioelectron 16:653–660
97. Anicet N, Bourdillon C, Moiroux J et al (1998) Simple electrochemical procedure for measuring the rates of electron transfer across liquid/liquid interfaces formed by coating graphite electrodes with thin layers of nitrobenzene. J Phys Chem B 102:9845–9854
98. Hodak J, Etchenique R, Calvo EJ et al (1997) Layer-by-layer self-assembly of glucose oxidase with a poly(allylamine)ferrocene redox mediator. Langmuir 13:2708–2716
99. Ma H, Hu N, Rusling J (2000) Electroactive myoglobin films grown layer-by-layer with poly (styrenesulfonate) on pyrolytic graphite electrodes. Langmuir 16:4969–4975
100. Lojou E, Bianco P (2003) Quartz crystal microbalance and voltammetry monitoring for layer-by-layer assembly of cytochrome c_3 and poly(ester sulfonic acid) films on gold and silver electrodes. J Electroanal Chem 557:37–47
101. Lojou E, Bianco P (2004) Buildup of polyelectrolyte–protein multilayer assemblies on gold electrodes. Role of the hydrophobic effect. Langmuir 20:748–755
102. Galeska I, Chattopadhyay D, Moussy F et al (2000) Calcification-resistant Nafion/Fe^{3+} assemblies for implantable biosensors. Biomacromolecules 1:202–207

103. Harris JJ, Stair JL, Bruening ML (2000) Layered polyelectrolyte films as selective, ultrathin barriers for anion transport. Chem Mater 12:1941–1946
104. Calvo EJ, Wolosiuk A (2002) Donnan permselectivity in layer-by-layer self-assembled redox polyelectrolyte thin films. J Am Chem Soc 124:8490–8497
105. De Longchamp DM, Hammond PT (2003) Fast ion conduction in layer-by-layer polymer films. Chem Mater 15:1165–1173
106. Bertoncello P, Ram MK, Notargiacomo A et al (2002) Fabrication and physico-chemical properties of Nafion Langmuir-Schaefer films. Phys Chem Chem Phys 4:4036–4043
107. Bertoncello P, Ugo P (2003) Preparation and voltammetric characterization of electrodes coated with Langmuir-Schaefer ultrathin films of Nafion®. J Braz Chem Soc 14:517–522
108. Moretto LM, Kohls T, Chauvin A et al (2008) Epifluorescence imaging of electrochemically switchable Langmuir–Blodgett films of Nafion. Langmuir 24:6367–6374
109. Moretto LM, Kohls T, Badocco D et al (2010) Electrochemiluminescence of $Ru(bpy)_2^{3+}$ loaded in Nafion Langmuir–Blodgett films: role of the interfacial ultrathin film. J Electroanal Chem 640:35–41
110. Bertoncello P, Dennany L (2007) Nafion–tris(2-2′-bipyridyl)ruthenium(II) ultrathin Langmuir–Schaefer films: redox catalysis and electrochemiluminescent properties. Anal Chem 79:7549–7553
111. Wring SA, Hart JP (1992) Chemically modified, carbon-based electrodes and their application as electrochemical sensors for the analysis of biologically important compounds. A review. Analyst 117:1215–1229
112. Ugo P, Moretto LM, Bertoncello P et al (1998) Determination of trace mercury in saltwaters at screen-printed electrodes modified with Sumichelate Q10R. Electroanalysis 10:1017–1021
113. Zumbrunnen HR, Anson FC (1983) Electrostatic binding of anions and cations to graphite electrodes coated with a polyelectrolyte containing both positive and negative fixed charges. J Electroanal Chem 152:111–124
114. Espenscheid MW, Martin CR (1985) Electroactive ion exchange polymers. J Electroanal Chem 188:73–84
115. Buyukserin F, Kohli P, Wirtz MC et al (2007) Electroactive nanotube membranes and redox-gating. Small 3:266–270
116. Fan FF, Bard AJ (1986) Polymer films on electrodes. J Electrochem Soc 133:301–304
117. Penner RM, Martin CR (1986) Electronically conductive composite polymer membranes. J Electrochem Soc 133:310–315
118. Nagasubramanian G, Di Stefano S, Moacanin J (1986) Electrochemical incorporation of poly (pyrrole) into Nafion and comparison of the electrochemical properties of Nafion-poly (pyrrole) and poly(pyrrole) films. J Phys Chem 90:4447–4451
119. Orata D, Buttry DA (1988) Virtues of composite structures in electrode modification. Preparation and properties of poly(anyline)/Nafion composite films. J Electroanal Chem 257:71–82
120. Hirai T, Kuwabata S, Yonegani H (1988) Electrochemical behaviours of polypyrrole, poly-3-methylthiophene, and polyaniline deposited on Nafion-coated electrodes. J Electrochem Soc 135:1132–1137
121. Shimidzu T, Ohtani A, Iyoda T et al (1987) Charge-controllable polypyrrole/polyelectrolyte composite membranes. Part II. Effect of incorporated anion size on the electrochemical oxidation-reduction process. J Electroanal Chem 224:123–135
122. Shimidzu T, Ohtani A, Iyoda T et al (1988) Charge-controllable polypyrrole/polyelectrolyte composite membranes. Part III. Electrochemical deionization system constructed by anion-exchangeable and cation-exchangeable polypyrrole electrodes. J Electroanal Chem 251:323–337
123. Zhou QX, Miller LL, Valentine JR (1989) Electrochemically controlled binding and release of protonated dimethyldopamine and other cations from poly(N-methyl-pyrrole)/polyanion composite redox polymers. J Electroanal Chem 261:147–164
124. Elliot CM, Kopelove AB, Albery WJ et al (1991) Nonaqueous electrochemistry of polypyrrole/polystyrenesulfonate composite films: voltammetric, coulometric, EPR, and a. c. impedance studies. J Phys Chem 95:1743–1747

125. Wang J, Sun Z, Lu Z (1991) Electrochemical behaviour of polypyrrole/Kodak AQ composite polymeric films. J Electroanal Chem 310:269–279
126. Gbatu TP, Ceylan O, Sutton KL et al (1999) Electrochemical control of solid phase microextraction using unique conducting polymer coated fibers. Anal Commun 36:203–205
127. Wu J, Yu X, Lord H et al (2000) Solid phase microextraction of inorganic anions based on polypyrrole film. Analyst 125:391–394
128. Wu J, Pawliszyn J (2001) Preparation and applications of polypyrrole films in solid-phase microextraction. J Chromatogr A 909:37–52
129. Liljegren G, Petterson J, Markides KE et al (2002) Electrochemical solid-phase microextraction of anions and cations using polypyrrole coatings and an integrated three-electrode device. Analyst 127:591–597
130. Deinhammer RS, Porter MD, Shimazu K (1995) Retention characteristics of polypyrrole as a stationary phase for the electrochemically modulated liquid chromatographic (EMLC) separations of dansyl amino acids. J Electroanal Chem 387:35–46
131. Scopece P, Moretto LM, Polizzi S et al (2006) Composite films of poly-(ester-sulphonated) and poly-(3-methylthiophene) for ion-exchange voltammetry in acetonitrile solutions. Electrochim Acta 51:2153–2160
132. Hanzlik J, Ugo P, Daniele S et al (1996) Ion-exchange voltammetry of tris(2,2′-bipyridyl) ruthenium(II), iron(II), osmium(II) and tris(2,2′-bipyrazyl) ruthenium(II) in acetonitrile solutions at poly(ester-sulphonate) coated electrodes. J Electroanal Chem 404:89–97
133. Wang J, Tuzhi P (1986) Selectivity and sensitivity improvements at perfluorinated ionomer/cellulose acetate bilayer electrodes. Anal Chem 58:3257–3261
134. Wang J, Tuzhi P, Golden T (1987) Amperometric detection of cationic neurotransmitters at Nafion-coated glassy carbon electrodes in flow streams. Anal Chim Acta 194:129–138
135. Brazell MP, Kasser RJ, Renner KJ et al (1987) Electrocoating carbon fiber microelectrodes with Nafion improves selectivity for electroactive neurotransmitters. J Neurosci Methods 22:167–172
136. Kristensen EW, Kuhr WG, Wightman RM (1987) Temporal characterization of perfluorinated ion exchange coated microvoltammetric electrodes for in vivo use. Anal Chem 59:1752–1757
137. Crespi F, Martin KF, Marsden CA (1988) Measurement of extracellular basal levels of serotonin in vivo using Nafion-coated carbon fibre electrodes combined with differential pulse voltammetry. Neuroscience 27:885–896
138. Ugo P, Ballarin B, Daniele S et al (1991) Determination of trace amounts of Eu^{3+} and Yb^{3+} ions at Nafion-coated thin mercury film electrodes. Anal Chim Acta 244:29–38
139. Ugo P, Ballarin B, Daniele S et al (1990) Electrochemistry of Yb^{3+} and Eu^{3+} at Nafion modified electrodes. J Electroanal Chem 291:187–199
140. Zhou J, Wang E (1991) Ion exchange of cationic drugs at a Nafion-coated electrode in flow-through analysis. Anal Chim Acta 249:489–494
141. Gorski W, Cox JA (1992) Stripping voltammetry with preconcentration through chemical reactions coupled to charge transfer in an ionomer-coated electrode: application to the determination of a nitrosoamine. Anal Chem 64:2706–2710
142. Boyd D, Barreira Rodriguez JR, Miranda Ordieres AJ et al (1994) Voltammetric study of salbutamol, fenoterol and metaproterenol at unmodified and Nafion-modified carbon paste electrodes. Analyst 119:1979–1984
143. Zen J-M, Ting Y-S (1996) Square-wave voltammetric stripping analysis of lead(II) at a Nafion®/copper-mercury film electrode. Anal Chim Acta 332:59–65
144. Hu Z, Seliskar CJ, Heineman WR (1998) PAN-incorporated Nafion-modified spectroscopic graphite electrodes for voltammetric stripping determination of lead. Anal Chim Acta 369:93–101
145. Brett CMA, Fungaro DA, Morgado JM et al (1999) Novel polymer-modified electrodes for batch injection sensors and application to environmental analysis. J Electroanal Chem 468:26–33
146. Lu T-H, Yang H-Y, Sun I-W (1999) Square-wave anodic stripping voltammetric determination of thallium(I) at a Nafion/mercury film modified electrode. Talanta 49:59–68

147. Zen J-M, Jou J-J, Kumar AS (1999) A sensitive voltammetric method for the determination of parathion insecticide. Anal Chim Acta 396:39–44
148. Yang H-Y, Sun I-W (2000) Cathodic stripping voltammetric determination of selenium(IV) at a Nafion coated mercury film electrode modified with 3,3′-diaminobenzidine. Electroanalysis 12:1476–1480
149. Brett CMA, Fungaro DA (2000) Poly(ester sulphonic acid) coated mercury thin film electrodes: characterization and application in batch injection analysis stripping voltammetry of heavy metal ions. Talanta 50:1223–1231
150. Cheng HL, Sun IW (2001) Square-wave voltammetric detection of apomorphine on a Nafion film modified glassy carbon electrode. Electroanalysis 13:1544–1546
151. Yi H, Wu K, Hu S et al (2001) Adsorption stripping voltammetry of phenol at Nafion-modified glassy carbon electrode in the presence of surfactants. Talanta 55:1205–1210
152. Collyer SD, Bradbury SE, Hatfiled JV et al (2001) A study of factors affecting the enhanced voltammetric stripping analysis of N-nitrosamines at sulfopolyester modified electrodes. Electroanalysis 13:332–337
153. Ugo P, Zangrando V, Moretto LM et al (2002) Ion-exchange voltammetry and electrocatalytic sensing capabilities of cytochrome *c* at polyestersulfonated ionomer coated glassy carbon electrodes. Biosens Bioelectron 17:479–487
154. Wang HY, Xu GB, Dong SJ (2003) Electrochemiluminescence sensor using tris(2,2′-bipyridyl)ruthenium(II) immobilized in Eastman-AQ55D–silica composite thin-films. Anal Chim Acta 480:285–290
155. Xu Q, Xu C, Wang Q et al (2003) Application of a single electrode, modified with polydiphenylamine and dodecyl sulfate, for the simultaneous amperometric determination of electro-inactive anions and cations in ion chromatography. J Chromatogr A 997:65–71
156. Rodríguez Gutiérrez JA, Petit Domínguez MD, Pinilla Macías JM (2004) Development of ionoselective electrochemical sensors by using the sol–gel process. Anal Chim Acta 524:339–346
157. Monterroso SCC, Carapuça HM, Duarte AC (2005) Ion-exchange and permselectivity properties of poly(sodium 4-styrenesulfonate) coatings on glassy carbon: application in the modification of mercury film electrodes for the direct voltammetric analysis of trace metals in estuarine waters. Talanta 65:644–653
158. Moretto LM, Bertoncello P, Vezzà P et al (2005) Electrochemistry of cytochrome *c* incorporated in Langmuir–Blodgett films of Nafion and Eastman AQ 55. Bioelectrochemistry 66:29–34
159. Buriez O, Moretto LM, Ugo P (2006) Ion-exchange volatmmetry of tris-(2,2′-bipyridine) nickel(II), cobalt(II), and Co(salen)at polyestersulfonated ionomer coated electrodes in acetonitrile: reactivity of the electrogenerated low-valent complexes. Electrochim Acta 52:958–964
160. Monterroso SCC, Carapuça HM, Duarte AC (2006) Mixed polyelectrolyte coatings on glassy carbon electrodes: ion-exchange, permselectivity properties and analytical application of poly-l-lysine–poly(sodium 4-styrenesulfonate)-coated mercury film electrodes for the detection of trace metals. Talanta 68:1655–1662
161. Riccardi CD, Yamanaka H, Josowicz M et al (2006) Label-free DNA detection based on modified conducting polypyrrole films at microelectrodes. Anal Chem 78:1139–1145
162. Lin YH, Cui XL (2006) Electrosynthesis, characterization, and application of novel hybrid materials based on carbon nanotube–polyaniline–nickel hexacyanoferrate nanocomposites. J Mater Chem 16:585–592
163. Bertoncello P, Dennany L, Forster RJ et al (2007) Nafion–Tris(2-2′-bipyridyl)ruthenium(II) Ultrathin Langmuir–Schaefer films: redox catalysis and electrochemiluminescent properties. Anal Chem 79:7549–7553
164. Huang M, Jiang H, Zhai J et al (2007) A simple route to incorporate redox mediator into carbon nanotubes/Nafion composite film and its application to determine NADH at low potential. Talanta 74:132–139

165. Zhang LB, Li J, Xu YH et al (2009) Solid-state electrochemiluminescence sensor based on the Nafion/poly(sodium 4-styrene sulfonate) composite film. Talanta 79:454–459
166. Silva CP, Carapuça HM, Rocha LS et al (2009) Evaluation of poly(sodium 4-styrenesulfonate) film coating in thin mercury film electrodes for lead determination. J Electroanal Chem 626:192–196
167. Çelebi MS, Özyörük H, Yıldız A et al (2009) Determination of Hg^{2+} on poly(vinylferrocenium) (PVF^{+})-modified platinum electrode. Talanta 78:405–409
168. Yuan Y, Yuan X, Chai Y et al (2010) A novel label-free electrochemical aptasensor for thrombin based on the $\{$nano-Au/thionine$\}_n$ multilayer films as redox probes. Anal Chim Acta 668:171–176
169. Xub H, Chenc J, Shia G, Jina L (2010) Nafion stabilized silver nanoparticles modified electrode and its application to Cr(VI) detection. J Electroanal Chem 652:60–65
170. Zhou C, Liu Z, Dai J et al (2010) Electrospun $Ru(bpy)_3{}^{2+}$-doped Nafion nanofibers for electrochemiluminescence sensing. Analyst 135:1004–1009
171. Taffa DH, Kathiresan M, Walder L et al (2010) Pore size and surface charge control in mesoporous TiO_2 using post-grafted SAMs. Phys Chem Chem Phys 12:1473–1482
172. Azad UP, Ganesan V (2010) Efficient sensing of nitrite by $Fe(bpy)_3{}^{2+}$ immobilized Nafion modified electrodes. Chem Commun 46:6156–6158
173. Cox JA, Kulesza PJ (1983) Stripping voltammetry of chromium(VI) at a poly(4-vinvlpyridine)-coated platium electrode. Anal Chim Acta 154:71–78
174. Cassidy JF, Tokuda K (1990) Preconcentration and voltammetric determination of copper ions in aqueous chloride solutions at a cross-linked poly(4-vinylpyridine)-coated electrode. J Electroanal Chem 285:287–294
175. Ugo P, Moretto LM, Mazzocchin GA (1995) Voltammetric determination of trace mercury in chloride media at glassy carbon electrodes modified with polycationic ionomers. Anal Chim Acta 305:74–82
176. Zen J-M, Chung M-J (1996) Square-wave voltammetric stripping analysis of bismuth(III) at a poly(4-vinylpyridine)/mercury film electrode. Anal Chim Acta 320:43–51
177. Moretto LM, Mazzocchin GA, Ugo P (1997) Electrochemical study on the ion-exchange voltammetric behaviour of Hg(II) at Tosflex®-coated glassy carbon electrode. J Electroanal Chem 427:113–121
178. Lu T-H, Sun I-W (1998) Anodic stripping voltammetric determination of thallium(III) using a Tosflex/mercury film electrode. Electroanalysis 10:1052–1056
179. Zen J-M, Chung H-G, Ilagovan G (1999) Selective voltammetric determination of lead(II) on partially quaternized poly(4-vinylpyridine)/mercury film electrodes. Electroanalysis 11:108–113
180. Yang H-Y, Chen W-Y, Sun I-W (1999) Anodic stripping voltammetric determination of bismuth(III) using a Tosflex-coated mercury film electrode. Talanta 50:977–984
181. Maizels M, Heineman WR, Seliskar CJ (2000) Graphite electrodes coated with poly (dimethyldiallylammonium)chloride network films cross-linked by gamma-irradiation. Electroanalysis 12:241–247
182. Lu T-H, Huang J-F, Sun I-W (2001) Perfluorinated anion-exchange polymer mercury film electrode for anodic stripping voltammetric determination of zinc(II): effect of model organic compounds. Anal Chim Acta 454:93–100
183. Cheng H-L, Kuei C-H, Sun I-W (2002) Electrochemical detection of O, O-diethyldithiophosphoric acid at a Tosflex film modified glassy carbon electrode. Electroanalysis 14:767–772
184. Prasad BB, Arora B (2003) Application of polymer-modified hanging mercury drop electrode in the indirect determination of certain β-lactam antibiotics by differential pulse, ion-exchange voltammetry. Electroanalysis 15:1212–1218
185. Rodríguez Gutiérrez JA, Petit Domínguez MD, Pinilla Macías JM (2004) Development of ionoselective electrochemical sensors by using the sol–gel process. Anal Chim Acta 524:339–346

186. Tsai DM, Kumar AS, Zen JM (2006) A highly stable and sensitive chemically modified screen-printed electrode for sulfide analysis. Anal Chim Acta 556:145–150
187. Silva CP, Carapuça HM (2006) Glassy carbon electrodes coated with poly(allylamine hydrochloride), PAH: characterization studies and application to ion-exchange voltammetry of trace lead(II) at combined PAH/mercury film electrodes. Electrochim Acta 52:1182–1190
188. Tsai M-C, Chen P-Y (2007) Electrochemical detection of 2-naphthol at a glassy carbon electrode modified with Tosflex film. Electroanalysis 19:1315–1321
189. Vago M, Tagliazucchi M, Williams FJ et al (2008) Electrodeposition of a palladium nanocatalyst by ion confinement in polyelectrolyte multilayers. Chem Commun 5746–5748
190. Hui T-W, Baker MD (2001) Ion exchange and electron transport at methyl viologen Y modified electrodes. J Phys Chem B 105:3204–3210
191. Li Y-J, Liu C-Y (2001) Silver-exchanged zeolite Y-modified electrodes: size selectivity for anions. J Electroanal Chem 517:117–120
192. Hui T-W, Baker MD (2002) Redox processes of methyl viologen cation radicals at zeolite Y-modified electrodes. J Phys Chem B 106:827–832
193. Zen JM, Wang HF, Kumar AS et al (2002) Preconcentration and electroanalysis of copper(II) in ammoniacal medium on nontronite/cellulose acetate modified electrodes. Electroanalysis 14:99–105
194. Zhang Y, Chen F, Shan W et al (2003) Fabrication of ultrathin nanozeolite film modified electrodes and their electrochemical behavior. Micropor Mesopor Mat 65:277–285
195. Tonle IK, Ngameni E, Walcarius A (2004) From clay – to organoclay-film modified electrodes: tuning charge selectivity in ion exchange voltammetry. Electrochim Acta 49:3435–3443
196. Yuan S, Chen W, Hu S (2004) Simultaneous determination of cadmium (II) and lead (II) with clay nanoparticles and anthraquinone complexly modified glassy carbon electrode. Talanta 64:922–928
197. Guo Z, Shen Y, Zhao F et al (2004) Electrochemical and electrogenerated chemiluminescence of clay nanoparticles/$Ru(bpy)_3^{2+}$ multilayer films on ITO electrodes. Analyst 129:657–663
198. Lin C-L, Tien P, Chau L-K (2004) Electrochemical behavior of an anion-exchanger modified electrode prepared by sol–gel processing of an organofunctional silicon alkoxide. Electrochim Acta 49:573–580
199. Walcarius A, Etienne M, Delacote C (2004) Uptake of inorganic Hg^{II} by organically modified silicates: influence of pH and chloride concentration on the binding pathways and electrochemical monitoring of the processes. Anal Chim Acta 508:87–98
200. Walcarius A, Ganesan V (2006) Ion-exchange properties and electrochemical characterization of quaternary ammonium-functionalized silica microspheres obtained by the surfactant template route. Langmuir 22:469–477
201. Letaief S, Tonle IK, Diaco T et al (2008) Nanohybrid materials from interlayer functionalization of kaolinite. Application to the electrochemical preconcentration of cyanide. Appl Clay Sci 42:95–101
202. Ganesan V, Walcarius A (2008) Ion exchange and ion exchange voltammetry with functionalized mesoporous silica materials. Mat Sci Eng B 149:123–132
203. Gan T, Hu C, Chen Z et al (2010) Direct electrochemical determination of methyl jasmonate in wheat spikelet at a nano-montmorillonite film modified electrode by derivative square wave voltammetry. J Agric Food Chem 58:8942–8947
204. Tcheumi HL, Tonle IK, Ngameni E et al (2010) Electrochemical analysis of methylparathion pesticide by a gemini surfactant-intercalated clay-modified electrode. Talanta 81:972–979

Chapter 12
Sulfonated Poly(Ether Ether Ketone) (SPEEK): A Promising Membrane Material for Polymer Electrolyte Fuel Cell

Amir-Al-Ahmed, Abdullah S. Sultan, and S.M. Javaid Zaidi

Abstract Nafion has been the material of choice for polymer electrolyte membrane fuel cells (PEMFCs), but during the last two decades, considerable efforts have been made to find an alternative with similar or better physicochemical properties and with lower manufacturing cost. Developments over the last two decades have resulted to some novel membrane materials with improved properties. Among the materials researched and developed, sulfonated poly(ether ether ketone) (SPEEK) has been the most promising and has the potential for commercialization. In this chapter, the properties of SPEEK and its characteristics are discussed.

12.1 Introduction

Proton exchange membrane (PEM) is an electrolyte used in both polymer electrolyte membrane fuel cells (PEMFCs) and direct methanol fuel cells (DMFCs) which allows protons (H^+) to conduct from the anode to the cathode. It is expected to provide a barrier against electrons, gas, and methanol transporting through the membrane. The pioneer of PEM is Nafion, a perfluorinated polymer with sulfonic acid groups as

Amir-Al-Ahmed
Center of Research Excellence in Renewable Energy,
King Fahd University of Petroleum & Minerals, Dhahran 31261, Saudi Arabia

A.S. Sultan
Center for Petroleum & Minerals, King Fahd University of Petroleum & Minerals,
Dhahran 31261, Saudi Arabia

S.M.J. Zaidi (✉)
Center of Research Excellence in Renewable Energy,
King Fahd University of Petroleum & Minerals, Dhahran 31261, Saudi Arabia

Department of Chemical Engineering, King Fahd University of Petroleum & Minerals,
Dhahran 31261, Saudi Arabia
e-mail: zaidismj@kfupm.edu.sa

Inamuddin and M. Luqman (eds.), *Ion Exchange Technology I: Theory and Materials*, DOI 10.1007/978-94-007-1700-8_12,

pendant, which was developed by Dupont in the 1960s. Nafion has excellent ionic properties; the most important advantage of Nafion is its excellent proton conductivity in the hydrated condition. Furthermore, it has good physical, chemical, and mechanical stability. However, the main drawbacks are its high cost, high methanol permeability (in the case of DMFC), and loss of proton conductivity at higher temperatures [1, 2]. Ideally, PEMs should possess certain characteristics such as high proton conductivity; low water and methanol permeability; excellent mechanical properties; good thermal and chemical stabilities; high resistance to dehydration; low swelling; high resistance to oxidation, reduction, and hydrolysis; and low cost. Furthermore, the polymeric membrane must be an electron insulator to prevent electron leakage, which leads to low efficiency. Commonly used PEM materials are, for example, Nafion 117 (proton conductivity 90–120 mS cm^{-1} at 80°C and 34–100% RH), Nafion 115 (41 mS cm^{-1} at 25°C), sulfonated poly(ether ether ketone) (SPEEK) (0.9 mS cm^{-1} at 20°C), sulfonated poly(benzimidazole) (SPBI) (10–40 mS cm^{-1} at 130–180°C), sulfonated poly(phosphazene) (40 mS cm^{-1} at 25°C), asymmetric-based acrylic (AMPS) (42 mS cm^{-1} at 90°C), poly(vinylidene fluoride) (PVDF) with styrene membrane (resistance 0.454 $\Omega\ cm^2$), PVDF with SiO_2 or SiO_2 gel (70 mS cm^{-1} (SiO_2) and 200 mS cm^{-1} (gel) at 25°C), etc. [1–6].

To get a replacement, membrane materials other than Nafion were tried for their low price and better properties. In some cases, blends and/or composites were fabricated to improve membrane properties. The poly(arylene ether ketone) family (Fig. 12.1) is of particular interest because it is an engineering thermoplastic material with excellent thermal and chemical stabilities, high mechanical strength, good insulating properties, and relatively low cost compared to Nafion. Several sulfonated derivatives of polymers such as poly(ether ether ketone), poly(sulfone), poly(arylene ether sulfone), poly(styrene), and poly(phenylene sulfide) have been developed for fuel cells [7–11]. Copolymers of sulfonated poly(arylene ether sulfone) and/or sulfonated poly(arylene ether ketone)s by direct copolymerization of bisphenol, disulfonated activated aromatic halide monomers, and the precursor-activated aromatic halide monomer for fuel cell membrane applications were synthesized [9, 12]. Sulfonated poly(arylene ether sulfone) copolymers were also synthesized by the McGrath group [13]. Four bisphenols (4,4′-bisphenol A; 4,4′-bisphenol AF; 4,4′-bisphenol; and hydroquinone) were investigated for the synthesis of novel copolymers with controlled degrees of sulfonation. These copolymers are promising candidates for high-temperature PEM fuel cells [9, 12, 14]. The poly (aryl ether ketones) family are thermostable polymers in which ether (E) and ketone (K) units connect phenylene rings, giving a range of polymers, such as PEK, PEEK, PEKEKK, etc. The proton conductivity and thermal and mechanical properties of sulfonated PEEK [15, 16] and its fuel cell performance in hydrogen–air and hydrogen–oxygen up to 110°C [16], as well as in DMFC, have been reported in recent years, and long-term tests have claimed lifetimes of up to 4,300 h at 50°C [17]. In this chapter, some properties of SPEEK are discussed in view of its applications in PEMFC/DMFC.

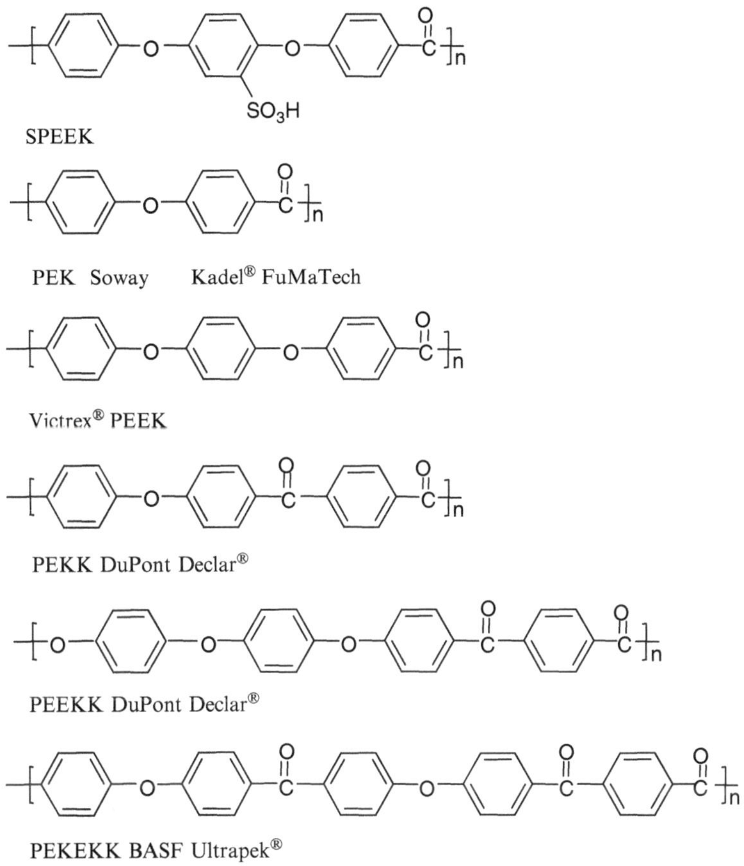

Fig. 12.1 Sulfonated poly(ether ether ketone) and representative membranes of the poly(ether ketone) family

12.2 Sulfonated Poly(Ether Ether Ketone) (SPEEK)-Based Membranes for Fuel Cells

12.2.1 Pure Sulfonated Poly(Ether Ether Ketone) (SPEEK)

Fuel cell membranes can be divided into two broad classes: (1) perfluorosulfonic acid (PFSA)-based membranes (e.g., Nafion), which have been the standard for a long time, and (2) membranes based on aromatic sulfonic acid, which have been the primary class of materials synthesized, as people search for alternatives to PFSAs. The latter include a wide range of polymers such as sulfonated poly (phosphazenes), sulfonated poly(sulfone), sulfonated poly(ether ether ketone), sulfonated poly(trifluorostyrene), and so on. Some particular advantages for the latter class include excellent mechanical properties, enabling the preparation of very thin low-resistance films and more resistance to swelling. As compare to other membranes, perfluorosulfonic acid membranes have almost all of the necessary properties for a good fuel cell membrane. Major drawbacks of the PFSA membranes are their high cost and their limited stability at temperatures substantially in excess of 100°C [15, 18, 19].

The first preparation of poly(ether ether ketone) was reported by Bonner [20], who obtained a polymer of low molecular weight by Friedel–Crafts polycondensation using aluminum chloride as the catalyst. Later, crystalline poly(aryl ether ketones) and poly(aryl ether ketone-co-sulfones) were also obtained by polycondensation of bis-4-halogenophenyl ketones (plus bis-4-halogenophenyl sulfones for copolymers) with potassium salt at 280–340°C using certain diarylsulfones as solvent for the reaction [21]. In the case of PEEK, it was observed that if the mol % of the ether ketone repeating unit decreases, the T_m and crystallizability from melt increases, and T_g decreases on the other side. Helmer-Metzmann et al. [22] proposed (in their patent) sulfonated poly(ether ether ketone) (SPEEK) membranes as proton conductors in a PEM fuel cell. They reported good chemical stability of the SPEEK membranes. They checked the degree of sulfonation (DS) and the ion (proton) exchange capacity (IEC) of SPEEK. A certain percentage of aqueous swelling of SPEEK layers is required to minimize the mass transport resistance in the case of composites and to increase the permeability of cations through these stabilization layers [23]. On the other hand, it has been found by Gates and Newman [24] that a high degree of swelling of the polymer network allows the components of the liquid membrane phase to permeate through the SPEEK layer into the surrounding aqueous phase. Wijers et al. [23] reported that the degree of swelling decreases with increasing ionic strength or decreasing chemical potential of the aqueous solution. At room temperature, PEEK is insoluble in common organic solvents and dissolves only in some concentrated acids such as sulfuric acid, methane sulfuric acid, or hydrofluoric acid, and it was also found that close to the melting point, PEEK dissolves in high-boiling-point esters, benzophenone, and diphenyl sulfones also [25]. Often, resins or fillers such as glass and carbon fibers are mixed with PEEK polymer for different applications, where it requires strength, high-temperature performance, and chemical resistance. For example- model components for medical, automotive, aerospace, chemical processing and microelectronics industries etc. Till 1985, its molecular characterization has remained limited to the melt or sulfuric acid solution's viscosities, for example, the average molecular weight ($\bar{M}_w$) and the radius of gyration ($\bar{R}_G$), and the second virile coefficients (A_2) of five melt viscous samples of PEEK were determined by light scattering (LS) in concentrated sulfuric acid [25]. Correlations between light scattering and molecular weights and melt viscosity in sulfuric acid or in a 50:50 phenol–trichlorobenzene (TCB) mixture were reported at 25°C and 115°C, respectively [25].

Poly(ether ether ketone) had been modified using various sulfonation agents such as sulfuric acid (concentrated/fuming), chlorosulfonic acid, or sulfur trioxide (or complexes thereof). This polymer consists of sequences of ether and carbonyl linkage between phenyl rings. The presence of adjacent ortho-directing ether groups confers highest reactivity to the four equivalent sites on the hydroquinone unit situated between the ether segments. The sulfonation of PEEK has been reported to be a second-order reaction, which takes place at the aromatic ring flanked by two ether links, due to the higher electron density of the ring [26]. Since the electron density of the other two aromatic rings in the repeat unit is relatively low due to the electron-attracting nature of the neighboring carbonyl group, one sulfonic acid group per repeat unit may be substituted. Postmodification reactions are moderately less favorable due to their lack of precise control over the degree of sulfonation, site specificity, and the possibility of side reactions, or degradation of the polymer backbone. It has been reported that sulfonation of PEEK with chlorosulfonic acid or fuming sulfuric acid causes mostly unexplored degradation of the polymer; therefore, concentrated sulfuric acid is typically used [27]. The sulfonation rate of PEEK in concentrated sulfuric acid can be controlled by changing the reaction time, temperature, and acid concentration to provide polymers with a sulfonation range of 30–100% without degradation and cross-linking reactions [28]. However, it has been shown that the sulfonation of PEEK in sulfuric acid cannot be used to produce truly random copolymers at sulfonation levels less than 30% because the dissolution and sulfonation occur in a heterogeneous environment [29]. Nevertheless, this area of PEM synthesis has received much attention and may be the source of merging products such as sulfonated Victrex poly(ether ether ketone) [30–32]. A novel approach was adopted by many researchers to obtain sulfonated aromatic copolymers, i.e., the copolymerization of sulfonated monomers. Direct copolymerizations of sulfonated monomers have become an alternative approach with some distinct advantages compared with the modification of a preformed polymer. This process gave closer control on the molecular design of the resulting copolymer [14]. It has been of interest to investigate the effect of sulfonation, for example, on the deactivated sites of the repeat units since one might expect enhanced stability and higher acidity from sulfonic acid groups which are attached to electron-deficient aromatic rings rather than from sulfonic acid groups bonded to electron-rich aromatic rings as in the case of postmodification [33]. The possibilities of controlling and/or increasing molecular weight to enhance durability are not feasible in the case of postreaction on an existing commercial product.

SPEEK is made up of highly branched, narrow channels for proton conduction and has more dead-end components than the wider, less-branched channels of Nafion (Fig. 12.2). The more extensive hydrophobic–hydrophilic interface region in SPEEK results in greater separation between sulfonic acid functional groups [34]. In sulfonated hydrocarbon polymers, the hydrocarbon backbones are less hydrophobic and the sulfonic acid functional groups are less acidic and polar. As a result, the water molecules of hydration may be completely dispersed in the nanostructure of the sulfonated hydrocarbon polymers. Both perfluorosulfonic acid (PFSA) and sulfonated hydrocarbon membranes have similar water uptakes at low water activities, whereas at high relative humidity (100%), PFSA membranes have a much higher water uptake due to the more polar character of sulfonic acid functional groups. Sulfonated aromatic polymers have different

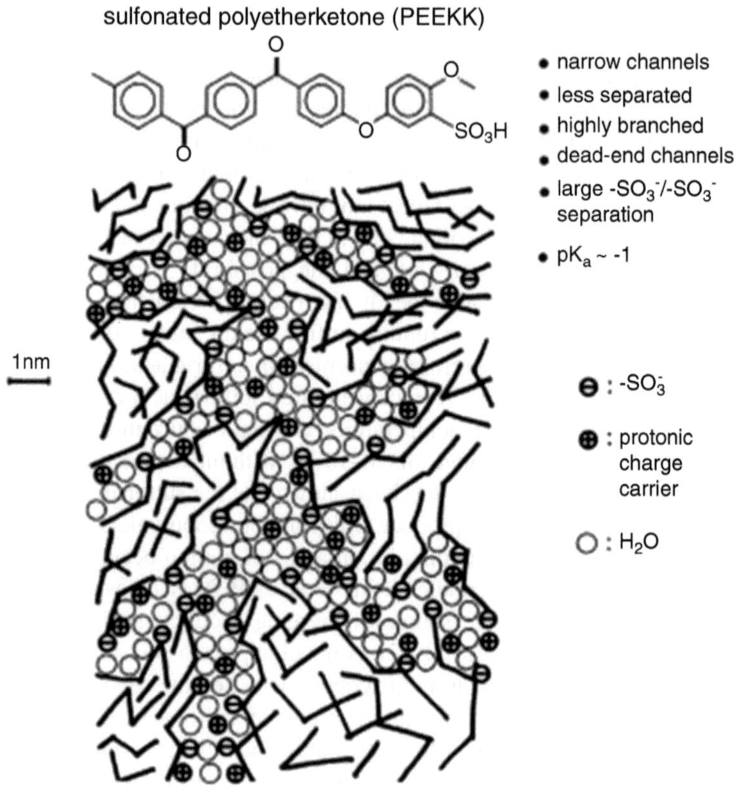

Fig. 12.2 Schematic structure of SPEEK (Reprinted from Ref. [34] with kind permission of © Elsevier (2001))

microstructures from those of PFSA membranes [35]. Sulfonated PEEK-WC membranes with a degree of sulfonation of 15–40% have been employed in PEM fuel cells. The membranes exhibit electrochemical performances comparable to Nafion membranes [36].

12.2.2 Sulfonated Poly(Ether Ether Ketone) (SPEEK)-Based Blends

Blending technology appears to be a more versatile approach for the development of new and better membranes as the properties of the membrane materials can be varied over a wide range. Polymer blends has been prepared by combining polymeric nitrogen-containing bases (N-base) with the polymeric sulfonic acids. The sulfonic acid groups interact with the N-base by forming hydrogen bonds with the basic N sites of the other polymer [37, 38]. The most advanced acid–base polymer blends are those on SPEEK or ortho-sulfone-sulfonated poly(ether sulfone) (SPSU) as the acidic component and poly(benzimidazole) (PBI) as the basic component. SPEEK-based blends are explored to obtain good mechanical properties, high proton conductivity, and optimized membrane characteristics. Matthew et al. [27] studied solubility and properties of PEEK in strong acids such as H_2SO_4 and HSO_3Cl by combining different techniques including steady and dynamic light scattering, viscometry, NMR, elemental analysis, and absorption spectra. They compared PEEK and poly(ether imide) (PEI) in terms of strength and stiffness, ductility, and glass transition temperature (T_g) (around 215°C) and found better chemical resistance properties in PEEK as compared to PEI [39], and both of them are cheaper than Nafion. Now, PEEK/PEI blends have been the subject of several fundamental investigations dealing with basic mechanical properties, miscibility, morphology, and the crystallization behavior of the blend. Gensler et al. [39] studied the effect of temperature on the deformation behavior of different amorphous PEEK/PEI thin films, where all pure and blend polymer materials showed localized shear deformation at temperatures well below $T_g \approx 143°C$. In case of pure PEI thin films and the blend containing 60% PEEK, they observed a transition from shear to disentanglement crazing with the increasing temperature toward Tg, such transition occurred also in bulk tensile bars of these materials, leading to fracture of the samples near T_g [39]. Rheological studies on these polymer solutions focused on characterizing their intrinsic properties through solution viscometry or melt properties. One of these studies is the derivation of the intermolecular hydrodynamic contribution toward the reduced viscosity of polyelectrolyte solution as a function of polymer concentration by separating the theoretically calculated intermolecular electrostatic contribution from the observed reduced viscosity. The resulting intermolecular part reflects the effect of the polyion conformation which increases with decreasing polymer concentration [40, 41]. Kerres et al. [42] developed novel acid–base polymer blend membranes consisting of SPEEK and sulfonated poly(sulfone) (sPSU) as the acidic components and poly(sulfone) (PSU), poly-4-vinylpyridine (P4VP), poly(benzimidazole) (PBI), and poly(ether imide) (PEI) as the basic components. The membranes showed good proton conductivities at ion exchange capacity, (IEC) = 1 and are thermally stable up to 270°C. Some of these membranes were tested in a H_2 fuel cell and showed impressive performance. When tested for direct methanol fuel cell (DMFC), these blend membranes showed low methanol permeabilities than Nafion® membranes [42]. SPEEK was found to be miscible with poly(ether sulfone) in N-methyl-2-pyrrolidinon (NMP) over a weight ratio of 0.02–0.040% of SPEEK in the total polymer blend, and partially miscible with PSU in NMP over a small range of weight ratio [43]. Thermal analysis of SPEEK–PEI and SPEEK–PC binary blend systems showed single- and double-glass transition temperatures, respectively. This result suggests that SPEEK and PEI are miscible with each other, but SPEEK and polycarbonate (PC) remain as individual components in the blends. Similarly, for SPEEK–PEI–PC ternary blends, phase separation regions showed two T_g's. Those miscibility results of the three systems were also proven to be in agreement with the values of the polymer–polymer

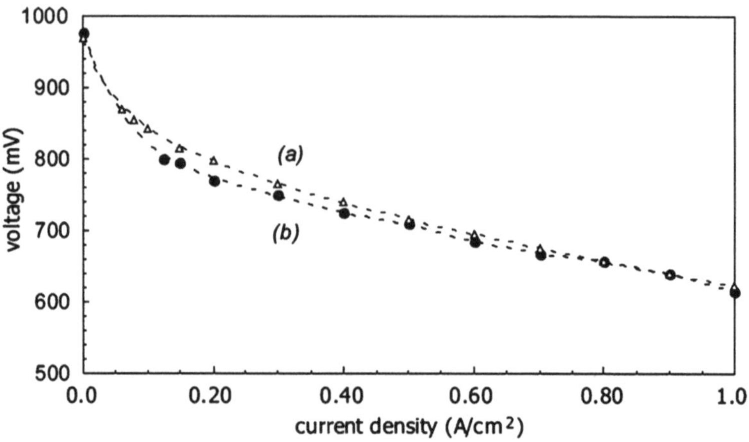

Fig. 12.3 Polarization characteristics of (**a**) SPEEK, (**b**) SPEEK-modified silica (10 wt% SiO2) (Reprinted from Ref. [48] with kind permission of © JOHN WILEY AND SONS (2002))

interaction parameter χ_{12}, which was obtained using the modified Flory–Huggins equation [44].

The proton conductivity of pure SPEEK membranes is not as good as that of Nafion®, so various inorganic materials such as SiO_2, ZrO_2, heteropolyacids, and phosphates were incorporated into SPEEK to improve its proton conductivity [1, 2]. Hydrated tungstophosphoric acid (HPA) has good proton conductivity (1.9×10^{-1} S/cm) and produce stable Keggin-type anions. When embedded into the polymer matrix, it increases the conducting properties of the membrane while retaining other desirable properties of the polymer films [45]. A different SiO_2/SPEEK composite membrane was also prepared, and its performance was compared to that of pure SPEEK in a single-cell assembly with an active surface area of 5 cm^2; one example is shown in Fig. 12.3 (electrodes were loaded with 1 mg Pt and 0.35 mg Nafion® per square cm) [46]. The ionic interaction between inorganic and organic components improved the mechanical properties of the membranes by increasing the interface region between inorganic and organic components. This also allows enhanced proton conductivity compared to the pure sulfonated polymer [46]. Zaidi et al. [47] prepared membranes by blending SPEEK and PBI and incorporating boron phosphate (BPO_4) into these blends. These acid–base blend membranes showed reduced methanol permeability. But due to the reduction in water uptake, the conductivities were lowered. They [47, 48] studied the membrane properties with different BPO_4 loading (10–40% wt) and found that the water uptake increased with the increasing BPO_4 content, reaching maximum at 30% and then drops down. With the addition of PBI into the SPEEK matrix, the amount of water uptake was decreased, resulting in the reduction of the swelling of the membranes, bringing up improvement in the mechanical properties of the

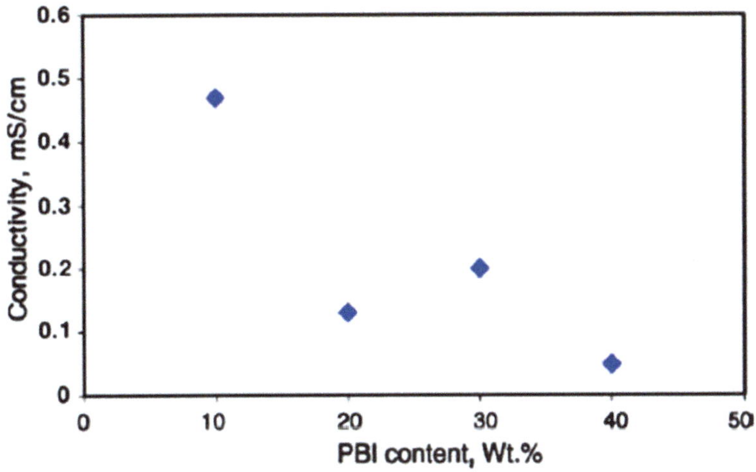

Fig. 12.4 Conductivity of SPEEK/PBI blends membranes with increasing PBI content (Reprinted from Ref. [49] with kind permission of © Elsevier (2005))

membranes. But the conductivity of the SPEEK-blended PBI was found to decrease with the increase of the percentage of PBI in the matrix. This is due to the decrease in the number of free SO_3H groups by the formation of ionic cross-linkings to the imidazole groups by proton transfer, leading to the increase in protonic resistance and decrease in conductivity [49]. Poly(ether sulfone)/sulfonated poly(ether ether ketone) (PES/SPEEK) blend membranes were investigated for methanol uptake, water uptake, permeability, and proton conductivity [50]. The blend of PES and SPEEK was prepared by mixing the two polymers in N-methyl-2-pyrroelidone (NMP) solution. It was found out that poly(ether sulfone) plays an important role in decreasing water uptake and methanol permeability while enhancing thermal stability. Composite membranes of SPEEK with various compounds of solid heteropolyacids (HPA) have also been prepared by Zaidi et al. [45]. These membranes were found to have high proton conductivity and good mechanical strength. But after prolonged exposure to water, the HPA were leached out, reducing the conductivity which results in the poor performance of the membranes. In order to tackle the problem of leaching out HPAs from the SPEEK polymer matrix, Ahmad et al. (2006) [51] synthesized novel composite proton-conducting solids – heteropolyacids loaded onto Y-zeolite or MCM-41 [52]. Various weight percentages of HPAs were loaded onto Y-zeolite and MCM-41 structures. The conductivity was found to increase with the increasing percentages of the HPA loading onto Y-zeolite and MCM-41, as shown in Fig. 12.4. The highest proton conductivity in the order of 10^{-2} S/cm was obtained at room temperature for a fully hydrated membrane containing 50 wt% HPAs. These materials were reported to combine the high thermal and structural stability of Y-zeolite and MCM-41 with high conductivity, making them one of the most promising solid proton conductors. These solid proton-conducting

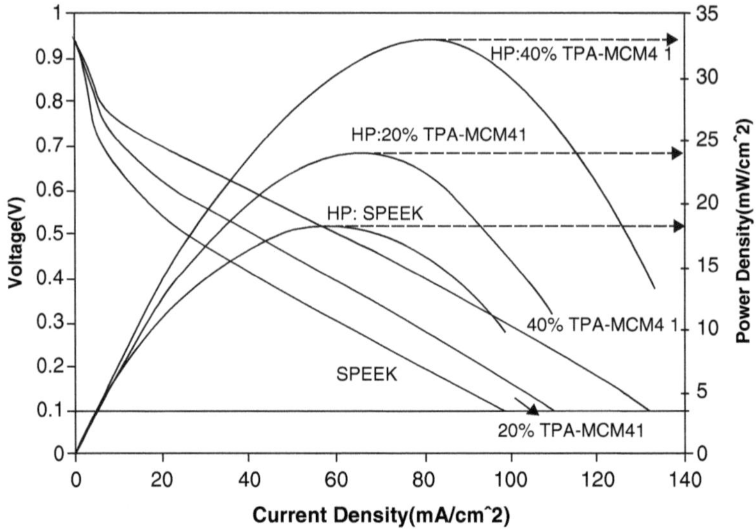

Fig. 12.5 Polarization curves and plots of power density vs. current density for different percentages of TPA-MCM41 loading into SPEEK matrix. [Operating conditions: Anode flow (H2) = 100 sccm, cathode flow (air) = 200 sccm, Tcell = Thumidifiers = 60°C, Pressure = 20 psi]. (Re. M.S.)

powders were embedded into the SPEEK matrix in order to increase its conductivity. The composite SPEEK membranes with heteropolyacids, impregnated into the pores of Y-zeolite or MCM-41, acting as proton conductors exhibited better proton conductivities at room temperatures, and this conductivity increased with increasing temperature [53]. Fuel cell performance of these composite membranes had been carried out in H_2/O_2 fuel cells. The polarization curves and plots of power density vs. current density for various percentages of TPA-MCM41 loading into the SPEEK matrix are shown in Fig. 12.5. Zaidi et al. [46] also studied the methanol transport behavior of novel composite membranes prepared for use in the DMFC using the potentiometric technique. The composite membranes were prepared by embedding different proportions (10–30 wt%) of an inorganic proton-conducting material (tungstophosphoric acid (TPA)/MCM-41) into the SPEEK polymer matrix. The methanol permeability and crossover flux increase with increasing loading of the solid proton-conducting material. The lowest permeability value of 5.7×10^{-9} cm^2s^{-1} was obtained for the composite membrane with 10 wt% of the solid proton-conducting material (40 wt% TPA and 60 wt% MCM-41). However, all the composite membranes showed higher selectivity (ratio between the proton conductivity and the methanol permeability) compared to the pure SPEEK membrane. In addition, the membranes are thermally stable up to 160°C. Using the potentiometric technique, potential values were recorded for the pure SPEEK membrane and also for the fabricated composite membranes. The methanol concentration history in reservoir B for the pure

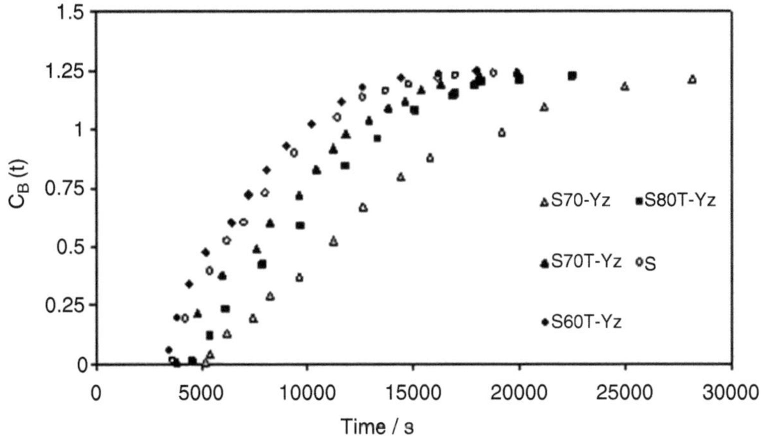

Fig. 12.6 Methanol concentration history in the reservoir B for the pure SPEEK and the composite membrane samples for 2.5 M initial methanol concentration in the reservoir A

SPEEK membrane and the fabricated composite membranes is shown in Fig. 12.6. As can be observed, the methanol amount in reservoir B is higher for the pure SPEEK membrane than the fabricated composite membranes, except for the sample S60T-Yz. From the experimental data, they observed that the lower inorganic loading gives better reduction in the methanol crossover. This is due to the voids created because of the incompatibility of the inorganic materials with the polymer matrix at higher inorganic loadings. It has been reported that sizing of the carbon molecular sieve (CMS) particles using polyvinylpyrrolidone (PVP) can be carried out to promote adhesion between the inorganic substrate and the polymer matrix which in turn reduced the interfacial voids. In general, these membranes showed lower permeability values than the pure SPEEK membrane. The membrane sample S60T-Yz showed the highest permeability value of 7.04×10^{-8} cm^2/s, while the membrane sample S70-Yz, which does not contain any TPA, showed the least permeability value of 1.64×10^{-8} cm^2/s. The TPA was incorporated to enhance the proton conductivity of the membranes [46–50]. Hasani-Sadrabadi et al. [54] prepared SPEEK with various degrees of sulfonation using H_2SO_4 and prepared blends with montmorillonite (MMT) clay. They fabricated membranes by solvent-casting methods and found increased ion exchange capacity and water uptake properties with the increased degree of sulfonation. They also reported that the methanol permeability of the membrane decreases to 5×10^{-8} cm^2/s by the addition of only 10% MMT clay. Chen et al. [55] prepared SPEEK-based membranes by grafting divinylbenzene (DVB) and ethyl styrenesulfonate (ETSS) onto the PEEK film. They compared properties of Nafion 212 with the hydrolyzed DVB and ETSS-grafted SPEEK and found better proton conductivity and mechanical properties when tested in a fuel cell, as evident in Fig. 12.7.

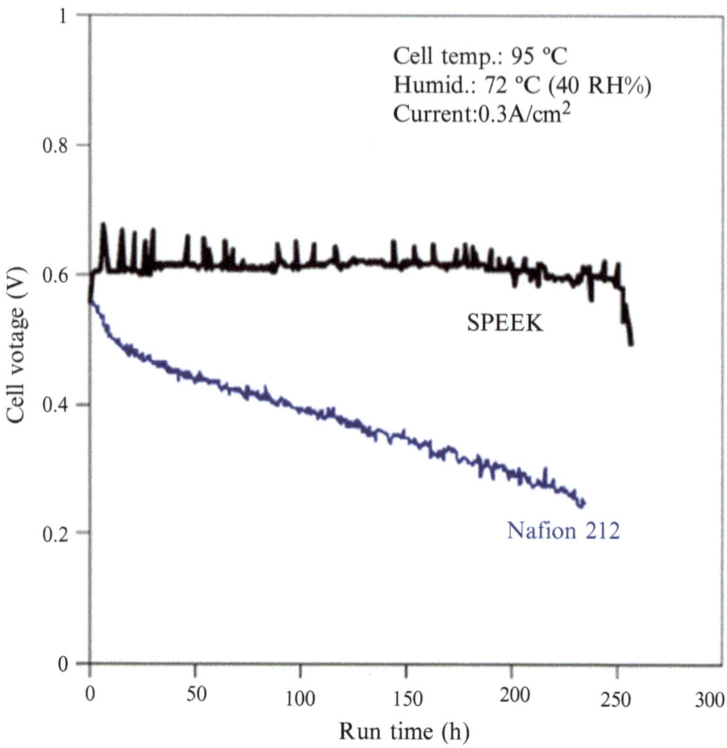

Fig. 12.7 Durability testing of the ssPEEK and Nafion-based fuel cells at a high temperature and low humidity condition 95°C and 40 RH% (Reproduced from Ref. [55] with kind permission of © Elsevier (2008))

12.3 Conclusion and Future Directions

Sulfonated aromatic high-performance polymers with good durability and high proton conductivity (in the wet state) show promises for low-cost alternatives to Nafion. However, the two main hurdles for polymeric membranes are the high proton conductivity at low water contents (e.g., under conditions of 120°C and 50% RH) and long-term durability under fuel cell conditions. Attempts have been made to increase the hydrophobicity and length of nonsulfonated properties as well as the hot water stability of these membranes by incorporating larger and rigid moieties in the structure. It is thought that the fluorine content in bisphenol 6F promotes adhesion and electrochemical compatibility with Nafion-based electrodes and reduces swelling. Blends of SPEEK showed good proton conductivities, which is close to that of Nafion under similar conditions. Some of the blends gave low methanol permeabilities (in the case of MDFC) lower than that of Nafion as well. As already mentioned, the cost of SPEEKs is much lower than that of Nafion; therefore, SPEEKs have the potential to replace Nafion and can accommodate itself as PEM materials for PEMFCs.

The future of PEM fuel cell strongly depends on the technological solution of the challenges in the fuel cell industry. Academic organizations and industries both are looking into the technological aspects and cost factors. Attempts can be made to examine the effect of asymmetric or composite structures on the performance of PEM. Considering that only a small amount of polymeric additive is required for surface coating of the membrane, the future direction of research and development efforts should be focused on the development of new methods of membrane surface coating and surface modification. In this context, the approach of surface-modifying macromolecules (SMM) might be interesting. Another novel approach is the use of an electrospun nanofiber membrane (ENM) for fuel cells. Polymeric materials having sulfonic acid groups have been fabricated into ENMs. The electrospinning process is capable of producing fibers in the submicron to nanoscale range.

Acknowledgement The authors would like to thank the Center of Research Excellence in Renewable Energy, King Fahd University of Petroleum and Minerals.

References

1. Zaidi SMJ (2009) Research trends in polymer electrolyte membranes for PEMFC. In: Zaidi SMJ, Matsmura T (eds) Polymer membranes for fuel cells. Springer, New York
2. Maiyalangan T, Pasupathi S (2010) Components of PEM fuel cell. In: Al-Ahmed A (ed) Current application of polymers and nano materials. Trans Tech, Publication Ltd., Switzerland
3. Lakshmanan B, Huang W, Olmeijer D, Weider JW et al (2003) Polyetheretherketone membrane for elevated temperature PEMFCs. Electrochem Solid-State Lett 6(12):A282–A285
4. Kim DS, Guiver MD (2009) Development of sulfonated poly(ether ether ketone)s for PEMFC and MDFC. In: Zaidi SMJ, Matsmura T (eds) Polymer membranes for fuel cells. Springer, New York
5. Jiang R, Kunz HR, Fenton JM et al (2005) Investigation of membrane property and fuel cell behavior with sulfonated poly(ether ether ketone) electrolyte: temperature and relative humidity effects. J Power Sour 150:120–128
6. Deb PC, Rajput LD, Hande VR, Sasane S, Kumar A et al (2007) Modification of sulfonated poly(ether ether ketone) with phenolic resin. Polym Adv Technol 18(6):419–426
7. Alberti G, Casciola M, Massinelli L, Bauer B et al (2001) Polymeric proton conducting membranes for medium temperature fuel cells (110–160°C). J Membr Sci 185:73–81
8. Genova-Dimitrova P, Baradie B, Foscallo D, Poinsignon C, Sanchez JY et al (2001) Ionomeric membranes for proton exchange membrane fuel cell (PEMFC): sulfonated polysulfone associated with phosphatoantimonic acid. J Membr Sci 185:59–71
9. Wang F, Hickner M, Kim YS, Zawodzinski TA, McGrath JE et al (2002) Direct polymerization of sulfonated poly(arylene ether sulfone) random (statistical) copolymers: candidates for new proton exchange membranes. J Membr Sci 197:231–242
10. Carretta N, Tricoli V, Picchioni F et al (2000) Ionomeric membranes based on partially sulfonated poly(styrene): synthesis, proton conduction and methanol permeation. J Membr Sci 166:189–197
11. Miyatake K, Shouji E, Yamamoto K, Tsuchida E et al (1997) Synthesis and proton conductivity of highly sulfonated poly(thiophenylene). Macromolecules 30:2941–2946
12. Kim YS, Wang F, Hickner M, Zawodzinski TA, McGrath JE et al (2003) Fabrication and characterization of heteropolyacid ($H_3PW_{12}O_{40}$)/directly polymerized sulfonated poly(arylene

ether sulfone) copolymer composite membranes for higher temperature fuel cell applications. J Membr Sci 212:263–282
13. Kim YS, Wang F, Hickner M, McCartney S, Hong YT, Harrison WT, Zawodzinski A, McGrath JE et al (2003) Influence of the bisphenol structure on the direct synthesis of sulfonated poly(arylene ether) copolymers. J Polym Sci Part A 41(14):2264–2276
14. Harrison WL (2002) Ph.D. Dissertation, Virginia Tech Polytechnic Institute and State University, Blacksburg
15. Rikukawa M, Sanui K et al (2000) Proton-conducting polymer electrolyte membranes based on hydrocarbon polymers. Prog Polym Sci 25:1463–1502
16. Bauer B, Jones DJ, Rozière J, Tchicaya L, Alberti G, Casciola M, Massinelli L, Peraio A, Besse S, Ramunni E et al (2000) Electrochemical characterisation of sulfonated polyetherketone membranes. J New Mater Elect Syst 2:93–98
17. Soczka-Guth T, Baurmeister J, Frank G, Knauf R WO 99/29763
18. Smitha B, Sridhar S, Khan AA et al (2005) Solid polymer electrolyte membrane for fuel cell application – a review. J Membr Sci 259:10–26
19. Steck AE (1995) Membrane materials in fuel cells. In: First international symposium on new materials for electrochemical systems, Montreal, pp 74–94
20. Bonner WH (1962) US Patent 3065205
21. Attwood TE, Dawson PC, Freman JL, Hoy LR, Ross JB, Stainland PA et al (1981) Synthesis and properties of polyarylether ketones. Polymer 22:1096–1103
22. Helmer-Metzman et al (1993) EP0574791.AZ, Hoeclist AG
23. Wijers MC, Jin M, Wessling M, Strathmann H et al (1998) Supported liquid membrane modification with sulfonated poly (ether-ether ketone): permeability, selectivity and stability. J Membr Sci 147:117–130
24. Gates CM, Newman J et al (2000) Equilibrium and diffusion of methanol and water in a nafion 117 membrane. AICHE J 46(10):2076–2085
25. Devaux J, Delimoy D, Daoust D, Legras R, Mercier P, Strazielle C, Nield E et al (1985) On the molecular weight determination of poly(aryl-ether-ether-ketone) (PEEK). Polymer 26:1994–2000
26. Xing P, Robertson GP, Guiver MD, Mikhailenko SD, Wang K, Kaliaguine S et al (2004) Synthesis and characterization of sulfonated poly(ether ether ketone) for proton exchange membranes. J Membr Sci 229:95–106
27. Bishop MT, Karasz FE, Russo PS, Langley KH et al (1985) Solubility and properties of a poly (aryl ether ketone) in strong acids. Macromolecules 18:86–93
28. Huang RYM, Shao P, Burns CM, Feng X et al (2001) Sulfonation of poly(ether ether ketone) (PEEK): kinetic study and characterization. J Appl Polym Sci 82:2651–2660
29. Bailly C, Williams DJ, Karasz FE, MacKnight WJ et al (1987) The sodium salts of sulphonated poly(aryl-ether-ether-ketone) (PEEK): preparation and characterization. Polymer 28:1009–1016
30. Rozière J, Jones DJ (2003) Non-fluorinated polymer materials for proton exchange membrane fuel cells. Annu Rev Mater Res 33:503–555
31. Lee HC, Hong HS, Kim YM, Choi SH, Hong MZ, Lee HS, Kim K et al (2004) Preparation and evaluation of sulfonated-fluorinated poly(arylene ether)s membranes for a proton exchange membrane fuel cell (PEMFC). Electrochim Acta 49:2315–2323
32. Kim DS, Liu B, Guiver MD et al (2006) Influence of silica content in sulfonated poly(arylene ether ether ketone ketone)(SPAEEKK) hybrid membranes on properties for fuel cell application. Polymer 47:7871–7880
33. Kim YS, Wang F, Hickner M, McCartney S, Hong YT, Zawodzinski TA, McGrath JE et al (2003) Effect of acidification treatment and morphological stability of sulfonated poly(arylene ether sulfone) copolymer proton-exchange membranes for fuel-cell use above 100°C. J Polym Sci Part B 41:2816–2828
34. Kreuer KD (2001) On the development of proton conducting membranes for hydrogen and methanol fuel cell. J Membr Sci 185:29–39

35. Li Q, He R, Jensen JO, Bjerrum NJ et al (2003) Approaches and recent development of polymer electrolyte for fuel cells operating above 100°C. J Chem Mater 15:4896–4915
36. Fontananova E, Trottac F, Jansen JC, Drioli E et al (2010) Preparation and characterization of new non-fluorinated polymeric and composite membranes for PEMFCs. J Membr Sci 348:326–336
37. Kerres J, Ullrich A, Meier F, Häring T et al (1999) Synthesis and characterization of novel acid-based polymer blends for application in membrane fuel cell. Solid State Ionics 125:243–249
38. Kerres JA (2001) Development of ionomers membranes for fuel cells. J Membr Sci 185:3–27
39. Gensler R, Plummer CJG, Kausch HH, Münstedt H et al (1997) Thin film and bulk deformation behavior of poly(ether ether ketone)/poly(ether imide) blends. J Mater Sci 32:3037–3042
40. Andrady AL, Nunez CM, Chiou B, Khan SA et al (2002) Rheology of concentrated solution of hyperbranched polymers. Polym Eng Sci 42(11):2065–2071
41. Koji N, Kaji K, Kanaya T, Fanjat N et al (2002) Theoretical calculation of reduced viscosity of polyelectrolyte solution. Polymer 43:1295–1300
42. Kerres J, Ullrich A, Meier F, Haring T et al (1999) Syntheses and characterization of novel acid base polymer blends for application in membrane fuel cell. Solid State Ionics 125:243–249
43. Bowen WR, Doneva TA, Yin H et al (2002) the effect of sulfonated poly (ether-ether ketone) additives on membrane formation and performance. Desalination 145:39–45
44. Chun YS, Kwon HS, Kim WN, Yoon HG et al (2000) Compatibility studies of sufonated poly (ether-ether ketone)- poly(ether-imide)- polycarbonate ternary blends. J Appl Polym Sci 78:2488–2494
45. Zaidi SMJ, Mikhailenko SD, Robertson GP, Guiver MD, Kaliaguine S et al (2000) Proton conducting composite membrane from polyether ether ketone and heteropolyacids for fuel cell. J Membr Sci 173:17–34
46. Zaidi SMJ (2003) Polymer sulfonation – a versatile route to prepare proton-conducting membrane material for advance technology. Arabian J Sci Eng 28(2B):183–194
47. Bauer B, Jones DJ, Roziere J, Tchicaya L, Alberti G, Casciola M, Massinelli L, Peraio A, Besse S, Ramunni E et al (2000) Electrochemical characterization of sulfonated polyetherketone membranes. J New Mater Elect Syst 3(2):93–98
48. Tchicaya-Bouckary L, Jones DJ, Rozière J et al (2002) Hybrid polyaryletherketone membranes for fuel cell application. Fuel Cells 2(1):40–45
49. Zaidi SMJ (2005) Preparation and characterization of composite membranes using blends of SPEEK/PBI with boron phosphate. Electrochem Acta 50:4771–4777
50. Wu H, Ma CM, Liu F, Chen C, Lee S, Chiang C et al (2006) Preparation and characterization of poly(ether sulfone)/sulfonated poly(ether ether ketone) blend membranes. Eur Polym J 42:1688–1695
51. Irfan MI, Zaidi SMJ, Rahman SU, Ahmad S et al (2006) Synthesis and proton conductivity of heteropolyacids loaded Y-zeolite as solid proton conductors for fuel cell application. Micropor Mesopor Mater 91:296–304
52. Irfan MI, Zaidi SMJ, Ahmad S et al (2008) Proton conducting composites of heteropolyacids loaded MCM-41. J Power Sources 157:35–44
53. Ahmad MI (2005) Synthesis and characterization of composite polymeric membranes for proton exchange membrane (PEM) fuel cell applications. M.S. Thesis, Chemical Engineering Department, King Fahd University of petroleum & Minerals
54. Hasani-Sadrabadi MM, Emami SH, Ghaffarian R, Moaddel H et al (2008) Nanocomposite membranes made from sulfonated poly(ether ether ketone) and montmorillonite clay for fuel cell application. Energy Fuel 22:2539–2542
55. Chen J, Asano M, Maekawa Y, Yoshida M et al (2008) Fuel cell performance of polyetheretherketone-based polymer electrolyte membranes prepared by two-step grafting method. J Membr Sci 319:1–4

Chapter 13
Preparation and Use of Organic-Inorganic Hybrid Ion Exchangers in Catalysis

Dilson Cardoso and Leandro Martins

Abstract Organic-inorganic hybrids constitute a new class of compounds in the research area of materials design. In this context, the preparation of hybrid organic-inorganic catalysts is presented, focusing on catalysts where an organic molecule is ionically bonded to the surface of an inorganic material. The application of this class of hybrid catalysts in some organic reactions is presented.

13.1 Introduction

Why use inorganic catalysts containing ionic organic guests to perform different catalyzed reactions? This topic has been written to show that these hybrid heterogeneous catalysts allow for more efficient catalytic processes, which are more cost-effectively derived from a higher productivity of the catalysts and from a lower temperature of reactor operation.

One of the major problems of homogeneously catalyzed reactions is the difficulty in recovering the catalyst from reaction products. Therefore, heterogeneous catalysts have a significant advantage over the homogeneous counterparts. There are several procedures attempting to "heterogenize" homogeneous catalysts in order to combine the advantages of the homogeneous catalyst with those of the heterogeneous [1]. Traditional heterogeneous catalysts are relatively limited in the nature of their active sites and thus the reactions that they can accomplish.

D. Cardoso (✉)
Chemical Engineering Department, Federal University of São Carlos – UFSCar, 676, São Carlos, SP 13565-905, Brazil
e-mail: dilson@ufscar.br

L. Martins
Institute of Chemistry, Universidade Estadual Paulista – UNESP, 355, Araraquara, SP 14800-900, Brazil
e-mail: leandro@iq.unesp.br

Inamuddin and M. Luqman (eds.), *Ion Exchange Technology I: Theory and Materials*, DOI 10.1007/978-94-007-1700-8_13,

However, soluble organic catalysts can catalyze a much larger variety of reaction types than traditional solid catalysts. In this context, the first attempt to heterogenize homogeneous catalysts occurred when Haag et al. [2] showed that transition metal complexes could be immobilized on polymer resins to be used in hydrogenation and oxidation reactions.

An important class of reactions is those catalyzed by liquid acids or bases, and there is a strong incentive to find solid acids [3] or bases [4] that can replace their liquid equivalents in large-scale reactions. An alternative is the use of ion exchange resins that have been available for many years [5]. Acid ion exchangers are generally based on a linked polymer with acid functionalities in the form of MSO_3H. Similarly, strongly basic ion exchangers contain basic groups such as the quaternary ammonium hydroxides, $-N(CH_3)_3^+OH^-$.

Nafion, a perfluorinated ion exchange polymer, known as containing a very strong Brønsted acid site, has been used in a wide range of catalyzed reactions such as alkylation, esterification, and transesterification [6]. Poly(vinylalcohol), another type of organic matrix, has been modified with different agents in order to improve their hydrophobic/hydrophilic properties. In a previous work, it was shown that these membranes can be successfully used as acid catalysts [6].

Attempts have also been made to modify porous inorganic materials owing to their superior thermal and mechanical properties [7]. Hybrid catalysts with organic groups attached to the support by silica functionalization techniques have been proposed [8].

However, it must be pointed out that not only the acid or basic sites are important in catalysis. Indeed, a solid surface may interact with reactants also as a function of its hydrophobicity [6]. A weak interaction hinders any catalytic step, while too strong interaction of reactants to the surface makes desorption too slow or impossible. Again, the hydrophilic-hydrophobic character of the surface may be manipulated by an appropriate choice of the exchangeable ions [7].

In the context of materials development for more efficient catalytic processes, inorganic-organic hybrid materials offer new and exciting opportunities. This topic focuses on organic-inorganic hybrid catalysts, with organic molecules ionically bonded to the surface of inorganic support materials. This topic does not aim to cover the properties of hybrids with covalently bonded molecules, because these materials have already been subject to a growing amount of research during the last few years [8].

13.2 Anionic Inorganic Supports: Zeolites and Mesoporous Molecular Sieves

The most typical inorganic matrices with anionic character are the zeolites, also known as molecular sieves due to its microporous system. These materials are formed by silica-alumina networks in which the aluminum atom has a tetrahedral configuration and, hence, generating negative framework charges [9]. These anions

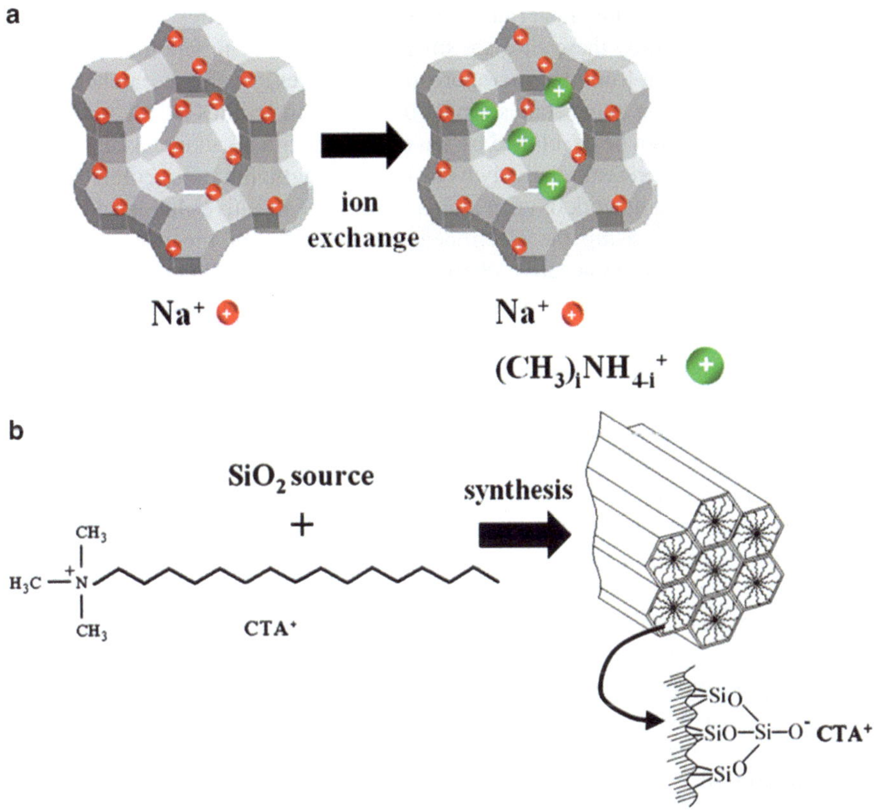

Fig. 13.1 Organic-inorganic hybrid anionic exchangers prepared by different procedures: (**a**) Ion exchange in FAU zeolites, (**b**) direct synthesis of MCM-41 structure

are compensated by mobile cations, which can be easily exchanged by others present in liquid solutions as depicted in Fig. 13.1a. More recently, Kubota and coauthors [10] discovered very important properties of new anionic matrices formed by direct synthesis of mesoporous silicas (Fig. 13.1b).

The preparation and properties of these two hybrid materials are described in more detail in the following items.

13.2.1 Anionic Inorganic Supports: Prepared by Ion Exchange

Ion exchange is an intrinsic property of zeolites. It is associated with the presence of cations that compensate the negative charge of structural AlO_4^- species. Various zeolite applications are directly or indirectly based on this property. Additionally, zeolites have attracted attention mainly due to their environmentally benign character and their potential use in fine chemical synthesis [11].

Zeolites are interesting materials not only by the properties of their active sites but also by the way these sites are situated inside their microporous structure. In an attempt to profit from the microporosity of zeolites, several researchers have devoted their attention to generate stronger basic sites in zeolites [11]. Usually, the basicity of zeolites can be developed by ion exchange with alkali metal cations [9]; however, these sites present relatively weak basic strength. Intrinsic zeolite basicity is associated with framework oxygen atoms bearing the negative charge and is thus of Lewis type. Decreasing the electronegative character of the non-framework compensating cations the negative charge over oxygen ions is enhanced, i.e., the basicity of alkali-exchanged zeolites increases with the compensating cation size in the order $Li^+ < Na^+ < K^+ < Rb^+ < Cs^+$. Indeed, this is a way of describing zeolite basicity. According to Barthomeuf [9], the electron density on the oxygen atom depends also on other factors, such as zeolite structure nature and cation location. In this alkali metal's sequence, francium cation (Fr^+) could ensure a higher basicity to zeolites because it has the lowest known electronegativity. However, this element is the second rarest naturally in occurrence and is a highly radioactive metal. An alternative to incorporate stronger basic sites in the zeolite pores is by supporting alkali compounds. Since cesium ions induce stronger basic sites than other alkali cations, researchers have focused significant attention on supported cesium catalysts.

A new basic zeolite catalyst, comprising ion-exchanged methylammonium [12] as well linear alkylammonium cations [13] in Faujasite zeolite (FAU) was found. These catalysts showed to be six times more active than cesium-FAU zeolites in Knoevenagel condensation of benzaldehyde with ethyl cyanoacetate (Reaction 1). This original result indicates that zeolites containing organic cations might have huge potential in various applications, which have not been properly explored. In addition, ammonium cation compounds are much cheaper than cesium. In this previous work [12], four different cations were evaluated, and interestingly, as occurs with the alkali metals, for the same ion exchange degree, the zeolite catalytic activity increases with the size of the organic compensating cation: $NH_4^+ < (CH_3)NH_3^+ < (CH_3)_3NH^+ < (CH_3)_4N^+$. Among the Me_i cations (where $0 < i < 4$, stands for the number methylammonium groups in the cation), the monomethyl cation, $(CH_3)NH_3^+$, showed the highest ion exchange degree and consequently ensured the highest catalytic activity to the FAU zeolite (Table 13.1).

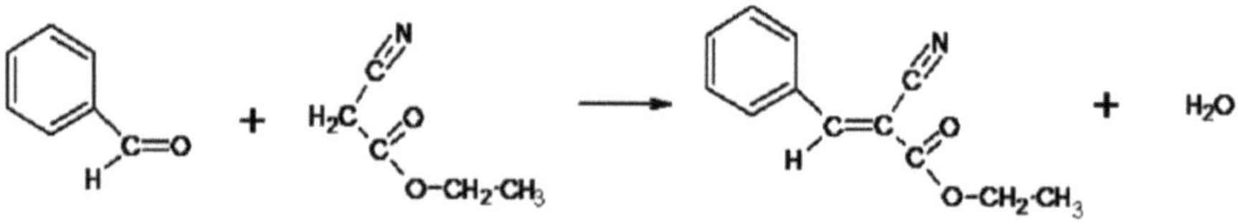

An alternative manner of increasing zeolite basicity is by incorporating aluminum atoms in the zeolite framework because, in this way, the negative framework charge density is increased [9]. FAU zeolites with different framework aluminum content and exchanged with methylammonium cations have been tested in the

Table 13.1 Activity of FAU zeolites in Knoevenagel condensation [14]

	Zeolite Si/Al = 2.5		Zeolite Si/Al = 1.4	
Catalyst	Conversion (%)	Selectivity (%)	Conversion (%)	Selectivity (%)
Na^+-FAU	8.2	100	16.6	99.4
Cs^+-FAU	20.7	99.3	50.4	98.5
$CH_3NH_3^+$-FAU	38.4	99.2	83.1	95.8
$(CH_3)_4N^+$-FAU	29.4	98.8	60.1	98.5

Scheme 13.1 Reactions taking place during propylene oxide methanolysis

$$CH_3OH + PrO \rightarrow CH_3O(PrO)H \quad (1^{st}\ \text{reaction})$$

$$CH_3O(PrO)H + PrO \rightarrow CH_3O(PrO)_2H \quad (2^{nd}\ \text{reaction})$$

$$CH_3O(PrO)_{n\text{-}1}H + PrO \rightarrow CH_3O(PrO)_nH \quad (\text{n reaction})$$

Table 13.2 Activity of FAU zeolites in propylene oxide methanolysis [14]

Catalyst	Propylene oxide conversion (%)	Selectivity to monomers (%)	Selectivity to dimers (%)
Na^+-FAU	39.4	45.2	19.1
Cs^+-FAU	41.2	46.1	17.1
$CH_3NH_3^+$-FAU	90.7	74.7	2.3

Knoevenagel condensation [14], and the material with higher aluminum content (Si/Al ratio of 1.4) showed a conversion up to three times superior compared with a zeolite having lower aluminum content (Si/Al ratio of 2.5). The main idea when developing these Me_i^+-FAU hybrid materials was to take advantage of the best properties of each component that forms the hybrid material, trying to eliminate the cesium high costs drawback and obtain catalysts with superior basicity.

Another class of reaction in organic synthesis over basic catalysts is the production of glycols and glycol ethers [14]. These products are versatile molecules and extremely important from an industrial point of view. There are several methods for the synthesis of propylene glycol ethers; however, the propylene oxide (PrO in Scheme 13.1) reaction with methanol is most suitable and commercialized. In the presence of basic zeolites, the C-O epoxide bond preferentially opens at the least sterically hindered position, leading mainly to secondary alcohols (1-methoxy-2-propanol – 1M2P). However, on the contrary, acid zeolites provide most primary alcohols (2-methoxy-1-propanol – 2M1P) [14].

Commonly, propylene oxide (PrO) methanolysis can be described by a sequence of consecutive reactions depicted in Scheme 13.1. The most preferable conditions would be obtained for the highest content of monopropylene glycol methyl ethers (first reaction), which is a mixture of isomers 1-methoxy- 2-propanol (1M2P) and 2-methoxy-1-propanol (2M1P). In this reaction, the zeolite $CH_3NH_3^+$-FAU, prepared by ion exchange with an alkylammonium cation, was found to be more active than Cs^+-FAU for the conversion of propylene oxide (Table 13.2). This indicates higher

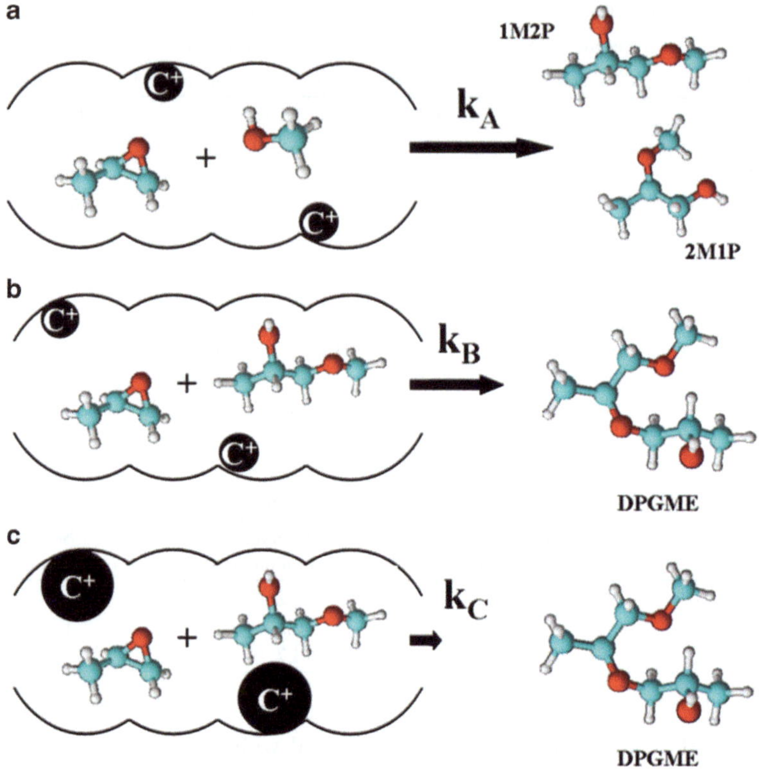

Fig. 13.2 Methanolysis of propylene oxide in FAU zeolite pores: (**a**) Formation of the monomer $CH_3O(PrO)H$ (1M2P and 2M1P), (**b**) of the dimer $CH_3O(PrO)_2H$ (DPGME) in the pores of the zeolite Na^+ or Cs^+-FAU, and (**c**) lowered formation of the bulky dimer in zeolite $Me_1{}^+$-FAU

basicity of the hybrid catalyst. Interestingly, selectivity toward the monomers 1M2P or 2M1P also showed to be strongly dependent on catalyst. Controlling selective production of 1M2P and 2M1P is a very important step in obtaining high-quality glycol ethers. The hybrid catalyst $CH_3NH_3{}^+$-FAU showed an excellent result for obtaining the desired selectivity toward 1M2P and 2M1P isomers [14].

Zeolites are also known by that the shape or topology of the internal pore structure that can strongly affect the selectivity, with which particular product molecules are formed, designated as shape selectivity [11]. Particularly in zeolite $CH_3NH_3{}^+$-FAU, pore narrowing is more severe, and thus cation volume is considerably increased after ion exchange of Na^+ from 4 to 49 Å^3, making more evident the shape selectivity toward formation of smaller molecules. This is an enormous advantage when using hybrid alkylammonium zeolites, thus the selectivity is easily controlled by ion exchange.

Generally, during propylene oxide methanolysis, dipropylene glycol methyl ether mixture of isomers (DPGME, Reaction B, Fig. 13.2) is also formed, and its content increases with advancement of the reaction. However, on zeolite $CH_3NH_3{}^+$-FAU good selectivity towards $CH_3O(PrO)H$ (Reaction A, Fig. 13.2) was obtained

Table 13.3 Estimated cation volume (V_C), maximum ion exchange percentage, and micropore volume of FAU zeolites [17]

Cation	V_C (Å^3)	% Expected exchange	% Experimental exchange	Micropore vol. (mL/g)
Na^+	3.6	100	100	0.25
Cs^+	20.2	100	70	0.20
NH_4^+	13.6	100	74	0.21
$CH_3NH_3^+$	49	100	71	0.19
$(CH_3)_3NH^+$	129	74	52	0.11
$(CH_3)_4N^+$	175	54	35	0.14
$CH_3CH_2NH_3^+$	94	81	71	0.10
$CH_3CH_2CH_2NH_3^+$	115	77	65	0.12

and the formation of bulkier products was suppressed. Actually, the micropore narrowing caused by ion exchange with a more voluminous cation hinders the second step of the methanolysis (Reaction C, Fig. 13.2) and leads to high selectivity toward the smaller monomers molecules.

Basic property of zeolites, as a consequence of ion exchange with inorganic cations, has been studied in a number of papers [9], but almost no attention had been given to zeolites ion-exchanged with organic cations. With exception of two works from the 1960s [15, 16], which studied uniquely ion exchange isotherms of alkylammonium cations, no other paper described the effect of organic cation in zeolite properties.

The exchangeable cations in FAU zeolite are usually found at four different crystallographic positions, namely, type I, I′, II, and III [12]. Type I exchange sites are located inside hexagonal prisms connecting sodalite cavities. Sites I′ are inside the sodalite cavity, with its face positioned toward the hexagonal prisms. Sites II are located in the six-member ring of the sodalite catalyst, facing the supercavity. And sites III are also in the supercavity, next to four-member rings in the sodalite cavity. Sites types I and I′ represent approximately 17% and 13% of aluminum atoms present in this zeolite, respectively [12]. Sites II and III correspond to 20% and 50% of ion exchange sites, respectively. These results, obtained from crystallographic data and theoretical models, provide an explanation for ion exchange results obtained with methylammonium cations summarized in Table 13.3 [17]. The maximum ion exchange obtained for cation with intermediate volumes (between 13.6 and 49 Å^3, in Table 13.3) is in agreement with the observations that the proportion of Na^+ ions located in the supercavity is around 70%.

In the ion exchange of FAU zeolites with alkylammonium cations, Table 13.3 shows that some organic cations should allow 100% ion exchange, but none of the alkylammonium cations used was able to exchange all sodium present in the zeolite. This limitation is attributed to two stereo-spatial barriers that may take place simultaneously, depending on cation volume. The first is due to the inaccessibility of internal region of hexagonal prisms and sodalite cavities to voluminous cations, thus restricting ion exchange to the supercavity. The second barrier is related to the lack of space between neighbor anions to voluminous cations

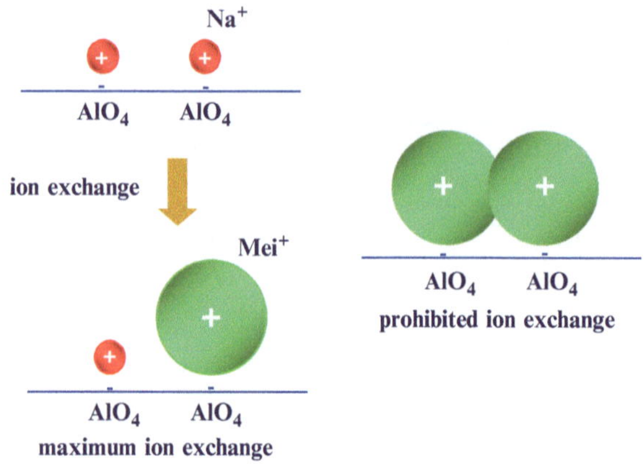

Fig. 13.3 Maximum ion exchange degree and prohibited ion exchange in the channels of zeolite FAU

(i.e., Me_3 and Me_4) because they have kinetic diameter larger than the distance between two negative neighboring charges, induced by the incorporation of aluminum atoms in the zeolite framework. Figure 13.3 illustrates better this second barrier that the ion exchange at neighboring aluminum atoms is difficult.

13.2.2 Anionic Inorganic Supports: Prepared by Direct Synthesis

In 2004, Kubota et al. [10] published their pioneering work about the use of Beta zeolite as catalyst without going through calcination, i.e., with pores still occluded with the cationic structure directing template, the tetraethylammonium cation. This hybrid cationic-anionic material showed to be an excellent catalyst in the base-catalyzed condensation reactions [18]. This novel result indicated that zeolites containing organic cations might have high potential of application, which motivated other researchers to test different zeolite structures enclosing entrapped cationic species prepared through ion exchange (Fig. 13.1a) [12] or synthesis procedures (Fig. 13.1b) [18].

Particularly during zeolites synthesis, organic-inorganic coulombic forces that exist between the cationic structure directing agents and the anionic structure of the zeolites play an important role in stabilizing zeolite phases. Actually, these forces are important for the nucleation and crystallization of high-silica zeolites [18]. During the nucleation process, the organic species arrange TO_4 tetrahedra ($T = Si$ and Al atoms) into a particular topology, providing the building block for a particular structure type [18]. This is a strong indication of the presence of organic cationic species, in concert with anionic ones such as siloxy $\equiv SiO^-$ or aluminoxy groups $[AlO_4]^-$. In these catalysts, the siloxy groups are very strong basic sites that are capable to catalyze several reactions. Figure 13.4 evidences this finding because as silicon fraction is increased in different zeolites and mesoporous materials, the same tendency is observed in benzaldehyde conversion. The catalyst with

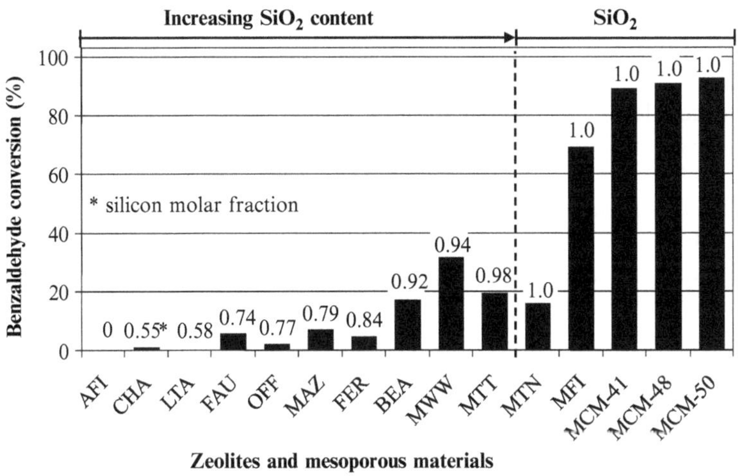

Fig. 13.4 Conversion in Knoevenagel condensation (Reaction 12-1), between ethyl cyanoacetate and benzaldehyde, as a function of molar fraction of silicon in molecular sieves containing entrapped organic cations (Reproduced from Ref. [18] with kind permission of © Elsevier (2002))

superior activity in this reaction is the siliceous MCM-41containing CTA cations (Fig. 13.1b).

Although the mesoporous materials designed as MCM-41 was discovered in 1992, many properties have already been explored, and the commercial application of some of them is likely underway. One important novel property of this molecular sieve was found in 2004: the as-synthesized MCM-41, denoted as [CTA^+]-MCM-41 (CTA stands for cetyltrimethylammonium cation), showed interesting and remarkable basic properties that is attributed to the presence of siloxy anions in combination with CTA^+ cations. This attractive catalyst has already been tested in three different reactions – Knoevenagel condensation [19–21], Michael addition [22], and cycloaddition reaction of CO_2 with epoxides [23] – and showed superior activity compared to other basic catalysts.

Because the MCM-41 catalyst channels are blocked with the ionic surfactant, the high activity in the investigated reactions is attributed to basic sites located at the pore mouth. The [CTA^+]-MCM-41 possesses a surface area of only 1 m^2/g, and therefore, the basic sites present inside the channels are inaccessible for the reactants [19]. This catalyst shows high catalytic activity even at low temperatures and has in addition an extra great advantage because it can be used as synthesized, without further modification.

The main goal in heterogenization of homogeneous catalysts is to utilize an organic molecule as the active site and the solid to provide possibilities of recovering and recycling the catalyst. The support material itself is usually not catalytically active. However, the use of ionic organic molecules (e.g., alkylammonium cations) changes dramatically the acidity/basicity character of these surfaces and consequently the purposes of their use. For example, in the case of as-synthesized MCM-41, the active sites are the siloxy anions $\equiv SiO^-$ that exist in combination with CTA^+ cations.

Frequently, the basicity of mesoporous silicas, like MCM-41, MCM-48, SBA-15 molecular sieves, is achieved by functionalizing its surface with compounds containing terminal amines [8]. The material is functionalized by anchoring organic bases at the silanol groups, thereby forming a covalent bond. Prior to functionalizing, the cationic surfactants present in the pores of the mesoporous support have to be removed in the first place in order to make the pores free and to generate the silanol groups, usually carried out by calcination under airflow. Another method of obtaining basic MCM-41 can be achieved by dispersing alkali metal oxides [24]. However, due to the high pH of the impregnating solution, the MCM-41 structure can be somewhat damaged. Cesium acetate is usually used for the impregnation of calcined MCM-41, followed by heating to generate the basic Cs_2O centers [24].

Since their first description in 1992, the mesoporous micelle templated silicas, like MCM-41, have become favored carriers for homogeneous catalysts. In fact, discovery of these materials is mainly responsible for the renewed interest in homogeneous catalysts "heterogenization." However, the main drawbacks of these methods for preparing functionalized supports are the high experimental expenditures.

13.3 Cationic Inorganic Supports: Layered Double Hydroxides or Hydrotalcites

Layered double hydroxides (LDH) are well known as anionic inorganic supports [25]. The ion exchange of interlayer cations (organic or inorganic) has received considerable attention due to their importance as functional materials. LDH has a positively charged hydroxide layer due to trivalent cations substituted for a part of divalent cations in the layer. The positive layer charge is compensated by intercalated anions. The chemical composition of LDH is represented as $[M^{2+}_{1-x}M^{3+}_x(OH)_2][A^{n-}_{x/n}\cdot yH_2O]$, where M^{2+} is a divalent cation, M^{3+} is a trivalent cation, and A^{n-} is an anion (inorganic and/or organic anions). The M^{3+} cation substitution parameter x varies between 0.20 and 0.33. The nature of the layer cations can be also changed, although most of the studies reported in the literature refer to systems with M^{2+} and M^{3+} cations in the layers [26].

The intercalation of organic compounds into LDH has been investigated because of the resulting compound is a nanostructured material with new functional properties. The system is constituted of sheets lying on top of each other in which coulombic forces maintain the chemical integrity and define an interlamellar distance filled up with an organic anion, forming a sandwich-like structure [25]. Exchange of interlamellar anions by organic ions was first reported by Miyata and Kumura [27]. LDH easily exchanges short- and long-chain alkyl sulfate anions or fatty acid anions. As a consequence of this ion exchange, the basal spacing of the LDH then increases with the fatty acid chain length.

Catalytic applications of LDH have for a long time focused on the organic condensation reactions carried out with mixed oxides obtained by calcinations of Mg/Al LDH. However, the versatile character of materials prepared from these LDH exhibiting catalytic activity has recently been highlighted and directed them to extended functions [28].

Several LDH containing organic cations or cationic complexes have been prepared; however, restricted examples of the catalytic application of these materials are known. One example is the intercalation of anionic complexes such as porphyrins or other supramolecular assemblies between the lamellar structures. In their cationic or anionic forms, the porphyrins display a very high affinity toward negatively or positively charged inorganic supports [29]. The charge density of the layers directly controls the concentration of the catalytic species. This concentration can be high, especially with high charge density supports such as layered zirconium phosphates or double layered hydroxides. Indeed, LDH display very high ionic exchange capacities compared to the cationic-exchange capacity of other supports. This might be interesting because, in the case of porphyrins adsorbed on clays, the guest-host interactions have shown improved chemical stability, including the prevention of rapid photodegradation.

Porphyrins are extensively known as oxidation catalysts, especially for their ability to catalyze the oxidation of various substrates under mild conditions, including inert molecules such as alkanes. As verified by Bonnet et al. [29], the ability of anionic porphyrins to exchange and to remain bonded to the LDH layers, even under thermal treatment up to 300°C, encourages the use of such heterogenized catalyst. Additionally, the intercalation process seems to favor the catalytic activity of iron porphyrins probably because the supported catalyst is less susceptible to inactivation by molecular aggregation and/or bimolecular self-destruction than the parent porphyrin in solution.

Additionally, there has been increasing interest in supporting metallophthalocyanines on LDH, owing to the fact that these macrocyclic complexes can function as catalysts for the oxidation of organic molecules in aqueous solution [24, 30]. Cobalt(II) phthalocyanine tetrasulfonate, for example, is active for the oxidation of a thiolate. Incorporation of the phthalocyanine complex into a MgAl LDH was found to significantly improve the catalyst reactivity and longevity for this reaction, compared to the homogeneous catalyst.

Another example is the intercalation into Zn/Al layered double hydroxides with sulfonato-salen metal complexes containing Fe or Mn metal centers and application as catalyst for the epoxidation of cyclohexene. In this case, the catalysts showed good activity and selectivity in the epoxidation reaction using oxygen as oxidant and at room temperature [31].

A major disadvantage of LDH is that many of the applications are largely restricted due to the inaccessibility to the inner surfaces of the layers. The most effective solution to this problem may be delamination into single layers as this maximizes the use of the layers. Nevertheless, the strong electrostatic interactions between layers and guest anions make the delamination difficult [32].

13.4 Concluding Remarks

The heterogenization of homogeneous catalysts is an area that has been subject to increasing attention during recent years, especially in the case of hybrid composites where organic molecules are ionically associated to the surface of an inorganic material. There are two main reasons for that: (1) the discovery of new basic catalysts with superior basicity compared to the traditional ones and (2) the possibility of choosing different ionic organic molecules allows tailoring of the heterogeneous catalyst according to the specific strength needs of the active sites.

While many of the materials traditionally used as catalyst are too expensive for industrial applications today (e.g., cesium containing zeolites), organic-inorganic hybrid ion exchangers can lower their prices to a point where the advantages of these versatile catalysts can lead to their use in real industrial conditions.

Acknowledgment The authors are grateful to the Fundação de Amparo a Pesquisa do Estado de São Paulo for financial support.

References

1. Jaenicke S, Chuah GK, Lin XH, Hu XC (2000) Organic-inorganic hybrid catalysts for acid- and base-catalyzed reactions. Micropor Mesopor Mat 35–36:143–153
2. Haag WO, Whitehurst DD (1968) Verfahren zur Durchführung organischer Umsetzungen, die durch einen Metallverbindungskatalysator heterogen katalysiert werden. Patent BE 721 686
3. Corma A (1995) Inorganic solid acids and their use in acid-catalyzed hydrocarbon reactions. Chem Rev 95:559–614
4. Hattori H (1995) Heterogeneous basic catalysis. Chem Rev 95:537–550
5. Bárbaro P, Liguori F (2009) Ion exchange resins: catalyst recovery and recycle. Chem Rev 109:515–529
6. Guerreiro L, Castanheiro JE, Fonseca IM, Martin-Aranda RM, Ramos AM, Vital J (2006) Transesterification of soybean oil over sulfonic acid functionalised polymeric membranes. Catal Today 118:166–171
7. Vaccari A (1998) Preparation and catalytic properties of cationic and anionic clays. Catal Today 41:53–71
8. Valkenberg MH, Hölderich WF (2002) Preparation and use of hybrid organic–inorganic catalysts. Catal Rev Sci Eng 44:321–374
9. Barthomeuf D (2003) Framework induced basicity in zeolites. Micropor Mesopor Mat 66:1–14
10. Kubota Y, Nishizaki Y, Ikeya H, Saeki M, Hida T, Kawazu S, Yoshida M, Fujii H, Sugi Y (2004) Organic-silicate hybrid catalysts based on various defined structures for Knoevenagel condensation. Micropor Mesopor Mat 70:135–149
11. Martins L, Cardoso D (2006) Aplicação catalítica de peneiras moleculares básicas micro e mesoporosas. Quím Nova 29:358–364
12. Martins L, Boldo RT, Cardoso D (2007) Ion-exchange and catalytic properties of methylammonium FAU zeolite. Micropor Mesopor Mat 98:166–173
13. Almeida KA, Martins L, Cardoso D (2010) Preparação e propriedades de zeólitas faujasita contendo cátions amônio. Quím Nova 33:1077–1081
14. Martins L, Hölderich W, Cardoso D (2008) Methylammonium FAU zeolite: investigation of the basic sites in base catalyzed reactions and its performance. J Catal 258:1–11

15. Theng BKG, Vansant E, Uytterhoeven JB (1968) Ion-exchange in synthetic zeolites. I. Ammonium and some of its alkyl derivatives in Linde sieves X and Y. Trans Faraday Soc 64:3370–3382
16. Barrer RM, Papadopoulos R, Rees LVC (1967) Exchange of sodium in clinoptilolite by organic cations. J Inorg Nucl Chem 29:2047–2063
17. Martins L, Vieira KM, Rios LM, Cardoso D (2008) Basic catalyzed Knoevenagel condensation by FAU zeolites exchanged with alkylammonium cations. Catal Today 133:706–710
18. Oliveira AC, Martins L, Cardoso D (2009) Basic catalytic properties of as-synthesized molecular sieves. Micropor Mesopor Mat 120:206–213
19. Martins L, Bonagamba TJ, Azevedo ER, Bargiela P, Cardoso D (2006) Surfactant containing Si-MCM-41: an efficient basic catalyst for the Knoevenagel condensation. Appl Catal A Gen 312:77–85
20. Martins L, Cardoso D (2007) Influence of surfactant chain length on basic catalytic properties of Si-MCM-41. Micropor Mesopor Mat 106:8–16
21. Martins L, Hölderich W, Hammer P, Cardoso D (2010) Preparation of different basic Si-MCM-41 catalysts and application in the Knoevenagel and Claisen-Schmidt condensation reactions. J Catal 271:220–227
22. Kubota Y, Ikeya H, Sugi Y, Yamada T, Tatsumi T (2006) Organic-inorganic hybrid catalysts based on ordered porous structures for Michael reaction. J Mol Catal A Chem 249:181–190
23. Srivastava R, Srinivas D, Ratnasamy P (2006) Syntheses of polycarbonate and polyurethane precursors utilizing CO_2 over highly efficient, solid as-synthesized MCM-41 catalyst. Tetrahedron Lett 47:4213–4217
24. Ernst S, Bongers T, Casel C, Munsch S (1999) Cesium-modified mesoporous molecular sieves as basic catalysts for Knoevenagel condensations. In: Kiricsi I, Pál-Borbély G, Nagy JB, Karge HG (eds) Porous materials in environmentally friendly processes, studies in surface science and catalysis, vol 125. Elsevier, Amsterdam
25. Khan AI, O'Hare D (2002) Intercalation chemistry of layered double hydroxides: recent developments and applications. J Mater Chem 12:3191–3198
26. Rives V, Ulibarri MA (1999) Layered double hydroxides (LDH) intercalated with metal coordination compounds and oxometalates. Coord Chem Rev 181:61–120
27. Miyata S, Kumura T (1973) Synthesis of new hydrotalcite-like compounds and their physicochemical properties. Chem Lett 2:843–848
28. Tichit D, Coq B (2003) Catalysis by hydrotalcites and related materials. CATTECH 7:206–217
29. Bonnet S, Forano C, Roy A, Besse JP (1996) Synthesis of hybrid organo-mineral materials: anionic tetraphenylporphyrins in layered double hydroxides. Chem Mater 8:1962–1968
30. Carrado KA, Forman JE, Botto RE, Winans RE (1993) Incorporation of phthalocyanines by cationic and anionic clays via ion-exchange and direct synthesis. Chem Mater 5:472–478
31. Bhattacharjee S, Anderson JA (2006) Comparison of the epoxidation of cyclohexene, dicyclopentadiene and 1,5-cyclooctadiene over LDH hosted Fe and Mn sulfonato-salen complexes. J Mol Catal A Chem 249:103–110
32. Liu Z, Ma R, Osada M, Iyi N, Ebina Y, Takada K, Sasaki T (2006) Synthesis, ion-exchange and delamination of Co-Al layered double hydroxide: assembly of the exfoliated nanosheet/polyanion composite films and magneto-optical studies. J Am Chem Soc 128:4872–4880

Chapter 14
Ion Exchange to Fabrication of Waveguides for Optical Telecommunication

Victor Anthony Garcia Rivera

Abstract Ion exchange technique in the process of design and fabrication of passive devices for telecommunication has been a procedure to increment the glass refraction index and has received special attention because it improves surface-mechanical properties of glass and, more importantly, creates a wave-guiding region in the glass. Today, glass waveguides are considered to be suitable candidates for passive and active device for the integrated optical (IO) such as couplers, multiplexers, demultiplexers, and others. Their importance is borne out by their compatibility with optical fibers, low cost, low propagation loss, and integration into the system. In this way, with increasing speed of data transmission, the importance of integrated photonic devices and circuits grows. Many research groups have focused in the development of optical amplifier based on Erbium-doped materials since their potential to realize broadband and inherently linear compatible with current technology. This chapter gives an introduction to the ion exchange technique in solid matter, mainly optical glasses. I shall present the main results and equations describing the traditional ion exchange processes.

14.1 Theory

The modern history of ion exchange goes back over a hundred years (H. S. Thompson and J. T. Way, 1850), during which there have been many scientific and technical developments in this field. Progress has accelerated recently as a result of the introduction of optical glasses. In 1972, Izawa and Nakagone [1] reported the first ion exchange waveguides by thallium ions in silicate glass containing oxides of sodium and potassium. This report marked the starting of the understanding the ion exchange process and the fabrication conditions of waveguides.

V.A.G. Rivera (✉)
Institute of Physics and National Institute of Optical and Photonic (INOF) of São Carlos, University of São Paulo, Caixa Postal 369, São Carlos, SP 13560-970, Brazil
e-mail: garcia@ursa.ifsc.usp.br

Inamuddin and M. Luqman (eds.), *Ion Exchange Technology I: Theory and Materials*, DOI 10.1007/978-94-007-1700-8_14,

The ion exchange reaction is defined as the reversible interchange of ions between a solid phase and a solution phase, that is, the ion exchange existence in the medium in which the exchange is made. If an ion exchanger $M^{-}A^{+}$, carrying cations A^{+} as the exchanger ions, is placed in an aqueous solution phase containing B^{+} cations, an ion exchange reaction takes place which may be represented by the following equation:

$$\underbrace{M^{-}A^{+}}_{solid} + \underbrace{B^{+}}_{solution} \leftrightarrow \underbrace{M^{-}B^{+}}_{solid} + \underbrace{A^{+}}_{solution}\ . \tag{14.1}$$

The equilibrium represented by Eq. 14.1 is an example of cation exchange where M^{-} is the insoluble fixed anionic complement of the ion exchanger $M^{-}A^{+}$, often called simply the fixed anion. The cations A^{+} and B^{+} are referred to as counter ions, while ions in the solution which bear the same charge as the fixed anion of the exchanger are called co-ions. This way, anions can be exchanged provided that an anion-receptive medium is employed. Analogous representation of an anion exchange reaction may be written:

$$\underbrace{M^{+}A^{-}}_{solid} + \underbrace{B^{-}}_{solution} \leftrightarrow \underbrace{M^{+}B^{-}}_{solid} + \underbrace{A^{-}}_{solution}\ . \tag{14.2}$$

Equation 14.2 illustrates the essential difference between ion exchange. The main fact is that electroneutrality is preserved at all times in both the exchange and solution phases. Important features characterizing an ideal exchange are:

- Controlled and effective ion exchange capacity
- Rapid rate of exchange
- Chemical stability
- Physical stability in terms of mechanical strength and resistance to attrition

Several cation pairs have been studied, and in most cases [2, 3], sodium ions in the substrate glass are exchanged for one of the cations, namely, Cs^{+}, Rb^{+}, Li^{+}, K^{+}, Ag^{+}, or Tl^{+}; of the various monovalent ions, each one has its advantages and drawbacks. In all the cases, surface indices change and thereby is possible fabricate single- and multimode waveguides for optical telecommunication system. However, for integrated-optic (IO) applications, it is desirable to fabricate a waveguide with low propagation loss and desired refractive index profile. Both depend upon the substrate glass, that is, the relatively slow progress in ion exchange technology can be attributed to this dependence. For example, losses are dominated by absorption, due to presence of foreign impurities and scattering contributions caused by glass inhomogeneity and surface imperfections. In the same way, dopants introduce additional scattering to the background losses caused by the multicomponent nature of the glass.

In the production process of waveguides (telecommunication optics) for a given substrate glass, such a systematic study entails the following distinct steps:

(a) Ion exchange equilibrium to determine the relationship between the melt composition and the glass-surface-ion concentration
(b) Diffuse equation with appropriate boundary conditions to predict the concentration profile
(c) Measurement of the diffusion profile
(d) Waveguide characterization propagation for the given index profile
(e) Optical characterization of the waveguide and correlation between the optical characteristic and the process parameters

The exchange capacity is possibly the most important characteristic of an ion exchange material since it is a measure of its capability to carry out useful ion exchange work [2–4], for example, the rate at which an ion exchange reaction proceeds is a complex function of several physical-chemical processes such that the overall reaction may be influenced by the separate or combined effects of concentration gradients in both phases, electrical charges gradient in both phases, ionic interaction in either phase, exchanger properties, and chemical reactions in either phase.

The basic step in the fabrication of ion exchange waveguides is the development of suitable glass substrate having sufficient alkali ion concentration (typically Na^+, K^+, and Ca^+), chemical durability, and activity ion solubility [5]. Thus, the exchange is realized by thermal movement of cations in salt bath, which can be assisted by an electric field external, so these cations are exchange to other cations with the same valence (e.g., Ag^+, Rb^+, Cs^+, Li^+, K^+, and Tl^+). We can wait that the dielectric permittivity increment proportionately to the dopant cationic concentration, and thus, planar/strip optical waveguides can be obtained. Being a diffusion-driven process, ion exchange produces a gradient like concentration profile. Also, due to differing polarizabilities and sizes between the exchange ions, it produces a refractive index change in the glass. Therefore, the prediction of the cation concentration is of great importance to design high-performance optical device [6].

Basically, the net index change (Δn) depends on the extent of the ion exchange dictated by the equilibrium at the melt-glass interface [6, 7], and the effective diffusion constant of cations species into of bulk glass and the time diffusion [8]. Penetrability of the glass can be thus considered as one of the figures of merit for application of ion exchange for the fabrication of photonics components.

The only major disadvantage of the ion exchange technology is that it requires thoughtful adjustment of the technology procedure appropriate for a certain combination of a chosen glass substrate [9] and the expected utilization of the waveguide. However, once the fabrication procedure is adjusted, this technology enables formation of enormous amount of low-cost waveguide with fully reproducible properties.

In the next section are shown theoretical approaches that describe the development of ion exchange for the production of devices for the optical telecommunications.

14.1.1 Standard Models

In case of the simultaneous diffusion of two species (e.g., cations A and B) with different diffusion coefficients (D_A and D_B) in a medium (glass) where fixed anions are present this process is governed by three partial differential equations of the second order: the Poison's and two drift-diffusion equations (one for each specific cation). In most cases, this system can be simplified under the approximation of charge neutrality, i.e., the total cation concentration is assumed to be almost equal that of anion everywhere. It is supported by the fact that a very small charge disequilibrium is enough to generate a strong electric field which dominates the drift-diffusion equations; this field moves the cations in such a way that the charge neutrality tends to be restored [10, 11]. This approximation was developed in the ref. [12] for the cases where the total flux of cations (J_0) is null. Thus, the following diffusion equation (without drift term) is written as:

$$\frac{\partial C_A}{\partial t} = \vec{\nabla}\left(\frac{D_A D_B}{D_A C_A + D_B C_B}\vec{\nabla} C_A\right), \tag{14.3}$$

where t is the time and C_i is the mole fraction of the species $i = A, B$. Each mole fraction is related with concentrations of both species through:

$$C_i = \frac{C_i}{C_A + C_B};\ i = A, B. \tag{14.4}$$

Note that C_B could be eliminated from Eq. 14.3 since $C_A + C_B = 1$.

A generalization of charge neutrality approximation is desirable to describe these kind of cases; for one-dimensional problems [13], the following was obtained:

$$\frac{\partial C_A}{\partial t} + \frac{D_A D_B}{(D_A C_A + D_B C_B)^2}\frac{J_{0x}}{C_0}\frac{\partial C_A}{\partial x} = \frac{\partial}{\partial x}\left(\frac{D_A D_B}{D_A C_A + D_B C_B}\frac{\partial C_A}{\partial x}\right), \tag{14.5}$$

where x is the Cartesian coordinate, C_0 the anion concentration, and J_{ox} the cation flux density along the x-axis.

A first approach was made in [14] by approximation of the total electric field (but not its divergence) by external one $(\overrightarrow{E}_{ext})$ in three dimensions, which leads to:

$$\frac{\partial C_A}{\partial t} + \frac{D_A D_B}{(D_A C_A + D_B C_B)^2}\frac{e\vec{E}_{ext}}{k_B T}\vec{\nabla} C_A = \frac{\partial}{\partial x}\left(\frac{D_A D_B}{D_A C_A + D_B C_B}\vec{\nabla}^2 C_A\right). \tag{14.6}$$

Here e, k_B, and T are the proton charge, the Boltzmann's constant, and the absolute temperature, respectively; however, this is a not a true generalization. Defining the drift term proportional to the field, $\overrightarrow{E}_{ext}$ is obtained:

$$\frac{e\overrightarrow{E}_{ext}}{k_B T} = \frac{\overline{J_0}}{D_A C_A + D_B C_B}. \tag{14.7}$$

Albert and Lit [15] claim that $\overrightarrow{E}_{ext}$ is the external applied field, but it will be shown later that charges located inside glass contribute to this field. Specifically, $\left(\overrightarrow{E}_{ext}\right)_x$ should be independent of the position in one-dimensional problems (like above J_{0x}), which makes clear that the drift terms of Eqs. 14.5 and 14.6 are different although their qualitative behavior is similar. On the other hand, both Refs. [16] and [17] have obtained the same multidimensional electro-diffusion equation:

$$\frac{\partial C_A}{\partial t} + \frac{D_A D_B}{(D_A C_A + D_B C_B)^2} \frac{\overline{J_0}}{C_0} \overrightarrow{\nabla} C_A = \overrightarrow{\nabla}\left(\frac{D_A D_B}{D_A C_A + D_B C_B} \overrightarrow{\nabla} C_A\right). \tag{14.8}$$

That depends on total flux density instead of $\overline{J_0}$ of the electric field. It can be regarded as a satisfactory extension of both Eqs. 14.3 and 14.5.

The influence conductivity change on the flux density has been formalized in Ref. [18] to explain why channel waveguides are buried more slowly than slab ones: they assumed the Ohm's law: $e\overrightarrow{J_0} = -\sigma\overrightarrow{\nabla} V$, where V is the electric potential and σ is the conductivity, this last can be related with the mole fraction by means of the following: $\sigma = \frac{e^2 C_0}{k_B T}(D_A C_A + D_B C_B)$. Finally we can have:

$$\overrightarrow{\nabla}\left((D_A C_A + D_B C_B)\overrightarrow{\nabla} V\right) = 0, \tag{14.9}$$

from Eq. 14.9 we can determine V if the concentration profile is known; next, $\overrightarrow{J_0}$ is trivially calculated from V.

The above referred works assumed both constant self-diffusion coefficients and ideal behaviors of cations. Nevertheless, the experimental profile of the field-assisted burying of slab waveguide is much more asymmetric (sharp front of profile and smooth rear tail) than predicted from constant self-diffusion coefficients. For that reason, a strong dependence of self-diffusion coefficients on cation composition is needed to get an accurate description [19]. Such dependence is called mixed alkali effect and is known [4, 20–22], although no theory has been universally accepted to explain effect.

The measurements realized by Wakabayashi and Messerschmidt et al. [21] indicate the dependence on concentration and a nonideal behavior.

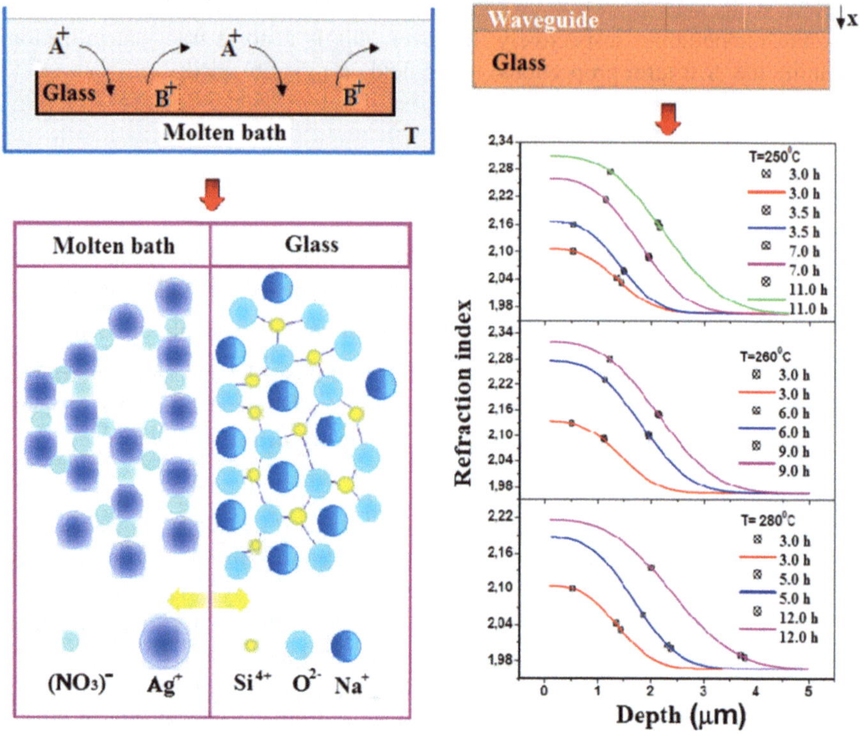

Fig. 14.1 Representation of ion exchange process. *Top left*: glass into molten salt bath A^+-B^+ to temperature T. *Down*: schematic of the ion exchange Ag^+-Na^+ to silicate glass between molten salts and the glass. *Top right*: stage final, planar waveguide with x depth. *Down*: refractive index in function of the depth; the better fitting is obtained from of Eq. 14.26 (Reproduced from Ref. [22] with kind permission of V.A.G. Rivera, Master thesis, IFGW-UNICAMP-SP-Brazil (2005))

14.1.2 Diffusion Equation

The diffusion of the ion A^+ (e.g., Ag^+) can be derived in the following form [14]:

$$\frac{\partial C_A}{\partial t} = \mu_A \vec{E}_{ext} \cdot \vec{\nabla} C_A \left(\frac{1}{1 - \alpha C_A} \right) + \frac{n}{1 - \alpha C_A} D_A \nabla^2 C_A, \tag{14.10}$$

where $\alpha = 1 - \frac{\mu_A}{\mu_B}$, μ_i is the mobility of ion, and

$$n = \frac{\partial \ln \overline{\alpha}_A}{\partial \ln C_A}. \tag{14.11}$$

Here, $\overline{\alpha}_A$ is the thermodynamic activity of the ion A^+. Equation 14.10 can be solved to obtain the concentration profile provided the initial condition and the boundary conditions are known. In Eq. 14.10, it has been assumed that the Nernst-Einstein relation is satisfied. One of the boundary conditions involves knowledge of the concentration of cation A^+ at the glass surface, which is a function of the melt concentration. Figure 14.1 shows the ion exchange process. The determination of this boundary condition was accomplished by ion exchange equilibrium studies reported earlier [4].

In addition, other important characteristic of the self-diffusion coefficient which is of interest in ion exchange is the mixed alkali or double alkali effect [22]. When a second alkali is added to a glass, a significant reduction in the self-diffusion coefficient of the original alkali ion occurs. While this reduction occurs whether the second alkali ion is smaller or larger than the original one, the magnitude of the change varies directly with the concentration and the size mismatch of the second alkali. In fact, the self-diffusion coefficient of an alkali ion is always considerably higher than that of the impurity alkali ion, and the diffusion curves of the two alkalis intersect at a certain concentration ratio, suggesting the alternation of the principal current carrying species.

As a result, the electrical conductivity passes through a minimum where the two diffusion curves intersect. Although such a concentration dependence of the self-diffusion coefficients has been measured in several glasses [22], the data for most glasses of practical interest for ion-exchanged glass waveguide applications are almost nonexistent. This is especially true for the Ag^+-Na^+ case as silver-rich glasses have not been fabricated and studied. The mixed alkali effect has considerable influence on the diffusion profile [3].

14.1.3 Diffusion Without External Electric Field

Equation 14.10 can be further simplified if there were no external electric field, $\vec{\nabla} \cdot \vec{E} = 0$, without the presence of space charge. Therefore, we can write the Eq. 14.10 in two spatial dimensions [23]:

$$\frac{\partial N_A}{\partial t} = \frac{\partial}{\partial x}\left(\frac{nD_A}{1-\alpha N_A}\frac{\partial N_A}{\partial x}\right) + \frac{\partial}{\partial y}\left(\frac{nD_A}{1-\alpha N_A}\frac{\partial N_A}{\partial y}\right), \tag{14.12}$$

where N_A is the concentration of ion, e.g., of silver in glass in mole fractions.

From Eq. 14.12, we can have the following cases:

- If $\alpha = 0$, we have the two cations with the same mobile. In the case of planar waveguide, Eq. 14.12 reduces the one-dimensional equation:

$$\frac{\partial N_A}{\partial t} = \frac{\partial}{\partial x}\left(\tilde{D}\frac{\partial N_A}{\partial x}\right) \tag{14.13}$$

where $\tilde{D}$ is the interdiffusion coefficient given by:

$$\tilde{D} = \frac{nD_A}{1-\alpha N_A} = \frac{nD_A D_B}{D_A N_A + D_B N_B}. \tag{14.14}$$

- For $N_A << 1$ as in the case of single-mode Ag^+-Na^+ waveguides, $D = nD_A$. In this case, the diffusion equation has the form:

$$\frac{\partial N_A}{\partial t} = nD_A \frac{\partial^2 N_A}{\partial x^2} \tag{14.15}$$

The solution of Eq. 14.15 is [24]:

$$N_A(x,t) = N_0 erfc\left(\frac{x}{W_0}\right) \tag{14.16}$$

where $W_0 = 2\sqrt{nD_A t}$ is the effective depth of diffusion and N_0 is the mole fraction the Ag^+ ion in the surface; see Fig. 14.1 (planar waveguide).

14.1.4 Diffusion with External Electric Field

Diffusion of an incoming ion can be enhanced by applying an electric field across the substrate. In this case, one has to solve Eq. 14.10, which is a difficult task at best in the two-dimensional case. However, in the case of one dimension (planar waveguides), approximate solutions can be obtained in certain conditions. For example, in the case of low concentrations ($N_A << 1$), space-charge effects can be neglected, and for planar waveguides, Eq. 14.10 reduces to:

$$\frac{\partial N_A}{\partial t} = nD_A \frac{\partial^2 N_A}{\partial x^2} - \mu_A E_{ext} \frac{\partial N_A}{\partial x}, \tag{14.17}$$

where $N_A(x,t)$ is calculated by Laplace-transform technique:

$$N_A = \frac{1}{2} N_0 \{erfc(x' - r) + \exp(4rx')erfc(x' + r)\}. \tag{14.18}$$

Here, $x' = x/W_0 = x/2\sqrt{nD_A t}$ is the normalized effective depth of diffusion without an external field and $r = \mu Et/W_0$.

It has been shown that the influence of the space charge leads to steeper diffusion fronts when compared to the solution of Eq. 14.18 [25]. For large values of r, the contribution of second term in Eq. 14.18 is negligible, and the diffusion profile can be approximated as:

$$N_A(x', r) \cong \frac{N_0}{2} \{erfc(x' - r)\} \quad r>2.5. \tag{14.19}$$

In the two-dimensional case, Eq. 14.10 has been solved, neglecting the concentration dependence of the interdiffusion coefficient and assuming that the diffusion

Table 14.1 Some results of ion exchange to obtained waveguides from different ions

Ion	Host matrix	Melt salts	Temperature (°C)	Δn	Ref.
Tl^+	Borosilicate	$TlNO_3 + KNO_3 + NaNO_3$	530	0.001	1
K^+	Soda lime	KNO_3	365	0.008	2
Ag^+	Aluminosilicate	$AgNO_3$	225–270	0.13	2
Rb^+	Soda lime	$RbNO_3$	520	0.015	3
Cs^+	Soda lime	$CsNO_3$	520	0.03	3
Li^+	Soda lime	$Li_2SO_4 + K_2SO_4$	520–620	0.012	31
Ag^+	Tellurite	$AgNO_3 + KNO_3 + NaNO_3$	280	0.13	35

term is small [25]. This solution is valid in many cases, nevertheless, in determined conditions they fail to describe the diffusion of a determined ion into the glass. In such a situation, Eq. 14.10 can be solved by numerical methods.

Abou-el-Lei and Cooper [13] have analyzed the problem of electric-field-induced ion exchange in detail and compared their results with the data from the viewpoint of strengthening of glass. Not much consideration was given to conditions characteristic of waveguide formation.

14.1.5 Boundary Conditions

Boundary conditions depend on the substance that is in contact with the glass. Three kinds of substances are usually studied: molten salts, metallic films as cation source, and masks. The standard theory of electrolytes [26] is applied to obtain the boundary conditions for both C_A and V, in such a way that the usual boundary conditions of results in a potential difference that arises between the glass and the salt or the metallic film, that is, a contact potential exists when cations can cross the boundary.

For example, the solution of Eq. 14.11 is shown in Eq. 14.26, which is obtained from of the initial and boundary conditions:

$$\begin{aligned} &N_A(x,0) = 0 \qquad \text{for } x>0 \\ &\left.\begin{aligned} N_A(\infty,t) &= 0 \\ N_A(0,t) &= N_0 \end{aligned}\right\} \text{for all } t \geq 0 \end{aligned} \tag{14.20}$$

14.2 Ion Exchange Process

Since Izawa and Nakagome [1] reported the first ion exchange waveguide made by $T1^+$-Na^+ exchange from a mixture of molten nitrate salts, molten bath has almost invariably been used as the source of exchanging ions for fabrication of glass waveguides. Several cation pairs have been studied. In most cases [1, 2, 27, 28], sodium ions in the substrate glass are exchanged for one of the cations, namely, Cs^+, Rb^+, Li^+, K^+, Ag^+, or Tl^+ (Table 14.1). Also, Cs^+-K^+ exchange has been reported for waveguide fabrication [10]. Table 14.1 shows the first published reports for each cation pair along with the ion exchange conditions, the salt bath, the substrate glass, and the accompanying surface-index change (Δn). It is necessary to say that of the various monovalent ions, each one has its advantages and drawbacks.

Several methods have been developed to produce optical waveguides in glasses. Among these methods, ion exchange [28–50] is the most known and explored technique to obtain waveguide candidates for IO.

14.2.1 Exchange from Molten Bath

In order to create a waveguide in glass systems, it is necessary to create an area where the refraction index is higher than that of the adjacent areas. In the process of ion exchange from of bath salt molten, variations can be obtained in the refractive index, absorption, luminescence, and others. In this sense, the most often used salts are nitrates of sodium, potassium, and silver or mixtures of them [21, 25]. For IO applications, it is desirable to fabricate waveguides with low propagation loss and desired refractive index profile. Still, both parameters depend upon the substrate glass. Resulting in a slow progress in the ion exchange technology for the fabrication of optical waveguides due to the dependence above mentioned.

Furthermore, molten salts (nitrates) of such cations are common sources of dopants, although a metallic film deposited on the glass surface can be also a source of cations. The exchange happens by thermal movement of also cations or it can be assisted by an electric field. The refractive index of the glass increases proportionally to the dopant cation concentration, and thus, slab optical waveguides can be obtained after a time determined of ion exchange. By selective masking of the glass surface (aluminum or others), ion exchange can be locally prevented or allowed to fabricate IO elements such as channel waveguides, splitters, multiplexers, and others optical planar devices. It must be stressed that even amplifiers, lasers, and frequency doubling have already been demonstrated in rare-earth-doped glasses.

On the other hand, large diffusion times are required for suitable waveguide to occur, limiting its use for single-mode guides. The resulting waveguides are birefringent [51] although they exhibit negligible depolarization [30] and small propagation losses. Li^+ is very mobile in glasses, and highly multimode guides with index profiles compatible with optical fibers can be produced in a few minutes [31]. The maximum index change of the order of 0.015 is achievable, but so far, no low-loss guides have been demonstrated.

Moreover, because of large mismatch in the ionic radius of the two ions (Table 14.1), large stress is introduced in the glass network, causing surface damage. In the case of Cs^+, an index change of the order of 0.04 has been achieved [25] in a special potassium-rich BGG2 glass with very low losses (<0.2 dB/cm). However, one disadvantage of using Cs^+ is that cesium salts react with most materials and diffusion must be carried out in crucibles made of special materials such as platinum, thereby adding to the cost.

The requirements for an ideal substrate glass have been listed in Ref. [31, 32]. The losses are dominated by absorption, primarily due to presence of foreign impurities, and scattering contributions caused by glass inhomogeneity and surface or geometric imperfections. By careful choice of the host glass, waveguide losses below 0.1 dB/cm can be obtained [27, 33, 34]. Scattering losses are generally introduced in the processing steps.

Dopants introduce additional Rayleigh scattering to the background losses caused by the multicomponent nature of the glass. Reduction of cations to metallic form, such as in the case of silver, as pointed out earlier, can give rise to additional absorption and scattering losses.

Hence, the refractive index profile depends on various process parameters in an intricate manner. The composition of the host glass,[1] the nature of the incoming ion and its concentration in the source, the temperature, the diffusion time [8], and the magnitude of the externally applied field all affect the index profile. Since the waveguide characteristics are determined by the index profile, it is necessary to establish a correlation between the process parameters and the index distribution before the process can be tailored to produce waveguides of desired characteristics through a careful choice and control of the process parameters.

The reproducibility of the fabrication process should be verified for a given diffusion time from the bath salt molten.

Due to the presence of sodium in the very glass composition, ion exchange at Na^+-Ag^+ was chosen [28, 35]. For example, the bath composition can be a mixture of $XAgNO_3$-$YNaNO_3$-$ZKNO_3$ (weight %, $X + Y + Z = 100$), which was performed at near (up-down transition temperature, T_g), in a time period ranging from a few minutes to hours in a special ion exchange equipment representative (see Fig. 14.2). The salt was contained in a crucible held in a vertical furnace, in which the temperature was controlled.

14.2.2 Equilibrium at the Melt-Substrate Interface

The simultaneous diffusion of two cation (or more) species (A and B) with different diffusion coefficients (D_A and D_B) in a melt-substrate interface occurring in order to maintain charge neutrality. This process can be represented by Eq. 14.1 or 14.2. It is supported by the fact that a very small charge disequilibrium is enough to generate a strong electric field which dominates the drift-diffusion equations; this field moves cations in such a way that the charge neutrality tends to be restored [11].

[1] The multicomponent optical glasses used for ion exchange are usually silicate glasses composed of SiO_2 and various other oxides. The oxides of monovalent cations such as Na_2O, K_2O, Li_2O, etc., are called the network modifiers, and it is believed that the basic structure of the glass does not change as a result of the binary ion exchange where one of the network modifiers is exchanged with an ion of higher polarizability.

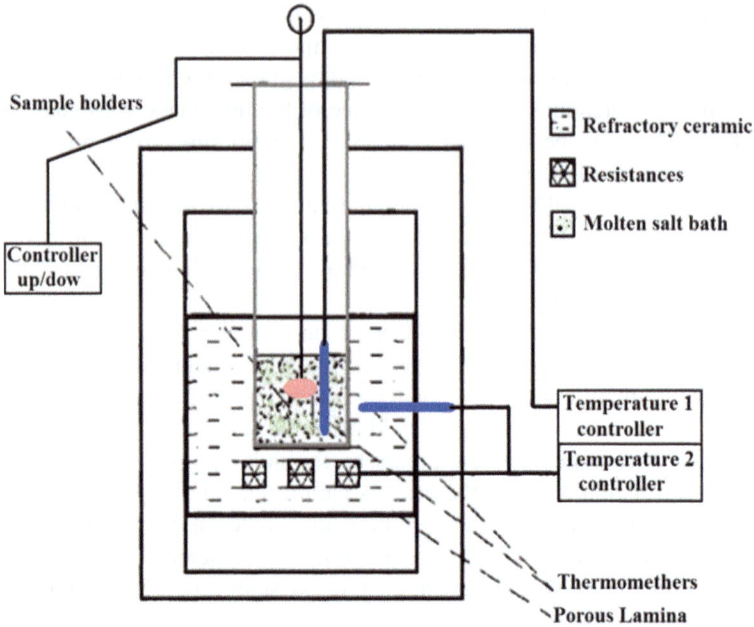

Fig. 14.2 Schematics of the experimental arrangement for the diffusion

This equilibrium may be limited by the following processes:

- Mass transfer of reactants
- Kinetics of the reaction at the interface
- Transport of cation in the glass phase

Transfer of cations in the melt takes place through of the diffusion and convection processes. Convection can be enhanced by stirring the melt. For diffusion to be a limiting process in the melt, the important parameter is $\sqrt{(N_B/N_A)(D_B/D_A)}$, where $N_A(N_B)$ denotes the concentrations in mole fractions of cation A in the melt (glass) and D_A the self-diffusion coefficient in the respective medium.

In addition, the ion exchange process consists basically the net index change Δn that depends on the extent of the ion exchange dictated by the equilibrium at the melt-glass interface and on the differences in the radii and polarizabilities of the two cation species involved (particularly in the binary exchange). While the contribution of the ionic polarizability is easy to understand, the role of the mismatch of the ionic size in the determination of the index change is not well understood.

The equilibrium state is governed by an equilibrium constant K defined as:

$$K = \frac{\overline{a}_A a_B}{\overline{a}_B a_A}, \tag{14.21}$$

where a's represent thermodynamic activities of the cations in the respective phase.

14.2.2.1 One Species of Cations

This occurs when the substrate containing a monovalent cation A into a molten salt solution containing the diffusing monovalent cation B. The ion B is driven into the glass by an interphase chemical potential gradient.

These cations move from salt to glass and vice versa to achieve the equilibrium, which happens when the cation electrochemical potentials are equal in both phases and they do not depend on the position. As cations are charged particles, each electrochemical potential can be split into an electrical and a chemical part:

$$e(V_s - V) = \bar{\mu}_A^0 C_A - \bar{\mu}_{As}^0 C_{As}. \tag{14.22}$$

The subindex "s" means the salt; no subindex means the glass. Note that $\bar{\mu}_A^0 C_A$, $\bar{\mu}_{As}^0 C_{As}$, V, and V_s depend on thc position. Let us suppose that the salt has less energetic states which results in a lower value of its chemical potential in bulk region; thus, if both phases are put in contact, glass cations tend to cross the interface. As cations cross, a negative charge builds up in the glass and a positive one also appears in the salt. The resultant electrostatic force keeps these charges confined in two thin surface layers. These charges cause a potential difference between the glass and the molten salt. Obviously, the cation migration stops when this difference balances the original energetic stop [36].

Likewise, the bulk value of the salt potential (V_s) is used instead of its value at the surface ($V_{s/F}$; the subindex "$/F$" represents the charged layers) because the conductivity of the molten salt is much higher than that of the glass; hence, the potential differences into the salt can be neglected. Furthermore, the charge density near the interface is not neglected, but the thickness of the charged region is. Finally, the boundary condition of ϕ is equal to that of V because $C_A = 0$:

$$\phi|_F = V_s + \frac{\bar{\mu}_{As}^0 - \bar{\mu}_A^0}{e}. \tag{14.23}$$

14.2.2.2 Two Species of Cations

When two species are present in the process of ion exchange, the electrochemical potentials also depend on the mole fractions of cations; therefore:

$$\begin{aligned} \mu_{As} &= \mu_{A|F}(A_s \leftrightarrow A) \\ \mu_{Bs} &= \mu_{B|F}(B_s \leftrightarrow B). \end{aligned} \tag{14.24}$$

We can assume the salt as homogeneous because its cation mobility is much higher than the one in glass; in addition, the salt convection continuously renews the salt at the boundary. So, we can obtain the following relation:

$$\bar{\mu}_{As} + \bar{\mu}_{B|F} = \bar{\mu}_{A|F} + \bar{\mu}_{Bs}(A_s + B \leftrightarrow A + B_s). \tag{14.25}$$

Nevertheless, as neither salt nor glass behaves ideally, we have [11]:

$$\left.\begin{aligned}\bar{\mu}_i &= \bar{\mu}_i{}^0 + k_B T \ln a_i \\ \bar{\mu}_{is} &= \bar{\mu}_{is}{}^0 + k_B T \ln a_{is}\end{aligned}\right\} i = A, B$$
$$\ln \frac{a_{As} a_B|_F}{a_{Bs} a_A|_F} = \frac{\bar{\mu}_A{}^0 - \bar{\mu}_{As}{}^0 - (\bar{\mu}_B{}^0 - \bar{\mu}_{Bs}{}^0)}{k_B T}. \tag{14.26}$$

Consequently, to relate mole fractions at the glass boundary ($C_{A|F}$ and $C_{B|F}$) with those of salt ($C_{A|s}$ and $C_{B|s}$), can be written [37]:

$$\frac{a_B}{a_A} = \left(\frac{C_B}{C_A}\right)^n, \tag{14.27}$$

where n is a constant to the ratio of the thermodynamic activities in glass.

Once mole fractions at the glass boundary are known, the potential differential is fixed by Eq. 14.21:

$$V_{|F} = V_s + \frac{\overline{\mu}^0_{As} - \overline{\mu}^0_A}{e} + \frac{k_B T}{e} \ln \frac{a_{As}}{a_A|_F} = V_s + \frac{\overline{\mu}^0_{Bs} - \overline{\mu}^0_B}{e} + \frac{k_B T}{e} \ln \frac{a_{Bs}}{a_B|_F} \tag{14.28}$$

and thus, the following Dirichlet condition for ϕ is obtained:

$$\begin{aligned}\phi_{|F} &= V_s + \frac{\overline{\mu}^0_{As} - \overline{\mu}^0_A}{e} + \frac{k_B T}{e} \ln \left[a_{As} \left(\frac{a_B|_F}{a_A|_F} + \frac{D_A C_A}{D_B C_B} \right) \right] \\ &= V_s + \frac{\overline{\mu}^0_{Bs} - \overline{\mu}^0_B}{e} + \frac{k_B T}{e} \ln \left[a_{Bs} \left(1 + \frac{D_A C_A}{D_B C_B} \right) \right].\end{aligned} \tag{14.29}$$

Note that $C_{A|F}$ is the same in all points of the glass surface in contact with the salt. Therefore, both $V_{|F}$ and $\phi_{|F}$ are also constant along the frontier between salt and glass but different from V_s.

14.2.2.3 Experimental Cases: Ag^+-Na^+, K^+-Na^+, and Cs^+-K^+

Most of the progresses in the ion exchange process for obtained (single- or multimode) waveguides at optical telecommunication are based in three basic cation pair systems. The main results are summarized as follows:

Ag^+-Na^+ Exchange: This is the most studied system. There are several advantages of choosing this cation pair: (1) Index change can be varied diluting the $AgNO_3$ melt with $NaNO_3$. No birefringence or depolarization is observed. Losses are also kept low. (2) Diffusion temperatures are the lowest, and diffusion times are also reasonable as silver has high mobility. With the control of diffusion time, it is possible to obtain single- or multimode guides. (3) Buried waveguides are easy to

fabricate by thermal diffusion as Ag^+ near the glass surface diffuses out in the second step ion exchange. (4) While the attempts to optimize the single-mode guides for low losses and fiber compatibility have given favorable results, the progress on the device end using Ag^+-Na^+ exchange is slow. This can be attributed to the lack enthusiasm, owing to the general belief that silver ions embedded in a glass matrix can exhibit optical losses and, therefore are prone to deterioration of their optical transmission. While this is true if pure $AgNO_3$ melts are used, single-mode structures with low silver contents are not likely to suffer from these drawbacks.

K^+-Na^+ Exchange: This system has by far been the most used for single-mode device fabrications. This preference may be attributed to two principal factors: (1) the index change with pure KNO_3 melt is highly compatible with single-mode fibers and (2) unlike Ag^+-Na^+ of the exchange process, no concentration control of the melt is required, thus simplifying the fabrication procedure. Yip et al. [38, 51] have reported an extensive study of K^+-Na^+ planar guides in soda-lime glass.

Cs^+-K^+ Exchange: This cation pair system has recently been adopted by several workers. The composition of the glass [32] is such that it allows optimization of the process parameters for single-mode waveguide fabrication. The process diffusion is performed from a molten mixture of $CsNO_3$ and CsCl.

Recent studies in ion exchange process have permitted a better understanding of the role of the processing conditions and the matrix glass in influencing the index profile of planar and buried single- and multimode waveguides; this allows fabrication of low-loss, fiber-compatible, reproducible. Buried channel waveguides have characteristics excellent for applications in photonics.

In the ion exchange process, it is possible to have the space-charge effect. This is created by the unequal ionic mobilities ($\mu_A/\mu_B \neq 1$) since the substrate glass is rich in cation *B*, $\mu_A < \mu_B$ when glass is immersed in the melt. Cation *B* migrates faster than *A*, causing space-charge region. However, as ion exchange proceeds, the depth of the cation *A*–rich region increases, and in this region, $\mu_A > \mu_B$ provided sufficient cations of species *B* have been replaced by *A*. As a result of this interplay, the ionic conductivity of the glass varies across the substrate, being the lowest in the space-charge region. This variation is time dependent, and the ionic current is not constant [39]. The extent to which space-charge effects modify the index profile depends upon the range of variation of μ_A and μ_B in the two extreme concentrations. In the case of Ag^+-Na^+ exchange, the effects are not severe, and the ionic current is time independent during the process. However, in the cases such as K^+-Na^+, it may be an important factor and therefore must be considered in the analysis.

Dielectric Mask

A mask is deposited on the glass in order that the cation crosses the glass only in the region desired, still if the sample is totally immersed in molten salt. If the mask is a dielectric like Al_2O_3 [23], SiO_2, Si_3N_4, polyimide [40], or photoresist [41], no charge transference is possible between cations and mask; thus:

$$\begin{aligned} \vec{J}_{A|F} \cdot \hat{e}_F &= 0 \\ \vec{J}_{B|F} \cdot \hat{e}_F &= 0 \end{aligned} . \quad (14.30)$$

It leads to the Newman conditions for C_A, V, and ϕ; thus:

$$\begin{aligned} \vec{\nabla} C_{A|F} \cdot \hat{e}_F &= 0 \\ \vec{\nabla} V_{|F} \cdot \hat{e}_F &= 0 \\ \vec{\nabla} \phi_{|F} \cdot \hat{e}_F &= 0 \end{aligned} . \quad (14.31)$$

Metallic Mask

Aluminum or titanium films are widely used as metallic masks [42]. Chromium masks have been also proposed for dry field-assisted ion exchange [43]. Their boundary conditions have been also modeled by Eq. 14.31, in spite of several known failures of the model [23, 44].

14.2.3 Type Waveguide: Formation Process

A waveguide is a structure which guides electromagnetic waves. These waveguides differ in their geometry which can confine energy in one dimension such as in slab waveguides or two dimensions as in fiber, planar, channel, or buried waveguides.[2]

A propagation mode in a waveguide is one solution of the wave equations or, in other words, the form of the wave. Due to the constraints of the boundary conditions, there are only limited frequencies and forms for the wave function which can propagate in the waveguide. The mode with the lowest cutoff frequency is the basic mode of the waveguide, and its cutoff frequency is the waveguide cutoff frequency.

Waveguides used at optical frequencies are typically dielectric waveguides (glass generally), structures in which a dielectric material with high permittivity, and thus high index of refraction, is surrounded by a material with lower permittivity.

[2] The wave propagation along the waveguide axis is described by the wave equation, where the wavelength depends on the waveguide structure as well as on the frequency. Along the width of the waveguide, the wave is confined in a standing wave pattern. The equation that describes the transverse wave form is more complicated and is derived in the case of electromagnetic waves from Maxwell's, along with boundary conditions that depend on the shape of the waveguide and the materials from which it is made. These equations have multiple solutions, called propagation modes (TE, transversal electric, and TM, transversal magnetic).

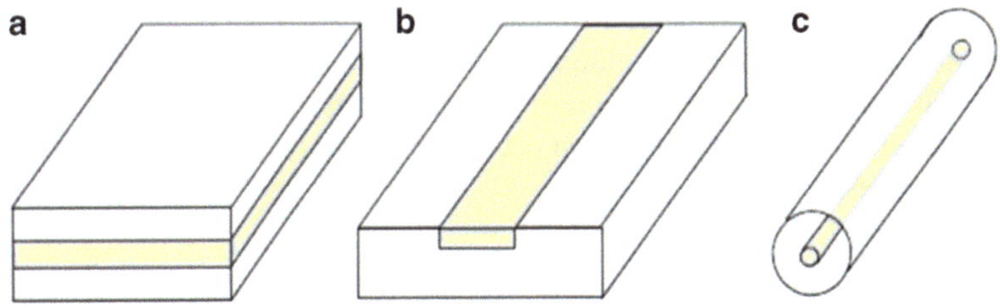

Fig. 14.3 Basic waveguide geometries: (**a**) planar, (**b**) channel, and (**c**) optical fiber

Practical rectangular-geometry optical waveguides are most easily understood as variants of the simple dielectric slab waveguide, also called planar waveguide (see Figs. 14.1 and 14.3). The slab waveguide consists of three layers of materials with different dielectric constants, extending infinitely in the directions parallel to their interfaces.

Light may be confined in the middle layer by total internal reflection. This occurs only if the dielectric index of the middle layer is larger than that of the surrounding layers. In practice, slab waveguides are not infinite in the direction parallel to the interface, but if the typical size of the interfaces is much larger than the depth of the layer, the slab waveguide model will be an excellent approximation.

Another type is the strip waveguide, which is basically a strip of the guiding layer confined between cladding layers. The simplest case is a rectangular waveguide, which is formed when the guiding layer of the slab waveguide is restricted in both transverse directions rather than just one. Rectangular waveguides are used in IO circuits, and in laser diodes. They are commonly used as the basis of such optical components as Mach-Zehnder interferometers and wavelength division multiplexers. The cavities of laser diodes are frequently constructed as rectangular optical waveguides. Optical waveguides with rectangular geometry are produced by a variety of means, usually by a planar process. In a channel waveguide, the light propagates within a rectangular channel (the dashed region in Fig. 14.3) which is embedded in a planar substrate. To confine light within the channel, it is necessary for the channel to have a refractive index greater than that of the substrate and, of course, greater than the refractive index of the upper medium, which is usually air or the same for buried waveguides. Because the substrate is planar, the technology associated with IO circuits is also called *planar lightwave circuits* (PLC).

Figure 14.3 shows the geometry of an optical fiber, which can be considered as a cylindrical channel waveguide.

The technology and fabrication methods associated with IO circuits and components are very varied; in addition, they depend on the substrate material with which the optical device is fabricated. Optical integration can expand in two directions: serial integration and parallel integration. In serial integration for optical communication devices, the different elements of the optical chip are consecutively interconnected: laser and driver, modulator and driver electronics, and detector and receiver electronics. In parallel integration, the chip is built by bars of amplifiers, bars of detectors, and wavelength (de)multiplexors. Also, a combination of these two architectures should incorporate optical cross-connects and add-drop modules.

The glass-based IO devices have the great advantage of the low cost of the starting material and the fabrication technique, mainly performed by an ionic exchange process.

14.3 Characterization

The important characterization of passive waveguides includes measurement of index (dopant) profile, propagation constants of the guided modes, attenuation, and mode field pattern. The results may be a valuable supplement to numerical modeling verification.

In addition, two models attributed to Huggins and Sun (HS) and Append have been successfully used to evaluate the index change due to ionic substitution in multicomponent glasses and are described below [45].

Ion exchange process is only possible when the process temperature is above T_g. In this sense, the thermal properties were measured by differential thermal analysis (DTA) technique. Powdered samples were characterized by DTA using α-alumina crucibles as reference and sample holder in the 50–1,100°C range, heating rates of 10°C/min, sample mass of orders mg, and N_2 flux of 50 ml/min, usually experimental conditions. This analysis determines the features temperatures (T_g, onset crystallization temperature on heating, T_x, peak crystallization temperature on heating, T_c, and melting point, T_m). These characteristic temperatures refer to the first crystallization and melting peaks on heating. The thermal stability against crystallization parameters of the glasses is calculated from equations proposed by Nascimento et al. [46]. Another similar technique is differential scanning calorimetry (DSC).

The purpose of this section is to show the most important methods used to characterize waveguides obtained through of the ion exchange process and the consequence this process into the glass matrix.

14.3.1 Index Change by Ion Substitution

The effective indices of the glass samples and planar waveguides are measured at transverse electric (TE) and magnetic (TM) polarizations with an m-line apparatus (Model 2010, Metricon) based in the prism coupling technique with 0.0005 resolution [28]. Through this equipment, it is possible to determine the guided modes and effective refractive index of the sample. In this way, the effective refractive index of the guided mode increases with enhancing exchange time [8], and the temperature will play a more important role in the ion diffusion.

In addition, the index change after ion exchange process is due to polarizability, volume, or stress-induced change into the system glass, which are associated with the substitution of the A ion for a B ion.

14.3.1.1 Polarizability and Volume Change

The refractive index of an isotropic medium such as glass can be expressed by the Clausius-Mossotti or Lorentz-Lorenz equation [45]. For a single-component oxide glass, such as SiO_2 (fused silica), the above equation is $\frac{n^2-1}{n^2+2} = \frac{R}{V}$, where $R = 1/(3\varepsilon_0)N\alpha$ is called molar refraction of the glass, α is the molecular polarizability, N is the Avogadro number, and V is the molar volume.

Thus, the ion-exchanged glass never attains the volume predicted by free expansion. If so, the difference between the initial and final volumes is limited due to stress in the glass when subjected to ion exchange process (an A ion for a B ion).

14.3.1.2 Stress-Induced Index Change

Ion exchange process involves exchange of two cations of different ionic radii at temperatures well below the stress relaxation temperature of the glass. As a result, the accompanying volume changes have to be accommodated only in the direction normal to the substrate since the surface is prevented from expanding (or contracting) laterally due to resistance to bending of the glass substrate.

The stress profile generally follows the concentration profiles. Moreover, since the stresses are anisotropic, the index change depends on the state of polarization of the optical electric field, that is, the resulting waveguides are birefringent [47].

Another factor which causes stress in ion-exchanged waveguides is the mismatch between the thermal expansion coefficients of the waveguide and the substrate materials.

However, this effect is very small [47] and can be neglected in comparison to the compositional stresses unless the exchange is carried out above the stress relaxation temperature as in the case of Li^+-Na^+ exchange in aluminum silicate glass [48].

14.3.2 Refractive Index Profile

There are several techniques for measurement of the index profile in waveguides. These include interferometry, reflectivity measurement, and the inverse Wentzel-Kramers-Brillouin method (WKB method) which relies on the mode-index data.

Interferometry may be the most accurate and direct technique [1]. However, the sample preparation is time consuming and laborious, and the technique is destructive. The two-dimensional index profile can also be determined by measuring reflectivity from polished faces with the aid of an optical multichannel detection system [14]. The index distribution is calculated from Fresnel's formula.

Another method which yields the index profile of the waveguide relies on the measurement of mode indices. This can be easily accomplished by the prism coupler technique with accuracies approaching 1×10^{-4} [14]. From the mode-index data, the index profile is derived using the inverse WKB method [49]. This method gives reasonably good results for multimode guides with ERFC[3] and exponential profiles. Moreover, the WKB approximation is known to be erroneous near the mode cutoffs [50].

The method may thus give reliable results for diffusion depths, but the value of surface-index change is always subject to error. In case of surface waveguides (such that prism coupling can be employed for measurement of mode indices) with the large number of guided modes, WKB method has been used extensively for characterization of waveguides [28].

There are basically two ways of calculating the normal modes from the fundamental parameters (refractive indices and thicknesses of the layers):

1. A classical theory based on geometrical optics
2. The direct resolution of Maxwell's equations whose analysis, as discussed below, shows a similar quantum pretty interesting

In the framework of geometrical optics, the confinement of light inside the film must be described by the total reflection at two interfaces (substrate-film and film-superstrate). Starting from this definition, there are two different types of waveguides: single-[4] and multimode guides.

14.3.3 Determination of Diffusion Coefficient $\overline{D}$ and Mobility μ

The diffusion process that occurred by ion exchange process was characterized by diffusion coefficient $\overline{D}$ calculated from the following equation:

$$\overline{D} = d^2/4t, \tag{14.32}$$

where d is a depth of the waveguide (μm) and t stands for time of the ion exchange process.

Besides the boundary conditions obtained by the ion exchange equilibrium reported earlier [4], complete solutions of diffusion Eq. 14.32 also require knowledge of interdiffusion $\overline{D}$ and mobility $\mu_{\mathrm{A-B}}$.

[3] ERFC is the complementary error function, commonly denoted erfc(z), and is an entire function defined by $erfc(z) = \frac{2}{\sqrt{\pi}} \int_z^\infty e^{-t^2} dt$.

[4] A single-mode waveguide is a waveguide designed to carry only a single ray of light (mode). This ray of light often contains a variety of different wavelengths. Although the ray travels parallel to the length of the fiber, it is often called the transverse mode since its electromagnetic vibrations occur perpendicular (transverse) to the length of the fiber.

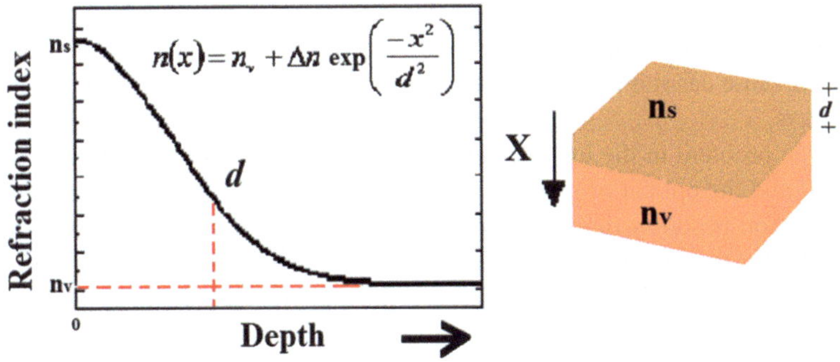

Fig. 14.4 Change of refraction index in depth waveguide function

In addition, field-assisted ion exchange can, in general, be used to decrease the processing time or, as in our case, to permit fabrication of buried channel waveguides. Then, the profile the refraction index can be written as:

$$n(x) = n_v + \Delta n \exp\left(\frac{-x^2}{d^2}\right). \tag{14.33}$$

This equation is very important to determine the characteristic of waveguides make. Then, Δn is the refraction index change, n_v is the glass refraction index, and d is the effective depth of the waveguide. Figure 14.4 shows the behavior described above.

From experimental measurement (e.g., prism coupler), it is possible to determine the parameter d to a time determined and, finally, the diffusion coefficient, Eq. 14.32. Moreover, $\overline{D}$ is temperature function described for Arrhenius:

$$D_e = D_0 \exp\left(\frac{E_D}{RT}\right), \tag{14.34}$$

where D_0 is a pre-exponential factor, E_D is the activation energy, T is the absolute temperature, and R is the gas constant (8.314 J/K mol). From this, we can determine the parameter better for the fabrication of channel or planar waveguide, single- or multimode, etc.

14.4 Conclusion

The correlation between the theory and experiment allows determination or relevant parameters such as the index profile, the electric mobility of A(B) ion in matrix glass, and the interdiffusion coefficient for given melt composition and other process parameters. The low diffusion coefficient is responsible for the longer time needed to fabricate a waveguide at low temperature. Furthermore, complex network in the glass makes the ion exchange process slow.

Consequently, after the diffusion profile was determined, a method was used to obtain the mode-index and the mode-field profiles. Therefore, it may be possible to obtain the same desired waveguide performance by a range of combination of these parameters.

A large problem in the ion exchange process is that the surface of glass will be corroded and the optical property of the fabricated waveguide will be deteriorated. The corrosion becomes more and more severe with increasing exchange time and temperature. One reason possible is the intrinsically weak stabilization of glass structure; weakly acidic property of glass is the main factor leading to the corrosion of glass surface.

Finally, several methods have been developed to produce optical waveguides in glasses, but process ion exchange is the most known and explored that allows integration of optoelectronic devices on the same chip and efficient waveguides to optical fiber coupling.

Acknowledgments The author would like to thank for the financial support by the Brazilian agencies Fapesp, CNPq, and CEPOF/INOF.

References

1. Izawa T, Nakagone H (1972) Optical waveguide formed by electrically induced migration of ions in glass plates. Appl Phys Lett 21:584–586
2. Giallorenzi TG, West EJ, Kirk R, Ginther R, Andreews RA (1973) Optical waveguides formed by thermal migration of ions in glass. Appl Opt 12:1240–1245
3. Neuman V, Parriaux O, Walpita LM (1979) Double alkali effect: influence on index profile of ion-exchanged waveguides. Electron Lett 15:704–706
4. Ramaswamy RV, Srivastava R, Chludzinski P (1988) Influence of Ag^+-Na^+ ion-exchange equilibrium on waveguide index profiles. IEEE J Quatum Elect 24:780–786
5. Jose G, Sorbello G, Taccheo S, Cianci E, Foglieti V, Laporta P (2003) Active waveguide devices by Ag-Na ion exchange on erbium-ytterbium doped phosphate glasses. J Non-Cryst Solids 322:256–261
6. Ramaswany RV, Srisvatava R, Chludzinski P, Anderson TJ (1988) Influence of Ag^+-Na^+ ion-exchange on waveguide index profile. IEEE J Quantum Elect 24:780–786
7. Garfinkel HM (1968) Ion-exchange equilibriums between glass and molten salts. J Phys Chem 72:4175–4181
8. Gato L, Srivastava R (1996) Time-dependent surface-index change in ion-exchanged waveguides. Opt Comm 123:483–486
9. Francini R, Giovenal R, Grassano UM, Laporta P, Taccheo S (2000) Spectroscopy of Er of Er-Yb-doped phosphate glasses. Opt Mater 13:417–425
10. Quaranta A, Gonella F (1995) On the role of local electric field correlation effects on the ionic interdiffusion in soda-lime glass. J Non-Cryst Solids 192:334–337
11. Prieto X, Srivastava R, Linares J, Montero C (1996) Prediction of space-charge density and space-charge field in thermally ion-exchanged planar surface waveguide. Opt Mater 5:145–151
12. Helferich F, Plesset MS (1958) Ion exchange kinetics. A nonlinear problem. J Chem Phys 28:418–424
13. Abou-el-Leil M, Cooper AR (1979) Analysis of field-assisted binary ion exchange. J Am Ceram Soc 62:390–395

14. Lilienhof HJ, Voges E, Ritter D, Pantschew B (1982) Field-induced index profiles of multimode ion-exchanged strip waveguides. IEEE J Quantum Elect 18:1877–1883
15. Albert J, Lit JWY (1990) Full modeling of field-assisted ion exchange for graded index buried channel optical waveguides. Appl Opt 29:2798–2804
16. Tervonen A (1990) A general model for fabrication processes of channel waveguides by ion exchange. J Appl Phys 67:2746–2752
17. Cantor AJ, Abou-el-leil M, Hobbs RH (1991) Theory of 2-D ion exchange in glass: optimization of microlens arrays. Appl Opt 30:2704–2713
18. Cheng D, Saarinen J, Saarikoski H, Tervonen A (1997) Simulation of field-assisted ion exchange for glass channel waveguide fabrication: effect of nonhomogeneous time-dependent electric conductivity. Opt Commun 137:233–238
19. Lupascu A, Kevorkian A, Bondet T, Saint-André F, Persegol D, Levy M (1996) Modeling ion exchange in glass with concentration-depend diffusion coefficients and mobilities. Opt Eng 35:1603–1610
20. Teray R, Hayami R (1975) Ionic diffusion in glasses. J Non-Cryst Solids 18:217–264
21. Wakabayashi H (1996) The relationship between kinetic and thermodynamic properties in mixed alkali glass. J Non-Cryst Solids 203:274–279
22. Day DE (1976) Mixed alkali glasses – their properties and uses. J Non-Cryst Solids 21:343–372
23. Walker RG, Wilkinson CDW, Wilkinson JAH (1983) Integrated optical waveguiding structures made by silver ion-exchange in glass. 1: the propagation characteristic of stripe ion-exchanged waveguides; a theoretical and experimental investigation. Appl Opt 22:1923–1928
24. Crank J (1956) Mathematics of diffusion. Oxford University Press, London
25. Inman JM, Bentley JL, Houde-Walter SN (1995) Modeling ion-exchanged glass photonics: the modified quasi-chemical diffusion coefficient. J Non-Cryst Solids 191:209–215
26. Guggenheim EA (1967) Thermodynamics, 5th edn. North-Holland, Amsterdam
27. Ross L, Lilienhof HJ, Holscher H, Schlaak HF, Bran Brandenburg A (1986) Improved substrate glass for planar waveguide by Cs-ion exchange. Technical Digest, Topical Meeting on Integrated and Guided-Wave Optics, Atlanta, ThBB2, pp 25–26
28. Rivera VAG, Chillcce EF, Rodriguez E, Cesar CL, Barbosa LC (2006) Planar waveguides by ion Exchange in Er^{3+}-doped tellurite glass. J Non-Cryst Solids 352:363–367
29. Stewart G, Millar CA, Laybourn PJR, Wilkinson CDW, De La Rue RM (1997) Planar optical waveguide formed by silver migration in glass. IEEE J Quantum Elect 13:192
30. Jackel JL, Lee KY, Favire FJ (1985) Do glass waveguide depolarize? J Lightwave Technol 3:818 820
31. Chartier GH, Jaussaud P, Oliveira AD, Pamaux O (1977) Fast fabrication method for thick and highly multimode optical waveguides. Electron Lett 13:763–764
32. Ross L, Fabricius N, Oeste H (1987) Single-mode integrated optical waveguides by ion-exchange in glass. Eur Fiber Opt Conf LAN 87:99–102
33. Eguchi RG, Maunders EA, Naik IK (1983) Fabrication of low loss waveguides in BK7 by ion exchange. Proc Soc Photo-Opt Instrum Eng 408:21–26
34. Lagu RK, Ramaswamy V (1984) Fabrication of single mode glass waveguides by electrolytic release of silver ions. Appl Phys Lett 45:117–118
35. Rivera VAG, Chillcce EF, Rodriguez E, Cesar CL, Barbosa LC (2006) Waveguides produced by ion-exchange in Er^{3+}-doped tellurite glass. Proc SPIE – Int Soc Opt Eng 6116:611610–611613
36. Prieto-Blanco X (2008) Electro-diffusion equations of monovalent cations in glass under charge neutrality approximation for optical waveguide fabrication. Opt Mater 31:418–428
37. Rothmund V, Kornfeld G, Anorg Z (1918) Der basenaustausch im permutet. Z Anorg Allg Chem 103:129
38. Albert J, Yip GL (1985) Refractive index profiles of planar waveguides made by ion-exchange in glass. Appl Opt 24:3692–3963

39. Najafi SI, Suchoski PG Jr, Ramaswamy RV (1986) Silver film-diffused glass waveguides diffusion process and optical properties. IEEE J Quantum Elect 22:2213–2218
40. Pantchev B, Danesh P (1997) Masking problem in the fabrication of optical waveguide structures in glass by double ion exchange. Jpn J Appl Phys 36:4320–4322
41. Poyhonen P, Honkanen S, Tervonen A, Tahkokorpi M, Albert J (1991) Planar 1/8 splitter in glass by photoresist masked silver film ion exchange. Electron Lett 27:1319–1320
42. Pantchev B, Danesh P, Nikolov Z (1993) Field-assisted ion exchange in glass: the effect of masking films. Appl Phys Lett 62(11):1212–1214
43. Honkanen S, Tervonen A, Von Bagh H, Salin A, Leppihalme M (1987) Fabrication of ion-exchanged channel waveguides directly into integrated circuit mask plates. Appl Phys Lett 51:296–298
44. Tervonen A, Honkanen S (1996) Feasibility of potassium-exchanged waveguides in BK7 glass for telecommunication devices. Appl Opt 35:6435–6437
45. Miliou AN, Srivastava R, Ramaswamy RV (1991) Modeling of the index change in K^+–Na^+ ion-exchanged glass. Appl Opt 30:674–681
46. Nascimento MLF, Souza LA, Ferreira EB, Zanotto ED (2005) Can glass stability parameters infer glass forming ability? J Non-Cryst Solids 351:3296–3308
47. Brandenburg A (1986) Stress in ion-exchanged glass waveguides. J Light Technol 4:1580–1593
48. French WG, Pearson AD (1970) Refractive index changes produced in glass by exchange. Am Ceram Soc Bull 49:974–977
49. White JM, Heidrich PF (1976) Optical waveguide refractive index profiles determined from measurement of mode indices: a simples analysis. Appl Opt 15:151–155
50. Srivastava R, Kao CK, Ramaswamy RV (1987) WKB analysis of planar surface waveguides with truncated index profiles. J Light Technol 5:1605–1608
51. Yip GL, Albert J (1985) Characterization of planar optical waveguides by K^+-ion exchange in glass. Opt Lett 10:151–153

Chapter 15
Network Simulation of the Electrical Response of Ion Exchange Membrane Systems

A.A. Moya

Abstract The steady-state, transient and small-signal ac responses of ion exchange membrane systems are studied by using the network simulation method. A network model for the Nernst-Planck and Poisson equations is used to describe the ionic transport processes through a cation-exchange membrane and the two diffusion boundary layers on both sides of the membrane. With this model and the electric circuit simulation programme PSpice, the steady-state, chronoamperometric, chronopotentiometric and small-signal ac responses are simulated. In this work, we analyse the influence of the fixed-charge concentration inside the membrane on (1) the steady-state current-voltage characteristic, (2) the ionic fluxes ratio describing the permselectivity of the membrane, (3) the chronoamperometric response, (4) the chronopotentiometric response and (5) the electrochemical impedance. Some of the results obtained for highly charged membranes can be compared with the analytical solutions in ideal membranes.

15.1 Introduction

Theoretical studies on ionic transport through ion exchange membranes have received considerable attention in the last years because of the interesting technological applications in fuel cells, batteries and electrodialysis for seawater desalination [1]. It is now well known that the ionic transport properties in ion exchange membrane systems can be obtained by analysing the response of the system to controlled electric potential or current perturbations. Chronopotentiometry and electrochemical impedance spectroscopy are examples of experimental techniques commonly used to characterizing the electrical properties of ion exchange membrane systems [2].

A.A. Moya
Departamento de Física, Universidad de Jaén, Edificio A-3,
Campus Las Lagunillas, Jaén 23071, Spain
e-mail: aamoya@ujaen.es

Inamuddin and M. Luqman (eds.), *Ion Exchange Technology I: Theory and Materials*, DOI 10.1007/978-94-007-1700-8_15,

The characterization of electrochemical systems by electrical response measurements requires the development of mathematical models for which the response can be determined theoretically and which serve as a basis for the interpretation of experimental data. For the case of ion exchange membranes, the theoretical treatment most commonly employed is based on a model in which the ionic transport through the system is governed by the Nernst-Planck and Poisson equations [3–5]. However, the nature of these equations is such that an exact analytical solution for the electrical response of ion exchange membrane system is almost impossible to obtain in most of the interesting physical situations.

Although the main electrical properties of ion exchange membrane systems can be analytically obtained using certain approximations, such as the consideration of ideal membranes, the electrical neutrality condition in the membrane and in the diffusion boundary layers or the Donnan equilibrium relations at the membrane-solution interfaces, the general and rigorous treatment of the problem requires the use of numerical methods. Excellent procedures have been developed to solve the Nernst-Planck and Poisson equations in ion exchange membrane systems [6–11]. However, they found serious difficulties in adapting to the different experimental conditions.

An alternative method to the classical numerical method is the network simulation method [12]. This simulation method basically consists in modelling a physicochemical process by means of a graphical representation analogous to circuit electrical diagrams which is analysed by means of an electric circuit simulation programme. Highly developed, commercially available software for circuit analysis can thus be employed to obtain the dynamic behaviour of the whole membrane system without having to deal with the solution of the governing differential equations. In this work, it is shown that the network simulation method can be used satisfactorily in the study of the electrical response of ion exchange membrane systems. Some of the reasons for this have been given in our previous papers [12–15], and they can be summarized as follows: (1) the network simulation method permits to impose any condition on the electric potential and the electric current; (2) steady-state, transient and small-signal ac responses can be dealt with by means of this approach and (3) it permits the study of any transport process through spatial regions where large gradients in the ionic concentrations and the electric potential occur. Moreover, it must be noted that the network model of the system is an appropriate electric circuit to model the diffusion-migration impedance of ion exchange membrane systems [16].

In this chapter, a network model for the Nernst-Planck and Poisson equations is used to describe the ionic transport processes through a cation-exchange membrane and the two diffusion boundary layers on both sides of the membrane. With this model and the electric circuit simulation programme PSpice, the steady-state, chronoamperometric, chronopotentiometric and small-signal ac responses are simulated. We analyse the influence of the membrane fixed-charge concentration on (1) the steady-state current-voltage characteristic, (2) the ionic fluxes ratio describing the permselectivity of the membrane, (3) the chronoamperometric response, (4) the chronopotentiometric response and (5) the electrochemical impedance. Some of the results obtained for highly charged membranes can be compared with the analytical solutions in ideal membranes.

15.2 Ionic Transport in Ion Exchange Membrane Systems

15.2.1 Mathematical Model

The system under study is constituted by a membrane that extends from $x = 0$ to $x = d'$ and two diffusion boundary layers adjacent to the membrane lying from $x = -\delta'$ to $x = 0$ and from $x = d'$ to $x = d' + \delta'$. The membrane is bathed by two bulk solutions with the same concentration $c^{0\prime}$ of a z:z electrolyte, and it will be assumed to have a negative fixed charge. Assuming that the ionic transport is one dimensional and perpendicular to the membrane/solution interface, with x the direction of transport, the equations determining the behaviour of the system are the laws of mass conservation or continuity equations:

$$\frac{\partial J_i{}' (x, t)}{\partial x} = -\frac{\partial c_i{}' (x, t)}{\partial t}, \quad i = 1, 2 \tag{15.1}$$

the Nernst-Planck flux equations written for dilute solutions:

$$J_i{}' (x, t) = -D_{ip}{}' \left[\frac{\partial c_i{}' (x, t)}{\partial x} + z_i \, c_i{}' (x, t) \frac{F}{RT} \frac{\partial \phi (x, t)}{\partial x}\right] \tag{15.2}$$

and the Poisson equation:

$$\frac{\partial D' (x, t)}{\partial x} = F \left[z_1 \, c_1{}' (x) + z_2 \, c_2{}' (x, t) - \theta' (x)\right] \tag{15.3}$$

where

$$\boldsymbol{D}' (x, t) = -\varepsilon' \frac{\partial \phi' (x, t)}{\partial x} \tag{15.4}$$

Here J_i', D_{ip}', $c_i'(x,t)$ and z_i denote the ionic flux, the diffusion coefficient, the molar concentration and the charge number ($z_1 = z$ and $z_2 = -z$) of ion i, respectively. In this work, we consider the ion diffusion coefficients to be different in the diffusion boundary layers and in the ion exchange membrane, and D_{iS}' and D_{iM}' stand for the diffusion coefficients of ion i in the solution S and membrane M phases, respectively. The electric potential is represented by $\phi'(x,t)$, the electric permittivity by ε' and the electric displacement by $\boldsymbol{D}'(x,t)$. The constants F, R and T have their usual meanings: Faraday constant, ideal gas constant and absolute temperature, respectively. $\theta'(x)$ is the fixed-charge concentration, which is presumed known and expressed in a general way as a function of position x.

On the other hand, since the total electric current density through the membrane system, I', is not a function of x [17], it can be written as

$$I'(t) = F\left[z_1 J_1'(-\delta', t) + z_2 J_2'(-\delta', t) + \frac{d\,\boldsymbol{D}'(-\delta', t)}{d\,t}\right] \tag{15.5}$$

In order to study the transient response of the ion exchange membrane system, the following boundary conditions are used:

$$c_1'(-\delta', t) = c_1'(d' + \delta', t) = c^{0'} \tag{15.6}$$

$$c_2'(-\delta', t) = c_2'(d' + \delta', t) = c^{0'} \tag{15.7}$$

$$\frac{d\boldsymbol{D}'(-\delta', t)}{dt} = I'(t) - z_1 J_1'(-\delta', t) - z_2 J_2'(-\delta', t) \tag{15.8a}$$

$$\phi'(-\delta', t) = \phi_A'(t) \tag{15.8b}$$

$$\phi(d' + \delta') = 0 \tag{15.9}$$

Equations 15.6–15.9 specify all the physical conditions to be imposed on the solution of the Nernst-Planck and Poisson equations. In particular, Eqs. 15.6 and 15.7 indicate that the system is electrically neutral at the outer boundaries of the diffusion boundary layers, because c_1' and c_2' have the constant concentration $c^{0\prime}$. Equation 15.8a, which is obtained from Eq. 15.5, is a boundary condition for the time evolution of the electric displacement, $\boldsymbol{D}'$, at $\xi = -\delta'$, and it is imposed when the total electric current, $I'(t)$, is the externally controlled variable. When the electric potential, $\phi_A'(t)$, is the externally applied perturbation, it is necessary to take into account the boundary condition given by Eq. 15.8b, instead of Eq. 15.8a; in this case, the electric current through the system must be evaluated from Eq. 15.5. Finally, Eq. 15.9 defines the reference level for the electric potential.

Since it is easier to numerically solve the above equations using groups of dimensionless variables, the following relations are defined:

$$\xi = \frac{x}{\lambda}\ ;\ \tau = \frac{D_a t}{\lambda^2}$$

$$c_i = \frac{c_i'}{c_a};\ J_i = \frac{\lambda J_i'}{D_a c_a};\ D_{ip} = \frac{D_{ip}'}{D_a};\ i = 1, 2$$

$$\phi = \frac{F\phi'}{RT};\ \varepsilon = \frac{RT\varepsilon'}{F^2 c_a \lambda^2};\ D = \frac{\boldsymbol{D}'}{Fc_a\lambda};\ \rho = \frac{\rho'}{Fc_a};\ \theta = \frac{\theta'}{c_a}$$

$$I = \frac{\lambda I'}{F D_a c_a}$$

$$c^0 = \frac{c^{0\prime}}{c_a};\ \delta = \frac{\delta'}{\lambda};\ d = \frac{d'}{\lambda};\ \phi_A = \frac{F {\phi_A}'}{RT}$$

where λ, D_a and c_a are scaling factors with the dimensions of length, diffusion coefficient and molar concentration, respectively. λ, D_a and c_a are taken as characteristic values of the system studied. In particular, we have taken $\varepsilon = 1$, and so the length λ is given by

$$\lambda = \sqrt{\frac{\varepsilon' RT}{F^2 c_a}}$$

and it can be considered as the Debye length in the system [18].

15.2.2 Network Model

The network model for the electrodiffusion process in a spatial region or compartment of width δ_k has been obtained elsewhere [13], and it is shown in Fig. 15.1a. In this figure, the network elements are as follows: R_{dik} is the resistor representing the diffusion of ion i in the compartment k; $GJ_{eik}(\pm)$ is the voltage-controlled current source modelling the electrical contribution to the ionic flux, minus and plus signs meaning the flux entering and leaving the compartment k, respectively; C_{dk} is the capacitor representing the nonstationary effects of the electrodiffusion process in the compartment k; R_{pk} is the resistor modelling the constitutive equation of the medium; GJ_{pk} is the voltage-controlled current source modelling the electric charge stores in the compartment k. The relation between those network elements and the parameters of the system is given by

$$R_{dik} = \frac{\delta_k}{2D_{ip}} \tag{15.10}$$

$$GJ_{eik}(\pm) = \pm D_{ip} z_i c_i(\xi_k \pm \frac{\delta_k}{2}) \frac{\phi(\xi_k) - \phi(\xi_k \pm \frac{\delta_k}{2})}{\delta_k/2}, \quad i = 1, 2 \tag{15.11}$$

$$C_{dk} = \delta_k \tag{15.12}$$

$$R_{pk} = \frac{\delta_k}{2\varepsilon} \tag{15.13}$$

$$GJ_{pk} = -\delta_k \left[z_1 c_1(\xi_k) + z_2 c_2(\xi_k) - \theta(\xi_k)\right] \tag{15.14}$$

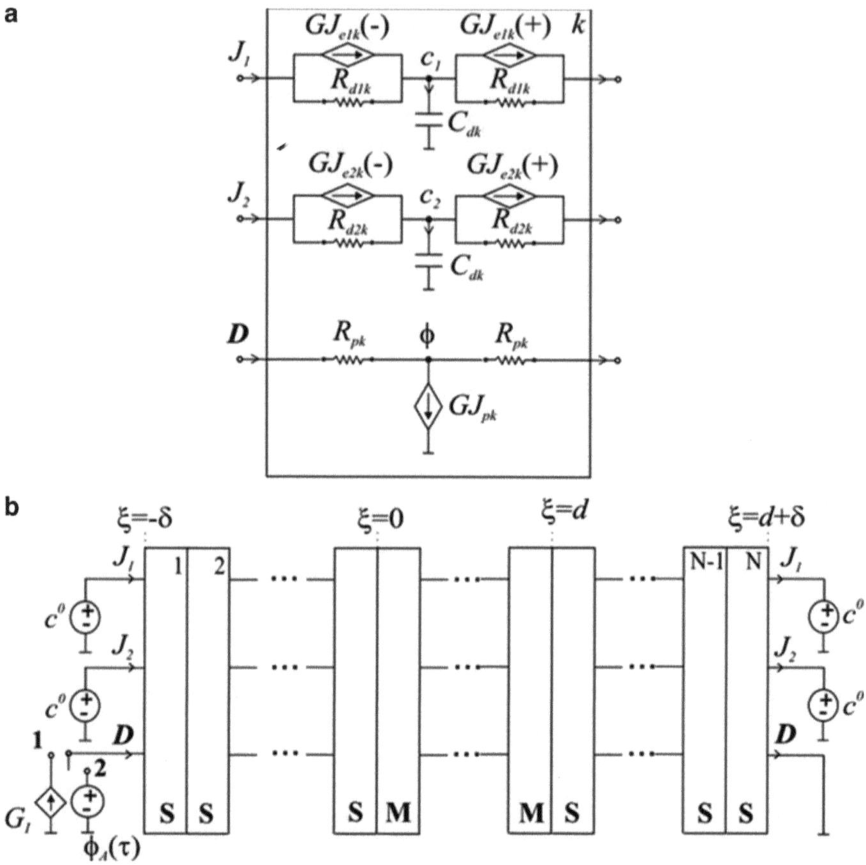

Fig. 15.1 (**a**) Network model for the stationary electrodiffusion in a volume element. (**b**) Network model for an ion exchange membrane system. Details of the structures of boxes 1-*N* are shown in (**a**). Letters *S* and *M* indicate solution and membrane phases, respectively

For network modelling purposes, a number N of circuit elements like those in Fig. 15.1a ($k = 1, \ldots, N$) must be connected in series to form a network model for the entire physical region undergoing an electrodiffusion process.

Figure 15.1b shows the network model for the membrane system. In this figure, the details of the structures of boxes 1,. . .,N are shown in Fig. 15.1a, and the letters S and M indicate solution ($D_{ip} = D_{iS}$) and membrane ($D_{ip} = D_{iM}$) phases, respectively. In the network model of Fig. 15.1b, the concentrations of the ionic species in the bathing solutions (Eqs. 15.6, 15.7) are represented by independent voltage sources of values c^0. The perturbation of electric current (Eq. 15.8a) is introduced from a current-controlled current source, G_I, which impose the value of the electric displacement at $\xi = -d$ from the value of the external electric current, I, when the switch is in position 1. The perturbation of electric potential (Eq. 15.8b) is represented by an independent voltage source of value $\phi_A(\tau)$, and it is imposed with the switch in position 2, while the origin of the electric potential Eq. 15.9 is introduced into the network model by short-circuiting the node ϕ at $\xi = d + \delta$.

15.3 Results and Discussion

The results for the steady-state, transient and small-signal ac electrical properties of ion exchange membrane systems are obtained by means of the simulation into the electric circuit simulation package PSpice [19], under dc, transient or ac conditions, of the network model depicted in Fig. 15.1b, with the appropriate numerical values for the system parameters.

In this work, we consider a homogeneous cation-exchange membrane system in which the negative fixed charge obeys the following equation:

$$\theta(\xi) = \begin{cases} X, 0 \leq \xi < d \\ 0, \quad -\delta < \xi < 0 \, and \, d < \xi < d + \delta \end{cases} \tag{15.15}$$

where X is the membrane fixed-charge concentration. Figure 15.2a shows a sketch of the membrane system under study.

We will now study the electrical responses of ion exchange membrane systems with various values of X, $z = 1$, $d = 1{,}000$, $\delta = 1{,}000$, $D_{1S} = D_{2S} = 100$, $D_{1M} = D_{2M} = 10$, $c^0 = 0.5$ and $\varepsilon = 1$. The chosen spatial grid contains $N = 480$ compartments, and the compartment thickness, δ_k, is given by

$$\delta_k = 33, \; k = 1, \ldots, 30 \; and \; 451, \ldots, 480$$

$$\delta_k = 0.6, \; k = 31, \ldots, 40; \; 201, \ldots, 210; \; 271, \ldots, 280 \; and \; 441, \ldots, 450$$

$$\delta_k = 0.05, \; k = 41, \ldots, 200 \; and \; 281, \ldots, 440$$

$$\delta_k = 16.33, \; k = 211, \ldots, 270$$

This spatial grid considers the presence of two membrane/solution interfaces and three volume phases. Moreover, it takes into account that the membrane/solution interfacial regions have a size of about 4λ in each phase [8]. Figure 15.2b shows a sketch of the characteristic zones used in the discretization of the membrane system under study. This spatial grid is symmetrical about the middle point of the membrane, $\xi = d/2 = 500$. In the left solution phase, we have chosen 30 compartments of width 33 from $\xi = -\delta/\lambda = -1{,}000$ to $\xi = -10$, 10 compartments of width 0.6 from $\xi = -10$ to $\xi = -4$ and 80 compartments of width 0.05 from $\xi = -4$ to $\xi = 0$. On the other hand, in the membrane phase, we have chosen 80 compartments of width 0.05 from $\xi = 0$ to $\xi = 4$, 10 compartments of width 0.6 from $\xi = 4$ to $\xi = 10$, and 30 compartments of width 16.33 from $\xi = 10$ to $\xi = d/2\lambda = 500$.

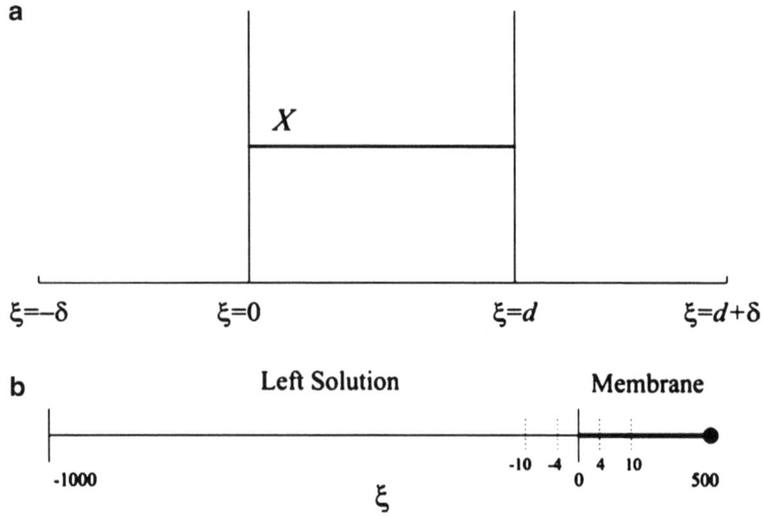

Fig. 15.2 (**a**) Schematic view of the ion exchange membrane system. (**b**) Schematic representation of the zones used in the discretization of the electrodiffusion region

15.3.1 Steady-State Current-Voltage Characteristics

Figure 15.3 gives the total electric current, I, through the system as a function of the applied electric potential, ϕ_A, for various values of the membrane fixed-charge concentration, namely, $X = 2$, 4 and 6. It is worth noting that, when the applied electric potential is positive, $\phi_A > 0$, the electric current is also positive, $I > 0$, and the anions move from the membrane to the solution at the left interface and from the solution to the membrane at the right interface. In this way, the left interface is reverse biased, while the right interface is forward biased [7]. In Fig. 15.3, two regions are observed: an ohmic region for small values of ϕ_A (underlimiting current regime) and one region in which the electric current varies very slowly with the voltage (overlimiting current regimes) for high values of ϕ_A. In the lower voltage range, the slope of the current-voltage curves increases as the fixed-charge concentration, X, increases. However, in the higher voltage region, this slope decreases as X increases. This behaviour is typical of negatively charged membranes [20–23]. In this way, a limiting value for the electric current is expected in highly charged membrane systems. The classical expression for the limiting electric current, I_L, which is obtained in a membrane system with an ideal permselectivity (i.e. zero coion flux), is given by [8]

$$I_L = \frac{2\,zD_{1S}c^0}{\delta} \qquad (15.16)$$

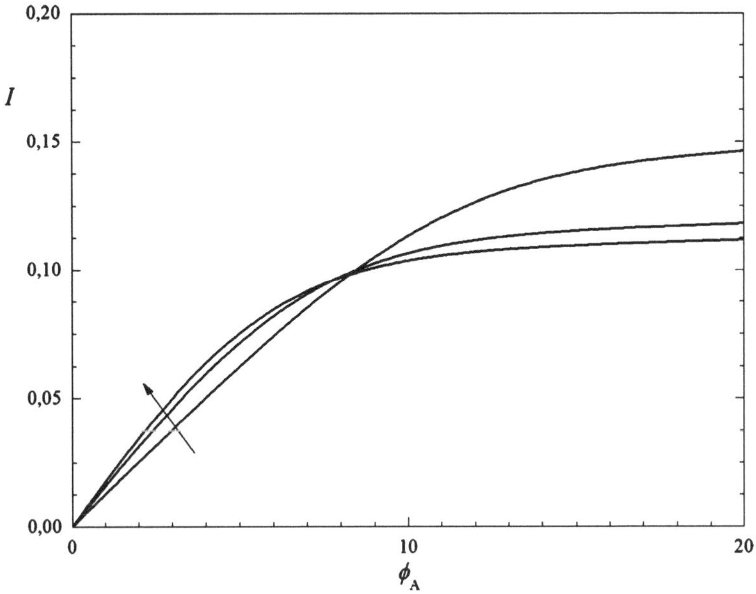

Fig. 15.3 Steady-state current-voltage characteristics of membrane systems with $z = 1, d = 1{,}000$, $\delta = 1{,}000, D_{1S} = D_{2S} = 100, D_{1M} = D_{2M} = 10, c^0 = 0.5, \varepsilon = 1$ and $X = 2, 4$ and 6. The *arrow* indicates increasing values of X

and for the systems here considered ($z = 1, D_{1S} = 100, \delta = 1{,}000, c^0 = 0.5$), one obtains $I_L = 0.1$.

15.3.2 Permselectivity

It is worth noting that the previous papers on steady-state ionic transport through ion exchange membranes have used the flux ratio, η, given by

$$\eta = -\frac{J_1}{J_2} \tag{15.17}$$

in order to characterize the permselectivity of the membrane [24]. Figure 15.4 shows the steady-state flux ratio-voltage characteristics under the same conditions given in Fig. 15.3. In Fig. 15.4, it can be seen that the flux ratio, η, is a function of the applied electric potential, ϕ_A. This fact is due to the concentration polarization in an ion exchange membrane system [24]. In Fig. 15.3, it is observed that the flux ratio, η, increases as the fixed-charge concentration of the membrane, X, increases, regardless of the value of the applied electric potential, ϕ_A. In this figure, it is also observed that all the curves show an appreciable curvature due to the

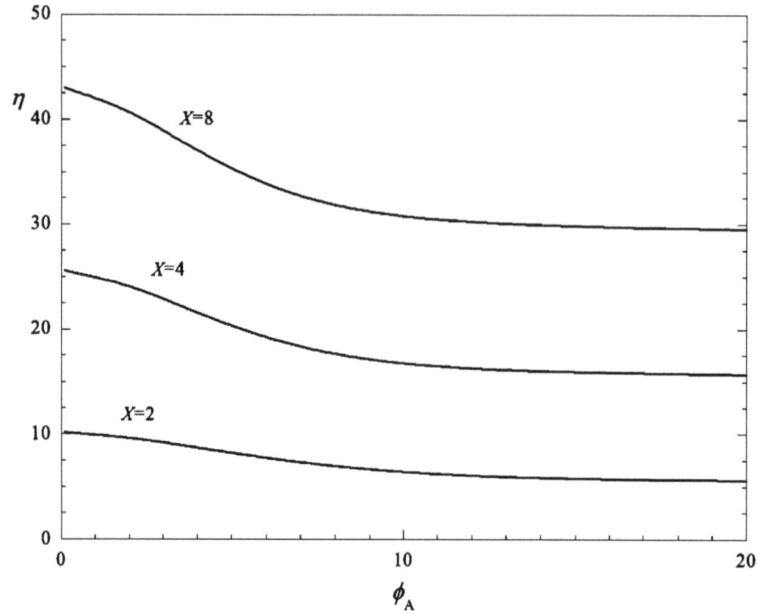

Fig. 15.4 Flux ratio vs. applied voltage for membrane systems with $z = 1, d = 1{,}000, \delta = 1{,}000$, $D_{IS} = D_{2S} = 100$, $D_{IM} = D_{2M} = 10$, $c^0 = 0.5$, $\varepsilon = 1$ and $X = 2$, 4 and 6

concentration polarization and this curvature is more pronounced for the higher values of X, i.e. for the more charged membranes, as expected [24].

15.3.3 Chronoamperometric Response

In this section, the transient response of the ion exchange membrane systems to an externally applied electric potential, $\phi_A(\tau)$, is reported. The initial conditions for the ionic concentrations and the electric potential correspond to the equilibrium values, and they are determined from a steady-state analysis for $\phi_A = 0$ ($I = 0$). The system is then perturbed with an electric signal given by

$$\phi_A(\tau) = \begin{cases} 0, & \tau = 0 \\ \phi_S, & \tau > 0 \end{cases} \tag{15.18}$$

where ϕ_S is the amplitude of the perturbing electric potential.

Figure 15.5 gives the total electric current, $I(\tau)$, for the amplitude of the perturbing electric potential $\phi_S = 10$, in charged membranes with the membrane fixed-charge concentration $X = 2$, 4 and 6. The current-time responses show an

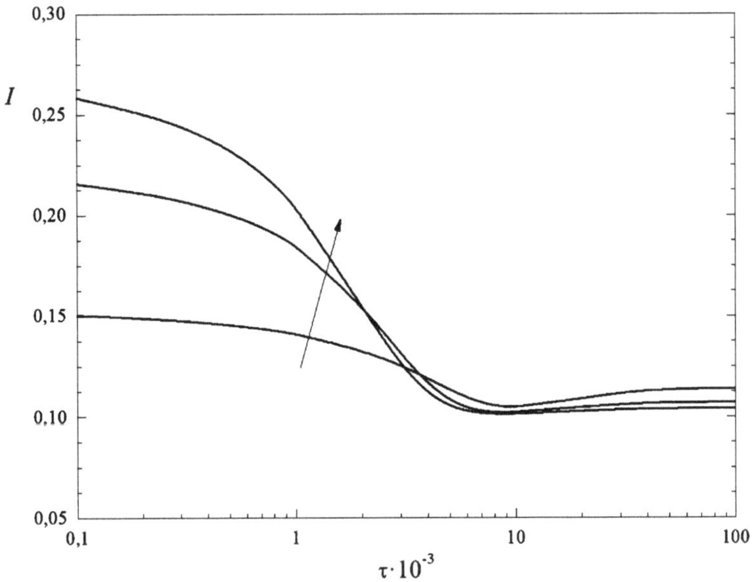

Fig. 15.5 Time evolution of the electric current in response to a step-function potential of amplitude $\phi_S = 10$, for membrane systems with $z = 1$, $d = 1{,}000$, $\delta = 1{,}000$, $D_{1S} = D_{2S} = 100$, $D_{1M} = D_{2M} = 10$, $c^0 = 0.5$, $\varepsilon = 1$ and $X = 2, 4$ and 6. The *arrow* indicates increasing values of X

initial ohmic current region, followed by a decay into the steady-state current region. It is interesting to note that the diffusional relaxation time of the ion exchange membrane system, τ_D, is

$$\tau_D = \frac{d^2}{D_{1M}} \tag{15.19}$$

while the electric relaxation time, τ_E, of the system can be estimated from that of an electroneutral membrane system as [13]

$$\tau_E = \varepsilon \left[\frac{d}{z^2 \left(D_{1M}\, c_+ + D_{2M}\, c_- \right)} + \frac{2\delta}{z^2\, c^0 \left(D_{1S} + D_{2S} \right)} \right] \tag{15.20}$$

where c_+ and c_- are the equilibrium concentrations of the cation and the anion, respectively, in the volume of that membrane and they can be evaluated by using the Donnan equilibrium relations [7]:

$$c_\pm = \pm \frac{X}{2} + \sqrt{\left(\frac{X}{2} \right)^2 + \left(c^0 \right)^2} \tag{15.21}$$

For the systems here considered ($z = 1$, $d = 1{,}000$, $\delta = 1{,}000$, $D_{IS} = D_{2S} = 100$, $D_{IM} = D_{2M} = 10$, $c^0 = 0.5$, $\varepsilon = 1$), one obtains $\tau_D = 100 \times 10^3$ and $\tau_E = 0.065 \times 10^3$ for $X = 2$ ($c_+ = 2.12$ and $c_- = 0.12$), $\tau_E = 0.044 \times 10^3$ for $X = 4$ ($c_+ = 4.062$ and $c_- = 0.062$) and $\tau_E = 0.036 \times 10^3$ for $X = 6$ ($c_+ = 6.041$ and $c_- = 0.041$). From these results, it is clearly inferred that the electric relaxation time of the system decreases as the membrane fixed-charge concentration, X, increases.

Now, the initial ohmic current through the system is given by

$$I(\tau = 0^+) = \frac{\varepsilon}{\tau_E}\,\phi_A \tag{15.22}$$

and for the systems here considered with $\phi_S = 10$, one obtains $I(\tau = 0^+) = 0.154$ for $X = 2$, 0.227 for $X = 4$ and 0.278 for $X = 6$. The results inferred from Fig. 15.4 for the initial value of the electric current is then in good agreement with the theoretical expectations.

15.3.4 Chronopotentiometric Response

In this section, the transient response of the ion exchange membrane systems to an externally applied electric current, $I(\tau)$, is reported. The initial conditions for the ionic concentrations and the electric potential correspond to the equilibrium values, and they are determined from a steady-state analysis for $\phi_A = 0$ ($I = 0$). Now, the system is perturbed with an electric signal given by

$$I(\tau) = \begin{cases} 0, & \tau = 0 \\ I_S, & \tau > 0 \end{cases} \tag{15.23}$$

where I_S is the amplitude of the perturbing electric potential.

The time evolution of the electric potential difference across the membrane systems, $\phi_M(\tau) = \phi(-\delta/\lambda,\tau)$, with $X = 2$, 4 and 6, for a value of the amplitude of the current in the overlimiting regime, $I_S = 0.11$, is shown in Fig. 15.6. In this figure, it can be observed that, after an electric potential vertical jump due to the ohmic resistance of the system, the electric potential difference increases and reaches a maximum, and then it evolves to the steady-state value. In this figure, it can also be seen that the maximum value of the electric potential increases as the membrane fixed-charge concentration, X, increases. Moreover, the time in which this maximum is reached increases as X increases. In Fig. 15.6, one observes that the transition time to the steady-state value of the membrane system potential increases as the membrane fixed-charge concentration, X, increases, according with the experimental results obtained by other authors in this field [25–30].

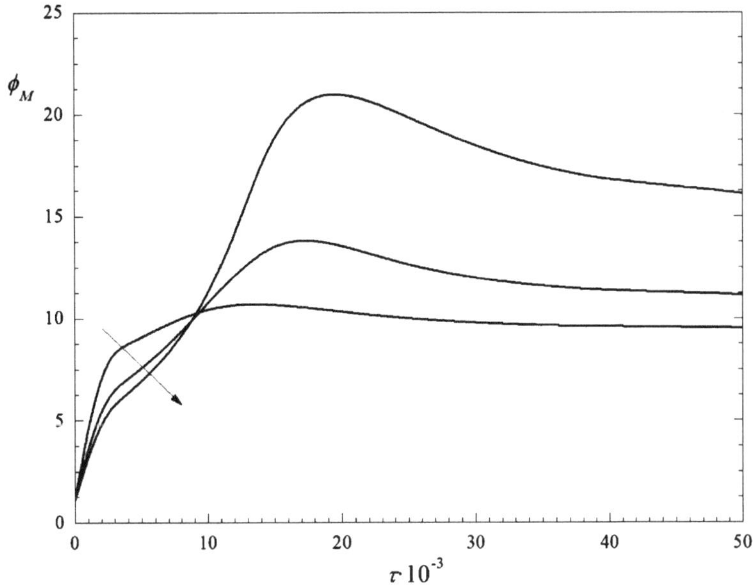

Fig. 15.6 Time evolution of the membrane system potential in response to a step-function current of amplitude $I_S = 0.11$, for membrane systems with $z = 1$, $d = 1{,}000$, $\delta = 1{,}000$, $D_{1S} = D_{2S} = 100$, $D_{1M} = D_{2M} = 10$, $c^0 = 0.5$, $\varepsilon = 1$ and $X = 2$, 4 and 6. The *arrow* indicates increasing values of X

15.3.5 Electrochemical Impedance

In order to study the electrochemical impedance, the system is perturbed around an equilibrium state with an electric potential given by

$$\phi_A(t) = \phi_0 \sin(\omega\tau) \tag{15.24}$$

where ϕ_0 is the amplitude of the perturbing sine electric potential and ω is the angular frequency, which can be written as a function of the conventional frequency, f:

$$\omega = 2\pi f \tag{15.25}$$

When a small-amplitude ac signal ($\phi_0 < <1$) is used, the membrane system potential can be written as

$$I(t) = I_0 \sin(\omega\tau + \varphi) \tag{15.26}$$

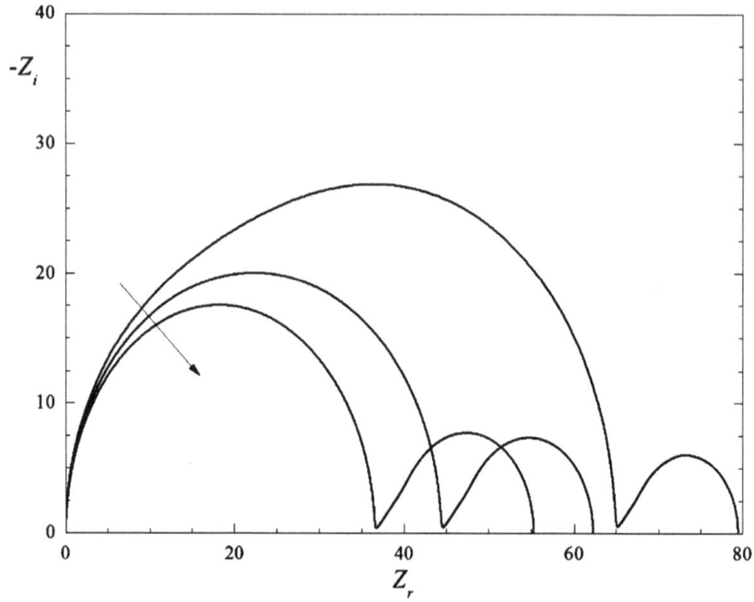

Fig. 15.7 Complex-plane impedance plot for membrane systems with $z = 1$, $d = 1{,}000$, $\delta = 1{,}000$, $D_{1S} = D_{2S} = 100$, $D_{1M} = D_{2M} = 10$, $c^0 = 0.5$, $\varepsilon = 1$ and $X = 2$, 4 and 6. The *arrow* indicates increasing values of X

where I_0 is the ac amplitude of the electric current and φ is the phase difference between the current and the perturbing voltage. Now, the electrochemical impedance, Z, is a complex quantity given by the following equation [31]:

$$Z(j\omega) = \frac{\phi_0}{I_0} e^{-j\varphi} = Z_r(\omega) + jZ_i(\omega) \tag{15.27}$$

Where $j = \sqrt{-1}$ is the imaginary unit and Z_r and Z_i are the real and imaginary parts of the impedance, respectively.

Figure 15.7 gives the complex-plane impedance plot of the charged membrane systems with $X = 2, 4$ and 6. In this graph, $-Z_i(\omega)$ is plotted against $Z_r(\omega)$ with the angular frequency, ω, as a parameter increasing from the right to the left of the plot. These impedance plots are typical of ion exchange membrane systems [32–35], and they show two regions: a geometric arc at high frequencies and a diffusional arc at low frequencies. In Fig. 15.7, it can be seen that the geometric arc is distorted and it can be considered as the combination of two different geometric arcs associated with the electric migration processes taking place in the bathing electrolyte solutions and in the membrane. Moreover, in Fig. 15.7, it is clearly observed that the diffusional arcs are Warburg-type impedances [36] since they present a 45° straight line at high frequencies and a semicircle at low frequencies. It must be noted that analytical solutions for the impedance are not possible in this case because of the presence of the electric double layers at the interfaces. However, the characteristic parameters of the impedance can be estimated from those of single systems [37, 38]. Thus, the theoretical characteristic frequency of a solution geometric arc in an electroneutral system is given by

$$f_{1S} = \frac{z^2\, c^0\, (D_{1S} + D_{2S})}{2\,\pi\,\varepsilon} \tag{15.28}$$

while the theoretical characteristic frequency of an homogeneous, electroneutral membrane geometric arc is

$$f_{1M} = \frac{z^2\, (D_{1M}\, c_+ + D_{2M}\, c_-)}{2\,\pi\,\varepsilon} \tag{15.29}$$

where c_+ and c_- are the equilibrium concentrations of the cation and the anion, respectively, in the volume of that membrane (see Eq. 15.21). Moreover, the width of a solution geometric semicircle is given by

$$R_{acS} = \frac{2\,\delta}{2\,\pi\,\varepsilon f_{1S}} \tag{15.30}$$

while the theoretical width of an homogeneous, electroneutral membrane geometric arc is

$$R_{acM} = \frac{d}{2\,\pi\,\varepsilon f_{1M}} \tag{15.31}$$

On the other hand, the theoretical peak frequency of a diffusional arc in an electroneutral layer, f_d, is given by the following relation [36]:

$$2\,\pi f_d = 2.54\, \frac{2\, D_{1S}\, D_{2S}}{(D_{1S} + D_{2S})\, \delta^2} \tag{15.32}$$

Moreover, in a membrane system with an ideal permselectivity, the width of the diffusional arc is [35]

$$R_d = \frac{D_{2S}}{D_{1S}}\, R_{acS} \tag{15.33}$$

From Eqs. 15.28, 15.30, 15.32 and 15.33 for the systems here considered ($z = 1$, $d = 1{,}000$, $\delta = 1{,}000$, $D_{1S} = D_{2S} = 100$, $D_{1M} = D_{2M} = 10$, $c^0 = 0.5$, $\varepsilon = 1$), one obtains $f_{1S} = 15.9$, $R_{acS} = 20$, $f_d = 4.03 \times 10^{-5}$ and $R_d = 20$. Moreover, from Eqs. 15.29 and 15.31, one obtains $f_{1M} = 3.565$ and $R_{acM} = 44.64$ for

Table 15.1 Numerical results obtained for the characteristic parameters of the impedance plots of membrane systems with $z = 1$, $d = 1{,}000$, $\delta = 1{,}000$, $D_{IS} = D_{2S} = 100$, $D_{IM} = D_{2M} = 10$, $c^0 = 0.5$, $\varepsilon = 1$ and various values of X

	$X = 2$	$X = 4$	$X = 6$
R_{DC}	79.46	62.22	55.27
$f_2 \times 10^5$	3.631	3.802	3.89
$-Z_{i2}$	6.035	7.412	7.793
R_{ac}	65.12	44.59	36.73
$R_{DC}-R_{ac}$	14.34	17.63	18.54
f_I	4.365	9.55	12.88

$X = 2$, $f_{IM} = 6.564$ and $R_{acM} = 24.25$ for $X = 4$ and $f_{IM} = 9.68$ and $R_{acM} = 16.44$ for $X = 6$.

On the other hand, Table 15.1 shows the results obtained for different parameters of the impedance plots for systems of charged membranes with $X = 2$, 4 and 6. The tabulated parameters are:

- The dc resistance of the system, $R_{DC} = Z_r(f = 0)$, i.e. the real part of the impedance at the limit of zero frequency.
- The peak frequency of the diffusional arc, f_2, and the imaginary part of the impedance at this frequency, $Z_{i2} = Z_i(f_2)$.
- The real part of the impedance at the intersection point between the geometric and diffusional arcs, R_{ac}. This parameter is obtained from the relative minimum value of the imaginary part of the impedance at low frequencies.
- The width of the diffusional arc, $R_{DC}-R_{ac}$.
- The peak frequency of the total geometric arc, f_I.

It must be noted that the values obtained for the total ohmic resistance of the systems, R_{ac}, are in good agreement with those theoretically estimated from Eqs. 15.30 and 15.31 as the sum of R_{acS} and R_{acM}, i.e. $R_{acS} + R_{acM} = 64.64$ for $X = 2$, 44.25 for $X = 4$ and 36.44 for $X = 6$. Moreover, the peak frequencies of the geometric arcs, f_I, take values in the range from f_{IM} to f_{IS}. Also, from Eq. 15.32, one obtains $f_d = 4.03 \times 10^{-5}$, and the results obtained for the peak frequencies of the diffusional arcs are then in good agreement with the theoretical expectations. However, the width of the diffusional arc for $X = 2$, $R_{DC}-R_{ac} = 14.34$, differs from the value obtained from Eq. 15.33, $R_d = 20$, since the membrane is weakly charged and it is not ideal.

From the data shown in Table 15.1, it can be also inferred that the influence of the membrane fixed-charge concentration, X, on the peak frequency of the diffusional arc, f_2, is very small. Moreover, it has been found that the dc resistance of the system, R_{DC}, decreases and the width, $R_{DC}-R_{ac}$, and the height, $-Z_{i2}$, of the diffusional arc increase as the membrane fixed-charge concentration, X, increases.

Acknowledgement I am indebted to Professor J. Horno for many helpful suggestions.

References

1. Xu T (2005) J Membr Sci 263:1
2. Szymczyk A (ed) (2008) Surface electrical phenomena in membranes and microchannels. Transworld Research Network, Trivandrum
3. Helfferich F (1962) Ion exchange. McGraw-Hill, New York
4. Lakshminarayanaiah N (1969) Transport phenomena in membranes. Academic, New York
5. Rubinstein I (1990) Electro-diffusion of ions. SIAM Studies in Applied Mathematics, Philadelphia
6. Bassignana IC, Reiss H (1983) J Phys Chem 87:136
7. Selvey C, Reiss H (1987) J Membr Sci 30:75
8. Manzanares JA, Murphy WD, Mafé S, Reiss H (1993) J Phys Chem 97:8524
9. Sokalski T, Lingenfelter P, Lewenstam A (2003) J Phys Chem B 107:2443
10. Volgin VM, Davydov AD (2005) J Membr Sci 259:110
11. Morf WE, Pretsch E, De Rooij NF (2007) J Electroanal Chem 602:43
12. Moya AA, Horno J (1999) J Phys Chem B 103:10791
13. Moleón JA, Moya AA (2008) J Electroanal Chem 613:23
14. Moleón JA, Moya AA (2009) J Electroanal Chem 633:306
15. Moya AA, Moleón JA (2010) J Electroanal Chem 647:53
16. Moya AA (2010) Electrochim Acta 55:2087
17. Buck RP (1984) J Membr Sci 17:1
18. Murphy WD, Manzanares JA, Mafé S, Reiss H (1992) J Phys Chem 96:9983
19. Tuinenga PW (1992) SPICE: a guide to circuit simulation and analysis using PSpice. Prentice-Hall, Englewood Cliffs
20. Rubinstein I, Shtilman L (1981) J Chem Soc Faraday Trans II 77:1595
21. Nikonenko VV, Zabolotskii VI, Gnusin NP (1989) Sov Electrochem 25:262
22. Listovnichii AV (1989) Sov Electrochem 25:1479
23. Sistat P, Pourcelly G (1999) J Electroanal Chem 460:53
24. Sokirko AV, Manzanares JA, Pellicer J (1994) J Colloid Interface Sci 168:32
25. Karlin YuV, Kropotov VN (1995) Russ J Electrochem 31:472
26. Sistat P, Pourcelly G (1997) J Membr Sci 123:121
27. Krol JJ, Wessling M, Strathmann H (1999) J Membr Sci 162:145
28. Ibanez R, Stamatialis DF, Wessling M (2004) J Membr Sci 239:119
29. Volodina E, Pismenkaya N, Nikonenko V, Larchet C, Pourcelly G (2005) J Colloid Interface Sci 285:247
30. Lee XT (2008) J Colloid Interface Sci 325:215
31. Barsoukov E, Macdonald JR (2005) Impedance spectroscopy: theory, experiment and applications. Wiley, New York
32. Urtenov MA-Kh, Kirillova EV, Seidova NM, Nikonenko VV (2007) J Phys Chem B 11:14208
33. Park J-S, Choi J-H, Yeon S-H, Moon S-H (2006) J Colloid Interface Sci 294:129
34. Park J-S, Choi J-H, Woo J-J, Moon S-H (2006) J Colloid Interface Sci 300:655
35. Sistat P, Kozmai A, Pismenskaya N, Larchet C, Pourcelly G, Nikonenko V (2008) Electrochim Acta 53:6380
36. Franceschetti DR, Macdonald JR, Buck RP (1991) J Electrochem Soc 138:1368
37. Freger V (2005) Electrochem Commun 7:957
38. Diard J-P, Glandut N, Montella C, Sánchez J-Y (2005) J Electroanal Chem 578:247

Chapter 16
Computer Modeling of Strong Acid Cation Exchangers on Styrene: Divinylbenzene Matrix

V.S. Soldatov and V.M. Zelenkovskii

Abstract This chapter is a review of the authors' works on the mathematical and computer modeling of ion exchangers on styrene–divinylbenzene matrix. A mathematical model based on the concept of influence of neighboring exchange sites on the properties of each other allowed explaining the dependence of selectivity and additive properties of the ion exchange system on the degree of ion exchange. The information of interaction between the nearest neighbors and water molecules in the vicinity of the functional groups as well of the structures on the fragments of the polymeric structure of the ion exchanger has been obtained using quantum chemistry nonempirical ab initio calculations. These data have been obtained for different degrees of hydration (0–10 water molecules per sulfonic group) and the following counter ions: H^+, Li^+, Na^+, K^+, Rb^+, Cs^+, Ag^+, Be^{2+}, Mg^{2+}, Ca^{2+}, Sr^{2+}, Ba^{2+}, Cd^{2+}, Ni^{2+}, and Pb^{2+}. The mathematical modeling as well as molecular mechanic and quantum chemistry calculations is consistent with the main concepts used in interpreting of ion exchange equilibria: a concept of nonregularity of the exchange sites in polymeric ion exchangers and competitive interaction of the counter ions with the fixed ion and water molecules.

16.1 Introduction

This chapter is devoted to interpreting of known regularities of ion exchange and physical chemical properties of polymeric ion exchangers by their comparison with the computer images of representative fragment of their structure. The visualization

V.S. Soldatov (✉)
Institute of Physical Organic Chemistry, Belarus National Academy of Sciences, 13 Surganov Str, Minsk 220072, Republic of Belarus

Lublin University of Technology, 40 Nadbystrzycka, 20-618 Lublin, Poland
e-mail: soldatov@ifoch.bas-net.by

V.M. Zelenkovskii
Institute of Physical Organic Chemistry, Belarus National Academy of Sciences, 13 Surganov Str, Minsk 220072, Republic of Belarus

Inamuddin and M. Luqman (eds.), *Ion Exchange Technology I: Theory and Materials*, DOI 10.1007/978-94-007-1700-8_16,

of microstate with a possibility of immediate evaluation of distances and angles between atoms, the forces of interaction between them, effective electrical charges on the atoms, and the degree of covalence of the bonds is the best way for understanding physical reasons for ion exchange selectivity.

The traditional methods of the interpretation are based on common chemical and physical knowledge, analysis of the macroscopic properties of ion exchange systems, spectroscopic studies, and different interpretive models. They also have, as their final goal, the construction of visually perceptible situation in the ion exchanger. This is achieved by the creation of the schematic drawing of imaginable structures and/or their verbal description. Such an interpretation may not be complete and unambiguously understandable. The knowledge of structure of the representative fragment of ion exchangers and the characteristics of interionic and ion–molecule interactions are necessary conditions for correct interpretation of its properties and sorption behavior. Computer modeling gives such a possibility. Its reliability is restricted by the perfectness of the computer programs used, characteristics of the hardware applied, the correctness and completeness of the initial database set, and the correctness of the choice of the microstate. The computer visualization of structure does not leave place for a double meaning interpreting of the structure and it is easily subjected to criticism.

This method is extensively used in physical chemistry of polyelectrolyte solutions (see, e.g., monograph [1] and reviews [2, 3]). The methods used in these publications are not suitable for ion exchangers with dense polymer network and high concentration of the charged fixed ionic groups and the counter ions. In spite of their obvious great potential, the methods of computer modeling for studies of ion exchange processes practically have not been used in this area. Except for our own publications starting from 1993, there are no other ones directly related to the computer modeling of ion exchangers. The closest papers containing results of ab initio calculations of the fragments of polymeric ion exchangers are Refs. [4–6] and [7–10]. In the first group of references, the structure of fragments of Nafion membrane has been calculated. In the second one, the monomeric units of sulfostyrene resin in Na^+ form have been modeled by one molecule of sodium benzene sulfonate in the presence of 0–3 water molecules [8], and the monomeric units of a carboxylic acid resin in Na^+ form have been modeled by the sodium salt of pivalic acid in the presence of 6 and 10 water molecules [10].

This chapter is a review of our works on the mathematical and computer modeling of sulfostyrene ion exchangers [11–22].

The computer modeling of polymeric ion exchangers has been performed on two levels: on the level of relatively large fragments of the polymer structure including its typical elements such as linear chains, cross-links, different order of the repetitive unit junction, etc., and on the level of one or several pairs of interacting ions in the presence of certain numbers of water molecules. In the first case, the methods of molecular mechanics have been used; in the second case, the methods of ab initio quantum chemical calculations have been applied.

It is the aim of this chapter to demonstrate potential ability of computer modeling on the example of the most spread and the best studied resins, and those are sulfonated copolymers of styrene and divinylbenzene. We wish to attract the attention of researchers to computer modeling as a powerful method for investigating ion exchange phenomena.

16.2 Mathematical and Molecular Mechanic Modeling of Ion Exchangers on Styrene: Divinylbenzene (ST–DVB) Matrix

Ion exchangers on the base of ST–DVB matrix are the best studied type of ion exchange resins. The experimental data obtained on these materials are the fundament for theoretical interpretation of ion exchange processes as a whole class of physical chemical phenomena. Therefore, this matrix has been chosen as a specific object for the mathematical and computer modeling. It is to note that the same method can be used for modeling of the other polymeric ion exchanger.

The interpretation of regularities of ion exchange equilibria implies clear answer at least to two questions:

- Why is one ion absorbed stronger than the other by the ion exchanger with a different functional group and matrix?
- Why does the selectivity of ion exchange depend on the loading of the sorbent with the exchanging ions?

The answer to the first question can be obtained from quantum chemical calculation of the structures of one or several pairs of ions in the presence of the water molecules. The answer to the second question implies assumption on the irregularity of the exchange centers of ion exchanger caused by mutual influence of the nearest neighboring groups on the energy of interaction of the counter ions and fixed ions.

In this section, we consider the possible reasons for nonequivalency of the fixed ionic groups in the structure of monofunctional ion exchanges on ST–DVB matrix. We do not consider such reasons for irregularity of the ion exchange structure as uncontrolled defects caused by the imperfectness of their method of preparation.

16.2.1 The Mathematical Model

The ion exchange equilibrium of monovalent ions is formulated as

$$\bar{I}_1 + I_2 \leftrightarrow I_1 + \bar{I}_2 \tag{16.1}$$

The over bar denotes the ion exchanger phase.

The model allows obtaining dependence of any additive property Y of the ion exchange system as a function of ionic composition of the ion exchanger expressed as equivalent fraction of ion 2, $\bar{x}$, and some structural parameters of the ion exchanger such as a total number of neighbors of the observed exchange site i and its local composition expressed as a number of ion 2 denoted as j: $Y = \sum Y(\bar{x}, i, j)$. Quantity Y for each $i(Y)$ can be obtained as a weight average of properties y of the microstates differing only by the local composition. The Y can be logarithm of the equilibrium or selectivity coefficient, Gibbs energy or enthalpy, the water uptake, molar volume, etc.:

$$Y_i = \sum_{j=o}^{j=i} y(i-j,j)p(i-j,j) \tag{16.2}$$

where $y(i - j, j)$ is property Y related to the microstate $i - j, j$; $p(p(i - j, j)$ is the probability of the existence of the exchange site having $i - j$ neighbors of type 1 and j neighbors of type 2. It is proportional to the molar fraction of each counter ion raised to a power equal to the number of the neighbors of a relative type and a number of their permutations. If the sum of the probabilities is 1, then

$$p(i-j,j) = \frac{i!}{(i-j)!j!} y(i-j,j)(1-\overline{x})^{(i-j)}\overline{x}^{j} \tag{16.3}$$

The overall property Y is proportional to the probability (mole fraction) of the i sites P_i and the $p(i - j, j)$:

$$Y_i = \sum_{j=o}^{i=n} P_i \sum_{j=o}^{j=i} \frac{i!}{(i-j)!j!} y(i-j,j)(1-\overline{x})^{(i-j)}\overline{x}^{j} \tag{16.4}$$

Equation 16.4 is a polynomial of power n. The number of neighbors in real systems is rarely more than 4. Thus, Eq. 16.5 represents the following set of equations:

$$Y_0 = y(0,0) \tag{16.5a}$$

$$Y_1 = y(1,0)(1-\overline{x}) + y(0,1)\overline{x} \tag{16.5b}$$

$$Y_2 = y(2,0)(1-\overline{x})^2 + 2y(1,1)(1-\overline{x})\overline{x} + y(0,2)(1-\overline{x})\overline{x}^2 \tag{16.5c}$$

$$Y_3 = y(3,0)(1-\overline{x})^3 + 3y(2,1)(1-\overline{x})^2\overline{x} + 3y(1,2)(1-\overline{x})\overline{x}^2 + y(0,3)\overline{x}^3 \tag{16.5d}$$

$$Y_4 = y(4,0)(1-\overline{x})^4 + 4y(3,1)(1-\overline{x})^3 x + 6y(2,2)(1-\overline{x})^2\overline{x}^2 + 4y(1,3)(1-\overline{x})\overline{x}^3 + y(0,4)\overline{x}^4 \tag{16.5e}$$

Quantities $y(i - j, j)$ are constants which can be found by fitting Eq. 16.5 to the experimental dependence $Y = Y(\overline{x})$ in each specific case.

The power of polynomial indicates the largest detectable number of neighbors of the ion exchange site in the ion exchanger.

Here we need to specify what *exchange sites* are accounted *neighbors*. The location of exchange site is defined as the center of the sulfur atom in the functional group. The neighbors are a functional group influencing the interaction of the observed counter ion with the fixed ion within the predetermined precision of our experiment. They are situated inside a sphere with the center in the exchange site with radius R_0 named the *radius of interaction*. Quantitative evaluation of the R_0 is given in the following section of this chapter.

16.2.2 The Molecular Mechanic Modeling of Ion Exchangers on Styrene: Divinylbenzene Matrix

The aim of this modeling is quantitative evaluation of the radius of interaction and the probabilities of existence of a certain number of neighbors in different fragments of the polymeric matrix of ion exchanger.

The polymer matrix is assumed to be a statistical mixture of elementary fragments with different structures. They are randomly integrated into a matrix, and a property of the integrated system can be found as a weight average of the properties of the fragments. The sulfonic group is assumed situated in *para* position relative to the chain. The following types of fragments were chosen to represent the polymer structure: linear chain (*L*), H- and T-shaped fragments, and small loops (rings – *R*). Any other structures can be obtained as combinations of these fragments. The choice of size of the elementary fragments is important. We accepted it as a junction of 8–10 repeating units, since with the larger fragments it would be impossible to represent the structure of highly cross-linked resins. The smaller fragments do not have statistical meaning. Each fragment can have a different internal structure due to difference in the order of junction between the monomer units (head to tail, *ht*; head to head, *hh*; tail to tail, *tt*), absence or presence of the functional group in the DVB ring, and the cross-bridge type (*para-* or *meta-*DVB fragment). According to this classification, each structure has been given an individual letter code as shown in Table 16.1.

The probabilities of finding of the functional group in the distance R from the observed group require calculating the geometric structure of the whole fragment, which implies knowledge of distances between any atoms in the system. The number of groups with distances between them less than R_0 can be determined.

The structure of the fragment was determined by finding the structure with the minimal energy as described in Ref. [11]. In each specific fragment, the probability of existence of i neighbors was found using the Boltzmann distribution equation:

$$p_i = \exp(-E_q/RT \Big/ \sum_{q=1}^{s} \exp(-E_s/RT) \quad (16.6)$$

where q and s denote the number of the accounted conformers in the fragment (1, 2, ...q, ...,s). The overall probability P_i can be found as the weight sum of the p_i values. No methods for evaluation of probability of these fragments themselves

Table 16.1 The letter codes and the $P_{i>2}$ for different fragments of the spatial polymer net

	No functional group in the DVB residue		The functional group in DVB ring is present	
Type of the fragment	The letter code	$P_{i>2}$	The letter code	$P_{i>2}$
Linear chain	*L(ht)*	0.046		
	L(hh)	0.461		
	L(tt)	0.217		
T – shaped fragment	*T(ht)*		*T(ht)**	
	T(hh)	0.000	*T(hh)**	0.144
	T(tt)		*T(tt)**	
H – shaped fragment	*H(ht)*	0.000	*H(ht)**	0.416
	H(hh)		*H(hh)**	
	H(tt)		*H(tt)**	
R – ring-shaped fragment[a]	*R(ht)N*		*R(ht)N**	
	R(ht)6	0.047	*R(ht)6**	0.373

[a]In place of *N* in each specific case should stay the number of monomer units in the ring
* In this cases the asterisk means that fragment is sulfonated

are known. Therefore, our further consideration will be restricted with analysis of values p_i only.

In order to evaluate the distances between sulfonic groups on the polymer chain, the calculation of structure of dimer of sodium *p*-styrene sulfonate in aqueous medium (10 water molecules per ionic pair) was done by nonempirical method SCF MO LCAO and program GAMESS [23] using basis set MINI [24]. Initial distances between the sulfur atoms were chosen by variation of torsion angles formed by four carbon atoms of the hydrocarbon chain in the fragment. Optimizing of the coordinates of all atoms in the system was done to find the total energy minimum. It appeared that such a molecule has three stable conformations as shown in Fig. 16.1. The sulfonic groups interact via bridges formed by water molecules; at a distance between 0.8 and 1.2 nm, the interaction disappears. Increasing the distance between the sulfonic groups above 1 nm leads to spontaneous separation of the sulfonic groups to a distance 1.15–1.16 nm. Their hydrate structure splits into two independent structures. At such distances, electrostatic interaction between sulfonic groups in the presence of water molecules is weaker than the conformation strength in the hydrocarbon chain determining the shape of the structure. At distances shorter than 1 nm, a decisive role belongs to electrostatic interaction in the common hydration structure; the sulfonic groups spontaneously approach each other until they form stable structures 2 and 1 with a lower energy for the latter structures. The least energetically favorable is structure 3.

Evaluation of the radius of interaction for the sulfonic ion exchanger was performed in the following way. It is known that dependences $Y = Y(\overline{x})$ for weakly cross-linked ion exchangers and soluble polyelectrolytes are rather simple and can be expressed by the polynomial with at most the second power [25]. In terms of the model, it means that the exchange sites in this case have two neighbors.

The calculated probabilities of occurrence of zero to five neighbors in different linear fragments of the sulfonic acid polyelectrolyte as a function of distance from the exchange site are presented in Fig. 16.2. It is seen that in the fragments with

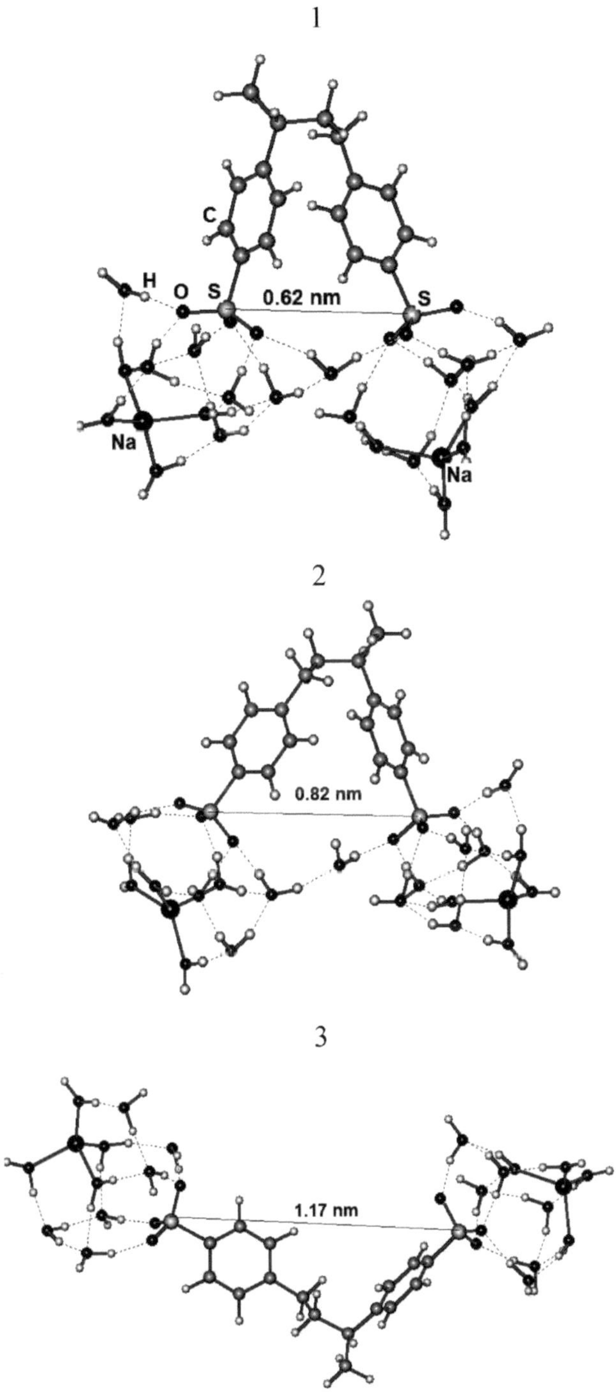

Fig. 16.1 Computer images of tree stable conformations of Na^{+} salt of a dimer of sulfostyrene hydrated with 20 water molecules, corresponding to minima of the energy

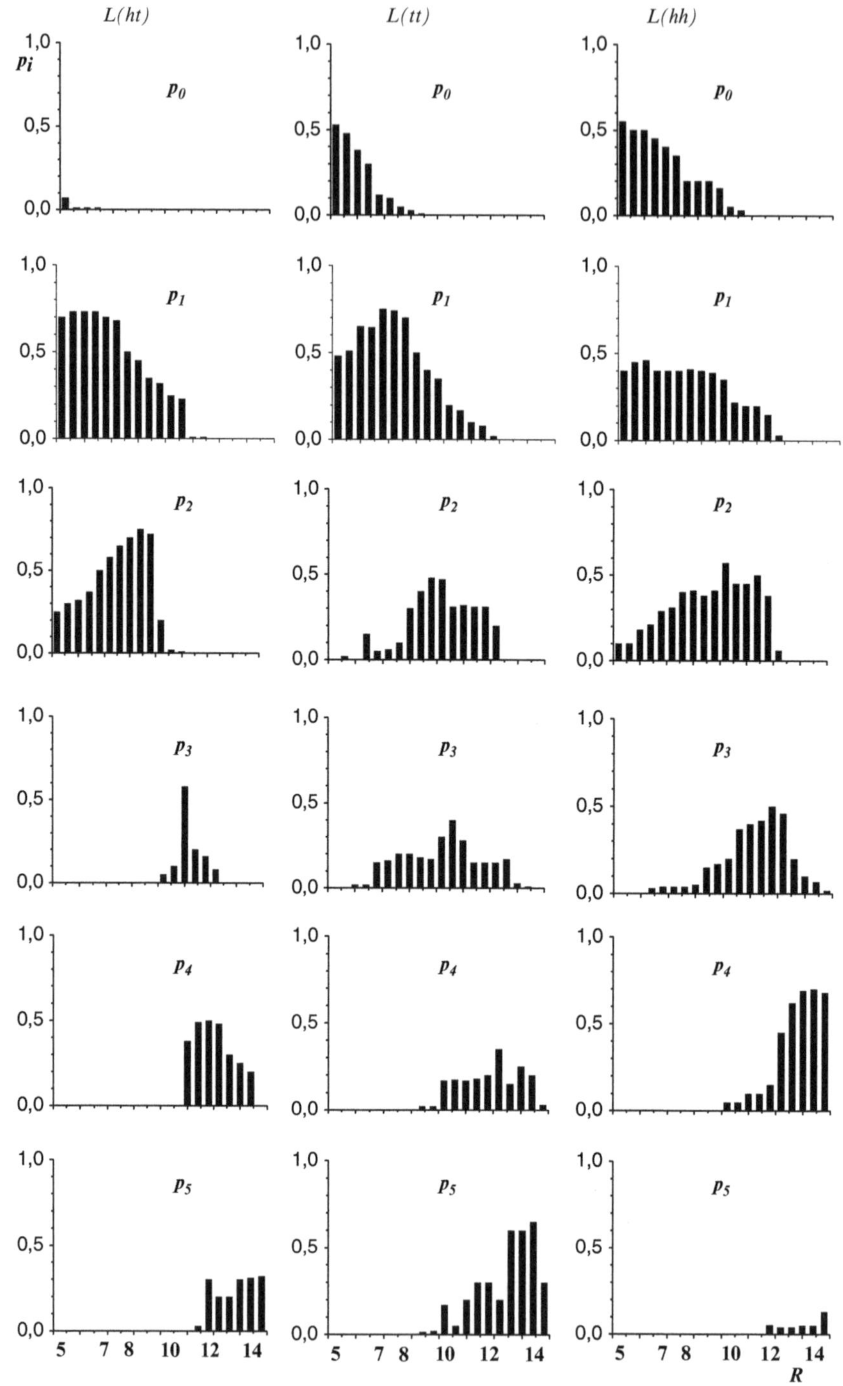

Fig. 16.2 Calculated probabilities of occurrence of zero to five neighbors in different linear fragments of the sulfonic acid polyelectrolyte as a function of distance from the exchange site

irregular structures (*hh* and *tt*), the probabilities of larger numbers of neighbors at a shorter distances from the exchange site are rather high. The dependences $Y = Y(\bar{x})$ of such polyelectrolytes can be very complicated, which contradicts to the experimental data. Therefore, it is reasonable to assume that in such systems, the predominant fragment in the polymer structure is linear chain consisting mainly of the *L(ht)* subfragments.

The probabilities of number of neighbors more than two for $R_O \leq 1$ nm for this and the other structures for the fragments containing 8–10 monomer units are given in Table 16.1. As seen from the probability histogram of *L(ht)* structure (Fig. 16.2), a significant probability of three and more neighbors appears when the distance from the observed exchange site is more than 1.0 nm. At the smaller distances, the exchange sites have two or one neighbors, and the radius of interaction may not be longer than 1.0–1.2 nm. If the R_O were assumed to be longer, then the site would have three or more neighbors, and the property composition dependence would be described with polynomial Eq. 16.5 of a power higher than 2, which was not observed in the experiments. Value $R_O = 1.0$ nm agrees well with the result of the quantum chemical calculations. Based on these considerations, we assumed the interaction radius to be equal to 1.0 nm. Calculation of the number of exchange sites in a sphere with $R = 1.0$ nm allows identifying the fragments responsible for the complexity of the dependence Y of the mole fraction of the exchanging ions in the sorbent. The property of primary importance for ion exchange systems is $Y = \log k(\bar{x})$, where k is the equilibrium coefficient defined as

$$k = \bar{x}(1 - x)/(x(1 - \bar{x})) \tag{16.7}$$

Examples of dependences of enthalpy of ion exchange ΔH and $\log k$ of $\bar{x}$ are given in Figs. 16.3 and 16.4. In order to determine correctly the power of polynomial, the k value should be determined in a sufficiently wide range of $\bar{x}$. The regions near $\bar{x} = 0$ and $\bar{x} = 1$ are especially important [12] as are the regions also seen from Figs. 16.3 and 16.4. Unfortunately, most of the published data may not be used for such analysis because the interval of the studied ionic composition is too narrow and the number of experimental points on dependences $\log k = f(\bar{x})$ is not large enough for the mathematical analysis. Only in separate cases (e.g., [26, 27]) do the quality of the data and the way of their presentation allow their treatment in terms of the model described.

It is seen from Fig. 16.3 that the $\log k = f(\bar{x})$ curves for resins with low cross-linkage (<8% DVB) are described by Eq. 16.5 of the second power and those for the higher cross-linkage need the polynomial of the third power.

In these cases, constants $y(i - j, j)$ denote logarithms of the equilibrium constants of elementary ion exchanges related to each microstate. For example, if $i = 3$, the ion exchange equilibrium is a superposition of four ideal elementary equilibria each related to one of the four possible microstates of the exchanging ion:

$$\mathrm{I}_1\,(3,0) + \mathrm{I}_2 \leftrightarrow \mathrm{I}_2(3,0) + \mathrm{I}_1, y(3,0) = \log K(3,0), K(3,0)$$
$$= \bar{x}(3,0)(1 - x)/(x(1 - (3,0)))$$

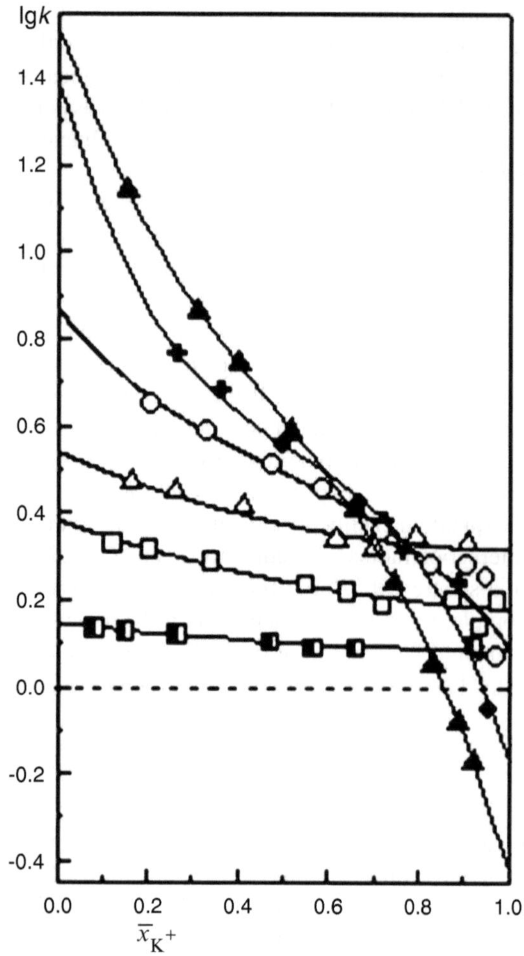

Fig. 16.3 Dependences of $\log k = f(\bar{x})$ of the degree of ion exchange H^+–K^+ on styrene–*m*-DVB ion exchanger KRSm with different percent of *m*-DVB: 1, 1%; 2, 5%; 3, 8%; 4, 12%; 5, 16%; and 6, 25%; t = 25°C. The points are experimental data. The *curves* are calculated from Eq. 16.5c (*curves* 1–3) and Eq. 16.5d (*curves* 3–6) (Reproduced from Ref. [11] with kind permission of © Elsevier (2004))

$$I_1(2,1) + I_2 \leftrightarrow I_2(2,1) + I_1, y(2,1) = \log K(2,1), K(2,1) = \bar{x}(2,1)(1-x)/(x(1-(2,1)))$$

$$I_1(1,2) + I_2 \leftrightarrow I_2(1,2) + I_1, y(1,2) = \log K(1,2), K(3,0) = \bar{x}(1,2)(1-x)/(x(1-(1,2)))$$

$$I_1(0,3) + I_2 \leftrightarrow I_2(0,3) + I_1, y(0,3) = \log K(0,3), K(0,3) = \bar{x}(0,3)(1-\mathrm{x})/(x(1-(0,3)))$$

The dependence of the equilibrium coefficient of $\bar{x}$ for %DVB ≥ 8 is described by Eq. 16.5 of the third power.

Complexity of such dependences may increase with increasing the number of neighbors *i*. If $I = 0$, the ion exchange is ideal and described by one equilibrium

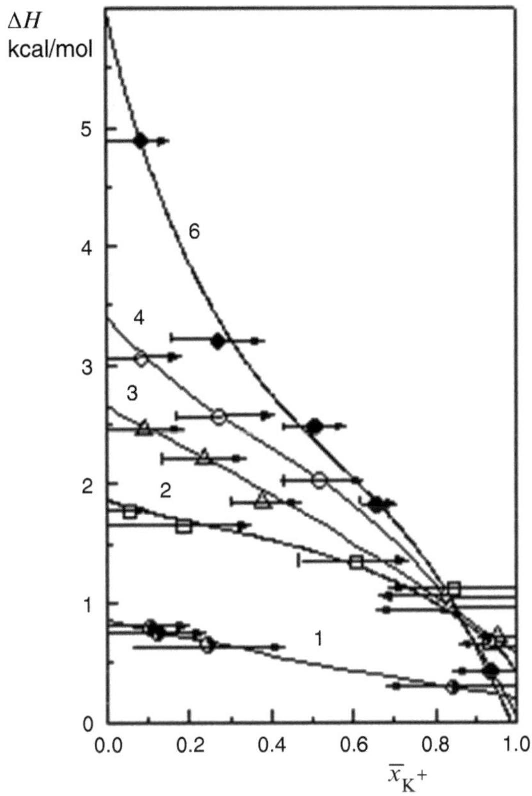

Fig. 16.4 Dependences of enthalpy of the degree of ion exchange H^+–K^+ on styrene–*m*-DVB ion exchanger KRSm with different percent of *m*-DVB. See captions to Fig. 16.3. The data were obtained by measuring heat effect of ion exchange. The beginning and the end of the arrows show the directing, starting, and end points of the process in the calorimetric measurements [26]

constant (polynomial of the zeroth power). If $i = 1$, the ion exchange can be named regular because it is described by linear equation similar to the property–composition function of regular solution. If $i = 2$, the ion exchange is described to be the polynomial of the second power, and dependence of property–composition may have one extremum but may not have an inflection point (see Figs. 16.3 and 16.4). Both extremums and inflections may occur if $i \geq 3$ (the case of large organic ions, [28]).

This model was applied for interpreting the selectivity, heat effects of ion exchange, and water uptake by resinous and liquid ion exchangers [29, 30].

The data obtained can be summarized in the following way: In the structures formed by regular linear chains, T and H fragments without the functional group in the cross bridge, and some of the ring structures, the probability of more than two neighbors is negligently small. This corresponds to ion exchangers with low cross-linkage demonstrating the simplest property–composition dependences. The fragments responsible for the complicated $Y = Y(x)$ functions are H structures with the functional group in the cross bridge. The calculations of the probabilities for the fragments with *hh* and *tt* junctions in all kind of fragments have shown [11] that the probabilities of higher than two neighbors sharply increase. The simplicity of the $Y = Y(x)$ dependence for the resins with a low cross-linkage and soluble polyelectrolytes shows that the head-to-tail junctions dominate. This means that the conditions of polymerization process to obtain the matrix polymers can affect the selectivity of ion exchangers with identical compositions.

In the case of sorption of large organic ions, complicated shape of function $\log k = f(x)$ is often observed both on the resinous and liquid ion exchangers. These dependences correspond to the presence of four numbers of sites and are well described by Eq. 16.5 for $i = 4$.

16.3 Quantum Chemical Modeling of the Sorption Complexes

The macroscopic model described above does not give an answer to the questions how in the sphere of interaction the neighboring exchange sites are distributed, in what distances the interacting ions and water molecules are located, and how they oriented to each others. The nature of the bonds between the interacting atoms is not specified either. The answer to these questions can give quantum chemical calculation of the structure of hydrated sorption complexes including one or several functional groups.

For these calculations, we used program pack GAMESS [23]. The full gradient optimizing of all geometric parameters for obtaining minimal potential energy of the system was done. All structural characteristics of our systems discussed further in this chapter are related to the minima on the potential energy surface. Preliminary calculations have shown that the most acceptable for our systems is nonempirical method SCF MO LCAO applying minimal basis set MINI [24]. It allows calculation on personal computers of large multiatomic systems (200–300 atoms). The calculations were performed by method RHF SCF MO LCAO. In the case of some heavy metal ions, for example, Ni and Cu in the main valence state, the doublet electronic state is realized; in these cases, the calculations were done by ROHF SCF MO LCAO method.

The quantum chemical calculations give structural, electronic, and energy characteristics of the system under consideration. They are coordinates of each atom, electrical charges on the atoms, and bond order of bonds between atoms (the value reflecting the degree of the bond covalence). These data allow reconstruction of the visual image of the sorption complex and calculation of the thermodynamic functions and spectra of the modeled substances.

The simplest calculations are for the structure of complexes formed by one counter ion and several water molecules. Some features of the interactions in the real ion exchange systems can be obtained by such calculations. At the same time, the interaction between several functional groups and the counter ions always occurring in the ion exchangers may not be foreseen from such calculations. Therefore, more reliable information on the interactions in the real ion exchangers can be obtained only for representative fragment of the structure containing several function groups with sufficient number of water molecules.

The following conventions were used in drawing and discussion of the computer images.

The atoms are considered chemically bound if the calculated distance between them is not larger than 25% of the sum of their radii. The drawing program is made so that in this case, it draws a bar between the atoms meaning the chemical bond, no matter ionic or covalent one.

The hydrogen bond is shown as the dotted lines drown if in the system – *X*. . .*H*. . .*Y* – the distances between atoms are in the range 0.18–0.35 nm and the angle *XHY* is 150–250°. The effective electrical charge (in electron units) is shown near the relative atom with the relative sign; its length in nanometers is placed near the bond; the order of bond is given in square parentheses.

The meaning of all above-mentioned terms correspond to IUPAC recommendations [31].

The totality of molecules, ions, and fragments of the polymer network in the considered fragment is named *hydration system*. The structure of atoms and ions bound directly to water molecules is named *hydration complex*. It does not include the water molecules bound by hydrogen bonds only with the other water molecules.

Complete information on structural and electronic characteristics of the hydrated fragments obtained in the calculations is too bulky for its presentation in this chapter. Therefore, we present only several examples and the most important characteristics of the molecular fragments. The information of the other characteristics – for example, valence angles, electrical charges on the atoms, and geometric and electronic characteristics of water molecules bound to the functional groups and/or the counter ion – is stored in electronic form and can be used for interpretation of the properties of ion exchangers. In this case, a situation occurs when we cannot use all information obtained and restrict our consideration only with its part according to our subjective idea of the importance of different factors affecting ion exchange selectivity.

16.3.1 Hydration Systems with Univalent Counterions

The ion exchanger was modeled by *p*-ethylbenzene sulfonic acid and the oligomers of polystyrene sulfonic acid and their salts with the degree of polymerization $p = 2$–4. As counter ions, Li^+, Na^+, K^+, Rb^+, Cs^+, Ag^+, and H^+ have been chosen. The structures of the hydration systems in the presence of $W = 0$–10 water molecules per sulfonic group have been calculated. One of the important questions to clarify is whether the structure of the hydration system of *p*-ethylbenzene sulfonic acid and its salts is identical to those for the oligomers and what is the minimal size of the fragment which can be used for modeling the ion exchanger. The structures of the hydration systems presented in this section have been calculated using MIDI(d) basis with accounting for electronic correlation MP2/MIDI(d). In several cases, 3-21G*, MIDI(d), MP2/3-21G*, and MP2/MIDI(d) bases have been used. The results obtained were practically the same as those for the used basis.

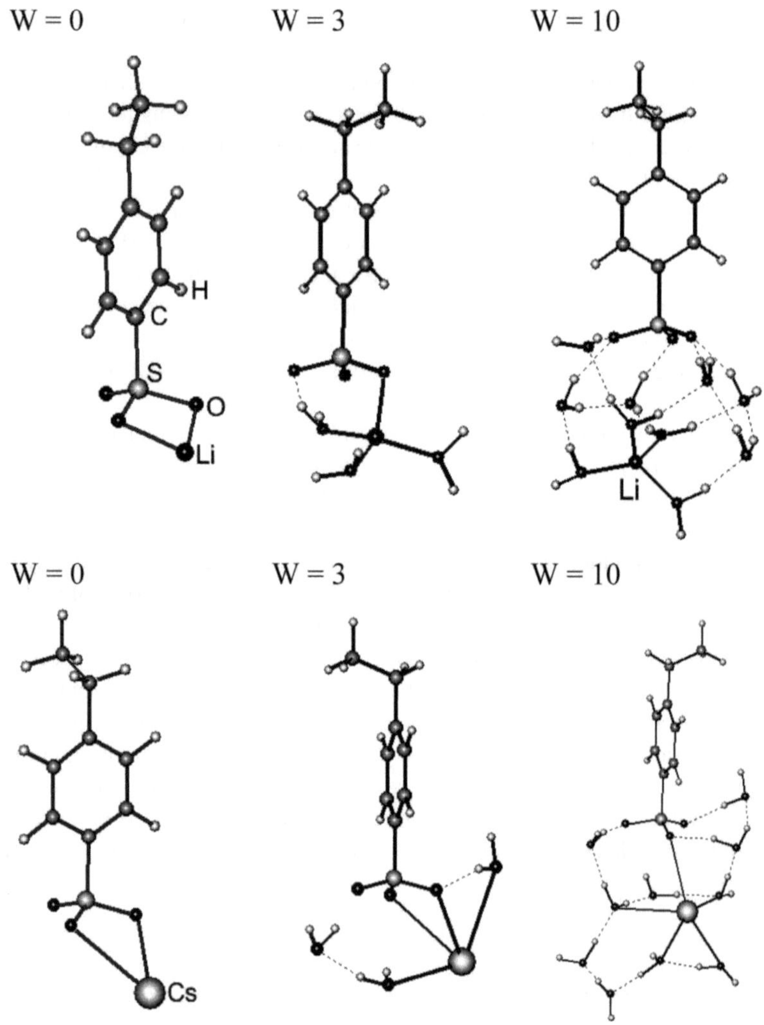

Fig. 16.5 Calculated structure of *p*-ethylbenzene sulfonates (*p*-EBS) of Li and Cs at different degrees of hydration. W = 0, 3, and 10 water molecules per sulfonic group

16.3.1.1 Alkali Metal Ions

The Models of Fragment Containing One Isolated Sulfonic Group

Alkali metal ions are expected to be the simplest case of such systems. The structures of sorption complexes at the different degrees of hydration of *p*-ethylbenzene sulfonates of Li^+ and Cs^+ are given as the examples in Fig. 16.5.

These examples reflect some common features in the structures of the hydrated complexes which also can be related to the lonely exchange groups present in the structure of the ion exchangers as it was described in Sect. 16.2.

In all cases, the cations in the waterless structure are bound to two oxygen atoms of sulfonic groups forming unsymmetrical structure. Zundel [32] obtained the IR spectral proof of the asymmetry of these groups, but he assumed that the cation is bound to one oxygen atom. Nevertheless, his experimental data are in agreement with the results of our calculations. The stepwise hydration of the Li^+ salt leads to complete separation of the ions while Cs^+ remains bound to one of the O atoms of the sulfonic group even in presence of 10 water molecules which corresponds to the swelling of ion exchanger with approximately 4% DVB. The calculations done for the other alkali metals show that Na^+ salts hydrated similar to that of Li^+ ones and K^+ and Rb^+ salts behave like Cesium salts of ethylbenzenesulfoacid (CsEBS).

In Refs. [33, 34], a difference in selectivity of Li^+ and Na^+–H^+ exchange on one side and K^+, Rb^+, and Cs^+–H^+ on the other side was observed and explained by the difference in the type of the hydration mode of these groups of ions. Li^+ and Na^+ are "structure-forming" ions ("positively hydrated"), and K^+, Rb^+, and Cs^+ are "structure-breaking" or "negatively" hydrated ions [35–37]. The quantum chemical calculations explain the nature of this phenomenon. In the complete hydrated ion exchangers with a moderate and high cross-linkage, Li^+ and Na^+ ions are not directly bound to the sulfonic group while K^+, Rb^+, and Cs^+ remained fixed on one of the oxygen atoms.

Figure 16.6 shows the location of the cation in the vicinity of the sulfonic group as three numbers expressing the distances between the centers of the cation and the centers of the oxygen atoms. It is seen from this diagram that distances of two oxygen atoms in the absence of water and at the first steps of hydration are practically equal to the sum of the ionic radii of oxygen and the cation which corresponds to existence of the bonds between them. The order of these bonds in this case (degree of covalence) is less than 0.05, and the bonds are purely electrostatic. Upon hydration, one of the bonds became broken. In the case of structure-forming ions (Li^+ and Na^+), all three bonds became broken when 10 water molecules are present in the system. In difference with that, the structure-breaking cations (K^+, Rb^+, Cs^+) do not move away from one of the $-SO_3^-$ oxygen atoms, indicating existence of the bond between them.

The state of water molecules in the hydration system can be expressed by the number of molecules with a certain type of binding to the ions in the hydration system. We classify and denote them as follows:

- W_R – bound only to the fixed ion
- W_{RR} – bound only to two fixed ions
- W_I – bound only to one counter ion
- W_{II} – bound only to two counter ions
- W_{RI} – bound both to the fixed ion and counter ion
- W_W – bound only to another water molecules

The total number of water molecules in the hydration system W is a sum of the molecules of all categories:

$$W = W_R + W_{RR} + W_I + W_{II} + W_{RI} + W_W$$

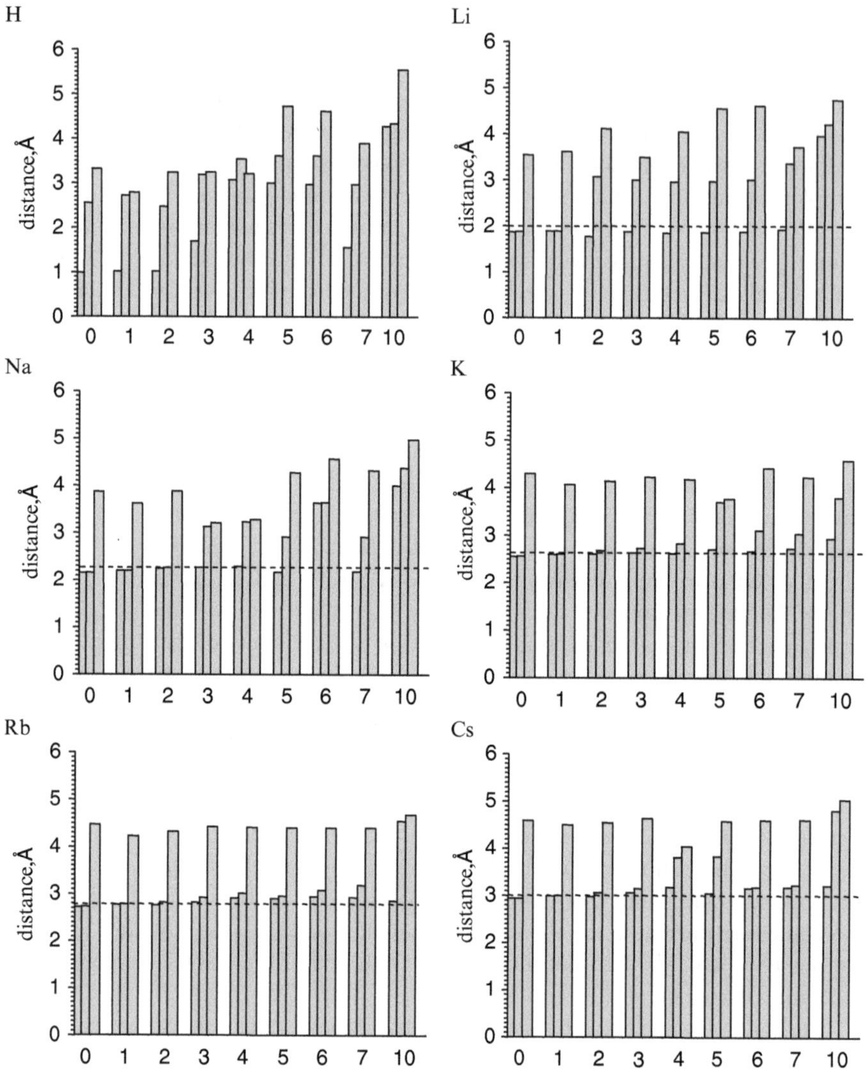

Fig. 16.6 Distances between counter ions and the oxygen atoms of sulfonic groups at the different hydration stages. The dashed lines designate the sums of ionic radii of Cat^{+}–O [38]. Each column of three corresponds to distance from the counter ion to one of the oxygen atoms of sulfonic groups

The number of water molecules in the hydration complex is

$$W_{\Sigma RI} = W - W_{W}$$

The other states of water molecules are not considered because they are not met in our specific cases.

Table 16.2 Distribution of water molecules according to the type of binding with ions for the fragments including one isolated sulfonic group

		The number of water molecules, W							
I^+	Type of H_2O	1	2	3	4	5	6	7	10
Li	W_R	1	1	1	2	3	3	4	4
	W_I	0	1	2	1	1	2	3	5
	W_{RI}	0	0	0	1	1	0	0	0
	W_W	0	0	0	0	0	1	0	1
Na	W_R	0	0	0	1	2	3	3	4
	W_I	0	0	0	0	2	1	4	5
	W_{RI}	1	2	3	3	1	2	0	0
	W_W	0	0	0	0	0	0	0	1
K	W_R	0	0	1	2	1	2	2	3
	W_I	0	0	1	2	1	2	2	5
	W_{RI}	1	2	1	0	3	2	2	1
	W_W	0	0	0	0	0	0	1	1
Rb	W_R	0	1	0	2	2	2	2	4
	W_I	0	0	1	2	2	2	3	4
	W_{RI}	1	1	2	0	1	1	1	0
	W_W	0	0	0	0	0	1	1	2
Cs	W_R	0	1	0	1	2	1	2	2
	W_I	0	0	1	1	2	3	3	4
	W_{RI}	1	1	2	2	1	1	1	2
	W_W	0	0	0	0	0	1	1	2

The data on the state of water molecules in the hydration systems at different degrees of hydration for the alkali metal salts of *p*-ethylbenzene acid is given in Table 16.2. They will be discussed together with the data on the hydration of the fragments containing several functional groups.

The Models of Fragments Containing Several Sulfonic Groups

As noted above in the real ion exchanger alongside with the isolated sulfonic groups, the groups having several neighbors are present. The linear fragments consisting of 2, 3, and 4 monomeric units were modeled with 10 water molecules per one sulfonic group. Their total composition can be expressed by the following formula: $H(CH_2CH_2C_6H_4SO_3^-)_fCH_3(fMe^+) + 10fH_2O$. The calculations were done for nonhydrated fragments (W = 0) and fragments with W = 10 for Li^+, Na^+, and K^+ cations.

The Nonhydrated Fragments

It appeared that the fragments in which the sulfonic groups are situated in *cis* positions relative the hydrocarbon chain (on the same side of it) are energetically favorable and their structures are similar. They are exemplified by Fig. 16.7.

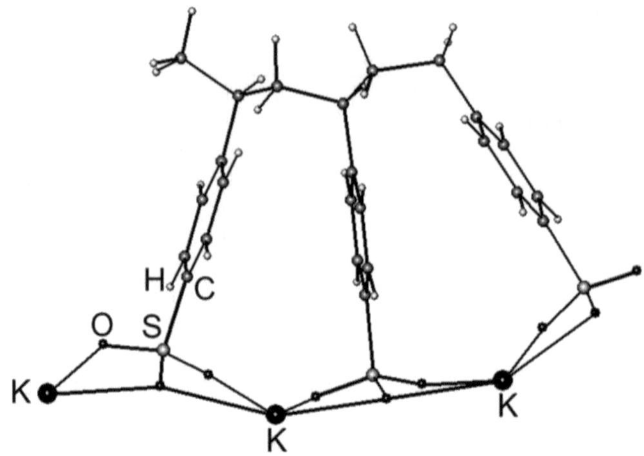

Fig. 16.7 Calculated structure of nonhydrated fragment of sulfostyrene ion exchanger in K^+ form

All cations of alkali metals except one in each kind of fragment are bound to the oxygen atoms of two neighboring groups. One cation is bound to one sulfonic group situated in the end of fragment, but that is an artifact caused by the specificity of the end position of the sulfonic group; it is not accounted at the further consideration. All cations are four coordinated; their Me–O bonds have approximately equal length, practically equal to the sum of the ionic radii. The sulfur atoms lie near the plain. The order of Me–O bonds in all cases is less than 0.05, indicating their pure electrostatic character. Increasing the number of repeated units in the fragment little affects the length of the Me–O bonds.

The Hydrated Fragments

Similar to the hydration the salts of *p*-ethylbenzenesulfoacid fragments, containing several functional groups, two types of structures are formed. The Li^+ and Na^+ forms are completely dissociate to the counter ions and fixed ions. The distances between the cation and nearest O atom of $-RSO_3^-$ are much longer than the sum of the ionic radii of Me^+ and O^- ions (for Li^+ 0.445 and 0.198 nm, respectively). The counter ions are bound only to water molecules independently of the size of the molecular fragment. The geometry of the fragments with Li^+ and Na^+ is similar and illustrated by Fig. 16.8.

Ions Li^+ and Na^+ are bound with four water molecules, the length of Me–O bonds is independent of the fragment size and equal to 0.185–0.188 nm for Li^+ and 0.214–0.217 nm for Na^+, which is close to the sum of the relative ionic radii. The sulfonic groups are situated under angle 16–49° relative to each other. The neighboring groups are pointed to different directions from the plain formed by the carbon atoms of the polymer chain. The *trans* configurations are possible, but they are less favorable energetically than the *cis* configurations.

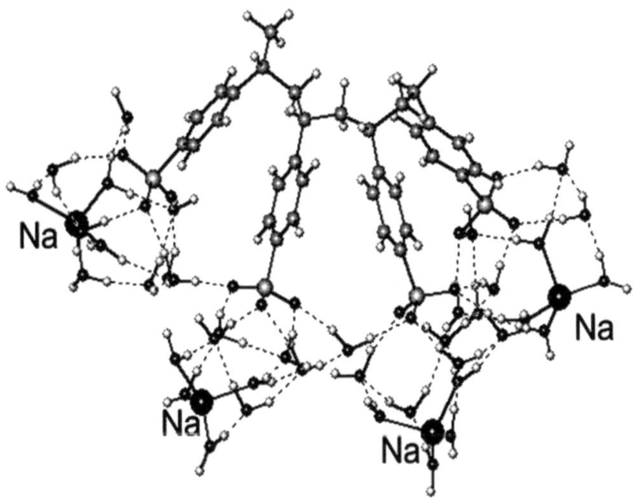

Fig. 16.8 Calculated structure of linear fragment of sulfostyrene ion exchanger in Na^+ form containing four sulfonic groups and 10 water molecules per sulfonic group

The distribution of water molecules on the type of bonding with the ions in the hydration systems is given in Table 16.3.

Hydration of K^+ forms of the fragments with 10 water molecules does not cause dissociation of the ion pairs independently of the size of the fragment (Figs. 16.9 and 16.10). Nevertheless, the complete dissociation occurs in the solution of the soluble potassium polystyrene sulfonates or in swollen ion exchanges with very high swelling. That follows from the osmotic coefficients of their solutions tending to the same value for polystyrene sulfonates of all alkali metals at infinite dilution [39] and the studies of the selectivity of ion exchange by low cross-linked ion [40]. In the ion exchangers whose swelling corresponds to 10 and less water molecules per functional groups in potassium form (approximately 4% DVB), K^+ is bound to one of O atoms of the sulfonic group. The length of the K^+–OSO_2 bond in the hydrated system independently of their size is 0.262–0.277 nm, which is very close to the length in the nonhydrated fragments 0.260–0.273 nm. The order of the bonds in all cases is below 0.05, indicating electrostatic nature of the bond.

The data in Table 16.3 can be summarized as follows: The number of water molecules per one functional group in the hydration system, $W_{\Sigma RI}$, for Li^+ and Na^+ forms is 9 per group and independent of the number of monomeric units in the fragment. These molecules are almost equally divided between those hydrating only the cation, $W_I = 4$, and hydrating only the sulfonic group, $W_R = 4.25$. The characteristics of hydration in the hydration system containing several functional groups are the same as those for Li and Na salts of *p*-ethylbenzene sulfonates modeling the polymeric fragment with one functional group. The distribution of water molecules in the fragments with K^+ ion is substantially different from that for Li^+ and Na^+.

Table 16.3 The distribution of water molecules on the type of bonding with the ions in the hydration systems with Li^+, Na^+, and K^+. The numbers of water molecules of a certain type are average per one functional group

Type of water molecule	Number of sulfonic groups in the fragment *p* 1	2	3	4
	Li⁺			
W_R	5	4.5	4.33	4.25
W_I	4	4	4	4
W_{RI}	0	0	0	0
W_{RR}	0	0.5	0.67	0.75
W_{II}	0	0	0	0
$W_{\sum RI}$	9	9	9	9
W_W	1	1	1	1
	Na⁺			
W_R	5	4.5	4.33	4.25
W_I	4	4	4	4
W_{RI}	0	0	0	0
W_{RR}	0	0.5	0.67	0.75
W_{II}	0	0	0	0
$W_{\sum RI}$	9	9	9	9
W_W	1	1	1	1
	K⁺			
W_R	5	3	2.67	3
W_I	4	2	1.67	2.75
W_{RI}	0	0	0.33	0.75
W_{RR}	0	0	0	0
W_{II}	0	1	1	0
$W_{\sum RI}$	9	6	5.67	6.5
W_W	1	4	4.33	3.5

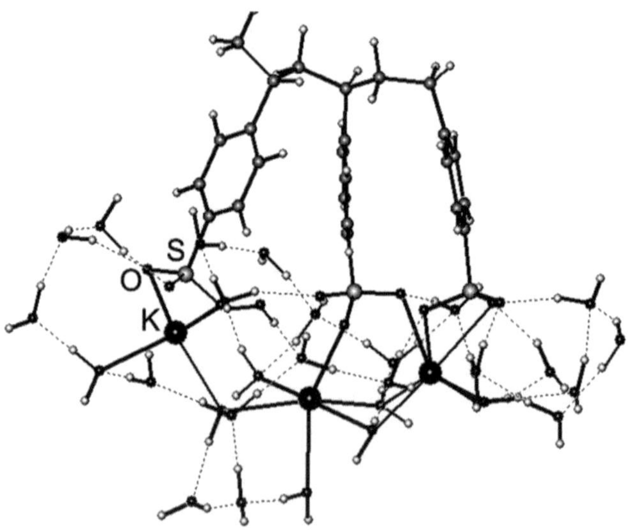

Fig. 16.9 Calculated structure of linear fragment sulfostyrene ion exchanger in K^+ form containing three sulfonic groups and 10 water molecules per sulfonic group

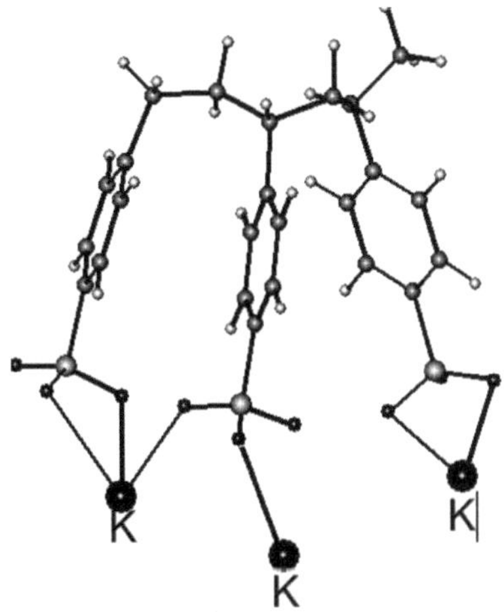

Fig. 16.10 Calculated structure of linear fragment sulfostyrene ion exchanger in K^+ form containing three sulfonic groups and 10 water molecules per sulfonic group. The water molecules are not shown

The total number of water molecules in the hydrate structure with one sulfonic group is 9, the same as for Li^+ and Na^+. But already in the fragment with two functional groups, it becomes 6 and remains close to that value for the bigger fragments. The number of water molecules joined to only cation and only anion is by one or two units smaller in the fragment with several functional groups than in the fragments with one group, which indicates that only fragments containing several functional groups may be chosen as representative fragment of the structure of ion exchanger and used for determining of the water state in the resin.

16.3.1.2 Silver Ion

Ion Ag^+ has crystallographic radius 0.115 nm, which is between Na^+ (0.095 nm) and K^+ (0.133 nm). Nevertheless, the selectivity of its sorption by sulfostyrene ion exchangers is much higher than that of the both Na^+ and K^+ [41]. It was also observed that judging from the IR of nonhydrated liquid ion exchanger of sulfonic type Ag^+ forms symmetric $–SO_3Ag$ structure [42]. In very general terms, such a behavior was explained by polarization interaction between silver ion and sulfonic groups. Quantum chemistry calculations allow clarifying the nature of chemical bonds between Ag^+ and $–SO_3^-$.

In Fig. 16.11, the structure of nonhydrated ion pair $–SO_3Ag$ calculated in MIDI(d) basis with accounts for electron correlation MP2/MIDI(d) is presented. The silver ion symmetrically interacts with all three oxygen atoms of sulfonic group and with the sulfur atom. It is probably due to interaction of the valence *d* electrons of silver with vacant *d* orbitals of the sulfur atom. The distances between Ag^+ and each of the atoms of sulfonic group are longer than the sum of the relative ionic radii, but instead of two ionic bonds, as in the cases of Na^+ and K^+, it forms four bonds with a large fraction of covalence (see bond orders in Fig. 16.11). The most covalent bond is formed with the sulfur atom.

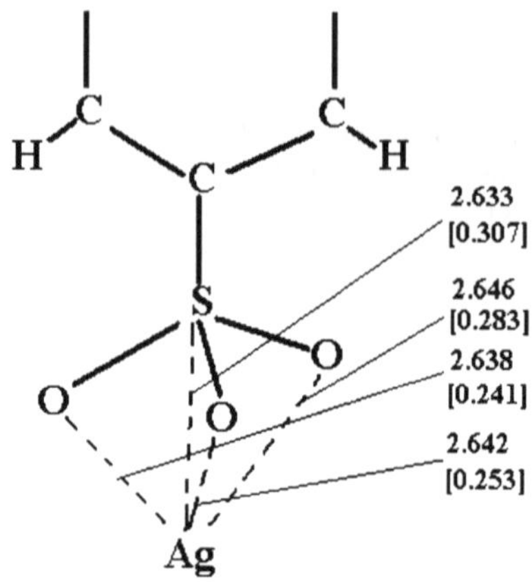

Fig. 16.11 The structure of nonhydrated ion pair $-SO_3Ag$. The distances between atoms and bond orders (in *square brackets*) are given on the figure

16.3.1.3 Hydrogen Ion

The state of proton in aqueous media was studied by theoretical and experimental methods [32, 43–48]. The resulting most common opinion followed from these studies is that that two predominant proton hydrate are formed. They are Zundel's proton $H_5O_2^+$ [32] and Eigen's proton $H_9O_4^+$ [44]. The latter is hydroxonium ion H_3O^+ bound to three water molecules. The state of proton in usually studied dilute aqueous solutions is different from that in the hydrated H^+ form of cation exchangers because in the latter case, the concentration of the protons is much higher and the counter ions fixed on the polymer matrix are not regularly distributed in the volume.

A special interest deserves the state of proton in dry and slightly hydrated resins used in the acid catalysis and sorption of base gases from the air.

The hydration of H^+ forms of sulfonic ion exchangers was studied by thermodynamic [42, 49] and IR spectroscopic [32, 42] methods. On the basis of these studies, assumptions on the state of the proton in the ion exchangers were made. Nevertheless, these assumptions are often contradictory and cannot be unambiguously related to specific atomic structures because of the restriction of the experimental methods used and complexity of the systems studied.

The fragments of the polymeric structure of the ion exchanger containing one, two, and three sulfonic groups were modeled.

The structures obtained by calculations are presented in Figs. 16.12–16.14 and characterized by Tables 16.4 and 16.5.

From the results of these calculations, we can estimate the number of protons present in different states:

- Connected to the oxygen atom of sulfonic group
- Forming hydroxonium ion (H_3O^+)
- Interacting simultaneously with two water molecules (ion $H_5O_2^+$)

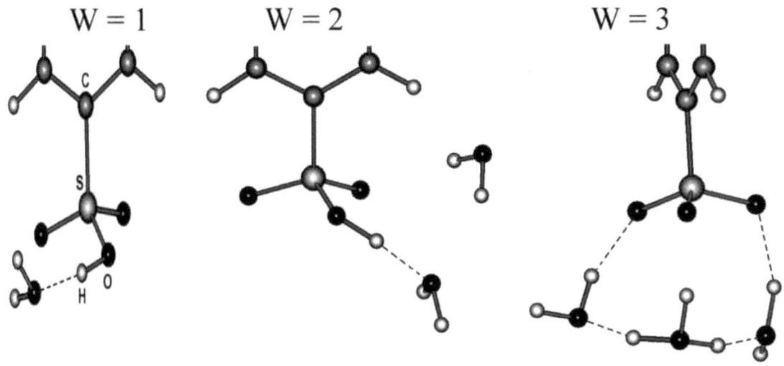

Fig. 16.12 Calculated structures of isolated functional groups of sulfostyrene ion exchanger in H^+ form at different stages of hydration

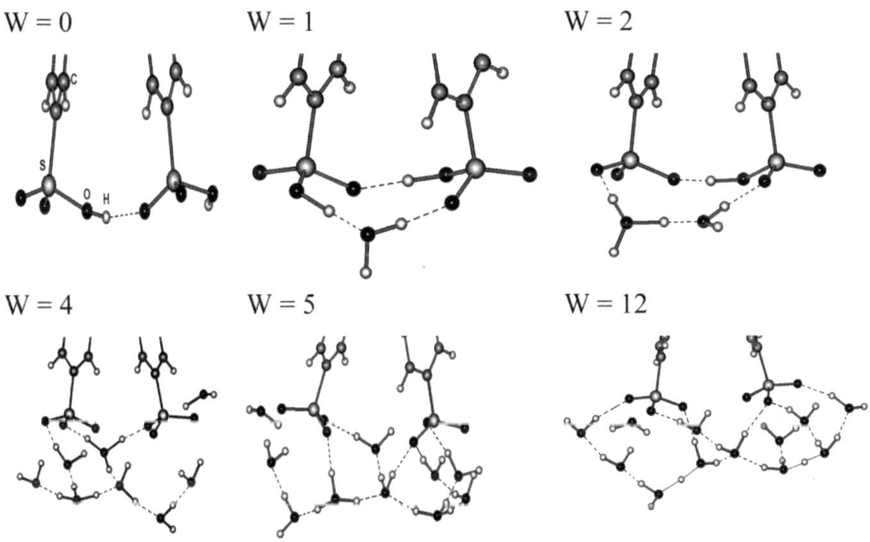

Fig. 16.13 Calculated structures of two adjacent functional groups of sulfostyrene ion exchanger in H^+ at different stages of hydration

It is important to note that unambiguous separation of specific protons from the hydration system is impossible because in the system with hydrogen bonds, the barrierless redistribution of the proton is possible. Nevertheless, the data presented in Tables 16.4 and 16.5 allow evaluation of the degree of hydration and the presence of the neighboring sulfonic groups on the concentration of the protons of each type. From Figs. 16.12–16.14, it is seen that in the studied hydrogen systems, formation of both H_3O^+ and $H_5O_2^+$ is possible.

The water molecules interact with the sulfonic groups in two ways: by formation of the hydrogen bonds and by simple electrostatic attraction. Assigning the latter to hydrogen bonds is impossible because the H–O–H angle is too small (130–150°).

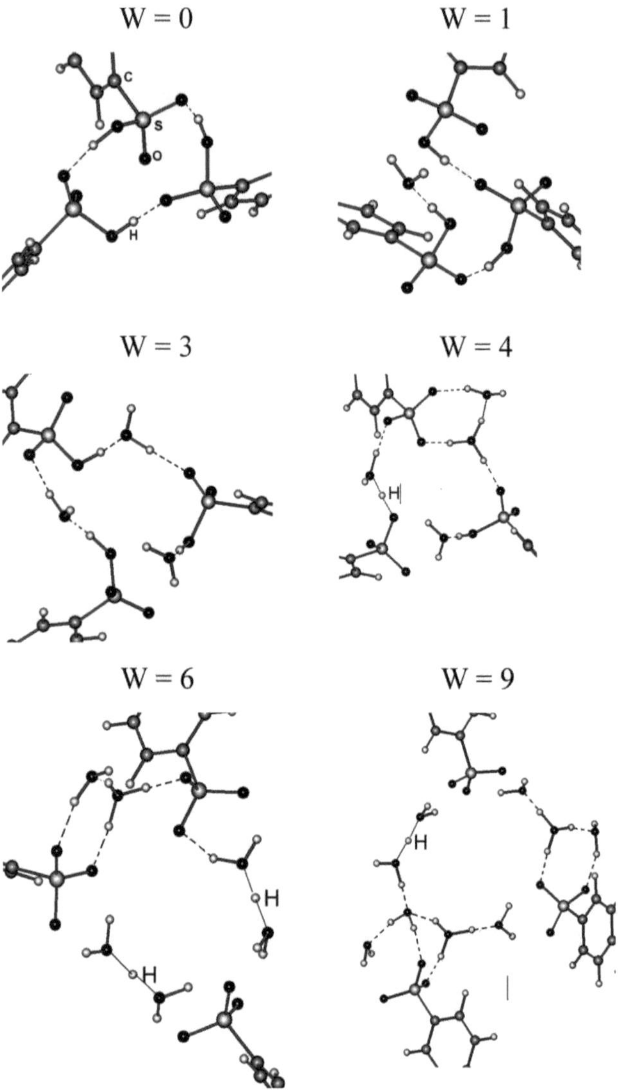

Fig. 16.14 Calculated structures of three adjacent functional groups of sulfostyrene ion exchanger in H^+ at different stages of hydration

It follows from the data obtained that dissociating of the sulfonic groups occurs when three water molecules are present at the isolated group, one water molecule per group in the case of two groups in the fragment and two water molecules per three sulfonic groups. The presence in the system of three water molecules per sulfonic group in all cases leads to complete dissociation of sulfonic group. Introduction into the system additional water molecules does not cause substantial changes in the state of ions. Ion $H_9O_4^+$ can be formally identified only in the system with 10 water molecules per sulfonic group.

Table 16.4 The structural and electronic characteristics of proton for the model of an isolated sulfonic group in the sulfostyrene ion exchanger

W	Complex of H^+	R_h[a]	R_{ho}	q_h	B_o
0	SO_3H	0.098	0.098	0.362	0.847
1	SO_3H	0.103	0.103	0.416	0.698
2	SO_3H	0.102	0.102	0.409	0.716
3	H_3O^+	0.108	0.288	0.442	0.592
4	H_3O^+	0.111	0.303	0.448	0.540
5	H_3O^+	0.113	0.300	0.452	0.491
6	H_3O^+	0.111	0.297	0.438	0.525
10	H_3O^+	0.110	0.428	0.435	0.550

[a]Here and further, the following designations are used: R_h distance from the acidic proton to the nearest oxygen atom of water molecule or sulfonic group; R_{ho} distance from the acid proton to the nearest oxygen atom of the sulfonic group only; q_h charge on the acid proton; and B_o bond order between the acid proton and the nearest oxygen atom

Table 16.5 The structural and electronic characteristics of proton for the model of three interacting sulfonic groups in the sulfostyrene ion exchanger

Number of functional group	Complex of H^+	R_s[a]	R_h	R_{hh}	R_{ho}	q_h	B_o
0	SO_3H	0.360	0.102	0.362	0.102	0.427	0.732
	SO_3H	0.366	0.102	0.286	0.102	0.422	0.720
	SO_3H	0.514	0.101	0.415	0.101	0.410	0.755
1	SO_3H	0.389	0.102	0.445	0.102	0.418	0.721
	SO_3H	0.468	0.105	0.392	0.105	0.441	0.631
	SO_3H	0.492	0.102	0.281	0.102	0.427	0.723
2	SO_3H	0.361	0.108	0.501	0.108	0.444	0.580
	H_3O^+	0.563	0.107	0.547	0.140	0.464	0.608
	H_3O^+	0.573	0.115	0.355	0.126	0.464	0.302
3	SO_3H	0.624	0.112	0.448	0.112	0.452	0.493
	SO_3H	0.691	0.109	0.395	0.109	0.447	0.546
	SO_3H	0.670	0.111	0.478	0.111	0.452	0.524
4	SO_3H	0.684	0.112	0.519	0.112	0.452	0.496
	H_3O^+	0.686	0.107	0.590	0.296	0.452	0.584
	$H_5O_2^+$	0.601	0.119	0.435	0.121	0.461	0.396
5	H_3O^+	0.617	0.113	0.617	0.128	0.466	0.480
	$H_5O_2^+$	0.754	0.119	0.577	0.354	0.469	0.389
	$H_5O_2^+$[b]	0.796	0.117	0.587	0.342	0.469	0.414
6	H_3O^+	0.645	0.106	0.564	0.143	0.444	0.635
	$H_5O_2^+$	0.744	0.119	0.763	0.354	0.467	0.394
	$H_5O_2^+$	0.826	0.119	0.639	0.364	0.471	0.397
9	H_3O^+	0.812	0.110	0.663	0.332	0.454	0.524
	$H_5O_2^+$	0.827	0.118	0.621	0.356	0.460	0.394
	H_3O^+	1.016	0.112	0.561	0.328	0.464	0.490
12	H_3O^+	0.834	0.114	0.622	0.341	0.455	0.455
	H_3O^+	0.835	0.113	0.745	0.338	0.466	0.482
	H_3O^+	0.982	0.113	0.613	0.359	0.460	0.472
15	H_3O^+	0.816	0.115	0.631	0.345	0.460	0.449
	H_3O^+	0.828	0.110	0.549	0.376	0.453	0.538
	H_3O^+	1.154	0.110	0.676	0.325	0.452	0.545

(continued)

Table 16.5 (continued)

Number of functional group	Complex of H^+	R_s [a]	R_h	R_{hh}	R_{ho}	q_h	B_o
18	H_3O^+	0.831	0.113	0.587	0.340	0.448	0.496
	H_3O^+	0.932	0.108	0.601	0.328	0.449	0.570
	$H_5O_2^+$	1.156	0.118	0.596	0.347	0.459	0.444
30	H_3O^+	0.872	0.112	0.745	0.299	0.444	0.507
	H_3O^+	1.202	0.111	0.476	0.374	0.436	0.502
	H_3O^+	1.229	0.108	0.924	0.526	0.432	0.580

[a]R_s – distance between the sulfur atoms
[b]In this case, one oxygen atom belongs to the sulfonic group

The data on location of the ions in hydration systems for the isolated sulfonic group and fragments with two, three, and four functional groups at different degrees of hydration have been obtained. In Tables 16.4 and 16.5, the data for one and three functional groups are presented.

It is to be noted that the methods for calculations of the electrical charges on atoms are rather conventional and depend strongly on the quantum chemical approximation used and local surrounding of the atom. More reliable characteristics of relative strength of the chemical bonds of the same type (in our case, the acid proton–oxygen atom) are their lengths and the orders.

The computed lengths and orders of these bonds indicate a substantial dependence of proton binding from the degree of hydration and the number of the neighboring functional groups. The longer the length of the bond and lower its order, the greater ability of the proton to act in the acid–base, ion exchange, and catalytic processes (i.e., its higher reactivity). It is seen from the tables that the lowest reactivity has the protons bound to the sulfonic anion. The highest reactivity has protons bound to ions H_3O^+ and $H_5O_2^+$ for the functional groups having two neighbors. It also follows from the tables that the probability of formation of $H_5O_2^+$ in the hydration process is substantially lower than that of H_3O^+.

Geometry of the structures with three and four hydrated monomeric units is presented in Figs. 16.14 and 16.15. It is seen from Fig. 16.14 and Table 16.5 that in the first case, formation of two ions H_3O^+ and one $H_5O_2^+$ occurs. In the fragment with four sulfonic groups, all four acidic protons are present as H_3O^+.

The state of water in fully hydrated fragments containing 1–4 monomeric units is characterized by Table 16.6. It is seen from the table that the greatest change in the water state in the fragments with different numbers of monomeric units occurs between $f = 1$ and 2. Further increase in the number of elements in the fragment gives less pronounced effect. Six water molecules are bound to the hydrated complex, if $f > 1$.

The data on quantum chemical calculation of the structure of hydration system for H^+ form can be summarized as follows: Two types of the proton hydrates are formed in these systems. They are H_3O^+ and $H_5O_2^+$. They interact with the sulfonic groups and the other water molecules. The separation of proton from sulfonic group accompanied by formation of its hydrate complexes occurs when three water molecules are present in the system with the isolated group, one water molecule per group in the system with two sulfonic groups and two water molecules per three sulfonic groups in the system.

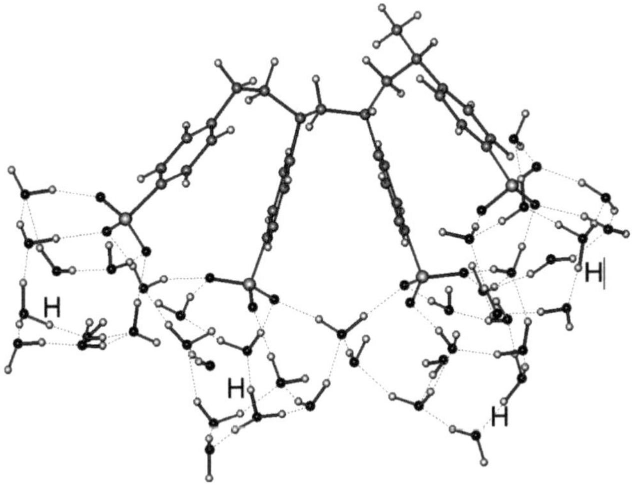

Fig. 16.15 Calculated structure of linear fragment sulfostyrene ion exchanger in H^+ form containing four sulfonic groups and 10 water molecules per sulfonic group

Table 16.6 Distribution of the water molecules according to the type of their bonding with elements of the ion exchanger structure. The hydrate systems contain 10 water molecules per the sulfonic groups. The values given are average per one sulfonic group

	Number of sulfonic group in fragment, *f*			
The water molecule type	1	2	3	4
W_R	5	3	3.5	3.67
W_I	2	1.5	1.5	1.0
W_{RI}	0	1.0	0.25	1.0
W_{RR}	0	0.5	0.75	0.33
W_{II}	0	0	0	0
$W_{\Sigma RI}$	7	6.0	6.0	6.0
W_W	3	4.0	4.0	4.0

16.3.2 Hydration Systems with Bivalent Ions

16.3.2.1 Cations of Alkaline Earth Metals

The structures of nonhydrated alkaline earth metals ionic forms of sulfostyrene ion exchangers for the fragments containing two and four monomeric units obtained in the calculations appeared similar. They are illustrated by Fig. 16.16. The cations form symmetric structures with two sulfonic groups having pure electrostatic bonds and practically equal length. They are compared with the sum of ionic radii in Table 16.7.

In the structures including 10 water molecules per sulfonic, the alkaline earth metal ions most often are bound with one sulfonic group by two bonds and by one bond with the neighboring group.

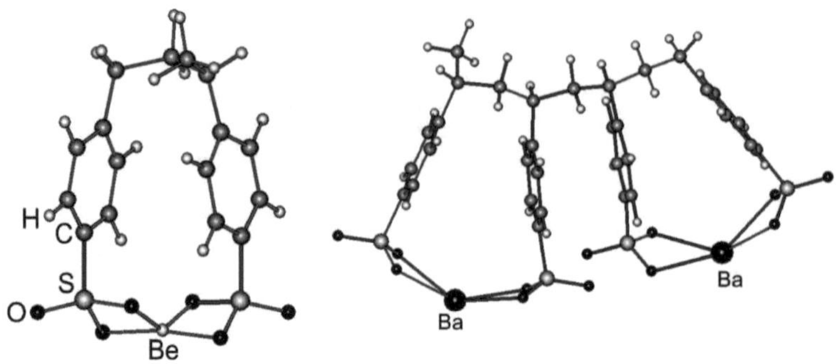

Fig. 16.16 Calculated structures of nonhydrated Be^{2+} and Ba^{2+} forms of sulfostyrene ion exchanger

Table 16.7 Characteristics of nonhydrated complexes Me^{2+}

Ion	Sum of ionic radii	Bond length	Bond order
Be^{2+}	0.174	0.170	0.140
		0.170	0.140
		0.177	0.137
		0.177	0.137
Mg^{2+}	0.210	0.196	0.098
		0.196	0.098
		0.199	0.103
		0.199	0.103
Ca^{2+}	0.236	0.233	0.058
		0.233	0.058
		0.234	0.056
		0.234	0.055
Sr^{2+}	0.256	0.251	<0.05
		0.251	<0.05
		0.251	<0.05
		9.251	<0.05
Ba^{2+}	0.274	0.271	<0.05
		0.271	<0.05
		0.271	<0.05
		0.271	<0.05

Ion Mg^{2+} coordinates six O atoms, three of which are O atoms of water. Ca^{2+}, Sr^{2+}, and Ba^{2+} form eight coordinated bonds, five of which are the bonds with water molecules. The charges on atoms are in the range 1.79–1.89, and the order of the bonds in all cases is less than 0.05, indicating their pure electrostatic character. As an example in Fig. 16.17, the structure of hydrated fragment in Ca form is given.

Classification of the water molecules according to the type of their bonding in the hydration system is given in Table 16.8. In difference with ions of alkali metals in this case, the water molecules bound simultaneously to the cation and anion, two cations or two anions are absent. The sorption complexes of each cation form its own closed system interacting with each other only via a chain of the water

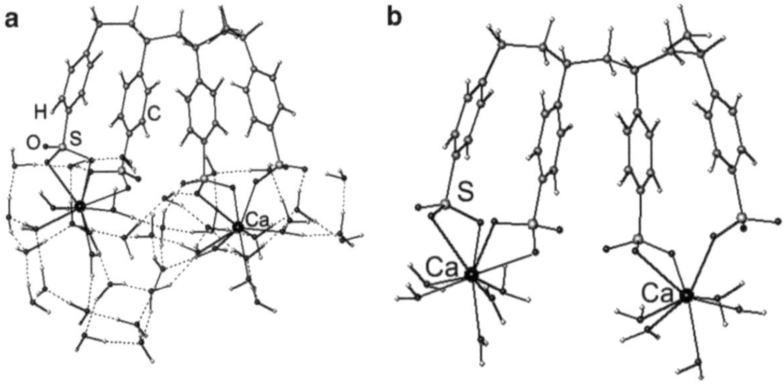

Fig. 16.17 The structure of the fragment of ion exchanger in Ca form including four functional groups and 40 water molecules: (**a**) all water molecules are shown; (**b**) only water molecules bound to the cations are shown

Table 16.8 Classification of the water molecules according to the type of their bonding in the hydration system for the fragment with four monomeric units. The numbers are given as the average in calculation per one sulfonic group

	Be^{2+}	Mg^{2+}	Ca^{2+}	Sr^{2+}	Ba^{2+}
W_R	2.25	2.0	2.25	2	1.75
W_I	3.0	1.5	2.5	2.5	2.5
W_{RI}	0	0	0	0	0
W_{RR}	0	0	0	0	0
W_{II}	0	0	0	0	0
$W_{\Sigma RI}$	5.5	3.5	4.75	4.5	4.25
W_W	4.5	6.5	5.25	5.5	5.75

molecules (see Fig. 16.17). This means that the cations are weakly interact with each other and the selectivity of ion exchange should not strongly depend on the loading of the ion exchanger with the bivalent counter ion. This correlates well with the data on selectivity of ion exchange Me^{2+}–H^{+} [50]. The interaction between the neighboring sorption complexes is weak which is reflected in a weak dependence of the selectivity coefficient on the ionic composition of the resin, Fig. 16.18. The selectivity coefficient was calculated as

$$k_x = \bar{x}_{Me}{}^{1/2} \cdot x_H / (x_{Me}{}^{1/2} \bar{x}_H) \qquad (16.8)$$

16.3.2.2 The Heavy Metal Ions

The structures of hydration systems of Cd^{2+}, Pb^{2+}, Ni^{2+}, and 20 water molecules computed for the fragments containing two sulfonic groups are presented in Fig. 16.19, compared with the structure of Ca^{2+} system.

They are rather individual and will be discussed separately. Some characteristics of these structures are presented in Tables 16.9 and 16.10.

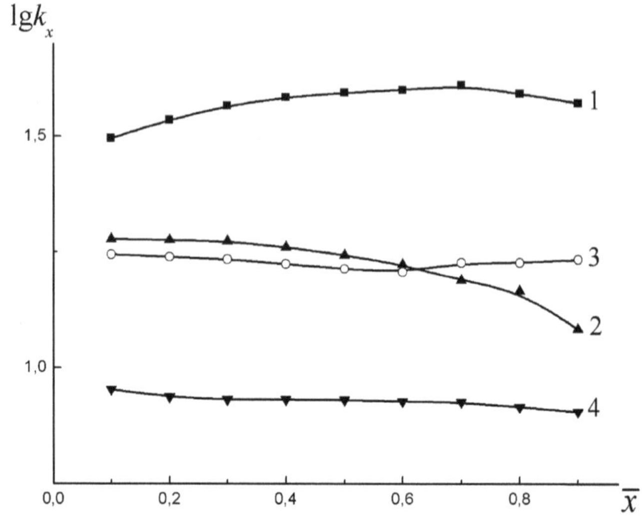

Fig. 16.18 Exchange Me^{2+}–H^+. $\log k_x = f(x_{Me})$ for the ion exchanger with 10.5% DVB, t = 20°C. (1) Ba^{2+}–H^+, (2) Sr^{2+}–H^+, (3) Ca^{2+}–H^+, (4) Mg^{2+}–H^+

Cd. The hydration system of this ion has the structure similar to that for Ca^{2+}. These ions have almost equal ionic radii. The lengths of Me–O bonds with sulfonic groups are almost the same and have electrostatic character. Nevertheless, selectivity of Ca sorption is significantly higher than Cd (Fig. 16.20, Table 16.10). The reason for that is the formation of stronger bonds of Cd with the water molecules. In spite of the fact that both ions have equal number of water molecules in the hydrates, the degree of their covalence is higher for Cd indicated by the higher order of these bonds (Table 16.10). Stronger interaction with water lowers the energy of the interaction with sulfonic group.

Pb. Ion Pb^{2+} is bound to two atoms of oxygen of the sulfonic groups and four water molecules. Its ionic radius is close to that of Ba^{2+}, and the selectivity of these two ions is the highest among the studied ions and close to each other. Nevertheless, the characters of the bonding of Pb^{2+} and Ba^{2+} in the ion exchanger are different. Ba^{2+} forms pure ionic bonds both with O atom of sulfonic groups and water. The latter are rather weak and do not prevent interaction of Ba^{2+} with sulfonic group. Ion Pb^{2+} forms strong bonds with O atom of sulfonic group having significant degrees of covalence and very strong bonds with water molecules with high degrees of covalence (Table 16.10). These two oppositely directed phenomena compensate and result in a similar selectivity sorption of Ba^{2+} and Pb^{2+} (Fig. 16.20, Table 16.10).

Ni. Ion Ni^{2+} in the hydrated structure does not form direct bonds with sulfonic group because of its strong interaction with five water molecules. Four of them form a flat square, and the fifth one is situated on a longer distance from this square almost symmetrically to its corners. The four molecules in the square form strong bonds with Ni^{2+} having a large fraction of covalence. Formation of such stable structure prevents Ni^{2+} from strong interaction with sulfonic group and results in low selectivity of its sorption.

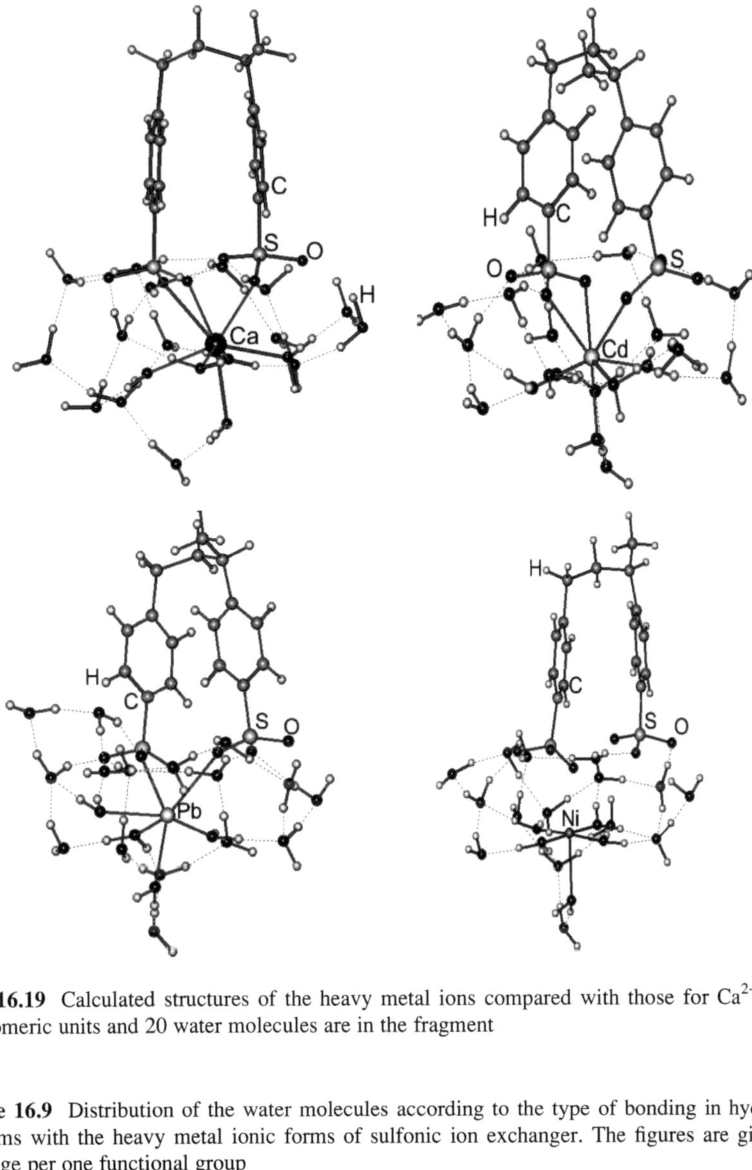

Fig. 16.19 Calculated structures of the heavy metal ions compared with those for Ca^{2+}. Two monomeric units and 20 water molecules are in the fragment

Table 16.9 Distribution of the water molecules according to the type of bonding in hydration systems with the heavy metal ionic forms of sulfonic ion exchanger. The figures are given as average per one functional group

Type of water molecule	Ca^{2+}	Cd^{2+}	Ni^{2+}	Pb^{2+}
W_R	2.5	2.5	3.5	3.0
W_I	2.5	2.5	2.5	2.0
W_{RI}	0	0	0	0
W_{RR}	0	0	0	0
W_{II}	0	0	0	0
$W_{\Sigma RI}$	5.0	5.0	6.0	5.0
W_W	5.0	5.0	4.0	5.0

Table 16.10 Characteristics of the hydration systems of the heavy metal ions compared to that for Ca^{2+} ion

Ion	Charge on the ion	Number of water molecules in the complex	Number of ligands of the cation	Number of water molecules bound to the cation	Number of bonds with O atoms of sulfonic group	Length of Me–O–SO_2 bonds, nm	Bond order Me–O–SO_2 bonds,	Length of Me–OH_2 bonds, in the hydrate, nm	Bond order of Me–OH_2 bonds, in the hydrate
Ca^{2+}	1.791	10	8	5	3	0.236	<0.05	0.234	<0.05
						0.251	<0.05	0.235	0.059
						0.271	<0.05	0.243	0.053
								0.244	0.057
								0.244	0.056
Cd^{2+}	1.684	10	8	5	3	0.229	0.05	0.228	0.086
						0.244	<0.05	0.230	0.080
						0.272	<0.05	0.236	0.079
								0.236	0.083
								0.240	0.083
Pb^{2+}	1.18	10	6	4	2	0.243	0.192	0.223	0.330
						0.288	0.062	0.232	0.301
								0.243	0.250
								0.249	0.206
Ni^{2+}	1.548	2	5	5	0	–	–	0.182	0.189
								0.183	0.178
								0.186	0.170
								0.189	0.165
								0.247	0.057

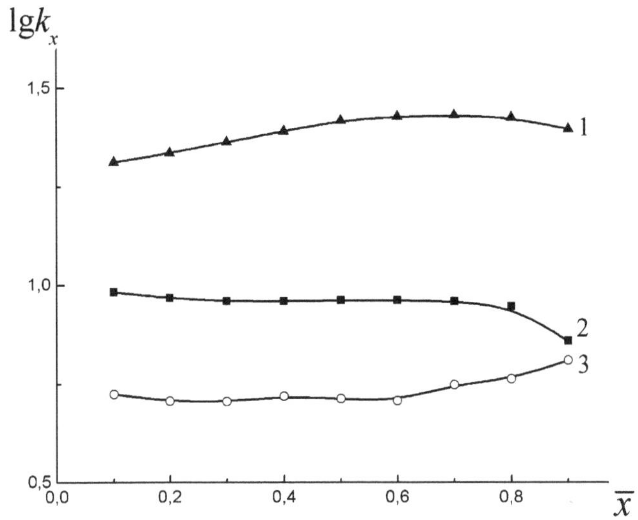

Fig. 16.20 Exchange Me^{2+}–H^+. logk_x = $f(x_{Me})$ for the ion exchanger with 10.5% DVB, t = 20°C. (1) Pb^{2+}–H^+, (2) Ni^{2+}–H^+, (3) Cd^{2+}–H^+

16.4 Summary

A mathematical model based on the concept of influence of neighboring exchange sites on the properties of each other allows explaining dependence of logarithm of the selectivity coefficient and the other additive properties of ion exchange systems on the degree of ion exchange. The higher the probability of the larger number of neighbors, the more complicated dependence is possible. The degree of polynomial describing these dependences is equal to the maximal number of the nearest neighbors. The computer modeling of sulfonic ion exchangers has shown that at low degree of cross-linkage, the maximal number of neighbors of the exchange site with inorganic counter ion is two and the dependences of property–ionic composition are simple and described by the polynomial of the second power. In the case of ion exchangers with high cross-linkage, these dependences can be more complicated and are described by the polynomial of the third power because the appearance of three neighbors is high enough to affect the shape of these dependences. Very complicated dependences described by polynomial of the forth power can occur for large organic ions where the probability of four neighbors can be high due to the large size of the counter ion. These predictions of the model are consistent with the experimental data on simple cation exchange on the resins with different percentages of divinylbenzene. The ion exchangers of the considered type may have significant fraction of isolated functional groups and groups with one, two, three, and more (less probable) neighboring groups. The information of interaction between the nearest neighbors and water molecules in the vicinity of the functional groups as well as of the structures of the fragments of the polymeric structure of the ion exchanger has been obtained using quantum chemistry nonempirical ab initio calculations.

The water molecules present in the hydration system have been classified as the molecules joined only to the counter ion, only to the oxygen atoms of the sulfonic group, both to the counter ion and the sulfonic group, only to the other water molecules, etc. Evaluation of the length of the bonds between the counter ion, functional group, and the water molecules and the degree of their covalence allowed obtaining detailed information on the state of the ions and water molecules in the ion exchanger. These data have been obtained for different degrees of hydration (0–10 water molecules per sulfonic group) for the following counter ions: H^+, Li^+, Na^+, K^+, Rb^+, Cs^+, Ag^+, Be^{2+}, Mg^{2+}, Ca^{2+}, Sr^{2+}, Ba^{2+}, Cd^{2+}, Ni^{2+}, and Pb^{2+}.

The following main results have been obtained. The dissociation of R-SO_3H groups occur when three water molecules are present at the isolated group, one water molecule per group in the case of two groups in the fragment and two water molecules per three sulfonic groups. Two types of the proton hydrates are predominantly formed in these systems; they are H_3O^+ and $H_5O_2{}^+$. The ions of alkali metals in the absence of water form asymmetric structures with sulfonic group in which one counter ion is bound to two oxygen atoms. Upon hydration, the structure of the sorption complex changes, and in the systems with 10 water molecules per sulfonic group, Li^+ and Na^+ ions are separated from the sulfonic groups and form direct bond only with the water molecules; ions K^+, Rb^+ and Cs^+ remained fixed on the sulfonic group with one direct Me^{2+}–O bond. This explains the difference in the ion exchange behavior of the water structure-forming and structure-breaking ions. The silver ion has abnormally high selectivity to sulfonic ion exchangers because it forms four bonds with large degree of covalence; especially strong bond is formed with the sulfur atom due to interaction of the vacant *d* orbitals of sulfur with Ag valence *d* electrons. Cations of alkaline earth metals form electrostatic bonds with the pairs of sulfonic groups. This complex weakly interacts with the neighboring exchange sites. In the state of complete hydration, these cations remained bound to the sulfonic groups with one or two Me–O bond except of Be^{2+} which can completely separate from the fixed anion. Ions of the heavy metal have rather low selectivity (comparable or lower than the alkaline earth metals) because of formation of strong hydrates in which Me^{2+}–O bonds have large fraction of covalence preventing their strong interaction with the sulfonic groups.

In conclusion, we note that mathematical modeling as well as molecular mechanic and quantum chemistry calculations of the structures of ion exchangers is consistent with the main concepts used in the interpreting ion exchange equilibria. They are a concept of nonregularity of the exchange sites in polymeric ion exchangers and competitive interaction of the counter ions with the fixed ion and water molecules.

Acknowledgment This work has been supported by Belarus foundation for the fundamental research, grants No: X09-091.

References

1. Frenkel D, Smit B (2001) Understanding molecular simulations. From algorithms to applications, vol 1, 2nd edn, Computational science series. Academic, San Diego
2. Holm C, Joanny JF, Kremer K et al (2004) Polyelectrolyte theory. Adv Polym Sci 166:67–111
3. Lyubartsev AP, Nordenskiöld L (2002) Computer simulations of polyelectrolytes. In: Tripathy SK, Kumar J, Nalwa HS (eds) Handbook of polyelectrolytes and their applications, vol 3. American Scientific, Los Angeles, Chapter 11
4. Okomoto Y (2004) An ab initio study of the side chain of Nafion. Chem Phys Lett 389:64–67
5. Warren DS, McQuillan AJ (2008) Infrared spectroscopic and DFT vibrational mode study of perfluoro(2-ethoxyethane)sulfonic acid (PES), a model Nafion chain molecule. J Phys Chem B 112:10535–10543
6. Kreuer KD, Paddison SJ, Spohr E et al (2004) Transport in proton conductors for fuel-cell applications: simulations, elementary reactions and phenomenology. Chem Rev 104:4637–4678
7. Butyrskaya EV, Shaposhnik VA, Butyrskii AM et al (2007) Interpretation of hypsochromic and bathochromic shifts of vibrational frequencies of a cation exchanger. J Anal Chem 62:930–934
8. Butyrskaya EV, Shaposhnik VA (2002) Interpretation of infrared spectra for ion exchange systems. Opt Spectrosc 202:370–374
9. Shaposhnik VA, Butyrskaya EV (2004) Computer simulations of cation exchange membrane structure. An elementary act of hydrogen ion transport. Russ J Electrochem 40:370–374
10. Nechaeva LS, Butyrskaya EV, Shaposhnik VA (2009) Structurally-group analysis of carboxylic cation exchanger. Sorp Chromatogr Process (Sorbzionnye i Chromatografycheskiye prozessy, Rus) 9:208–213
11. Soldatov VS, Zelenkovskii VM, Bezyazychnaya TV (2003) Structure and properties of spatial polyelectrolytes on styrene-divinylbenzene matrixes. React Funct Polym 54:63–85
12. Soldatov VS (1993) Mathematical modeling of ion exchange equilibria on resinous ion exchangers. React Polym 19:105–121
13. Soldatov VS, Gogolinsky VI, Zelenkovskii VM et al (1996) Computer model of sulfonated styrene-divinylbenzene copolymers. In: Creig JA (ed) Ion exchange. Development and applications. Royal Society of Chemistry, Cambridge
14. Soldatov VS, Bezyazychnaya TV, Zelenkovskii VM et al (1998) Effect of irregularities in the region of crosslinking junction on the structure of network polystyrene. Polym Sci B 40:268–271
15. Soldatov VS, Bezyazychnaya TV, Gogolinskii VI et al (1999) Influence of cross-linking on selectivity of cation and anion exchange on styrene-divinylbenzene matrix. Russ J Phys Chem (Zhurn Fiz Khim) 73:1080–1084
16. Soldatov VS, Bezyazychnaya TV, Gogolinskii VI et al (1999) Evaluation of the spatial distribution of sulfonic groups in the ring fragments of polystyrene networks. Polym Sci A 41:1085–1089
17. Zelenkovskii VM, Bezyazychnaya TV, Soldatov VS (2007) Hydration of strongly basic anionites: ab initio calculations. Polym Sci B 49:30–35
18. Zelenkovskii VM, Bezyazychnaya TV, Soldatov VS (2004) Quantum-chemical modeling of interactions between adjacent functional groups in cationites. Polym Sci A 46:1085–1089
19. Soldatov VS, Zelenkovskii VM, Bezyazychnaya TV (2004) Computer modeling of the hydration of the functional group in ion exchangers. In: Cox M (ed), Ion Exchange technology for today and tomorrow (Proceedings of IEX'04). University of Hertforthire, Cambridge
20. Soldatov VS, Zelenkovskii VM, Bez'yazychnaya TV (2001) The relative positions of ions in hydrate structures of p-ethyl benzene sulfonates. Dokl NAN Belarusi 45:75–80 (Russ)
21. Zelenkovskii VM, Soldatov VS, Bezyazychnaya TV (2004) Non-empirical calculations of p-ethylbenzenesulfo-acid salts structure. Vestsi Natsyyanal'nai Akad Navuk Belarusi Ser Khim Navuk 4:24–27 (Russ)

22. Zelenkovskii VM, Bezyazychnaya TV, Soldatov VS (2010) Quantum chemical calculations of vibrational spectra of para-ethylbenzene sulfinic acid. J Appl Spektrosk 77:189–193
23. Schmidt MW, Baldridge KK, Boatz JA et al (1993) General atomic and molecular electronic structure systems. J Comp Chem 14:1347–1363
24. Huzinaga S, Andzelm J, Klobukowski M (1984) Gaussian basis sets for molecular calculations. Elsevier, Amsterdam
25. Reichenberg D, McCaley DJ (1955) Properties of ion exchange resin relation to their structure. VII Cation exchange equilibria on sulfonated polysterene resins of varying degree of crosslinking. J Chem Soc 77:2741–2749
26. Soldatov VS, Marzynkevich RV, Pokrovskaya AI et al (1976) Comparative characteristics of sulfostyrene resins with different isomers of divinylbenzene. Russ J Phys Chem (Zhurn Fiz Khim) 50:480–484
27. Soldatov VS, Marzynkevich RV, Pokrovskaya AI (1969) The features of ion exchange on the resins with high cross-linkage. Russ J Phys Chem (Zhurn Fiz Khim) 43:2889–2893
28. Samsonov GV, Trostianskaya EB, Elkin GE (1969) Ion exchange. Sorption of organic substances. Nauka, Lenigrad
29. Arkhangelskij LK, Materova EA, Mikhailova SS et al (1968) Investigation of the relation between the apparent constants of equilibria, the composition of the ion exchanger and the water uptake for the regions of the large relative contents of one of the counter ions. In: Thermodynamics of ion exchange (Termodinamika ionnogo obmena, Rus). Nauka i Technica, Minsk
30. Soldatov VS, Kuvaeva ZI, Bychkova VA et al (1998) Ion-exchange equilibria of aliphatic amino acid cations on a liquid sulphonic ion exchanger: comparison with ion exchange resins. React Funct Polym 38:237–247
31. IUPAC (1999) Glossary of terms used in theoretical organic chemistry. Pure Appl Chem 71:1919–1981
32. Zundel G (2000) Hydrogen bonds with large proton polarizability and proton transfer processes in electrochemistry and biology. Adv Chem Phys 111:1–217
33. Starobinets GL, Soldatov VS (1963) On thermodynamics of ion exchange on sulfonated copolymers of styrene and divinylbenzene. Russ J Phys Chem (Zhurn Fiz Khim) 37:294–299
34. Soldatov VS (1972) Simple ion exchange equilibria (Prostye ionoobmennie ravnkvesia, in Russian). Nauka i Technika, Minsk
35. Samoilov OY (1957) A new approach to the study of hydration of ions in aqueous solution. Discuss Faraday Soc 24:141–146
36. Samoilov OY (1965) Structure of aqueous electrolyte solutions and the hydration of ions. Consultants Bureau, New York
37. Markus Y (2009) Effect of ions on the structure of water: structure making and breaking. Chem Rev 109:1346–1370
38. CRC (1990) In: Weast RC (ed) Handbook of chemistry and physics, CRC Press, Boca Raton, pp P. F–187
39. Dolar D (1974) Thermodynamic properties of polyelectrolyte solutions. In: Selengu E, Mandel M, Straus UP (eds) Polyelectrolytes. Reidel, Dordrecht/Holland
40. Soldano P, Larson QV, Myers CE (1955) The osmotic approach to ion-exchange equilibrium II. Cation exchangers. J Am Chem Soc 77:1339–1344
41. Bonner OD, Rhett V (1953) Equilibrium studies of the solver-sodium-hydrogen system on Dowex 50. J Am Chem Soc 57:254–256
42. Hogfeldt E, Chiarizia R, Danesi PR et al (1981) Structure and ion exchange properties of dinonyl naphthalene sulfonic acids and its salts. Chem Scripta 18:13–43
43. Soldatov VS, Starobinets GL (1964) On thermodynamics of ion exchange on sulfonated copolymers of styrene and divinylbenzene. II. The effect of cross-linkage, ionic radii and temperature on the equilibria. Russ J Phys Chem (Zhurn Fiz Khim) 38:681–685
44. Eigen M (1964) Proton transfer acid–base catalysis, and enzymatic hydrolysis. Part I: elementary processes. Angew Chem Int Ed 3:1–19

45. Chandra A, Tuckerman ME, Marx D (2007) Connecting salvation shall structure to proton transport kinetics in hydrogen-bonded networks via populations analysis. Phys Rev Lett 99:145901.1–145901.4
46. Dey TJF, Schmitt UW, Voth GA (2000) The mechanism of hydrated proton transport in water. J Am Chem Soc 122:12027–12208
47. Wu Y, Voth GA (2003) A computer simulation study of the hydrated proton in a synthetic proton channel. Biophys J 85:864–875
48. Marx D, Tuckerman ME, Hutter J et al (1999) The nature of the hydrated excess proton in water. Nature (London) 397:601–604
49. Glueckauff E, Kitt GP (1955) A theoretical treatment of cation exchangers III Hydrations of cations in polystyrene sulphonates. Proc R Soc A 228:322–341
50. Krylova AA, Soldatov VS, Starobinets GL (1966) On thermodynamics of ion exchange on sulfonated copolymers of styrene and divinylbenzene. IX. Selectivity and hydration of bivalent ions. Rus J Phys Chem (Zhurn Fiz Khim) 40:1203–1206

Index

Inamuddin and M. Luqman (eds.), *Ion Exchange Technology I: Theory and Materials*, DOI 10.1007/978-94-007-1700-8,

FSC
www.fsc.org
MIX
Papier aus verantwortungsvollen Quellen
Paper from responsible sources
FSC® C105338

If you have any concerns about our products,
you can contact us on
ProductSafety@springernature.com

In case Publisher is established outside the EU,
the EU authorized representative is:
Springer Nature Customer Service Center GmbH
Europaplatz 3, 69115 Heidelberg, Germany

Printed by Libri Plureos GmbH
in Hamburg, Germany